ELECTRIC CIRCUITS AND SIGNALS

ELECTRIC CIRCUITS AND SIGNALS

Nassir H. Sabah
American University of Beirut, Lebanon

CRC Press
Taylor & Francis Group
Boca Raton   London   New York

CRC Press is an imprint of the
Taylor & Francis Group, an **Informa** business

MATLAB® is a trademark of The MathWorks, Inc. and is used with permission. The Mathworks does not warrant the accuracy of the text or exercises in this book. This book's use or discussion of MATLAB® software or related products does not constitute endorsement or sponsorship by The MathWorks of a particular pedagogical approach or particular use of the MATLAB® software.

The author would like to thank Cadence Design Systems, Inc. for allowing Taylor and Franics Group to distribute OrCAD software with this book.

CRC Press
Taylor & Francis Group
6000 Broken Sound Parkway NW, Suite 300
Boca Raton, FL 33487-2742

© 2008 by Taylor & Francis Group, LLC
CRC Press is an imprint of Taylor & Francis Group, an Informa business

Library of Congress Cataloging-in-Publication Data

Sabah, Nassir H.
Electric circuits and signals / by Nassir H. Sabah.
p. cm.
Includes bibliographical references and index.
ISBN 1-4200-4589-X (alk. paper)
1. Electric circuits--Textbooks. I. Title.

TK454.S28 2007
621.319'2--dc22

2006049275

Visit the Taylor & Francis Web site at
http://www.taylorandfrancis.com

and the CRC Press Web site at
http://www.crcpress.com

To Gharid

In loving appreciation of her support and forbearance

And in fond remembrance of

J.T. Allanson

My first teacher of electric circuits

Let the main object of this, our didactic, be as follows: to seek and to find a method of instruction by which teachers may teach less, but learners may learn more.

John Comenius (1592–1670)

Contents

The Author

Nassir Sabah is professor of electrical and computer engineering at the American University of Beirut, Lebanon. He received his B.Sc. (hons. Class I) and M.Sc. in electrical engineering from the University of Birmingham, U.K., and his Ph.D. in biophysical sciences from the State University of New York (SUNY/Buffalo). He has served as Chairman of the Department of Electrical Engineering, Director of the Institute of Computer Studies, and Dean of the Faculty of Engineering and Architecture at the American University of Beirut. In these capacities, he was responsible for the development of programs, curricula, and courses in electrical, biomedical, communications, and computer engineering.

Professor Sabah has extensive professional experience in the fields of electrical engineering, electronics, and computer systems and more than 30 years teaching experience in electric circuits, electronics, neuroengineering, and biomedical engineering. He has over 100 technical publications, mainly in neurophysiology, biophysics, and biomedical instrumentation.

He has served on several committees and panels in Lebanon and elsewhere in the region. He is a Fellow of the IET, U.K., and a member of American Society of Engineering Education.

Preface

This book is intended for one or two courses on basic electric circuits as well as the "Circuits and Signals" course, CE-CSG, described in the Body of Knowledge of the IEEE/ACM. It differs from other textbooks in its coverage, organization, and presentation, which strongly emphasize fundamentals and stresses creative problem solving. By emphasis on fundamentals I mean: (1) being comprehensive enough to include all the basic material that I believe should be included, (2) giving adequate explanation or justification of assertions or mathematical expressions (within the scope of the book, of course), and (3) helping the student understand the logic of circuit behavior and gain insight into the meaning, significance, and interrelations of the concepts involved. Insight is like an x-ray of the mind: it penetrates to the heart of a matter or problem. The main concepts are explicitly highlighted throughout the book for special emphasis.

There is no doubt that integration, generalization, and emphasis on fundamentals are conducive to a thorough understanding of the material. A thorough understanding is what distinguishes quality engineering education from mere technical training and is a prerequisite for creative problem solving, providing a solid foundation for life-long learning.

The book stresses creative problem solving, in contrast to problem solving based on following set procedures without really understanding their rationale. Needless to say, creative problem solving is the most important conceptual step leading to a superior engineering design. In addition to emphasis on fundamentals, creative problem solving is fostered by encouraging the student to look at a given problem in different ways, particularly fresh and original ways. This is done through solving a relatively large number of carefully selected problems, many of them using different methods, with in-depth explanations and insightful comments and interpretations. A prologue on solving circuit problems follows this preface.

I have always tried to impress upon my students that a course on electric circuits involves much more than learning set procedures and techniques or writing down equations and solving for the variables involved. The factual knowledge acquired in these courses is much less important than the insight gained into concepts and their interrelations, and the problem-solving approach that is learned. Factual knowledge is liable to be forgotten or become obsolete, whereas a grasp of fundamentals and the pursuit of creative problem solving become integral to the mindset of a true engineering professional. This book reflects such a philosophy.

No special preparation is assumed on the part of the student beyond trigonometry, elementary calculus, and some basic physics. Discussion starts from almost level zero in circuit theory. Mathematics alone is not relied upon to present concepts and relations. Qualitative and physical explanations and interpretations are emphasized whenever possible.

Organization

The progression of material is logical and coherent. The Foreword highlights some general aspects of electric circuits: historical landmarks, the nature of electric circuits, and their relevance. Basic concepts are introduced for the dc state (Chapter 1 through Chapter 4), followed by the sinusoidal steady state (Chapter 5) and transformers (Chapter 6). The sinusoidal steady state is applied to power relations (Chapter 7) and to balanced three-phase systems (Chapter 8). It is extended to periodic inputs (Chapter 9) and to frequency response (Chapter 10). Energy storage elements and their series and parallel connections are then discussed (Chapter 11), followed by natural responses and convolution (Chapter 12). More general transient responses are considered next (Chapter 13). Discussion of basic electric circuits concludes with two-port circuits (Chapter 14) and the Laplace transform (Chapter 15). Because it is optional in CE-CSG, the Laplace transform is postponed to this stage.

The signals part is limited to continuous signals, as required in CE-CSG. It formally begins with the Fourier transform (Chapter 16), continues with basic signal-processing operations (Chapter 17), and ends with signal processing using operational amplifiers (Chapter 18). The last chapter (Chapter 19) is on electric circuit analogs of mechanical, fluid, and thermal systems as well as the ionic system of the cell membrane.

Every chapter begins with a brief overview, followed by learning objectives. These are divided into two groups, inspired by Bloom's taxonomy on cognitive learning: the "to be familiar with" group and the "to understand" group, to which the student must pay particular attention. Learning outcomes are stated at the end of every chapter together with a summary of the main concepts and results.

Good use is made of the companion CD. Almost half as much material as is in the book is included on the CD in the form of supplementary topics (having the section number prefixed by ST) and supplementary examples (prefixed by SE). The CD also includes some appendices (prefixed by S). A given topic is included on the CD, rather than in the text, if it is not basic enough, in the sense that it may be somewhat advanced, or it may not introduce additional fundamental concepts, or its outcomes could be realized using more basic alternatives.

The book is written in a concise, get-to-the-point style, but not at the expense of elaborating important concepts at a slow enough pace.

Pedagogy

Specific features of the book's pedagogy are as follows:

1. Extensive use is made of duality in order to provide a more integrated treatment of capacitive and inductive circuits. A consequence of this is the rightful emphasis on flux linkage in inductive circuits as the dual of electric charge in capacitive circuits.

2. The substitution theorem is discussed and applied as a very useful tool for solving some types of problems.

3. Practical, real-life material is included whenever appropriate, mainly in the form of "Application Windows" that apply theory and provide additional motivation for students.

4. The computer aids emphasized in the book are MATLAB® (MATLAB® is a registered trademark of The Mathworks, Inc., Natick, Massachusetts) and PSpice® with Capture, which is part of OrCAD 15.7 Demo (OrCAD®, PSpice®, SPECTRA® for OrCAD, and Cadence® are registered trademarks of Cadence Design Systems, Inc., San Jose, California). MATLAB is used more extensively than in comparable introductory texts as an aid in solving circuit problems and in plotting results. PSpice simulations based on schematic capture are presented in some detail and are integrated within the discussion. The PSpice simulation examples are included on the companion CD, together with OrCAD 15.7 Demo software, compatible with Microsoft® Windows® (Microsoft® and Windows® are registered trademarks of Microsoft Corporation, Redmond, Washington). Instructions on installing this software and running the PSpice simulation examples are given in Appendix SD on the companion CD.

5. Answers for all problems are included on the companion CD.

Solutions Manual and Classroom Presentations

A Solutions Manual for all problems and exercises is available with its own companion CD that includes classroom presentations. These are Microsoft® Office Word (Microsoft® Office Word is a registered trademark of Microsoft corporation, Redmond, Washington) files for each chapter that present, in the form of bulleted text with figures, the main ideas discussed in the chapter, as well as the examples illustrating these ideas. The files are intended for projection in the classroom by instructors and used as a basis for explaining the chapter material.

There are three advantages to having the files in Word: i) the files can be easily modified by instructors if they so wish, ii) the top and bottom margins can be hidden for smooth scrolling from one end of the file to the other, as if the file is a single page, and iii) windows can also be opened that allow reference to a figure, say, on a preceding page.

Acknowledgments

I am deeply indebted to my students over the years for the motivation to better explain, justify, and present key concepts in electric circuits. In fact, it was these pursuits that impelled me to write this book. I am very grateful to my family, friends, and colleagues for their encouragement. I would also like to express my sincere appreciation of the efforts of the staff of CRC Press in producing and promoting this book, particularly Nora Konopka, Publisher, Engineering and Environmental Sciences, for her admirable professionalism and her considerate support. I would also like to thank Tamara Chehayeb Makarem for her contribution to the design of the front cover of the book.

Prologue

Solving Circuit Problems

The essence of engineering is solving practical problems through design. But not all solutions are created equal. In the parade of solutions, the more creative ones stand head and shoulders above the mundane variety. A creative solution is not only effective, but is typically original, neat, and elegantly simple as well. In the engineering balance sheet, simplicity is generally synonymous with low cost and reliability. Creative solutions are pertinent not only to new problems but to old ones as well (as in the proverbial "better mousetrap").

Consider, for example, a 1-m-high water tank that is to be emptied periodically. A novice engineer may immediately think of a solenoid valve at the outlet controlled by a timer (Figure 1a). To prevent overflow, our novice engineer may decide to install a float valve at the inlet. (He or she could do worse and opt for a solenoid valve at the inlet controlled by a level detector, as illustrated in the figure). Now consider a more creative solution that consists simply of an open, bent tube that discharges directly out of the tank (Figure 1b). Timing is controlled by the rate of inflow, which could be varied by a simple valve like that of a water tap. When the water level reaches the top of the bend in the tube, the tank automatically empties under atmospheric pressure all the way down to the tube opening (siphon effect). The cycle then repeats.

Both systems empty the tank periodically at a variable rate. If this is all that is required, the solution in Figure 1b is the clear winner. It is neat, simple, and reliable. It may not be immediately obvious, which calls for some creative thinking, guided by the following considerations:

1. The specifications imposed no restriction on the type of system to be used. The novice engineer may have unconsciously restricted the system to be electrically

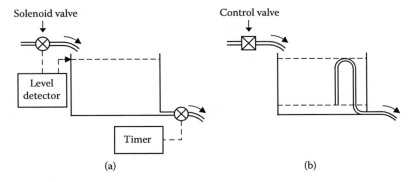

(a)   (b)

FIGURE 1
Periodic emptying of water tank. (a) Electrical system, (b) system based on siphon effect.

operated. Such self-imposed restrictions are common among inexperienced problem solvers and hinder effective problem solving.

2. Knowledge of fundamentals is required. It is not at all intuitive that atmospheric pressure, if unopposed, can support the weight of a column of water approximately 10 m high.

There are courses in the engineering curricula, such as mechanics and basic structures, that lend themselves to promoting problem-solving skills. But nothing, in my opinion, beats electric circuits in this respect because they offer so much variety and versatility, often to the dismay of students. Circuit problems are valuable in training students to think logically, systematically, and creatively, and to be able to reduce an apparently complex problem to its bare-bone essentials. It is indeed a big loss if circuit courses are not adequately exploited to foster creative problem-solving skills. I have always tried to impress on my students that it is not sufficient to solve a problem and get the right answer. The method of solution should be the most efficient one can think of, involving minimum time and effort (the journey is at least as important as the destination). That is where the real fun in problem solving is, and that is what one should train oneself to do.

Books have been written on creative problem solving in engineering, and several textbooks on electric circuits give helpful procedures for problem solving. Here, I would like to offer some advice based on my experience with students in solving circuit problems.

1. *Resist the urge to charge ahead with whatever method of solution that first comes to mind.* Understandably, this may be motivated by a desire to "do something" without wasting time. Unfortunately, more often than not, this is a self-defeating strategy because the method that first comes to mind may well be the most tedious, time-consuming, and error-prone.

   A case in point that I find myself constantly grappling with is the tendency of some students to become "fixated" on the node-voltage and mesh-current methods of circuit analysis once they learn that these are powerful and systematic methods that, at least in principle, can always be applied to solve circuit problems. The trouble is, solving circuit problems by these methods becomes largely routine, and too much reliance on these "safe" methods stifles the imagination and prevents the student from seeking alternative, more creative solutions. My advice is to keep such routine methods as a last resort (like having no choice but to read the manual).

2. *Always take time to think about the problem first.* Thinking is most productive when guided by some well-defined goals such as:

   a. Formulate the problem in some precise, systematic, and organized manner. Mathematical expressions, tabulation of given data and required results, and redrawing circuit diagrams to make them more meaningful are generally helpful.

   b. Try to reduce a given problem to a simpler one; this is where insight and knowledge of fundamentals are essential. Reasonable assumptions may be made, and their validity verified by checking the consistency of the results obtained.

3. *In thinking about a problem, do not be constrained by preconceived notions.* For example, students usually assume that the only logical and organized way of solving a problem is to start with the given input and determine the output. In some cases, however, it is far quicker and less error-prone to assume a certain output, then

work backwards to determine the input that gives this assumed output, and finally scale the assumed output in accordance with the actual input.

4. *Seek alternative solutions or alternative steps in a given solution.* Additional insight into the problem may be gained in this way, and the alternatives may prove to be better ways of solving the problem. Even if they are not, they can be useful for checking purposes.

5. *Decide on a solution strategy.* If you have thought about the problem and have come up with a number of alternatives, choose the alternative that seems to be the most direct and which requires the fewest number of steps. The next best alternative may be used for checking purposes.

6. *Check and recheck your results.* Other than checking your calculations and using alternative methods, it is good practice to get into the habit of making some simplifying assumptions in order to get some rough ballpark figures that you can compare your results to. This is especially valuable in solving problems using computer aids, as it enables you to judge whether or not the computer results make sense. It may also help you get additional insight into the problem. PSpice simulations are a good habit to acquire in order to verify your results.

7. *Make the most out of solving problems.* Solving circuit problems should never be a matter of just writing equations and solving for unknowns. You would be foregoing a valuable opportunity to foster your problem-solving skills and understand the logic of circuit behavior. After solving a problem you should stop and ask yourself about the concepts that the problem illustrated, and make sure you understand these concepts. The best test for understanding a concept is to be able to apply it in a context that is different from that in which it was first encountered.

8. *Persevere.* When you feel almost ready to give up, remember the verse:

> Problems worthy of attack
> Prove their worth by hitting back.

Foreword

Historical Landmarks, Nature, and Relevance of Electric Circuits

Historical Landmarks of Electric Circuits

The end of the 18th century marked a watershed in the history of science. Why? Because of a famous controversy over muscular contraction of frogs' legs.

The Greeks discovered around 600 B.C. that if amber resin was rubbed against cat fur, or glass against silk, the amber or glass attracted bits of various materials. *Amber* in Greek is *elektron*, which is the name later given to the atomic particle discovered in 1897 by the English physicist Joseph John Thompson (1856–1940). Following the Greeks, static electricity was produced by rubbing together many different substances. By the late 18th century, quite sophisticated machines were developed in which materials were continuously rubbed together by hand rotation of a movable member against a fixed member. Cumulated electrostatic charges were produced on a large scale in this manner, resulting in sparking between metallic objects spaced some distance apart.

Considerable experimental work was conducted on static electricity in the 18th century, which led to the notion that electricity was some kind of fluid that could flow through certain bodies but was held by others. The American experimenter Benjamin Franklin (1706–1790) argued that rubbing two bodies together resulted in excess fluid in one body and deficiency in the other, a view that is the forerunner of conservation of charge. He postulated that one body becomes "electricised positively," the other negatively, and that electricity flowed from positive to negative, a convention that is still adhered to today. In 1752, he demonstrated through his famous kite experiment that lightning was a giant electric spark.

It was known that exposure to static electricity caused muscular contraction, or electric shock, which could also be delivered by certain fish, such as the torpedo fish. As muscular contraction was believed to be due to a "nervous fluid," or animal spirit, that courses through nerves to contracting muscles, there was speculation that muscular contraction and nervous function were electrical in nature.

One day in April 1786, the Italian anatomist and physician Luigi Galvani (1737–1798) performed a routine dissection of the legs of a frog and the associated nerve. When one of his assistants accidentally touched the nerve with his scalpel at the instant that a spark was generated by an electrostatic machine nearby, the muscles of the frog's legs contracted strongly. Galvani was struck by the novelty of this phenomenon and set out to investigate it in detail. After a series of experiments under different conditions, he determined that if the leg muscles were in contact with one metal, such as iron, while the nerve was in contact with a different metal, such as brass, and the two metals were brought into contact, the muscles contracted. He postulated that the cause of the contraction was "animal electricity" and that the two metals, the nerve, and the muscles formed a circuit that discharged the electric fluid in the muscles. He did not attach much importance to the fact that using a single metal to form the connection between nerve and muscle resulted in weak or no contractions.

The publication of Galvani's results and conclusions in 1791 was received with great interest, and his experiments were repeated by many investigators, one of whom was a prominent Italian physicist Alessandro Volta (1745–1827). Volta at first accepted Galvani's explanation but soon began to have his doubts. After a series of experiments, he proposed in 1792 that the contractions were due to "metallic electricity" generated by contact between two dissimilar metals and that the nerve and muscles merely detected this electricity. A vigorous controversy ensued that divided the European scientific community into two opposing camps. Galvani and his supporters proved the existence of animal electricity to their satisfaction when they succeeded in eliciting contractions in a frog's leg by direct contact between the muscles and a freshly cut nerve without any metals at all. Volta retorted by showing that electricity could be produced by two dissimilar conductors separated by a conducting solution, and explained muscular contraction without metals on the basis that the nerve and muscle were dissimilar conductors separated by a conducting liquid. In reaching this conclusion, Volta constructed the first electric battery, referred to as "Volta's pile" at the time, which he described in a letter, made public in 1800, to the president of the Royal Society in London. In one form of the pile, a number of units were connected in series, each unit consisting of a zinc plate and a silver plate immersed in brine. Volta noted that this arrangement produced electricity continuously and not just momentarily, as when static electricity is discharged.

The introduction of the electric battery was a momentous event. Static electricity could not be adequately controlled, which allowed only qualitative deductions, with some notable exceptions such as the work of the French physicist Charles A. Coulomb (1726–1806), who formulated in 1785 the inverse-square law for the force of attraction or repulsion between charged bodies. The electric battery allowed for the first time a continuous and controlled flow of current, thereby greatly accelerating research and development on electricity. In 1800, the English chemists William Nicholson (1753–1815) and Anthony Carlisle (1768–1840) used Volta's pile to electrolyze water into hydrogen and oxygen, thereby launching electrochemistry. In 1809, the English chemist Sir Humphry Davy (1778–1829) produced the first electric light by striking an arc between two charcoal electrodes connected to a Voltaic pile.

The following is a chronology of some important events in the history of physics and mathematics that are related to electric circuits:

1782 — The French mathematician Pierre-Simon Laplace (1749–1827) used a form of what is now referred to as the Laplace transform to obtain an integral solution of a general linear, second-order, partial differential equation. In 1785, he used a form that is closer to the present-day Laplace transform to approximate functions of very large numbers. Laplace-transform-like integrals were mentioned by Euler and Lagrange before Laplace.

1806 — Swiss-born French amateur mathematician Jean Robert Argand (1768–1822) interpreted complex numbers in geometrical terms, and associated multiplication by $i = \sqrt{-1}$ with a 90° rotation in the complex plane.

1807 — French mathematician and physicist Jean Baptiste Joseph Fourier (1768–1830), in his work on the propagation of heat in solids, expanded a function over an interval as a trigonometric series, in what is now called Fourier series. The expansion was criticized at the time by prominent mathematicians as "lacking in rigor."

1819 — Danish physicist and philosopher Hans Oersted (1777–1851) discovered that an electric current deflected a compass needle.

1820 — French mathematician and physicist André Ampère (1775–1836) extended Oersted's work and derived the right-hand-screw-rule of force on a current-carrying conductor. In 1823 he postulated that magnetism in iron is due to tiny electric currents.

1827 — German mathematics and physics teacher and investigator Georg Simon Ohm (1789–1854) promulgated what is known as Ohm's law.

1831 — British physicist and chemist Michael Faraday (1791–1867) discovered electromagnetic induction. In 1834 he announced what is known as Faraday's law.

1831 — American scientist Joseph Henry (1797–1878) discovered self-inductance and noted its analogy to mechanical momentum.

1834 — German physicist Heinrich Friedrich Emil Lenz (1804–1865) generalized the law of magnetic induction. A year earlier, he determined that the resistance of metals changes with temperature.

1840 — British physicist James Prescott Joule (1818–1889) discovered what became known as Joule's law, relating the heat produced by an electric current in a wire to the resistance of the wire and the magnitude of the current.

1842 — German physicist Julius Robert von Mayer (1814–1878) enunciated the Principle of Conservation of Energy, based mainly on philosophical reasoning and physiological considerations. In 1847, Joule asserted conservation of energy from his experiments on the mechanical equivalent of heat. About the same time, Helmholtz deduced "conservation of forces" from Newton's law of motion.

1845 — The German physicist Gustav Robert Kirchhoff (1824–1887) announced what are now known as Kirchhoff's laws.

1853 — The German physicist, anatomist, and physiologist Hermann Ludwig Ferdinand von Helmholtz (1821–1894) enunciated the Principle of Superposition. In the same paper, he proclaimed what is now almost universally referred to as Thevenin's theorem, rediscovered in 1883 by the French telegraph engineer Léon Charles Thévenin (1857–1926). Thevenin's theorem is less commonly known as Helmholtz's theorem.

1885 — Serbian-American electrical engineer and inventor Nikola Tesla (1856–1943) sold to Westinghouse Electric Company in Pittsburgh the patent rights to his polyphase system of alternating-current dynamos, transformers, and motors. The polyphase ac system was eventually adopted as the standard for power generation, transmission, and distribution, in preference to Edison's dc system.

1890s — English electrical engineer Oliver Heaviside (1850–1925) adapted complex numbers to the study of electric circuits, developed operational calculus, and elaborated techniques for applying the Laplace transform to the solution of differential equations.

1926 — American electrical engineer at Bell Telephone Laboratories Edward Lawry Norton (1898–1983) pronounced in a paper what is known as Norton's equivalent circuit.

1938 — American mathematician and physicist at Bell Telephone Laboratories Hendrik Wade Bode (1905–1982) developed the asymptotic phase and magnitude plots that bear his name.

1952 — Dutch electrical engineer Bernard T.H. Tellegen (1900–1990) enunciated what is known as Tellegen's theorem.

Nature of Electric Circuits

Electrical engineering and its allied fields are concerned with devices, equipment, and systems that depend for their operation on phenomena associated with electric charges. This is true, for example, of computer, communications, control, electric power, and instrumentation systems as well as biomedical equipment and household appliances. A magnetic field is associated with moving electric charges, whereas an electric field is associated with electric charges at rest or in motion. Hence, electrical problems are fundamentally problems in electromagnetic fields. Field problems are generally awkward or difficult to solve, however, so that an alternative, simpler approach would be most advantageous.

The success of the engineering approach has depended to a large extent on the ability of engineers to devise appropriate models of the system under consideration. To be useful, these models should be relatively simple and tractable yet adequately representative, under some desired conditions, of those features of the system that are most relevant to the task at hand. The validity of such models is verified through experiments on the real system to ensure that they are sufficiently accurate for practical purposes under given conditions. This implies that different models may have to be used for the same system under different conditions. For example, the model of a transistor or a transformer at low frequencies has to be modified to represent the same transistor or transformer at high frequencies. The circuit approach is a prime example of successful modeling that has been found to be applicable to a large class of electromagnetic field problems.

In essence, three essential attributes of the electromagnetic field, namely, power dissipation, energy stored in the electric field, and energy stored in the magnetic field, are modeled by three basic **circuit parameters**: *resistance, capacitance,* and *inductance,* respectively. These basic circuit parameters can be physically realized by means of three **circuit elements**: *resistors, capacitors,* and *inductors,* respectively. However, it was found that combinations of these circuit elements are extremely useful not just for modeling but also for **signal processing**, that is, modifying, in a prescribed manner, the magnitude and time course of current or voltage signals. The operation of almost all electronic equipment depends on the generation and processing of electric signals using the aforementioned basic circuit elements in conjunction with other passive devices, such as transformers and diodes, and active devices such as transistors in the form of discrete components or integrated circuits (ICs). Hence, electric circuits and elements not only serve as models of electromagnetic field phenomena but are also extensively employed on their own merit for some highly useful purposes.

Circuit theory is concerned with the behavior of interconnected circuit elements. It has two main subdivisions: circuit analysis and circuit synthesis or design. In circuit analysis, the behavior of a given combination of circuit elements is determined. The objective of circuit synthesis is to design circuits that behave in a desired manner.

Relevance of Electric Circuits

Circuit theory, in terms of both circuit analysis and synthesis, is an important foundation of electrical engineering and its allied fields. Practically all electrical and electronic "hardware" involves electric circuit elements or is described in terms of circuit parameters in some form or other. The ubiquitous printed circuit that is found in all kinds of electronic, computer, control, or communications equipment is almost invariably replete with resistors, capacitors, and ICs. Even in the case of an IC composed entirely of transistors, equivalent circuits involving resistance, capacitance, and dependent sources are used to model transistor behavior. The interconnections between transistor elements on the IC may have to be modeled by resistance and capacitance. Power systems, whether in the form of electric machinery, power transmission, or distribution, invariably involve electric circuit elements or models. In all cases, the understanding of the behavior of signal-processing or modeling elements is based on circuit analysis, and the modification of system behavior in some desired manner involves circuit design.

Because of its importance, circuit theory has been highly developed over the years. This made it a valuable tool for analysis and design of many types of nonelectrical systems that may be modeled by electric circuits. Examples of such systems are mechanical, fluid, thermal, and ionic systems (Chapter 19). It is not unusual in the analysis and design of these systems to work in terms of the equivalent electric circuit model, making use of the powerful circuit analysis and synthesis techniques and computational tools available for electric circuits.

Units, Symbols, Acronyms, and Abbreviations

The International System of Units

The International System of Units, referred to as the SI system (its acronym in French), is almost universally used for scientific and engineering purposes and is adopted throughout this book. It is a system of physical units in which the fundamental quantities are the seven listed in Table 1, together with their corresponding units and symbols of these units.

In addition, circuit theory uses many units that are derived from these fundamental units. Table 2 lists the more common of these derived units together with their relations to other units and their expressions in terms of other units.

In many practical problems, the SI units defined below are either too large or too small. Standard prefixes in powers of ten are applied in order to bring the numeral preceding the power of ten to a convenient value, generally between 1 and 10. Table 3 lists the prefixes associated with the SI system of units. The prefixes *centi, deci, deka,* and *hekto* are

TABLE 1

SI System of Units

Fundamental Quantity	Unit	Symbol
Length	meter	m
Time	second	s
Mass	kilogram	kg
Electric current	ampere	A
Temperature	degree Kelvin	K
Luminous intensity	candela	cd
Amount of substance	mole	mol

TABLE 2

SI-Derived Units

Quantity	Unit	Symbol	Relation to Other Units	Expressions
Frequency	hertz	Hz	—	s^{-1}
Angular frequency	radians per second	rad/s	$2\pi \times$ (frequency)	s^{-1}
Energy or work	joule	J	Force × distance	N.m
Power	watt	W	Energy/time	J/s or A.V
Electric charge	coulomb	C	Current × time	A.s
Electric potential difference (voltage)	volt	V	Power/current	V
Electric resistance	ohm	Ω	Voltage/current	V/A
Electric conductance	siemens (or mho)	S	Current/voltage	A/V
Electric capacitance	farad	F	Charge/voltage	A.s/V
Magnetic flux	weber	Wb	Voltage × time	V.s
Magnetic flux linkage	weber-turn	Wb-turns	(magnetic flux) × (number of turns)	V.s
Inductance	henry	H	Flux linkage/current	V.s/A

TABLE 3

Power of Ten Prefixes Used with the SI System

Prefix	Symbol	Power
atto	a	10^{-18}
femto	f	10^{-15}
pico	p	10^{-12}
nano	n	10^{-9}
micro	μ	10^{-6}
milli	m	10^{-3}
centi	c	10^{-2}
deci	d	10^{-1}
deka	da	10^{1}
hecto	h	10^{2}
kilo	k	10^{3}
mega	M	10^{6}
giga	G	10^{9}
tera	T	10^{12}
peta	P	10^{15}
exa	E	10^{18}

not used with electrical quantities. The remaining prefixes progress in powers of ten that are divisible by three.

Current and Voltage Symbols

- Capital letter with capital subscript denotes dc, or average, quantity. Example: V_O.
- Capital letter with lowercase subscript denotes rms value of an alternating quantity, its Fourier transform, or its Laplace transform. In some cases, the capital subscript is used, as when referring to a circuit element or when the arguments s or $j\omega$ are explicitly used. Examples: I_o, $V_i(j\omega)$, $I_C(s)$, $V_{Th}(s)$, $V_{SRC}(j\omega)$.
- Lowercase letter with capital subscript denotes a total instantaneous quantity. Example: v_{SRC}.
- Lowercase letter with lowercase subscript denotes a small signal of zero average value. Example: i_y.
- Double subscript in a voltage symbol denotes a voltage drop from the node or terminal designated by the first subscript to the node or terminal designated by the second subscript. Example: v_{ab}.
- Double subscript in a current symbol denotes a current flowing from the node or terminal designated by the first subscript to the node or terminal designated by the second subscript. Example: i_{ab}.
- Boldface, not italicized, denotes a phasor. Example: $\mathbf{V_b}$.

Main Acronyms and Abbreviations

ac	Alternating current
ADC	Analog-to-digital converter
BW	3-dB bandwidth
CCCS	Current-controlled current source
CCVS	Current-controlled voltage source
CMR	Common-mode rejection ratio expressed in decibels
CMRR	Common-mode rejection ratio
dB	Decibel
dc	Direct current
emf	Electromotive force
FSD	Full-scale deflection
FSE	Fourier series expansion
FT	Fourier transform
IA	Instrumentation amplifier
IC	Integrated circuit
i-f	Intermediate frequency
IFT	Inverse Fourier transform
ILT	Inverse Laplace transform
Im	Imaginary part of a complex quantity
KCL	Kirchhoff's current law
KVL	Kirchhoff's voltage law
LHS	Left-hand side (of an equation)
LTI	Linear, time-invariant
mmf	Magnetomotive force
NEC	Norton's equivalent circuit
p.f.	Power factor
PFE	Partial fraction expansion
Re	Real part of a complex quantity
rf	Radio frequency
RHS	Right-hand side (of an equation)
rms	Root mean square
SI	Système International — International System of Units
SR	Slew rate
TEC	Thevenin's equivalent circuit
VA	Volt-ampere
VAR	Volt-ampere reactive
VCCS	Voltage-controlled current source
VCVS	Voltage-controlled voltage source

1

Circuit Variables and Elements

Overview

Physically, an electric circuit is an interconnection of circuit elements of various types, such as sources of electric energy, resistors, capacitors, and inductors. In the presence of sources, current flows in different parts of the circuit, and voltages appear at the terminals of circuit elements. The circuit is described by a circuit diagram that specifies how the circuit elements are interconnected and on which are shown the values assigned to these elements. For each type of circuit element, voltage and current are related in a particular way that is characteristic of the given type. In the case of ideal resistors, for example, the voltage across a resistor and the current through it are related by Ohm's law.

A discussion of electric circuits logically begins with the definitions of the two basic circuit variables, namely current and voltage, and their relation to electric energy and power. It is emphasized in this discussion that the two fundamental laws of conservation of charge and conservation of energy must be satisfied in any valid electric circuit. Ideal circuit elements are also considered in this introductory chapter. Voltage sources, current sources, and resistors are presented in some detail. Capacitors and inductors are only briefly introduced at this stage in preparation for Chapter 5 on the sinusoidal steady state. A discussion of their general behavior is postponed to Chapter 11. The chapter ends with concluding remarks on the assumptions made in basic circuit theory.

Learning Objectives

- To be familiar with:
  - The general nature of electric resistance in metals
  - The nature of the assumptions made in basic circuit theory
- To understand:
  - The basic circuit concepts of current, voltage, and power
  - How power absorbed or power delivered by a circuit element is related to the assigned positive directions of current through the element and voltage drop or rise across the element
  - The attributes of independent and dependent, ideal voltage sources and ideal current sources

- The attributes of an ideal resistor, Ohm's law, and the power dissipated in a resistor
- The attributes of an ideal capacitor and an ideal inductor, their voltage–current relations, and the energy they store

1.1 Electric Current

Definition: *Current is the rate of flow of electric charge.*

To illustrate this definition and derive an expression for current, consider a stream of charged particles moving with a velocity u m/s, each particle carrying a positive charge of e coulombs (Figure 1.1.1). For simplicity, it is assumed that the stream is of uniform cross-sectional area A m^2, and that the concentration of particles and their velocity are uniform throughout. In an infinitesimal time dt, the charge dq that crosses a given plane xx′ is that contained in the disk of volume $Audt$ m^3. Multiplying this volume by the concentration n of the particles per cubic meter, and by e, gives $dq = Aunedt$. According to the preceding definition, the current i equals dq/dt. Hence,

$$i = \frac{dq}{dt} = Aune \qquad (1.1.1)$$

When charge is expressed in coulombs and time in seconds, current is in amperes (A). The following should be noted:

1. The dimensions of the quantities in Equation 1.1.1 are:

$$A = \frac{C}{s} = m^2 \cdot \frac{m}{s} \cdot \frac{particles}{m^3} \cdot \frac{C}{particle}$$

Quantities on the RHS cancel out between numerators and denominators, leaving C/s. It is helpful to check the dimensions of all defining relations so as to gain a better appreciation of the quantities involved.

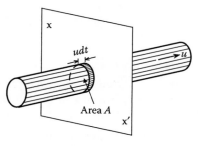

FIGURE 1.1.1
Flow of electric charges in a conducting medium.

2. Unlike charge, which has only magnitude and sign, current is, in general, a vector quantity that has both magnitude and direction. By convention, and for purely historical reasons, *the direction of current is considered as that of motion of positive electric charges.* In Equation 1.1.1, e is positive, so that i has the same sign as u. If the particles are negatively charged, e is negative, and the direction of i is opposite that of u.

3. In the presence of an electric field, positive and negative charges experience forces in opposite directions. As a result, positive charges move in the direction of the electric field, whereas negative charges move in the opposite direction. However, the direction of current is the same for both positive and negative charges because, in the case of negative charges, the signs of both e and u in Equation 1.1.1 change simultaneously.

Current Carriers

In metals, current carriers are predominantly conduction electrons, which have sufficient energy to detach from their parent atoms, becoming free electrons within the crystal. These electrons are able to acquire a mean drift velocity in the presence of an applied electric field, as explained more fully in Section 1.7. In some solids, such as semiconductors, current can also be carried by what are effectively positive charges, or holes. In conducting (or electrolytic) solutions, conduction is by means of positively charged ions (cations) or negatively charged ions (anions). In gases, current carriers could be positively charged ions, negatively charged ions, or electrons.

Example 1.1.1: Drift Velocity in a Copper Wire
A copper wire 2 mm in diameter has 8.4×10^{28} conduction electrons/m³ and carries a current of 10 A. It is required to determine the mean drift velocity of these electrons, assuming the charge per electron is -1.6×10^{-19} C .

SOLUTION

The cross-sectional area A of the wire is $A = \frac{\pi}{4} d^2 = \frac{\pi}{4}(2 \times 10^{-3})^2 = 3.14 \times 10^{-6}$ m². Assuming that the current is positive and uniformly distributed across the cross-section of the wire, it follows from Equation 1.1.1 that

$u = \dfrac{10}{3.14 \times 10^{-6} \times 8.4 \times 10^{28}(-1.6 \times 10^{-19})} = -2.37 \times 10^{-4}$ m/s. The negative sign of velocity signifies that the electrons move in a direction opposite of that of the current.

EXERCISE 1.1.1

A fuse is rated at 1 A. What is the minimum steady rate of flow of electrons per hour that will cause the fuse to blow? Is the direction of current relevant?

Answer: 2.25×10^{22} electrons/h. No, because the fuse blows as a result of the heating effect of the current, which is independent of the direction of current (Section 1.7).

EXERCISE 1.1.2

In a one-dimensional flow of current through a semiconductor, holes move in the positive x-direction at a steady rate of 5×10^{18} holes/min and electrons move in the negative x-direction at a steady rate of 2.5×10^{18} electrons/min. Determine the total current in milliamperes in: (a) the positive x-direction and (b) the negative x-direction. (c) What would be the current if the holes and electrons are moving in the same direction? Note that the positive charge of a hole has the same magnitude as the charge of an electron.

Answer: (a) 20 mA; (b) –20 mA; (c) 20/3 mA in the direction of movement of holes.

1.2   Voltage

Definition: *The voltage, or electric potential difference, between two points is the difference in electric potential energy, per unit positive charge, between these points.*

To illustrate this definition and derive an expression for voltage, consider two conductors A and B separated in the x-direction, A having an excess positive charge $+q$ and B having an excess negative charge $-q$ (Figure 1.2.1). An electric potential difference, or voltage, v is established between the two conductors, with A at a positive electric potential with respect to B, and with an electric field ξ_x in the x-direction, from A to B. Suppose that a positive charge $+dq$ is moved from B to A so that the charge on A becomes $(+q+dq)$ and the charge on B becomes $(-q-dq)$. Work dw is done in moving dq against the attraction of $-q$ on B and the repulsion of $+q$ on A, the work being stored as potential energy in the electric field. As dq is moved, v changes, eventually becoming $v+dv$. Had the voltage remained constant at v while dq was moved, $dw = v\,dq$. Had it become $v+dv$ at the beginning of the movement, $dw = (v+dv)dq$. Hence, $v\,dq < dw < (v+dv)dq$. Neglecting second-order infinitesimals, $dw = v\,dq$, or

$$v = \frac{dw}{dq} \tag{1.2.1}$$

The following should be noted:

1. The unit of voltage is the volt (V). If a charge of +1 C is moved to a region whose voltage is 1 V higher, the increase in electric potential energy is 1 J.

2. Similar to potential energy, in general, voltage does not have an absolute zero reference. It is always measured with respect to some arbitrary zero reference.

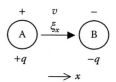

FIGURE 1.2.1
Electric field and potential difference between two charged conductors.

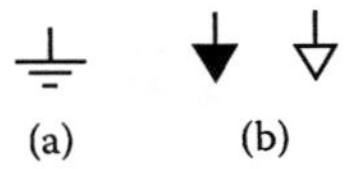

(a)   (b)

FIGURE 1.2.2
(a) Ground symbol, (b) voltage reference symbols.

This is quite acceptable because voltages with respect to the same arbitrary reference are of interest in practice.

3. The Earth is such a huge conductor that its voltage is not affected by any current flow to or from Earth. Hence, it is common practice to assign a zero voltage to Earth and to measure voltages with respect to this zero reference, just as altitude is measured with respect to a zero at sea level. A conductor that is electrically connected to Earth is said to be **grounded**; the circuit symbol for a ground connection is shown in Figure 1.2.2a.

4. Electric circuits may or may not be grounded. Portable equipment, for example, is not normally connected to ground. However, voltages in a circuit are generally specified with respect to a particular reference terminal or point in the circuit, which may or may not be grounded. A PSpice simulation does not run unless a ground is specified *and* assigned a voltage of zero. A circuit that is not grounded at any point is said to be **floating**. Figure 1.2.2b shows commonly used symbols for a voltage reference other than ground.

EXERCISE 1.2.1

Electrons are emitted from a heated metal plate A at a constant rate of 6.25×10^{14} electrons/s, with zero kinetic energy. They are accelerated toward a parallel metal plate B that is separated from A by 5 mm. Plates A and B are connected to an external power supply that maintains B at a constant voltage of $+10$ V with respect to A. (a) How much potential energy does an electron gain or lose in going from A to B? (b) What happens to this potential energy? (c) If the mass of the electron is 9.1×10^{-31} kg, what is the velocity of the electron when it arrives at B?

Answer: (a) Loses potential energy of 1.6×10^{-18} J; (b) it is converted to kinetic energy; (c) 1.88×10^{6} m/s.

1.3 Electric Power and Energy

Definition: *Power, p, is the rate at which energy, w, is delivered or absorbed,*

that is, $p = \dfrac{dw}{dt}$.

Using Equation 1.1.1 and Equation 1.2.1, it follows that:

$$p = \frac{dw}{dt} = \frac{dw}{dq}\frac{dq}{dt} = vi \qquad (1.3.1)$$

If *i* is expressed in amperes and *v* in volts, *p* is in watts (W), or J/s.

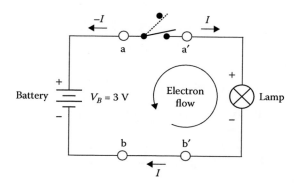

FIGURE 1.3.1
Flashlight circuit.

To interpret Equation 1.3.1, consider a simple flashlight circuit consisting of a 3 V battery, a lamp that draws a current of 0.25 A from the battery, and a switch that makes or breaks the connection between the lamp and battery (Figure 1.3.1). When the switch is closed, the battery impresses a voltage of 3 V across the metallic filament of the lamp, with terminal a′ at a positive voltage with respect to terminal b′, as indicated by the plus and minus signs in the figure. The conduction electrons in the filament, being negatively charged, have a higher electric potential energy at b′ than at a′. The force of the electric field will therefore drive them down an electric potential energy difference from b′ to a′ through the filament. This loss of electric potential energy is converted to heat in the filament, causing it to glow and emit light.

In the battery, the chemical reactions of the battery expend energy to move the electrons up an electric potential energy difference from a to b, so they can again flow down an electric potential energy difference from b′ to a′ through the filament, and so on. As long as the switch is closed, electrons flow continuously in the circuit in the counterclockwise direction in Figure 1.3.1. Current always flows in a continuous loop, as illustrated in the figure, in accordance with conservation of charge. The battery acts like a pump that circulates water by raising it to a higher level so that it can perform useful work while falling back under gravity.

In circuit terms, the current I is conventionally taken as the flow of positive charge, so I is considered to flow in the direction opposite that of the flow of electrons, as indicated in Figure 1.3.1. If the 3 V battery voltage is denoted by V_B and the 0.25 A current by I, then $P = V_B I = (3 \text{ V}) \times (0.25 \text{ A}) = 0.75 \text{ W}$. This is the power delivered by the battery and absorbed by the lamp, in accordance with conservation of energy in the whole circuit. The distinction between power delivery and power absorption in terms of the direction of power flow is explained in the next section.

It is common practice to describe a nominally steady, or constant, voltage or current as dc, although dc stands for direct current. dc quantities, such as the battery voltage in Figure 1.3.1 and the resulting current, are denoted by capital letters, with capital subscripts, if required. Lowercase letters with capital subscripts denote instantaneous values, which generally vary with time (see Units, Symbols, Acronyms, and Abbreviations). Thus, p, v, and i in Equation 1.3.1 are instantaneous quantities.

Example 1.3.1: Flashlight Circuit

If the current is 0.25 A in the circuit of Figure 1.3.1, how many electrons, on average, pass through the lamp per second? What is the electric potential energy of an electron at a relative to that at b? How much energy does the battery deliver in 1 h?

SOLUTION

A current of +0.25 A in the direction shown corresponds to an electron flow through the lamp of −0.25 C/s. Because the charge of the electron is −1.6×10⁻¹⁹ C, it follows that

$$\frac{-0.25 \text{ C}}{\text{s}} \Big/ \frac{-1.6\times10^{-19} \text{ C}}{\text{electron}} = 15.6\times10^{17} \text{ electrons/s flow through the lamp, on average.}$$

The electric potential energy difference is the charge times the voltage. Considering the electric potential energy at b to be at zero reference, the electric potential energy of an electron at a is $(-1.6\times10^{-19} \text{ C})\times(3 \text{ V}) = -4.8\times10^{-19}$ J. Because it is negatively charged, the electron has greater potential energy at b than at a.

Because $P = 0.75$ W, the energy delivered in 1 h is $(0.75 \text{ W})\times(3600 \text{ s/h}) = 2700$ J/h.

EXERCISE 1.3.1

Consider the case described in Exercise 1.2.1. (a) What is the total kinetic energy of the electrons that arrive at B during 1 second? (b) The accelerated electrons are collected at B, where they flow through plate B. What is the magnitude and direction of current in the space between the plates? (c) What happens to the kinetic energy of the electrons once they are collected at B? (d) How much power is expended by the power supply to keep the voltage between A and B at 10 V? (e) How is this power related to the kinetic energy given up by the electrons?

Answer: (a) 1 mJ; (b) 100 μA, directed from B to A; (c) converted to heat; (d) the power supply has to remove charge from B because of electrons being collected at a rate of 100 μA. In doing so, it expends a power of 1 mW; (e) the power is equal to the rate at which kinetic energy is given up per second.

Example 1.3.2: Discharge of Battery

An A-size, 1.5 V battery is rated at 3 Ah. During continuous use, with the battery supplying a current of 100 mA, the battery voltage stays substantially constant at 1.5 V for the first 20 hours. During the next 10 hours, the voltage drops linearly to 1.25 V, whereas the current drops linearly to 80 mA (Figure 1.3.2). At this point, the battery is no longer considered useful. What was the useful ampere-hour capacity of the battery, and how much energy was delivered by the battery during the 30 hours?

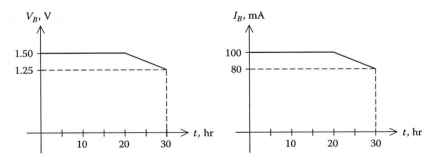

FIGURE 1.3.2
Figure for Example 1.3.2.

SOLUTION

During the first 20 hours, the battery delivers $0.1 \times 20 = 2$ Ah. During the next 10 hours, the current drops linearly with time. The average current during this interval is 90 mA, or 0.09 A. The battery delivers $0.09 \times 10 = 0.9$ Ah during this interval. The battery therefore delivers 2.9 Ah during the 30 hours of operation. In practice, the ampere-hours delivered by a battery depend on the discharge current, on whether the discharge is continuous or intermittent, and if intermittent, on the relative duration of the on and off periods, that is, the *duty cycle*. The rated capacity in ampere-hours quoted by the manufacturer is for specified discharge conditions. Note that, although 1 Ah is 3600 C, battery capacities are specified in ampere-hours rather than coulombs.

To determine the energy delivered by the battery, we again have to consider the first 20 hours and the following 10 hours separately. During the first 20 hours, the battery voltage is constant at 1.5 V, and the current is constant at 0.1 A. The battery delivers a constant power of $1.5 \times 0.1 = 0.15$ W. The energy delivered up to a time t hours is $0.15 \times 3600 \times t = 540t$ J. By the end of this period, the energy delivered is $540 \times 20 = 10{,}800$ J, or 10.8 kJ.

During the interval $20 \le t \le 30$ h, $v = 2 - 0.025t$ V, and $i = 0.14 - 2 \times 10^{-3}t$ A. The instantaneous power p is $p = vi = (2 - 0.025t) \times (0.14 - 2 \times 10^{-3}t) = 0.28 - 0.0075t + 5 \times 10^{-5}t^2$ W, where t is in hours. The energy delivered is:

$$w = \int_{20}^{30} p \, dt = \left[0.28t - \frac{0.0075}{2}t^2 + \frac{5 \times 10^{-5}}{3}t^3 \right]_{20}^{30} = 1.24 \text{ Wh}$$

To convert to joules, this has to be multiplied by 3600 s/h, which gives 4.46 kJ. The total energy delivered by the battery is therefore $10.8 + 4.46 = 15.26$ kJ.

Conservation of Power

Concept: *Conservation of energy implies conservation of power.*

It may be argued qualitatively that energy cannot be conserved in a given system only over some finite interval of time. If this were the case, there would be a net surplus of energy during part of the given interval and an equal net deficiency of energy during the rest of the interval. In principle, it would be possible to activate the system when there is a net surplus of energy and deactivate it when energy deficiency occurs. Energy can then be extracted from the system at no energy cost, in violation of conservation of energy. Hence, energy must be conserved at every instant, which means that power must be conserved. A mathematically based argument is given in Section ST1.1.

1.4   Assigned Positive Directions

In circuit analysis, the actual polarity of voltage across every circuit element, or the direction of current through every circuit element, are generally not known beforehand. Hence, positive directions of unknown voltages or currents are assumed arbitrarily, subject to restrictions imposed by the voltage–current relations of the circuit elements, as will be discussed later. These assumed directions are the **assigned positive directions**. The assigned positive direction of current is indicated by an arrow, whereas the assigned positive direction of voltage is indicated by + and − signs at the points between which the voltage assumes its value. If, after solving the circuit problem and obtaining the numerical values of currents and voltages, the *numerical value* of a current or voltage is found to be positive, the actual direction of the given quantity is the same as the assigned positive direction. If the numerical value is negative, the actual direction is opposite that of the assigned positive direction.

Negative values of current or voltage are therefore perfectly natural in circuit analysis. If $I = 0.25$ A is the current that flows clockwise around the circuit in Figure 1.3.1, then $-I = -0.25$ A flows counterclockwise around the circuit. V_B is assigned as a voltage drop from a to b and could also be denoted as V_{ab}, where, by convention, the *voltage drop* is in the direction from the first subscript to the second. The numerical value of V_{ab} is +3 V. V_{ab} is also a *voltage rise* from b to a. As a voltage drop, $V_{ba} = -V_{ab} = -3V$.

Because currents and voltages in a circuit have direction, then power must also have a direction. This is implicit in Equation 1.3.1 and may be expressed as follows:

> **Concept:** *If current i is in the direction of a voltage drop v, then the power p = vi represents power absorbed by the element through which i flows. Conversely, if i is in the direction of a voltage rise v, then the power p represents power delivered by the element through which i flows.*

This is true of the flashlight circuit of Figure 1.3.1. In going through the lamp from a′ to b′, in the direction of I, there is a voltage drop of 3 V, and power is absorbed. In going through the battery from b to a in the direction of I, there is a voltage rise of 3 V, and power is delivered. In general, when positive charges flow through a voltage drop across a given element, they lose electric potential energy, which means power is being absorbed by the element. Conversely, when positive charges flow through a voltage rise across a given element, they gain electric potential energy; so power is delivered.

Negative power delivered is power that is actually absorbed, and negative power absorbed is power that is actually delivered, as illustrated by Example 1.4.1. Power that is actually absorbed is either dissipated as heat, as in a resistor, or partly converted to another form of energy, as in a lamp or electric motor, or stored as electric or magnetic energy in a capacitor or inductor, respectively.

Example 1.4.1: Assigned Positive Directions of Current and Voltage

Assume that in the case of the flashlight discussed earlier, the battery is contained in a box having unmarked terminals, so we do not know which terminal is positive and which is negative. Suppose we decide to assign the positive direction of V_B as shown in Figure 1.4.1a, that is, assume that a given battery terminal a that is connected to the lamp terminal a′ is positive with respect to the battery terminal b that is connected to the lamp terminal b′. It would then be reasonable to assign the positive direction of I as before, that is, flowing into the lamp at a′.

Suppose that when the current I is measured in the assigned positive direction by means of an instrument called an *ammeter* (Section 7.5, Chapter 7), it is found to be 0.25 A in the direction opposite to that assumed, which means that $I = -0.25$ A. This would seem to indicate that the positive terminal of the battery is b and its negative terminal is a. We confirm this by measuring the battery voltage V_B in the assigned positive direction by means of an instrument called a *voltmeter* (Section 7.5, Chapter 7). So $V_B = -3$ V. From Figure 1.4.1a, the assigned positive direction of I is in the direction of a voltage drop V_B in the lamp and a voltage rise V_B in the battery. Hence, according to Equation 1.3.1, the

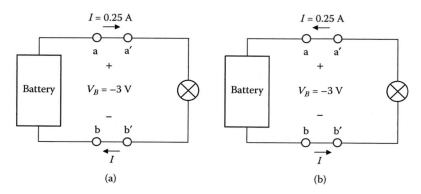

FIGURE 1.4.1
Figure for Example 1.4.1.

power absorbed by the lamp, and that delivered by the battery, is $P = VI = (-3)(-0.25) = 0.75$ W, as before.

Suppose we had assigned the positive directions of V_B and I as shown in Figure 1.4.1b, so I is now in the direction of a voltage rise V_B in the lamp and a voltage drop V_B in the battery. The same aforementioned measurements, in the assigned positive directions, would give $I = 0.25$ A, and $V_B = -3$ V. Equation 1.3.1 now gives $P = VI = (-3)(0.25) = -0.75$ W as the power *delivered* by the lamp and *absorbed* by the battery. The negative sign means that a power of 0.75 W is, in fact, absorbed by the lamp and delivered by the battery, as before.

It is seen that Equation 1.3.1 gives the same value of power absorbed by the lamp or delivered by the battery, *irrespective of the assigned positive directions of v and i,* as long as Equation 1.3.1 is interpreted correctly in terms of actual values, whether measured or calculated.

EXERCISE 1.4.1

An electrically operated car has a 12 V battery rated at 2000 Ah. When cruising at a constant speed of 30 km/h, the car motors draw a total current of 15 A. During charging, a current of 20 A is supplied to the battery. Determine how much power flows, and in what direction, during: (a) cruising; (b) charging of the battery.

Answer: (a) Battery delivers 180 W, motors absorb 180 W; (b) battery absorbs 240 W, charging source delivers 240 W.

Application Window 1.4.1: Electroplating

In electroplating, a metal is deposited on a metallic object by electrolysis, which involves passing an electric current through an electrolytic solution containing ions of the metal to be deposited. The purpose may be decorative, or protective, or to impart some desirable surface characteristic. The object is made the cathode, whereas the anode is usually made up of the metal to be deposited.

The process is illustrated diagrammatically in Figure 1.4.2. During passage of current, a reduction reaction takes place at the cathode. Metal ions in solution gain electrons from the cathode and become metal atoms that are deposited on the object; (metal ion)$^{+z} + zq \rightarrow$ (metal atom), where z is the valence of the metal ion and q is the charge of an electron. An oxidation reaction takes place at the anode, whereby metal ions lose electrons to the anode, becoming metal ions in solution: (metal atom) $\rightarrow$ (metal ion)$^{+z} + zq$.

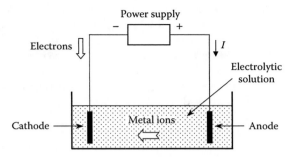

FIGURE 1.4.2
Figure for Application Window 1.4.1.

Under the influence of the power supply connected in the external circuit between the anode and cathode, electrons flow in the power supply from the positive terminal to the negative terminal, up a potential energy gradient, because the cathode is at a negative voltage with respect to the anode. In terms of the conventional current of positive charges, current flows in the opposite direction. The current loop is completed in the solution by flow of positively charged metal ions from the anode to the cathode and by the flow of negatively charged ions in the opposite direction.

The deposition of a single metal atom requires zq C. The number of atoms in a gram-atomic weight of the metal is given by Avogadro's number (6.025×10^{23}). Hence, the deposition of 1 gram-atomic weight of the metal requires $zq \times 6.025 \times 10^{23}$ C. To make this independent of z, a gram-equivalent weight is defined as a gram atomic weight divided by the valence. The deposition of 1 gram-equivalent weight of any metal therefore requires $|q| \times 6.025 \times 10^{23} = 1.602 \times 10^{-19} \times 6.025 \times 10^{23} \cong 96,500$ C. This quantity is **Faraday's constant**.

Suppose that the metal to be deposited is silver, having a valence of 1 and a gram-atomic weight of 0.1079 kg. A current of 10 A deposits in 1 hour, a mass m of silver, given by:

$$m = \left(\frac{10 \text{ C}}{\text{s}} \right) \times \left(\frac{3600 \text{ s}}{\text{h}} \right) \times \left(\frac{\text{gram-equivalent weight}}{96,500 \text{ C}} \right) \times \left(\frac{\text{gram-atomic weight}}{\text{gram-equivalent weight}} \right) \times$$

$$\left(\frac{0.1079 \text{ kg}}{\text{gram-atomic weight}} \right) = 0.0403 \text{ kg/h} \equiv 40.3 \text{ g/h}.$$

1.5 Active and Passive Circuit Elements

Definition: *Active devices can generate electrical energy through conversion from another source of energy, whereas passive devices cannot.*

Examples of active devices are: (1) batteries, which convert chemical energy to electrical energy, (2) solar panels, which convert solar energy to electrical energy, (3) electromechanical generators, which convert mechanical energy to electrical energy, and (4) electronic devices such as transistors, which accept input signals at a low power level and output them at a higher power level, the difference coming from a dc bias source. Active devices are represented in electric circuits by voltage or current sources (Section 1.6). Examples of passive circuit elements are resistors, capacitors, and inductors.

Note that active devices were not defined as those that deliver energy, nor passive devices as those that absorb energy. This is because some active devices can, under certain conditions, absorb energy. A rechargeable battery, for example, absorbs electrical energy during recharging; an electromechanical generator can absorb electrical energy and act as a motor. Similarly, capacitors and inductors can deliver electrical energy that has been previously stored in them.

1.6 Voltage and Current Sources

> **Definition:** *An ideal voltage source maintains a specified voltage across its terminals irrespective of the current through the source.*

The following should be noted:

1. The voltage–current characteristic of an ideal voltage source is shown in Figure 1.6.1a, where v_{SRC} is the specified source voltage. It should be emphasized that this is not a plot of v_{SRC} against time but against the current i through the source. As a function of time, v_{SRC} could be constant, as for a dc source such as a battery or a dc power supply. It could be a sinusoidal function of time, as for an alternator, or it could be any specified function of time, such as a rectangular waveform, a triangular waveform, or any other waveform provided by a function generator.
2. The current i through the voltage source is not determined exclusively by the source but depends on the rest of the circuit to which the source is connected.

In an **independent voltage source**, v_{SRC} is specified independently of any other voltage or current in the circuit. The source is denoted by the circle symbol with + and − signs that indicate the assigned positive direction of v_{SRC} (Figure 1.6.1b). In a **dependent voltage source**, denoted by the diamond symbol (Figure 1.6.1c), v_{SRC} depends on another voltage or current in the circuit. It is designated as a **voltage-controlled voltage source** (VCVS) or a **current-controlled voltage source** (CCVS) according to whether v_{SRC} depends on another voltage or current, respectively. This dependency is usually a direct proportionality of the form $v_{SRC} = \alpha v_\phi$ or $v_{SRC} = \rho i_\phi$, where v_ϕ and i_ϕ are, respectively, a voltage or a current in the circuit, and α and ρ are constants.

The case of an ideal current source is exactly analogous to that of an ideal voltage source, with the roles of current and voltage interchanged.

> **Definition:** *An ideal current source maintains a specified current through it irrespective of the voltage across its terminals.*

The voltage–current characteristic of an ideal current source is shown in Figure 1.6.2a, where i_{SRC} is the specified source current. For a given i_{SRC}, v depends on the rest of the circuit to which the current source is connected.

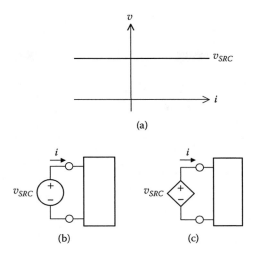

FIGURE 1.6.1
Ideal voltage sources.

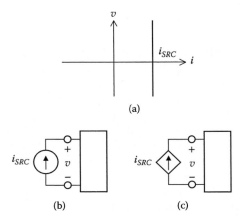

FIGURE 1.6.2
Ideal current sources.

In an **independent current source**, i_{SRC} is specified independently of any other voltage or current in the circuit. The source is denoted by the circle symbol, with an arrow that indicates the assigned positive direction of i_{SRC} (Figure 1.6.2b). i_{SRC} could be a constant or it could be any specified function of time, such as a sinusoidal function.

In a **dependent current source**, denoted by the diamond symbol (Figure 1.6.2c), i_{SRC} depends on another voltage or current in the circuit. It is designated as a **voltage-controlled current source** (VCCS) or a **current-controlled current source** (CCCS) according to whether i_{SRC} depends on another voltage or current, respectively. This dependency is usually a direct proportionality of the form $i_{SRC} = \sigma v_\phi$ or $i_{SRC} = \beta i_\phi$, where v_ϕ and i_ϕ are, respectively, a voltage or a current in the circuit, and σ and β are constants.

In a linear circuit, dependent sources can only be some linear function of other voltages or currents in the circuit. For example, a dependent voltage source can be a linear combination of a VCVS and CCVS: $v_{SRC} = \alpha v_\phi + \rho i_\phi$, but $v_{SRC} = \alpha v_\phi^2$ is not allowed.

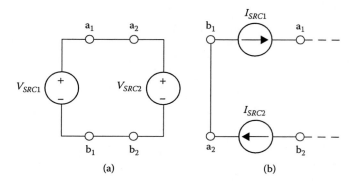

FIGURE 1.6.3
Figure for Example 1.6.1.

Interconnection of Sources

> **Concept:** *A connection of sources that violates conservation of energy or conservation of charge is invalid and is not allowed in electric circuits.*

This is elaborated in Example 1.6.1.

Example 1.6.1: Invalid and Valid Source Connections

The connection of two ideal voltage sources in parallel (Figure 1.6.3a) is not valid when $V_{SRC1} \neq V_{SRC2}$, because it violates conservation of energy. To show this, let a charge q be moved around the circuit, say clockwise, starting with terminal b_1 and assuming that V_{SRC1} and V_{SRC2} have positive numerical values. In taking q through V_{SRC1}, from b_1 to a_1, the work done on the charge is qV_{SRC1}, and the electric potential energy increases by this amount. In taking q through V_{SRC2}, from a_2 to b_2, the work done by the charge is qV_{SRC2}, and the electric potential energy decreases by this amount. In going around the circuit, therefore, there is a net gain or loss of energy of $q|V_{SRC1} - V_{SRC2}|$, in violation of conservation of energy, unless $V_{SRC1} = V_{SRC2}$.

The connection of two current sources in series as in Figure 1.6.3b is not valid when $I_{SRC1} \neq I_{SRC2}$, because it violates conservation of charge. If we consider the connection a_2b_1 between the two sources, then the charge entering this connection per second is I_{SRC2}, whereas the charge leaving this connection per second is I_{SRC1}. There is a net gain or loss of charge of $|I_{SRC1} - I_{SRC2}|$ per second, in violation of conservation of charge, unless $I_{SRC1} = I_{SRC2}$.

Consider an ideal voltage source of 12 V to be connected to an ideal current source of 2 A, as in Figure 1.6.4. Such a connection is valid. The current source forces a current of 2 A through the voltage source, and the voltage source impresses a voltage of 12 V across the current source. Because the current through the voltage source is in the direction of a voltage rise across the source, the voltage source delivers a power $P = (12\ \text{V}) \times (2\ \text{A}) = 24\ \text{W}$. The current through the current source is in the direction of a voltage drop across the source, so this source absorbs the same power $P = (12\ \text{V}) \times (2\ \text{A}) = 24\ \text{W}$. If the polarity of either source is reversed, the current source delivers power, and the voltage source absorbs power.

EXERCISE 1.6.1

A 10 V independent source K is connected to a CCVS H and to a VCCS M as in Figure 1.6.5. (a) How much is I and what determines it? (b) What is the voltage across

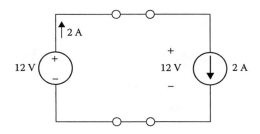

FIGURE 1.6.4
Figure for Example 1.6.1.

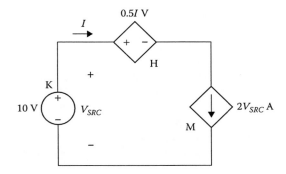

FIGURE 1.6.5
Figure for Exercise 1.6.1.

source H? (c) What is the power delivered or absorbed by sources K, H, and M? (d) What is the voltage across source M?

Answer: (a) I is determined by the current source and equals 20 A; (b) 10 V; (c) source K delivers 200 W, and source H absorbs 200 W. By conservation of power, source M neither absorbs nor delivers any power; (d) 0 V.

Simulation Example 1.6.2: Interconnected Sources

A 20 V independent voltage source, a 10 A independent current source, and a VCCS are connected as shown in Figure 1.6.6. It is required to determine the power absorbed or delivered by each source and to simulate the circuit.

SOLUTION
Each source has one terminal connected to point a and the other terminal connected to point b. The voltage V_{ab} therefore appears across the three sources. This voltage is 20 V, as determined by the voltage source. Considering the independent current source, the source current is in the direction of a voltage rise V_{ab}. It follows that this source delivers a power $P_3 = (20) \times (10) = 200$ W. The current of the dependent current source is $I_2 = 0.8V_{ab}$ = 16 A. This current is in the direction of a voltage drop V_{ab}. Hence, this source absorbs a power $P_2 = (20) \times (16) = 320$ W.

To determine the power absorbed or delivered by the 20 V source, we can invoke conservation of power. Instead, we will determine the current I_1 through this source from conservation of charge. Charge enters junction a from the 10 A source at the rate of 10 C/s and leaves this junction through the dependent current source at the rate of 16 C/s. Conservation of charge requires that charge enters junction a from the 20V source at the

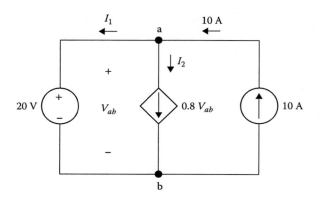

FIGURE 1.6.6
Figure for Example 1.6.2.

rate of 6 C/s. That is, the current from this source toward junction a is 6 A. Because this current is in a direction opposite to the assigned positive direction of I_1, then $I_1 = -6$ A. Moreover, I_1 is in the direction of a voltage drop V_{ab}. Hence, the 20 V source absorbs a power $P_1 = (20) \times (-6) = -120$ W. This means that the source actually delivers 120 W. The total power delivered by the two independent sources is 320 W, and the power absorbed by the dependent source is 320 W, as required by conservation of power.

SIMULATION

Appendix SD.2 explains the basic mechanics of PSpice simulation and provides a convenient reference for some general features of PSpice. In the schematic of Figure 1.6.7, the independent sources are VDC and IDC from the source library, and the VCVS is the G source from the analog library. Dependent sources are simulated as two-port (that is, four-terminal) devices. They have an input (or control port) and an output (or source port). Note the polarities of the current and the controlling voltage. To enter the value of the dependent source, click on its symbol to display the Property Editor window. Move the cursor to the cell under Gain and enter 0.8. To display this value, click on the Display tab, with the cell containing 0.8 selected. Choose Value Only under Display Format.

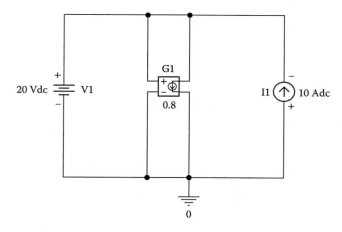

FIGURE 1.6.7
Figure for Example 1.6.2.

In the simulation profile, choose Bias Point/General Settings and run the simulation by selecting PSpice/Run. In the Capture page, press the V button in the lowest toolbar to display the voltages, press the I button to display the currents, and press the W button to display power. Note the signs of current, power absorbed, and power delivered in PSpice (Appendix SD.2). Compare with the preceding solution.

Comments on Sources

1. Dependent sources are most commonly encountered in the equivalent circuits of electronic devices such as transistors and operational amplifiers (Chapter 18). Voltage sources are more commonly encountered than current sources, the latter usually being derived from the former using electronic devices.

2. Ideal sources are circuit abstractions. An ideal voltage source maintains a specified voltage across its terminals and can deliver or absorb any current. An ideal current source supplies a specified current irrespective of the magnitude or polarity of the voltage across the source terminals. Practical voltage sources deviate from the ideal in that the voltage across the source terminals normally decreases with increasing current delivered by the source. Similarly, practical current sources deviate from the ideal in that the current they deliver decreases with increasing voltage across the current source.

3. Practical voltage sources have limitations on the current or power that they can supply, which are referred to as **current rating** or **power rating** of the source. Similarly, practical current sources have limitations on the voltage and power that they can handle.

4. Some practical sources cannot absorb power. Dry batteries and mercury batteries, for example, are not rechargeable and are damaged by trying to force a reverse current through them.

5. A CCVS whose controlling current is the source current itself violates the definition of an ideal voltage source in that the source voltage is not independent of the source current. Similarly a VCCS whose controlling voltage is the voltage across the source itself violates the definition of an ideal current source. However, these very special sources can be replaced by resistors (Fig. 4.2.2) and should properly be regarded as such.

1.7   The Resistor

The Nature of Resistance

As far as electrical conduction is concerned, materials fall into a broad spectrum, at one end of which are the good conductors, which pass a substantial current when subjected to an applied voltage; at the other end are the good insulators, which pass an insignificant current. Examples of good conductors are metals such as silver (the best conductor), copper, and aluminum, which are the most widely used commercially. Examples of good insulators are mineral oil, quartz, and many types of ceramics and plastics. A rigorous discussion of the nature of electrical conduction involves some advanced concepts of solid-state physics and is well beyond the scope of the present discussion. It is sufficient for our purposes to consider a highly simplified view of electrical conduction in metals.

> **Concept:** *Electrical resistance is due to impediments to the flow of conduction electrons under the influence of an applied electric field.*

These impediments can be considered to be due to 'collisions' between conduction electrons accelerated by the applied electric field and metal atoms of the crystal, which vibrate about their rest positions with an amplitude that increases with temperature. As a result of these collisions, (1) electrons lose kinetic energy. But because their velocities increase between collisions, the net effect is that they acquire a mean **drift velocity** that is superimposed on their much larger random thermal velocities; and (2) the metal atoms gain kinetic energy so they vibrate with a larger amplitude. This is reflected as an increase in the temperature of the metal, referred to as **Joule heating**.

The electrical **resistance** R of a conductor is the ratio of the voltage v applied to the conductor to the resulting current i in the conductor. Thus,

$$R = \frac{v}{i} \tag{1.7.1}$$

When v is expressed in volts and i in amperes, R is in ohms and is denoted by the capital omega symbol Ω.

The reciprocal of the resistance R is the **conductance** G:

$$G = \frac{i}{v} \tag{1.7.2}$$

When v is expressed in volts and i in amperes, G is in siemens, denoted by the symbol S. The mho (ohm read backwards) is sometimes used as the unit of conductance and is denoted by the inverted omega symbol $\mho$.

Ohm's Law

It is found that for metals at a given temperature, R is constant over a wide range of v and i. Equation 1.7.1 may be written as

$$v = Ri \tag{1.7.3}$$

which is known as **Ohm's law**.

The circuit element that represents the property of resistance is the ideal resistor. A resistor that obeys Ohm's law has a linear v–i characteristic whose slope is R, as expressed by Equation 1.7.3 and illustrated in Figure 1.7.1a. It is described as a linear resistor and denoted by the symbol shown in Figure 1.7.1b. Because such a resistor dissipates power, the assigned positive direction of current i is *always* in the direction of the voltage drop v across the resistor. With v and i positive quantities, R in Equation 1.7.3 is a positive quantity for a passive resistor.

The following should be noted:

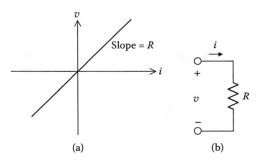

FIGURE 1.7.1
Ideal resistor.

1. The linearity expressed by Equation 1.7.3 implies that v is proportional to i for all values of v and i, positive or negative. A linear resistor is therefore **bilateral**; that is, its resistance is the same for either direction of current through it.

2. A connection of zero resistance, or infinite conductance, is a **short circuit**, whereas a connection of infinite resistance, or zero conductance, is an **open circuit** because there will be no current flow.

3. It follows that a voltage source should not be short-circuited by directly connecting its terminals by a connection of very low resistance. Under these conditions, if v is finite and $R \to 0$, then $i \to \infty$ in accordance with Equation 1.7.1 or Equation 1.7.2. In practice, the current through the source is limited by the internal resistance of the source; nevertheless, unless specially protected, a voltage source is likely to be damaged by a short circuit. Similarly, a current source should not be open-circuited, that is, its terminals left open. Under these conditions, if i is finite and $R \to \infty$ (or $G \to 0$), then $v \to \infty$.

4. Electronic devices and circuits may present a negative resistance. Whereas a positive resistance dissipates energy, and is passive, a negative resistance does the opposite, that is, it delivers electric energy and is therefore an active element.

Power Dissipated

Substituting from Equation 1.7.2 or Equation 1.7.3 in Equation 1.3.1 gives the power absorbed by a resistor:

$$p = vi = Ri^2 = \frac{v^2}{R} = \frac{i^2}{G} = Gv^2 \tag{1.7.4}$$

When v is in volts, i in amperes, R in ohms, and G in siemens, the power dissipated is in watts.

In practice, some units are conveniently used together. For example, if current i is in milliamperes (mA) and resistance R is in kilo-ohms (kΩ), then from Equation 1.7.3,

$$v = (R \text{ k}\Omega)\frac{10^3 \Omega}{\text{k}\Omega} \times (i \text{ mA})\frac{10^{-3} \text{A}}{\text{mA}} = Ri \text{ V}$$

Thus, multiplying current in milliamperes by resistance in kilo-ohms gives the voltage drop in volts. Similarly, multiplying current in microamperes by resistance in mega-ohms gives the voltage drop in volts.

Also, if current i is in mA and resistance R is in kΩ, then from Equation 1.7.4, power p is in mW:

$$p = \left[(i \text{ mA}) \frac{10^{-3} \text{A}}{\text{mA}} \right]^2 \times (R \text{ k}\Omega) \frac{10^3 \Omega}{\text{k}\Omega} = i^2 R \ \text{ mW}$$

Similarly, if current is in microamperes and resistance is in mega-ohms, power is in microwatts.

EXERCISE 1.7.1

A 220 V electric heater is rated at 2 kW and has two identical heating elements. Determine the heater current, resistance, and conductance when (a) one element is switched on; (b) both elements are switched on. How does this compare with part a?

Answer: (a) 4.55 A, 48.4 Ω, 20.7 mS; (b) 9.09 A, 24.2 Ω, 41.3 mS; resistance is halved, current and conductance are doubled.

Temperature Effect

Based on the preceding discussion, it is to be expected that the resistance of a metal increases with temperature because of the increased amplitude of vibration of metal ions. As a result, the probability of collision increases, the mean time between collisions decreases, the drift velocity and current decrease, and the resistance increases. The variation of the resistance of a metal with temperature is given by

$$R_2 = R_1[1 + \alpha_m(T_2 - T_1)] \tag{1.7.5}$$

where R_1 is the resistance of the metal at temperature T_1°C, R_2 is its resistance at a higher temperature T_2°C, and α_m is the temperature coefficient of resistance of the given metal. α_m is always positive for pure metals, and is approximately 0.004/°C for good conductors such as copper and aluminum. It is as low as $\pm$ 0.00001/°C for constantan, an alloy of copper and nickel, with a trace of manganese, which is especially formulated to have a low temperature coefficient. Some nonmetallic materials have a high temperature coefficient that may be positive or negative (Problem P1.3.11).

Application Window 1.7.1: Incandescent Lamp

Incandescent lamps have filaments made of tungsten, a metal that has a high melting point (3400°C) and a low vapor pressure so that it does not rapidly vaporize under vacuum at the high temperatures that cause the filament to glow (about 2500°C). A 220 V lamp that is rated at 100 W has an operating current of (100 W)/(220 V) = 0.45 A and a resistance of $((220)^2/100) = 484 \ \Omega$, at 2500°C. Assuming the temperature coefficient of tungsten to be

0.0045/°C, then from Equation 1.7.5, the resistance R_1 at room temperature (20°C) is given by $R_1 = \dfrac{484}{1+0.0045(2500-20)} \cong 40\ \Omega$. This means that when voltage is applied to a cold filament, the initial current is about 5.5 A, which is more than 12 times the operating current. Hence, there is a relatively large "inrush" current in a cold filament that produces a substantial thermal shock and must be allowed for in the design of the lamp and the switch that is used to turn the lamp on.

Application Window 1.7.2: Current Ratings of Conductors

A copper conductor of 1.8 mm diameter has a current rating, or current-carrying capacity (also known as **ampacity**) of about 18 A and a resistance of 6.6 mΩ/m at room temperature. If a 50 m length of this wire is used in an electrical installation, and is carrying 15 A, it is required to determine: (a) the heat generated by the wire and (b) the voltage drop along the wire.

SOLUTION

(a) The total resistance of 50 m length of the wire is 50 × 6.6 = 330 mΩ ≡ 0.33 Ω. The heat generated is $I^2R = 0.33 \times (15)^2 \cong 75$ W.

(b) The voltage drop along the wire is, from Ohm's law: $RI = 0.33 \times 15 \cong 5$ V.

Although the preceding calculations are quite simple applications of heating in a conductor, and of Ohm's law, they have important practical implications. Conductors supplying electric power have specified current ratings that are based primarily on the heating effect of current in the conductor. The effect of excessive heating of the conductor is twofold: (1) If the temperature rise is sufficient to bring any surrounding flammable material to its ignition point, fire would result. (2) For every type of insulation, there is a maximum safe temperature beyond which the insulation is permanently damaged and loses its functional properties. Proper insulation presents a very high resistance that effectively isolates "live" conductors at different voltages from one another and from metallic enclosures. When this insulation is damaged, large, uncontrolled currents can flow, which pose a fire hazard, and metallic enclosures can acquire a dangerously high voltage, which poses a risk of electric shock (Section ST8.6, Chapter 8). For these reasons, electric conductors in installations are protected from excessive current by circuit protection devices, such as fuses or circuit breakers.

The current rating of a conductor, if not exceeded, ensures that the temperature of the conductor remains within safe limits. However, even though the 1.8 mm conductor is adequate for the 15 A load, the length used may cause an excessive voltage drop, as calculated in (b), depending on the voltage in the installation and the type of load supplied by the conductor. An excessive voltage drop in the installation wiring may reduce the voltage supplied to loads to less than the rated value, which is undesirable. A 5% drop in the voltage supplied to an incandescent lamp reduces the light output by about 15% and makes the light more reddish. Fluorescent lights are not affected in this manner, but a reduced supply voltage may cause starting problems and may shorten the life of the lamp. The effect of a 5% drop in the supply voltage V to a heating load of resistance R is to reduce the heat power generated by about 10%, as $(0.95\ V)^2/R \cong 0.9\ V^2/R$. A reduced voltage supply to an electric motor reduces the maximum torque produced by the motor and increases the motor current for a given power output. Hence, to reduce the voltage drop in the aforementioned case, a conductor of larger cross-sectional area, and hence lower resistance, is used (Section 2.4, Chapter 2). Electrical codes specify a maximum recommended voltage drop for a given type of load.

It may be noted that in wiring installations, voltages and currents are almost invariably sinusoidal alternating current (ac), not dc. However, as discussed in Section 5.4 of Chapter 5, the aforementioned relations derived involving voltage, current, and power apply to sinusoidal quantities when root-mean-square (rms) values of voltage and current are used instead of dc values. Hence, voltages and currents in this application window and Application Window 1.7.1 are to be understood as ac rms values.

Simulation Example 1.7.1: Interconnected Circuit Elements

Figure 1.7.2 shows six interconnected circuit elements. The positive directions of current through each element and the voltage across each element are arbitrarily assigned as indicated. Measurements on elements 1, 2, and 4 to 6 gave the values entered in Table 1.7.1. It is required to: (1) determine the power delivered or absorbed by element 3, (2) deduce the values of the current in element 3 and the voltage across it, and (3) simulate the circuit.

SOLUTION

Table 1.7.2 is formed by having the first two rows under the title row identical to those of Table 1.7.1 and adding a column for element 3. Entries are made for each of the elements 1, 2, 4, 5, and 6 according to the following steps:

1. From the assigned positive directions of v and i in Figure 1.7.2, it is ascertained whether these directions correspond to power delivery or absorption. If i is in the direction of a voltage rise v, which corresponds to power delivery, in accordance with Equation 1.3.1, the product vi for the element concerned is entered in row 3 under the title row for the given element in Table 1.7.2. If i is in the direction of a voltage drop, which corresponds to power absorption, the product vi is entered in row 4. For element 1, for example, i_1 is in the direction of a voltage rise v_1, and the product $v_1 i_1$ is entered in row 3 for element 1. For element 6, i_6 is in the direction of a voltage drop v_6, and the product $v_6 i_6$ is entered in row 4 for element 6.

2. The numerical value of the product $p = vi$ is calculated from the values given in Table 1.7.1 and entered under the corresponding vi. If this value is negative in row 3, it denotes actual power absorbed, so it is entered as a positive quantity in row 4. Note that numerical values were not substituted until *after* it was ascertained whether or not $p = vi$ represented power delivered or absorbed according to the assigned positive directions of v and i.

The positive values of power in Table 1.7.2 indicate the power that is actually being delivered or absorbed by each element. These values are boldfaced for clarity.

Power must be conserved in the whole circuit at every instant. This may be expressed as:

$$\sum \text{Actual power delivered} = \sum \text{Actual power absorbed} \qquad (1.7.6)$$

where positive values are summed on each side.

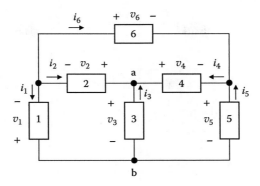

FIGURE 1.7.2
Figure for Example 1.7.1.

TABLE 1.7.1

Currents and Voltages in Figure 1.7.2

Element	1	2	4	5	6
i (A)	−3	4	4	5	−1
v (V)	−12	−4	−8	16	−4

TABLE 1.7.2

Power in Circuit Elements of Figure 1.7.2

Element	1	2	4	5	6	3
i (A)	−3	4	4	5	−1	−8
v (V)	−12	−4	−8	16	−4	8
Power delivered (W)	$v_1 i_1$ +36	$v_2 i_2$ −16	$v_4 i_4$ −32	$v_5 i_5$ +80		$v_3 i_3$ −64
Power absorbed (W)		+16	+32		$v_6 i_6$ +4	+64

Adding the positive quantities in columns 2 to 6 for power delivered gives 116 W for the total actual power delivered by all the elements except element 3. Similarly, adding the positive quantities in columns 2 to 6 for power absorbed gives 52 W for the total actual power absorbed by all the elements except element 3. For conservation of energy, it follows that element 3 must actually absorb 64 W, so that −64 is entered for this element in row 3 and +64 in row 4. Power is therefore conserved in the circuit, in accordance with the preceding summations.

The current i_3 could be determined from conservation of charge. Consider junction a between elements 2, 3, and 4 in Figure 1.7.2. Current $i_2 = +4$ A, which means that charge flows through element 2 toward this junction at the rate of 4 C/s. Similarly, $i_4 = +4$ A means that charge flows through element 4 toward junction a also at the rate of 4 C/s. Conservation of charge requires that charge leave junction a through element 3 at a rate equal to the sum of the two preceding rates, that is, at a rate of 8 C/s. Otherwise, a net charge would appear at junction a from nowhere, or would leave this junction to nowhere. That is, charge would not be conserved. The current through element 3, in the direction from a to b should therefore be 8 A. Because this current is in the opposite direction to the assigned positive direction of i_3, it follows that $i_3 = −8$ A.

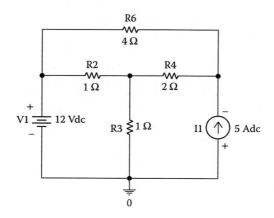

FIGURE 1.7.3
Figure for Example 1.7.1.

To determine v_3, we note that the assigned positive direction of i_3 is in the direction of a voltage rise v_3. These directions correspond to power being delivered, as indicated by the product $v_3 i_3$ entered in row 3. It was found earlier that the power actually absorbed by element 3 is +64 W, or that delivered by element 3 is −64 W. Dividing this power by i_3 gives $v_3 = +8$ V.

SIMULATION
Power-absorbing elements are simulated by resistors, whereas power-delivering elements are simulated by dc sources. The schematic is shown in Figure 1.7.3. Select Bias Point for Analysis type and run the simulation. By pressing the W button in the Capture page, the power dissipated in each resistor is found to be: R2:16 W, R3:64 W, R4:32 W, R6:4 W, the total power absorbed being 116 W. The power delivered by the voltage source is 36 W, whereas that delivered by the current source is 80 W. The total power delivered is 116 W, in accordance with conservation of power.

1.8   The Capacitor

> **Concept:** *The fundamental attribute of a capacitor is its ability to store energy in the electric field resulting from separated positive and negative electric charges.*

As a prototypical capacitor, consider a device consisting of two metal plates, each having an area of A m^2 separated by a dielectric material of permittivity ε F/m and thickness d m (Figure 1.8.1). If the capacitor is momentarily connected to a source of voltage v, it acquires a charge $+q$ on the plate connected to the positive terminal of the source and a charge $-q$ on the plate connected to the negative terminal. The charge on the capacitor is proportional to the voltage across it and is given by:

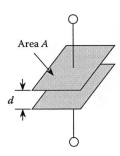

Area A

d

FIGURE 1.8.1
Parallel plate capacitor.

$$q = Cv \tag{1.8.1}$$

where C is the capacitance of the capacitor. If q is in coulombs and v is in volts, C is in farads (F). In an ideal capacitor, C is constant, so q and v are linearly related, and there is no power dissipation. This implies that insulation of the dielectric is perfect; that is, there is no leakage of charge between the plates of the capacitor, and charge is retained indefinitely.

The capacitance of the parallel-plate capacitor shown in Figure 1.8.1 can be readily calculated from elementary principles of electrostatics, neglecting fringe effects at the edges of the plates. From Gauss's law, the electric displacement is $D = \frac{q}{A}$ C/m^2. The electric field is $\xi = \frac{D}{\varepsilon}$ V/m, where ε is the permittivity in F/m. The voltage is $v = \xi d$ V.

Combining these relations gives: $q = \frac{\varepsilon A}{d} v$. Comparing with Equation 1.8.1 gives the following equation for the parallel plate capacitor:

$$C = \frac{\varepsilon A}{d} \tag{1.8.2}$$

The capacitance depends on the geometry of the capacitor and the permittivity of the dielectric. Irrespective of the physical structure of the capacitor, capacitance generally increases as the permittivity and the surface area of the conducting surfaces are increased, and as the separation between these surfaces is reduced.

EXERCISE 1.8.1

A capacitor consists of two metallic disks of 5 cm diameter separated by 1 mm, the space between the disks being filled with an insulator having a relative permittivity of 5,000. How many electrons have to be moved from one disk to the other so that the voltage between them becomes 50 V? Assume the permittivity of free space is 8.85 × 10^{-12} F/m.

Answer: 2.7×10^{13} electrons.

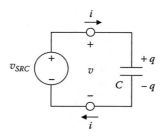

FIGURE 1.8.2
Capacitor symbol and assigned positive directions.

v–i Relation

The symbol for a capacitor is shown in Figure 1.8.2 together with the assigned positive direction of voltage drop v across the capacitor. The v–i relation is obtained from Equation 1.8.1 and $q = \int i\, dt$ as:

$$v = \frac{1}{C} \int i\, dt, \text{ or } i = C\frac{dv}{dt} \tag{1.8.3}$$

The signs in Equation 1.8.3 require more careful consideration than in the case of a resistor because even when v and i have positive values, dv/dt can have a negative value, but C must be a positive quantity for a passive capacitor. The proper sign can be readily assigned based on the direction of power flow discussed in Section 1.4, and expressed as the following **passive sign convention**:

> **Concept:** *If the assigned positive direction of current through a passive circuit element is in the direction of a voltage drop across the element, a positive sign is used in the v–i relation of that circuit element. If the assigned positive direction of current through the element is in the direction of a voltage rise across the element, a negative sign is used in the v–i relation of that circuit element.*

In the case of a resistor, the assigned positive direction of current is in the direction of a voltage drop (Figure 1.7.1b), and Ohm's law (Equation 1.7.1) is written with a positive sign so that R is positive. In the case of a capacitor, the assigned positive direction of current in Figure 1.8.2 is also in the direction of a voltage drop, and the v–i relation of Equation 1.8.3 is written with a positive sign. If $\dfrac{dv}{dt} > 0$, the charge on the capacitor increases with time, so current must be flowing into the positive terminal of the capacitor and out of the negative terminal. According to the power flow convention, power is absorbed by the capacitor and stored as electric energy. The values of i and $\dfrac{dv}{dt}$ in Equation 1.8.3 are both positive, so C is positive. If $\dfrac{dv}{dt} < 0$, the charge on the capacitor decreases with time, so current must be flowing out of the positive terminal of the capacitor and into the negative terminal. The values of i and $\dfrac{dv}{dt}$ in Equation 1.8.3 are both negative, so C remains a positive quantity. The *actual* current is now in the direction of a voltage rise across the capacitor. This means, according to the power flow convention, that the capacitor is now returning to the rest of the circuit energy that had been stored in it.

If the assigned positive direction of i or v is reversed in Figure 1.8.2, so that i is in the direction of a voltage rise across the capacitor, then according to the preceding argument, Equation 1.8.3 must be written with a negative sign, so that C is positive for positive or negative $\dfrac{dv}{dt}$.

Because the insulation in an ideal capacitor is perfect, the current in a capacitor cannot be a conduction current, like that through a resistor. Rather, it is a **displacement current**. To appreciate the nature of this current, assume that the current i flowing into the positive terminal of the capacitor increases the positive charge on the positive plate of the capacitor by an amount δq in a time δt, where $\delta q = i\delta t$. The positive charge δq displaces, through electrostatic repulsion, an equal positive charge δq from the negative plate. The displaced positive charge δq flows out of the negative terminal, thereby completing the current path through the capacitor. At the end of the time interval δq, the charge on the positive plate of the capacitor is $(+q+\delta q)$ and that at the negative plate is $(-q-\delta q)$.

Steady Voltage and Current

> **Concept:** *When a dc voltage is applied to an ideal capacitor, the resulting current through the capacitor is zero, and the capacitor behaves as an open circuit as far as the dc voltage is concerned.*

For a dc voltage, $\dfrac{dV}{dt} = 0$, and the capacitor charge does not change with time. Hence, $I = 0$, in accordance with Equation 1.8.3, which means that the capacitor behaves as an open circuit as far as the dc voltage is concerned.

EXERCISE 1.8.2

The voltage waveform shown in Figure 1.8.3 is applied to a 1 µF capacitor. Determine: (a) the capacitor current as a function of time; (b) the charge on the capacitor at $t = 1.5$ ms, assuming that the capacitor had an initial charge of 5 µC.

Answer: (a) 5 mA, $0 < t < 1$ ms; 0, $1 < t < 2$ ms; 10 mA, $2 < t < 2.5$ ms; 0, $t > 2.5$ ms; (b) 10 µC.

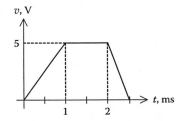

FIGURE 1.8.3
Figure for Exercise 1.8.2.

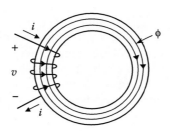

FIGURE 1.9.1
Coil wound on a toroidal core.

Stored Energy

The energy stored in the electric field of a capacitor is equal to the work done in separating the charges $+q$ and $-q$. From Equation 1.2.1, $w = \int_0^q v\,dq$. For a capacitor, $q = Cv$ and $dq = Cdv$. Substituting in the expression for w:

$$w = C\int_0^v v\,dv = \frac{1}{2}Cv^2 = \frac{1}{2}\frac{q^2}{C} = \frac{1}{2}qv \qquad (1.8.4)$$

1.9   The Inductor

> **Concept:** *The fundamental attribute of an inductor is its ability to store energy in the magnetic field associated with current.*

As a prototypical inductor, consider a coil of N turns that is tightly wound on a toroidal core of magnetic material that could have a circular or rectangular cross-section (Figure 1.9.1). Magnetic flux ϕ is associated with current through the coil. Because of the toroidal shape and the tight winding, the flux is confined to the core and links all the turns of the coil, neglecting the finite thickness of the insulation of the wire and the wire diameter. The flux linkage λ then simply equals $N\phi$. The inductance L of the coil is defined as the flux linkage per unit current of the coil, or:

$$\lambda = Li \qquad (1.9.1)$$

When i is in amperes and λ is in weber-turns (Wb-turns), L is in henries (H). In an ideal inductor, L is constant, so λ and i are linearly related and there is no power dissipation. The inductance of the toroidal coil of Figure 1.9.1 is derived in Example 1.9.1.

v–i Relation

The symbol for an inductor is shown in Figure 1.9.2 together with the assigned positive direction of voltage drop v across the inductor. If the flux changes with time, a voltage is induced in the coil whose magnitude is given by Faraday's law:

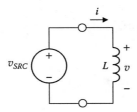

FIGURE 1.9.2
Inductor symbol and assigned positive directions.

$$v = \frac{d\lambda}{dt} \qquad (1.9.2)$$

Substituting from Equation 1.9.1,

$$v = L\frac{di}{dt}, \text{ or } i = \frac{1}{L}\int v\,dt \qquad (1.9.3)$$

According to Lenz's law, the polarity of v due to a change in i is such that it opposes the change in i. If $\frac{di}{dt} > 0$, v opposes the increase in i by being a voltage drop in the direction of i, thereby resisting the flow of current. Power is absorbed by the inductor and stored as energy in the magnetic field. If $\frac{di}{dt} < 0$, v opposes the decrease in i by being a voltage rise in the direction of i, thereby adding to the source voltage and aiding the flow of current. The inductor now returns to the rest of the circuit the energy that had previously been stored. In both cases, L is a positive quantity in Equation 1.9.3. Again, the passive sign convention applies, so that if the assigned positive direction of i is in the direction of a voltage rise across the inductor, Equation 1.9.3 is written with a negative sign, and L remains a positive quantity.

Example 1.9.1: Inductance of Toroidal Coil

Consider the toroidal core shown in Figure 1.9.1. Let the radius of the toroid be a m, the cross-sectional area of the core be A m², and its relative permeability be μ_r. The cross-sectional dimensions of the core are assumed to be small compared to a, so that the magnetic field intensity H can be considered constant across the cross-section of the core. A coil of N turns and negligible diameter and thickness of insulation is tightly wound around the core so that all the magnetic flux lies entirely in the core. It is required to determine the inductance of the toroidal coil.

SOLUTION
Consider a circle of radius r that lies inside the core. From Ampere's circuital law: $H \times 2\pi r = NI$, where I is the current in the coil, so that:

$$H = \frac{NI}{2\pi r} \qquad (1.9.4)$$

The magnetic flux density B is:

$$B = \mu_r \mu_0 H = \mu_r \mu_0 \frac{NI}{2\pi r} \tag{1.9.5}$$

where μ_0 is the permeability of free space ($4\pi \times 10^{-7}$ H/m). Because the cross-sectional dimensions of the core are small compared to the radius of the toroid, we may replace r with a so that the flux in the core is:

$$\phi = B.A = \mu_r \mu_0 \frac{A}{2\pi a} NI \tag{1.9.6}$$

The flux linkage is:

$$\lambda = N\phi = \mu_r \mu_0 \frac{A}{2\pi a} N^2 I \tag{1.9.7}$$

It follows from Equation 1.9.1 that the inductance of the coil is:

$$L = \frac{\lambda}{I} = \mu_r \mu_0 \frac{AN^2}{2\pi a} \tag{1.9.8}$$

Steady Voltage and Current

> Concept: *When a dc current flows through an ideal inductor, the voltage across the inductor is zero, and the inductor behaves as a short circuit as far as the dc current is concerned.*

When the current is dc, $\frac{dI}{dt} = 0$, and the flux does not change with time. Hence, $V = 0$, in accordance with Equation 1.9.3, which means that the inductor behaves as a short circuit.

EXERCISE 1.9.1

A current waveform as in Figure 1.8.3 with volts replaced by amperes, is applied to a 1 µH inductor. Determine: (a) the inductor voltage as a function of time; (b) the flux linkage in the inductor at $t = 1.5$ ms, assuming that the inductor had an initial flux linkage of 5 µWb-turns.

Answer: (a) 5 mV, $0 < t < 1$ ms; 0, $1 < t < 2$ ms; 10 mV, $2 < t < 2.5$ ms; 0, $t > 2.5$ ms; (b) 10 µWb-turns. (Note the similarity with Exercise 1.8.2.)

Stored Energy

Assume that the current i in the inductor, and hence the flux linkage λ, are positive and increasing with time. The power input to the inductor is: $vi = i\frac{d\lambda}{dt}$. The work done in increasing λ from 0 at $t = 0$ to λ at time t, which equals the energy stored in the magnetic field, is:

$$w = \int_0^t vi\,dt = \int_0^\lambda i\,d\lambda = \int_0^i i(L\,di) = \frac{1}{2}Li^2 = \frac{1}{2}\lambda i = \frac{1}{2}\frac{\lambda^2}{L} \qquad (1.9.10)$$

1.10 Concluding Remarks

1. Fundamentally, as mentioned in the Foreword, the three basic circuit parameters of resistance, capacitance, and inductance model respectively, three essential attributes of the electromagnetic field, namely: power dissipation, energy stored in the electric field, and energy stored in the magnetic field.

2. Electromagnetic fields are inherently distributed. Consider, for example, a current-carrying conductor, such as the wiring or interconnections in electronic equipment, or a power line that could be hundreds of kilometers long. The current is carried by moving electric charges that are distributed along the conductor. The power dissipation, as well as the electric and magnetic fields, are also distributed along the conductor. Strictly speaking, the electromagnetic fields associated with the electric signals of voltage and current in any electrical system travel as waves at speeds that may approach that of light in vacuum. However, it is found that in many cases, the wave nature of electric signals can be neglected. This is tantamount to ignoring the distributed nature of the power dissipation, the energy stored in the electric field, and that stored in the magnetic field. Each of these entities is then separately lumped and represented by discrete circuit elements, namely, resistors, capacitors, and inductors. It is argued in Section ST1.2 that the **lumped-parameter** assumption is valid as long as the physical dimensions of the system under consideration are small compared to the smallest wavelength of the voltage or current signals in the system. Thus, for a signal having a frequency of 1 GHz (gigahertz), that is, 10^9 cycles per second, which is in the range of frequencies used in some communications systems, the wavelength is approximately 30 cm, assuming the velocity of the electromagnetic waves is that of light in vacuum (3×10^8 m/s). If the physical dimensions of the system do not exceed, say, one tenth of the wavelength, that is, 3 cm, the lumped-parameter representation may be considered to be a reasonably good approximation. On the other hand, if the physical dimensions of the system are, say, 10 cm, the wave nature of signal propagation cannot be neglected, and the lumped-parameter representation breaks down at this frequency. The situation is very different at low frequencies such as a power frequency of 50 Hz. The wavelength at this frequency is 6×10^6 m, or 6,000 km. Even a transmission line that is 600 km long may be adequately represented at this frequency by lumped circuit parameters.

3. In basic circuit theory it is assumed that the circuit parameters — R, C, and L — are constant; that is, they do not vary with current or voltage, nor with time, which makes the corresponding v–i, v–q, or λ–i relations linear. The system is designated as **linear, time-invariant** (LTI). A linear system obeys superposition, as discussed in Section 16.6 of Chapter 16.

4. Like ideal sources, ideal resistors, capacitors, and inductors are abstractions. Not only are their values constant, but each is assumed to model solely the field attribute it represents. Thus, ideal resistors only dissipate energy; they do not store electric or magnetic energy. Ideal capacitors only store electric energy; they do not store magnetic energy, nor do they dissipate energy. Ideal inductors only store magnetic energy; they do not store electric energy, nor do they dissipate energy. Practical circuit elements depart from this ideal to varying degrees, depending on their physical realization and operating conditions. They all have limitations on the maximum current or voltage they can handle, which must be taken into consideration in circuit design by ensuring that the specified ratings are not exceeded. The values of practical circuit elements may be considered constant over a limited operating range and may change over a prolonged period. For example, the heat generated by an appreciable current in a resistor changes its temperature and, hence, its resistance. A magnetic field, and hence some inductance, is associated with the current through a resistor. A capacitance is associated with the electric field due to the voltage difference between the ends of a current-carrying resistor. Similarly, energy is dissipated by the finite resistance of the coil of an inductor, and a capacitance is associated with the voltage drops between the ends of a coil as well as between adjacent turns of the coil. The dielectric of a capacitor is never perfect and dissipates charge and energy, and an inductance is associated with the leads of a capacitor. These parasitic effects are generally negligible at low frequencies but become significant at high-enough frequencies.

5. Ideal circuit elements are *basic*, in the sense that an ideal circuit element cannot be modeled in terms of other ideal circuit elements. On the other hand, practical circuit elements are modeled in terms of a combination of ideal circuit elements. For example, a coil is modeled at low frequencies by a combination of an ideal inductor and an ideal resistor.

6. As is true of macroscopic systems in general, energy and charge must be conserved in all electric circuits.

Summary of Main Concepts and Results

- Current is the rate of flow of electric charge.
- The voltage, or electric potential difference, between two points is the difference in electric potential energy, per unit positive charge, between these points.
- Power is the rate at which energy is delivered or absorbed.
- Conservation of energy implies conservation of power.
- If current i is in the direction of a voltage drop v, then the power $p = vi$ represents power absorbed by the element through which i flows. Conversely, if i is in the direction of a voltage rise v, then the power p represents power delivered by the element through which i flows.
- Active devices can generate electrical energy through conversion from another form of energy, whereas passive devices cannot.
- An ideal voltage source maintains a specified voltage across its terminals, irrespective of the current through the source.

- An ideal current source maintains a specified current through it, irrespective of the voltage across its terminals.
- A connection of sources that violates conservation of energy or conservation of charge is invalid and is not allowed in electric circuits.
- Electrical resistance is due to impediments to the flow of conduction electrons under the influence of an applied electric field.
- An ideal resistor obeys Ohm's law, $v = Ri$, where the resistance R is constant, and the power dissipated is $p = Ri^2 = v^2/R$.
- Conductance G is the reciprocal of resistance. In terms of conductance, dissipated power is $p = Gv^2 = i^2/G$.
- According to the passive sign convention, if the assigned positive direction of current through a passive circuit element is in the direction of a voltage drop across the element, a positive sign is used in the v–i relation of that circuit element. If the assigned positive direction of current through the element is in the direction of a voltage rise across the element, a negative sign is used in the v–i relation of that circuit element.
- The fundamental attribute of a capacitor is its ability to store energy in the electric field resulting from separated positive and negative electric charges.
- In an ideal capacitor $q = Cv$, where the capacitance C is constant; there is no power dissipation and the energy stored in the electric field is $qv/2$.
- Under dc conditions, an ideal capacitor behaves as an open circuit.
- The fundamental attribute of an inductor is its ability to store energy in the magnetic field associated with the current.
- In an ideal inductor $\lambda = Li$, where the inductance L is constant; there is no power dissipation and the energy stored in the magnetic field is $\lambda i/2$.
- Under dc conditions, an ideal inductor behaves as a short circuit.

Learning Outcomes

- Understand the concepts of current, voltage, power, and energy, and their inter-relations.
- Understand the basic attributes of voltage sources, current sources, resistors, capacitors, and inductors and their voltage–current relations.

Supplementary Topics on CD

ST1.1 Conservation of power: Gives a mathematical justification that conservation of energy implies conservation of power.

ST1.2 Validity of lumped-parameter assumption: Derives the condition for validity of this assumption.

Problems and Exercises

P1.1   Current, Voltage, and Power

P1.1.1 A room is lit by two 100 W lamps. How many kilowatt-hours of energy do the lamps consume in a month if the lamps are lit an average of 6 h/day? Energy consumption in electrical installations is usually metered in kilowatt-hours rather than kilojoules.

P1.1.2 The voltage drop across a certain device, and the current through it, in the direction of voltage drop, are shown in Figure P1.1.2. Determine: (a) the charge q through the device at the end of each 1 s interval from $t = 0$ to $t = 6$ s; (b) the instantaneous power p during the aforementioned intervals; and (c) the total energy consumed by the device.

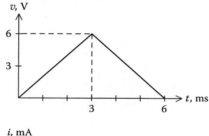

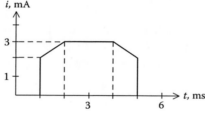

FIGURE P1.1.2

P1.1.3 The voltage drop across a certain device, and the current through it, in the direction of voltage drop, are given by:

$$v = 2t + 1 \text{ V}, \quad i = 4 - 2t \text{ mA}, \quad 0 < t < 4 \text{ s}$$

At what instant of time is the instantaneous power absorbed by the device: (a) maximum? (b) zero? (c) What is the energy delivered to the element at $t = 2$ s and at $t = 4$ s?

P1.1.4 The voltage drop across a certain device, and the current through it, in the direction of voltage drop, are given by:

$$v = \sin 2\pi t \text{ V}, \quad i = \cos 2\pi t \text{ A}$$

Determine: (a) the direction of power flow during successive quarter cycles; (b) the maximum instantaneous power absorbed or delivered by the device; (c) the minimum instantaneous power absorbed or delivered by the device; and (d) the average power over one cycle.

P1.1.5 The voltage drop across a certain device, and the current through it, in the direction of voltage drop, are given by:

$$v = 3\sin(100\pi t - 45°) \text{ V}, \quad i = 4\cos(100\pi t) \text{ A}$$

Sketch the instantaneous power delivered to the element. Determine: (a) the average power delivered to the element; (b) the maximum instantaneous power; and (c) the minimum instantaneous power.

P1.1.6 The voltage drop v V across a certain device, and the current i A through it, in the direction of voltage drop, are related by:

$$i = 8 - 2v^2, \quad 0 \le v \le 2 \text{ V}$$

$$i = 0, \quad v \ge 2 \text{ V}$$

(a) Determine the power absorbed by the load when $v = 1$ V and when $v = 2$ V; (b) at what value of v is the instantaneous power a maximum? (c) If $v(t) = 2e^{-t}$ V, $t \ge 0$ s, what is the total charge that passes through the device from $t = 0$ to $t = 2$ s?

P1.1.7 Seven circuit elements are interconnected as shown in Figure P1.1.7. The assigned positive directions of voltage drops and currents are indicated. Based on these assigned positive directions, which elements absorb power and which deliver power? If the values of the currents and voltages are as listed in Table P1.1.7, which elements actually absorb power and which deliver power? How much is the total power absorbed, and how much is the total power delivered? Are they equal? Verify your results with PSpice; simulate power-absorbing elements with resistors and power-delivering elements with sources.

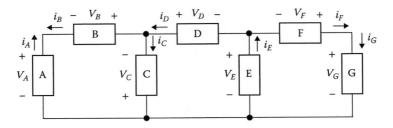

FIGURE P1.1.7

TABLE P1.1.7

Currents and Voltages in Figure P1.1.7

Element	A	B	C	D	E	F	G
Voltage, V	5	−3	−2	5	−3	7	4
Current, A	3	−3	1	−2	−1	1	1

P1.2   Sources

Use conservation of power to solve Problem P1.2.1 to Problem P1.2.6. Verify your results with PSpice. Substitute resistors of appropriate value for power-absorbing elements shown as rectangles.

P1.2.1   Determine I_{SRC} and I_A in Figure P1.2.1, given that element A absorbs 40 W.

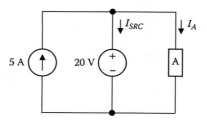

FIGURE P1.2.1

P1.2.2   Determine V_{SRC} and V_A in Figure P1.2.2, given that element A absorbs 240 W.

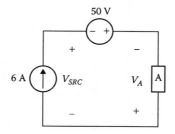

FIGURE P1.2.2

P1.2.3   Determine I_{SRC} in Figure P1.2.3.

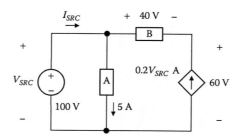

FIGURE P1.2.3

P1.2.4 Determine V_{SRC} in Figure P1.2.4.

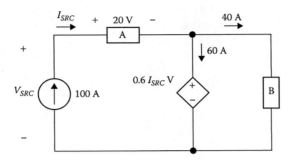

FIGURE P1.2.4

P1.2.5 Determine V_B in Figure P1.2.5.

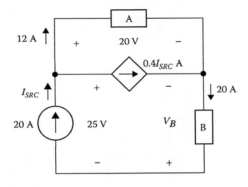

FIGURE P1.2.5

P1.2.6 Determine I_B in Figure P1.2.6.

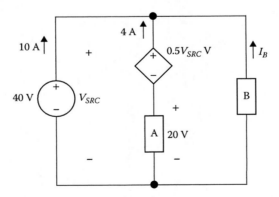

FIGURE P1.2.6

P1.2.7 A and B are identical circuit elements connected as in Figure P1.2.7. The v–i relation of either element is: $v = 2i^2$ V, where i is in amperes. Determine i, the power absorbed by each element, and that delivered by the source. Note that power is conserved although the elements are nonlinear, because conservation of energy applies to all systems, linear or nonlinear.

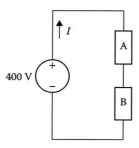

FIGURE P1.2.7

P1.2.8 The elements of Problem P1.2.7 are connected as in Figure P1.2.8. Determine I_{SRC}, the power absorbed by each element, and that delivered by the source.

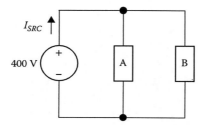

FIGURE P1.2.8

P1.2.9 A and B are identical circuit elements connected as in Figure P1.2.9. The v–i relation of either element is: $i = 2v^2$ A, where v is in volts. Determine V_{SRC}, the power absorbed by each element, and that delivered by the source.

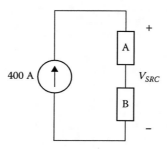

FIGURE P1.2.9

P1.2.10 The elements of P1.2.9 are connected as in Figure P1.2.10. Determine V_{SRC}, the power absorbed by each element, and that delivered by the source.

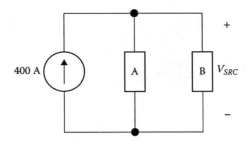

FIGURE P1.2.10

P1.3 Resistors

P1.3.1 Given a 220 V, 10 kW oven, determine: (a) the current that the oven draws from the mains supply, (b) the resistance of the oven, and (c) its conductance.

P1.3.2 The resistance of a copper power line is 60 Ω at 20°C when not carrying any current. Its resistance, when carrying its rated current, is 70 Ω. Find the temperature of the conductor under these conditions, assuming that the temperature coefficient of copper is 0.0039/°C.

P1.3.3 A 1.5 MΩ resistor is rated at 1/2 W. What is the maximum voltage that can be applied to the resistor without exceeding its power rating?

P1.3.4 An ideal ammeter has zero resistance; when it measures current, the voltage across it is zero. An ideal voltmeter has zero conductance; when it measures voltage, the current through it is zero. If a practical ammeter has a resistance of 1 mΩ what is the voltage across it when it reads 10 A? If a practical voltmeter has a conductance of 0.1 μS, what is the current through it when it reads 100 V?

P1.3.5 A *pn* junction diode has an exponential *i–v* relation of the form: $i = 10^{-9}(e^{20v} - 1)$ A, where v is in volts. Determine the diode current for: (a) $V = 0.7$ V and (b) $V = -0.7$ V. Note that the *i–v* relation is highly asymmetrical.

P1.3.6 The triangular voltage waveform of Figure P1.3.6 is applied to a 100 Ω resistor. Determine: (a) the resistor current; (b) instantaneous power dissipation in the resistor; and (c) the average power dissipation. Is the statement "the average power in a resistor is the product of the average voltage across the resistor and the average current through the resistor" valid? Justify your answer.

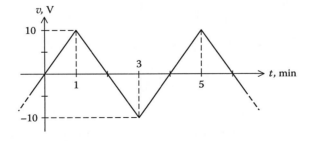

FIGURE P1.3.6

P1.3.7   The current through a resistor is $4\sin100\pi t$ A when the voltage across it is $60\sin100\pi t$ V. Find the resistance and the average power dissipated.

P1.3.8   A voltage $v(t) = 10\cos100\pi t$ V is applied across a 10 Ω resistor. (a) Sketch p. (b) Find the average power dissipated in the resistor and the energy dissipated during a half-cycle of the supply waveform.

P1.3.9   The voltage shown in Figure P1.3.9 is applied across a 5 Ω resistor. (a) Determine p, $0 \le t \le 1$ min; (b) the energy dissipated in the resistor at $t = 3$ min.

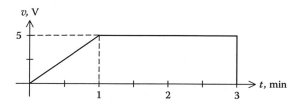

FIGURE P1.3.9

P1.3.10 The v–i relation of a resistor is $v = 5i^2$ V, where i is in amperes. If $i(t) = \cos\omega t + \cos2\omega t$ A, determine v. Note the presence of frequencies in the output that are not in the input because of the nonlinearity.

P1.3.11 **Thermistors** are highly temperature-sensitive resistors that are used for temperature measurement in many applications, including electronic thermometers. They are made of a ceramic material that is usually a mixture of oxides, sulfates, or silicates of various elements such as nickel, copper, magnesium, manganese, cobalt, aluminum, and titanium. The resistance of a thermistor decreases with temperature and is given by: $R_T = R_{T0}e^{\beta\left(\frac{1}{T} - \frac{1}{T_0}\right)}$, where R_T is the resistance at an absolute temperature T K, R_{T0} is the resistance at a reference absolute temperature T_0 K. β, the temperature coefficient of the material, is usually in the range 2,500 to 5,000 K. Using MATLAB (Appendix SD.1), or another appropriate program, plot $\dfrac{R_T}{R_{300}}$ over the temperature range 200 to 400 K for β = 2,500 K and β = 5,000 K. Compare with $\dfrac{R_T}{R_{300}}$ for copper, assuming that Equation 1.7.5 applies, with $\alpha_m = 0.0039/°C$. Use a logarithmic scale for $\dfrac{R_T}{R_{300}}$.

P1.4   Capacitors

P1.4.1   A parallel plate capacitor consists of two plates separated by 1 mm. Each plate is 5 cm square. A dielectric of relative permittivity, or **dielectric constant**, of 10 partially fills the space between the plates for a distance of 3 cm (Figure P1.4.1). Determine the capacitance, ignoring edge effects.

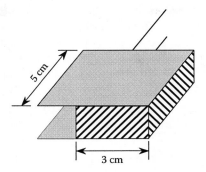

FIGURE P1.4.1

P1.4.2 Any physical structure capable of separating electric charges, and hence storing energy in the electric field, may be modeled by means of a capacitance. Membranes of living cells are composed of fatty molecules oriented back-to-back across the membrane so as to form a thin sheet that is essentially two molecules thick. The inside and the outside media of cells are weak electrolytic solutions. As a good insulator separating two relatively good conductors, the cell membrane exhibits a surprisingly large capacitance of about $1\ \mu F/cm^2$, due to its thinness. The electric charges separated by the membrane are ions, mainly Na^+, K^+, and Cl^-. The membrane is not an ideal capacitor, because the separation of ions is not perfect; ions can move from one side of the membrane to the other through channels in the membrane. Assuming the membrane thickness is 8 nm, determine the effective permittivity of the membrane.

P1.4.3 A capacitor consists of two thin concentric cylinders l m long, the space between the two cylinders being filled with a dielectric of permittivity ε F/m. The radii of the inner and outer cylinders are a and b m, respectively. Apply Equation 1.8.2 to a cylindrical shell of radius r and thickness dr, and then integrate from a to b to show that the capacitance is: $\dfrac{2\pi\varepsilon l}{\ln(b/a)}$, neglecting end effects.

P1.4.4 A **transducer** is a device that converts one form of energy to another. A capacitive transducer for measuring displacement is illustrated in longitudinal section in Figure P1.4.4. It consists of two concentric metal cylinders 5 cm long. The

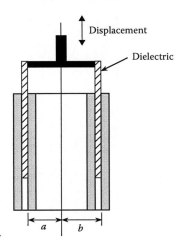

FIGURE P1.4.4

outer radius of the inner cylinder is $a = 1$ cm, and the inner diameter of the outer cylinder is $b = 1.2$ cm. The displacement to be measured causes a cylinder of dielectric material to move axially between the metal cylinders. The distance is determined by measuring the capacitance between the two cylinders. Using the result of Problem P1.4.3, determine the range of capacitance to be measured, assuming a relative dielectric constant of 100. $\varepsilon_0 = 8.85 \times 10^{12}$ (F/m).

P1.4.5 The voltage across a 2 mF capacitor is 10 V, $t \leq 0$, and $Ate^{-10t} + Be^{-10t}$ V, $t \geq 0$, where t is in seconds. If the capacitor current is 200 mA at $t = 0$, determine A and B.

P1.4.6 The capacitance of a linear time-varying capacitor is given by: $C(t) = C_0(1 - e^{-\alpha t})$ If the voltage across the capacitor is $v(t) = V_0 t^2$, determine the capacitor current.

P1.4.7 A series of current pulses of 10 mA amplitude and 2 ms duration is applied to an initially uncharged, ideal 5 μF capacitor. How many pulses are required to charge the capacitor to 20 V?

P1.4.8 A 2 μF capacitor is initially uncharged. A current pulse of amplitude 100 μA and 200 ms duration is applied at $t = 0$. Express the capacitor voltage as a function of time.

P1.4.9 A 2 μF capacitor is initially uncharged. A current impulse of 100 μC strength is applied at $t = 0$, and another impulse of 100 μC strength is applied at $t = 200$ ms. Express the capacitor voltage as a function of time.

 (Refer to Section 11.3, Chapter 11, for a discussion of the impulse function. Note that the current input in this problem is the time derivative of that of the preceding problem. Because the given capacitor is a linear circuit element, the response is also the time derivative of the response of the preceding problem. It is sometimes easier to derive the response to the time derivative or integral of a given input and then integrate or differentiate the response accordingly.)

P1.4.10 Repeat the preceding two problems assuming the 2 μF capacitor was initially charged to 10 V.

P1.4.11 The current waveform of Figure P1.4.11 is applied to a 0.5 μF capacitor that is initially uncharged. (a) Derive expressions for the voltage across the capacitor during the time intervals: $0 \leq t \leq 10$ μs, $10 \leq t \leq 40$ μs, $40 \leq t \leq 60$ μs, $60 \leq t \leq 80$ μs, and $t > 80$ μs. (b) What is the charge on the capacitor at $t = 10$ μs and at $t = 50$ μs? (c) What is the energy stored in the capacitor at $t = 80$ μs? (d) How do the expressions for the voltage across the capacitor derived in (a) change if the capacitor was initially charged to 0.5 V?

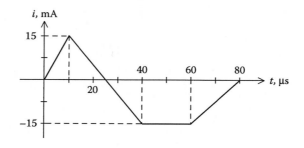

FIGURE P1.4.11

P1.4.12 The triangular voltage pulse of Figure P1.4.12 is applied to a 0.1 µF capacitor that is initially uncharged. Plot as a function of time: (a) the charge on the capacitor; (b) the energy stored in the capacitor, as derived from Equation 1.8.4; (c) the capacitor current; and (d) the instantaneous power input to the capacitor. How is the energy stored in (b) related to the instantaneous power in (d)?

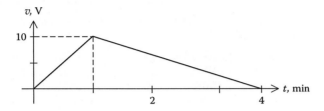

FIGURE P1.4.12

P.1.4.13 Repeat Problem P1.4.12 assuming the voltage is the first half-cycle of the waveform $v_C(t) = 10\sin500t$ V, where t is in seconds; that is, $v_C(t) = 10\sin500t$ V, $0 \le t \le (\pi/500)$ s, and $v_C(t) = 0$ elsewhere.

P1.4.14 The voltage applied to an initially uncharged 5 µF capacitor is $v_C = 10te^{-5t}$ V, where t is in ms and $t \ge 0$. Derive the expressions for the capacitor current i_C, the energy stored in the capacitor, and the instantaneous power input to the capacitor. Plot these quantities as well as v_C, using MATLAB or some other appropriate program.

P1.4.15 Let the current applied to an initially uncharged 5 µF capacitor be $i_C = 10te^{-5t}$ mA, where t is in ms and $t \ge 0$. (a) Derive the expression for capacitor voltage v_C using integration by parts. (b) Determine the voltage at the instant the current is a maximum. (c) Verify your result using MATLAB's int(y,t,a,b) command (Appendix SD.1).

P1.4.16 The charge on a nonlinear capacitor is given by $q = 0.5v^2$ C, when v is in volts. If the voltage across the capacitor is $v = 1+t$ V, determine the additional energy stored in the capacitor from $t = 0$ to $t = 3$ s.

P1.5   Inductors

P1.5.1 A current of 10 mA through a coil of 100 turns produces a flux of 10^6 Wb in the coil. Assuming that all the flux links all the turns, determine the inductance of the coil.

P1.5.2 The current through a 2 mH inductor is 10 A, $t \le 0$, and $Ate^{-10t} + Be^{-10t}, t \ge 0$, where t is in seconds. If the voltage across the inductor is 200 mV at $t = 0$, determine A and B.

P1.5.3 The inductance of a linear time-varying inductor is given by: $L(t) = L_0(1 - e^{-\alpha t})$. If the inductor current is $i(t) = I_0t^2$, determine the voltage across the inductor.

P1.5.4 A series of voltage pulses of 10 mV amplitude and 2 ms duration are applied to an ideal 5 µH inductor, with no initial energy storage. How many pulses are required to bring the inductor current to 20 A?

P1.5.5 A 2 μH inductor has no initial energy storage. A voltage pulse of amplitude 100 μV and 200 ms duration is applied at $t = 0$. Express the inductor current as a function of time.

P1.5.6 A 2 μH inductor has no initial energy storage. A voltage impulse of 100 μVs strength is applied at $t = 0$, and another impulse of -100 μVs strength is applied at $t = 200$ ms. Express the inductor current as a function of time. (Refer to P1.4.9).

P1.5.7 Repeat the preceding two problems assuming the 2 μH inductor had an initial current of 10 A.

P1.5.8 A voltage waveform is described by Figure P1.4.11, with the voltage in millivolts replacing the current in milliamperes. The voltage waveform is applied to a 0.5 μH inductor that has no initial energy storage. (a) Derive expressions for the inductor current during the time intervals: $0 \leq t \leq 10$ μs, $10 \leq t \leq 40$ μs, $40 \leq t \leq 60$ μs, $60 \leq t \leq 80$ μs, and $t > 80$ μs. (b) What is the flux linkage in the inductor at $t = 10$ μs and at $t = 50$ μs? (c) What is the energy stored in the inductor at $t = 80$ μs? (d) How do the expressions for the inductor current derived in part a change if the initial current is 0.5 A?

P1.5.9 A triangular current waveform is described by Figure P1.4.13, with the current in amperes replacing the voltage in volts. The triangular current pulse is applied to a 0.1 μH inductor that has no initial energy storage. Plot as a function of time: (a) the flux linkage in the inductor; (b) the energy stored in the inductor as derived from Equation 1.9.10; (c) the voltage across the inductor; and (d) the instantaneous power input to the inductor. How is the energy stored in (b) related to the instantaneous power in (d)?

P.1.5.10 Repeat Problem P1.5.9 assuming the current is $i_L(t) = 10\sin500t$ A, where t is in seconds.

P1.5.11 The current through a 5 μH inductor that has no initial energy storage is $i_L(t) = 10te^{-5t}$ A, where t is in ms and $t \geq 0$. Using MATLAB, plot on the same graph: $i_L(t)$, the voltage $v_L(t)$ across the inductor, the energy stored in the inductor, and the instantaneous power input to the inductor.

P1.5.12 Let the voltage applied to a 5 μH inductor that has no initial energy storage be $v_L = 10te^{-5t}$ mV, where t is in ms and $t \geq 0$. (a) Derive the expression for inductor current i_L using integration by parts. (b) Determine the current at the instant the voltage is a maximum. (c) Verify your result using MATLAB's int(y, t, a, b) command (Appendix SD.1).

P1.5.13 The flux linkage in a nonlinear inductor is given by $\lambda = 0.5i^2$ Wb-turns, when i is in amperes. If the inductor current is $i = 1 + t$ A, determine the additional energy stored in the inductor from $t = 0$ to $t = 3$ s.

P1.6   Miscellaneous

P1.6.1 A tube 10 cm long and 1 cm² in cross-sectional area contains a 0.1 M solution of $CuCl_2$ (cupric chloride). A voltage is applied between the two ends of the tube so that the velocity of Cu^{++} is 1 mm/min and that of Cl^- is 2 mm/min. Calculate the current through the tube.

P1.6.2 Look up typical power ratings of some of the following appliances and determine how many kilowatt-hours they would consume per month, based on normal, average usage: dishwasher, microwave oven, toaster, blender, washing machine, clothes dryer, room air conditioner, circulation fan, vacuum cleaner, hair dryer, electric iron, soldering iron, electric shaver, television set, and a small portable radio.

2

Basic Circuit Connections and Laws

Overview

There are two basic circuit laws, known as Kirchhoff's laws, which are expressions of conservation of charge and conservation of energy. They must be obeyed, therefore, by electric circuits under all conditions, independently of the circuit configuration and the circuit elements involved.

Kirchhoff's laws are first applied in this chapter to analyze some simple circuits, namely, resistive voltage dividers and current dividers. These simple circuits feature the two basic circuit connections: the series and parallel connections. Circuit elements connected in series or in parallel can be combined with other circuit elements in series or in parallel combinations to obtain circuits of any desired complexity. When resistors are connected in series or in parallel, the individual resistances or conductances combine according to some simple rules to give an equivalent resistance or conductance. The concept of equivalence is fundamental to circuit analysis and is applied to the Δ-Y transformation and transformation between voltage and current sources.

The voltage divider supplying a load at a reduced voltage is considered at the end of the chapter to illustrate a common and important feature of engineering design in general, namely, the trade-off between conflicting performance requirements. The example also illustrates the practical effects of tolerance of resistance values due to inevitable variations in component values during manufacturing.

Learning Objectives

- To be familiar with:
  - The definitions of terms used in circuit analysis, such as nodes, essential nodes, branches, essential branches, loops, and meshes
- To understand:
  - The nature and significance of Kirchhoff's current and voltage laws
  - The basic characteristics of voltage divider and current divider circuits
  - How resistances and conductances combine in series and in parallel
  - The equivalence between nonideal voltage and current sources
  - The fundamental concept of the equivalence of two circuits at a specified pair of terminals

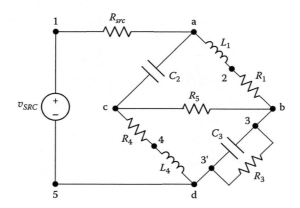

FIGURE 2.1.1
Bridge circuit.

- The trade-off in the design of a resistive voltage divider supplying a load at a reduced voltage

2.1 Circuit Terminology

A **node** is the junction of two or more circuit elements. An **essential node** is the junction of three or more circuit elements. Normally, an essential node is indicated on circuit diagrams by a filled circle.

A **path** is a set of one or more adjoining circuit elements that may be traversed in succession without passing through the same node more than once. A path generally has an initial node at its beginning and a final node at its end. If the initial and final nodes are the same, the path is closed and becomes a **loop**. A **mesh** is a loop that does not enclose any other loop.

A **branch** is a path that connects two nodes. An **essential branch** is a branch that connects two essential nodes without passing through an essential node.

Figure 2.1.1 shows a circuit configuration known as a bridge circuit. The nodes labeled a, b, c, and d are essential nodes, whereas nodes 1, 2, and 4 are not essential nodes. Nodes 3 and b are one and the same, as are nodes 3′, d, and 5 because no circuit element is connected between them, only a connection of zero resistance that is used for convenience of illustration. R_{src}, v_{SRC}, L_1, R_1, L_4, R_4, C_3, and R_3 taken individually are branches. The combinations R_{src}-v_{SRC}, L_1-R_1, L_4-R_4, and the individual branches C_3 and R_3 are essential branches. The closed paths d-5-1-a-b-d and d-5-1-a-c-b-d are loops. The loops d-5-1-a-c-d, a-b-c-a, and d-c-b-d (going through C_3) are meshes. C_3 and R_3 can also be considered to form a mesh.

2.2 Kirchhoff's Laws

Kirchhoff's Current Law (KCL)

> Statement: *At any instant of time, the sum of currents entering a node is equal to the sum of currents leaving the node.*

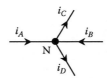

FIGURE 2.2.1
Kirchhoff's current law.

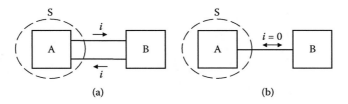

(a)   (b)

FIGURE 2.2.2
KCL applied to a surface S.

At node N in Figure 2.2.1, for example, KCL gives: $i_A + i_B = i_C + i_D$. KCL is a direct expression of conservation of current, which follows from conservation of charge, just as conservation of power follows from conservation of energy (Section 1.3, Chapter 1). We have already used conservation of charge in Chapter 1.

An alternative statement of KCL is: *at any instant of time, the algebraic sum of all the currents at any node is zero*. To apply this form, it is necessary to assign opposite signs to currents entering a node and to currents leaving a node. In Figure 2.2.1, for example, i_A and i_B may be assigned a positive sign, so that i_C and i_D will have to be assigned a negative sign. Then, according to this statement of KCL: $i_A + i_B - i_C - i_D = 0$, which is identical to the previous statement.

KCL may be applied not just to a node but also to interconnected circuits or parts of a circuit. In Figure 2.2.2a, for example, KCL may be applied to a surface S that encloses circuit A to conclude that, at any instant of time, the current in the upper connection between circuits A and B is equal in magnitude but flows in the opposite direction to the current in the lower connection. This is because the net current entering or leaving S must be zero because A and B, consisting of whole circuit elements, do not accumulate any net charge (see Problem P2.1.7). It also follows that if two circuits or parts of a circuit are connected by a single connection (Figure 2.2.2b), then no current flows in that connection because there is no return path for an equal but opposite current.

Comment

Evidently, conservation of charge applies to the actual directions of currents. It may be wondered if it also applies to assigned positive directions of current. The answer is yes, because if the value of a given current in the assigned positive direction is negative, then its direction is already opposite that of the actual current, and KCL would apply to both the actual and assigned positive directions of currents. For example, if in Figure 2.2.1 the actual currents are $i_A = 2$ A, $i_B = 3$ A, $i_C = 1$ A, and $i_D = 4$ A, KCL takes the form: $i_A + i_B = i_C + i_D$. If $i_A = -2$ A, then in order for its value to be negative, it would have to be assigned a direction opposite that of its actual direction. It would therefore be directed away from N in Figure 2.2.1, and KCL takes the form $i_B = i_A + i_C + i_D$. Substituting $i_A = -2$ A gives the same KCL equation.

EXERCISE 2.2.1

Consider KCL for node N in Figure 2.2.1 in the form $i_A + i_B = i_C + i_D$ based on the assigned positive directions of currents. Suppose that the following numerical values are obtained: $i_A = -2$ A, $i_C = -1$ A, and $i_B = 4$ A. Determine i_D from $i_A + i_B = i_C + i_D$, then redraw Figure 2.2.1 based on the actual directions of currents and verify KCL.

Answer: $i_D = 3$ A.

Kirchhoff's Voltage Law (KVL)

Statement: *At any instant of time, the sum of voltage rises around any loop is equal to the sum of voltage drops around the loop.*

An equivalent statement of KVL is: *at any instant of time, the algebraic sum of the voltages around any loop is zero*, because voltage drops and voltage rises have opposite signs. In Figure 2.2.3, for example, if the loop is traversed clockwise, then v_1, v_2, and v_3 are voltage rises, whereas v_4 and v_5 are voltage drops. According to the first statement of KVL, $v_1 + v_2 + v_3 = v_4 + v_5$, whereas according to the second statement, $v_1 + v_2 + v_3 - v_4 - v_5 = 0$, which is evidently an equivalent statement.

Concept: *KCL and KVL together are an expression of conservation of energy.*

Assume, for the sake of argument, that the sum of the voltage rises around the loop in Figure 2.2.3 is less than the sum of the voltage drops; that is, $(v_1 + v_2 + v_3) < (v_4 + v_5)$. Then, in taking a charge $+q$ clockwise around the loop, the work done on the charge is $q(v_1 + v_2 + v_3)$. But the charge can do more work, $q(v_4 + v_5)$, in going through the voltage drops. It would therefore be possible to continuously extract a net amount of energy from the circuit, simply by taking the charge $+q$ around the loop, in violation of conservation of energy. Note that taking the charge $+q$ around a loop, which necessarily involves passing through nodes or essential nodes between the circuit elements, implies that the charge $+q$ is conserved at these nodes; that is, KCL applies. Hence, KCL and KVL *together* are an expression of conservation of energy. This is in fact embodied in Tellegen's theorem (Section ST7.1, Chapter 7).

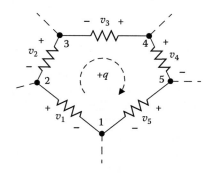

FIGURE 2.2.3
Kirchhoff's voltage law.

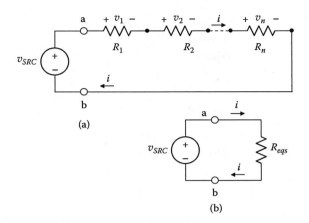

FIGURE 2.3.1
Series connection of resistors.

2.3   Voltage Division and Series Connection of Resistors

Definition: *In a series connection of elements, the same current flows through all the elements.*

Figure 2.3.1a illustrates a series connection of n resistors and a voltage source where the same current i flows through all the elements, and the voltage drops in the resistors are assigned in the direction of i. Because the same current flows through all the elements, KCL is automatically satisfied at every node. KVL gives

$$v_{SRC} = v_1 + v_2 + \ldots + v_n \qquad (2.3.1)$$

Applying Ohm's law to the individual resistors,

$$v_{SRC} = R_1 i + R_2 i + \ldots + R_n i \qquad (2.3.2)$$

Let R_m denote one of the resistors R_1 to R_n and v_m be the voltage drop across R_m so that $v_m = R_m i$. Dividing this relation by Equation 2.3.2, i cancels out, giving:

$$\frac{v_m}{v_{SRC}} = \frac{R_m}{R_1 + R_2 + \ldots + R_n} , \; m = 1, 2, \ldots, n \qquad (2.3.3)$$

According to Equation 2.3.3, v_{SRC} divides across the string of resistors in the ratio of each individual resistance to the total resistance. Thus, in the case of $v_{SRC} = 6$ V and three resistors $R_1 = 1\ \Omega$, $R_2 = 2\ \Omega$, and $R_3 = 3\ \Omega$, the voltages across the resistors are $v_1 = 1$ V, $v_2 = 2$ V, and $v_3 = 3$ V.

It also follows that the voltages across any two resistors are in the ratio of the corresponding resistances and in the inverse ratio of the conductances. For example:

$$\frac{v_1}{v_m} = \frac{R_1}{R_m} = \frac{G_m}{G_1} \tag{2.3.4}$$

Equation 2.3.2 may be expressed as:

$$v_{SRC} = (R_1 + R_2 + ... + R_n)i \tag{2.3.5}$$

An equivalent series resistor R_{eqs} is defined such that if v_{SRC} is applied across R_{eqs}, the resulting current is i (Figure 2.3.1b):

$$v_{SRC} = R_{eqs}i \tag{2.3.6}$$

Comparing Equation 2.3.5 and Equation 2.3.6, it follows that:

$$R_{eqs} = R_1 + R_2 + ... + R_n \tag{2.3.7}$$

For example, in the case of the 1, 2, and 3 Ω resistors connected in series, $R_{eqs} = 6~\Omega$. Note that because of the summation of positive quantities in Equation 2.3.7, R_{eqs} is larger than the largest of the individual resistances. If all the n resistances are equal, $R_{eqs} = nR$.

If the resistors in Figure 2.3.1a are represented by their conductances, then every resistance in Equation 2.3.7 is replaced by its reciprocal conductance, so that

$$R_{eqs} = \frac{1}{G_{eqs}} = \frac{1}{G_1} + \frac{1}{G_2} + ... + \frac{1}{G_n} \tag{2.3.8}$$

Thus, the conductances of the three series-connected 1, 2, and 3 Ω resistors are, respectively, 1, $\frac{1}{2}$, and $\frac{1}{3}$ S. The sum of their reciprocals is 6 S^{-1}, and $G_{eqs} = \frac{1}{6}$ S, which is, of course, the reciprocal of R_{eqs}.

In summary:

> **Concept:** *In a series connection of resistors, the resistances, or the reciprocals of the conductances, add.*

2.4 Current Division and Parallel Connection of Resistors

> **Definition:** *In a parallel connection of elements, the same voltage is applied across all the elements.*

Figure 2.4.1a illustrates a parallel connection of n resistors, where the same voltage v_{SRC} is applied to all the resistors, and the currents are assigned in the direction of the voltage drop v_{SRC}. In going around each mesh there is a voltage drop of v_{SRC} and a voltage rise of v_{SRC}, so that KVL is automatically satisfied. KCL at node a or node b gives:

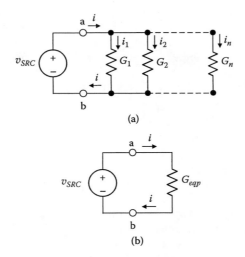

FIGURE 2.4.1
Parallel connection of resistors.

$$i = i_1 + i_2 + \dots + i_n \tag{2.4.1}$$

Applying Ohm's law to the individual resistors, in terms of their conductances:

$$i = G_1 v_{SRC} + G_2 v_{SRC} + \dots + G_n v_{SRC} \tag{2.4.2}$$

Let G_m denote one of the conductances G_1 to G_n and let i_m be the current through G_m so that $i_m = G_m v_{SRC}$. Dividing this relation by Equation 2.4.2, v_{SRC} cancels out, giving:

$$\frac{i_m}{i} = \frac{G_m}{G_1 + G_2 + \dots + G_n} , \quad m = 1, 2, \dots, n \tag{2.4.3}$$

According to Equation 2.4.3, i divides between the paralleled resistors in the ratio of each individual conductance to the total conductance. Moreover, the currents through any two resistors are in the ratio of the corresponding conductances, or the inverse ratio of the resistances. Thus:

$$\frac{i_1}{i_m} = \frac{G_1}{G_m} = \frac{R_m}{R_1} \tag{2.4.4}$$

A case that is often encountered is that of two resistors in parallel (Figure 2.4.2). From Equation 2.4.3 and Equation 2.4.4:

$$\frac{i_1}{i} = \frac{G_1}{G_1 + G_2} = \frac{R_2}{R_1 + R_2} \tag{2.4.5}$$

$$\frac{i_2}{i} = \frac{G_2}{G_1 + G_2} = \frac{R_1}{R_1 + R_2} \tag{2.4.6}$$

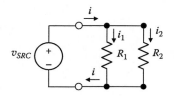

FIGURE 2.4.2
Two-resistor current divider.

$$\frac{i_1}{i_2} = \frac{G_1}{G_2} = \frac{R_2}{R_1} \tag{2.4.7}$$

The currents divide in direct proportion to the conductances, or in inverse proportion to the resistances, the smaller resistance drawing a larger current. If $v_{SRC} = 6$ V, $R_1 = 1$ Ω, and $R_2 = 2$ Ω, then $i_1 = 6$ A, $i_2 = 3$ A, and $i = 9$ A.

Equation 2.4.2 may be expressed as:

$$i = (G_1 + G_2 + \ldots + G_n)v_{SRC} \tag{2.4.8}$$

An equivalent parallel conductance G_{eqp} is defined such that if the parallel combination is replaced by G_{eqp}, then the same voltage v_{SRC} applied between terminals ab produces the same current i (Figure 2.4.1b):

$$i = G_{eqp}v_{ab} \tag{2.4.9}$$

Comparing Equation 2.4.8 and Equation 2.4.9, it follows that

$$G_{eqp} = G_1 + G_2 + \ldots + G_n \tag{2.4.10}$$

For example, if three resistors of 1, 2, and 3 S are connected in parallel, $G_{eqp} = 6$ S.

If the resistors in Figure 2.4.1a are represented by their resistances, then every conductance in Equation 2.4.10 is replaced by its reciprocal resistance, so that:

$$G_{eqp} = \frac{1}{R_{eqp}} = \frac{1}{R_1} + \frac{1}{R_2} + \ldots + \frac{1}{R_n} \tag{2.4.11}$$

Thus, the resistances of the three parallel-connected 1, 2, and 3 S resistors are, respectively, 1, $\frac{1}{2}$, and $\frac{1}{3}$ Ω. The sum of their reciprocals is 6 Ω⁻¹, and $R_{eqp} = \frac{1}{6}$ Ω, which is the reciprocal of G_{eqp}. Note that if R_m is the smallest resistance in the parallel combination, it follows from Equation 2.4.11 that $\frac{1}{R_{eqp}} > \frac{1}{R_m}$, so that $R_{eqp} < R_m$. That is, the parallel resistance is smaller than the smallest resistance of the paralleled resistors. If all the resistances are equal, $R_{eqp} = R/n$.

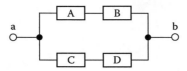

FIGURE 2.4.3
Figure for Exercise 2.4.1.

In summary:

> **Concept:** *In a parallel connection of resistors, the conductances, or the reciprocals of the resistances, add.*

According to Equation 2.4.11, R_{eqp} is the product of the n resistances divided by the sum of their products $(n-1)$ at a time. In the case of three resistors in parallel,

$$R_{eqp} = \frac{R_1 R_2 R_3}{R_1 R_2 + R_2 R_3 + R_3 R_1} \tag{2.4.12}$$

For two resistors,

$$R_{eqp} = \frac{R_1 R_2}{R_1 + R_2} \tag{2.4.13}$$

Note that Ohm's law applied to Equation 2.4.7 and to Equation 2.4.5, or Equation 2.4.6 using Equation 2.4.13, gives: $v_{SRC} = R_1 i_1 = R_2 i_2 = R_{eqp} i$.

EXERCISE 2.4.1

Determine the equivalent circuit parameter if elements A to D in Figure 2.4.3 are, respectively: (a) resistances of value 2, 3, 6, and 9 Ω; (b) conductances of value 3, 6, 10, and 15 S. (c) If a voltage of 12 V is applied between terminals ab, determine the voltage across each element and the current through it, for both cases (a) and (b).

Answer: (a) 3.75 Ω; (b) 8 S. (c) In case a, the current is 2.4 A in the AB branch and 0.8 A in the CD branch; the voltages are 4.8 V across A and C, and 7.2 V across B and D; in case b, the current is 24 A in the AB branch and 72 A in the CD branch; the voltages are 8 V across A, 4 V across B, 7.2 V across C, and 4.8 V across D.

Example 2.4.1: Resistivity

An instructive and useful problem is the calculation of the resistance R between opposite ends of a block of material of length L units and uniform cross-sectional area A square units (Figure 2.4.4). For simplicity, a block of rectangular cross-section is shown.

SOLUTION

Let the block be divided into a number of unit cubes, each having a resistance ρ between opposite faces. A strip of length L units and cross-sectional area of one square unit will have L cubes in series and a resistance ρL end-to-end. Because the block consists of A such strips in parallel:

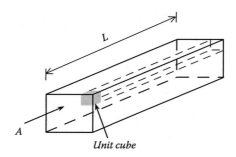

FIGURE 2.4.4
Figure for Example 2.4.1.

$$R = \rho \frac{L}{A} \qquad (2.4.14)$$

ρ is the **resistivity**, or **specific resistance**, of the material. It is an intrinsic property of the material and is independent of the size or shape of a given sample, just as density is independent of size and shape. According to Equation 2.4.14, the unit of resistivity is Ω- (unit length). For copper, ρ is approximately 1.7×10^{-6} Ωcm at room temperature, or 1.7×10^{-8} Ωm. The reciprocal of resistivity is **conductivity**, usually denoted by σ.

Application Window 2.4.1: Strain Gauges

In a **strain gauge**, a metallic wire or conductor is stretched within its elastic limit, and the change in its resistance measured. If the conductor is bonded to a surface, the change in resistance is a measure of the strain $\Delta L/L$. If the conductor is unbonded, the change in resistance is a measure of the applied force or pressure. When the conductor elongates under stress, its cross-sectional area decreases, but its volume remains constant, as long as the elastic limit is not exceeded. Because the conductor is usually of uniform cross-section, Equation 2.4.14 becomes $R = \rho \dfrac{L^2}{\Lambda}$, where $\Lambda = LA$ is the volume. Taking logarithms, $\ln R = \ln \rho + 2\ln L - \ln \Lambda$, and then taking differentials, $\dfrac{\Delta R}{R} = \dfrac{\Delta \rho}{\rho} + 2\dfrac{\Delta L}{L}$. Dividing both sides by $\Delta L/L$:

$$\mathrm{GF} = \frac{\Delta R / R}{\Delta L / L} = 2 + \frac{\Delta \rho / \rho}{\Delta L / L} \qquad (2.4.15)$$

where GF is the **gauge factor**. In the case of many metals, $\Delta \rho \cong 0$, so that GF $\cong 2$. In some metallic alloys, the distortion of the crystalline structure due to the applied stress increases ρ, which increases GF to about 3.5. ρ in semiconductors is much more sensitive to stress, and $\Delta \rho$ could be positive or negative. The $\dfrac{\Delta \rho / \rho}{\Delta L / L}$ term in Equation 2.4.15 becomes dominant, giving $|\mathrm{GF}| \cong 120$.

Example 2.4.2: Series-Parallel Connection of Lamps

100 lamps rated at 12 V, 6 W each are to be connected across a 240 V supply such that the rated voltage of 12 V is applied to each lamp. Design an appropriate arrangement of the lamps, and calculate the equivalent resistance and the total current drawn from the supply.

SOLUTION

Because $\dfrac{240 \text{ V}}{12 \text{ V}} = 20$, then 20 lamps may be connected in series across the supply. The voltage across each lamp will be the rated voltage of 12 V, assuming the lamp resistances are all equal. To accommodate 100 lamps, 5 such series combinations will have to be paralleled across the supply. The resistance of each lamp is $R = \dfrac{v^2}{p} = \dfrac{(12)^2}{6} = 24\ \Omega$ at the normal operating temperature. The equivalent series resistance of 20 lamps is $24 \times 20 = 480\ \Omega$. The equivalent parallel resistance of five of these series combinations is $\dfrac{480}{5} = 96\ \Omega$.

The lamp current at 12 V is $\dfrac{6 \text{ W}}{12 \text{ V}} = 0.5$ A, which is also the current in each series combination. The total current drawn from the supply is $5 \times 0.5 = 2.5$ A.

As a check, we may calculate P, the total power supplied. From the total current, $P = 240 \times 2.5 = 600$ W. From the total resistance, $P = (5 \times 0.5)^2 \times 96 = \dfrac{(240)^2}{96} = 600$ W. From the total number of lamps and their individual ratings, $P = 100 \times 6$ W $= 600$ W.

2.5 Δ-Y Transformation

A useful application of series and parallel connection of resistances is the Δ-Y transformation: given three resistances R_a, R_b, and R_c connected in Δ (Figure 2.5.1a), it is required to determine the resistances R_1, R_2, and R_3 connected in Y (Figure 2.5.1b) that will make the two circuits equivalent between terminals a, b, and c. This means that the resistance seen between any two terminals is the same for the two circuits, the third terminal being connected in the same way in both cases. The inverse problem is also required: given three resistances R_1, R_2, and R_3 connected in Y, it is required to determine R_a, R_b, and R_c of the equivalent Δ.

Let us equate the resistances between corresponding pairs of terminals in the two circuits, with the third terminal left open. In Figure 2.5.1a, the resistance R_{ab} between terminals ab, with terminal c open, is R_c in parallel with R_a in series with R_b. That is, $R_{ab} = \dfrac{R_c(R_a + R_b)}{R_a + R_b + R_c}$. The resistance between terminals ab in Figure 2.5.1b, with the terminal c open, is $R_1 + R_2$. Equating the resistances in both cases:

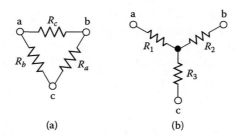

FIGURE 2.5.1
Δ and Y circuits.

$$R_1 + R_2 = \frac{R_c\left(R_a + R_b\right)}{R_a + R_b + R_c} \tag{2.5.1}$$

Similarly, equating the resistances between terminals bc, with terminal a open,

$$R_2 + R_3 = \frac{R_a\left(R_b + R_c\right)}{R_a + R_b + R_c} \tag{2.5.2}$$

Equating the resistances between terminals ac, with terminal b open,

$$R_1 + R_3 = \frac{R_b\left(R_a + R_c\right)}{R_a + R_b + R_c} \tag{2.5.3}$$

Equation 2.5.1 to Equation 2.5.3 are three independent equations. If R_a, R_b, and R_c of the Δ-circuit are given, these equations can be solved for R_1, R_2, and R_3 of the equivalent Y-circuit to give:

$$R_1 = \frac{R_b R_c}{R_a + R_b + R_c} \tag{2.5.4}$$

$$R_2 = \frac{R_a R_c}{R_a + R_b + R_c} \tag{2.5.5}$$

$$R_3 = \frac{R_a R_b}{R_a + R_b + R_c} \tag{2.5.6}$$

Conversely, if the resistances R_1, R_2, and R_3 of the Y-circuit are given, Equation 2.5.1 to Equation 2.5.3 can be solved for R_a, R_b, and R_c of the equivalent Δ-circuit to give:

$$R_a = \frac{R_1 R_2 + R_2 R_3 + R_3 R_1}{R_1} \tag{2.5.7}$$

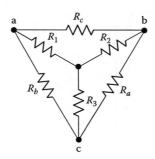

FIGURE 2.5.2
Δ-Y transformation.

$$R_b = \frac{R_1 R_2 + R_2 R_3 + R_3 R_1}{R_2} \tag{2.5.8}$$

$$R_c = \frac{R_1 R_2 + R_2 R_3 + R_3 R_1}{R_3} \tag{2.5.9}$$

To help remember the preceding relations, the two equivalent circuits are superimposed (Figure 2.5.2). Each Y-resistance is the product of the Δ-resistances on either side, divided by the sum of the three Δ-resistances. Conversely, each Δ-resistance is the product of Y-resistances, two at a time, divided by the Y-resistance that is opposite the given Δ-resistance.

When the three resistances in either configuration are equal, the preceding relations reduce to:

$$R_\Delta = 3R_Y \text{ and } R_Y = R_\Delta/3 \tag{2.5.10}$$

To help remember these relations, note that the Δ-circuit is more of a parallel circuit, whereas the Y-circuit is more of a series circuit. Because resistances in series give a larger resistance whereas resistances in parallel give a smaller resistance, then if the two circuits are to be equivalent, the Δ-circuit should have the larger resistances.

A common application of the Δ-Y transformation is to replace a Δ connection in a circuit by its equivalent Y, using Equation 2.5.4 to Equation 2.5.6, which results in a simpler circuit, as illustrated by Example 2.5.1. The Δ-Y transformation is also extensively used in three-phase systems (Chapter 8).

EXERCISE 2.5.1

Three 6 Ω resistors connected in Δ are shown in Figure 2.5.3. Determine: (a) the equivalent Y circuit; (b) the resistance between terminals ab in both circuits with terminal c short-circuited to a. (c) Assuming that a 12 V source is connected between terminals ab with terminal c open, calculate the power dissipated in both circuits. Are the results in part b and part c as expected?

Answer: (a) R_Y = 2 Ω; (b) 3 Ω; (c) 36 W. Yes, the two circuits are equivalent at corresponding terminals, with the third terminal connected in the same way in both cases. Because the *v–i* relation is the same at terminals ab, the power delivered, and hence the power dissipated in the two circuits, must be the same.

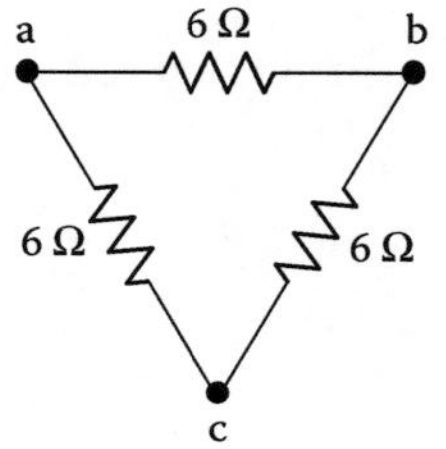

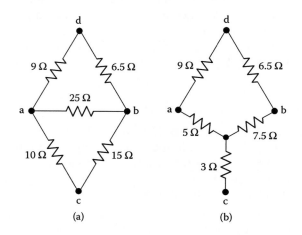

FIGURE 2.5.3
Figure for Exercise 2.5.1.

Simulation Example 2.5.1: Δ-Y Transformation

It is required to obtain the resistance seen between terminals cd in the circuit of Figure 2.5.4a analytically and by simulation.

ANALYSIS

The lower abc Δ is transformed to its equivalent Y (Figure 2.5.4b, as in Figure 2.5.2).

This gives: $R_1 = \dfrac{25 \times 10}{50} = 5\ \Omega$, $R_2 = \dfrac{25 \times 15}{50} = 7.5\ \Omega$, and $R_3 = \dfrac{10 \times 15}{50} = 3\ \Omega$. The 5 Ω resistance is then added to the 9 Ω to give 14 Ω, and the 7.5 Ω resistance is added to the 6.5 Ω to give 14 Ω. The two 14 Ω resistors in parallel give 7 Ω, which is added to the 3 Ω to give 10 Ω between terminals cd.

SIMULATION

The five resistors of Figure 2.5.4a are entered in the schematic. To determine the resistance between terminals cd, a 1 V dc voltage source is applied between these terminals and a Bias Point simulation run. By pressing the I button, the source current is found to be 100 mA, which gives a resistance of 10 Ω between terminals cd. Note that in PSpice, the source current is directed through the source from the + to the − terminal, in the direction of a voltage drop. Press the W button and note that the source absorbs a power of −100 mW. The total power dissipated in the resistors is also 100 mW.

FIGURE 2.5.4
Figure for Example 2.5.1.

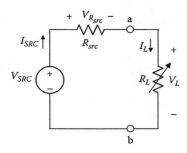

FIGURE 2.6.1
Voltage source having a source resistance.

2.6 Source Equivalence and Transformation

A practical voltage source departs from the ideal in that the voltage at its terminals decreases with increasing current supplied by the source. If the variation is linear, then at least in the case of dc sources, this effect may be simulated by adding a resistance R_{src} in series with an ideal voltage source element (Figure 2.6.1). This **source resistance**, or internal resistance of the source, is included with the source, so that the source terminals become a and b. R_L, having an arrow across it, is a variable resistance representing a load that can draw different currents from the source.

From KCL: $I_L = I_{SRC}$ at nodes a or b. In going clockwise around the circuit, KVL gives:

$$V_{SRC} = V_{R_{src}} + V_L \qquad (2.6.1)$$

Ohm's law gives $V_{R_{src}} = R_{src}I_{SRC} = R_{src}I_L$. Substituting in Equation 2.6.1:

$$V_L = V_{SRC} - R_{src}I_L \qquad (2.6.2)$$

It is seen from Equation 2.6.2 that as I_L increases, the voltage V_L at the source terminals decreases. When $I_L = 0$, $V_L = V_{SRC}$. Thus, V_{SRC} is the **open-circuit voltage** of the source. When I_L equals a specified, or rated, value, V_L drops to the value given by Equation 2.6.2. This variation in V_L, due to the loading effect of I_L, is the **voltage regulation** of the voltage source. It is equal, in percentage, to $\dfrac{V_{SRC} - V_L}{V_L} \times 100$. The voltage regulation is due to a nonzero R_{src} and is zero for $R_{src} = 0$.

The corresponding case of a current source is shown in Figure 2.6.2. KCL at node a or b gives:

$$I_{SRC} = I_{R_{src}} + I_L \qquad (2.6.3)$$

KVL is automatically satisfied because the same voltage V_L is across all the circuit elements. Multiplying both sides of Equation 2.6.3 by R_{src} and substituting $V_L = R_{src}I_{R_{src}}$:

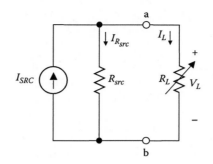

FIGURE 2.6.2
Current source having a source resistance.

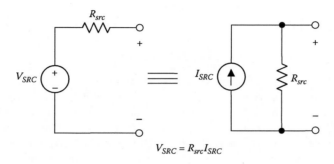

$$V_{SRC} = R_{src}I_{SRC}$$

FIGURE 2.6.3
Source transformation.

$$V_L = R_{src}I_{SRC} - R_{src}I_L \tag{2.6.4}$$

When $V_L = 0$, $I_L = I_{SRC}$. As V_L increases, more of I_{SRC} is diverted to R_{src}, so I_L decreases. **Current regulation** occurs, analogous to voltage regulation in a voltage source.

It is seen that Equation 2.6.2 and Equation 2.6.4 are identical if $V_{SRC} = R_{src}I_{SRC}$, with R_{src} being the same in both cases. Under these conditions, if the two sources are hidden from view, and only the variation of V_L with I_L is observed, it would be impossible to tell which source is connected to R_L. In other words, the two sources are *equivalent as far as the load is concerned.*

It is concluded, therefore, that a voltage source V_{SRC} in series with a resistance R_{src} is

equivalent at its terminals to a current source $I_{SRC} = \dfrac{V_{SRC}}{R_{src}}$ in parallel with the same resis-

tance R_{src}. Conversely, a current source I_{SRC} in parallel with a resistance R_{src} (or a conduc-

tance $G_{src} = \dfrac{1}{R_{src}}$) is equivalent at its terminals to a voltage source $V_{SRC} = R_{SRC}I_{src}$ in series

with the same resistance R_{src}. Source transformation is summarized in Figure 2.6.3 and applies to dependent sources as well. By its very nature, source transformation preserves the polarities of voltage and current at the source terminals.

The concept of equivalence of two circuits at a pair of terminals is of fundamental importance and may be generalized as follows:

Concept: *Two circuits are said to be equivalent at a specified pair of terminals if the voltage–current relation at these terminals is identical for both circuits.*

We have used this concept in the case of Δ-Y transformation, because having the same resistance between a specified pair of terminals is a consequence of having the same voltage–current relation. It should be emphasized that the equivalence applies only at the specified terminals and does not apply, in general, to the rest of the circuit, as illustrated by Example 2.6.1.

Example 2.6.1: Equivalent Sources

Given a voltage source having an open-circuit voltage of 12 V and a source resistance of 0.5 Ω (Figure 2.6.4a), it is required to derive the equivalent current source.

SOLUTION

According to the preceding discussion, the equivalent current source has a source current

of $\dfrac{12\ \text{V}}{0.5\ \Omega} = 24$ A and a source resistance of 0.5 Ω (Figure 2.6.4b).

If both sources are connected to a 5.5 Ω load, for example, I_L for the voltage source case

is, from Figure 2.6.4a: $I_L = \dfrac{12\ \text{V}}{(0.5 + 5.5)\ \Omega} = 2$ A and $V_L = 2 \times 5.5 = 11$ V. In the case of the

current source (Figure 2.6.4b), the resistance of the parallel combination is

$\dfrac{0.5 \times 5.5}{0.5 + 5.5} = \dfrac{5.5}{12}\ \Omega; V_L = \dfrac{5.5}{12} \times 24 = 11$ V, as in the case of the voltage source. However, the

power delivered by the ideal voltage source is $12 \times 2 = 24$ W, whereas the power delivered by the ideal current source is $11 \times 24 = 264$ W. The power absorbed by R_{src} is $0.5 \times (2)^2 =$

2 W in the case of the voltage source and $\dfrac{(11)^2}{0.5} = 242$ W in the case of the current source.

Moreover, the current in the 0.5 Ω resistor is not the same in both cases. It is 2A in Figure 2.6.4a and 22 A in Figure 2.6.4b. This emphasizes that equivalence between two circuits holds only at the specified terminals and does not apply inside the circuits containing equivalent sources. The difference between the power delivered by the ideal source element and that dissipated in R_{src} is 22 W, the power absorbed by the load. It is the same in both cases because the voltage and current at the load terminals are the same.

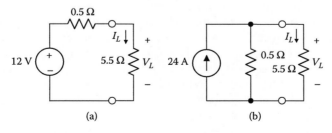

(a)   (b)

FIGURE 2.6.4
Figure for Example 2.6.1.

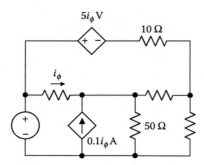

FIGURE 2.6.5
Figure for Exercise 2.6.1.

EXERCISE 2.6.1

In the circuit of Figure 2.6.5, transform: (a) the dependent current source to a voltage source, and (b) the dependent voltage source to a current source. Can the independent voltage source be transformed to a current source?

Answer: (a) CCVS of $5i_\phi$ V in series with 50 Ω; (b) CCCS of $0.5i_\phi$ A in parallel with 10 Ω. No, because it is not in series with a resistor only.

Before ending this section, we wish to emphasize that according to Equation 2.6.2, the full voltage V_{SRC} is available at the terminals of a voltage source only when $I_L = 0$; that is, when the source terminals are open-circuited. Similarly, the full current I_{SRC} is available at the terminals of a current source only when $V_L = 0$; that is, when the source terminals are short-circuited. Hence:

> **Concept:** *The ideal termination for a voltage source is an open circuit, whereas the ideal termination for a current source is a short circuit.*

2.7   Reduced-Voltage Supply

The simple resistive voltage divider is often used to supply a load at a reduced voltage. A common example is the supply of the base of a discrete bipolar junction transistor in the common-emitter configuration, or of the gate of a discrete metal-oxide semiconductor transistor in the common-source configuration. Another example is given in Example 2.7.2. In this section we will investigate the design considerations for such a voltage divider.

The basic circuit, shown in Figure 2.7.1, consists of a voltage source V_{SRC}, R_1, and R_2, with a load connected across R_2. Normally, V_{SRC} is given together with the nominal load voltage V_L and a range of variation of the load current I_L. It is desired to determine R_1 and R_2 such that the variation in V_L remains within specified limits, as I_L changes over the given range.

From KCL at either essential node:

$$I_1 = I_2 + I_L \qquad (2.7.1)$$

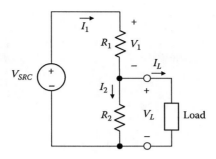

FIGURE 2.7.1
Voltage divider supplying a load.

If the voltage drop across R_1 is denoted by V_1, the voltage drop across R_2 being V_L, then applying KVL clockwise around the mesh composed of V_{SRC}, R_1, and R_2:

$$V_{SRC} = V_1 + V_L \qquad (2.7.2)$$

From Ohm's law,

$$V_1 = R_1 I_1 \quad \text{and} \quad V_L = R_2 I_2 \qquad (2.7.3)$$

I_1, I_2, and V_1 may be eliminated from the preceding equations to give:

$$V_L = \frac{R_2}{R_1 + R_2}\left(V_{SRC} - R_1 I_L\right) \qquad (2.7.4)$$

$$= \frac{R_2}{R_1 + R_2}V_{SRC} - (R_1 \| R_2)I_L \qquad (2.7.5)$$

where $R_1 \| R_2 = \dfrac{R_1 R_2}{R_1 + R_2}$ is the parallel resistance of R_1 and R_2. Comparing Equation 2.7.5 with Equation 2.6.2, it follows that the voltage divider appears to the load as a source of open-circuit voltage $\dfrac{R_2}{R_1 + R_2}V_{SRC}$, when $I_L = 0$, and source resistance $R_1 \| R_2$ (Figure 2.7.2).

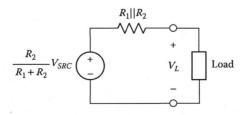

FIGURE 2.7.2
Voltage divider as source.

This is an important interpretation to which we will return in Section 4.1 (Chapter 4), in connection with Thevenin's equivalent circuit.

Given V_{SRC}, V_L, and I_L, Equation 2.7.5 provides a relation between R_1 and R_2, but is not sufficient for uniquely determining their values. The additional information needed for this purpose comes from specifying the voltage variation $\Delta V_L = V_{L2} - V_{L1}$ for $\Delta I_L = I_{L2} - I_{L1}$. Substituting V_{L1} and I_{L1} first in Equation 2.7.5 and then V_{L2} and I_{L2}, and subtracting, eliminates the term in V_{SRC} and gives:

$$\Delta V_L = -(R_1 \| R_2)\Delta I_L \qquad (2.7.6)$$

Equation 2.7.6 also follows from Equation 2.7.5 by taking differentials, with V_{SRC} constant. It is readily interpreted with reference to Figure 2.7.2. When $\dfrac{R_2}{R_1 + R_2} V_{SRC}$ is constant and I_L changes by ΔI_L, ΔV_L is due to the voltage drop in $R_1 \| R_2$, which is $(R_1 \| R_2) \Delta I_L$. The minus sign in Equation 2.7.6 signifies that V_L decreases with I_L.

It is desirable, in practice, to have a small ΔV_L; that is, a small variation in V_L with I_L. It follows from Equation 2.7.6 that $R_1 \| R_2$ should be small in this case. If $R_1 \| R_2$ is small, it is seen from Equation 2.7.5 that the second term $(R_1 \| R_2)I_L$ will be small compared to the first term $\dfrac{R_2}{R_1 + R_2} V_{SRC}$, which means that $V_L \cong \dfrac{R_2}{R_1 + R_2} V_{SRC}$. Under these conditions, $\dfrac{R_2}{R_1 + R_2}$ is fixed for given V_{SRC} and V_L. To reduce $R_1 \| R_2$ while keeping $\dfrac{R_2}{R_1 + R_2}$ constant, both R_1 and R_2 should be reduced. But reducing R_1 and R_2 increases the current drain from the supply and generally increases the power dissipated in the voltage divider (Section ST2.1). This is undesirable, particularly in battery-operated equipment. Hence a trade-off must be made in the design of the resistive voltage divider between good voltage regulation, that is, a small ΔV_L, on the one hand, and current drain from the supply and power dissipation in R_1 and R_2 on the other hand (Example 2.7.1). For these reasons, the resistive voltage divider is impractical where there is a substantial variation in load current. However, if the load current is small, the resistive voltage divider is quite satisfactory, as illustrated by Example 2.7.2. This is also true of the base supply of low-power bipolar junction transistors and of the gate supply of metal-oxide-semiconductor field-effect transistors, where the load current of the voltage divider is practically zero in the latter case.

EXERCISE 2.7.1

Assume that in the voltage divider of Figure 2.7.1, $V_{SRC} = 45$ V, $R_1 = 60\ \Omega$, $R_2 = 40\ \Omega$, and the load is a 120 Ω resistor. Determine: (a) the current in each resistor; (b) the voltage across the 120 Ω load; (c) the open-circuit voltage and source resistance of the voltage divider (Equation 2.7.5); and (d) the power dissipated in R_1 and R_2.

Answer: (a) $I_1 = 0.5$ A, $I_2 = 375$ mA, $I_L = 125$ mA; (b) 15 V; (c) 18 V, 24 Ω; (d) $P_1 = 15$ W, $P_2 = 5.625$ W.

Design Example 2.7.1: Voltage Regulation and Power Dissipation in a Voltage Divider

A 3 V, 0.1 A dc load is to be supplied from a 12 V dc supply. The load current may vary by ± 20% of its nominal value of 0.1 A. The corresponding variation in the load voltage should not exceed ± 5% of the nominal value of 3 V. It is required to determine suitable values of R_1 and R_2.

SOLUTION

According to Equation 2.7.5, I_L has its largest value when V_L has its smallest value, and conversely. At $I_L = 0.8 \times 0.1 = 0.08$ A, $V_L = 1.05 \times 3 = 3.15$ V, and at $I_L = 1.2 \times 0.1 = 0.12$ A, $V_L = 0.95 \times 3 = 2.85$ V. Substituting these values in Equation 2.7.5 gives two relations between R_1 and R_2:

$$3.15 = 12\frac{R_2}{R_1 + R_2} - 0.08\frac{R_1 R_2}{R_1 + R_2} \quad \text{and} \quad 2.85 = 12\frac{R_2}{R_1 + R_2} - 0.12\frac{R_1 R_2}{R_1 + R_2}$$

From these, $R_1 = 24\ \Omega$ and $R_2 = 10.9\ \Omega$. Note that an easy way to solve these equations is to consider $\dfrac{R_2}{R_1 + R_2}$ as one variable and $\dfrac{R_1 R_2}{R_1 + R_2}$ as another variable.

Let us calculate P, the total power dissipated in R_1 and R_2 under these conditions. With reference to Figure 2.7.1, $P = \dfrac{(V_L)^2}{R_2} + \dfrac{(V_{SRC} - V_L)^2}{R_1}$. When $V_L = 3.15$ V, $P = 4.17$ W, and when $V_L = 2.85$ V, $P = 4.23$ W. P may not seem like an excessive amount of power. But if it is noted that the power consumed by the load does not exceed 0.34 W, when $V_L = 2.85$ V and $I_L = 0.12$ A, it is seen that $\dfrac{0.34}{0.34 + 4.23} \times 100 \cong 7.5\%$ of the power supplied by V_{SRC} goes to the load under these conditions, the rest is dissipated in R_1 and R_2. The power efficiency is therefore quite poor.

Suppose that in order to dissipate less power in R_1 and R_2, R_1 is increased by a factor of 3 to 72 Ω. From Equation 2.7.5: $3 = (12 - 0.1 \times 72)\dfrac{R_2}{72 + R_2}$, which gives $R_2 = 120\ \Omega$. With these values of R_1 and R_2, $V_L = 3$ V when $I_L = 0.1$ A, and the power dissipated in R_1 and R_2 is now $\dfrac{(3)^2}{120} + \dfrac{(9)^2}{72} = 1.2$ W. From Equation 2.7.6, $\Delta V_L = -\dfrac{72 \times 120}{192} \times 0.04 = -1.8$ V. Thus, as I_L changes from 0.08 A to 0.12 A, V_L changes from $3 + \dfrac{1.8}{2} = 3.9$ V to $3 - \dfrac{1.8}{2} = 2.1$ V, a variation of ±30% of the nominal value of 3 V. Such a variation will most likely not be acceptable.

If the load current is constant, as may occur in the case of a lamp or a heater, for example, R_2 may be dispensed with altogether and only R_1 used. If $I_L = 0.1$ A and $V_L = 3$ V, then $R_1 = \dfrac{12 - 3}{0.1} = 90\ \Omega$ and the power dissipated is $(0.1)^2 \times 90 = 0.9$ W. R_1 is used in this case to drop the supply voltage by 9 V at 0.1 A.

Design Example 2.7.2: Measurement of Partial Pressure of Oxygen

The partial pressure of oxygen in the blood is measured as a direct indication of lung function, and as an indirect measurement of glucose concentration. Approximately 0.7 V is applied between two suitable electrodes, the resulting current being directly proportional to the partial pressure of oxygen in solution. For the partial pressures normally of interest, the range of current is 0 to 1.5 μA. It is required to determine suitable values of R_1 and R_2, assuming $V_{SRC} = 12$ V and $V_L = 0.7$ V.

SOLUTION

A good guide for selecting R_1, according to Equation 2.7.4, is to have the largest value of

$R_1 I_L$ small compared to V_{SRC}, say, one tenth of V_{SRC}. Then, $R_1 = \dfrac{0.1 \times 12 \text{ V}}{1.5 \text{ μA}} = 800$ kΩ. The

nearest smaller standard value of a 5% tolerance resistor is 750 kΩ (Appendix SE.1). Using this value in Equation 2.7.5, with $V_L = 0.7$ V and a mid-range value of $I_L = 0.75$ μA, gives $R_2 = 48.9$ kΩ. The nearest smaller standard value of a 5% tolerance resistor is 47 kΩ. These values give $V_L = 0.708$ V at $I_L = 0$ and $V_L = 0.641$ V at $I_L = 1.5$ μA. The variation in V_L is about ± 5% with respect to the nominal value of 0.7 V, which is quite acceptable.

With $R_1 \| R_2 = 44.23$ kΩ, the power dissipated in R_1 and R_2 is $\dfrac{(0.7)^2}{44.23} + \dfrac{(12-0.7)^2}{750} = 0.18$ mW,

which is small because I_L is small, so that R_1 and R_2 are relatively large.

Simulation Example 2.7.3: Sensitivity Analysis of Resistive Voltage Divider

Given a voltage divider having $V_{SRC} = 12$ V dc, $R_1 = 1$ kΩ and $R_2 = 3$ kΩ supplying a load. If V_{SRC} can vary by ± 5%, and the tolerance on the values of R_1 and R_2 is also ± 5%, it is required to determine the largest variation in load voltage at a load current of 1 mA, due to variations in V_{SRC} and resistor tolerances.

SOLUTION

PSpice can be used to perform a sensitivity analysis that gives the variation of each specified output variable for small changes in every circuit parameter or source. We can determine the sensitivity of V_L to changes in circuit parameters at a load current of 1 mA by connecting a 1 mA current source as a load, as shown in the schematic of Figure 2.7.3. The output node is named VL using net alias (Appendix SD.2). To perform the sensitivity analysis, select Bias Point under Analysis type in the Simulation Settings window, then check the Perform Sensitivity analysis box under Output File Options. After running PSpice, the results of the sensitivity analysis displayed in the output file may be tabulated as follows:

DC Sensitivities of Output V(VL)

Element Name	Element Value	Element Sensitivity (Volts/Unit)	Normalized Sensitivity (Volts/Percent)
R_R1	1.000E+03	−2.813E−03	−2.813E−02
R_R2	3.000E+03	6.875E−04	2.063E−02
V_V1	1.200E+01	7.500E−01	9.000E−02
I_I1	1.000E−03	−7.500E+02	−7.500E−03

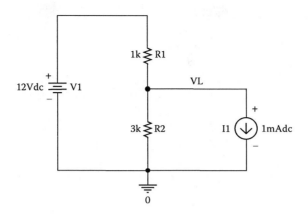

FIGURE 2.7.3
Figure for Example 2.7.3.

Considering R_1, the third column indicates that an increase (or decrease) of 1 Ω in R_1 decreases (or increases) V_L by 2.813×10^{-3} V. The fourth column indicates that an increase (or decrease) of 1% in R_1 decreases (or increases) V_L by 2.813×10^{-3} V. Similar interpretations apply to the other rows.

PSpice can also be used to determine the maximum and minimum values of V_L due to resistor tolerances, assuming V_{SRC} remains constant. We will use a resistor R_3 of 8.25 kΩ and 5% tolerance for the load, instead of the 1 mA current source. This value is chosen because with the given values, Equation 2.7.4 gives: $V_L = \dfrac{3}{1+3}(12 - 1 \times 1) = 8.25$ V. Divided by 1 mA, this gives a resistance of 8.25 kΩ. To perform the simulation, first enter the tolerance value of 5% in the TOLERANCE column of the Property Editor spreadsheet of each resistor. In the Simulation Settings window, select DC Sweep under Analysis type. With the Primary Sweep box checked, and the voltage source selected as the Sweep Variable, enter the name of the voltage source, V1, in the Name field. Under Sweep type, with Linear selected, enter 12 for the Start value, 12 for the End Value, and 1 for Increment. Check the Monte Carlo/Worst Case box under Options. Press the Worst Case/Sensitivity button and enter V(VL) in the Output Variable field. Press the More Settings button and choose the 'maximum value (MAX)' from the pull-down menu in the Find field, and Hi under Worst-Case direction. Press the OK buttons and run the simulation. Choose the Worst Case All Devices from the Available Selections window and press the OK button. The output file shows:

```
RUN       MAXIMUM VALUE
WORST CASE ALL DEVICES
          8.5031 at V_V1 = 12
          (103.07% of Nominal)
```

To find the minimum worst-case value, repeat the aforementioned procedure except that under More Settings, choose the 'minimum value (MIN)' from the pull-down menu in the Find field and select Low under Worst-Case direction. The output file shows:

```
RUN       MINIMUM VALUE
WORST CASE ALL DEVICES
          7.9873 at V_V1 = 12
          (96.815% of Nominal)
```

Summary of Main Concepts and Results

- According to KCL, at any instant of time, the sum of the currents entering a node is equal to the sum of the currents leaving the node. KCL is an expression of conservation of charge.
- According to KVL, at any instant of time, the sum of the voltage rises around any loop is equal to the sum of the voltage drops around the loop. KVL and KCL together are an expression of conservation of energy.
- In a series connection of elements, the same current flows through all the elements.
- In a series connection of resistors, the resistances, or the reciprocals of the conductances, add.
- In a parallel connection of elements, the same voltage is applied across all the elements.
- In a parallel connection of resistors, the conductances, or the reciprocals of the resistances, add.
- A Δ-connection in a circuit may be replaced, terminal for terminal, by the equivalent Y-circuit, and conversely.
- A voltage source v_{SRC} in series with a resistance R_{src} is equivalent at its terminals to a current source $i_{SRC} = v_{SRC}/R_{src}$ in parallel with the same resistance R_{src}. Conversely, a current source i_{SRC} in parallel with a resistance R_{src} is equivalent at its terminals to a voltage source $v_{src} = R_{src}I_{SRC}$ in series with a resistance R_{src}.
- Two circuits are said to be equivalent at a specified pair of terminals if the voltage–current relation at these terminals is identical for both circuits.
- The ideal termination of a voltage source is an open circuit, whereas the ideal termination for a current source is a short circuit.
- In the design of a voltage divider, there is inherently a trade-off between good voltage regulation on the one hand and current drain and power dissipation on the other hand.

Learning Outcomes

- Apply KCL at a node and KVL around a loop.
- Derive the equivalent series resistance and conductance as well as the equivalent parallel resistance and conductance.
- Design a resistive voltage divider to supply a load at a nominal voltage and over a given range of load current, within a specified variation in the load voltage.

Supplementary Topics and Examples on CD

ST2.1 Resistive voltage divider: Derives an expression for R_1 in terms of percentage variation in load voltage and current and investigates power dissipation.

SE2.1 Redrawing of circuits: Illustrates systematic redrawing of a circuit to facilitate analysis.

SE2.2 Successive source transformation and combination: Analyzes a circuit by successively transforming and combining sources.

SE2.3 Source transformation and KCL: Illustrates the correct interpretation of the current through a voltage source after a source transformation, and the effect of a resistor in series with a current source.

SE2.4 Ohm's law at a point: Derives Ohm's law at a point, $J = \sigma\xi$.

SE2.5 Modification of circuit parameters by dependent sources: Demonstrates how dependent sources can be used to make circuit elements appear to have larger or smaller values.

SE2.6 Simulation of variable source and negative resistance: Illustrates the DC sweep capability of PSpice and the production of a negative resistance using a dependent source.

Problems and Exercises

P2.1   Kirchhoff's Laws

P2.1.1   Use KCL and KVL to solve Problem P1.2.1 to Problem P1.2.6 of Chapter 1.

P2.1.2   Determine I_{SRC} in Figure P2.1.2, and the power absorbed or delivered by each source.

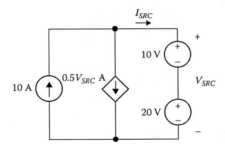

FIGURE P2.1.2

P2.1.3   Determine the voltage across each current source and the current through each voltage source in Figure P2.1.3.

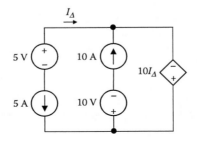

FIGURE P2.1.3

P2.1.4 Determine the voltage across each current source and the current through each voltage source in Figure P2.1.4.

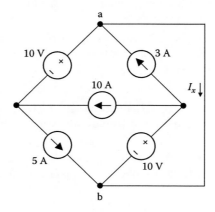

FIGURE P2.1.4

P2.1.5 Determine V_x in Figure P2.1.5, and the power absorbed or delivered by each source.

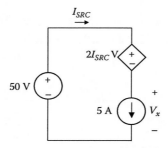

FIGURE P2.1.5

P2.1.6 Determine the voltage across each current source and the current through each voltage source in Figure P2.1.6.

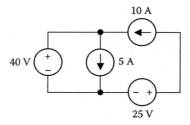

FIGURE P2.1.6

P2.1.7 Consider a surface S that encloses whole elements of a circuit, except for the negatively charged plate of a capacitor. Only a current i enters this surface. Is charge conserved? Does KCL apply?

P2.2 Series and Parallel Connections

After solving each of Problem P2.2.1 to Problem P2.2.6 analytically, simulate it with PSpice using a current source (or a voltage source) and determine R_{eq} and G_{eq} from the source voltage (or source current).

P2.2.1 Determine R_{eq} between terminals ab in Figure P2.2.1.

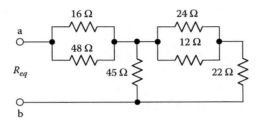

FIGURE P2.2.1

P2.2.2 Determine G_{eq} between terminals ab in Figure P2.2.1 if each resistance is replaced by a conductance having the same numerical value in S.

P2.2.3 Determine R_{eq} between terminals ab in Figure P2.2.3 when terminals cd are: (a) open-circuited and (b) short-circuited.

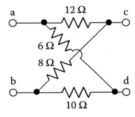

FIGURE P2.2.3

P2.2.4 Determine G_{eq} between terminals ab in Figure P2.2.3 if each resistance is replaced by a conductance having the same numerical value in S, and with terminals cd: (a) open circuited and (b) short circuited.

P2.2.5 Determine G_{eq} between terminals ab in Figure P2.2.5.

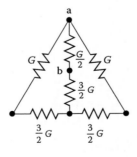

FIGURE P2.2.5

P2.2.6 Determine R_{eq} between terminals ab in Figure P2.2.6.

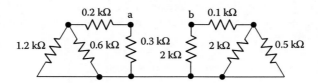

FIGURE P2.2.6

P2.3   Source Transformation, Current and Voltage Division

P2.3.1 In Figure P2.3.1, transform the voltage source and the 40 Ω resistance to the equivalent current source and combine with the dependent current source to find V_O. How much is I_{SRC}?

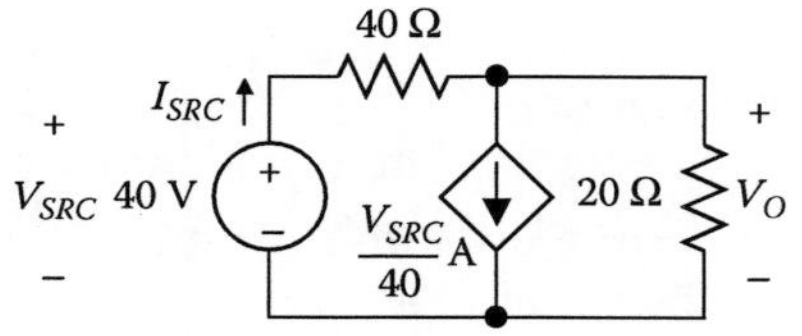

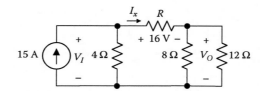

FIGURE P2.3.1

P2.3.2 Determine R, V_O, V_I, and I_x in Figure P2.3.2. How is V_O related to V_I? How is I_x related to the source current?

FIGURE P2.3.2

P2.3.3 Find V_x and the currents in the 12 Ω and 6 Ω resistors in Figure P2.3.3.

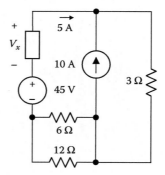

FIGURE P2.3.3

P2.3.4 Determine V_O in Figure P2.3.4 using the following methods: (a) voltage division; (b) transform the voltage source and the 20 Ω resistance to the equivalent current source and apply current division and Ohm's law.

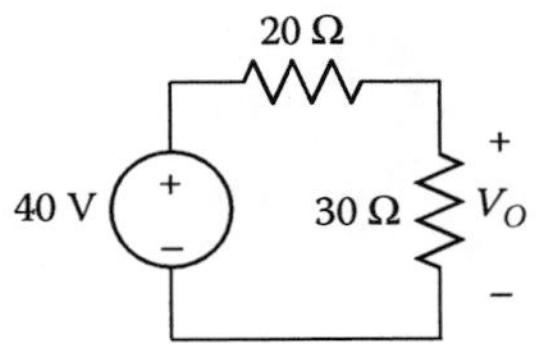

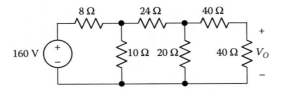

FIGURE P2.3.4

P2.3.5 Determine V_O by redrawing the ladder network of Figure P2.3.5 as a cascade of three voltage dividers.

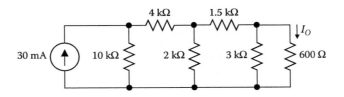

FIGURE P2.3.5

P2.3.6 Determine I_O by redrawing the ladder network of Figure P2.3.6 as a cascade of three current dividers.

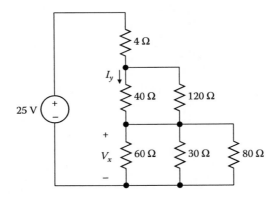

FIGURE P2.3.6

P2.3.7 Determine V_x and I_y in Figure P2.3.7.

FIGURE P2.3.7

P2.4   Solution of Circuit Problems Based on Kirchhoff's Laws, Source Transformation, and Current and Voltage Division

After solving each of problems P2.4.1 to P2.4.13 analytically, simulate it with PSpice and verify your results.

P2.4.1   Determine V_O in Figure P2.4.1 using two methods: (a) applying KCL at node a and invoking KVL and Ohm's law, and (b) transforming the voltage source to an equivalent current source that is combined with the 4 mA source.

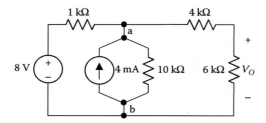

FIGURE P2.4.1

P2.4.2   Determine V_O in Figure P2.4.2 using two methods: (a) applying KCL at node a and invoking KVL and Ohm's law, and (b) using two successive source transformations that result in a voltage source of known source resistance, which can then be combined with the 15 V source.

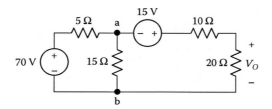

FIGURE P2.4.2

P2.4.3   Determine I_{SRC} in Figure P2.4.3 so that no current flows in R_L.

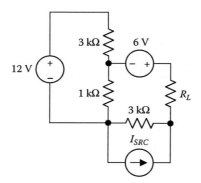

FIGURE P2.4.3

P2.4.4 Determine V_O in Figure P2.4.4.

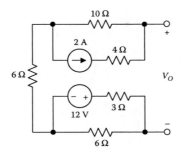

FIGURE P2.4.4

P2.4.5 Determine V_O in Figure P2.4.5, assuming $\alpha = 3$.

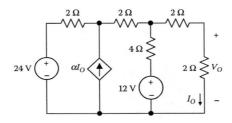

FIGURE P2.4.5

P2.4.6 Determine I_x in Figure P2.4.6.

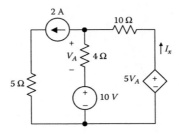

FIGURE P2.4.6

P2.4.7 Determine I_x, V_A, and V_B in Figure P2.4.7.

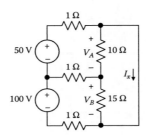

FIGURE P2.4.7

P2.4.8 Determine V_x in Figure P2.4.8.

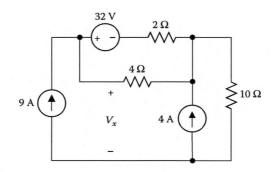

FIGURE P2.4.8

P2.4.9 Determine V_x and V_y in Figure P2.4.9.

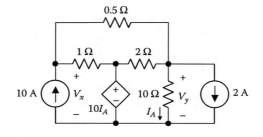

FIGURE P2.4.9

P2.4.10 Determine I_x in Figure P2.4.10.

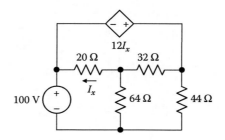

FIGURE P2.4.10

P2.4.11 Determine I_x in Figure P2.4.11.

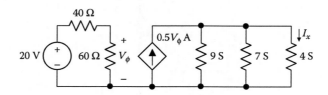

FIGURE P2.4.11

P2.4.12 Find V_{SRC}, V_x, V_y, I_{SRC}, and the current in each resistor in Figure P2.4.12.

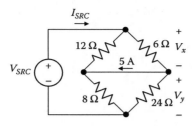

FIGURE P2.4.12

P2.4.13 Determine the input resistance $\dfrac{V_x}{I_x}$ seen at terminals ab in Figure P2.4.13 using two methods: (a) applying KCL at node a and (b) transforming the current source to an equivalent resistance.

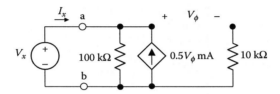

FIGURE P2.4.13

P2.4.14 A simple **ohmmeter** for measuring resistance consists of a battery, an ammeter, and a variable resistance R_0 (Figure P2.4.14). Initially, the instrument is zeroed by short-circuiting the terminals ab and adjusting R_0 so that the reading is zero on the ammeter scale calibrated in units of resistance. This corresponds to maximum current through the ammeter. When an unknown resistance R_x is connected between terminals ab, the current is reduced and the scale is calibrated so that the value of R_x is indicated. If the battery voltage is 3 V, and a current of 1 mA gives full-scale deflection of the meter, what value of R_x would reduce the ammeter current to 0.1 mA? Note that a better way of measuring resistance would be to have a current source pass a known current though R_x and measure the resulting voltage across R_x. What precaution will have to be taken, in case the resistance terminals are open circuited?

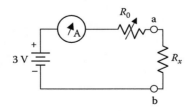

FIGURE P2.4.14

P2.4.15 In the ohmmeter circuit of Figure P2.4.15, the full-scale deflection of the ammeter is 100 μA. R_0 is adjusted so that the scale reading is 0 Ω when terminals ab are shorted together. What is the value of R_0? What value of R_x results in a current of 75 μA through the meter?

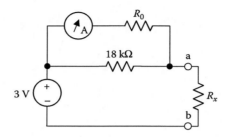

FIGURE P2.4.15

P1.4.16 Determine V_O in Figure P2.4.16 using two methods: (a) applying KCL at node a in terms of three unknown currents and invoking KVL and Ohm's law to evaluate the current through the 24 Ω resistor; (b) voltage division; and (c) successive source transformations to end up with a voltage source of known source resistance in series with the 24 Ω resistor.

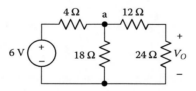

FIGURE P2.4.16

P2.4.17 Determine V_x and V_{SRC} in Figure P2.4.17 if I_{SRC} = 12 A.

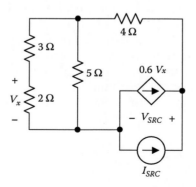

FIGURE P2.4.17

P2.4.18 Determine R and I_x in Figure P2.4.18, using two methods: (a) current division and (b) transforming the current source to an equivalent voltage source.

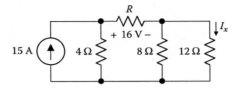

FIGURE P2.4.18

P2.4.19 Determine I_x in Figure P2.4.19.

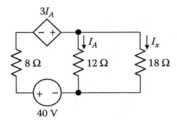

FIGURE P2.4.19

P2.4.20 Determine I_x in Figure P2.4.20.

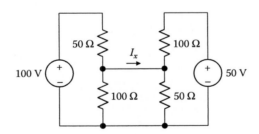

FIGURE P2.4.20

P2.4.21 Determine V_x in Figure P2.4.21.

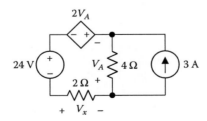

FIGURE P2.4.21

3

Basic Analysis of Resistive Circuits

Overview

Given a resistive circuit having a number of independent and dependent sources in its branches, the behavior of the circuit is completely governed by Kirchhoff's current law (KCL), Kirchhoff's voltage law (KVL), and the voltage–current relations of the branches. These provide the exact number of independent equations required to solve for all the unknown circuit variables. However, in more complicated circuits, having to solve for a large number of unknowns is tedious. Consequently, systematic procedures have been developed that reduce the number of equations to a minimum, and also facilitate their writing with the least likelihood of error.

The two basic systematic methods are those of node-voltage analysis and mesh-current analysis. The general procedures for these methods are explained and justified in terms of KCL, KVL, and the voltage–current relations of the branches. It is also shown how these methods can be applied in special situations that depart from the standard conditions for using these methods, including those involving dependent sources.

Another basic method of analysis is based on superposition. Superposition is fundamentally a defining property of linear, time-invariant (LTI) systems and involves applying independent sources one at a time and then obtaining the desired response as the algebraic sum of the responses to each source acting alone. Again, special considerations apply to dependent sources.

Learning Objectives

- To be familiar with:
  - The general procedure for node-voltage and mesh-current analyses
  - The general procedure for applying superposition
- To understand:
  - The underlying principles behind the systematic methods of node-voltage analysis and mesh-current analysis
  - How these systematic methods can be applied in particular situations that depart from standard conditions
  - The underlying principle behind applying superposition to circuit analysis

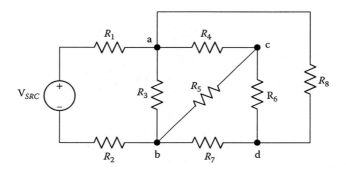

FIGURE 3.1.1
Circuit for illustrating the number of independent equations.

- How dependent sources can be dealt with in applying superposition
- How scaling can be usefully applied when a circuit is excited by a single independent source

3.1   Number of Independent Circuit Equations

To illustrate the number of independent equations required to analyze a circuit, Figure 3.1.1 is used as an example. The circuit shown has 4 essential nodes, 7 essential branches, and 4 meshes. There are 14 circuit variables; that is, a current and a voltage for each essential branch. Hence, 14 simultaneous equations have to be written to solve for the unknown variables. Seven equations are provided by the v–i relations for the branches. The number of equations provided by KCL and KVL is governed by the following relation between the number of essential branches B, the number of essential nodes N, and the number of meshes or independent loops L:

$$B = L + (N - 1) \tag{3.1.1}$$

Equation 3.1.1, proved in Section ST3.1, is quite general and applies to any circuit. $(N - 1)$ in this equation is the number of independent essential nodes. In Figure 3.1.1, KCL applied to the $(N - 1)$ independent essential nodes gives three equations whereas KVL applied to the L independent loops, or meshes, gives another four equations, thus providing the additional seven independent equations required to solve for all the variables.

3.2   Node-Voltage Analysis

> **Concept:** *In node-voltage analysis, the unknown node voltages are assigned in such a manner that KVL is automatically satisfied. Equations based on KCL are then written for each independent essential node directly in terms of Ohm's law.*

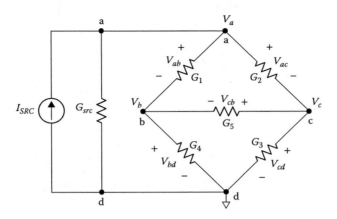

FIGURE 3.2.1
Node-voltage method.

Consider the bridge circuit of Figure 3.2.1, excited by a current source, with the resistors represented by conductances. One of the essential nodes, such as d, is arbitrarily chosen as the reference node, and the voltages of the other nodes are expressed with respect to this node. Thus, V_a, V_b, and V_c are, respectively, the voltage drops from nodes a, b, and c to d. The assignment of node voltages in this manner automatically satisfies KVL. To verify this, consider a mesh such as acb and express the voltage drops across the circuit elements in this mesh in terms of the assigned node voltages:

$$V_{ac} = V_a - V_c$$

$$V_{cb} = V_c - V_b$$

$$V_{ab} = V_a - V_b \text{ or } -V_{ab} = -V_a + V_b$$

When these equations are added, the node voltages on the RHS cancel out, giving: $V_{ac} + V_{cb} - V_{ab} = 0$, which is KVL for mesh acb. The same is true of any other mesh or loop in the circuit.

The next step is to write KCL equations for each of the nodes a, b, and c. Considering node a, the total current leaving this node through G_1, G_2, and G_{src} is: $G_2(V_a - V_c) + G_1(V_a - V_b) + G_{src}V_a$. This current must equal the source current I_{SRC} entering the node. Combining the coefficients of V_a, V_b, and V_c gives for KCL at node a:

$$(G_{src} + G_1 + G_2)V_a - G_1V_b - G_2V_c = I_{SRC} \tag{3.2.1}$$

As for nodes b and c, there is no source current entering these nodes. The current leaving node b through the conductances connected to this node is $G_1(V_b - V_a) + G_5(V_b - V_c) + G_4V_b$. Combining coefficients of the variables gives for KCL at node b:

$$-G_1V_a + (G_1 + G_4 + G_5)V_b - G_5V_c = 0 \tag{3.2.2}$$

The current leaving node c through the conductances is: $G_2(V_c - V_a) + G_5(V_c - V_b) + G_3V_c$. Combining coefficients of the variables gives for KCL at node c:

$$-G_2V_a - G_5V_b + (G_2 + G_3 + G_5)V_c = 0 \qquad\qquad (3.2.3)$$

Comparing Equation 3.2.1 to Equation 3.2.3, a definite pattern emerges for writing the node-voltage equation for any node n, which may be summarized as follows:

Procedure

1. The voltage of node n is multiplied by the sum of all the conductances connected directly to this node. This sum is the **self-conductance** of node n.

2. The voltage of every other node is multiplied by the conductance connected directly between node n and the given node. This is the **mutual conductance** between the two nodes. If there is no such conductance, the coefficient is zero. The sign of a nonzero coefficient is always negative because the current flowing away from node n toward the given node is proportional to the voltage of node n minus that of the given node.

3. The LHS of the node-voltage equation for node n is the sum of the terms from the preceding steps, ordered as the unknown node voltages. This sum is the total current leaving node n through the conductances connected to this node.

4. The RHS of the equation is equal to any source current entering node n.

Once the node voltages are known, branch currents can be determined from these and the branch conductances.
The following should be noted:

1. If a particular branch current is required, it is advantageous to take the reference node as one of the terminal nodes of that branch because the branch current is then the product of the branch conductance and a single node voltage rather than the difference between two node voltages.

2. Ideal resistors are bilateral; that is, the resistance is the same for both directions of current (Section 1.7, Chapter 1). This means that the mutual conductance terms in the equations of any two given nodes are the same. For example, the current flowing from node b toward node c is $G_5(V_b - V_c)$ in Figure 3.2.1, whereas the current flowing from node c toward node b is $G_5(V_c - V_b)$. The coefficient of V_c in the node-voltage equation for node b, which is $-G_5$ in Equation 3.2.2, is the same as the coefficient of V_b in the node-voltage equation for node c (Equation 3.2.3). When ordered in a matrix, or array, the conductance coefficients are symmetrical with respect to the diagonal, in the absence of dependent sources. This is a useful check on the node-voltage equations.

3. The number of independent node-voltage equations for a given circuit is one less than the number of essential nodes. KCL for the reference node is not an independent equation because it can be obtained by adding the KCL equations of all the other nodes (Exercise 3.2.1).

4. Node-voltage analysis applies to planar as well as nonplanar circuits, the latter being circuits that cannot be drawn in two dimensions without crossover connections (Section ST3.2). PSpice uses node-voltage analysis for circuit simulation.

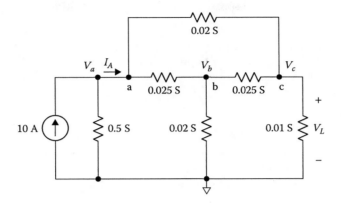

FIGURE 3.2.2
Figure for Example 3.2.1.

EXERCISE 3.2.1
Write KCL for nodes a, b, and c in Figure 3.2.1 in terms of the branch currents, then add the equations to obtain KCL for node d, so as to show that KCL for the reference node is not an independent equation.

Example 3.2.1: Node-Voltage Analysis

Given the circuit shown in Figure 3.2.2, it is required to determine I_A and V_L using node-voltage analysis.

SOLUTION
For direct application of node-voltage analysis, it is convenient to transform any voltage source in series with a resistor to its equivalent current source (Sections 2.6, Chapter 2) and represent resistors by their conductances. The lower node is chosen as the reference because V_L is with respect to this node. Following the procedure outlined earlier, the node-voltage equations for nodes a, b, and c, may be written directly:

$$(0.5 + 0.025 + 0.02)V_a - 0.025V_b - 0.02V_c = 10$$

$$-0.025V_a + (0.02 + 0.025 + 0.025)V_b - 0.025V_c = 0$$

$$-0.02V_a - 0.025V_b + (0.01 + 0.02 + 0.025)V_c = 0$$

The conductance coefficients are symmetrical with respect to the diagonal. Simultaneous equations may be conveniently solved using MATLAB (Appendix SD.1). The solution is: $V_a = 19.3$ V; $V_b = 11.2$ V; and $V_c = 12.1$ V.

I_A may be determined as the sum of the currents that flow into the 0.02 S and 0.025 S resistors connected to node a. That is, $I_A = 0.025 (V_a - V_b) + 0.02 (V_a - V_c) = 0.346$ A. Alternatively, $I_A = 10 - 0.5V_a = 0.346$ A. It is seen that $V_L = V_c = 12.1$ V.

EXERCISE 3.2.2

Assume that in the circuit of Figure 3.2.1, $I_{SRC} = 4$ A, $G_{src} = 0.05$ S, $G_1 = 0.05$ S, $G_2 = 0.025$ S, $G_3 = 0.05$ S, $G_4 = 0.025$ S, and $G_5 = 0.01$ S. Determine V_a, V_b, V_c, and V_{bc} using node-voltage analysis.

Answer: $V_a = 47.5$ V, $V_b = 30$ V, $V_c = 17.5$ V, and $V_{bc} = 12.5$ V.

3.3   Special Considerations in Node-Voltage Analysis

Dependent Sources

Dependent current sources in node-voltage analysis are treated exactly like independent sources. The node-voltage equations for the circuit of Figure 3.3.1 are:

$$(G_1 + G_2)V_a - G_2V_b = I_{SRC} + \beta V_L$$

$$-G_2V_a + (G_2 + G_3)V_b = -\beta V_L \qquad (3.3.1)$$

where the current due to the dependent source flows into node a and out of node b. Because $V_L = V_b$, the term βV_L may be moved to the LHS of Equation 3.3.1 to give:

$$(G_1 + G_2)V_a - (G_2 + \beta)V_b = I_{SRC}$$

$$-G_2V_a + (G_2 + G_3 + \beta) V_b = 0 \qquad (3.3.2)$$

Whereas the coefficients of V_a and V_b on the LHS of Equations 3.3.1 are symmetrical with respect to the diagonal, this symmetry is destroyed in Equations 3.3.2 when the term due to the dependent source is moved to the LHS.

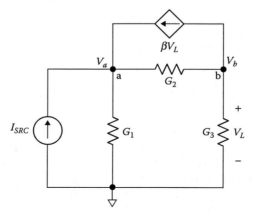

FIGURE 3.3.1
Dependent source in node-voltage analysis.

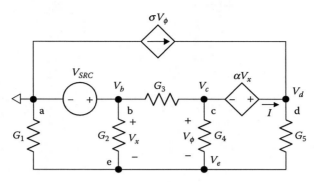

FIGURE 3.3.2
Nontransformable sources in node-voltage analysis.

EXERCISE 3.3.1

Assume that in the circuit of Figure 3.3.1, I_{SRC} = 5 A, G_1 = 0.1 S, G_2 = 0.4 S, G_3 = 0.1 S, and β = 0.2 A /V. Determine V_a and V_b using node-voltage analysis.

Answer: V_a = 31.8 V and V_b = 18.2 V.

Nontransformable Voltage Sources

Suppose that the circuit of Figure 3.3.2 is to be analyzed by the node-voltage method. Neither voltage source has a resistance directly in series with it, so it cannot be transformed to an equivalent current source. However, if we choose node a as a reference, $V_b = V_{SRC}$, leaving three unknown node voltages: V_c, V_d, and V_e.

The node equation for node e is:

$$-G_2V_b - G_4V_c - G_5V_d + (G_1 + G_2 + G_3 + G_4 + G_5)\, V_e = 0 \qquad (3.3.3)$$

When it comes to writing the node-voltage equations for nodes c and d, a problem arises in that the current through the dependent voltage source αV_x is not known. An unknown current I is introduced, which is arbitrarily assigned a direction from node c to node d. The equation for node c is:

$$-G_3V_b + (G_3 + G_4)\, V_c - G_4V_e = -I \qquad (3.3.4)$$

The equation for node d is:

$$G_5V_d - G_5V_e = I + \sigma V_\phi \qquad (3.3.5)$$

I is eliminated by adding Equation 3.3.4 and Equation 3.3.5 together:

$$-G_3V_b + (G_3 + G_4)V_c + G_5V_d - (G_4 + G_5)\, V_e = \sigma V_\phi \qquad (3.3.6)$$

The third equation is the voltage relation for the dependent voltage source:

$$V_d - V_c = \alpha V_x = \alpha \left(V_b - V_e \right) \tag{3.3.7}$$

Substituting $V_b = V_{SRC}$ and $V_\phi = V_c - V_e$, and rearranging the variables, gives three equations that may be solved for V_c, V_d, and V_e:

$$-G_4 V_c - G_5 V_d + \left(G_1 + G_2 + G_4 + G_5 \right) V_e = G_2 V_{SRC} \tag{3.3.8}$$

$$\left(G_3 + G_4 - \sigma \right) V_c + G_5 V_d - \left(G_4 + G_5 - \sigma \right) V_e = G_3 V_{SRC} \tag{3.3.9}$$

$$-V_c + V_d + \alpha V_e = \alpha V_{SRC} \tag{3.3.10}$$

Equation 3.3.6 may be interpreted as a node-voltage equation of a *supernode* that combines nodes c and d, and which may be written directly according to some rules. The concept of a supernode, which is simply a way of avoiding having to introduce the unknown current I, is discussed in Section ST3.3.

EXERCISE 3.3.2

Given the following values for the circuit elements in Figure 3.3.2: $V_{SRC} = 20$ V, $G_1 = 1$ S, $G_2 = 0.5$ S, $G_3 = 0.2$ S, $G_4 = 0.2$ S, $G_5 = 0.1$ S, $\sigma = 0.1$ A / V, and $\alpha = 0.5$, determine the node voltages.

Answer: $V_b = 20$V, $V_c = 12.5$ V, $V_d = 18.5$ V, and $V_e = 7.97$ V.

Change of Reference Node

Suppose that node e in Figure 3.3.2 is grounded, so that voltages with respect to this node are required. If we apply node-voltage analysis with node e as a reference, then we have to assign an unknown current to source V_{SRC}, which can be eliminated by adding the node-voltage equations for nodes a and b. Using node a as a reference avoids this added complication. If voltages are required with respect to node e, then all we have to do is subtract the value of V_e from all the node voltages determined with node a as a reference. For the values of Exercise 3.3.2, this gives: $V_a = -7.97$ V, $V_b = 12.0$ V, $V_c = 4.53$ V, $V_d = 10.5$ V, and $V_e = 0$. The justification is simply that the branch voltages, which are the basic quantities uniquely associated with the branch currents, depend on the difference between the node voltages at the ends of a given branch and are not changed by adding the same constant voltage to all the node voltages.

3.4   Mesh-Current Analysis

Concept: *In mesh-current analysis, the unknown mesh currents are assigned in such a manner that KCL is automatically satisfied. Equations based on KVL are then written for each mesh directly in terms of Ohm's law.*

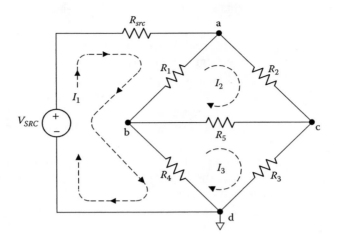

FIGURE 3.4.1
Mesh-current method.

Consider the same bridge circuit of Figure 3.2.1, redrawn in Figure 3.4.1 with the current source replaced by a voltage source and the resistors represented by resistances. Mesh currents are assigned to each mesh in the same sense, usually clockwise. The assignment of currents in this manner automatically satisfies KCL. To show this, consider the current flowing toward node a, for example. This current is I_1 through R_{src} and $(I_2 - I_1)$ through R_1. The current flowing away from the node through R_2 is I_2. Equating these currents gives $I_2 = I_2$, in accordance with KCL. The same is true at every other node.

KVL is then written for each mesh. Considering mesh 1, the total voltage drop across R_{src}, R_1, and R_4 in the direction of I_1 is: $R_{src}I_1 + R_1(I_1 - I_2) + R_4(I_1 - I_3)$. This must equal the voltage rise V_{SRC} in the mesh. Combining the coefficients of I_1, I_2, and I_3 gives KVL for mesh 1:

$$\left(R_{src} + R_1 + R_4\right)I_1 - R_1I_2 - R_4I_3 = V_{SR} \tag{3.4.1}$$

As for mesh 2 and mesh 3, they have no source voltage. The total voltage drop is: $R_1(I_2 - I_1) + R_2I_2 + R_5(I_2 - I_3)$ for mesh 2, and $R_4(I_3 - I_1) + R_5(I_3 - I_2) + R_3I_3$ for mesh 3. Combining coefficients of the variables as before, gives KVL for mesh 2 as:

$$-R_1I_1 + \left(R_1 + R_2 + R_5\right)I_2 - R_5I_3 = 0 \tag{3.4.2}$$

Similarly, KVL for mesh 3 becomes:

$$-R_4I_1 - R_5I_2 + \left(R_3 + R_4 + R_5\right)I_3 = 0 \tag{3.4.3}$$

Comparing Equation 3.4.1 to Equation 3.4.3, a definite pattern emerges for writing the mesh-current equation for any mesh m, which may be summarized as follows:

Procedure

1. The current of mesh m is multiplied by the sum of all the resistances around the mesh. This sum is the **self-resistance** of mesh m.
2. The current of every other mesh is multiplied by the common resistance between the given mesh and mesh m. This is the **mutual resistance** between the two meshes. If there is no such resistance, the coefficient is zero. The sign of a nonzero coefficient is always negative because the current of the given mesh produces a voltage rise in mesh m.
3. The LHS of the mesh-current equation for mesh m is the sum of the terms from the preceding steps, ordered as the unknown mesh currents. This sum is the total voltage drop across the resistances in mesh m.
4. The RHS of the equation is equal to the voltage rise due to any source voltage in mesh m.

Once the mesh currents are known, branch currents and branch voltages can be determined. The following should be noted:

1. As in the case of the node-voltage method, the matrix, or array, of resistances is symmetrical with respect to the diagonal in the absence of dependent sources and for similar reasons. For example, R_5 contributes a voltage rise $R_5 I_3$ in mesh 2, and a voltage rise $R_5 I_2$ in mesh 3. Hence, $-R_5$ is the coefficient of I_3 in mesh 2 and of I_2 in mesh 3.
2. The number of independent mesh-current equations for a given circuit equals the number of meshes.
3. Whether one uses the node-voltage or the mesh-current method in a particular problem may depend on the number of equations that have to be solved in each case, in accordance with Equation 3.1.1.
4. The mesh-current method is not as general as the node-voltage method in that the procedure outlined earlier does not apply to nonplanar circuits in which the concept of a mesh becomes ambiguous. However, the loop-current method, in which loop currents rather than mesh currents are used, can be applied to non-planar circuits (Section ST3.4).

Example 3.4.1: Mesh-Current Analysis

Given the same circuit of Figure 3.2.2, it is required to determine I_{SRC} and V_L using mesh-current analysis.

SOLUTION

The circuit is redrawn in Figure 3.4.2 showing the mesh currents. The 10 A source in combination with the 0.5 S is transformed to a voltage source of 20 V in series with 2 Ω. Following the procedure outlined earlier, the mesh-current equations for meshes 1, 2, and 3 are:

$$(2 + 40 + 50) I_1 - 40 I_2 - 50 I_3 = 20$$

$$-40 I_1 + 130 I_2 - 40 I_3 = 0$$

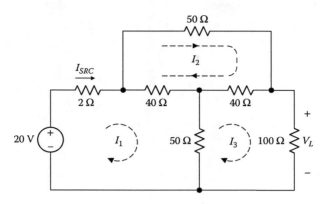

FIGURE 3.4.2
Figure for Example 3.4.1.

$$-50I_1 - 40I_2 + 190I_3 = 0$$

The resistance coefficients are symmetrical with respect to the diagonal. Solving these equations gives: $I_1 = 0.346$ A; $I_2 = 0.144$ A; $I_3 = 0.121$ A. Hence, $V_L = 100 \times I_3 = 12.1$ V, and $I_1 = I_{SRC} = I_A$, as in Example 3.2.1.

EXERCISE 3.4.1

Assume that in the circuit of Figure 3.4.1, $V_{SRC} = 80$ V, $R_{src} = 20$ Ω, $R_1 = 20$ Ω, $R_2 = 40$ Ω, $R_3 = 20$ Ω, $R_4 = 40$ Ω, and $R_5 = 100$ Ω. (a) Determine I_1, I_2, I_3, and V_{bc} using mesh-current analysis.

Answer: $I_1 = 1.625$ A, $I_2 = 0.750$ A, $I_3 = 0.875$ A, and $V_{bc} = 12.5$ V.

3.5 Special Considerations in Mesh-Current Analysis

Dependent Sources

Figure 3.5.1 illustrates a circuit with a CCVS ρI_L. The mesh-current equations are:

$$(R_1 + R_2)I_1 - R_2I_2 = V_{SRC} + \rho I_L$$

$$-R_2I_1 + (R_2 + R_3)I_2 = -\rho I_L \tag{3.5.1}$$

Substituting $I_L = I_2$ and collecting terms in I_2:

$$(R_1 + R_2)I_1 - (R_2 + \rho)I_2 = V_{SRC}$$

$$-R_2I_1 + (R_2 + R_3 + \rho)I_2 = 0 \tag{3.5.2}$$

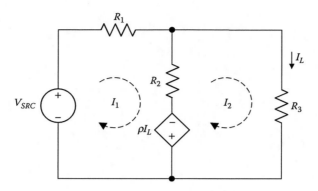

FIGURE 3.5.1
Dependent source in mesh-current analysis.

Whereas the matrix of coefficients is symmetrical about the diagonal in Equation 3.5.1, it is no longer so in Equation 3.5.2 after the substitutions for ρI_L are made.

EXERCISE 3.5.1

Assume that in Figure 3.5.1 $V_{SRC} = 50$ V, $R_1 = 10\ \Omega$, $R_2 = 25\ \Omega$, $R_3 = 100\ \Omega$, and $\rho = 5$ V/A. Determine I_1 and I_2.

Answer: $I_1 = 1.71$ A, $I_2 = 0.329$ A.

Nontransformable Current Sources

Consider the circuit of Figure 3.5.2. For mesh 1,

$$R_1I_1 = -V_1 - \rho I_\phi = -V_1 - \rho I_3$$

or,
$$R_1I_1 + \rho I_3 = -V_1 \tag{3.5.3}$$

where V_1 is an assumed voltage drop across the current source I_{SRC} and $I_\phi = I_3$

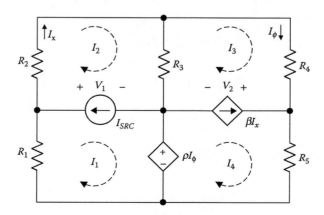

FIGURE 3.5.2
Nontransformable sources in mesh-current analysis.

For mesh 2,

$$(R_2 + R_3)I_2 - R_3I_3 = V_1 \tag{3.5.4}$$

For mesh 3,

$$-R_3I_2 + (R_3 + R_4)I_3 = -V_2 \tag{3.5.5}$$

where V_2 is an assumed voltage drop across the current source βI_x.
   For mesh 4,

$$R_5I_4 = V_2 + \rho I_\phi = V_2 + \rho I_3$$

or, $\qquad\qquad -\rho I_3 + R_5I_4 = V_2 \tag{3.5.6}$

Equation 3.5.3 to Equation 3.5.6 involve the four mesh currents plus two additional unknowns, V_1 and V_2. Two additional equations are required, which are derived from the relations between the two current sources and the mesh currents as follows:

$$I_2 - I_1 = I_{SRC} \tag{3.5.7}$$

$$I_4 - I_3 = \beta I_x = \beta I_2 \tag{3.5.8}$$

V_1 is eliminated by adding together Equation 3.5.3 and Equation 3.5.4:

$$R_1I_1 + (R_2 + R_3)I_2 - (R_3 - \rho)I_3 = 0 \tag{3.5.9}$$

Similarly, V_2 is eliminated by adding together Equation 3.5.5 and Equation 3.5.6:

$$-R_3I_2 + (R_3 + R_4 - \rho)I_3 + R_5I_4 = 0 \tag{3.5.10}$$

Equation 3.5.7 to Equation 3.5.10 can be solved for I_1, I_2, I_3, and I_4.
   Equation 3.5.9 and Equation 3.5.10 may be interpreted as mesh equations of a *supermesh* that combines two meshes in question, and which may be written directly according to some rules. The concept of a supermesh, which is simply a way of avoiding having to introduce an unknown voltage across a current source, is discussed in Section ST3.5.

EXERCISE 3.5.2

   Assume that in Figure 3.5.2, I_{SRC} = 5 A, R_1 = 10 Ω, R_2 = 20 Ω, R_3 = 40 Ω, R_4 = 10 Ω, R_5 = 20 Ω, β = 2, and ρ = 5 V/A. Determine I_1, I_2, I_3, and I_4.
   Answer: I_1 = − 4.29 A, I_2 = 0.714 A, I_3 = 0, and I_4 = 1.43 A.

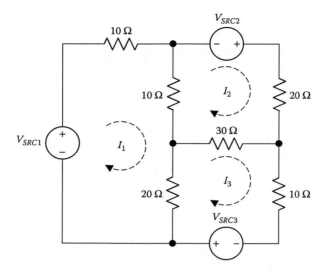

FIGURE 3.6.1
Superposition.

3.6 Superposition

Concept: *In an LTI circuit excited by more than one independent source, any voltage or current response is the algebraic sum of individual components due to each independent source acting alone, with all the other sources set to zero.*

To justify this, we will consider, for simplicity and without loss of generality, a three-mesh circuit, such as that of Figure 3.6.1. The mesh-current equations are:

$$40I_1 - 10I_2 - 20I_3 = V_{SRC1}$$

$$-10I_1 + 60I_2 - 30I_3 = V_{SRC2}$$

$$-20I_1 - 30I_2 + 60I_3 = V_{SRC3} \qquad (3.6.1)$$

Solving these equations by any of the standard methods, such as the method of determinants (Appendix SA) or simple elimination of variables, gives:

$$I_1 = \frac{9}{220}V_{SRC1} + \frac{1}{55}V_{SRC2} + \frac{1}{44}V_{SRC3}$$

$$I_2 = \frac{1}{55}V_{SRC1} + \frac{1}{33}V_{SRC2} + \frac{7}{330}V_{SRC3}$$

$$I_3 = \frac{1}{44}V_{SRC1} + \frac{7}{330}V_{SRC2} + \frac{23}{660}V_{SRC3} \qquad (3.6.2)$$

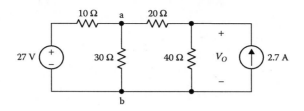

FIGURE 3.6.2
Figure for Example 3.6.1.

It is seen that I_1, I_2, or I_3 is the sum of three components, each of which is due to one of the sources acting alone with the other two sources set to zero. Although the above equations (3.6.2) were derived for a particular circuit, they apply, in general, to any LTI circuit excited by more than one source. In applying superposition, however, it must be clear how to set sources to zero, as described by the following concept:

> **Concept:** *A voltage source is set to zero by replacing the ideal voltage source element with a short circuit. A current source is set to zero by replacing the ideal current source element with an open circuit.*

The justification is that for an ideal voltage source, V_{SRC} is independent of source current (Figure 1.6.1a, Chapter 1). If $V_{SRC} = 0$, this means that the source will pass any current with zero voltage across the source, which is characteristic of a short circuit. Any source resistance in series with the ideal voltage-source element is not affected by setting the ideal voltage source to zero and is retained. Similarly, in the case of a current source, I_{SRC} is independent of voltage across the source (Figure 1.6.2a, Chapter 1). If $I_{SRC} = 0$, this means that the source will not pass any current, irrespective of the voltage across the source, which is characteristic of an open circuit. Any source resistance in parallel with the ideal current-source element is not affected by setting the ideal current source to zero and is retained.

Example 3.6.1: Superposition with Independent Sources

It is required to determine V_O in the circuit of Figure 3.6.2 using superposition.

SOLUTION
If the current source is replaced by an open circuit, the resistance between terminals ab is $30\|60 = 20\ \Omega$. Hence, $V_{ab} = \dfrac{20}{30} \times 27 = 18$ V and $V_{O1} = \dfrac{40}{60} \times 18 = 12$ V, where V_{O1} is the component of V_O due to the voltage source acting alone.

If the voltage source is replaced by a short circuit, the resistance between terminals ab is $10\|30 = 7.5\ \Omega$. The resistance in parallel with the 40 Ω is 27.5 Ω, and the resistance across the current source is $(27.5)\|40 = \dfrac{440}{27}\ \Omega$. Hence, $V_{O2} = \dfrac{440}{27} \times 2.7 = 44$ V, where V_{O2} is the component of V_O due to the current source acting alone.

By superposition, $V_O = 12 + 44 = 56$ V.

Dependent Sources

Only independent sources were considered in the preceding discussion. In the presence of dependent sources, superposition can be applied in one of two ways, depending on which method is easier to apply:

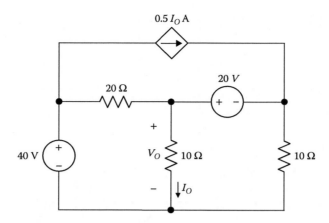

FIGURE 3.6.3
Figure for Example 3.6.2.

1. If we consider that the V_{SRC}'s or I_{SRC}'s on the RHS of the mesh-current or node voltage equations, respectively, are due to independent sources only, then the effect of dependent sources is to modify the resistance coefficients on the LHS, as in Equation 3.5.2, for example. The implication is that when applying superposition, dependent sources should remain unaltered. This is the conventional procedure.

2. The dependent source may be replaced by an independent source that is assigned an unknown value. Superposition is applied and a relation derived for the desired circuit variable in terms of the unknown value. This relation can then be used with the dependence relation of the source to solve the problem. In some problems, this is the quicker solution.

Example 3.6.2: Superposition with Dependent Sources

Given the circuit of Figure 3.6.3, it is required to find V_O using superposition.

SOLUTION

(a) The dependent source is left unaltered.

In this case, each of the two voltage sources is replaced by a short circuit, one at a time, and the two components of V_O are determined.

If the 20 V source is replaced by a short circuit, the circuit becomes as shown in Figure 3.6.4a. Let V_{O1} be the component of V_O and I_{O1} be the current in the leftmost 10 Ω resistor, where $V_{O1} = 10I_{O1}$. The current in the rightmost 10 Ω resistor is also I_{O1} because the same voltage V_{O1} is across this resistor. The current flowing away from node b through the two 10 Ω resistors is $2I_{O1}$. Because $0.5I_{O1}$ flows toward node b from the dependent source, it follows from KCL that a current $1.5I_{O1}$ flows toward node b through the 20 Ω resistor. Applying KVL to the mesh abca, $40 = 20 \times 1.5I_{O1} + V_{O1} = 4V_{O1}$ or $V_{O1} = 10$ V.

When the 40 V source is replaced by a short circuit, the circuit becomes as shown in Figure 3.6.4b. Let V_{O2} be the component of V_O and I_{O2} be the current in the leftmost 10 Ω resistor, where $V_{O2} = 10I_{O2}$. Because V_{O2} is also across the 20 Ω resistor, the current through this resistor is $0.5I_{O2}$. From KCL at node b, the current through the 20 V source is $1.5I_{O2}$ as shown. From KCL at node d, the current in the rightmost 10 Ω resistor is I_{O2} directed upward. Hence, the voltage across this resistor is also V_{O2} in the polarity shown. Applying KVL to the mesh bcd gives $20 = 2V_{O2}$ or $V_{O2} = 10$ V. From superposition, $V_O = V_{O1} + V_{O2} = 20$ V.

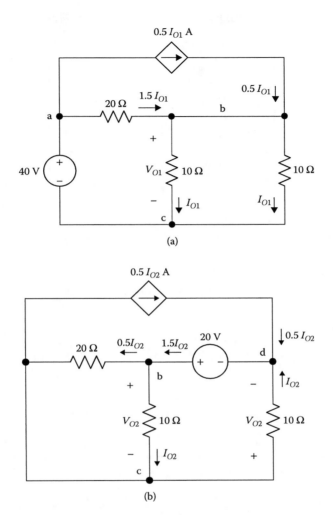

FIGURE 3.6.4
Figure for Example 3.6.2.

The way this problem was solved should be carefully noted. The solution proceeded step by step, based on a direct application of Ohm's law, KCL, and KVL, in a logical and organized manner. The relations that follow from KCL and Ohm's law could be written directly on the circuit diagram. This is a good habit to acquire. We did not have to use node-voltage or mesh-current analysis in this relatively simple circuit.

(b) The dependent source is treated as an independent source.

The dependent source is assigned a value, say I_x, as an independent source (Figure 3.6.5) and superposition is applied with each of the three sources acting alone. When the 40 V source is applied alone, with the 20 V source replaced by a short circuit and I_x by an open circuit, $V_{O1} = \dfrac{5}{25} \times 40 = 8$ V. When the 20 V source is applied alone, with the 40 V source replaced by a short circuit and I_x by an open circuit, $V_{O2} = \dfrac{20/3}{10 + 20/3} \times 20 = 8$ V. When I_x is applied alone, with the two voltage sources replaced by short circuits, $V_{O3} =$

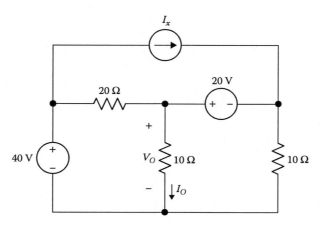

FIGURE 3.6.5
Figure for Example 3.6.2.

$(10 \| 10 \| 20)I_x$. To calculate the parallel resistance, the two 10 Ω resistors in parallel give 5 Ω. This resistance in parallel with 20 Ω is 4Ω, so that $V_{O3} = 4I_x$. It follows that $V_O = V_{O1} + V_{O2} + V_{O3} = 16 + 4I_x$. From the original circuit (Figure 3.6.3), $I_x = 0.5I_O = 0.05V_O$, so that $4I_x = 0.2V_O$. Substituting for I_x gives: $V_O = \dfrac{16}{0.8} = 20$ V, as before.

Power with Superposition

> **Concept:** *In a circuit excited by more than one source, the total power dissipated in a given resistor is NOT the sum of the powers due to each source acting alone, with all the other sources set to zero.*

The reason is that the power dissipated in a given resistor is proportional to the square of the current through the resistor, or the square of the voltage across it, and the sum of the squares of a set of quantities is not equal to the square of the sum of these quantities. Thus, in Example 3.6.2, part (a), $V_{O1} = V_{O2} = 10$ V. The sum of the powers due to these components is $\dfrac{(V_{O1})^2}{10} + \dfrac{(V_{O2})^2}{10} = 20$ W. The true power dissipated is $\dfrac{(V_{O1} + V_{O2})^2}{10} = 40$ W.

EXERCISE 3.6.1

Determine V_O in Figure 3.6.6 using superposition: (a) without altering the dependent source, and (b) considering the dependent source as an independent source. Determine the power dissipated in the 10 Ω output resistor due to each independent source acting alone and with both sources connected.

Answer: 8 V, 1.6 W, 1.6 W, 6.4 W.

Two other useful deductions can be made from equations (3.6.2) concerning scaling of inputs and excitation by dependent sources only.

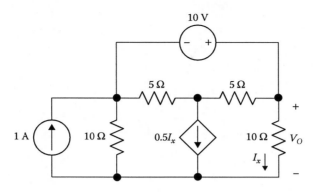

FIGURE 3.6.6
Figure for Exercise 3.6.1.

Scaling of Input

> **Concept:** *In a circuit excited by a single independent source, multiplying the excitation by a constant K multiplies all the voltage and current responses by the same constant.*

If only a single excitation, say V_{SRC1} (with $V_{SRC2} = V_{SRC3} = 0$), is applied to the circuit of Figure 3.6.1, equations (3.6.2) give:

$$I_1 = \frac{9}{220} V_{SRC1}, \quad I_2 = \frac{1}{55} V_{SRC1}, \quad I_3 = \frac{1}{44} V_{SRC1} \qquad (3.6.3)$$

If V_{SRC1} is multiplied by K, then I_1, I_2, and I_3 are also multiplied by K. This fact may be usefully exploited in some problems by working backward. That is, rather than determine the output for a given input, a convenient output is assumed and the input that produces this output is determined. The desired output is then obtained by simple scaling according to the given input, as illustrated by Example 3.6.3. Dependent sources are allowed because they affect only the resistance coefficients in the circuit equations (see Problem P3.2.17).

Example 3.6.3: Scaling Applied to a Ladder Circuit
It is required to determine I_O in the ladder circuit of Figure 3.6.7.

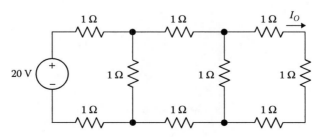

FIGURE 3.6.7
Figure for Example 3.6.3.

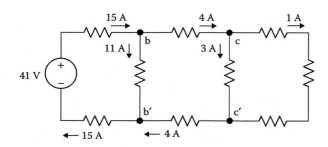

FIGURE 3.6.8
Figure for Example 3.6.3.

SOLUTION
The quickest way to solve this problem is to assume a convenient value for I_O and work backward toward the source. If $I_O = 1$ A (Figure 3.6.8), $V_{cc'} = 3$ V and the current in the cc' branch is 3 A. The current in the bc and c' b' branches is 4 A and $V_{bb'} = 11$ V. That makes the source current 15 A and the source voltage 41 V. But the given source is 20 V

so the actual value of I_O is $1 \times \dfrac{20}{41} = 0.49$ A.

Excitation by Dependent Sources

It would appear from equations (3.6.2) that if the circuit has dependent sources only, then all the V_{SRC}'s are zero and hence all the responses are zero. In Figure 3.6.9, for exam-

ple, $V_{ab} = \dfrac{3 \times 6}{9} \times 3V_\phi = 6V_\phi$ and $V_\phi = \dfrac{V_{ab}}{2}$. Substituting for V_{ab} gives: $2V_\phi = 6V_\phi$, which is

impossible unless $V_\phi = 0$. This means that the source current is zero and all responses in the circuit are zero. It should be pointed out, however, that in some cases, dependent sources can make the circuit unstable (Section 16.6, Chapter 16) and the response theoretically increases with time without limit. Hence, a more accurate statement is:

> **Concept:** *In a stable LTI circuit containing dependent sources and no independent sources, all circuit responses are zero.*

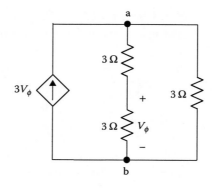

FIGURE 3.6.9
Excitation by dependent source.

It would appear that if the dependent source in Figure 3.6.9 is V_ϕ instead of $3V_\phi$, then $2V_\phi = 2V_\phi$, which makes the circuit responses indeterminate. In practice, however, neither the value of the source, nor the values of the resistances, are exact, so that this situation does not arise.

Superposition and Linearity

Before ending this chapter, it should be pointed out that superposition is a defining property of a linear system, as stated formally in Section 16.6 of Chapter 16. In Chapter 1 we characterized linearity of the passive circuit elements on the basis that $v = Ri$ for a resistor, $q = Cv$ for a capacitor, and $\lambda = Li$ for an inductor, where R, L, and C do not vary with the variables v, i, q, or λ. These linear relations obey superposition. In the case of a linear resistor R, for example, if a current i_1 produces a voltage $v_1 = Ri_1$, and a current i_2 produces a voltage $v_2 = Ri_2$, then a current $2i_1 + 3i_2$ produces a voltage $R(2i_1 + 3i_2) = 2v_1 + 3v_2$, as long as the relation $v = Ri$ applies, with R constant.

Summary of Main Concepts and Results

- $B = L + (N - 1)$, where B is the number of essential branches, N is the number of essential nodes, and L is the number of meshes or independent loops.
- In node-voltage analysis, the unknown node voltages are assigned in such a manner that KVL is automatically satisfied. Equations based on KCL are then written for each independent node directly in terms of the self-conductance of the node, its mutual conductances with respect to the other nodes, and any source current at the node. The sign of the mutual conductance term is always negative.
- In mesh-current analysis, the unknown mesh currents are assigned in such a manner that KCL is automatically satisfied. Equations based on KVL are then written for each mesh directly in terms of the self-resistance of the mesh, its mutual resistances with respect to other meshes, and any source voltage in the mesh. The sign of the mutual resistance term in the mesh current method is always negative.
- In both node-voltage and mesh-current analyses, dependent sources affect the coefficients of the circuit variables and destroy their symmetry with respect to the diagonal.
- In both node-voltage and mesh-current analyses, nontransformable sources are dealt with by assigning them an unknown circuit variable, which is then eliminated by adding the equations for the two nodes, or meshes, adjacent to the nontransformable source.
- In an LTI circuit excited by more than one independent source, any voltage or current response is the sum of individual components due to each independent source acting alone, with all the other independent sources set to zero.
- A voltage source is set to zero by replacing the ideal voltage source element with a short circuit. A current source is set to zero by replacing the ideal current source element by an open circuit.
- In a circuit containing dependent sources, superposition can be applied by leaving the dependent sources unaltered or by replacing them with independent sources of unknown value.

- In a circuit excited by more than one source, the total power dissipated in a given resistor is *not* the sum of the powers due to each source acting alone with all the other sources set to zero.
- In a circuit excited by a single independent source, multiplying the excitation by a constant K multiplies all the voltage and current responses by the same constant.
- In a stable LTI circuit containing dependent sources and no independent sources, all circuit responses are zero.

Learning Outcomes

- Analyze resistive circuits by the node-voltage method, mesh-current method, or superposition.

Supplementary Topics and Examples on CD

ST3.1 Number of independent equations: Derives the relation between the number of branches, the number of independent nodes, and the number of meshes.

ST3.2 Nonplanar circuits: Illustrates a nonplanar circuit.

ST3.3 Supernodes: Discusses the concept of a supernode.

ST3.4 Loop-current method: Explains the loop-current method of analysis.

ST3.5 Supermeshes: Discusses the concept of a supermesh.

SE3.1 Node-voltage analysis and simulation: Analyzes by the node-voltage method a circuit having nontransformable voltage sources and simulates it.

SE3.2 Superposition and simulation: Analyzes a bridge circuit by superposition and simulates it.

Problems and Exercises

P3.1   Node-Voltage and Mesh-Current Analysis

In the following problems, use MATLAB to solve simultaneous equations, whenever appropriate. Verify the solutions with PSpice simulation.

P3.1.1   In the circuit of Figure P3.1.1, write a KCL equation in terms of V_{ab} and the source voltages. Then determine I_1 and I_2. How do you interpret the result?

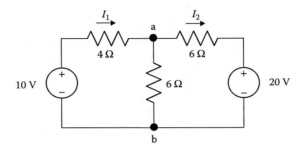

FIGURE P3.1.1

P3.1.2   Determine I_{SRC1} and I_{SRC2} in Figure P3.1.2 using the same method as in Problem P3.1.1. Note that the current through the 15 Ω resistor is independent of the other resistances in the circuit.

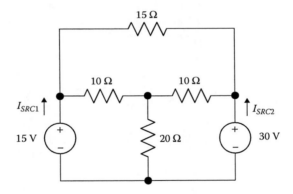

FIGURE P3.1.2

P3.1.3   Determine I_{SRC1} and I_{SRC2} in Figure P3.1.2 using mesh-current analysis.

P3.1.4   In the circuit of Figure P3.1.4, write a KVL equation around mesh abc in terms of I_{ab} and the source currents. Then determine V_{ac} and V_{bc}. How do you interpret the result?

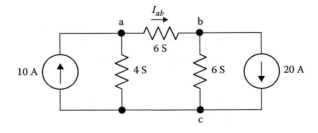

FIGURE P3.1.4

P3.1.5 Determine V_{SRC1} and V_{SRC2} in Figure P3.1.5 using the same method as in Problem P3.1.4. Note that the current through the 15 S resistor is independent of the other conductances in the circuit.

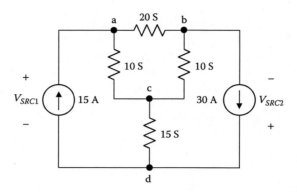

FIGURE P3.1.5

P3.1.6 Determine V_{SRC1} and V_{SRC2} in Figure P3.1.5 using node-voltage analysis.

P3.1.7 Determine V_O in Figure P3.1.7 by applying KCL at the middle node using the single variable V_O. How do you interpret this result?

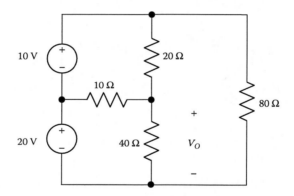

FIGURE P3.1.7

P3.1.8 Determine V_O in Figure P3.1.7 using mesh-current analysis.

P3.1.9 Determine I_O in Figure P3.1.9 by applying KVL around the mesh abc using the single variable I_O. How do you interpret this result?

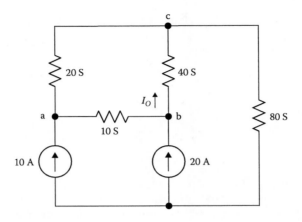

FIGURE P3.1.9

P3.1.10 Determine I_O in Figure P3.1.9 using node-voltage analysis.

P3.1.11 Determine V_O in Figure P3.1.11 using node-voltage analysis. Do not transform the voltage source. Consider that the current entering node a due to this source is $0.25 (10 - V_a)$ A.

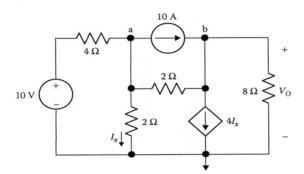

FIGURE P3.1.11

P3.1.12 Determine V_O in Figure P3.1.11 using mesh-current analysis. Redraw the circuit so as to show clearly how the 10 A source could be transformed to a voltage source.

P3.1.13 Determine I_O in Figure P3.1.13 using node-voltage analysis.

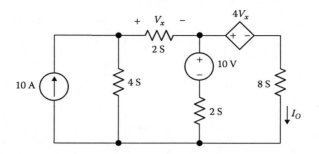

FIGURE P3.1.13

P3.1.14 Determine I_O in Figure P3.1.13 using mesh-current analysis. Do not transform the current source. Consider that the voltage-rise due to this source is $0.25\,(10 - I_1)\,$V, where I_1 is the current in the direction of the voltage drop V_x.

P3.1.15 Determine V_O in Figure P3.1.15 using node-voltage analysis.

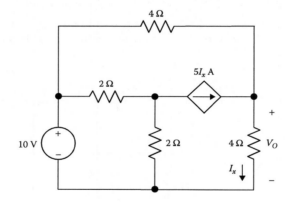

FIGURE P3.1.15

P3.1.16 Determine V_O in Figure P3.1.15 using mesh-current analysis.

P3.1.17 Determine I_O in Figure P3.1.17 using node-voltage analysis.

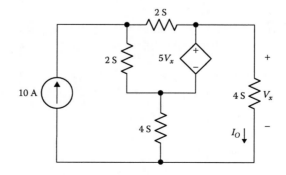

FIGURE P3.1.17

P3.1.18 Determine I_O in Figure P3.1.17 using mesh-current analysis.

P3.1.19 Determine V_O in Figure P3.1.19 using node-voltage analysis. Note that the 2 Ω resistors in series with the current sources do not affect V_O.

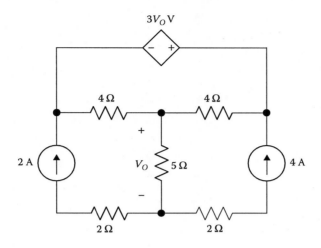

FIGURE P3.1.19

P3.1.20 Determine V_O in Figure P3.1.19 using mesh-current analysis. Note that V_O follows immediately from the assignment of mesh currents.

P3.1.21 Determine I_O in Figure P3.1.21 using node-voltage analysis. Note that I_O follows immediately from the assignment of node voltages.

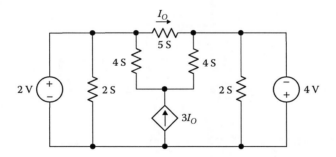

FIGURE P3.1.21

P3.1.22 Determine I_O in Figure P3.1.21 and the source currents in the direction of voltage-rise through the source using mesh-current analysis. Note that the 2 S resistors can be ignored in the mesh-current equations, but not in finding the currents in the voltage sources.

P3.1.23 Determine I_O in Figure P3.1.23 using node-voltage analysis.

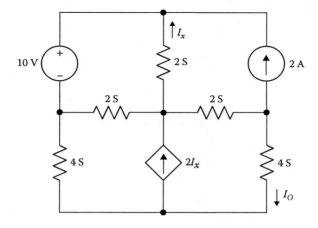

FIGURE P3.1.23

P3.1.24 Determine I_O in Figure P3.1.23 using mesh-current analysis.

P3.1.25 Determine V_O in Figure P3.1.25 using node-voltage analysis.

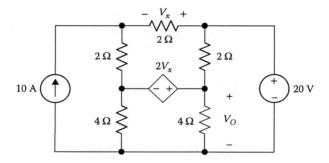

FIGURE P3.1.25

P3.1.26 Determine V_O in Figure P3.1.25 using mesh-current analysis.

P3.1.27 Determine V_O in the nonplanar circuit of Figure P3.1.27 using node-voltage analysis.

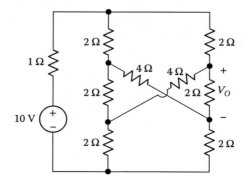

FIGURE P3.1.27

P3.2   Superposition and Scaling

P3.2.1 Determine V_{ab} in Figure P3.1.1 using superposition.

P3.2.2 Determine V_O in Figure P3.2.2 using superposition.

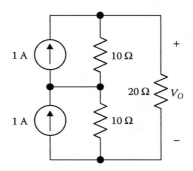

FIGURE P3.2.2

P3.2.3 Determine I_{SRC1} and I_{SRC2} in Figure P3.1.2 using superposition.

P3.2.4 Determine V_{ac} and V_{bc} in Figure P3.1.4 using superposition.

P3.2.5 Determine V_{SRC1} and V_{SRC2} in Figure P3.1.5 using Δ-Y transformation and superposition.

P3.2.6 Determine V_O in Figure P3.1.7 using superposition.

P3.2.7 Determine I_O in Figure P3.1.9 using superposition.

P3.2.8 Determine V_O in Figure P3.1.11 using superposition and leaving the dependent source unaltered.

P3.2.9 Repeat Problem P3.2.8 considering the dependent source as an independent source and determining the voltage across the 2 Ω resistor through which I_x flows.

P3.2.10 Determine I_O in Figure P3.1.13 using superposition and leaving the dependent source unaltered. Calculate the power dissipated in the 8 S resistor.

P3.2.11 Repeat Problem P3.2.10 considering the dependent source as an independent source and determining the current through the 2 S resistor across which V_x is taken.

P3.2.12 Determine V_O in Figure P3.1.19 using superposition and calculate the power dissipated in the 5 Ω resistor.

P3.2.13 Determine I_O in Figure P3.1.21 using superposition and calculate the power dissipated in the 5 S resistor.

P3.2.14 Determine I_O in Figure P3.1.23 using superposition and calculate the power dissipated in the 4 S resistor through which I_O flows.

P3.2.15 Determine V_O in Figure P3.1.25 using superposition.

P3.2.16 Repeat Problem P3.2.15 considering the dependent source as an independent source and determining the current through the 2 Ω resistor across which V_x is taken.

P3.2.17 Determine I_O in Figure P3.2.17 using scaling, assuming all resistances are 1 Ω. Note that because the controlling current is on the output side, it is convenient to work backward from the output.

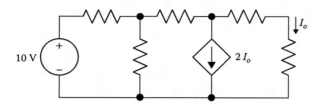

FIGURE P3.2.17

P3.2.18 Figure P3.2.18 shows an example of an *R*-2*R* ladder circuit used in digital-to-analog converters. The resistance looking to the right of every node is 2*R*, so that the current at each node divides equally between the series path and the shunt path. As a result, the current in the 2*R* branches is successively divided by two as one moves away from the source. If I_O = 1 mA and *R* = 1 kΩ, determine V_{SRC}.

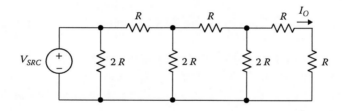

FIGURE P3.2.18

4

Circuit Simplification

Overview

The node-voltage and mesh-current methods of analysis discussed in the preceding chapter are general and powerful methods that can be used, at least in principle, to analyze any well-behaved electric circuit. They are, however, essentially "brute force" and mechanistic methods that may run counter to an important guiding principle in solving circuit problems, namely, to seek the simplest solution, thereby saving time and effort. Such simple solutions may be creative, deriving from particular insight into circuit behavior, or they may be based on circuit theorems and methods that reduce the circuit to a simpler one. We have already encountered one such a method, superposition, which essentially simplifies the circuit by considering one independent source at a time. The Δ-Y transformation and source transformations can also simplify a circuit, as demonstrated by the examples in Chapter 2.

The first part of this chapter is devoted to Thevenin's theorem and the substitution theorem. Thevenin's theorem is perhaps the most important and widely used theorem in circuit theory. It replaces a given circuit, between two specified terminals, by its simplest possible equivalent circuit. The substitution theorem is a simple, yet very useful, theorem that is based on Thevenin's theorem. In the second part of the chapter, we explore several general methods of reducing circuit complexity, based on source rearrangement, removal of redundant elements, and exploitation of symmetry.

The methods discussed in this chapter emphasize that much time and effort can be saved by first trying to simplify the circuit as much as possible based on applying circuit principles and gaining insight into how the circuit behaves.

Learning Objectives

- To be familiar with:
  - The general approach of analyzing a circuit by reducing it to a simpler circuit
- To understand:
  - The meaning and significance of Thevenin's equivalent circuit and the substitution theorem
  - How circuit complexity can be reduced by removing redundant elements, rearranging sources, or exploiting symmetry

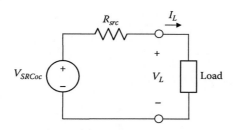

FIGURE 4.1.1
Thevenin's equivalent circuit of voltage divider.

4.1   Equivalent Circuits

Thevenin's Equivalent Circuit

To appreciate the essence of Thevenin's equivalent circuit (TEC), recall that for a voltage divider supplying a load, the *v–i* relation at the load terminals is given by Equation 2.7.5 (Chapter 2), which may be expressed as

$$V_L = V_{SRCoc} - R_{src}I_L \tag{4.1.1}$$

where V_{SRCoc} is the open-circuit voltage of an equivalent source and R_{src} is the effective source resistance (Figure 4.1.1). This circuit is, in fact, TEC of the voltage divider as seen from the load terminals.

One may wonder if this equivalent circuit applies only to the voltage divider. In fact, this result is quite general and can be expressed as follows:

> **Concept:** *In an LTI resistive circuit, the v–i characteristic at any specified pair of terminals is that of an ideal voltage source in series with a source resistance.*

To justify this, we consider the representative, generalized three-mesh circuit of Figure 3.6.1 (Chapter 3), redrawn in Figure 4.1.2 with the 10 Ω resistance in mesh 3 considered as a load resistance R_L connected to terminals ab of the circuit, and I_3 designated as I_L. The mesh current equations are:

$$40I_1 - 10I_2 - 20I_L = V_{SRC1}$$

$$-10I_1 + 60I_2 - 30I_L = V_{SRC2}$$

$$-20I_1 - 30I_2 + 50I_L = V_{SRC3} - V_L \tag{4.1.2}$$

where V_L is considered as a voltage drop in mesh 3 and is included on the RHS of the equation for this mesh as $-V_L$. Solving for I_L and rearranging,

$$V_L = \left[\frac{15V_{SRC1} + 14V_{SRC2}}{23} + V_{SRC3} \right] - \frac{430}{23}I_L \tag{4.1.3}$$

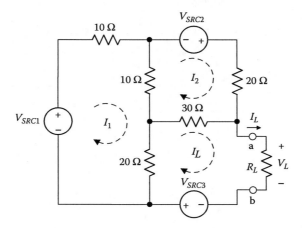

FIGURE 4.1.2
Thevenin's equivalent circuit for a three-mesh circuit.

It is seen that Equation 4.1.3 is indeed of the form of Equation 4.1.1, where V_{SRCoc} equals the bracketed terms and $R_{src} = 430/23$. Because the circuit of Figure 4.1.2 is quite arbitrary, it is concluded that the *v–i* relation for any given circuit at specified terminals ab is the same, in general, as that of an ideal voltage source V_{SRCoc} in series with a resistance R_{src}. The $V_{SRCoc}R_{src}$ circuit is **Thevenin's equivalent circuit** (TEC) of the given circuit at terminals ab and is the simplest possible equivalent circuit because it consists of just an ideal source and a resistor. It is customary to refer to V_{SRCoc} as the **Thevenin voltage** V_{Th}, and to R_{src} as the **Thevenin resistance** R_{Th}.

The next step is to formulate a general procedure for determining V_{Th} and R_{Th} for any given circuit at a specified pair of terminals. It is seen from the preceding discussion, and by setting I_L to zero in Equation 4.1.1, that V_{Th} is simply the voltage at the specified terminals when these terminals are open circuited. Thus, if terminals ab in Figure 4.1.2 are open-circuited, $I_L = 0$ and the mesh-current equations for the circuit become:

$$40I_1 - 10I_2 = V_{SRC1}$$

$$-10I_1 + 60I_2 = V_{SRC2} \tag{4.1.4}$$

Solving for I_1 and I_2, we obtain $I_1 = \dfrac{0.6V_{SRC1} + 0.1V_{SRC2}}{23}$ and $I_2 = \dfrac{0.1V_{SRC1} + 0.4V_{SRC2}}{23}$.
From KVL, $V_L = 30I_2 + 20I_1 + V_{SRC3}$. This gives:

$$V_{Th} = \frac{15}{23}V_{SRC1} + \frac{14}{23}V_{SRC2} + V_{SRC3} \tag{4.1.5}$$

which is the open-circuit voltage determined in Equation 4.1.3.

R_{Th} can be determined in one of two ways. The first follows from TEC (Figure 4.1.3a) when terminals ab are short-circuited, which gives: $R_{Th} = \dfrac{V_{Th}}{I_{SC}}$. Thus, R_{Th} readily follows from V_{Th} and I_{SC}. In the preceding example, if terminals ab are short-circuited, $V_L = 0$.

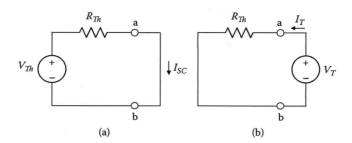

FIGURE 4.1.3
Determining R_{Th}.

Solving the mesh-current equations for I_L gives: $I_L = I_{SC} = \dfrac{15V_{SRC1} + 14V_{SRC2} + 23V_{SRC3}}{430}$.

Hence, $R_{Th} = \dfrac{430}{23}$, in accordance with Equation 4.1.3. Note that whereas V_{Th} and I_{SC} depend on the source voltages, as expected, R_{Th} does not.

The second method of determining R_{Th} can be understood with reference to Figure 4.1.3b. If $V_{Th} = 0$, then the resistance looking into terminals ab is R_{Th}. To set V_{Th} to zero, the three independent sources V_{SRC1}, V_{SRC2}, and V_{SRC3} should be set to zero, according to Equation 4.1.5. Once this is done, R_{Th} can be determined in many cases simply through series–parallel combinations of resistances in the circuit. In general, R_{Th} is determined by applying a test source, such as a voltage source V_T, to terminals ab and finding the resulting current I_T as a function of V_T. The resistance looking into terminals ab is V_T/I_T. In the preceding example, if the sources are replaced by short circuits, R_{Th} can be determined using Δ-Y transformation (Exercise 4.1.1). Alternatively, we can set the source voltages to zero in Equation 4.1.3 and substitute $V_T = V_L$ and $I_T = I_L$, which gives $R_{Th} = 430/23$. In summary:

Procedure

The derivation of TEC of a given circuit between a specified pair of terminals involves, in general, the following steps:

1. Determine the open-circuit voltage V_{Th} at the specified terminals.
2. Determine the short-circuit current I_{SC} at the specified terminals.
3. Set independent sources to zero and determine the resistance R_{Th} looking into the specified terminals. R_{Th} is then determined either directly, or by applying a test voltage and finding the test current.

Because $V_{Th} = R_{Th}I_{SC}$, only two of the three quantities in this relation need be determined through the preceding steps. However, it is useful for checking purposes to determine all three independently.
The following should be noted:

1. Dependent sources affect the resistance values in the circuit. They should not be set to zero in determining R_{Th}, as this will change the circuit. In step 1 and step 2 of the procedure, and when V_T is applied in step 3, superposition may be used and the dependent sources may be replaced by independent sources, as described in Section 3.6 (Chapter 3).

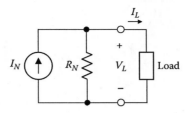

FIGURE 4.1.4
Norton's equivalent circuit.

2. As emphasized in Section 2.6 (Chapter 2), equivalence only applies at the specified terminals and does not apply, in general, to the rest of the circuit, as illustrated by Example 4.1.1.

3. In determining V_{Th}, I_{SC}, or R_{Th}, a simpler circuit is considered, which can facilitate analysis quite considerably (Example SE4.1 and Example SE4.2).

EXERCISE 4.1.1

Determine R_{Th} in Figure 4.1.2 using Δ-Y transformation.

Norton's Equivalent Circuit

When the source V_{Th}, in conjunction with R_{Th}, is transformed to an equivalent current source, Norton's equivalent circuit (NEC) is obtained (Figure 4.1.4). It follows from source transformation that:

$$R_N = R_{src}$$

$$I_N = I_{SC} = \frac{V_{Th}}{R_{src}} \qquad (4.1.6)$$

It may be noted that a circuit can have a TEC but not an NEC, and conversely. For example, TEC of an ideal voltage source is the source itself, with $R_{src} = 0$. This makes $I_N \rightarrow \infty$, so NEC does not exist. Similarly, NEC of an ideal current source is the source itself, with $R_{src} \rightarrow \infty$. This makes $V_{Th} \rightarrow \infty$, so TEC does not exist. Other examples are given in problems at the end of the chapter.

Example 4.1.1: TEC and NEC

It is required to determine TEC and NEC between terminals ab in Figure 4.1.5.

SOLUTION
This circuit is of the form of the equivalent circuit of the *h*-parameter, two-port circuit equations (Section 14.3, Chapter 14). Terminals ab are already open-circuited. $V_O = -20I_y \times 0.025 = -0.5I_y$, where I_y is in mA, so that the 25 Ω resistor is expressed in kΩ. However,

$I_y = \dfrac{5 - 3V_O}{2}$. Solving for V_O, $V_{Th} = V_O = -5$ V. $I_y = 10$ mA. If terminals ab are short-circuited

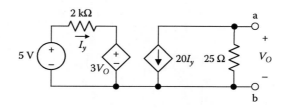

FIGURE 4.1.5
Figure for Example 4.1.1.

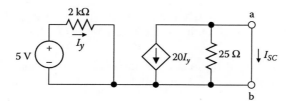

FIGURE 4.1.6
Figure for Example 4.1.1.

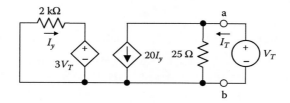

FIGURE 4.1.7
Figure for Example 4.1.1.

(Figure 4.1.6), $V_O = 0$, $I_y = \dfrac{5}{2} = 2.5$ mA, and $I_{SC} = -20I_y = -50$ mA. Hence, $R_{Th} = \dfrac{-5}{-50} \equiv 100\,\Omega$. If V_T is applied, with the 5 V source replaced by a short circuit (Figure 4.1.7), I_y

$$= -\frac{3V_T}{2} = -1.5V_T\,.$$ From KCL at node a, $I_T = \dfrac{V_T}{0.025} + 20I_y = 40V_T - 30V_T = 10V_T$. Hence,

$$R_{Th} = \frac{V_T}{I_T} = 0.1 \text{ k}\Omega \equiv 100 \ \Omega.$$

TEC and NEC at terminals ab are shown in Figure 4.1.8. It should be noted that the equivalence applies only to the *V–I* relations at terminals ab. For example, with terminals ab open-circuited, the Thevenin source does not dissipate any power, whereas the Norton source dissipates $(0.05)^2 \times 100 = 0.25$ W. The power delivered by the 5 V source in the given circuit is 50 mW and that delivered by the dependent source is 1 W. When a load is connected across terminals ab, the same power is delivered to the load in the three cases because V_L and I_L are the same. Example 4.1.1 is considered further in Example SE4.1, where TEC is obtained in two steps, first without the 25 Ω connected, and then by adding this resistor.

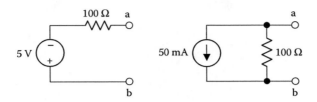

FIGURE 4.1.8
Figure for Example 4.1.1.

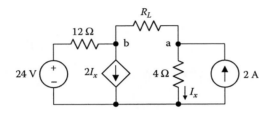

FIGURE 4.1.9
Figure for Exercise 4.1.2.

An advantage of using TEC is evident from this example. Suppose it is required to calculate the load current for a number of values of load resistance, say 900 Ω and 1900 Ω. Rather than determine the load current in each case from the original circuit, which may involve tedious calculations, TEC greatly eases this task because it is independent of the load resistance. Thus, the load current for a 900 Ω load is $5/(100+900)=5$ mA, and that for a 1900 Ω load is $5/(100+1900)=2.5$ mA.

EXERCISE 4.1.2

Determine TEC and NEC looking into terminals ab in Figure 4.1.9.
Answer: $V_{Th} = 32$ V, $R_{Th} = 40$ Ω, $I_N = 0.8$ A.

Simulation Example 4.1.2: Derivation of TEC Using PSpice
It is required to obtain TEC between terminals ab in Figure 4.1.10.

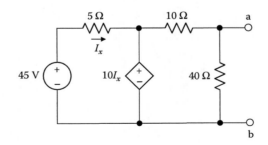

FIGURE 4.1.10
Figure for Example 4.1.2.

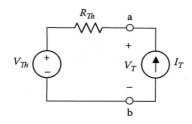

FIGURE 4.1.11
Figure for Example 4.1.2.

ANALYSIS

From KVL, $45 = 5I_x + 10I_x$, so that $I_x = 3$ A, irrespective of whether terminals ab are open-circuited or short-circuited. On open circuit, $V_{Th} = \dfrac{40}{50} \times 10I_x = 24$ V. On short circuit, $I_{SC} = \dfrac{10I_x}{10} = 3$ A. It follows that $R_{Th} = 8\ \Omega$.

SIMULATION

To obtain TEC using PSpice, one could perform two simulations to determine any two of the three quantities V_{Th}, I_{SC}, and R_{Th}. A more efficient method is to apply a test current source I_T at terminals ab, *without altering the circuit*. Replacing the circuit between terminals ab by its TEC, it follows from Figure 4.1.11 that $V_T = V_{Th} + R_{Th}I_T$. Hence, if I_T is swept over a range of values to obtain the linear V_T vs. I_T plot, the slope of the line is R_{Th} and its voltage intercept is V_{Th}.

After entering the schematic, with a current source IDC connected across the 40 Ω resistor, select DC Sweep under Analysis type in the Simulation Settings window of the Simulation Profile. Choose Current source under Sweep variable and enter the source name, such as I1. Enter 0 for Start value, 1 for End value, and 0.1 for Increment. After the simulation is run, select in the Schematic page Trace/Add Trace then choose V(I1:-) in the Add Traces dialog box. Alternatively, a voltage marker may be used (Appendix 5D.2) The line plot shown in Figure 4.1.12 is displayed. The intercept on the voltage axis is 24 V and the slope is 8 Ω.

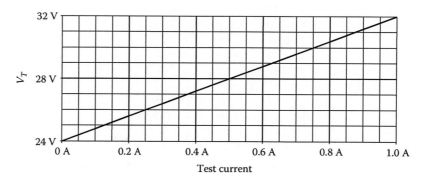

FIGURE 4.1.12
Figure for Example 4.1.2.

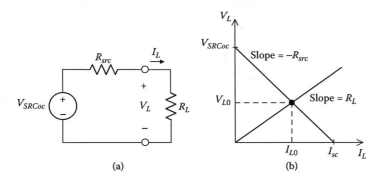

FIGURE 4.1.13
Figure for graphical analysis.

Graphical Analysis

Before leaving TEC, it is useful for a later discussion to introduce graphical analysis based on Equation 4.1.1 and Figure 4.1.13a. The plot of V_L vs. I_L is a straight line of slope $-R_{src}$, voltage intercept V_{SRCoc}, and current intercept $I_{SC} = V_{SRCoc}/R_{src}$ (Figure 4.1.13b). Because this line depends entirely on the source, it is referred to as the **source characteristic** and is given by Equation 4.1.1. V_L and I_L are also related by Ohm's law for the load, $V_L = R_L I_L$. When plotted on the same graph, this plot is a straight line of slope R_L passing through the origin. It is referred to as the **load line**. The intersection point of the load line and the source characteristic gives V_{L0} and I_{L0} for particular values of V_{SRCoc}, R_{src}, and R_L. This is because both the source characteristic and the load line equation are satisfied at the intersection point.

Graphical analysis applies to both linear and nonlinear circuits and is extensively used with diode and transistor circuits. It is very useful for visualizing the behavior of the circuit as the variables change. The load line construction is elaborated further in Section ST4.1.

4.2 Substitution Theorem

> **Statement:** *A resistor having a voltage V across it, or a current I through it, can be replaced by an ideal, independent voltage source V, or an ideal, independent current source I, without disturbing the rest of the circuit.*

Although this may appear counterintuitive at first sight, it follows readily from TEC. Consider a resistor connected to a circuit N that is represented by its TEC at terminals ab (Figure 4.2.1a). Then,

$$I = \frac{1}{R_{Th}}\ (V_{Th} - V) \tag{4.2.1}$$

or

$$V = V_{Th} - R_{Th}I \tag{4.2.2}$$

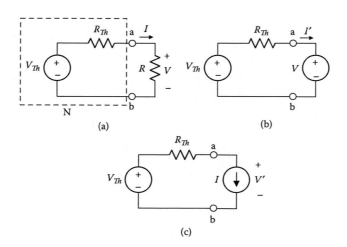

FIGURE 4.2.1
Substitution theorem.

If R is replaced by an ideal voltage source V (Figure 4.2.1b), $V_{ab} = V$, as before, and $I' = \dfrac{1}{R_{Th}}\ (V_{Th} - V)\ = I$ (Equation 4.2.1). The rest of the circuit is therefore undisturbed because the voltage and current at terminals ab are still V and I. If R is replaced by an ideal current source I (Figure 4.2.1c), the current in the circuit remains unchanged, and $V' = V_{Th} - R_{Th}I = V$ (Equation 4.2.2). Again, the rest of the circuit is undisturbed.

Evidently, the same reasoning holds if R were a dependent source, or a branch that includes combinations of resistances and sources, or part of a circuit. We already substituted an independent source for a dependent source in the second superposition method in Section 3.6 (Chapter 3).

Source Absorption Theorem

This is a weaker form of the substitution theorem, which may be stated as follows:

> **Statement:** *A resistor having a voltage V across it AND a current I through it, can be replaced by a CCVS having V = RI or a VCCS having I = V/R. Conversely, a CCVS having V = ρI can be replaced by a resistance ρ, and a VCCS having I = σV can be replaced by a conductance σ, where V is the voltage across the source and I is the current through it.*

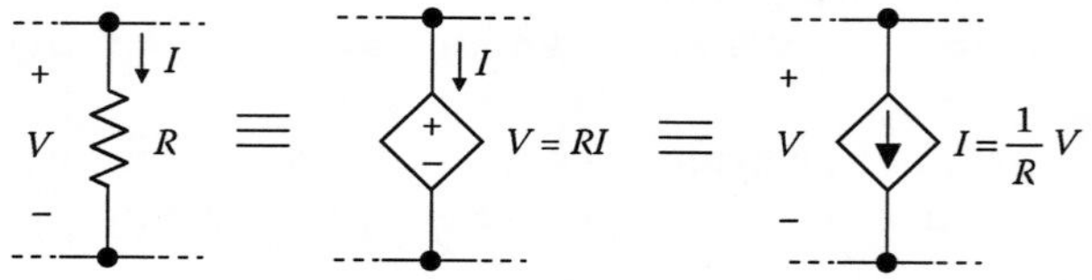

$$V\ \lessgtr R\ \equiv\ \text{(CCVS)}\ V = RI\ \equiv\ V\ \text{(VCCS)}\ I = \frac{1}{R}V$$

FIGURE 4.2.2
Source absorption.

These equivalence relations are summarized in Figure 4.2.2 and follow from the fact that the terminal voltages and currents are the same in the three cases.

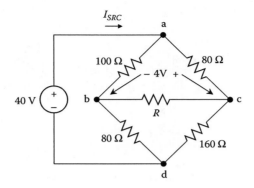

FIGURE 4.2.3
Figure for Example 4.2.1.

The substitution theorem is illustrated by Example 4.2.1, and can be very usefully applied in many cases, as illustrated by Exercise 4.3.1, Example 6.3.2 (Chapter 6), the universal active filter (Figure 18.4.7, Chapter 18), and Problem P18.3.1 to Problem P18.3.3 (Chapter 18) on active filters.

Example 4.2.1: Determination of Unknown Current Using Substitution Theorem

Given a bridge circuit (Figure 4.2.3) with an unknown resistance R connected at the bridge output. The circuit diagram of the bridge is known but the bridge is inaccessible except for R. With only a voltmeter available that measures a voltage of 4 V across R, of the polarity indicated, it is required to determine the source current I_{SRC}.

SOLUTION
According to the substitution theorem, R may be replaced by an ideal voltage source of 4 V. I_{SRC} can then be determined by deriving TEC of the bridge between terminals a and d in Figure 4.2.3. When the bridge is disconnected from the 40 V source at terminals a and d, the voltage V_{ad} between these terminals is $V_{ad} = V_{ab} - V_{db} = 4\left(\dfrac{100}{180} - \dfrac{80}{240}\right) = \dfrac{8}{9}$ V $= V_{Th}$. When the 4 V source is set to zero, the resistance between terminals a and d is

$\dfrac{80 \times 100}{180} + \dfrac{80 \times 160}{240} = \dfrac{880}{9}$ Ω. TEC between terminals a and d is a voltage source of $\dfrac{8}{9}$ V,

in series with a resistance $\dfrac{880}{9}$ Ω. It follows that $I_{SRC} = \dfrac{40 - 8/9}{880/9} = 0.4$ A.

EXERCISE 4.2.1

Suppose that instead of measuring the voltage across R, an ammeter included in series with R reads 50 mA, the current being directed from node c to node b (Figure 4.2.3). Determine the source current.

Answer: 0.4 A.

4.3   Source Rearrangement

> **Concept:** *Sources can be rearranged in a circuit so as to facilitate analysis of the circuit without affecting circuit responses, as long as KCL and KVL remain satisfied.*

In Figure 4.3.1a, for example, V_{SRC} can be replaced by two voltage sources as shown in Figure 4.3.1b. Because the same voltage V_{SRC} still appears at the terminals of N_1 and N_2, the currents I_1 and I_2 are unaltered.

Similarly, I_{SRC} in Figure 4.3.2a can be replaced by current sources as shown in Figure 4.3.2b. The same current I_{SRC} still flows out of N_1 and into N_2. The terminal voltages V_1 and V_2 are therefore unaltered.

A useful interpretation of Figure 4.3.1 can be made by assuming that I_1 in Figure 4.3.1a changes by ΔI_1 because of some disturbance. From KCL, the current in the V_{SRC} source changes by this same amount. I_2 does not change because the voltage at the terminals of N_2 remains equal to V_{SRC}. In other words, the ideal voltage source in shunt with both circuits essentially uncouples the two circuits, while applying the same voltage to both. The zero source resistance behaves as a short circuit, as far as current changes are concerned, and prevents such changes in one circuit from being transmitted to the other circuit. Similarly, a change of ΔV_1 in Figure 4.3.2a changes the voltage across the source by this same amount but is not transmitted to N_2 because I_{SRC} remains the same. The ideal current source in series with the two circuits in Figure 4.3.2a uncouples the two circuits while applying the same current to both. The infinite source resistance behaves as an open circuit, as far as voltage changes are concerned, and prevents such changes in one circuit from being transmitted to the other circuit.

Example 4.3.1: Source Rearrangement
It is required to determine I_x in Figure 4.3.3 using source rearrangement.

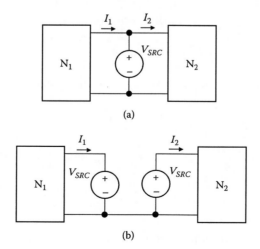

(a)

(b)

FIGURE 4.3.1
Rearrangement of voltage source.

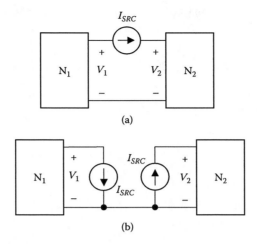

FIGURE 4.3.2
Rearrangement of current source.

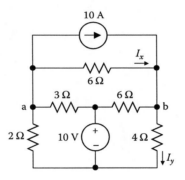

FIGURE 4.3.3
Figure for Example 4.3.1.

SOLUTION

The voltage source and current source may each be split into two sources as shown in Figure 4.3.4. In Figure 4.3.3, a current of 10 A enters node b from the source and a current of 10 A leaves node a. The same conditions are preserved in Figure 4.3.4. A source current of 10 A both enters and leaves node c in Figure 4.3.4, so that the net source current at this node is zero, as in Figure 4.3.3. I_x will be determined in two ways:

Method 1: The first method is to derive TEC between nodes b and a. The open-circuit voltages between nodes a and c and between nodes b and c will be determined by superposition.

When the 6 Ω resistor is removed, and the 10 A sources set to zero, $V_{ac} = 10 \times 2/5 = 4$ V and $V_{bc} = 10 \times 4/10 = 4$ V. Hence, nodes a and b will be at the same voltage, and this component of V_{ba} is zero. If the 10 V sources are set to zero, $V_{ac} = -10 \times 6/5 = -12$ V and $V_{bc} = 10 \times 24/10 = 24$ V. It follows that $V_{ba} = V_{Th} = 36$ V. With all sources set to zero, $R_{Th} = (4 \| 6) + (2 \| 3) = 3.6$ Ω. Hence $I_x = -36/9.6 = -3.75$ A.

Method 2: The second method is based on the substitution theorem and source transformation. The 6 Ω resistor is replaced by a current source I_x, in accordance with the substitution theorem, which adds I_x to the two 10 A sources. The 10 V source on the RHS in

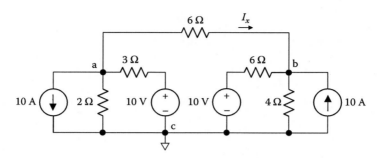

FIGURE 4.3.4
Figure for Example 4.3.1.

Figure 4.3.4 is transformed to a 5/3 A current source. The parallel resistance of 6 Ω and

4 Ω is 12/5 Ω, and $V_{bc} = \dfrac{12}{5}\left(\dfrac{5}{3} + 10 + I_x\right) = 28 + \dfrac{12}{5}I_x$. Similarly, the 10 V source on the LHS

in Figure 4.3.4 is transformed to 10/3 A current source in parallel with 6/5 Ω, and

$V_{ac} = \dfrac{6}{5}\left(\dfrac{10}{3} - 10 - I_x\right) = -8 - \dfrac{6}{5}I_x$. From the circuit without the 6 Ω resistor replaced by a

current source, $V_{ac} = V_{bc} + 6I_x$. Substituting for V_{bc} gives: $I_x = -3.75$A.

EXERCISE 4.3.1

Having found $I_x = -3.75$ A in Figure 4.3.3, determine I_y using the substitution theorem.

Answer: 4.75 A.

4.4   Removal of Redundant Elements

Concept: *Redundant elements can be removed from a circuit without affecting the circuit responses of interest.*

Examples of redundant elements are:

1. A resistor in series with an ideal current source. When this series resistor is removed, that is, replaced by a short circuit, the source current is unchanged. The voltage across the source decreases by an amount equal to the voltage drop across the resistor, but the rest of the circuit is not disturbed.

2. A resistor in parallel with an ideal voltage source. When this parallel resistor is removed, that is, replaced by an open circuit, the source voltage is unchanged. The current through the source decreases by an amount equal to the current through the resistor, but the rest of the circuit is not disturbed.

3. Resistors that do not carry any current. These may be replaced by an open circuit or a short circuit without disturbing the rest of the circuit.

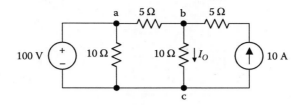

FIGURE 4.4.1
Figure for Example 4.4.1.

4. Inductors and capacitors in a circuit under dc operating conditions. The inductor is replaced by a short circuit and the capacitor by an open circuit, as discussed in Section 1.8 and Section 1.9 of Chapter 1 and implemented in electronic circuit analysis.

Example 4.4.1: Redundant Elements

It is required to determine I_O in the circuit of Figure 4.4.1.

SOLUTION
The 10 Ω resistor in parallel with the voltage source and the 5 Ω resistor in series with the current source are redundant and can be removed without affecting I_O. The circuit reduces to that of Figure 4.4.2. KCL at node b may be written as $10 + \dfrac{100 - 10I_O}{5} = I_O$, where

$10I_O$ is the voltage across the 10 Ω resistor and $\dfrac{100 - 10I_O}{5}$ is the current flowing toward node b through the 5 Ω resistor. Solving for I_O gives I_O = 10A. The two resistors do not affect I_O. In Figure 4.4.2, the current through the voltage source is zero and the voltage across the current source is 100 V. In Figure 4.4.1, the current through the voltage source is 10 A and the voltage across the current source is 150 V.

EXERCISE 4.4.1

Determine V_y in Figure 4.4.3 after eliminating all redundant elements.
Answer: 1 V.

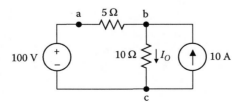

FIGURE 4.4.2
Figure for Example 4.4.1.

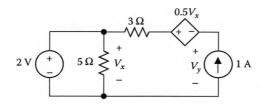

FIGURE 4.4.3
Figure for Exercise 4.4.1.

4.5   Exploitation of Symmetry

Concept: *In circuits possessing symmetry, the circuit can be simplified by removing elements that do not carry current, or by connecting together nodes that are at the same voltage.*

Exploitation of symmetry is illustrated by Example 4.5.1.

Example 4.5.1: Exploitation of Symmetry

Given a grid of twelve 1 Ω resistors connected as shown in Figure 4.5.1, a 21 V source is connected, successively, between nodes: (1) 1 and 7, (2) 1 and 9, and (3) 1 and 5. It is required to determine the source current in each case. The method of solution described is a good example of a creative problem-solving approach in that reasoning based on circuit principles is used to reduce the circuit to a much simpler one that can be analyzed readily with little effort.

SOLUTION

1. With the source connected between nodes 1 and 7, the source can be split into two 10.5 V in series, with the midpoint grounded (Figure 4.5.2a). Because the

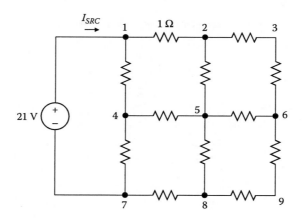

FIGURE 4.5.1
Figure for Example 4.5.1.

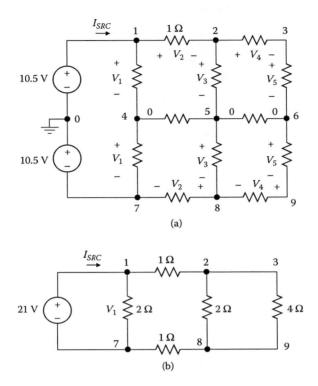

FIGURE 4.5.2
Figure for Example 4.5.1.

voltage of node 1 is +10.5 V and that of node 7 is –10.5 V with respect to ground, it is clear from symmetry that node 4 is at a ground voltage of zero. Similarly, it follows from symmetry that node 2 is at a voltage $10.5 - V_2$, whereas node 8 is at $10.5 + V_2$, where V_2 is a positive voltage drop. This means that the voltages of node 2 and node 8 are equal in magnitude but opposite in sign. Node 5 is therefore at zero voltage. Because node 2 and node 6 are connected by a resistance of 2 Ω, as are node 6 and node 8, then node 6 is also at a node voltage of zero. Hence, the resistors between node 4 and node 5, and between node 5 and node 6, do not carry any current. They could just as well be replaced by open circuits or short circuits. If they are replaced by open circuits, the circuit reduces to that shown in Figure 4.5.2b. The series–parallel combination of resistances evaluates to 1.25 Ω across the source, so that $I_{SRC} = 16.8$ A. Alternatively, it may be argued that the circuit of Figure 4.5.2a is symmetrical about a horizontal line passing through nodes 4, 5, 6, and the ground between the split sources. Voltages above this line are positive, whereas voltages below this line are negative. Hence, voltages along the line are zero.

2. If the 21 V source is connected between nodes 1 and 9 (Figure 4.5.1), the circuit becomes symmetrical about the diagonal and could be split into two halves along the diagonal. Figure 4.5.3a shows the source applied to one half-circuit. The half-circuit can be reduced to that shown in Figure 4.5.3b, from which it follows that the current drawn by the half circuit is 7 A, so that $I_{src} = 14$ A.

3. When the 21 V source is connected between node 1 and node 5 (Figure 4.5.4a), the currents are symmetrical about the diagonal through node 1, node 5, and node

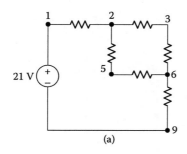

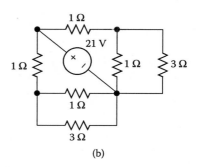

FIGURE 4.5.3
Figure for Example 4.5.1.

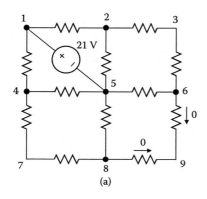

FIGURE 4.5.4
Figure for Example 4.5.1.

9. Thus, the current from node 1 to node 2 equals that from node 1 to node 4, the current from node 3 to node 6 equals that from node 7 to node 8, and the current from node 6 to node 9 equals that from node 8 to node 9. The currents at node 9 are therefore equal and sum to zero. Hence, each of them must be zero. It follows that the two resistors connected to node 9 do not carry current and can be removed. The resulting circuit becomes as shown in Figure 4.5.4b. The resistance on either side of the source is $7/4 \, \Omega$. The effective resistance across the source is $7/8 \, \Omega$, so that $I_{SRC} = 24$ A.

EXERCISE 4.5.1

(a) Verify that if the resistors between node 4 and node 5, and between node 5 and node 6, in Figure 4.5.1 are replaced by short circuits, $I_{SRC} = 16.8$ A.

(b) Determine the current distribution in the various resistors of Figure 4.5.3a.

(c) Repeat part b for the circuit of Figure 4.5.4a.

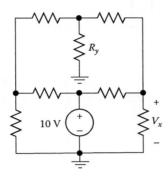

FIGURE 4.5.5
Figure for Exercise 4.5.2.

EXERCISE 4.5.2

All resistors in Figure 4.5.5 are 6 Ω, except R_y. Determine V_x assuming: (a) R_y is an open circuit; (b) $R_y = 3 \Omega$.

Answer: (a) 5 V; (b) 4 V.

Summary of Main Concepts and Results

- The v–i characteristic at any specified pair of terminals in an LTI resistive circuit is that of an ideal voltage source in series with a source resistance, which constitutes Thevenin's equivalent circuit.
- Norton's equivalent circuit is the current-source equivalent of Thevenin's equivalent circuit.
- The derivation of TEC of a given circuit between a specified pair of terminals involves, in general, the following steps:
  1. Determining the open-circuit voltage V_{Th} at the specified terminals.
  2. Determining the short-circuit current I_{SC} at the specified terminals.
  3. Setting independent sources to zero and determining the resistance R_{Th} looking into the specified terminals. R_{Th} is determined either directly, or by applying a test voltage (or current) and finding the test current (or voltage).
- Graphical analysis is useful for at least visualizing the behavior of a given circuit as variables change.
- According to the substitution theorem, a resistor having a voltage V across it, or a current I through it, can be replaced by an ideal, independent voltage source V, or an ideal, independent current source I, without disturbing the rest of the circuit.
- According to the source absorption theorem, a resistor having a voltage V across it and a current I through it can be replaced by a CCVS having $V = RI$ or a VCCS having $I = \dfrac{V}{R}$. Conversely, a CCVS having $V = \rho I$ can be replaced by a resistance ρ, and a VCCS having $I = \sigma V$ can be replaced by a conductance σ, where V is the voltage across the source and I is the current through it.

- Sources can be rearranged in a circuit, so as to facilitate analysis of the circuit, without affecting circuit responses as long as KCL and KVL remain satisfied.
- Redundant elements can be removed from a circuit without affecting the circuit responses of interest.
- In circuits possessing symmetry, the circuit can be simplified by removing elements that do not carry current, or by connecting together nodes that are at the same voltage.

Learning Outcomes

- Derive TEC and NEC between specified terminals
- Apply the substitution theorem and the source absorption theorem
- Simplify a circuit by rearranging sources, removing redundant elements, and exploiting symmetry

Supplementary Topics and Examples on CD

ST4.1 Graphical analysis and load lines: Elaborates the load-line construction, particularly as applied in electronic circuit analysis.

SE4.1 Alternative derivation of TEC: Considers an alternative and instructive derivation of TEC of Example 4.1.1.

SE4.2 Analysis of bridge circuit: Analyzes a bridge circuit by three methods: using TEC, scaling, and Δ-Y transformation.

SE4.3 Linear circuit with nonlinear load: Derives the current in a nonlinear resistor using TEC.

Problems and Exercises

P4.1   Thevenin's and Norton's Equivalent Circuits

Verify your results with PSpice simulation.

P4.1.1 Derive TEC between terminals ab in Figure P4.1.1.

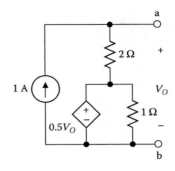

FIGURE P4.1.1

P4.1.2 Derive NEC between terminals ab in Figure P4.1.2.

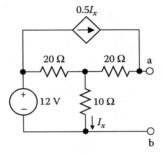

FIGURE P4.1.2

P4.1.3 Derive TEC between terminals ab in Figure P4.1.3, given $\rho = 30$.

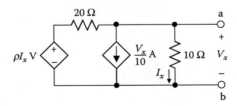

FIGURE P4.1.3

P4.1.4 Repeat Problem P4.1.3, given $\rho = 70$.

P4.1.5 Derive TEC between terminals ab in Figure P4.1.5 for a given V_{SRC} assuming:
(a) $\alpha = 1$ and (b) $\alpha = 2$.

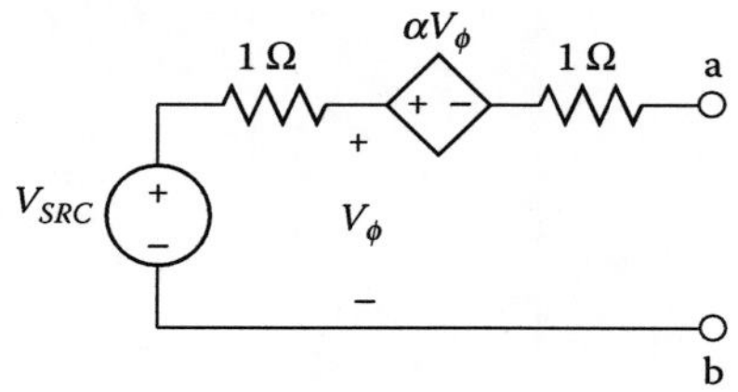

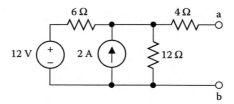

FIGURE P4.1.5

P4.1.6 Derive TEC between terminals ab in Figure P4.1.6.

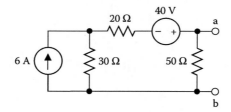

FIGURE P4.1.6

P4.1.7 Derive NEC between terminals ab in Figure P4.1.7.

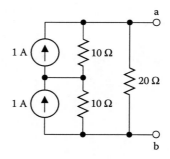

FIGURE P4.1.7

P4.1.8 Derive TEC between terminals ab in Figure P4.1.8.

FIGURE P4.1.8

P4.1.9 Derive TEC between terminals ab in Figure P4.1.9.

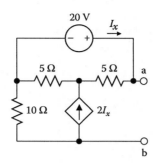

FIGURE P4.1.9

P4.1.10 Derive NEC between the short-circuited terminals ab in Figure P2.1.4 (Chapter 2).

P4.1.11 Determine V_O in Figure P3.1.7 (Chapter 3) using TEC.

P4.1.12 Determine I_O in Figure P3.1.9 (Chapter 3) using NEC.

P4.1.13 Determine V_O in Figure P3.1.11 (Chapter 3) using TEC.

P4.1.14 Determine I_O in Figure P3.1.13 (Chapter 3)using NEC.

P4.1.15 Determine V_O in Figure P3.1.15 (Chapter 3) using TEC.

P4.1.16 Determine I_O in Figure P3.1.17 (Chapter 3) using NEC.

P4.1.17 Determine V_O in Figure P3.1.19 (Chapter 3) using NEC.

P4.1.18 Determine I_O in Figure P3.1.21 (Chapter 3) using TEC.

P4.1.19 Determine I_O in Figure P3.1.23 (Chapter 3)using NEC.

P4.1.20 Determine V_O in Figure P3.1.25 (Chapter 3) using TEC.

P4.2 Substitution Theorem

Verify your results with PSpice simulation.

P4.2.1 Given that the current in the 15 Ω resistor in Figure P3.1.2 (Chapter 3) is 1 A flowing toward source 1, use the substitution theorem to find the current in the 20 Ω resistor.

P4.2.2 Given that $V_{ab} = 0.9$ V in Figure P3.1.5 (Chapter 3), use the substitution theorem to find V_{cd}.

P4.2.3 If the voltage drop across the dependent source in Figure P3.1.15 (Chapter 3) is 12.5 V in the direction of current, use the substitution theorem to find V_O.

P4.2.4 If the current through the dependent source in Figure P3.1.17 (Chapter 3) is 12.5 A in the direction of the voltage drop, use the substitution theorem to find I_O.

P4.2.5 If the voltage drop across the dependent source in Figure P3.1.23 (Chapter 3) is 15.5 V in the direction of current, use the substitution theorem to find I_O.

P4.2.6 If the current through the dependent source in Figure P3.1.25 (Chapter 3) is 10 A in the direction of the voltage drop, use the substitution theorem to find V_O.

P4.3   Circuit Reduction

P4.3.1 Given two circuits N_1 and N_2 connected by a VCCS σV_1, as indicated in Figure P4.3.1a, show that the source can be replaced by a resistance $1/\sigma$ connected to N_1 and a VCCS σV_1 connected to N_2 (Figure P4.3.1b).

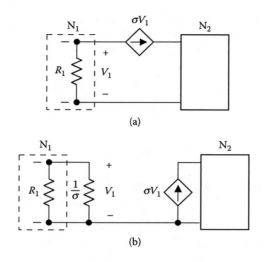

(a)

(b)

FIGURE P4.3.1

P4.3.2 Given two circuits N_1 and N_2 connected by a CCVS ρI_1, as indicated in Figure P4.3.2a, show that the source can be replaced by a resistance ρ connected to N_1 and a CCVS ρI_1 connected to N_2 (Figure P4.3.2b).

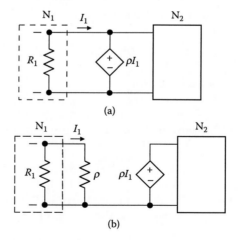

(a)

(b)

FIGURE P4.3.2

P4.3.3 Given two circuits N_1 and N_2 connected by a CCCS βI_1, as indicated in Figure P4.3.3a, show that this is equivalent to connecting the source to N_2 and increasing R_1 to $(1+\beta)R_1$ (Figure P4.3.4b).

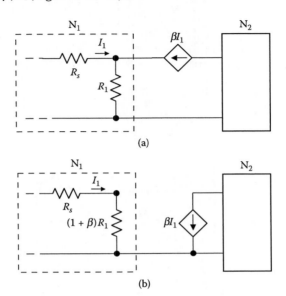

(a)

(b)

FIGURE P4.3.3

P4.3.4 Given circuits N_1 and N_2 that are connected by a VCVS αV_1, as indicated in Figure P4.3.4a, show that if this voltage source is replaced by a short circuit, currents in N_1 and N_2 remain unchanged if: (a) resistances and voltage sources in N_1 are multiplied by $(1+\alpha)$ (Figure P4.3.4b) or (b) resistances and voltage sources in N_2 are divided by $(1+\alpha)$ (Figure P4.3.4c). Current sources remain unchanged.

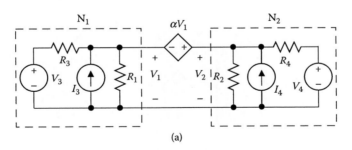

(a)

FIGURE P4.3.4

Continued.

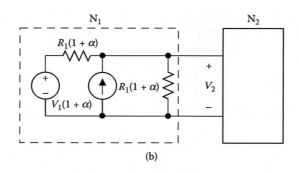

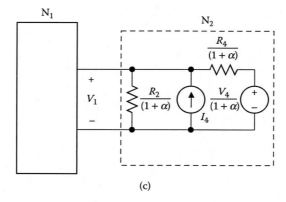

FIGURE P4.3.4

P4.3.5 Circuits N_1 and N_2 are connected by a CCCS βI_1, as indicated in Figure P4.3.5, where N_1 and N_2 are represented as in Problem P4.3.4. Show that if the current source is replaced by an open circuit, voltages in N_1 and N_2 remain unchanged if: (1) conductances and current sources in N_1 are multiplied by $(1+\beta)$ or (2) conductances and current sources in N_2 are divided by $(1+\beta)$. Voltage sources remain unchanged.

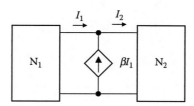

FIGURE P4.3.5

P4.3.6 Deduce from Figure 4.5.1, with the source connected between nodes 1 and 9, that the circuit may be split at node 5 and the horizontal and vertical branches cross connected, without changing the currents in any of the branches.

P4.3.7 Determine the voltage drop from a to b in Figure P4.3.7, assuming that all resistances are 1 Ω.

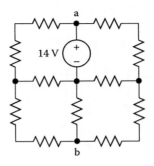

FIGURE P4.3.7

P4.3.8 Determine the voltage drop from a to b in Figure P4.3.8, assuming that all resistances are 1 Ω.

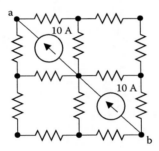

FIGURE P4.3.8

P4.3.9 Determine the voltage drop from a to b in Figure P4.3.9, assuming that all resistances are 1 Ω and all sources are 5 A.

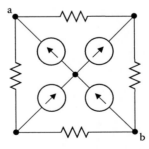

FIGURE P4.3.9

P4.3.10 Determine the resistance between terminals a and b in Figure P4.3.10.

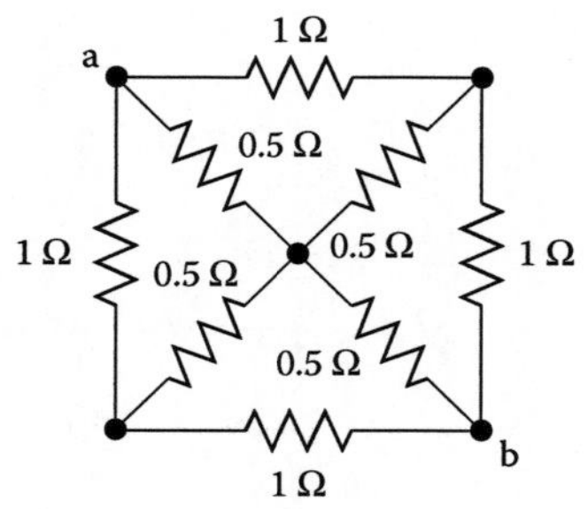

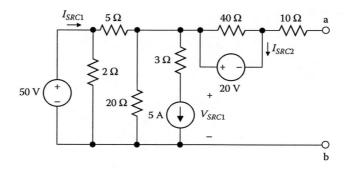

FIGURE P4.3.10

P4.3.11 Determine TEC between terminals ab in Figure P4.3.11 as well as I_{SRC1}, I_{SRC2}, and V_{SRC1}.

FIGURE P4.3.11

5

Sinusoidal Steady State

Overview

The analysis of electric circuits thus far was restricted to dc conditions. We will now consider circuits in the sinusoidal steady state; that is, when circuit excitation and responses are all sinusoidal functions of time that differ, in general, in amplitude and phase. Whereas only resistors are considered in dc circuit analysis, inductors and capacitors are included in sinusoidal steady-state analysis.

Sinusoidal steady-state analysis is of considerable importance. It is the basis for analyzing power systems, the responses to periodic functions, and the frequency response of circuits. Central to sinusoidal steady-state analysis are phasor notation and the concept of impedance. These are of fundamental importance because they reduce circuit differential equations to algebraic equations involving complex quantities. As will be shown, this allows a simple and direct generalization to the sinusoidal steady state of all the circuit relations and theorems discussed for dc conditions. The circuit is transformed to the frequency domain by representing voltages and currents as phasors and expressing the relation between phasor branch voltages and currents in terms of impedances. This same frequency domain representation also applies to periodic inputs (Chapter 9), to frequency responses (Chapter 10), and to the Fourier transform (Chapter 16). Moreover, the concept of impedance is readily extended by the Laplace transform method (Chapter 15) to the general case in which circuit excitations and responses are arbitrary functions of time.

Learning Objectives

- To be familiar with:
  - The definitions of impedance, reactance, admittance, and susceptance
  - The general procedure for analyzing circuits in the frequency domain
- To understand:
  - The interpretation of the circuit response to a complex sinusoidal excitation
  - The nature of phasor notation and the properties of phasors
  - Phasor relations of voltage and current for resistors, inductors, and capacitors and the representation of these relations in phasor diagrams
  - How the concept of impedance, together with phasor notation, transforms a differential equation of a circuit variable to an algebraic relation for the purpose of deriving the steady-state sinusoidal response

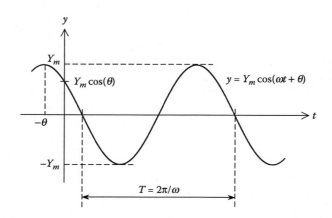

FIGURE 5.1.1
Sinusoidal function.

5.1 Sinusoidal Function

A voltage or current that varies sinusoidally with time may be represented as:

$$y = Y_m \cos(\omega t + \theta) \tag{5.1.1}$$

where Y_m is the amplitude, ω is the angular frequency, and θ is the phase angle (Figure 5.1.1). The time interval between successive repetitions of the same value of y is the **period** T. The full range of values of the function over a period is a **cycle**. The frequency f of repetitions of the function is:

$$f = \frac{1}{T} = \frac{\omega}{2\pi} \tag{5.1.2}$$

where T is in seconds, f is in cycles per second, or hertz (Hz), and ω is in radians/s. Voltages and currents that vary sinusoidally with time are designated as ac quantities, where ac stands for alternating current.

> **Concept:** *A characteristic property of the sinusoidal function is that it is invariant in nature under linear operations, such as addition, subtraction, differentiation, and integration.*

In other words, linear operations may change the amplitude and phase of a sinusoidal function but they do not change its general shape or its frequency. On the other hand, if a square wave, for example, is differentiated or integrated, it becomes a different function altogether.

EXERCISE 5.1.1

A sinusoidal function of the form of Equation 5.1.1 has a frequency of 100 Hz, and a phase angle of 30°. (a) What is the duration of 10 cycles? (b) At what value of t is the first zero of the 11th cycle?

Answer: (a) 100 ms; (b) 101.67 ms.

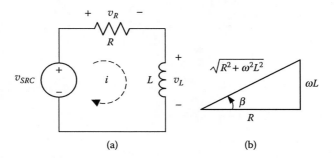

FIGURE 5.2.1
Response of *RL* circuit to sinusoidal excitation.

5.2   Response to Complex Sinusoidal Excitation

Response of *RL* Circuit to Sinusoidal Excitation

Consider a series *RL* circuit supplied from a voltage source $v_{SRC} = V_m\cos(\omega t + \theta)$, as in Figure 5.2.1a. From KVL, $v_{SRC} = v_R + v_L$, where $v_R = Ri$ and $v_L = Ldi/dt$. Substituting for these terms:

$$L\frac{di}{dt} + Ri = V_m\cos\left(\omega t + \theta\right) \qquad (5.2.1)$$

This is a linear, first-order differential equation with a forcing function $V_m\cos(\omega t + \theta)$ on the RHS. The complete solution is the sum of the following two components:

1. A *transient* component that is the solution to the equation $L\dfrac{di}{dt} + Ri = 0$, and which

    dies out with time, as discussed in Section 12.1 (Chapter 12). A steady state is assumed to prevail only after the transient component has become insignificant.

2. A *steady-state* component i_{SS} that satisfies Equation 5.2.1. As mentioned earlier, the linear operations on the LHS of Equation 5.2.1 affect the amplitude and phase of i_{SS} without affecting the frequency. Hence, we may consider i_{SS} to be of the form:

$$i_{SS} = I_m\cos\left(\omega t + \theta - \alpha\right) \qquad (5.2.2)$$

    where I_m and α are unknowns to be determined so as to satisfy Equation 5.2.1.

Substituting i_{SS} from Equation 5.2.2 in Equation 5.2.1:

$$I_m\left[-\omega L\sin\left(\omega t + \theta - \alpha\right) + R\cos\left(\omega t + \theta - \alpha\right)\right] = V_m\cos\left(\omega t + \theta\right) \qquad (5.2.3)$$

If the LHS of Equation 5.2.3 is multiplied and divided by $\sqrt{R^2 + \omega^2 L^2}$, it becomes:

$$I_m\sqrt{R^2+\omega^2L^2}\left[-\frac{\omega L}{\sqrt{R^2+\omega^2L^2}}\sin\left(\omega t+\theta-\alpha\right)+\frac{R}{\sqrt{R^2+\omega^2L^2}}\cos\left(\omega t+\theta-\alpha\right)\right] \quad (5.2.4)$$

Let β be the angle whose sine is $\dfrac{\omega L}{\sqrt{R^2+\omega^2L^2}}$ and whose cosine is therefore $\dfrac{R}{\sqrt{R^2+\omega^2L^2}}$ (Figure 5.2.1b). Equation 5.2.4 becomes:

$$I_m\sqrt{R^2+\omega^2L^2}\left[-\sin\beta\sin\left(\omega t+\theta-\alpha\right)+\cos\beta\cos\left(\omega t+\theta-\alpha\right)\right]=V_m\cos\left(\omega t+\theta\right)$$

or,

$$I_m\sqrt{R^2+\omega^2L^2}\left[\cos\left(\omega t+\theta+\beta-\alpha\right)\right]=V_m\cos\left(\omega t+\theta\right) \quad (5.2.5)$$

To equalize both sides of Equation 5.2.5 under all conditions, we must have $I_m=\dfrac{V_m}{\sqrt{R^2+\omega^2L^2}}$ and $\beta=\alpha$. It follows that:

$$i_{SS}=\frac{V_m}{\sqrt{R^2+\omega^2L^2}}\cos\left(\omega t+\theta-\alpha\right),\ \tan\alpha=\frac{\omega L}{R} \quad (5.2.6)$$

The steady-state current i_{SS} lags the applied excitation by an angle α. Both α *and* the amplitude of i_{SS} depend on ω, L, and R.

Response of *RL* Circuit to Complex Sinusoidal Excitation

We will determine, next, the steady-state current i_{SS} in response to an excitation $v_{SRC}=V_me^{j(\omega t+\theta)}$, where $j=\sqrt{-1}$. Although complex excitation is not physically realizable, it leads naturally to phasor notation and the concept of impedance, as will be demonstrated shortly. From Euler's formula:

$$v_{SRC}=V_me^{j(\omega t+\theta)}=V_m\left[\cos\left(\omega t+\theta\right)+j\sin\left(\omega t+\theta\right)\right] \quad (5.2.7)$$

Because the circuit is linear, superposition applies, and $i_{SS}=i_{SS1}+i_{SS2}$, where i_{SS1} is the steady-state response to $V_m\cos(\omega t+\theta)$, as given by Equation 5.2.6, and i_{SS2} is the steady-state response to $jV_m\sin(\omega t+\theta)$. The excitation $jV_m\sin(\omega t+\theta)$ may be written as $jV_m\cos\left(\omega t+\theta-\dfrac{\pi}{2}\right)$. Hence, i_{SS2} can be obtained from i_{SS1} by replacing θ by $\left(\theta-\dfrac{\pi}{2}\right)$ and multiplying V_m by j. This gives:

$$i_{SS}=\frac{V_m}{\sqrt{R^2+\omega^2L^2}}\left[\cos\left(\omega t+\theta-\alpha\right)+j\cos\left(\omega t+\theta-\alpha-\frac{\pi}{2}\right)\right]$$

$$= \frac{V_m}{\sqrt{R^2 + \omega^2 L^2}} \left[\cos(\omega t + \theta - \alpha) + j\sin(\omega t + \theta - \alpha) \right]$$

$$= \frac{V_m}{\sqrt{R^2 + \omega^2 L^2}} e^{j(\omega t + \theta - \alpha)}, \quad \tan\alpha = \frac{\omega L}{R} \tag{5.2.8}$$

This result expresses the following important concept:

> **Concept:** *When a complex sinusoidal excitation v_{SRC} is applied to an LTI circuit, the response is a complex sinusoidal function whose real part is the response to the real part of the excitation, $V_m\cos(\omega t + \theta)$, applied alone, and whose imaginary part is the response to the imaginary part of the excitation, $V_m\sin(\omega t + \theta)$, applied alone.*

In other words, the real and imaginary parts retain their separate identities in linear operations, without any mutual interaction. This is not true of nonlinear operations (Section ST5.1). The complex excitation may therefore be regarded as a means of applying both a cosinusoidal excitation and a sinusoidal excitation together and obtaining the two responses simultaneously.

EXERCISE 5.2.1

A complex sinusoidal voltage excitation having an amplitude of 2 V, a frequency of 400 Hz, and a phase angle of 45° is applied to an inductor having an inductance of 100 mH and a resistance of 10 Ω. Determine the steady-state inductor current.

Answer: $7.95\cos(800\pi t - 42.7°) + j7.95\sin(800\pi t - 42.7°)$ mA.

5.3 Phasor Notation

The complex sinusoid $Y_m e^{j(\omega t + \theta)}$ may be drawn in the complex plane, or Argand diagram (Appendix SB). At $t = 0$, $Y_m e^{j\theta}$ is a line OP of length Y_m and angle θ (Figure 5.3.1a). As t increases, the line OP rotates in the counterclockwise direction at an angular frequency ω. Its projection on the real axis traces the function $Y_m\cos(\omega t + \theta)$, whereas its projection on the imaginary axis traces the function $Y_m\sin(\omega t + \theta)$.

In a linear circuit, all circuit variables, such as $V_m e^{j(\omega t + \theta)}$ and $I_m e^{j(\omega t + \theta - \alpha)}$ mentioned earlier, have the same frequency. Their representations on an Argand diagram rotate at the same angular frequency ω, so that the relative phases between them are preserved. The rotation can be frozen at $t = 0$ without loss of the information contained in the magnitudes and relative phase angles (Figure 5.3.1b).

> **Definition:** *A phasor is a quantity such as $V_m e^{j\theta}$ or $V_m e^{j(\theta - \alpha)}$, representing a complex sinusoidal function of time, but with the time variation suppressed.*

Phasors are written in boldface and expressed as a magnitude and phase angle:

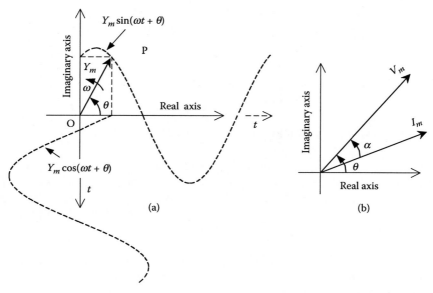

FIGURE 5.3.1
Sinusoidal functions in the complex plane.

$$\mathbf{V} = V_m e^{j\theta} = V_m \angle \theta \qquad (5.3.1)$$

Freezing the rotation of $V_m e^{j(\omega t + \theta)}$ and $I_m e^{j(\omega t + \theta - \alpha)}$ need not be done at $t = 0$. If it is done at $t = -\theta/\omega$, for example, phasor $\mathbf{V}$ will lie on the real axis, and the phase angle of phasor $\mathbf{I}$ is $-\alpha$. In a phasor diagram, the relative phase angles, and not the absolute phase angles, are generally of significance.

Properties of Phasors

Because phasors have magnitude and direction, they are similar to vectors in many respects, but have a real part $Y_r = Y_m \cos\theta$, and an imaginary part $Y_i = Y_m \sin\theta$. It follows that $Y_m = \sqrt{Y_r^2 + Y_i^2}$.

Multiplying a phasor by a real quantity K multiplies its magnitude by K, without changing its phase angle. Two phasors $\mathbf{Y}_1 = Y_1 \angle \theta_1$ and $\mathbf{Y}_2 = Y_2 \angle \theta_2$ may be added by drawing $\mathbf{Y}_2$ such that its origin lies at the tip of $\mathbf{Y}_1$ (Figure 5.3.2a). The sum $\mathbf{Y}_1 + \mathbf{Y}_2$ is the phasor whose origin is that of $\mathbf{Y}_1$ and whose tip is that of $\mathbf{Y}_2$. The sum of $\mathbf{Y}_1$ and $\mathbf{Y}_2$ may also be obtained by applying the parallelogram rule, as in Figure 5.3.2a. The real part of $\mathbf{Y}_1 + \mathbf{Y}_2$ is $Y_1 \cos\theta_1 + Y_2 \cos\theta_2$, and its imaginary part is $Y_1 \sin\theta_1 + Y_2 \sin\theta_2$. It follows that the magnitude of $\mathbf{Y}_1 + \mathbf{Y}_2$ is

$$|\mathbf{Y}_1 + \mathbf{Y}_2| = \sqrt{Y_1^2 + Y_2^2 + 2Y_1 Y_2 \cos(\theta_2 - \theta_1)} \qquad (5.3.2)$$

The phase angle of $\mathbf{Y}_1 + \mathbf{Y}_2$ is

$$\angle(\mathbf{Y}_1 + \mathbf{Y}_2) = \tan^{-1} \frac{Y_1 \sin\theta_1 + Y_2 \sin\theta_2}{Y_1 \cos\theta_1 + Y_2 \cos\theta_2} \qquad (5.3.3)$$

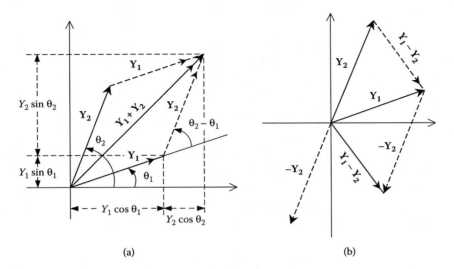

FIGURE 5.3.2
Phasor addition and subtraction.

The phasor difference $\mathbf{Y}_1 - \mathbf{Y}_2$ is obtained by adding $\mathbf{Y}_1$ and $-\mathbf{Y}_2$, where $-\mathbf{Y}_2$ is a phasor of the same magnitude as $\mathbf{Y}_2$ but having a phase angle $(\theta_2 + \pi)$ (Figure 5.3.2b). Alternatively, the phasor $\mathbf{Y}_1 - \mathbf{Y}_2$ may be obtained as the phasor whose origin lies at the tip of $\mathbf{Y}_2$ and whose tip lies at the tip of $\mathbf{Y}_1$. Then, $\mathbf{Y}_1 = \mathbf{Y}_2 + (\mathbf{Y}_1 - \mathbf{Y}_2)$.

A phasor $\mathbf{Y} = Y\angle\theta$ may be multiplied by a complex quantity $A\angle\alpha$ as follows:

$$Ye^{j\theta} \times Ae^{j\alpha} = AYe^{j(\theta + \alpha)} \tag{5.3.4}$$

The product is a phasor of magnitude AY and phase angle $(\theta + \alpha)$. A phasor $\mathbf{Y} = Y\angle\theta$ may be divided by a complex quantity $A\angle\alpha$ as follows:

$$\frac{Ye^{j\theta}}{Ae^{j\alpha}} = \frac{Y}{A}e^{j(\theta - \alpha)} \tag{5.3.5}$$

The quotient is a phasor of magnitude $\dfrac{Y}{A}$ and phase angle $(\theta - \alpha)$.

A special case is multiplication and division by j. In the complex plane, j is an imaginary quantity of unit magnitude and a phase angle of $\pi/2$:

$$j = \cos\frac{\pi}{2} + j\sin\frac{\pi}{2} = 1 \times e^{j\frac{\pi}{2}} \tag{5.3.6}$$

Multiplying a phasor by j rotates the phasor through an angle $\pi/2$ counterclockwise without changing its magnitude. Dividing a phasor by j or, conversely, multiplying it by $-j$ because $\dfrac{1}{j} = \dfrac{j}{j^2} = -j$ rotates the phasor through an angle $\pi/2$ clockwise without changing its magnitude.

EXERCISE 5.3.1

Determine the product and quotient of $(12 + j5)$ and $(3 + j4)$ by working in rectangular coordinates and in polar coordinates.

Answer: Product: $16 + j63 = 65\angle 75.8°$; quotient: $2.24 - j1.32 = 2.6\angle -30.5°$.

Example 5.3.1: Magnitude, Real Part, and Imaginary Part of a Complex Fraction

It is required to determine the magnitude, real and imaginary parts of $Y = \dfrac{a + jb}{c + jd}$.

SOLUTION

Let us rationalize Y, that is, make its denominator real, by multiplying numerator and denominator by the complex conjugate of the denominator, $c - jd$. Thus,

$$Y = \frac{a + jb}{c + jd} \times \frac{c - jd}{c - jd} = \frac{ac + bd}{c^2 + d^2} + j\frac{bc - ad}{c^2 + d^2} \tag{5.3.7}$$

The real part of Y is $\dfrac{ac + bd}{c^2 + d^2}$ and its imaginary part is $\dfrac{bc - ad}{c^2 + d^2}$.

The magnitude of Y may be obtained as the square root of the sum of the squares of the real and imaginary parts. An easier way is to convert the numerator and denominator to polar coordinates. Thus: $Y = \dfrac{\sqrt{a^2 + b^2}\angle \tan^{-1}(b/a)}{\sqrt{c^2 + d^2}\angle \tan^{-1}(d/c)}$. The magnitude of Y is therefore $\dfrac{\sqrt{a^2 + b^2}}{\sqrt{c^2 + d^2}}$ and its phase angle is $\tan^{-1}(b/a) - \tan^{-1}(d/c)$. It should be noted that, whereas the magnitude of Y is the magnitude of the numerator divided by that of the denominator, the real part of Y is *not* the real part of the numerator divided by that of the denominator. Nor is the imaginary part of Y the imaginary part of the numerator divided by that of the denominator.

EXERCISE 5.3.2

Show that the square root of the sum of the squares of the real and imaginary parts of the

RHS of Equation 5.3.7 reduces to $\dfrac{\sqrt{a^2 + b^2}}{\sqrt{c^2 + d^2}}$.

5.4   Phasor Relations of Circuit Elements

Phasor Relations for a Resistor

If the current through a resistor is $I_m e^{j(\omega t + \theta)}$ A, the voltage across the resistor is $RI_m e^{j(\omega t + \theta)}$ V. In phasor notation (Figure 5.4.1a):

$$\mathbf{V} = R\mathbf{I} \text{ or } \mathbf{I} = G\,\mathbf{V} \tag{5.4.1}$$

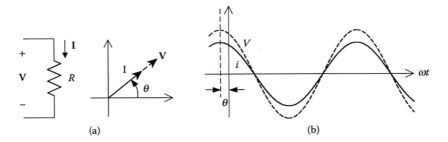

FIGURE 5.4.1
Phasor and time relations for a resistor.

According to the interpretation of complex sinusoidal excitation, a voltage $V_m\cos(\omega t + \theta)$ V produces a current $\dfrac{V_m}{R}\cos(\omega t + \theta)$ A that is in phase (Figure 5.4.1b).

The energy dissipated in a resistor R over a period is:

$$w = \int_0^{\frac{2\pi}{\omega}} RI_m^2\cos^2(\omega t + \theta)dt = \frac{RI_m^2}{2}\left(\frac{2\pi}{\omega}\right) \tag{5.4.2}$$

The average power dissipated is:

$$P = \frac{w}{(2\pi/\omega)} = \frac{RI_m^2}{2} \tag{5.4.3}$$

The operations leading to Equation 5.4.3 involve squaring the current $I_m\cos(\omega t + \theta)$, then taking the mean value over a period, which is $I_m^2/2$. Similarly, the mean square of the voltage is $V_m^2/2$. The square roots of these quantities are the **root-mean-square** (rms) values. Thus,

$$I_{rms} = \frac{I_m}{\sqrt{2}} \quad \text{and} \quad V_{rms} = \frac{V_m}{\sqrt{2}} \tag{5.4.4}$$

The power dissipated may be expressed in terms of rms values as follows:

$$P = RI_{rms}^2 = \frac{V_{rms}^2}{R} \text{ W} \tag{5.4.5}$$

Because the power dissipated in R by a dc current I is $P = RI^2 = \dfrac{V^2}{R}$ W, it follows that:

Concept: *A current of rms value I_{rms}, or a voltage of rms value V_{rms}, dissipates the same power in a given resistor as a dc current, or a dc voltage, of the same value.*

The rms value is also known as the **effective** value.

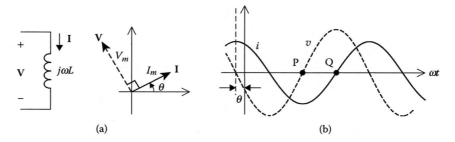

FIGURE 5.4.2
Phasor and time relations for an inductor.

Phasor Relations for an Inductor

If the inductor current is $I_m e^{j(\omega t + \theta)}$ A, the inductor voltage is:

$$v = L\frac{di}{dt} = j\omega L I_m e^{j(\omega t + \theta)} = \omega L I_m e^{j\left(\omega t + \theta + \frac{\pi}{2}\right)} \text{ V} \qquad (5.4.6)$$

In phasor notation, $\mathbf{I} = I_m \angle \theta$ A, and $\mathbf{V} = j\omega L I_m \angle \theta = \omega L I_m \angle\left(\theta + \frac{\pi}{2}\right)$ V, or:

$$\mathbf{V} = j\omega L \mathbf{I}, \text{ or } \mathbf{I} = \frac{1}{j\omega L}\mathbf{V} \qquad (5.4.7)$$

The magnitude of $\mathbf{V}$ is ωL times that of $\mathbf{I}$, and the phase angle of $\mathbf{V}$ is $\pi/2$ plus the phase angle of $\mathbf{I}$ (Figure 5.4.2a). According to the interpretation of complex sinusoidal excitation, if the inductor current is $i = I_m\cos(\omega t + \theta)$ A, the inductor voltage is,

$v = \omega L I_m \cos\left(\omega t + \theta + \frac{\pi}{2}\right) = -\omega L I_m \sin(\omega t + \theta)$ V. The voltage leads the current by 90°, or

the current lags the voltage by 90° (Figure 5.4.2b). This may be ascertained by comparing corresponding points on the two waveforms that are 90° apart, such as P and Q. Because P occurs before Q, v leads i.

Phasor Relations for a Capacitor

If the capacitor voltage is $V_m e^{j(\omega t + \theta)}$ V, the capacitor current is:

$$i = C\frac{dv}{dt} = j\omega C V_m e^{j(\omega t + \theta)} = \omega C V_m e^{j\left(\omega t + \theta + \frac{\pi}{2}\right)} \text{ A} \qquad (5.4.8)$$

In phasor notation, $\mathbf{V} = V_m \angle \theta$ A, and $\mathbf{I} = j\omega C V_m \angle \theta = \omega C V_m \angle\left(\theta + \frac{\pi}{2}\right)$ V, or:

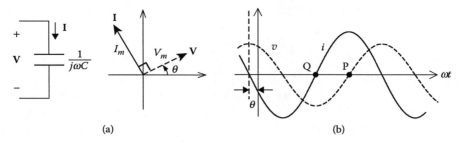

FIGURE 5.4.3
Phasor and time relations for a capacitor.

$$I = j\omega C V \text{ or } V = \frac{1}{j\omega C}I \tag{5.4.9}$$

The magnitude of V is $\frac{1}{\omega C}$ that of I, and the phase angle of V is that of I minus $\pi/2$ (Figure 5.4.3a). According to the interpretation of complex sinusoidal excitation, if the voltage across the capacitor is $v = V_m \cos(\omega t + \theta)$V, the current through the capacitor is $i = \omega C V_m \cos\left(\omega t + \theta + \frac{\pi}{2}\right) = -\omega C V_m \sin(\omega t + \theta)$ A. The voltage lags the current by 90°, or the current leads the voltage by 90° (Figure 5.4.3b).

The phase relations between voltage and current for resistors, inductors, and capacitors can be rationalized as follows:

> **Concept:** *The sinusoidal voltage and current for an ideal resistor are in phase because such a resistor is purely dissipative. They are in phase quadrature for ideal energy storage elements because these elements are nondissipative.*

In the case of a resistor, it is seen from Equation 5.4.2 that the squared cosine term in the power expression gives a nonzero average over a cycle. In the case of inductors and capacitors, the corresponding term is the product of $\sin(\omega t + \theta)$ and $\cos(\omega t + \theta)$, which is $\frac{1}{2}\sin 2(\omega t + \theta)$. This term averages to zero over a cycle. Power in circuit elements is discussed in detail in Section 7.1 (Chapter 7).

In comparing Equation 5.4.7 and Equation 5.4.9 with the corresponding v–i relations in the time domain, an important concept emerges that underlies the usefulness of phasor notation for steady-state sinusoidal analysis:

> **Concept:** *In phasor notation, differentiation in time is replaced by multiplication by $j\omega$, and integration in time is replaced by division by $j\omega$. Thus, differential and integral relations are transformed to algebraic relations in $j\omega$ for steady-state sinusoidal analysis ONLY.*

EXERCISE 5.4.1

The voltage applied to a 10 Ω resistor is $80\angle 35°$ V peak, the frequency being 50 Hz. Determine: (a) the expression for the current in the time domain, assuming the voltage is of the form of Equation 5.1.1; and (b) the power dissipated in the resistor.

Answer: (a) $8\cos(100\pi t + 35)$ A; (b) 320 W.

EXERCISE 5.4.2

The current through a series combination of a 10 mH inductor and a 50 μF capacitor is $15\angle -75°$ mA rms. Assuming the frequency is 200 Hz and the time variation is of the form of Equation 5.1.1, determine the expression in the time domain of the voltage across: (a) the inductor and (b) the capacitor.

Answer: (a) $60\pi\cos(400\pi t + 15°)$ mV rms; (b) $\dfrac{750}{\pi}\sin(400\pi t - 165°)$ mV rms.

5.5   Impedance and Reactance

Let $V_m\angle\theta_v$ be a voltage phasor and $I_m\angle\theta_i$ be a related current phasor, representing, for example, the voltage and current between any two nodes in a circuit, and let $\theta = \theta_v - \theta_i$. Then:

Definition: *Impedance Z is the ratio of the voltage phasor $V_m\angle\theta_v$ to the related current phasor $I_m\angle\theta_i$:*

$$\frac{\mathbf{V}}{\mathbf{I}} = Z = \frac{V_m}{I_m}\angle\theta \qquad\qquad (5.5.1)$$

When **V** is in volts and **I** is in amperes, Z is in ohms. Because Z is, in general, complex, it can be expressed as (Figure 5.5.1):

$$Z = R + jX \qquad\qquad (5.5.2)$$

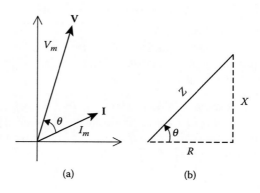

(a)   (b)

FIGURE 5.5.1
Definition of impedance.

The real part of Z is the resistance because, for an ideal resistor, $\mathbf{V}$ and $\mathbf{I}$ are in phase so that their ratio is real and equal to R (Equation 5.5.1), which means that $X = 0$ for an ideal resistor. The imaginary part X is the **reactance** and its unit is the ohm, just like resistance and impedance. Although Z is in general complex, it is not a phasor because it is not a complex sinusoidal function of time in which the time variation has been suppressed. To emphasize this, Z is drawn in Figure 5.5.1b on a separate Argand diagram from $\mathbf{V}$ and $\mathbf{I}$.

Because ideal inductors and capacitors do not dissipate power, $R = 0$ for these elements.

To determine X for an inductor, we note that according to Equation 5.4.7, $Z = \dfrac{\mathbf{V}}{\mathbf{I}} = j\omega L$, so

that $X = \omega L$. For a capacitor, Equation 5.4.9 gives $Z = \dfrac{1}{j\omega C} = -\dfrac{j}{\omega C}$, so that $X = -\dfrac{1}{\omega C}$. When

$X = 0$, $\mathbf{V}$ and $\mathbf{I}$ are in phase. When $X \neq 0$, Z is complex and $\mathbf{V}$ and $\mathbf{I}$ are out of phase. Hence:

> **Concept:** *Reactance is due to energy storage elements; that is, capacitors and inductors. Whereas sinusoidal voltages and currents are all in phase in a purely resistive circuit, energy storage elements introduce phase differences between sinusoidal voltages and currents.*

The following should be noted concerning reactance:

1. Being the imaginary part of impedance, reactance is always multiplied by j.
2. Because j appears in the $\mathbf{V}$–$\mathbf{I}$ relations as a result of differentiating or integrating expressions involving $e^{j\omega t}$, j is always associated with ω in the expression for impedance. Hence, reactance is always a function of frequency. Properties of reactance are discussed in Section ST5.2.

The reciprocal of impedance is the **admittance** Y and is expressed by the following equation:

$$\frac{\mathbf{I}}{\mathbf{V}} = Y = G + jB \tag{5.5.3}$$

where B is the **susceptance**. Like G, B and Y are in siemens. It follows from Equation 5.4.1, Equation 5.4.7, and Equation 5.4.9 that a resistor has an admittance G and zero susceptance;

an ideal inductor has an admittance $\dfrac{1}{j\omega L} = \dfrac{-j}{\omega L}$ and a susceptance $-1/\omega L$; an ideal capacitor

has an admittance $j\omega C$ and a susceptance ωC. Table 5.5.1 lists the circuit properties of the three circuit elements. The term **immittance** is used to describe a quantity that could be either an impedance or an admittance.

EXERCISE 5.5.1

Find the reactance and susceptance of the inductor and capacitor of Exercise 5.4.2.

Answer: (a) Inductor: $4\pi\,\Omega$, $-\dfrac{1}{4\pi}$ S; (b) $-\dfrac{50}{\pi}\,\Omega$, $\dfrac{\pi}{50}$ S.

TABLE 5.5.1

Circuit Properties of Circuit Elements

Circuit Property	Resistor	Inductor	Capacitor
Reactance (X)	0	ωL	$\dfrac{-1}{\omega C}$
Impedance (Z)	R	$j\omega L$	$\dfrac{1}{j\omega C} = \dfrac{-j}{\omega C}$
Admittance (Y)	$G = \dfrac{1}{R}$	$\dfrac{1}{j\omega L} = \dfrac{-j}{\omega L}$	$j\omega C$
Susceptance (B)	0	$\dfrac{-1}{\omega L}$	ωC

5.6   Representation in the Frequency Domain

The *RL* Circuit

To see how phasor notation greatly facilitates derivation of the sinusoidal steady-state response, let us again consider the *RL* circuit of Figure 5.2.1. The applied voltage $v_{SRC} = V_m \cos(\omega t + \theta)$ is a phasor $\mathbf{V_{SRC}} = V_m \angle \theta$. The current is an unknown phasor $\mathbf{I}$. In phasor notation, the term Ri in Equation 5.2.1 is $R\mathbf{I}$, according to Equation 5.4.1, and the term $L\dfrac{di}{dt}$ is $j\omega L\mathbf{I}$, according to Equation 5.4.7. Equation 5.2.1 becomes:

$$R\mathbf{I} + j\omega L\mathbf{I} = \mathbf{V_{SRC}} \tag{5.6.1}$$

or

$$\left(R + j\omega L\right)\mathbf{I} = V_m \angle \theta \tag{5.6.2}$$

The term $(R + j\omega L)$ is a complex quantity whose magnitude is $\sqrt{R^2 + \omega^2 L^2}$ and whose phase angle is $\alpha = \tan^{-1}\dfrac{\omega L}{R}$ (Figure 5.2.1). It follows from Equation 5.6.2 that:

$$\mathbf{I} = \frac{V_m \angle \theta}{\sqrt{R^2 + \omega^2 L^2} \angle \alpha} = \frac{V_m}{\sqrt{R^2 + \omega^2 L^2}} \angle(\theta - \alpha) \tag{5.6.3}$$

According to Equation 5.6.3, the magnitude of the current is $\dfrac{V_m}{\sqrt{R^2 + \omega^2 L^2}}$ and its phase angle is $(\theta - \alpha)$. Its angular frequency is ω, the same as that of the applied excitation. Its expression in the time domain is therefore $\dfrac{V_m}{\sqrt{R^2 + \omega^2 L^2}} \cos(\omega t + \theta - \alpha)$, the same as Equation

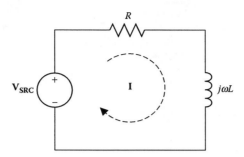

FIGURE 5.6.1
Frequency domain representation of *RL* circuit.

5.2.6. Note that because the applied excitation was expressed in time as a cosine function, the current is also expressed in time as a cosine function that lags the voltage by an angle α. If the voltage was expressed in time as $V_m\sin(\omega t + \theta)$, the current would be

$\dfrac{V_m}{\sqrt{R^2 + \omega^2 L^2}}\sin(\omega t + \theta - \alpha)$ so as to preserve the correct phase relations.

It would be advantageous to derive Equation 5.6.1 directly from the circuit, without having to write the differential equation and convert it term by term to phasor notation. This can be done readily by expressing excitations and responses as phasors and representing R and L by their impedances. The *RL* circuit then becomes as shown in Figure 5.6.1. The circuit is said to be represented in the **frequency domain**. It then follows from KVL that $\mathbf{V}_{SRC} = R\mathbf{I} + j\omega L\mathbf{I}$, as in Equation 5.6.1.

An important deduction can be made from the simple phasor analysis presented so far. Equation 5.6.1 is, in fact, KVL in the frequency domain, and is the exact counterpart of KVL as we have used it under dc conditions. This is not surprising because KVL fundamentally applies to instantaneous values of voltage drops and voltage rises. It would, therefore, apply when these values are expressed as sinusoidal functions of time or as complex exponential functions of time or as phasors (Exercise 5.6.1); similarly for KCL. Recall that circuit analysis under dc conditions is based on KCL, KVL, and Ohm's law, which expresses the *V–I* relation for a resistor. We have expressed in the preceding sections, the *v–i* relations for the three basic elements in terms of voltage phasors, current phasors, and impedances, in a form analogous to Ohm's law. We have just argued that KVL and KCL apply to phasor voltages and currents. The inevitable conclusion is that all the circuit relations, theorems, etc. that we have derived or applied under dc conditions carry over directly to the frequency domain. In fact, $(R + j\omega L)$ in Equation 5.6.2 is the equivalent series impedance Z_{eqs} of the two impedances R and $j\omega L$, and $(R + j\omega L)\mathbf{I}$ is the phasor voltage drop across the branch made up of R and L in series. We may thus state:

> **Concept:** *All circuit relations and theorems that apply to resistive circuits under dc conditions apply for sinusoidal steady-state analysis in the frequency domain to circuits that include resistance, inductance, and capacitance, with voltages and currents represented as phasors and impedances of circuit elements replacing resistance.*

Specifically, this applies to series and parallel connections of impedances, Δ-Y transformation, voltage division and current division, source transformation, node-voltage analysis, mesh-current analysis, Thevenin's and Norton's equivalent circuits, the substitution theorem, and the circuit simplification techniques discussed in the preceding chapter. In fact, we may assert that for the purpose of steady-state analysis, the dc steady state is a special case of the sinusoidal steady state with the frequency set to zero. Inductors are replaced by short circuits and capacitors by open circuits. The following exercises and examples illustrate some of these points.

EXERCISE 5.6.1

(a) Show that for the series RL circuit, KVL in terms of instantaneous values

is $-\omega L I_m \sin(\omega t + \theta - \alpha) + R I_m \cos(\omega t + \theta - \alpha) = V_m \cos(\omega t + \theta)$, where $I_m = \dfrac{V_m}{\sqrt{R^2 + \omega^2 L^2}}$

and $\alpha = \tan^{-1}(\omega L / R)$.

(b) Derive KVL when $v_{SRC} = V_m \sin(\omega t + \theta)$.

(c) Derive KVL when $v_{SRC} = V_m \cos(\omega t + \theta) + j V_m \sin(\omega t + \theta)$ and deduce that the complex exponential form satisfies KVL.

(d) Eliminate $e^{j\omega t}$ from KVL in the complex exponential form to express KVL in the frequency domain, as in Equation 5.6.2.

Example 5.6.1: Voltage Division and Current Division

Given the circuit of Figure 5.6.2, it is required to determine $\mathbf{V_L}$, $\mathbf{I_L}$, $\mathbf{I_1}$, and $\mathbf{I_2}$, assuming $\omega = 100$ rad/s.

SOLUTION

The reactance of the capacitor is $-\dfrac{1}{\omega C} = -\dfrac{1}{100 \times 10^{-3}} = -10\ \Omega$; hence, $Z_2 = 10 - j10\ \Omega$. The

reactance of the inductor is $\omega L = 100 \times 0.2 = 20\ \Omega$; hence, $Z_L = 20 + j20\ \Omega$. Z_2 in parallel

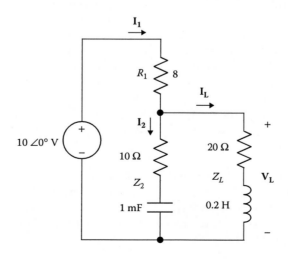

FIGURE 5.6.2
Figure for Example 5.6.1.

with Z_L is $\dfrac{(20+j20)(10-j10)}{(20+j20)+(10-j10)} = \dfrac{40}{3+j} = 4(3-j)$ Ω. Hence, $\mathbf{I_1} = \dfrac{10\angle 0°}{8+4(3-j)} = \dfrac{10\angle 0°}{20-j4} =$

$\dfrac{10\angle 0°}{20.4\angle -11.3°} = 0.49\angle 11.3°$ A.

From voltage division, $V_L = \dfrac{Z_L \parallel Z_2}{R_1 + Z_L \parallel Z_2} \times \mathbf{V_{SRC}} = \dfrac{4(3-j)}{8+4(3-j)} 10\angle 0° : \dfrac{3-j}{5-j} 10\angle 0° =$

$\dfrac{\sqrt{10}\angle -18.4°}{\sqrt{26}\angle -11.3°} 10\angle 0° = 6.2\angle -7.1°$ V.

From current division, $\mathbf{I_L} = \dfrac{10-10j}{(20+j20)+(10-10j)} \mathbf{I_1} = \dfrac{1-j}{3+j} \times \dfrac{5}{10-j2} = \dfrac{5(1-j)}{4(8+j)} =$

$\dfrac{5(1-j)(8-j)}{4(8+j)(8-j)} = \dfrac{7-j9}{52} = 0.22\angle -52.1°$ A.

From KCL, $\mathbf{I_2} = \mathbf{I_1} - \mathbf{I_L} = \dfrac{5}{2(5-j)} - \dfrac{5(1-j)}{4(8+j)} = \dfrac{5(3+j2)}{41-j3} = \dfrac{117+j91}{338} = 0.44\angle 37.9°$ A.

It should be carefully noted that in the frequency domain, currents and voltages add or subtract as phasors. Thus, in evaluating $\mathbf{I_2} = \mathbf{I_1} - \mathbf{I_L}$, we have $|\mathbf{I_2}| = |\mathbf{I_1} - \mathbf{I_L}|$, where the RHS is evaluated as the difference of two phasors, each having a magnitude and phase angle. It is incorrect to consider that $|\mathbf{I_2}| = |\mathbf{I_1}| - |\mathbf{I_L}|$. Similarly, the voltages across the 20 Ω resistor and the inductor, for example, add as phasors and not as magnitudes.

Under dc conditions, the current in the Z_2 branch is zero, because this branch consists of a resistor in series with a capacitor. Hence, this branch may be removed. The voltage drop across the inductor is zero. The circuit reduces to a simple resistive voltage divider of 8 Ω in series with 20 Ω. It follows that $V_L = \dfrac{20}{28} \times 10 = 7.14$ V and $I_1 = I_L = \dfrac{10}{28} = 0.36$ A.

EXERCISE 5.6.2

Find the impedance and admittance between terminals ab in Figure 5.6.3 at $\omega = 1$ krad/s.

Answer: $j25$ Ω; $j40$ mS.

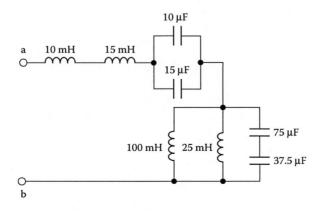

10 μF

a 10 mH 15 mH

15 μF

75 μF

100 mH 25 mH

37.5 μF

b

FIGURE 5.6.3
Figure for Exercise 5.6.2.

Example 5.6.2: Equivalent Parallel Circuit of Series *RL* Circuit

It is required to convert the series *RL* circuit to its equivalent parallel circuit.

SOLUTION

The impedance of the series *RL* circuit is $Z_s = R_s + j\omega L_s$. Its admittance is $Y_s = \dfrac{1}{Z_s} = \dfrac{1}{R_s + j\omega L_s} =$

$\dfrac{1}{R_s + j\omega L_s} \times \dfrac{R_s - j\omega L_s}{R_s - j\omega L_s} = \dfrac{R_s - j\omega L_s}{R_s^2 + \omega^2 L_s^2} = \dfrac{R_s}{R_s^2 + \omega^2 L_s^2} - \dfrac{j\omega L_s}{R_s^2 + \omega^2 L_s^2}$. The equivalent parallel circuit

will have $Y_p = G_p + jB_p = Y_s$. Equating real and imaginary parts:

$$G_p = \frac{R_s}{R_s^2 + \omega^2 L_s^2} \quad \text{and} \quad B_p = -\frac{\omega L_s}{R_s^2 + \omega^2 L_s^2} \tag{5.6.4}$$

Both G_p and B_p are frequency dependent. At $\omega = 0$, $G_p = \dfrac{1}{R_s}$ and $B_p = 0$, as expected. For

an ideal inductor, $R_s = 0 = G_p$ and $B_p = -1/\omega L_s$, also as expected. G_p can also be derived from equality of power dissipation in R_s and G_p, because no power is dissipated in L_s or B_p, and the two circuits are equivalent at their respective terminals. The power dissipated

in R_s is $|I|^2 \dfrac{R_s}{2} = \dfrac{V_m^2}{R_s^2 + \omega^2 L_s^2} \dfrac{R_s}{2}$, whereas the power dissipated in G_p is $V_m^2 G_p/2$. Equating

these two quantities gives the same value of G_p.

EXERCISE 5.6.3

Given a 40 µH inductor and a 100 nF capacitor: (a) At what frequency is the imped-ance of the series combination zero? (b) At what frequency is the admittance of the series combination zero? (c) At what frequency is the admittance of the parallel combination zero? (d) At what frequency is the impedance of the parallel combi-nation zero?

Answer: (a) 500 krad/s; (b) 0 and ∞; (c) 500 krad/s; and (d) 0 and ∞.

EXERCISE 5.6.4

An inductor of 50 mH inductance and 10 Ω resistance is connected in parallel with a 40 Ω resistor. The parallel combination is connected in series with a 25 µF capacitor across a 12∠0° V supply. Determine the inductor voltage and current and the total power dissipated in the circuit, given that the supply frequency is 1 krad/s.

Answer: 2(1 + j5) V, 0.2 A; 1.5 W.

EXERCISE 5.6.5

A 5 Ω resistor is connected in series with a 25 µF capacitor. This series combination is connected in parallel with a coil of 50 mH inductance and 5 Ω resistance. The parallel combination is supplied by a current source of $\sqrt{2}\angle 45°$ A. Determine the inductor current and the capacitor voltage, given that the supply frequency is 1 krad/s.

Answer: $0.5 - j4$ A, $200 - j20$ V.

EXERCISE 5.6.6
Repeat Exercise 5.6.4 and Exercise 5.6.5 using node-voltage analysis.

Simulation Example 5.6.3: Sinusoidal Steady State Using Node-Voltage Analysis

It is required to analyze the circuit of Figure 5.6.4 by the node-voltage method and to simulate it, assuming $\omega = 1$ rad/s. The circuit configuration is the same as that of Figure 3.3.2 (Chapter 3).

SOLUTION

The admittance of the 20 H inductor is $-\dfrac{j}{20}$ Ω, whereas that of the capacitor is $j0.05$ Ω.

Following the usual procedure for writing the node-voltage equations, but using admittances instead of conductances, the node-voltage equation for node e is $j0.05V_b - j0.05V_c$ $-0.1V_d + (0.1 - j0.05 + j0.05 + 0.1)V_c = 0$. Substituting $V_b = 10$ gives:

$$-j0.05V_c - 0.1V_d + 0.2V_e = -j0.5$$

For node c: $-0.2V_b + (0.2 + j0.05)V_c - j0.05V_e = -I$, and for node d: $0.1V_d - 0.1V_e = I + 0.1V_\phi$. Adding these two equations to eliminate **I** and substituting $V_b = 10$ and $V_\phi = V_c - V_e$ gives:

$$(0.1 + j0.05)V_c + 0.1V_d - j0.05V_e = 2$$

For the dependent voltage source, $V_d - V_c = 2V_x = 2(V_b - V_e)$, or:

$$-V_c + V_d + 2V_e = 20$$

Solving these three equations using MATLAB gives: $V_c = 6.7568 - j0.5405$ V, $V_d = 13.2432 + j0.5405$ V, and $V_e = V_c = 6.7568 - j0.5405$ V.

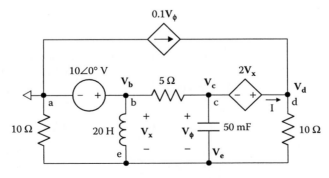

FIGURE 5.6.4
Figure for Example 5.6.3.

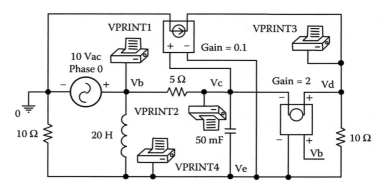

FIGURE 5.6.5
Figure for Example 5.6.3.

SIMULATION
The schematic is shown in Figure 5.6.5. For the sinusoidal steady-state, source VAC is used at a single frequency. A magnitude is entered, which is the rms value, and a phase, if required. One-terminal printers are used for measuring voltages with respect to ground. For ac measurements, a Y must be entered under AC in the Property Editor spreadsheet of each printer. A Y is also entered under REAL and IMAG to express complex values in rectangular form. The nodes are labeled Vb, Vc, Vd, and Ve using net aliases, as explained in Appendix SD.2.

For the simulation, choose AC Sweep/Noise under Analysis type, which allows the frequency to be varied over a certain range. For sinusoidal steady-state analysis, a single frequency is used. Under AC Sweep Type, select Linear, a Start Frequency of $f =$

$\dfrac{1}{2\pi} = 0.159155$ Hz, the same End Frequency, and a Total Points of 1. After the simulation is run, the part of the output file entitled AC analysis gives: $\mathbf{V_b} = 10 + j0$ V, as constrained by the source, $\mathbf{V_c} = 6.757 - j0.5405$ V, $\mathbf{V_d} = 13.24 + j0.5405$ V, and $\mathbf{V_e} = 6.757 - j0.5405$ V.

EXERCISE 5.6.7

Suppose that in Figure 5.6.4, node e is grounded, so that node voltages are required with respect to this node, instead of node a. Determine the node voltages with respect to ground.

Answer: $\mathbf{V_a} = -6.7568 + j0.5405$ V, $\mathbf{V_b} = 3.2432 + j0.5405$ V, $\mathbf{V_c} = 0$, $\mathbf{V_d} = 6.4864 + j1.0810$, $\mathbf{V_e} = 0$.

EXERCISE 5.6.8

Determine $\mathbf{V_L}$ in Figure 5.6.6 using node-voltage analysis and assuming $\mathbf{I_{SRC}} = 1\angle 45°$ A, $Y_1 = 1 - j0.5$ S, $Y_2 = 0.5$ S, $Y_3 = 1 + j0.5$ S, and $\beta = 0.5$ A/V.

Answer: $\mathbf{V_a} = 0.325 + j0.672$ V, $\mathbf{V_b} = 0.116 + j0.139$ V.

Simulation Example 5.6.4: Sinusoidal Steady State Using Mesh-Current Analysis

It is required to analyze the circuit of Figure 5.6.7 by the mesh-current method and to simulate it, assuming $\omega = 10$ rad/s. The circuit configuration is the same as that of Figure 3.5.2 (Chapter 3).

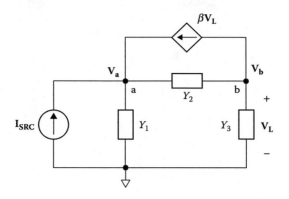

FIGURE 5.6.6
Figure for Exercise 5.6.8.

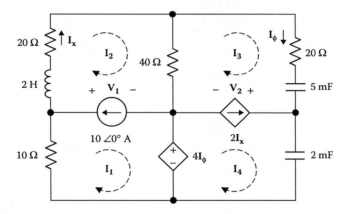

FIGURE 5.6.7
Figure for Example 5.6.4.

SOLUTION

The impedance of the 2 H inductor is $j20$ Ω, that of the 5 mF capacitor is

$\dfrac{-j}{10 \times 5 \times 10^{-3}} = -j20$ Ω, and that of the 2 mF capacitor is $-j50$ Ω. The mesh-current equation

for mesh 1 is: $10I_1 = -V_1 - 4I_\phi$, that for mesh 2 is: $(60 + j20)I_2 - 40I_3 = V_1$. Adding and substituting $I_\phi = I_3$:

$$10I_1 + (60 + j20)I_2 - 36I_3 = 0$$

The mesh-current equation for mesh 3 is $(60 - j20)I_3 - 40I_2 = -V_2$, that for mesh 4 is: $-j50I_4 = V_2 + 4I_\phi$. Adding and substituting $I_\phi = I_3$:

$$-40I_2 + (56 - j20)I_3 - j50I_4 = 0$$

For the independent current source:

$$-I_1 + I_2 = 10$$

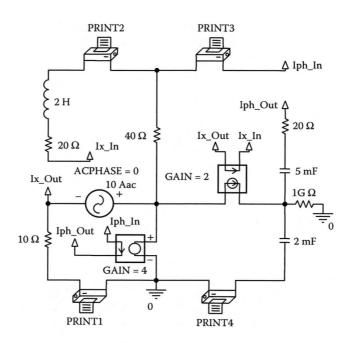

FIGURE 5.6.8
Example 5.6.4.

For the dependent current source: $I_4 - I_3 = 2I_x$ or

$$-2I_2 - I_3 + I_4 = 0$$

Solving using MATLAB gives $I_1 = -8.9443 + j0.2038$ A, $I_2 = 1.0557 + j0.2038$ A, $I_3 = -0.8383 + j0.9828$ A, and $I_4 = 1.2730 + j1.3904$ A.

SIMULATION

The schematic is shown in Figure 5.6.8. Several points should be noted about this simulation. Current printers are included in each mesh to measure the mesh currents in the clockwise sense. To avoid cluttering the drawing with the connections of the dependent sources, power connectors are used (Appendix SD.2) that are labeled in the correct polarities. Note the polarity of the independent current source, whereby current flows inside the source from the positive terminal to the negative terminal. A very large resistor of 1 GΩ connected to ground has been added at the node between the two capacitors because, if omitted, PSpice will give an error that this node is floating, that is, isolated from ground, because of the ideal capacitors and the infinite source resistance of the current source. Adding a large resistor avoids this problem without significantly affecting the simulation.

The simulation is run as described in Example 5.6.3, the frequency being 1.59155 Hz. The output file gives, under AC analysis, the current in each printer as: $I_1 = -8.9443 + j0.2038$ A, $I_2 = 1.0557 + j0.2038$ A, $I_3 = -0.8383 + j0.9828$ A, and $I_4 = 1.2730 + j1.3904$ A.

EXERCISE 5.6.9

Repeat Exercise 5.6.4 and Exercise 5.6.5 using mesh-current analysis.

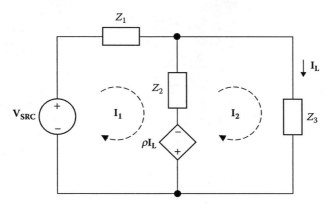

FIGURE 5.6.9
Figure for Exercise 5.6.10.

EXERCISE 5.6.10

Determine I_L in Figure 5.6.9 using mesh-current analysis and assuming $V_{SRC} = 10\angle{-30°}$ V, $Z_1 = 10 - j5$ Ω, $Z_2 = 5$ Ω, $Z_3 = 10 + j5$ Ω, and $\rho = 5$ V/A.

Answer: $I_1 = 0.7334 - j0.1395$ A, $I_2 = 0.1644 - j0.0760$ A.

Simulation Example 5.6.5: TEC in the Sinusoidal Steady State

It is required to obtain TEC between terminals ab in Figure 5.6.10, analytically and by simulation. The same circuit under dc conditions was simulated in Simulation Example 4.1.2 (Chapter 4), where it was found that $V_{Th} = 24$ V and $R_{Th} = 8$ Ω.

SOLUTION

The two shunt branches across ab are first removed and then added later. On open circuit, no current flows through the 10 Ω resistor. Because I_x flows through a CCVS of $10I_x$, this voltage source can be replaced with a 10 Ω resistor, in accordance with the source absorp-

tion theorem. Hence, $I_x = \dfrac{45}{15 + j10}$, and $V'_{Th} = 10I_x = \dfrac{90}{3 + j2} = \dfrac{90}{13}(3 - j2)$ V. On short circuit

$10I_x = 10I_{SC}$, so that $I_{SC} = I_x$. It follows that $I'_{SC} = \dfrac{9}{3 + j2}$, and $Z'_{Th} = \dfrac{10}{13}(3 - j2)(3 + j2) = 10$ Ω.

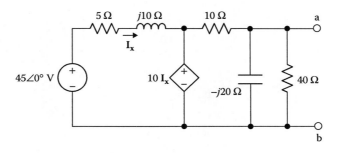

FIGURE 5.6.10
Figure for Example 5.6.5.

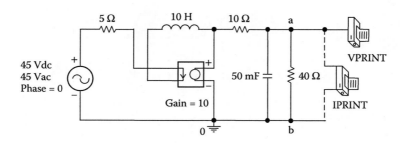

FIGURE 5.6.11
Figure for Example 5.6.5.

The impedance of 40 Ω in parallel with $-j20$ Ω is $\dfrac{40(-j20)}{40 - j20} = -j8(2 + j) = 8(1 - j2)\Omega$. It

follows that $\mathbf{V_{Th}} = \dfrac{8(1 - j2)}{10 + 8(1 - j2)} \times \dfrac{90}{13}(3 - j2) = 18.54 \angle -55.5°$ V. $Z_{Th} = (10 \| 40 \| (j20) =$

$1/[(1/10) + (1/40) + (1/-j20)] = 7.43\angle -21.8°$ Ω.

SIMULATION

The circuit will be simulated for both ac and dc as a demonstration that in an LTI circuit, signals of different frequencies do not interact. The schematic is shown in Figure 5.6.11, assuming $\omega = 1$ rad/s. The voltage source is assigned both ac and dc values by entering these values in the Property Editor spreadsheet of the source. PSpice performs both types of analysis simultaneously. In the Property Editor spreadsheet of each printer, a Y should be entered under AC, DC, MAG, and PHASE.

Two simulations are run, first with terminals ab open-circuited, then with these terminals short-circuited. The results of these simulations are as follows:

dc: Open-circuit voltage = 24 V, short-circuit current = 3 A

ac: Open-circuit voltage = $18.54\angle - 55.49°$ V, short-circuit current = $2.496\angle - 33.69°$ A.

The dc values can also be obtained by pressing the V and I buttons in the capture window.

It follows that $R_{Th} = 8$ Ω in the dc case and $Z_{Th} = 7.43\angle - 21.8°\Omega$. These values may be confirmed by applying a test current source between terminals ab and measuring the source voltage, with the independent voltage source replaced by a short circuit, as shown in Figure 5.6.12.

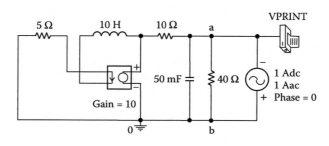

FIGURE 5.6.12
Figure for Example 5.6.5.

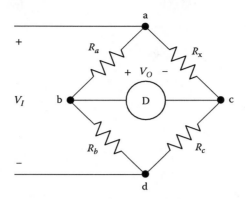

FIGURE 5.6.13
Application Window 5.6.1.

Application Window 5.6.1: Bridge Measurements

Bridge circuits are commonly used for accurate measurements of circuit parameters. The basic principle is illustrated by the Wheatstone bridge for dc measurement of resistance (Figure 5.6.13). When the bridge is balanced, $V_{bd} = V_{cd}$, so that $V_O = 0$, and the detector D does not draw any current. From voltage division, $V_{bd} = \dfrac{R_b}{R_a + R_b} V_I$ and $V_{cd} = \dfrac{R_c}{R_x + R_c} V_I$. Equating V_{bd} and V_{cd} gives:

$$\frac{R_x}{R_c} = \frac{R_a}{R_b} \quad \text{or} \quad R_x = \frac{R_a}{R_b} R_c \tag{5.6.5}$$

For accurate measurement of an unknown resistance R_x, R_c is a reference resistor of accurately known resistance value, and R_a and R_b are resistors of the same type whose values are known and can be varied in small-enough steps. R_a and R_b are varied until bridge balance is achieved. R_x is then determined from Equation 5.6.5. Two factors underlie the accuracy of this measurement:

1. A zero indication can be much more accurately determined than a nonzero value of current or voltage because the scale around zero can be expanded to almost any desired degree of accuracy. A sensitive, current-measuring device, known as a **galvanometer**, is normally used to indicate bridge balance. A method of measurement based on an indication of zero is referred to as a **null method**.

2. The ratio of two resistors of the same type, such as R_a and R_b, is much more accurate than the absolute value of either one because systematic, that is, nonrandom, errors in R_a and R_b tend to cancel out.

Capacitance can be accurately measured in an analogous manner using an ac bridge excited by an ac supply at an appropriate frequency. An unknown capacitance C_x is compared with a reference capacitor C_c using known, variable resistors R_a and R_b. Substituting $1/j\omega C_x$ for R_x, and $1/j\omega C_c$ for R_c, gives for bridge balance: $\dfrac{C_c}{C_x} = \dfrac{R_a}{R_b}$. Inductors are usually compared to accurately known capacitors because of the difficulty of constructing

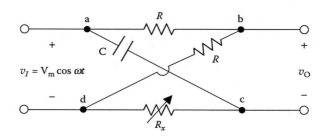

FIGURE 5.7.1
Figure for Example 5.7.1.

reference inductors. ac bridges are discussed in Example 5.7.2 and in problems at the end
of the chapter.

5.7 Phasor Diagrams

Phasor diagrams showing various voltage and current phasors in a given circuit are useful
for illustrating the interrelations between the various variables involved, particularly when
some circuit variable is varied, as illustrated by the following examples.

Example 5.7.1: Phase Shifter

Given the circuit of Figure 5.7.1, it is required to determine how the output voltage v_O
changes as R_x is varied from 0 to infinity, assuming that no current is drawn at the output.

SOLUTION
The lattice configuration of Figure 5.7.1 may be redrawn as a bridge configuration, for
easier visualization, and represented in the frequency domain as shown in Figure 5.7.2.

Because $V_m \angle 0° = R_x I + j X_c I$, where $X_c = -1/\omega C$, and the phasor $j X_c I$ lags the phasor $R_x I$
by 90°, then point Q joining these two phasors lies on the perimeter of a semicircle of
diameter V_m (Figure 5.7.3). If a phasor $-R_x I$ is drawn from point T at the tip of the phasor

$\dfrac{V_m}{2} \angle 0°$, then $\mathbf{V_O} = \dfrac{V_m}{2} - R_x I$ is the phasor from the origin O to S at the tip of $-R_x I$.

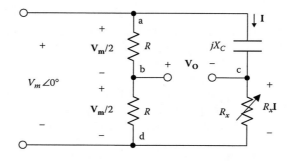

FIGURE 5.7.2
Figure for Example 5.7.1.

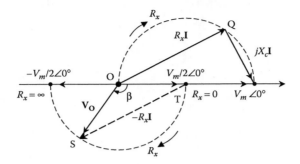

FIGURE 5.7.3
Figure for Example 5.7.1.

When $R_x = 0$, $\mathbf{V_O} = \dfrac{V_m}{2} \angle 0°$ and $v_O = \dfrac{V_m}{2} \cos\omega t$. Q coincides with O, and S coincides with T. As R_x increases, Q moves clockwise around the perimeter of its semicircle. Likewise, S moves clockwise around the perimeter of a semicircle because $-R_x\mathbf{I}$ is always parallel to $R_x\mathbf{I}$ and of the same magnitude. As $R_x \to \infty$, $\mathbf{I} \to 0$. Both plates of the capacitor will be at the same potential, and $\mathbf{V_O} = -\dfrac{V_m}{2} \angle 0°$. S would then lie at the tip of the phasor $-\dfrac{V_m}{2} \angle 0°$.

At any R_x, $\mathbf{V_O} = \dfrac{V_m}{2} \angle - \beta$ and $v_O = \dfrac{V_m}{2} \cos(\omega t - \beta)$.

It is seen from the geometry that $\sin(\beta/2) = \dfrac{R_x I_m/2}{V_m/2}$, where $I_m = \dfrac{V_m}{\sqrt{R_x^2 + X_c^2}}$. Substituting

for I_m gives: $\sin(\beta/2) = \dfrac{R_x}{\sqrt{R_x^2 + X_c^2}}$.

The circuit may be used to shift the phase of the output with respect to the input without altering the magnitude.

Example 5.7.2: ac Bridge
Given the ac bridge circuit of Figure 5.7.4, it is required to determine the balance condition and to draw a phasor diagram around balance.

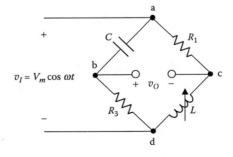

FIGURE 5.7.4
Figure for Example 5.7.2.

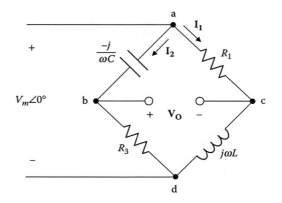

FIGURE 5.7.5
Figure for Example 5.7.2.

SOLUTION
Figure 5.7.5 shows the circuit in the frequency domain. From voltage division, $\mathbf{V}_{bd}$

$$=\left(\frac{R_3}{R_3 - j/\omega C}\right) V_m \angle 0° \text{ and } \mathbf{V}_{cd} = \left(\frac{j\omega L}{R_1 + j\omega L}\right) V_m \angle 0°. \text{ At balance, } \mathbf{V}_{bd} = \mathbf{V}_{cd}. \text{ The two}$$

expressions are equated to give: $R_3(R_1 + j\omega L) = j\omega L(R_3 - j/\omega C)$. The imaginary terms $j\omega L R_3$ are the same on both sides and cancel out. Equating the real terms gives for the balance condition: $L_b = CR_1R_3$, where L_b denotes the value of L at bridge balance.

The phasor diagram at bridge balance is shown in Figure 5.7.6. $\mathbf{I}_1$ is drawn having a

magnitude $\dfrac{V_m}{\sqrt{R_1^2 + \omega^2 L_b^2}}$ and a phase angle $\alpha = -\tan^{-1} \angle(\omega L_b / R_1)$. $\mathbf{I}_2$ is drawn having a mag-

nitude $\dfrac{V_m}{\sqrt{R_3^2 + (1/\omega^2 C^2)}}$ and a phase angle $\beta = \tan^{-1} \angle(1/\omega C R_3)$. At balance, $\mathbf{V}_{bd} = R_3\mathbf{I}_2 =$

$\mathbf{V}_{cd} = j\omega L_1 \mathbf{I}_1$. That is, $\mathbf{I}_2$ leads $\mathbf{I}_1$ by 90°, or $\alpha + \beta = 90°$. The phasor OQ represents $\mathbf{V}_{bd} =$

$\mathbf{V}_{cd}$. The phasor QS represents $\mathbf{V}_{ab} = -\dfrac{j}{\omega C}\mathbf{I}_2 = \mathbf{V}_{ac} = R_3\mathbf{I}_1$. Q lies on the perimeter of a

semicircle of diameter V_m.

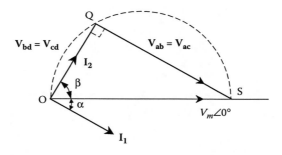

FIGURE 5.7.6
Figure for Example 5.7.2.

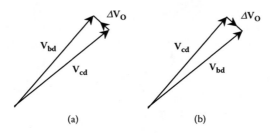

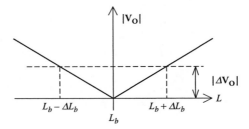

FIGURE 5.7.7
Figure for Example 5.7.2.

FIGURE 5.7.8
Figure for Example 5.7.2.

$$\mathbf{V}_{cd} = \frac{j\omega L}{R_1 + j\omega L}V_m\angle 0°.$$ If L is increased to $L_b + \Delta L_b$, the phasor $\mathbf{V}_{cd}$ increases slightly in magnitude and rotates clockwise with respect to $\mathbf{V}_{bd}$, which does not change. The output $\Delta \mathbf{V_O} = \mathbf{V}_{bd} - \mathbf{V}_{cd}$ will be as shown in Figure 5.7.7a. Conversely, if L is decreased to $L_b - \Delta L_b$, $\mathbf{I}_1$ increases in magnitude and decreases in phase angle. The phasor $\mathbf{V}_{cd}$ decreases slightly in magnitude and rotates counterclockwise with respect to $\mathbf{V}_{bd}$, which does not change. The output $\Delta \mathbf{V_O} = \mathbf{V}_{bd} - \mathbf{V}_{cd}$ will be as shown in Figure 5.7.7b. It is seen that $\Delta \mathbf{V_O}$ changes phase by 180° on either side of bridge balance. This change can also be shown analytically. Thus, around balance, $\Delta \mathbf{V_O} = \mathbf{V}_{bd} - \mathbf{V}_{cd} =$

$$\left(\frac{R_3}{R_3 - j/\omega C} - \frac{j\omega L}{R_1 + j\omega L}\right)V_m\angle 0° = \frac{R_1R_3 - L/C}{(R_1R_3 + L/C) + j(\omega LR_3 - R_1/\omega C)}V_m\angle 0°.$$ As L goes

through balance, the numerator changes sign but the denominator does not.

The change in sign of the output on either side of balance is true of all bridges, whether dc or ac, and has practical implications. In dc bridges, the detector gives opposite indications on either side of balance, so it is possible to distinguish between these two cases. In ac bridges, many types of detectors respond to magnitude only, and would not allow distinguishing the two sides of balance, as illustrated in Figure 5.7.8. A phase-sensitive detector is required for this purpose (Application Window 6.3.1, Chapter 6).

Summary of Main Concepts and Results

- A characteristic property of the sinusoidal function is that it is invariant in nature under linear operations, such as addition, subtraction, differentiation, and integration.

- When a complex sinusoidal excitation v_{SRC} is applied to an LTI circuit, the response is a complex sinusoidal function whose real part is the response to the real part of the excitation, $V_m\cos(\omega t + \theta)$, applied alone, and whose imaginary part is the response to the imaginary part of the excitation, $V_m\sin(\omega t + \theta)$, applied alone.

- A phasor is a quantity such as $V_m e^{j\theta}$ or $I_m e^{j(\theta - \alpha)}$ representing a complex sinusoidal function of time, but with the time variation suppressed.

- A current of rms value I_{rms}, or a voltage of rms value V_{rms}, dissipates the same power in a given resistor as a dc current, or a dc voltage, of the same value.

- The sinusoidal voltage and current for an ideal resistor are in phase because such a resistor is purely dissipative. They are in phase quadrature for ideal energy storage elements because these elements are nondissipative.

- In phasor notation, differentiation in time is replaced by multiplication by $j\omega$ and integration in time is replaced by division by $j\omega$. Thus, differential and integral relations are transformed to algebraic relations in $j\omega$ for steady-state sinusoidal analysis only.

- Impedance Z is the ratio of the voltage phasor $V_m\angle\theta_v$ to the related current phasor $I_m\angle\theta_i$. The real part of impedance is resistance and its imaginary part is reactance.

- Reactance is due to energy storage elements, that is, capacitors and inductors. Whereas sinusoidal voltages and currents are all in phase in a purely resistive circuit, energy storage elements introduce phase differences between sinusoidal voltages and currents.

- All circuit relations and theorems that apply to resistive circuits under dc conditions apply for sinusoidal steady-state analysis in the frequency domain to circuits that include resistance, inductance, and capacitance, with voltages and currents represented as phasors and impedances of circuit elements replacing resistance.

Learning Outcomes

- Represent circuits in the frequency domain in terms of phasors and impedances or admittances.

- Apply circuit relations and theorems in the frequency domain in order to derive steady-state sinusoidal responses.

Supplementary Topics and Examples on CD

ST5.1 Response of nonlinear resistor to complex excitation: Shows that the real and imaginary parts interact in a nonlinear system, and that the frequencies in the output are not the same as those in the input.

ST5.2 Properties of reactance: Discusses some basic properties of reactance.

SE5.1 *Balance in a bridged-T circuit:* Derives the condition for zero output in a bridged-T circuit of impedances.

SE5.2 *Series and shunt coupling:* Analyzes two parallel *RLC* circuits coupled by a series impedance and two series *RLC* circuits coupled by a shunt impedance.

SE5.3 *Impedance calculation using* MATLAB: Demonstrates how MATLAB can be used to derive the equivalent impedance of several branches connected in series and in parallel.

Problems and Exercises

P5.1 Phasors

P5.1.1 Given $\mathbf{A} = 10\angle15°$, $\mathbf{B} = 20\angle120°$, and $\mathbf{C} = 5\angle-45°$, determine the phasors resulting from the following operations: (a) $\mathbf{A} + \mathbf{B} + \mathbf{C}$, (b) $\mathbf{A} - \mathbf{B} + \mathbf{C}$, (c) $\mathbf{A} + \mathbf{B} - \mathbf{C}$, and (d) $\mathbf{A} - \mathbf{B} - \mathbf{C}$. Express the result in rectangular and polar forms.

P5.1.2 Given $\mathbf{A} = 3 + j5$, $\mathbf{B} = 10 - j8$, and $\mathbf{C} = j12$, determine the phasors resulting from the following operations: (a) $\mathbf{A} \times \mathbf{B} \times \mathbf{C}$, (b) $(\mathbf{A} \times \mathbf{B})/\mathbf{C}$, (c) $(\mathbf{A}/\mathbf{B}) \times \mathbf{C}$, and (d) $(\mathbf{A}/\mathbf{B})/\mathbf{C}$. Express the result in rectangular and polar forms.

P5.1.3 Given $\mathbf{A} = 5 + j10$, determine the phasor that is $\mathbf{A}$ raised to the fourth power.

P5.1.4 Given $\mathbf{A} = 24 + j32$, determine the phasor that is the cube root of $\mathbf{A}$.

P5.1.5 Using phasors, determine the steady-state y that satisfies the differential equation:

$$6\frac{dy}{dt} + 3y + 2\int_0^t y\,dt = 5\cos4t + 10\sin4t$$

Express y as a cosine time function.

P5.1.6 Given the sinusoidal time function v of Figure P5.1.6. Express v as a function of time and as a phasor.

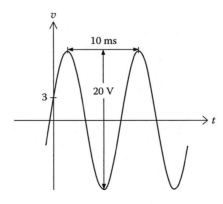

FIGURE P5.1.6

P5.2   Impedance and Admittance

P5.2.1   Given $i = 8\sqrt{2}\cos(2500\pi t - 45°)$ A and $i_1 = 2\cos 2500\pi t$ A in Figure P5.2.1, determine: (a) v and i_2 in the time and frequency domains; (b) Z, if composed of: (i) two series elements, or (ii) two parallel elements.

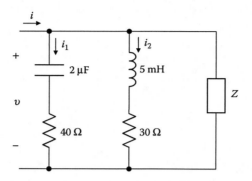

FIGURE P5.2.1

P5.2.2   Determine the input admittance Y_i in Figure P5.2.2. Verify with MATLAB and simulate with PSpice.

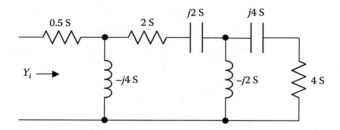

FIGURE P5.2.2

P5.2.3   Determine the input impedance Z_i in Figure P5.2.3. Simulate with PSpice.

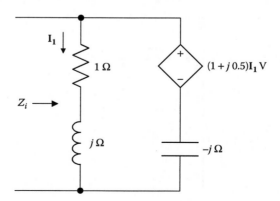

FIGURE P5.2.3

P5.2.4 Determine the input admittance Y_i in Figure P5.2.4.

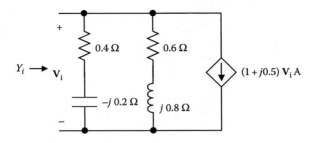

FIGURE P5.2.4

P5.2.5 Determine the input impedance Z_i in Figure P5.2.5.

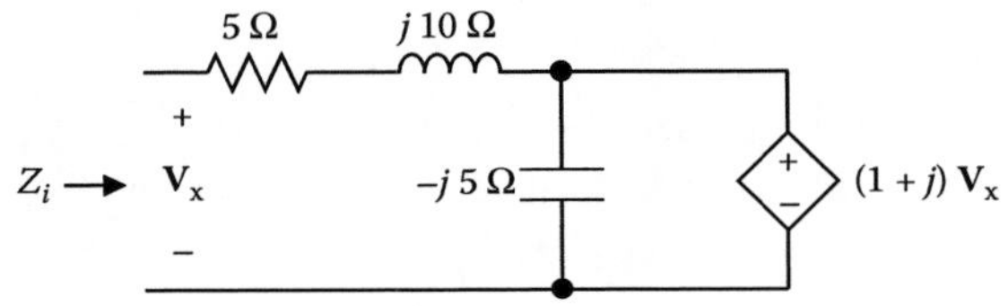

FIGURE P5.2.5

P5.2.6 Determine C in Figure P5.2.6 so that the impedance Z_i is purely resistive, assuming $\omega = 2000\pi$ rad/s. Draw a phasor diagram showing the relations between the currents in the two branches under these conditions.

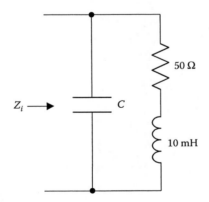

FIGURE P5.2.6

P5.2.7 Z_i and Y_i have equal values at all frequencies for the circuit shown in Figure P5.2.7. If $L = 1$ H, find R and C.

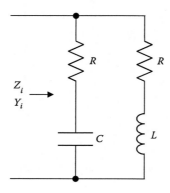

FIGURE P5.2.7

P5.2.8 In Exercise 5.6.5 determine the inductor current by transforming the current source and the capacitive branch to an equivalent voltage source. Then determine the capacitor current from KCL. Draw a phasor diagram showing the currents.

P5.3   Sinusoidal State Analysis

In the following problems, use MATLAB to solve simultaneous equations whenever appropriate and verify the solutions with PSpice simulation.

P5.3.1   Determine v_L in Figure P5.3.1. Express it both as a time function and as a phasor.

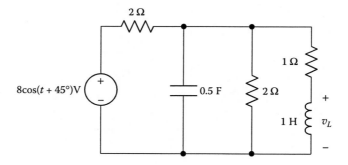

FIGURE P5.3.1

P5.3.2   Determine v_O in Figure P5.3.2. Express it both as a time function and as a phasor, assuming $\omega = 10^4$ rad/s.

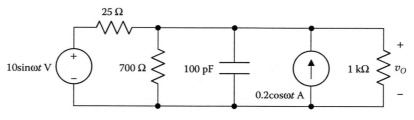

FIGURE P5.3.2

P5.3.3 Determine I_1 and I_2 in Figure P5.3.3.

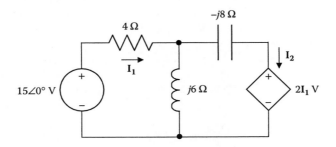

FIGURE P5.3.3

P5.3.4 Given that $v_{SRC1} = 125\sin(5 \times 10^4 t + 30°)$ V and $v_{SRC2} = 150\cos5 \times 10^4 t$ V in Figure P5.3.4, determine i_x and i_y.

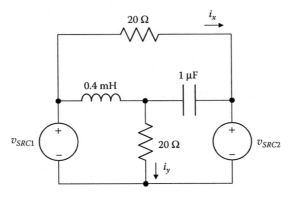

FIGURE P5.3.4

P5.3.5 Determine C in the circuit of Figure P5.3.5 so that v_O lags v_I by 90°, assuming $\omega = 400$ rad/s. Draw a phasor diagram showing the currents and voltages for each element and indicate how v_O and v_I are related to the phasors of the individual elements.

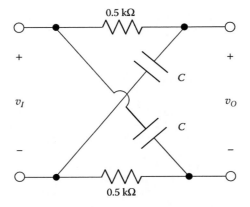

FIGURE P5.3.5

P5.3.6 In Figure P5.3.6, C is adjusted so that the reading of the ac ammeter does not change when the switch is alternately opened and closed. Show that under these conditions: $1 - 2\omega^2 LC = 0$. Draw a phasor diagram showing how the current magnitude remains the same. Note that this circuit allows using a capacitor to determine the value of a large inductance.

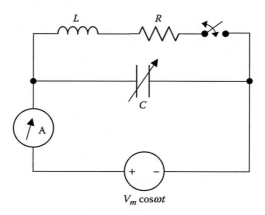

FIGURE P5.3.6

P5.3.7 Use Equation SE5.1.1 to show that $\mathbf{V_O} = 0$ in the bridged-T circuit of Figure P5.3.7 when $\omega^2 LC = 1$.

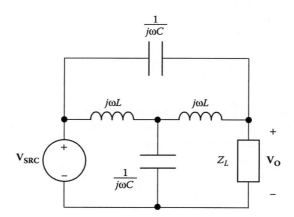

FIGURE P5.3.7

P5.3.8 Determine $\mathbf{I_O}$ in Figure P5.3.8 using superposition.

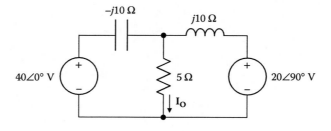

FIGURE P5.3.8

P5.3.9 Determine V_O in Figure P5.3.9 using superposition.

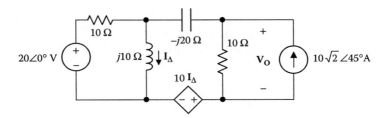

FIGURE P5.3.9

P5.3.10 Determine I_{SRC} in Figure P5.3.10 using Δ-Y transformation.

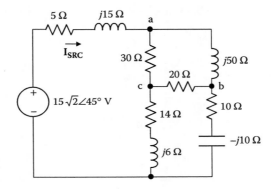

FIGURE P5.3.10

P5.3.11 Derive the T-circuit that is equivalent to the π-circuit shown in Figure P5.3.11 at: (a) $\omega = 2$ Mrad/s; and (b) $\omega = 1$ Mrad/s.

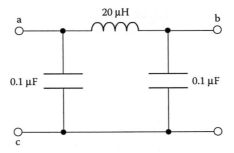

FIGURE P5.3.11

P5.3.12 Determine $\mathbf{V}_c$ and $\mathbf{I}_L$ in the circuit of Figure P5.3.12.

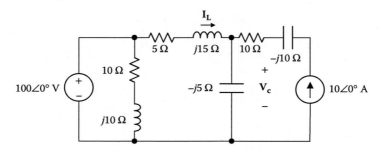

FIGURE P5.3.12

P5.3.13 Determine $\mathbf{I}_x$ in Figure P5.3.13 after eliminating all redundant elements.

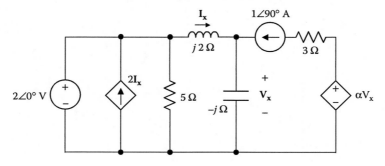

FIGURE P5.3.13

P5.3.14 Show that $\dfrac{\mathbf{I_O}}{\mathbf{V_I}} = \dfrac{1 - \omega^2 \tau^2}{2R(1 + j\omega\tau)}$ in Figure P5.3.14, where $\tau = CR$. Consider each T-circuit separately and use voltage or current division to find each component of $\mathbf{I_O}$. The circuit is known as a **twin-T notch filter**.

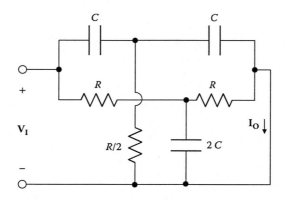

FIGURE P5.3.14

P5.3.15 Determine v_O in Figure P5.3.15, given that $v_{SRC} = 10\cos(2t - 30°)$V and $i_{SRC} = 5\cos(2t + 30°)$A.

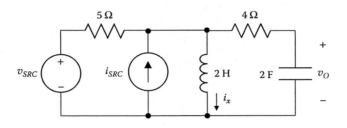

FIGURE P5.3.15

P5.3.16 Determine i_O in Figure P5.3.16 given that $v_{SRC1} = 10\cos(10^4t + 45°)$ V and $v_{SRC2} = 10\cos(10^4t\ 45°)$ V.

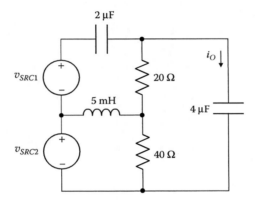

FIGURE P5.3.16

P3.5.17 Determine $\mathbf{I_O}$ in Figure P5.3.17 using repeated voltage division.

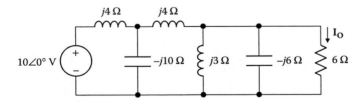

FIGURE P5.3.17

P5.3.18 Determine $\mathbf{V_x}$ in Figure P5.3.18 given that: $Z_1 = Z_2 = Z_3 = 2 + j2\,\Omega$ and $Z_a = Z_b = Z_c = 4 + j4\,\Omega$.

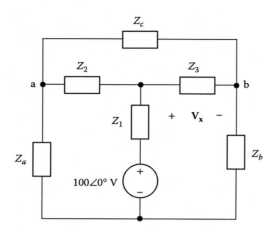

FIGURE P5.3.18

P5.3.19 Determine $\mathbf{V_O}$ in Figure P5.3.19.

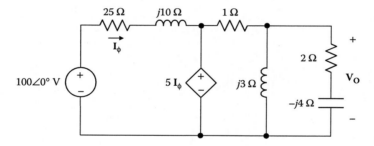

FIGURE P5.3.19

P5.3.20 Determine i_x in Figure P5.5.20.

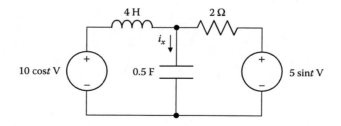

FIGURE P5.3.20

P5.3.21 Determine $\mathbf{V_O}$ in Figure P5.3.21.

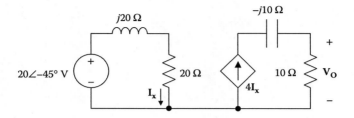

FIGURE P5.3.21

P5.3.22 Determine $\mathbf{I_{SRC}}$ and $\mathbf{V_L}$ in Figure P5.3.22 using the node-voltage method.

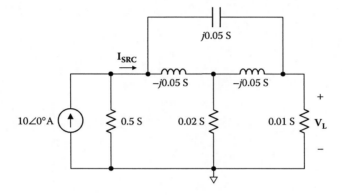

FIGURE P5.3.22

P5.3.23 Repeat Problem P5.3.22 using the mesh-current method.

P5.3.24 Determine $\mathbf{V_O}$ in Figure P5.3.24 by the mesh-current or node-voltage method, whichever is more advantageous.

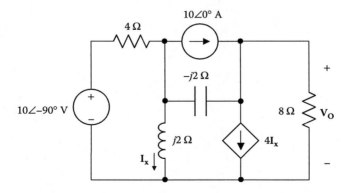

FIGURE P5.3.24

P5.3.25 Determine v_O in Figure P5.3.25 by the mesh-current or node-voltage method, whichever is more advantageous, assuming $v_{SRC} = 100\cos(10^3t - 30°)$ V.

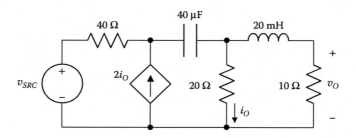

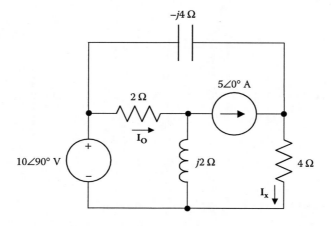

FIGURE P5.3.25

P5.3.26 Determine $\mathbf{I_O}$ in Figure P5.3.26 using the mesh-current method.

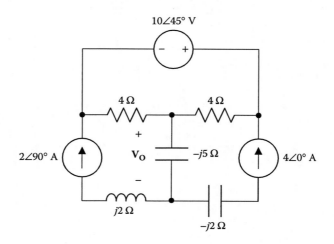

FIGURE P5.3.26

P5.3.27 Repeat Problem P5.3.26 using the node-voltage method.

P5.3.28 Determine $\mathbf{V_O}$ in Figure P5.3.28 using the mesh-current method.

FIGURE P5.3.28

P5.3.29 Repeat Problem P5.3.28 using the node-voltage method.

P5.3.30 Determine I_O in Figure P5.3.30 using the mesh-current method.

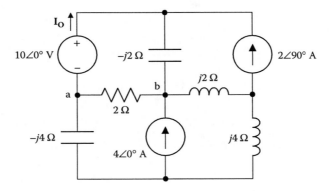

FIGURE P5.3.30

P5.3.31 Repeat Problem P5.3.30 using the node-voltage method. Use node a as a reference. Repeat using node b as a reference and check by shifting voltages.

P5.3.32 Determine I_1, I_2, and V_{ab} in Figure P5.3.32, given that $V_{SRC} = 50\angle 60°$ V. Draw a phasor diagram showing these quantities.

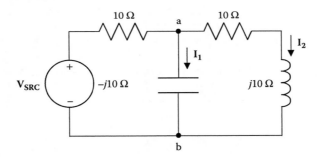

FIGURE P5.3.32

P5.3.33 Determine V_O in Figure P5.3.33.

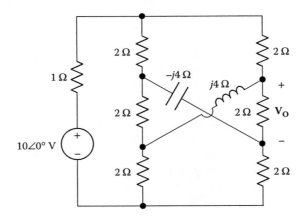

FIGURE P5.3.33

P5.3.34 Figure P5.3.34 shows a **Wien bridge** that can be used for measuring the frequency of a sinusoidal source. Show that at balance: $\omega_2 C_3 C_4 R_3 R_4 = 1$ and

$$R_1\left(R_3 + R_4 \frac{C_4}{C_3}\right) = R_2 R_4.$$

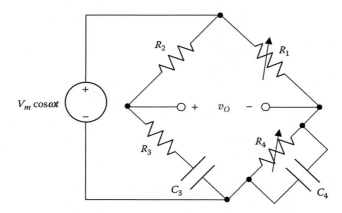

FIGURE P5.3.34

P5.3.35 Figure P5.3.35 shows a **Maxwell bridge** that can be used for measuring the inductance and resistance of a coil in terms of known capacitance and resistance values. Show that at bridge balance: $R_2 = \dfrac{R_1 R_3}{R_4}$ and $L_2 = C_4 R_1 R_3$. Note that the first condition is a dc-balance condition, the same as in the Wheatstone bridge.

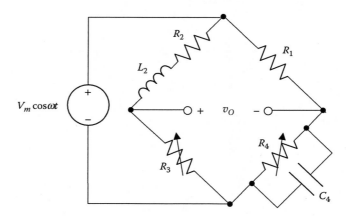

FIGURE P5.3.35

P5.3.36 Figure P5.3.36 shows an **Anderson bridge** that may also be used for measuring the inductance and resistance of a coil in terms of known capacitance and resistance values. Its advantage over the Maxwell bridge is that dc and ac balance may be separately established. Show that at bridge balance:

$$R_2 = \frac{R_1 R_3}{R_4} \quad \text{and} \quad L_2 = C_4 R_1\left[R + (R + R_4)\frac{R_3}{R_4}\right].$$

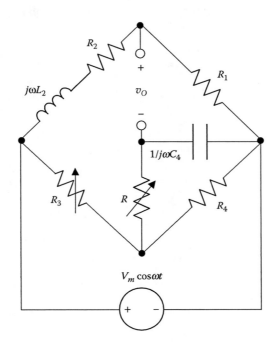

FIGURE P5.3.36

P5.3.37 Determine NEC between terminals ab in Figure P5.3.37.

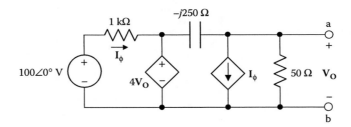

FIGURE P5.3.37

P5.3.38 Determine TEC between terminals ab in Figure P5.3.38.

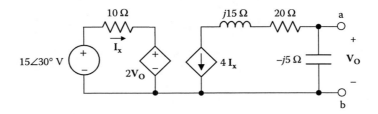

FIGURE P5.3.38

6

Linear and Ideal Transformers

Overview

The preceding chapter was devoted to steady-state sinusoidal analysis of circuits that included resistors, capacitors, and inductors. The present chapter extends steady-state sinusoidal analysis to circuits that include magnetic coupling between current-carrying coils. This type of coupling is described in terms of a fourth circuit parameter, namely, the mutual inductance between the coupled coils. The mutual inductance introduces a voltage-drop term or a voltage-rise term in the KVL equation, just like the other circuit parameters R, L, and C. Once the mutual inductance term is defined, in both magnitude and sign, the same circuit analysis techniques previously considered can be applied.

The idealization of magnetic coupling leads to the concept of an ideal transformer, which is of great practical and theoretical importance. Whereas resistors, capacitors, and inductors are one-port passive devices, an ideal transformer is a two-port passive device, having two input terminals, or input port, and two output terminals, or output port. It is a basic element because it cannot be represented in terms of the three passive one-port elements alone. A nonideal transformer, on the other hand, can be represented by an equivalent circuit consisting of inductors, resistors, and capacitors. Practical transformers, like capacitors, are generally close to the ideal under normal operating conditions. They are extensively used to transform voltages, currents, and impedances. Transformer imperfections, or departures from the ideal, set the limits for the normal operating conditions of the transformer.

Learning Objectives

- To be familiar with:
  - Terms describing magnetic systems
  - Some practical aspects of small inductors and transformers
- To understand:
  - How magnetic coupling between given coils is accounted for by the mutual inductance between the coils
  - Underlying principles for analyzing linear transformers and representing them by means of an equivalent circuit
  - The basic characteristics of the ideal transformer and autotransformer

- How to reflect circuits from one side of an ideal transformer to the other
- The effects of transformer imperfections

6.1   Mutual Inductance

Figure 6.1.1 shows two coils in air, wound on a former made from nonmagnetic material such as bakelite or polystyrene. The *B–H* relation for any magnetic medium may be expressed as $B = \mu_r \mu_0 H$, where μ_r is the relative permeability of the medium and μ_0 is the permeability of free space. If μ_r is constant, the material is linear, and if $\mu_r \cong 1$, as for air, the material is described as nonmagnetic. Some commonly used magnetic materials have μ_r as high as several tens of thousands, and are generally nonlinear because μ_r varies with *B* or *H* (Section 6.5).

Let the current i_1 in coil 1 be time-varying, whereas coil 2 is open-circuited to begin with. A voltage v_1 is induced in coil 1, in accordance with Faraday's law:

$$v_1 = \frac{d\lambda_1}{dt} = \frac{N_1 d\phi_{1e}}{dt} = L_1 \frac{di_1}{dt} \tag{6.1.1}$$

where ϕ_{1e} is an effective flux of coil 1 associated with i_1, which, if it links all N_1 turns, gives λ_1. Thus, $\lambda_1 = N_1 \phi_{1e} = L_1 i_1$. The reason we have to consider an effective flux is that in the case of cores of low permeability, not all of the magnetic flux links all the turns of the coil, as illustrated in Figure 6.1.1. Note that ϕ_{1e} is determined by N_1 and the time integral of v.

Let ϕ_{21e} be the fraction of the time-varying flux ϕ_{1e} that links coil 2. Again, ϕ_{21e} is an effective flux that, if multiplied by N_2, gives the flux linkage λ_{21} in coil 2 due to i_1. A voltage v_{21} is induced in this coil in accordance with Faraday's law:

$$v_{21} = \frac{d\lambda_{21}}{dt} = \frac{N_2 d\phi_{21e}}{dt} = M_{21} \frac{di_1}{dt} \tag{6.1.2}$$

where $\lambda_{21} = N_2 \phi_{21e}$. The quantity M_{21} is defined as the flux linking coil 2 per unit current in coil 1. Thus:

$$M_{21} = \frac{\lambda_{21}}{i_1} = \frac{N_2 \phi_{21e}}{i_1} \tag{6.1.3}$$

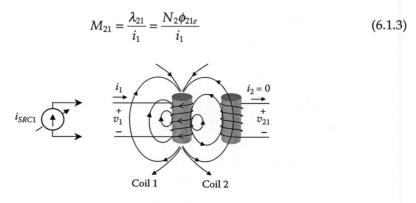

FIGURE 6.1.1
Magnetically coupled coils in air.

If a time-varying current i_2 is applied to coil 2, with coil 1 open-circuited, then following the same argument, we have, analogous to Equation 6.1.1 to Equation 6.1.3:

$$v_2 = \frac{d\lambda_2}{dt} = \frac{N_2 d\phi_{2e}}{dt} = L_2 \frac{di_2}{dt} \tag{6.1.4}$$

$$v_{12} = \frac{d\lambda_{12}}{dt} = \frac{N_1 d\phi_{12e}}{dt} = M_{12} \frac{di_2}{dt} \tag{6.1.5}$$

$$M_{12} = \frac{\lambda_{12}}{i_2} = \frac{N_1 \phi_{12e}}{i_2} \tag{6.1.6}$$

where M_{12} is the flux linking coil 1 per unit current i_2 in coil 2. Equation 6.1.4 to Equation 6.1.6 are the same as Equation 6.1.1 to Equation 6.1.3, but with the 1 and 2 subscripts interchanged.

Let us determine the energy expended in establishing steady currents I_1 and I_2, starting from zero. It is convenient to assume that I_1 and I_2 are established by variable current sources in two steps: (1) i_1 is first increased from zero to I_1 with $I_2 = 0$; and (2) i_2 is then increased from zero to I_2 with $i_1 = I_1$. For the sense of winding of coil 1 in Figure 6.1.1, the flux associated with i_1 is downward in this coil, according to the right-hand rule. While i_1 is increasing, the induced voltage $v_1 = L_1 \frac{di_1}{dt}$ in coil 1, opposes the increase in i_1, in accordance with Lenz's law, by being a voltage drop across L_1 in the direction of i_1. This voltage is concurrently a voltage rise across the current source, so that the total energy w_{11} delivered by the source is:

$$w_{11} = \int_0^t L_1 \frac{di_1}{dt} i_1 dt = L_1 \int_0^{I_1} i_1 di_1 = \frac{1}{2} L_1 I_1^2 \tag{6.1.7}$$

The second step is illustrated in Figure 6.1.2. With a constant current $i_1 = I_1$, the voltage induced in coil 2 is that due to increasing i_2 and the total energy supplied by the current source i_{SRC2} in establishing I_2 is $\frac{1}{2} L_2 I_2^2$, as in Equation 6.1.7. However, as i_2 increases, it induces a voltage $v_{12} = M_{12} \frac{di_2}{dt}$ in coil 1. The sense of winding of coil 2 and the direction of i_2 are such that the flux associated with i_2 is also downward in coil 1. The effect of

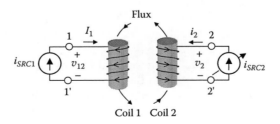

FIGURE 6.1.2
Current build-up in two magnetically coupled coils.

increasing i_2 is therefore the same as that of increasing i_1, so that v_{12} is of the same polarity as v_1 in Figure 6.1.1, and opposes the current in coil 1. The current source i_{SRC1} therefore has to deliver additional energy to maintain I_1. This energy is:

$$w_{12} = \int_0^t v_{12}I_1dt = \int_0^t M_{12}\frac{di_2}{dt}I_1dt = M_{12}I_1\int_0^{I_2}di_2 = M_{12}I_1I_2 \tag{6.1.8}$$

The total energy expended in establishing I_1 and I_2 is:

$$w_1 = \frac{1}{2}L_1I_1^2 + \frac{1}{2}L_2I_2^2 + M_{12}I_1I_2 \tag{6.1.9}$$

Suppose that I_1 and I_2 are established in the reverse order, that is, first I_2 is established with $i_1 = 0$, then i_1 is increased to I_1 with $i_2 = I_2$. Following the same preceding argument, the total energy expended in establishing I_1 and I_2 is:

$$w_2 = \frac{1}{2}L_1I_1^2 + \frac{1}{2}L_2I_2^2 + M_{21}I_1I_2 \tag{6.1.10}$$

w_1 and w_2 are equal because in a lossless, linear system, the total energy expended must depend only on the final values of I_1 and I_2 and not on the time course of i_1 and i_2 that led to these values. Otherwise, it would be possible, at least in principle, to extract energy from the system at no energy cost, in violation of conservation of energy. Suppose, for example, that $w_1 < w_2$. Then I_1 and I_2 may be established from zero with expenditure of energy w_1. In principle, I_1 and I_2 may be reduced to zero by reversing the steps that led to w_2, with recovery of more energy than was expended, which is clearly impossible. It follows that:

$$M_{12} = M_{21} = M \tag{6.1.11}$$

M is the **mutual inductance** between the two coils and is a constant in linear systems. In contrast, the individual inductances L_1 and L_2 are **self-inductances**.

> **Definition:** *The mutual inductance of two magnetically coupled coils is the flux linkage in one coil per unit current in the other coil. It is independent of which coil carries the current.*

If either the polarity of i_{SRC2} or the sense of winding of coil 2 is reversed in Figure 6.1.2, the flux due to i_2 becomes upward in coil 1. The polarity of v_{12} is reversed and becomes a voltage drop across the current source i_{SRC1}. Energy is therefore returned to the source and the sign of the energy term involving M becomes negative in Equation 6.1.9 and Equation 6.1.10. M, however, is *always* a positive quantity.

Because I_1 and I_2 are arbitrary values, they might just as well be replaced by instantaneous values i_1 and i_2. The energy stored in the magnetic field in building up the currents in two magnetically coupled coils to i_1 and i_2 starting from zero, may therefore be expressed, in general, as:

$$w = \frac{1}{2}L_1i_1^2 + \frac{1}{2}L_2i_2^2 \pm Mi_1i_2 \tag{6.1.12}$$

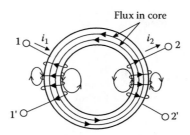

Flux in core

FIGURE 6.1.3
Coupling through a high-permeability core.

When the system is nonlinear and involves hysteresis, it is shown in Section 6.5 that the time course according to which i_1 and i_2 reach their final values affects the energy expended, the energy difference being dissipated as heat in the magnetic material.

Coupling through a High-Permeability Core

Suppose that coils 1 and 2, instead of being in air, are wound on a toroidal core of high, constant permeability (Figure 6.1.3). Because of the shape and high permeability of the core, the flux in the core follows a circular path that is *confined entirely to the core and links all the turns of both coils*. In other words, the effective flux that links all the turns of either coil is the same as the actual flux. The flux that links only one coil is the **leakage flux**. The leakage flux exists in the air space between the coils and the core, as shown, and generally does not link all the turns of the coil.

The flux relations assume a simpler form than in the case of a core of low permeability. If i_1 is the current in coil 1, with $i_2 = 0$, the total flux linkage λ_1 of coil 1 may be expressed as:

$$\lambda_1 = N_1\phi_{11e} + N_1\phi_{21} \tag{6.1.13}$$

where $N_1\phi_{21}$ is the flux linkage due to the flux ϕ_{21} in the core, and $N_1\phi_{11e}$ is the flux linkage due to the leakage flux. The self-inductance L_1 of coil 1 is:

$$L_1 = \frac{\lambda_1}{i_1} = \frac{N_1\left(\phi_{11e} + \phi_{21}\right)}{i_1} \tag{6.1.14}$$

Replacing M_{21} in Equation 6.1.3 by M, and ϕ_{21e} by ϕ_{21}, the mutual inductance becomes:

$$M = \frac{\lambda_{21}}{i_1} = \frac{N_2\phi_{21}}{i_1} \tag{6.1.15}$$

If i_2 is the current in coil 2, with $i_1 = 0$, the relations corresponding to Equation 6.1.13 to Equation 6.1.15 are obtained by repeating the earlier arguments, resulting in interchanging the 1 and 2 subscripts. Thus:

$$\lambda_2 = N_2\phi_{22e} + N_2\phi_{12} \tag{6.1.16}$$

$$L_2 = \frac{\lambda_2}{i_2} = \frac{N_2\left(\phi_{22e} + \phi_{12}\right)}{i_2} \tag{6.1.17}$$

$$M = \frac{\lambda_{12}}{i_2} = \frac{N_1\phi_{12}}{i_2} \qquad (6.1.18)$$

Coupling Coefficient

Multiplying together Equation 6.1.15 and Equation 6.1.18:

$$M^2 = \frac{N_2\phi_{21}N_1\phi_{12}}{i_1 i_2} \qquad (6.1.19)$$

Dividing by the product L_1L_2 from Equation 6.1.14 and Equation 6.1.17:

$$\frac{M^2}{L_1L_2} = \frac{\phi_{21}}{\phi_{11e} + \phi_{21}} \times \frac{\phi_{12}}{\phi_{22e} + \phi_{12}} \qquad (6.1.20)$$

The expression $\dfrac{\phi_{21}}{\phi_{11e} + \phi_{21}}$ is a measure of how effectively coil 1 is coupled to the core. If the coupling is perfect, $\phi_{11e} = 0$, and the expression is unity; if there is no coupling, $\phi_{21} = 0$, and the expression is zero. The same is true for the expression $\dfrac{\phi_{12}}{\phi_{22e} + \phi_{12}}$. The product on the RHS of Equation 6.1.20 is therefore a measure of how well the two coils are magnetically coupled together through the core. This product assumes values between zero for no coupling, and unity for perfect coupling. Equation 6.1.20 may be written as:

$$\frac{M^2}{L_1L_2} = k^2 \quad \text{or} \quad M = k\sqrt{L_1L_2} \qquad (6.1.21)$$

> **Definition:** *The coupling coefficient k of two magnetically coupled coils is defined as* $\dfrac{M}{\sqrt{L_1L_2}}$ *and is a measure of how tightly the two coils are coupled through the core. It assumes values in the range of 0 to unity, where k = 0 denotes no coupling, and k = 1 denotes perfect coupling.*

Although Equation 6.1.20 was derived for the case of a high-permeability core, for simplicity, it applies to the case of a low-permeability core with ϕ_{21e} and ϕ_{12e} replacing ϕ_{21} and ϕ_{12}, respectively. The coupling coefficient is defined by Equation 6.1.21 irrespective of the permeability of the core.

EXERCISE 6.1.1

Show that if two coils are perfectly coupled, the stored magnetic energy (Equation 6.1.12) is $w = \dfrac{1}{2}\left(\sqrt{L_i}\,i_1 \pm \sqrt{L_2}\,i_2\right)^2$.

Electric Circuit Analogy

A useful analogy can be made between a magnetic system and an electric circuit. This analogy, developed in some detail in Section ST6.1 is summarized in Table 6.1.1. Magnetic flux is

TABLE 6.1.1

Electric Circuit Analogy

Electric Circuit	Magnetic System
Current	Flux (ϕ)
Voltage excitation (emf)	mmf (Ni)
Resistance	Reluctance (mmf/flux)
Conductance	Permeance (flux/mmf)
Conductivity	Permeability

analogous to electric current, and magnetomotive force (mmf), which equals Ni, is analogous to voltage excitation, also known as electromotive force (emf). The concept of mmf is fundamental and will be used below. In fact, Ampere's circuital law relates the line integral of magnetic field intensity H over a closed path to the mmf: $\oint_l H dl = Ni$, where N is also the number of times the current is crossed in going around the closed path. Flux is the product of permeance and mmf, just as current is the product of conductance and voltage.

6.2 The Linear Transformer

> **Definition:** *A transformer consists of two or more coils that are magnetically coupled relatively tightly. In a linear transformer, permeability is constant, so that B and H, or ϕ and i, are linearly related.*

Consider a linear transformer consisting of two coils coupled through a high-permeability core, as illustrated in Figure 6.2.1, one coil being connected to a source of excitation v_{SRC}, the other to a load R_L. In transformer terminology, the coil connected to the source of excitation is the **primary winding**, whereas the coil connected to the load is the **secondary winding**. Let the primary and secondary currents be i_1 and i_2, the assigned positive directions being as shown. Because of linearity, the flux in the core is the superposition of ϕ_{21}, associated with i_1 alone, and ϕ_{12}, associated with i_2 alone, the direction of flux in each case being in accordance with the right-hand rule. The sense of the windings and the assigned positive directions of currents in Figure 6.2.1 are such that ϕ_{21} and ϕ_{12} are in opposite directions and therefore subtract. Because of high permeability, both ϕ_{21} and ϕ_{12} link all the turns N_1 of coil 1 and N_2 of coil 2.

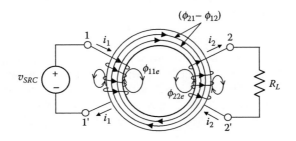

FIGURE 6.2.1
Linear transformer.

KVL for the primary circuit is $R_1 i_1 + N_1 \dfrac{d}{dt}(\phi_{11e} + \phi_{21} - \phi_{12}) = v_{SRC}$, where R_1 is the resistance of coil 1. From Equation 6.1.14, $\dfrac{d}{dt} N_1(\phi_{11e} + \phi_{21}) = L_1 \dfrac{di_1}{dt}$, and from Equation 6.1.18, $\dfrac{d}{dt}(N_1 \phi_{12}) = M \dfrac{di_2}{dt}$. Substituting:

$$R_1 i_1 + L_1 \frac{di_1}{dt} - M \frac{di_2}{dt} = v_{SRC} \tag{6.2.1}$$

KVL for the secondary circuit is: $(R_2 + R_L)i_2 + N_2 \dfrac{d}{dt}(\phi_{22e} + \phi_{12} - \phi_{21}) = 0$, where R_2 is the resistance of coil 2. Substituting from Equation 6.1.15 and Equation 6.1.17:

$$(R_2 + R_L)i_2 + L_2 \frac{di_2}{dt} - M \frac{di_1}{dt} = 0 \tag{6.2.2}$$

Once v_{SRC} and the circuit parameters R_1, R_2, R_L, L_1, L_2, and M are known, i_1 and i_2 are determined from Equation 6.2.1 and Equation 6.2.2. For the sinusoidal steady state, these equations are represented in the frequency domain and solved as two simultaneous equations in the current phasors $\mathbf{I_1}$ and $\mathbf{I_2}$, as discussed below. For arbitrary time variation, including initial values of i_1 and i_2, it is convenient to apply the Laplace transform method, as discussed in Section 15.5.

Sign of Mutual Inductance Term

It is clear from the preceding discussion that if the sense of one winding is reversed, for example as shown for coil 1 in Figure 6.1.3, or if the assigned positive direction of i_1 or i_2 is reversed, ϕ_{21} and ϕ_{12} will add, which changes the sign of the mutual inductance term relative to the self-inductance terms. This should not pose a problem if the sense of the windings is explicitly shown, as in Figure 6.2.1. It is inconvenient, however, to have to show the sense of the windings on circuit diagrams, so the relative sense of the windings is indicated by placing a dot on one terminal of each winding according to the following convention.

> **Dot Convention:** *One terminal of each coil is marked with a dot so that currents entering (or leaving) the marked terminals in each coil are associated with flux in the same direction in both coils.*

In Figure 6.2.1, for example, we may arbitrarily place a dot on terminal 1 of coil 1. Because i_1 entering at this terminal is associated with flux in the core in the clockwise direction, and i_2 entering terminal 2 is also associated with flux in the same direction, terminal 2 is dotted. Alternatively, terminals 1' and 2' may be dotted. If the sense of winding of either coil is reversed, as in Figure 6.1.3, then the dotted terminals will be 1 and 2', or 1' and 2. In Figure 6.1.2, terminals 1 and 2, or 1' and 2', will be dotted.

An alternative interpretation of the dot markings, which follows from the aforementioned, is that the *polarities of induced voltages in both coils are the same, relative to the dot markings*. In Figure 6.1.2, for example, when i_2 is increasing, v_2 opposes the increase in i_2 and v_{12} opposes I_1. Both voltages oppose currents entering at the dotted terminals, and the polarities of these voltages make the dotted terminal positive with respect to the unmarked terminal in both coils. Once the terminals are marked with dots, the sign of the M term readily follows:

> **Sign of M Term:** *If the assigned positive directions of currents are such that these currents both flow in or both flow out at the dotted terminals, the sign of the mutual inductance term ($M di_1/dt$ or $M di_2/dt$) for either coil is the same as that of the self-inductance term for that coil ($L_1 di_1/dt$ or $L_2 di_2/dt$). Otherwise, the sign of the mutual inductance term for either coil is opposite that of the self-inductance term for that coil.*

The justification is simply that if the assigned positive directions of coil currents are such that both currents flow into or out of the dotted terminals, i_1 and i_2 produce flux in the same direction in the core. This means that the $M \dfrac{di_2}{dt}$ voltage induced by i_2 in coil 1 is of the same polarity as the $L_1 \dfrac{di_1}{dt}$ voltage induced by i_1 in coil 1, so that these two terms have the same sign in the voltage relations of coil 1. Similarly, the $M \dfrac{di_1}{dt}$ voltage induced by i_1 in coil 2 is of the same polarity as the $L_2 \dfrac{di_2}{dt}$ voltage induced by i_2 in coil 2, so that these two terms have the same sign in the voltage relations of coil 2.

Frequency-Domain Representation

Equation 6.2.1 and Equation 6.2.2 are expressed in the frequency domain by replacing the time-varying currents with the corresponding phasors and replacing differentiation by $j\omega$. Thus:

$$(R_1 + j\omega L_1)\, \mathbf{I}_1 - j\omega M \mathbf{I}_2 = \mathbf{V}_{\text{SRC}} \tag{6.2.3}$$

and,

$$-j\omega M \mathbf{I}_1 + (R_2 + R_L + j\omega L_2)\, \mathbf{I}_2 = 0 \tag{6.2.4}$$

Figure 6.2.2 shows the linear transformer in the frequency domain. The dotted terminals are indicated, together with the self-inductance of each coil and the mutual inductance between them. Note that if the voltage induced by $\mathbf{I}_2$ in L_2, for example, is written as $-j\omega L_2 \mathbf{I}_2$, in accordance with the passive sign convention, the total voltage rise due to $\mathbf{I}_2$, and in the direction of $\mathbf{I}_2$, is $-j\omega L_2 \mathbf{I}_2 - (R_2 + R_L)\mathbf{I}_2$. The sign of the mutal inductance term is opposite that of $-j\omega L_2 \mathbf{I}_2$, so that this term is now $j\omega M \mathbf{I}_1$. KVL for the secondary circuit becomes $+j\omega M \mathbf{I}_1 - (R_2 + R_L + j\omega L_2)\mathbf{I}_2 = 0$, which gives the same equation as Equation 6.2.4.

T-Equivalent Circuit

The form of Equation 6.2.3 and Equation 6.2.4 allows a very convenient representation of a linear transformer in terms of a T-equivalent circuit, as shown in Figure 6.2.3. The mesh currents are $\mathbf{I}_1$ and $\mathbf{I}_2$ and the mesh-current equations are identical to Equation 6.2.3 and

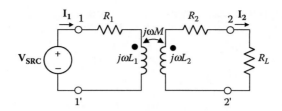

FIGURE 6.2.2
Linear transformer in the frequency domain.

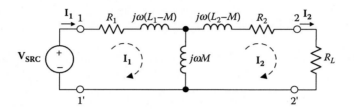

FIGURE 6.2.3
T-equivalent circuit.

Equation 6.2.4. Hence, the T-equivalent circuit is indeed equivalent to the linear transformer and can replace it in any given circuit, terminal for terminal. This applies, of course, to both the frequency and time domains.

If the dots on either coil are reversed, the sign of M is reversed. Figure 6.2.4 summarizes the T-equivalent circuit for the two possible dot markings, not including coil resistances. Note that the inductance of the shunt branch in Figure 6.2.4b is negative. This is acceptable because, as emphasized earlier, an equivalent circuit duplicates the v–i relations at the terminals and need not have a direct physical correspondence with the system it represents. The reactance of a negative inductance is negative, like a capacitive reactance. Its magnitude, however, increases with frequency, unlike a capacitive reactance, which decreases with frequency. It should be emphasized that the T-equivalent circuit is independent of the assigned positive directions of currents. This is further illustrated by Example 6.2.1. A π-equivalent circuit can be derived using Δ-Y transformation (Problem P6.2.9).

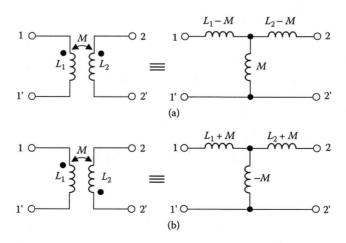

FIGURE 6.2.4
T-equivalent circuits for different dot markings.

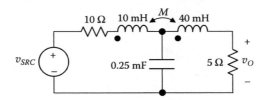

FIGURE 6.2.5
Figure for Example 6.2.1.

Example 6.2.1: Mesh-Current Analysis of Circuit Including Coupled Coils

Given the circuit of Figure 6.2.5 in which $v_{SRC} = 100\cos800t$ and $k = 0.25$, it is required to determine the steady-state value of v_O.

SOLUTION

$\omega L_1 = 800 \times 10^{-2} = 8$ Ω; $\omega L_2 = 800 \times 4 \times 10^{-2} = 32$ Ω; $M = 0.25\sqrt{10^{-2} \times 4 \times 10^{-2}} = 5 \times 10^{-3}$ H;

$\omega M = 800 \times 5 \times 10^{-3} = 4$ Ω; and $\dfrac{1}{\omega C} = \dfrac{1}{800 \times 0.25 \times 10^{-3}} = 5 \Omega$. The circuit in the frequency

domain is shown in Figure 6.2.6.

In writing the mesh-current equation for mesh 1, the total voltage drop in this mesh due to I_1 equals, as usual, I_1 multiplied by the total self-impedance of this mesh, that is, $(10 + j8 - j5)I_1$. I_2 introduces, as usual, a voltage rise of Z_cI_2 in mesh 1, where $Z_c = -j5$ Ω is the common impedance between meshes 1 and 2, plus a $j\omega MI_2$ term due to the magnetic coupling between the coils in the two meshes. Because both I_1 and I_2 enter at the dotted terminals, the sign of the $j\omega MI_2$ term is the same as that of the $j\omega L_1$ term in the equation of mesh 1 and the same as that of the $j\omega L_2$ term in the equation for mesh 2. The mesh-current equation for mesh 1 is, therefore:

$$(10 + j8 - j5)I_1 - (-j4 - j5)I_2 = 100 \qquad (6.2.5)$$

Similarly, the mesh-current equation for mesh 2 is:

$$-(-j4 - j5)I_1 + (5 + j32 - j5)I_2 = 0 \qquad (6.2.6)$$

Solving for I_2 gives: $I_2 = -3.0636 - j0.5375$ A, so that $V_O = 5I_2 = -15.3 - j2.69 = 15.55\angle-170.0°$V, and $v_O = 15.55\cos(800t - 170.0°)$ V.

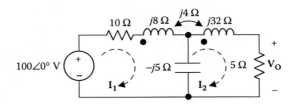

FIGURE 6.2.6
Figure for Example 6.2.1.

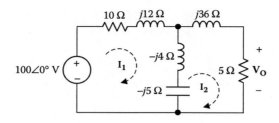

FIGURE 6.2.7
Figure for Example 6.2.1.

Equation 6.2.5 and Equation 6.2.6 could just as well be derived using the T-equivalent circuit (Figure 6.2.7). If L_1 in Figure 6.2.6 is rotated clockwise and L_2 is rotated counterclockwise so as to bring them to the upright position, the dots will be as in the transformer of Figure 6.2.4b. The appropriate T-equivalent circuit is therefore that having series branches $L_1 + M$ and $L_2 + M$ and a shunt branch $-M$.

If the assigned positive direction of I_2 in Figure 6.2.6 is made counterclockwise and this current is denoted by I_2', then I_2' flows into the unmarked terminal of coil 2. The sign of the mutual inductance term is now opposite that of the self-inductance term for each coil. The $Z_c I_2'$ term in the mutual impedance between the two meshes becomes positive because the voltage drop due to I_2' flowing in Z_c is also a voltage drop in mesh 1. Equation 6.2.5 and Equation 6.2.6 become:

$$(10 + j8 - j5)I_1 + (-j4 - j5)I_2' = 100 \tag{6.2.7}$$

and,

$$+(-j4 - j5)I_1 + (5 + j32 - j5)I_2' = 0 \tag{6.2.8}$$

Now $I_2' = -I_2$ and $V_O = -R_2 I_2' = R_2 I_2$ as before. The T-equivalent circuit is the same as in Figure 6.4.6b.

If the dot on coil 2 is reversed with the mesh currents still counterclockwise, the sign of the mutual inductance term is again the same as that of the self-inductance terms. Equation 6.2.7 and Equation 6.2.8 become:

$$(10 + j8 - j5)I_1 + (j4 - j5)I_2' = 100 \tag{6.2.9}$$

and,

$$(j4 - j5)I_1 + (5 + j32 - j5)I_2' = 0 \tag{6.2.10}$$

When L_1 and L_2 are rotated so as to bring them to the upright position, it is seen that dot marks are those of the transformer of Figure 6.2.4a. Using the corresponding T-equivalent circuit gives the same Equation 6.2.9 and Equation 6.2.10.

Under dc conditions, the flux does not vary with time and no voltage is induced in either coil due to current in the coil itself or in the other coil. The inductances behave as short circuits. Moreover, the capacitor acts as an open circuit. If $V_{SR°C} = 100$ V, it follows from voltage division that $V_O = \dfrac{5}{15} \times 100 = \dfrac{100}{3}$ V.

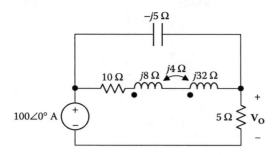

FIGURE 6.2.8
Figure for Exercise 6.2.1.

EXERCISE 6.2.1

Determine v_O if the capacitor in Example 6.2.1 is connected as shown in Figure 6.2.8.
Answer: $v_O = 73.5\cos(800t + 48.6°)$ V.

Application Window 6.2.1: Marking of Coil Terminals

Given two magnetically coupled coils, without any indication of the sense of the winding of the two coils, how does one determine experimentally which terminal on each coil should be marked with a dot?

SOLUTION
A low-voltage dc source, such as a battery, a switch, and a device that indicates voltage polarity, such as a moving-coil voltmeter or an oscilloscope, are connected to the coils as shown in Figure 6.2.9. One terminal of one coil, say a_1, which will be connected to the positive terminal of the battery when the switch is closed, is marked with a dot. When the switch is closed, the indicating device gives a momentary deflection as the flux linking the coils builds up to a steady value. If this deflection is such that a_2 goes positive with respect to b_2, then a_2 should be marked with a dot because it goes positive simultaneously with a_1. If a_2 momentarily goes negative, then b_2 should be marked with a dot.

When the switch is opened after having been closed for a sufficiently long time, the induced voltage in coil 1 will be such that it tends to keep the current flowing in accordance with Lenz's law. Hence, b_1 goes positive with respect to a_1. The indicating device will give a momentary deflection that is opposite to what it gave upon closure of the switch.

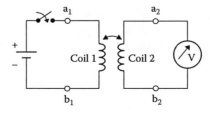

FIGURE 6.2.9
Figure for Application Window 6.2.1.

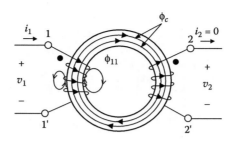

FIGURE 6.3.1
Transformer with open-circuited secondary.

6.3   The Ideal Transformer

To begin with, some fundamental cause-and-effect relations should be emphasized.

> **Concept:** *When a time-varying voltage v is applied to a coil, then neglecting the coil resistance, flux linkage λ is established in the coil in accordance with Faraday's law v = dλ/dt, irrespective of the parameters of the coil and of the characteristics of the medium in which the magnetic flux flows. On the other hand, the coil current is determined by the inductance of the coil, which in turn depends on the coil and on the characteristics of the medium in which the magnetic flux flows.*

In Figure 6.3.1, a voltage v_1 is impressed across coil 1 (the primary winding of a transformer), with coil 2 (the secondary winding) open-circuited. We may consider, for simplicity but without loss of generality, that v_1 is sinusoidal. Let the relative permeability of the core μ_r become infinite. Because inductance increases with permeability (Equation 1.9.8, Chapter 1), the coil inductance L_1 also becomes infinite, which means that $i_1 = 0$ because a coil of infinite inductance, and hence infinite impedance, draws no current. In fact, L_1 is the sum of two components (Equation 6.1.14), L_c due to the flux in the core ϕ_c, which equals ϕ_{21} in this case, and L_{leak} due to the leakage flux in air. It is L_c that becomes infinite, whereas L_{leak} remains finite. If $i_1 = 0$, the mmf acting on the leakage path is zero, so the leakage flux is zero, because flux = permeance × mmf. If $i_1 = 0$, the mmf acting on the core is also zero, but because the core is assumed to be of infinite permeability, and hence of infinite permeance, ϕ_c is finite, as required by Faraday's law. Moreover, because $i_1 = 0$, $p = v_1 i_1 = 0$, so no work is done in establishing ϕ_c, and hence no magnetic energy is stored in the core. With $i_1 = 0$, the voltage drop due to coil resistance or leakage flux is

zero, so that $v_1 = N_1 \dfrac{d\phi_c}{dt}$; ϕ_c also induces a voltage v_2 in coil 2 such that $v_2 = N_2 \dfrac{d\phi_c}{dt}$. Dividing these two equations gives:

$$\frac{v_1}{v_2} = \frac{N_1}{N_2} \qquad\qquad (6.3.1)$$

That $i_1 \to 0$ as $\mu_r \to \infty$ can also be concluded from the analysis of Example 1.9.1 of Chapter 1 for a core of finite permeability. According to Equation 1.9.6 (Chapter 1):

$$\phi_c = \mu_r \mu_0 \frac{A}{2\pi a} N_1 i_1 \tag{6.3.2}$$

When $\mu_r \to \infty$, then $i_1 \to 0$ if ϕ_c is to remain finite, as required by $v_1 = N_1 \dfrac{d\phi_c}{dt}$. It should

be noted that Equation 6.3.1 may be expressed as: $\dfrac{v_1}{N_1} = \dfrac{v_2}{N_2}$, which emphasizes that since

ϕ_c is common to both coils, the volts per turn are the same for both coils.

Next, let a resistor be connected across coil 2 so that i_2 flows in the direction indicated in Figure 6.3.1. If we assume for the moment that μ_r is finite, ϕ_c is related to i_1 and i_2 by

Ampere's circuital law, which now takes the form: $H_c = \dfrac{1}{2\pi a}(N_1 i_1 - N_2 i_2)$ because the mag-

netic field due to i_2 opposes that due to i_1. Multiplying the RHS of this equation by μ_r gives B_c, and multiplying by the cross-sectional area A gives ϕ_c:

$$\phi_c = \mu_r \mu_0 \frac{A}{2\pi a}(N_1 i_1 - N_2 i_2) \tag{6.3.3}$$

If we now let $\mu_r \to \infty$, then in order for ϕ_c to remain finite, $(N_1 i_1 - N_2 i_2) = 0$. The physical interpretation is that if the permeability of the core is very high, then only a negligibly small net mmf, $(N_1 i_1 - N_2 i_2)$, is associated with a finite ϕ_c. This is exactly analogous to a finite current flowing through a connection of very high conductance, the voltage across the conductance being negligible. If $(N_1 i_1 - N_2 i_2) = 0$, it follows that:

$$\frac{i_1}{i_2} = \frac{N_2}{N_1} \tag{6.3.4}$$

With i_1 and i_2 flowing, Equation 6.3.1 no longer strictly applies because there will now be a voltage drop due to coil resistance, and there will be an mmf acting on the leakage path to produce a leakage flux and hence a voltage drop. Thus, in order for both Equation 6.3.1 and Equation 6.3.4 to apply, we have to assume negligible resistance and perfect coupling between the coils and the core.

Equation 6.3.1 and Equation 6.3.4 define the v–i relations for an ideal transformer. They were derived for the dot markings and assigned positive directions of voltages and currents indicated in Figure 6.3.1. For other combinations, the sign of the RHS of these equations may be negative, in accordance with the interpretation of the dot convention.

Thus, if the assigned positive direction of v_1 and v_2 do not conform to the dots, then

$\dfrac{v_1}{v_2} = -\dfrac{N_1}{N_2}$. Similarly, if i_1 and i_2 both enter or leave at the dotted terminals, $N_1 i_1 + N_2 i_2 = 0$,

so that $\dfrac{i_1}{i_2} = -\dfrac{N_2}{N_1}$. The four possibilities are illustrated in Figure 6.3.2. Note that if the

assigned positive direction of the voltage across a given winding in Figure 6.3.2 is such that the dotted terminal is positive, the volts per turn for that winding can be considered positive; if not, the volts per turn for that winding is negative, as in the case of v_2 in Figure 6.3.2c and Figure 6.3.2d. By comparing $v_1 i_1$ and $v_2 i_2$, with due regard to the direction of power flow, it can be readily verified (Exercise 6.3.1) that the instantaneous power input equals the instantaneous power output, neglecting power losses in the core (Section 6.5).

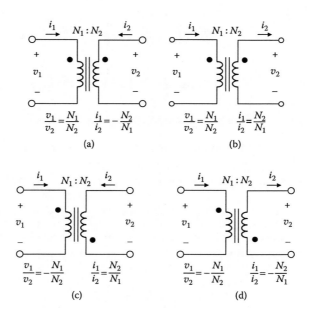

FIGURE 6.3.2
Signs of voltage and current ratios.

In summary, an ideal transformer can be defined as follows:

> **Definition:** *An ideal transformer is a two-port device that neither dissipates nor stores energy, and whose input–output v–i relations are given by:*

$$\frac{v_1}{v_2} = \pm\frac{N_1}{N_2} \text{ and } \frac{i_1}{i_2} = \pm\frac{N_2}{N_1}.$$

No power dissipation implies zero resistance and no core losses. No energy storage implies infinite core permeability. The voltage relation implies perfect coupling. Because the transformer does not itself generate power, it is a passive device. The symbol of an ideal transformer includes two parallel lines between the windings, as shown in Figure 6.3.2.

Finally, it should be noted that from Equation 1.9.8 (Chapter 1): $L_1 = \mu_r\mu_0 \dfrac{AN_1^2}{2\pi a}$ and $L_2 = \mu_r\mu_0 \dfrac{AN_2^2}{2\pi a}$, so that:

$$\frac{L_1}{L_2} = \left(\frac{N_1}{N_2}\right)^2 \tag{6.3.5}$$

independently of μ_r. Thus, although each of L_1 and L_2 becomes infinite as μ_r becomes infinite, their ratio remains finite and equal to the square of the turns ratio. It may be noted that Equation 6.3.5 also applies when L_1 and L_2 are finite provided the leakage flux is zero, or the permeances of the two leakage paths are equal (Section ST6.1).

EXERCISE 6.3.1
Verify that in all the cases shown in Figure 6.3.2, the instantaneous power absorbed on the primary side of the transformer is equal to the instantaneous power delivered by the secondary side.

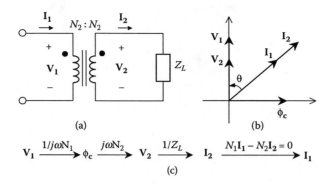

FIGURE 6.3.3
Phasor relations of ideal transformer.

Phasor Relations

Figure 6.3.3a shows an ideal transformer connected to a load impedance Z_L, with the corresponding phasor diagram depicted in Figure 6.3.3b. Figure 6.3.3c is a flow diagram of the causal relationships. A voltage $\mathbf{V_1}$ applied to the primary winding establishes a flux

ϕ_c in the core such that $\phi_c = \dfrac{1}{j\omega N_1}\mathbf{V_1}$ and lags $\mathbf{V_1}$ by 90°. ϕ_c induces a voltage $\mathbf{V_2}$ in the

secondary winding such that $\mathbf{V_2} = +j\omega N_2\phi_c$, assuming the assigned positive directions and dots shown. Because $\mathbf{V_2}$ leads ϕ_c by 90°, it is in phase with $\mathbf{V_1}$. $\mathbf{V_2}$ causes a current $\mathbf{I_2}$ in the secondary circuit, which lags $\mathbf{V_2}$ by an angle θ, assuming Z_L is inductive. To have zero mmf in the core, a current $\mathbf{I_1}$ flows in the primary winding such that $N_1\mathbf{I_1} - N_2\mathbf{I_2} = 0$.

Reflected Impedance

The impedance Z_{Lp} of Z_L *reflected* to the primary side can be determined from the **V–I**

ratios. For the dots and assigned positive directions of Figure 6.3.3a, $\dfrac{\mathbf{V_1}}{\mathbf{V_2}} = \dfrac{N_1}{N_2}$ and

$\dfrac{\mathbf{I_1}}{\mathbf{I_2}} = \dfrac{N_2}{N_1}$. Dividing these two equations, $\dfrac{\mathbf{V_1}}{\mathbf{I_1}}\dfrac{\mathbf{I_2}}{\mathbf{V_2}} = \left(\dfrac{N_1}{N_2}\right)^2$. Substituting $\dfrac{\mathbf{V_2}}{\mathbf{I_2}} = Z_L$:

$$Z_{Lp} = \frac{\mathbf{V_1}}{\mathbf{I_1}} = Z_L\left(\frac{N_1}{N_2}\right)^2 \tag{6.3.6}$$

Note that: (1) because of squaring of the turns ratio, Equation 6.3.6 is valid for any of the configurations of Figure 6.3.2a to Figure 6.3.2d (Exercise 6.3.2). (2) An open circuit is reflected as an open circuit, and a short circuit is reflected as a short circuit.

EXERCISE 6.3.2

Verify that Equation 6.3.6 is valid for any of the configurations of Figure 6.3.2a to Figure 6.3.2d.

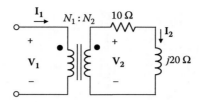

FIGURE 6.3.4
Figure for Example 6.3.1.

General Transformer Equation

It is of interest to derive this well-known equation for the sinusoidal steady state. From the relation $v_1 = N_1 \dfrac{d\phi_c}{dt}$ in phasor notation, $|\mathbf{V}_1| = \omega N_1 |\phi_c|$. If ϕ_m denotes the *peak* value of ϕ_c, and $V_{1\text{rms}}$, denotes the rms value of $\mathbf{V}_1$, then $V_{1\text{rms}} = \dfrac{2\pi f}{\sqrt{2}} N_1 \phi_m$. Substituting numerical values:

$$V_{1\text{rms}} = 4.44 f N_1 \phi_m \tag{6.3.7}$$

Example 6.3.1: Ideal Transformer Circuit

Assume that in Figure 6.3.4 $v_1 = 120\cos1000t$ V, $N_1{:}N_2 = 1{:}2$, and that Z_L consists of a resistance of 10 Ω in series with a 20 mH inductor. It is required to determine the primary current and the secondary current and voltage.

SOLUTION

$\mathbf{V}_2 = \dfrac{N_2}{N_1}\mathbf{V}_1 = 240\angle 0°$ V; hence $\mathbf{I}_2 = \dfrac{240\angle 0°}{10 + j20} = \dfrac{24}{\sqrt{5}}\angle{-63.4°}$ A. It follows that $\mathbf{I}_1 = \dfrac{N_2}{N_1}\mathbf{I}_2$

$= \dfrac{48}{\sqrt{5}}\angle{-63.4°}$ A.

The Ideal Autotransformer

This is a special type of an ideal transformer in which the primary and secondary windings are connected together in a particular way. Figure 6.3.5a shows a two-winding transformer having N_1 and N_2 turns on the primary and secondary windings, respectively, the load current being i_2. Assuming $N_2 > N_1$, the input voltage v_1 is stepped up to an output voltage v_2. An autotransformer that accomplishes the same purpose is shown in Figure 6.3.5b. In the autotransformer, the input voltage is part of the output voltage, so that the input voltage need only be boosted by an additional voltage $v_2' = v_2 - v_1$ to produce v_2. Hence, the number of turns of the output winding is $N_2' < N_2$. Moreover, the current in the common winding is $i_1 - i_2$. This winding therefore carries a reduced current and can have a conductor of smaller cross-sectional area. Both of these considerations make the autotransformer smaller and lighter. In addition, terminal a in Figure 6.3.5b could be connected to a slider over a bare part of the winding, so that a variable turns ratio is obtained. The principal disadvantage of the autotransformer is that the input and output sides are not electrically isolated because of the conduction path between them.

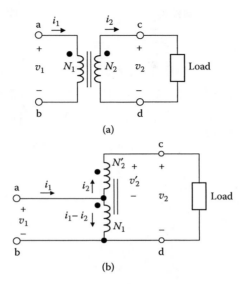

FIGURE 6.3.5
Ideal autotransformer.

The voltage and current relations for the ideal autotransformer follow directly from those of the two-winding ideal transformer by substituting $N_2 = N_1 + N'_2$. Thus:

$$\frac{v_2}{v_1} = \frac{N_1 + N'_2}{N_1} = 1 + \frac{N'_2}{N_1} \tag{6.3.8}$$

and

$$\frac{i_1}{i_2} = \frac{N_1 + N'_2}{N_1} = 1 + \frac{N'_2}{N_1} \tag{6.3.9}$$

If the dot on either winding is reversed, the effective turns ratio becomes $\left(1 - \dfrac{N'_2}{N_1}\right)$ (Exercise 6.3.3).

EXERCISE 6.3.3
Show that if the dot marking on either winding of an autotransformer is reversed, the effective turns ratio becomes $\left(1 - \dfrac{N'_2}{N_1}\right)$. Deduce that this relation, as well as Equation 6.3.8, may be derived from the fact that the voltage per turn is the same for all windings in a given transformer, taking into account the polarity of this voltage with respect to the dots.

EXERCISE 6.3.4
Show that in Figure 6.3.6 $\mathbf{I_{SRC}} = \dfrac{1}{Z_{src} + b^2 Z_L} \mathbf{V_{SRC}}$ and $\mathbf{V_L} = -\dfrac{b Z_L}{Z_{src} + b^2 Z_L} \mathbf{V_{SRC}}$, where $b = \dfrac{N'_1}{N_2} - 1$. How do you interpret $\mathbf{I_{SRC}}$ and $\mathbf{V_L}$ when $b = 0$?

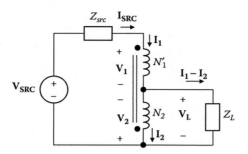

FIGURE 6.3.6
Figure for Exercise 6.3.4.

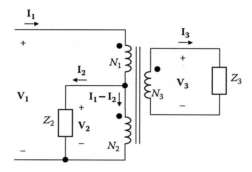

FIGURE 6.3.7
Figure for Example 6.3.2.

Example 6.3.2: Three-Winding Transformer

Given a three-winding ideal transformer with loads Z_2 and Z_3 connected as shown in Figure 6.3.7, it is required to determine the input impedance.

SOLUTION

Method 1: Because the voltage induced per turn is the same for all windings, $\dfrac{V_2}{V_1} = \dfrac{N_2}{N_1 + N_2}$

and $\dfrac{V_3}{V_1} = \dfrac{N_3}{N_1 + N_2}$, where the assigned positive directions of V_1, V_2, and V_3 are all in

accordance with the dots so that the volts per turn are positive for all the windings. Because

the net mmf in the core must be zero, $N_1 I_1 + N_2(I_1 - I_2) - N_3 I_3 = 0$, where $I_2 = \dfrac{1}{Z_2} V_2$ and

$I_3 = \dfrac{1}{Z_3} V_3$. Note that I_1 and $(I_1 - I_2)$ enter at the dotted terminals of their respective

windings. Their mmfs are in the same sense and may be assigned a positive sign. I_3 leaves
its winding at the dotted terminals, so its mmf is assigned a negative sign.
    Eliminating I_2, I_3, V_2, and V_3 from these equations gives:

$$\frac{V_1}{I_1} = \frac{(N_1 + N_2)^2}{N_2^2 / Z_2 + N_3^2 / Z_3} \tag{6.3.10}$$

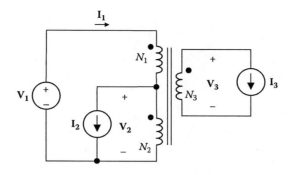

FIGURE 6.3.8
Figure for Example 6.3.2.

If $Z_2 \to \infty$, the impedance reflected to the primary side is $\dfrac{(N_1 + N_2)^2}{N_3^2} Z_3$, as for a two-winding transformer of turns ratio $\dfrac{N_1 + N_2}{N_3}$. If $Z_3 \to \infty$, the reflected impedance is

$\dfrac{(N_1 + N_2)^2}{N_2^2} Z_2$, as for an autotransformer of turns N_1 and N_2. In fact, these reflected imped-

ances in parallel give the reflected impedance of Equation 6.3.10.

Method 2: Let the impedances be replaced by current sources $\mathbf{I}_2$ and $\mathbf{I}_3$ in accordance

with the substitution theorem, as shown in Figure 6.3.8, where $\mathbf{I}_2 = \dfrac{1}{Z_2} \mathbf{V}_2$ and $\mathbf{I}_3 = \dfrac{1}{Z_3} \mathbf{V}_3$.

We now apply superposition. If $\mathbf{I}_3$ is applied alone, the primary current is $\dfrac{N_3}{N_1 + N_2} \mathbf{I}_3$. If

$\mathbf{I}_2$ is applied alone, the primary current is $\dfrac{N_2}{N_1 + N_2} \mathbf{I}_2$. With both sources applied, the

primary current is $\mathbf{I}_1 = \dfrac{N_2}{N_1 + N_2} \mathbf{I}_2 + \dfrac{N_3}{N_1 + N_2} \mathbf{I}_3$. Substituting for the currents in terms of

voltages and impedances, $\mathbf{I}_1 = \dfrac{N_2}{N_1 + N_2} \dfrac{1}{Z_2} \mathbf{V}_2 + \dfrac{N_3}{N_1 + N_2} \dfrac{1}{Z_3} \mathbf{V}_3$. But $\mathbf{V}_3 = \dfrac{N_3}{N_1 + N_2} \mathbf{V}_1$ and

$\mathbf{V}_2 = \dfrac{N_2}{N_1 + N_2} \mathbf{V}_1$. Substituting for $\mathbf{V}_2$ and $\mathbf{V}_3$ gives, $\mathbf{I}_1 = \dfrac{1}{(N_1 + N_2)^2} \left(\dfrac{N_2^2}{Z_2} + \dfrac{N_3^2}{Z_3} \right) \mathbf{V}_1$. The

reflected impedance $\mathbf{V}_1/\mathbf{I}_1$ is the same as that obtained in Equation 6.3.10.

Note that superposition strictly applies to voltage sources and current sources only. It cannot be applied directly to impedances, such as Z_2 and Z_3 in this example. However, the substitution theorem allowed us to apply superposition after replacing these imped-ances with sources.

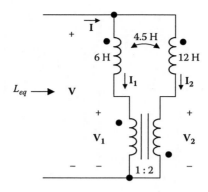

FIGURE 6.3.9
Figure for Example 6.3.3.

EXERCISE 6.3.5
Show that if the dot on any of the windings in Figure 6.3.7 is reversed, the sign of the corresponding N is reversed.

Simulation Example 6.3.3: Linear and ideal Transformers in Series

It is required to determine L_{eq} in Figure 6.3.9, assuming $\omega = 1\,\text{rad/s}$, and to simulate the circuit.

SOLUTION
We can assume that a current $\mathbf{I}$ flows, and determine $\mathbf{V}$. According to the dots of the ideal transformer, $\mathbf{I}_1 = 2\mathbf{I}_2$ and $\mathbf{V}_2 = -2\mathbf{V}_1$.
   Applying KVL to the primary side of the two transformers:

$$j6(2\mathbf{I}_2) + j4.5(\mathbf{I}_2) + \mathbf{V}_1 = \mathbf{V}$$

or
$$j16.5\mathbf{I}_2 + \mathbf{V}_1 = \mathbf{V} \qquad (6.3.11)$$

   Applying KVL to the secondary side of the two transformers:

$$j12(\mathbf{I}_2) + j4.5(2\mathbf{I}_2) - 2\mathbf{V}_1 = \mathbf{V}$$

or
$$j21\mathbf{I}_2 - 2\mathbf{V}_1 = \mathbf{V} \qquad (6.3.12)$$

   Eliminating $\mathbf{V}_1$ between Equation 6.3.11 and Equation 6.3.12 gives $\mathbf{I}_2 = -j\,\mathbf{V}/18$. Hence,

$$L_{eq} = \frac{1}{3j}\frac{\mathbf{V}}{\mathbf{I}_2} = 6\,\text{H}.$$ We also obtain $\mathbf{V}_1 = \mathbf{V}/12$.

SIMULATION
The schematic is shown in Figure 6.3.10. Coupled coils are modeled in PSpice by the part XFRM_Linear in the Analog library. Using the Property Editor spreadsheet, enter 6H under

L1_VALUE, 12H under L2_VALUE, and $k = \dfrac{4.5}{\sqrt{6 \times 12}} = 0.53033$ under COUPLING. As

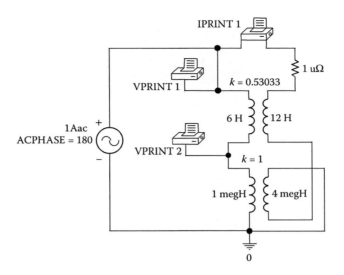

FIGURE 6.3.10
Figure for Example 6.3.3.

entered, the dots are assumed to be at the upper terminals of both coils. To model the ideal transformer, we will use two coupled coils having $k = 1$ and values of L1 and L2 whose impedances are much larger than those in the rest of the circuit, but with $\sqrt{\dfrac{L_1}{L_2}} = \dfrac{N_1}{N_2}$.

For the ideal transformer, enter 1MH under L1_VALUE, 4MH under L2_VALUE, and $k = 1$ under COUPLING. The connections to the secondary of the ideal transformer are reversed so as to account for the reversed dots on this side. Alternatively, k may be set equal to –1. A 1 μΩ resistance that is too small to affect the results is inserted in series with the secondaries because PSpice does not allow loops consisting of ideal inductors without any resistance.

A current source IAC of 1 A default value is used so that L_{eq} may be determined directly from the voltage across the source. To have the same direction as **I** in Figure 6.3.9, either the source symbol has to be rotated through 180° or 180 entered under ACPHASE in the Property Editor spreadsheet of the source. Two voltage printers measure **V** and $\mathbf{V_1}$ and a current printer measures $\mathbf{I_2}$.

In the simulation settings, choose AC Analysis/Noise Analysis type and select a Start Frequency of $f = \dfrac{1}{2\pi} = 0.159155$, the same End Frequency, and a Total Points of 1. After the simulation is run, the part of the output file concerned with AC analysis gives $\mathbf{I_2} = 1/3$ A, $\mathbf{V} = j6$ V, and $\mathbf{V_1} = j0.5$ V, in accordance with the analysis results.

Application Window 6.3.1: Linear Variable Differential Transformer

The linear variable differential transformer (LVDT) is a useful device for measuring displacement. It consists of a transformer having two secondary windings on either side of the primary winding (Figure 6.3.11a) and connected in series opposition, as indicated by the dot marks in Figure 6.3.11b. The transformer has a movable magnetic core that moves with the linear displacement to be measured.

The voltage $\mathbf{V_P}$ applied to the primary winding establishes flux in the core. If the core is centrally located between the two secondary windings, the induced voltages $\mathbf{V_{21}}$ and

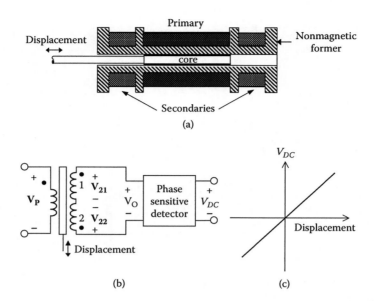

FIGURE 6.3.11
Figure for Application Window 6.3.1.

V_{22} are equal and opposite, and the voltage V_O is zero. If the core is moved to one side or the other of the central location, more flux will link the coil toward which the core has been moved and less flux will link the other coil because the core increases the permeance of the magnetic path of the coil toward which it moves. The voltages induced in the two coils are no longer equal, and a voltage $V_O = V_{21} - V_{22}$ is produced. With appropriate design, the magnitude of V_O is proportional to the displacement from the central location. Assuming that V_{21} and V_{22} have the same phase angle, say θ, the phase angle of V_O is θ or $180° + \theta$, depending on whether $V_{21} > V_{22}$, or $V_{21} < V_{22}$, respectively, as in an ac bridge (Example 5.7.2, Chapter 5). A phase-sensitive detector produces a dc output that is zero when the core is centrally located between the two secondary windings and that varies linearly, positively or negatively, with displacement on either side of the central location (Figure 6.3.11c).

6.4   Reflection of Circuits

Concept: *Circuits involving ideal transformers can be conveniently analyzed by reflecting the circuit on the primary side to the secondary side, or conversely.*

Consider Figure 6.4.1. KCL for node q is:

$$I_2 + I_L - \frac{1}{Z_L}(V_2 - V_y) = 0 \tag{6.4.1}$$

Substituting $V_2 = aV_1$ and $I_2 = \dfrac{1}{a} I_1$, Equation 6.4.1 may be expressed as:

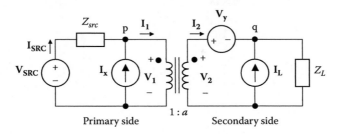

FIGURE 6.4.1
Circuit illustrating reflection.

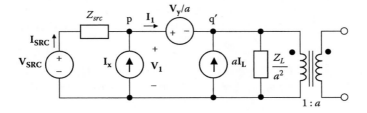

FIGURE 6.4.2
Secondary circuit reflected to the primary side.

$$\mathbf{I}_1 + a\mathbf{I}_\mathrm{L} - \frac{1}{(Z_L / a^2)}(\mathbf{V}_1 - \mathbf{V}_y') = 0 \tag{6.4.2}$$

where $\mathbf{V}_y' = \mathbf{V}_y/a$. This is KCL for a node q' in Figure 6.4.2. The voltage $\mathbf{V}_1$ at node p and the current $\mathbf{I}_1$ leaving this node are unaltered. The ideal transformer now appears at the extreme right with its secondary open-circuited. Because it is not serving any useful function, it can be omitted from the circuit. In Figure 6.4.2, the secondary circuit is reflected to the primary side, element by element. If $a > 1$, voltages are *stepped down* by a and currents are *stepped up* by a. Impedances are stepped down, like voltages, but by a factor a^2.

In a similar manner, we may write KCL for node p in Figure 6.4.1 as:

$$-\mathbf{I}_1 + \mathbf{I}_x + \frac{1}{Z_{src}}(\mathbf{V}_{SRC} - \mathbf{V}_1) = 0 \tag{6.4.3}$$

Substituting $\mathbf{V}_1 = \dfrac{1}{a}\,\mathbf{V}_2$ and $\mathbf{I}_1 = a\mathbf{I}_2$, Equation 6.4.3 may be expressed as:

$$-\mathbf{I}_2 + \frac{1}{a}\mathbf{I}_x + \frac{1}{a^2 Z_{src}}(\mathbf{V}_{SRC}' - \mathbf{V}_2) = 0 \tag{6.4.4}$$

where $\mathbf{V}_{SRC}' = a\mathbf{V}_{SRC}$. This is KCL for a node p' on the secondary side in Figure 6.4.3. The voltage and currents at node q are unaltered. The ideal transformer now appears at the extreme left of the figure with its primary open-circuited. Because it is not serving any useful function, it can be omitted from the circuit.

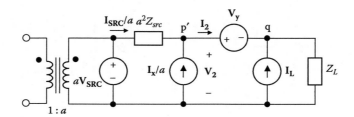

FIGURE 6.4.3
Primary circuit reflected to the secondary side.

In Figure 6.4.3, the primary circuit is reflected to the secondary side, element by element. Voltages are stepped up by a and currents are stepped down by a. Impedances are stepped up, like voltages, but by a factor a^2. The following should be noted concerning reflection of circuits:

1. In reflecting circuits from one side to the other, the order of the elements, left to right or right to left, must be preserved. Otherwise, the circuit is altered.

2. Admittance is reflected as the reciprocal of impedance.

3 Controlling currents or voltages of dependent sources are reflected like any other voltage or current.

4. When the dots are reversed, a negative value of a is used.

Some of these points are illustrated by Example SE6.4.

EXERCISE 6.4.1
Show that if in Figure 6.4.1 the connections of one of the windings are reversed, which reverses the dot on one of the coils, the effect on the circuits of Figure 6.4.2 and Figure 6.4.3 is the same as reversing the sign of a.

6.5   Transformer Imperfections

Practical transformers depart from the ideal in the following respects: (1) finite inductance of the windings; (2) finite leakage flux; (3) power losses because of finite resistance of the windings and core losses; and (4) capacitances between primary and secondary windings as well as between turns and layers of the same winding. The capacitances arise because of voltage differences and consequent energy storage in the electric field. In this section, these imperfections are accounted for by appropriate circuit representations.

The finite resistances of the windings are accounted for, at least at low frequencies, by adding them at the terminals of the respective winding, as was done with R_1 and R_2 in Figure 6.2.2. The distributed capacitance between windings and between turns and layers of the same winding is accounted for, in an approximate manner, by lumped capacitances connected across the terminals of the primary and secondary windings, and between these windings. Core losses will be considered later in this section. Ignoring the aforementioned imperfections, we are left with a linear transformer consisting of two coupled, lossless coils of inductances L_1 and L_2. The transformer is shown represented in the frequency

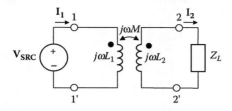

FIGURE 6.5.1
Idealized linear transformer in the frequency domain.

domain in Figure 6.5.1, connected to a source $\mathbf{V}_{SRC}$ on the primary side and to a load Z_L on the secondary side. The governing equations are Equation 6.2.3 and Equation 6.2.4, with $R_1 = 0 = R_2$ and with R_L replaced by Z_L:

$$j\omega L_1 \mathbf{I}_1 - j\omega M \mathbf{I}_2 = \mathbf{V}_{SRC} \qquad (6.5.1)$$

and,

$$-j\omega M \mathbf{I}_1 + (Z_L + j\omega L_2)\mathbf{I}_2 = 0 \qquad (6.5.2)$$

Any circuit that represents the effect of finite inductances of windings and finite leakage fluxes must also be described by these equations.

Finite Inductance of Windings

We will consider the coils to be perfectly coupled to begin with, that is, $M^2 = L_1 L_2$. Substituting for $\mathbf{I}_2$ from Equation 6.5.2 in Equation 6.5.1:

$$\frac{\mathbf{V}_{SRC}}{\mathbf{I}_1} = \frac{j\omega L_1 Z_L + \omega^2 (M^2 - L_1 L_2)}{Z_L + j\omega L_2} = \frac{L_1}{L_2} \frac{j\omega L_2 Z_L}{Z_L + j\omega L_2} \qquad (6.5.3)$$

With $M^2 = L_1 L_2$, it follows that the impedance seen by the source is that of $j\omega L_2$ in parallel with Z_L, reflected to the primary side of an ideal transformer having an inductance ratio of $L_1{:}L_2$, or a turns ratio $\sqrt{L_1} : \sqrt{L_2}$ (Equation 6.3.5 and Figure 6.5.2a). The impedance $j\omega L_2$ reflected to the primary side becomes $j\omega L_1$ (Figure 6.5.2b). It is seen that the effect of finite inductances of an otherwise ideal transformer is to introduce a shunt impedance $j\omega L_1$ on the primary side, or a shunt impedance $j\omega L_2$ on the secondary side.

Finite Leakage Flux

The next step is to modify the circuit of Figure 6.5.2b so as to include the effect of leakage. Because equivalence applies for any Z_L, we assume that $Z_L = 0$. Substituting $(M^2 - L_1 L_2) = L_1 L_2 (k^2 - 1)$ in Equation 6.5.3:

$$\frac{\mathbf{V}_{SRC}}{\mathbf{I}_1} = j\omega L_1 (1 - k^2) \qquad (6.5.4)$$

When terminals 22′ in Figure 6.5.2b are short-circuited by having $Z_L = 0$, $j\omega L_1$ is short-circuited. To satisfy Equation 6.5.4, a series impedance $j\omega L_1(1 - k^2)$ must be inserted in

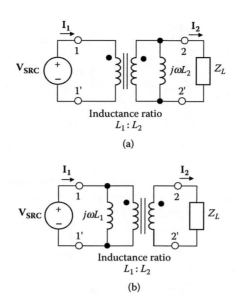

FIGURE 6.5.2
Finite inductance of windings.

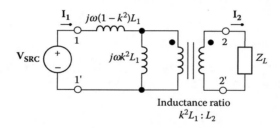

FIGURE 6.5.3
Finite inductance of windings and leakage flux.

series at terminal 1, as shown in Figure 6.5.3. However, when terminals 22′ are open-circuited, $I_2 = 0$, and Equation 6.5.1 gives $\dfrac{V_{SRC}}{I_1} = j\omega L_1$. To satisfy this condition, the shunt impedance must be $j\omega L_1 k^2$ instead of $j\omega L_1$, as shown. Moreover, L_2 in Figure 6.5.2a must now be reflected as $k^2 L_1$ in order to satisfy this same condition, which implies that the inductance ratio is $k^2 L_1 : L_2$. That the circuit of Figure 6.5.3, in fact, satisfies Equation 6.5.1 and Equation 6.5.2 is left as Exercise 6.5.1. The transformer equivalent circuit is developed in more detail in Section ST6.2 and several alternative versions derived.

The circuit of Figure 6.5.3 has important practical implications:

> **Concept:** *The performance of a transformer is limited at low frequencies by the reactance of the windings and at high frequencies by the leakage reactance.*

At low frequencies, the leakage impedance $j\omega L_1(1 - k^2)$ is negligible, but the impedance $j\omega L_2$ appears in parallel with Z_L. If the transformer is not to affect adversely the behavior of the circuit, $j\omega L_2 \gg Z_L$. At high frequencies, the shunting effect of $j\omega L_2$ can be neglected.

The leakage impedance $j\omega L_1(1-k^2)$ referred to the secondary side becomes $j\omega L_2\left(\dfrac{1}{k^2}-1\right)$ and appears in series with the load. If the transformer is not to affect adversely the behavior of the circuit, this impedance should be small compared with Z_L.

EXERCISE 6.5.1
Verify that Equation 6.5.1 and Equation 6.5.2 apply to the circuit of Figure 6.5.3.

Core Losses

Core losses are of two types: eddy-current and hysteresis. Eddy-current losses occur in any core made of electrically conducting material. Suppose that the secondary in Figure 6.3.1 consists of a single, closed turn of wire of resistance R. A voltage of rms value $j\omega\phi_{crms}$ is induced in this turn and causes a current $\dfrac{j\omega}{R}\phi_{crms}$, assuming the inductance of the turn is negligible with respect to its resistance R. The RI^2 power loss is $\omega^2\phi_{crms}^2/R$. But a similar situation occurs throughout the body of the core because any closed path inside the core may be considered to act like the closed turn of wire, as illustrated in Figure 6.5.4a. The resulting *eddy currents* circulate in the core. Apart from the power loss and resultant heating, the flux of these eddy currents ($\phi_e(t)$) in Figure 6.5.4a opposes, and hence decreases, the flux in the core. To reduce these currents, either a magnetic material of high resistivity is used, or the core is made of thin laminations that are insulated from one another and stacked together so that the flux is in a direction parallel to the plane of the laminations (Figure 6.5.4b). The induced currents are confined by the insulation to within each lamination. This effectively reduces the cross-sectional area of the loop that encloses the flux, and which can give rise to eddy currents. Laminations are usually made of iron alloys such as silicon steel, in which the added silicon increases the resistivity.

Iron, steel, nickel, cobalt, and their alloys are **ferromagnetic** materials, characterized by high permeability and nonlinearity that includes **hysteresis**. In general, hysteresis arises

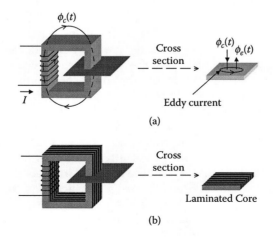

FIGURE 6.5.4
Eddy-current loss.

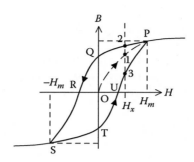

FIGURE 6.5.5
Hysteresis loop.

when an effect lags behind its cause. As a result, the state of a system depends on its previous history, that is, the manner in which this state was reached. In the case of ferromagnetic materials, if H in an unmagnetized specimen is increased from zero, B increases along the curve OP in Figure 6.5.5. The flattening of the curve at high values of H is described as **magnetic saturation**. If at point P, H is reduced, B lags behind H. When H is reduced to zero at Q, B retains the value of the positive intercept on the B-axis. To reduce B to zero at R requires a negative H. Making H more negative still brings the operating point to S. If H is now increased back to H_m, B changes along the lower part of the curve, STUP. The loop that is traced by a cyclic variation in H is a hysteresis loop. The area enclosed by the hysteresis loop represents power loss that appears as heat in the core (Section ST6.3). It is seen that at any particular value H_x, for example, B can take on different values corresponding to points 1, 2, or 3, or intermediate values, depending on how H_x is reached.

Both eddy-current loss and hysteresis loss are a function of the magnitude of ϕ_c. The voltage across the branch $j\omega k^2 L_1$ in Figure 6.5.3 can be considered proportional to ϕ_c. Core losses are sometimes accounted for by adding a resistance R_c in parallel with this branch.

Application Window 6.5.1: Construction of Small Inductors and Transformers

Small inductors and transformers at power and audio frequencies have laminated cores that are usually rectangular (Figure 6.5.6a) or of the shell type (Figure 6.5.6b). The windings are wound on formers made from insulating material and having a rectangular or preferably square cross–section. The formers are fitted around one or both sides of the rectangular core or around the central limb of the shell-type core. The shell-type core is compact, rugged, and naturally conforms to the two-sided distribution of the magnetic field of the windings.

Both primary and secondary windings are wound, using insulated wire, usually one on top of the other, with the lower voltage winding generally closer to the core, to reduce the stress on the insulation. Having the primary and secondary windings on separate limbs markedly reduces the coupling between them but reduces the capacitance between the two windings and practically eliminates the possibility of a direct short circuit between the two windings in the event of insulation breakdown. This type of construction is seldom used except for special applications, such as safety isolation transformers, where it is essential to have the primary and secondary sides isolated even in the event of insulation breakdown. Alternatively, isolation could be obtained by having a grounded metallic screen between primary and secondary windings. The screen also prevents coupling between primary and secondary windings through the interwinding capacitance.

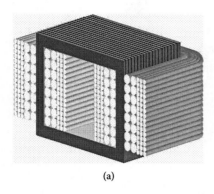
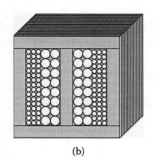

(a)   (b)

FIGURE 6.5.6
Cores of small transformers.

The cores of high frequency and pulse transformers are often made of high-resistivity material to reduce eddy currents. Examples of high-resistivity magnetic materials are the **ferrites**, which are composed of metallic oxides. They have high magnetic permeability and can be made into cores of various shapes, including toroidal. Ferrites made of manganese and zinc oxides or nickel and zinc oxides are linear, at least under normal operating conditions, whereas those containing iron oxide are generally nonlinear.

Applications of Transformers

Transformers are widely used for a variety of purposes. They are used for stepping supply voltages up or down, for measurement of ac currents and voltages, for impedance matching, for generating polyphase supplies, for coupling in tuned amplifiers, for oscillators, and in a host of electronic circuits. A combination transformer/inductor, or ballast, is commonly used with fluorescent-type lamps (Application Window 15.6.1). An inductor having an air gap is used in rectifier circuits supplying a relatively large dc current.

Summary of Main Concepts and Results

- The mutual inductance of two magnetically coupled coils is the flux linkage in one coil per unit current in the other coil. It is independent of which coil carries the current.

- The coupling coefficient k of two magnetically coupled coils is defined as $\dfrac{M}{\sqrt{L_1 L_2}}$

  and is a measure of how tightly the two coils are coupled. It assumes values in the range of 0 to unity, where $k = 0$ denotes no coupling and $k = 1$ denotes perfect coupling.

- A transformer consists of two or more coils that are magnetically coupled relatively tightly. In a linear transformer, permeability is constant, so that B and H, or ϕ and i, are linearly related.

- According to the dot convention, one terminal of each coil is marked with a dot so that currents entering (or leaving) the marked terminals in each coil are associated with flux in the same direction in both coils. An alternative interpretation of the dots, which follows from the preceding statement, is that the polarities of induced voltages in both coils are the same, relative to the dots.

- If the assigned positive directions of currents are such that these currents both flow in or both flow out at the dotted terminals, the sign of the mutual inductance term for either coil in the KVL equation is the same as that of the self-inductance term for that coil. Otherwise, the sign of the mutual inductance term for either coil is opposite that of the self-inductance term for that coil.

- When a time-varying voltage v is applied to a coil, then neglecting the coil resistance, flux linkage λ is established in the coil in accordance with Faraday's law $v = d\lambda/dt$, irrespective of the parameters of the coil and of characteristics of the medium in which the magnetic flux flows. The coil current is determined by the inductance of the coil, which in turn depends on the parameters of coil and on the characteristics of the medium in which the magnetic flux flows.

- An ideal transformer is a two-port device that neither dissipates nor stores energy and whose input–output v–i relations are given by: $\dfrac{v_1}{v_2} = \pm\dfrac{N_1}{N_2}$ and $\dfrac{i_1}{i_2} = \pm\dfrac{N_2}{N_1}$.

- In a multiwinding ideal transformer: (1) the magnitude of the voltage induced per turn is the same for all windings, and (2) the net mmf in the core must be zero, that is, $\displaystyle\sum_{r=1}^{m} N_r i_r = 0$ over all the m windings, taking into account the relative senses of the individual volts per turn and mmfs.

- An autotransformer is a specially connected transformer in which the voltage on one side is added algebraically to that of a winding connected between the input and output. For the same input and output currents and voltages, it is more economical than an equivalent two-winding transformer but does not provide isolation between the circuits on either side. It can also be made to provide a variable turns ratio.

- Circuit elements may be reflected from one side of a transformer to the other, element by element. In doing so, voltages are multiplied by a ratio equal to the number of turns on the winding to whose side the reflection is made divided by the number of turns on the winding from whose side the reflection is made. Currents are divided by this ratio, and impedances are multiplied by the square of this ratio.

- The performance of a transformer is limited at low frequencies by the reactance of the windings and at high frequencies by the leakage reactance.

- There are two types of core losses in a transformer: (1) eddy current losses due to currents induced in a conducting core, and (2) hysteresis losses because B lags H.

Learning Outcomes

- Analyze circuits that include linear transformers or ideal transformers.

Supplementary Topics and Examples on CD

ST6.1 Analogy between electric circuits and magnetic circuits: Discusses a useful analogy between electric and magnetic circuits, and elaborates on the relation between permeance and inductance.

ST6.2 Transformer equivalent circuits: Derives several versions of the equivalent circuit of a transformer.

ST6.3 Power loss due to hysteresis: Shows that the power loss due to hysteresis is proportional to the area enclosed by the hysteresis loop.

SE6.1 Input admittance using substitution theorem: Applies the substitution theorem and source rearrangement to convert an admittance connected between the primary and secondary windings of an ideal transformer to admittances connected across these windings.

SE6.2 Series and parallel connection of magnetically coupled coils: Derives and interprets the equivalent inductance of two magnetically coupled coils connected in series or in parallel.

SE6.3 Series–parallel connection of transformer windings: Analyzes a circuit that includes two identical ideal transformers having their primary windings connected in series and secondary windings connected in parallel.

SE6.4 Reflection of circuits with dependent source and reversed dot markings: Illustrates how a circuit is reflected from one side of a transformer to the other under these conditions.

Problems and Exercises

P6.1   Mutual Inductance and Dot Convention

P6.1.1   The terminal of one coil in Figure P6.1.1 is marked with a dot. Mark one terminal of the other coils with a dot and connect the coils in series for maximum total inductance.

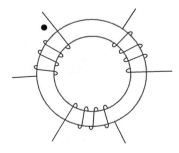

FIGURE P6.1.1

P6.1.2   Two coils are coupled through a high-permeability core. When $i_1 = 4$ A, $\phi_{11e} = 0.1$ mWb and $\phi_{21} = 0.4$ mWb. When $i_2 = 3$ A, $\phi_{12} = 0.6$ mWb. If $N_2 = 1000$ turns and $L_2 = 400$ mH, find N_1, L_1, M, and ϕ_{22e}. What is the total energy stored in the magnetic circuit? What is the dot convention with respect to i_1 and i_2 implied by the signs of the fluxes?

P6.1.3 Two coils having $N_1 = 800$ turns and $N_2 = 500$ turns are coupled through a high-permeability core. A current i_1 in coil 1 produces $\phi_{11e} = 500$ µWb and $\phi_{21} = 400$ µWb, whereas a current $2i_1$ in coil 2 produces $\phi_{22e} = 1400$ µWb. (a) What ϕ_{12} is produced by $2i_1$ in coil 2? (b) What is the coefficient of coupling? (c) If the permeance of the core is 50 nWb/A-turn, what is the mutual inductance? (d) What is the inductance of each coil?

P6.1.4 Given two magnetically coupled coils, if the current in one coil is $10\sin(1000t)$ mA, the voltage induced in the other coil is $32\cos(1000t)$ V. When the two coils are connected in series, the largest measured inductance is 16.4 H. If the inductance of one coil is 3.6 H, determine the inductance of the other coil and the coefficient of coupling.

P6.1.5 Determine the frequency at which the current i in Figure P6.1.5 has the same magnitude when the connections of one coil are reversed. Find for both connections: (a) the peak current (b) the peak stored magnetic energy if $v = 100\cos(1000t)$ V.

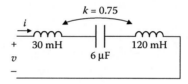

FIGURE P6.1.5

P6.1.6 Given three coils 1, 2, and 3 having the following inductances: $L_1 = 30$ mH, $L_2 = 50$ mH, $L_3 = 60$ mH, $M_{12} = 20$ mH, $M_{23} = 30$ mH, and $M_{13} = 40$ mH, determine the total series inductance when the coils are connected in series as in Figure P6.1.6a, P6.1.6b, and P6.1.6c.

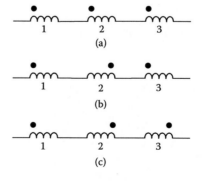

FIGURE P6.1.6

P6.1.7 If $I_1 = 2$ A in Figure P6.1.7, find the value of I_2 that minimizes the stored energy.

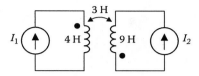

FIGURE P6.1.7

P6.1.8 Consider two perfectly coupled coils of turns N_1 and N_2 and inductances L_1 and L_2, respectively, wound on a linear core of high permeability. The two coils carry the same dc current I, the sense of the windings being such that the mmfs are additive. Determine: (a) the fluxes in the core due to each coil acting alone and the total flux, (b) the total flux linkage, and (c) deduce that the effective inductance is $L_1 + L_2 + 2M$. Note that: (1) mmfs add, (2) fluxes add as long as the core is linear, and (3) flux linkages do not add.

P6.1.9 Figure P6.1.9 diagrammatically illustrates in longitudinal section two coils that are uniformly wound around a toroidal core, one on top of the other. It is assumed that coil 2 is tightly wound around the core, the diameter of the wire of coil 2 and the thickness of its insulation being negligible. Hence, this coil has negligible leakage flux. Coil 1 is separated from the core, and has a leakage flux in the space between coils 1 and 2. This leakage flux follows a circular path around the core and links all the turns of coil 1 because it is inside this coil. It does not link coil 2 because it lies outside this coil. Assume that the cross-sectional dimensions of the core and windings are small compared to the radius a of the toroid. Let the relative permeability of the core be μ_r, and the cross-sectional areas of the core and coil 1 be A_c and A_1, respectively. Show that the inductances

are given by: $L_2 = \dfrac{\mu_0 \mu_r N_2^2 A_c}{2\pi a}$, $L_1 = \dfrac{\mu_0 N_1^2}{2\pi a}[A_1 + A_c(\mu_r - 1)]$ $L_{11} = \dfrac{\mu_0 N_1^2}{2\pi a}(A_1 - A_c)$,

and $M = \dfrac{\mu_0 \mu_r N_1 N_2 A_c}{2\pi a}$. Deduce that $k = 1 / \sqrt{1 + [(A_1 / A_c) - 1] / \mu_r}$. Verify that $L_{11} = (1 - k^2) / L_2$.

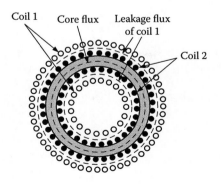

Coil 1   Core flux   Leakage flux of coil 1

Coil 2

FIGURE P6.1.9

P6.1.10 Figure P6.1.10 shows a mutual inductance bridge for comparing an unknown mutual inductance with a known one. Show that at bridge balance ($v_O = 0$):

$$\frac{M_1}{M_2} = \frac{R_1}{R_2} \text{ and } \frac{M_1}{M_2} = \frac{L_1}{L_2}.$$

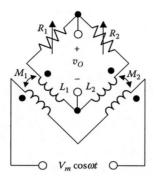

FIGURE P6.1.10

P6.1.11 Figure P6.1.11 shows a **Heaviside bridge** for measuring the mutual inductance in terms of the inductance L_1 of one of the coils and a known inductance L_3.

Show that at bridge balance ($v_O = 0$): $\dfrac{R_1}{R_2} = \dfrac{R_3}{R_4}$ and $M = \dfrac{L_3 R_2 - L_1 R_4}{R_2 + R_4}$.

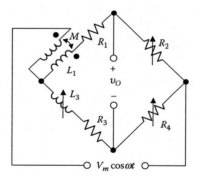

FIGURE P6.1.11

P6.1.12 Figure P6.1.12 shows a **Heydweiller bridge** for measuring the mutual inductance in terms of the inductance L_1 of one of the coils and a known capacitance C_2. Show that at bridge balance ($v_O = 0$): $M = \dfrac{L_1 R_4}{R_2 + R_4}$ and $M = C_2 R_1 R_4$.

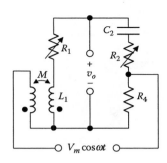

FIGURE P6.1.12

P6.2   Linear Transformers

P6.2.1   The open-circuit voltage ratio of a linear transformer is $\dfrac{V_2}{V_1} = \dfrac{1}{4}$, and the short-circuit current ratio is $\dfrac{I_2}{I_1} = 1$. If the coils are perfectly coupled, the mutual inductance is 8 H. Determine the inductance of each coil and the coefficient of coupling.

P6.2.2   A linear transformer has a primary inductance of 45 mH, a secondary inductance of 20 mH, and a coupling coefficient of 0.8. If a load of $20 - j50\ \Omega$ is connected to the secondary, what is the impedance reflected to the primary side, assuming $\omega = 1$ krad/s? What is the input impedance at the primary terminals of the linear transformer? What will be the reflected impedance and the input impedance if the transformer were ideal and of turns ratio equal to the square root of the inductance ratio? Simulate with PSpice.

P6.2.3   The linear transformer of Problem P6.2.2 has an open-circuited secondary and a primary current that is the periodic triangular waveform of Figure P6.2.3. Sketch the waveforms of the voltages across the primary and secondary windings.

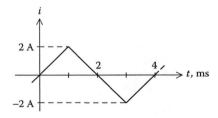

FIGURE P6.2.3

P6.2.4 C_1 in Figure P6.2.4 is initially charged to 6 V, and C_2 is uncharged. The switch is closed at $t = 0$. Calculate the total energy dissipated in the resistor.

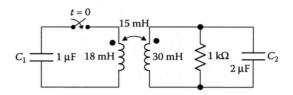

FIGURE P6.2.4

P6.2.5 Determine the input impedance in the circuit of Figure P6.2.5 assuming $\omega = 10^6$ rad/s.

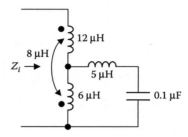

FIGURE P6.2.5

P6.2.6 Determine k in Figure P6.2.6 so that: (1) the input impedance Z_1 is purely resistive; (2) I_1 is maximum.

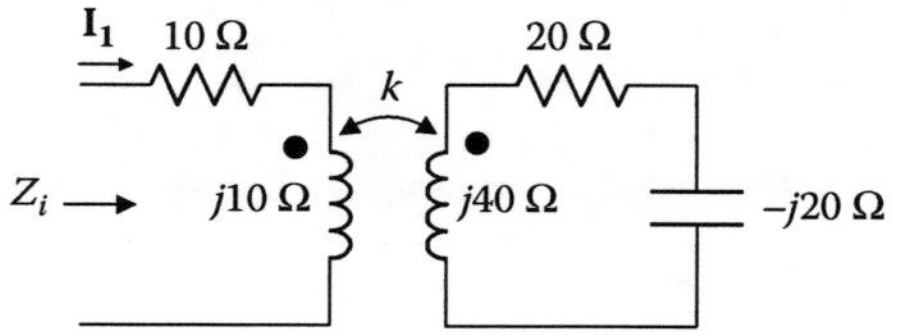

FIGURE P6.2.6

P6.2.7 A linear transformer has a primary inductance of 10 H, a secondary inductance of 2 H, and a coefficient of coupling of 0.9. The transformer is supplied at 50 V rms and connected to a load of 500 Ω. Compare the load voltages at 50 Hz and at 1 kHz. What can you conclude about the relative effects of the shunt inductance and the leakage inductance at the two frequencies?

P6.2.8 Consider the T-equivalent circuit with the assigned positive directions of voltages and currents as shown in Figure P6.2.8a. Write the mesh-current equations and solve for I_1 and I_2 in terms of V_1 and V_2. Express the result as two node-voltage equations and deduce the π-equivalent circuit of Figure P6.2.8b, where $L_a = \dfrac{L_1 L_2 - M^2}{L_2 - M}$, $L_b = \dfrac{L_1 L_2 - M^2}{L_1 - M}$, and $L_c = \dfrac{L_1 L_2 - M^2}{M}$. What happens if the coils are perfectly coupled?

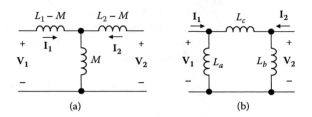

FIGURE P6.2.8

P6.2.9 Verify that the π-equivalent circuit of a linear transformer (Figure P6.2.8b) may be derived from the T-equivalent circuit by applying a Y-Δ transformation.

P6.2.10 The linear transformer of Figure P6.2.10 has $L_1 = 40$ mH, $L_2 = 60$ mH, and $M = 20$ mH. If $i_1 = 0.5\cos(500t)$ A, determine i_2, v_1, and v_2 both as time functions and as phasors. What is the power dissipated in the circuit?

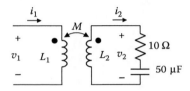

FIGURE P6.2.10

P6.2.11 Interpret Equation SE6.2.5 in terms of the T-equivalent circuit.

P6.2.12 Interpret Equation SE6.2.11 in terms of the T-equivalent circuit.

P6.2.13 Derive Equation SE6.2.10 by using the equivalent circuit of Figure 6.5.3 in the circuit of Figure SE6.2.2.

P6.3 Analysis of Transformer Circuits

P6.3.1 Determine k in Figure P6.3.1 so that no current flows in Z_x.

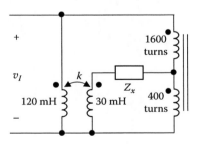

FIGURE P6.3.1

P6.3.2 Determine $\mathbf{I}_1$ and $\mathbf{I}_2$ in Figure P6.3.2 using the following methods: (a) branch current and induced voltages; (b) T-equivalent circuit; (c) node-voltage equations in terms of $\mathbf{V}_1$ and $\mathbf{V}_2$ using the π-equivalent circuit in Problem P6.2.8. Assume $\omega = 1$ rad/s. Simulate with PSpice.

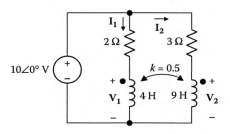

FIGURE P6.3.2

P6.3.3 Determine i_O in Figure P6.3.3, given that $v_{SRC} = 200\sin(1000t)$ V. Simulate with PSpice.

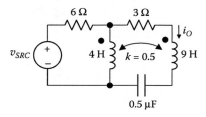

FIGURE P6.3.3

P6.3.4 Determine Z_x in Figure P6.3.4 so that $\mathbf{V_O} = 0$.

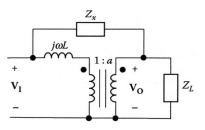

FIGURE P6.3.4

P6.3.5 In Figure P6.3.5, determine I_1, I_2, I_O, V_1, V_2, the power delivered to the 12 Ω resistor, and the impedance seen by the ideal voltage source, given that $V_{SRC} = 10\angle 30°$ V rms.

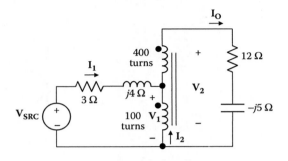

FIGURE P6.3.5

P6.3.6 Determine I_O in Figure P6.3.6 given: $\omega L_1 = 16$ Ω, $\omega L_2 = 12$ Ω, $\omega L_3 = 8$ Ω, $\omega M_{12} = 12$ Ω, $\omega M_{23} = 10$ Ω, and $\omega M_{31} = 8$ Ω.

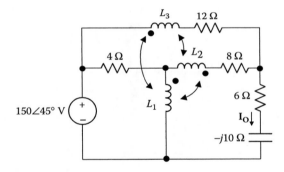

FIGURE P6.3.6

P6.3.7 Determine V_x and V_y in Figure P6.3.7, assuming $f = 50$ Hz.

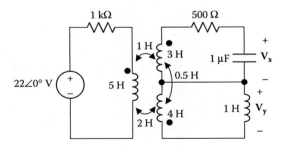

FIGURE P6.3.7

P6.3.8   Determine $\mathbf{I_x}$ in Figure P6.3.8. Simulate with PSpice.

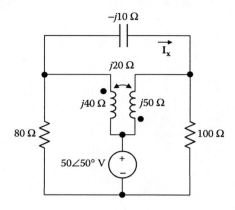

FIGURE P6.3.8

P6.3.9   Determine $\mathbf{V_O}$ in Figure P6.3.9. Simulate with PSpice.

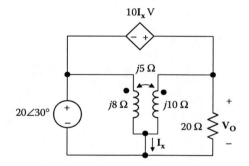

FIGURE P6.3.9

P6.3.10 Determine i_O in Figure P6.3.10, given that $v_{SRC} = 100\sin(100\pi t)$ V.

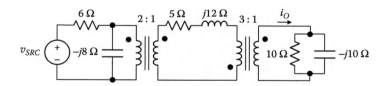

FIGURE P6.3.10

P6.3.11 Determine v_O in Figure P6.3.11 assuming $\omega = 10^5$ rad/s.

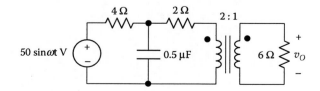

FIGURE P6.3.11

P6.3.12 Determine $\mathbf{V_O}$ in Figure P6.3.12 given that $\mathbf{V_{SRC}} = 50\angle45°\text{V}$.

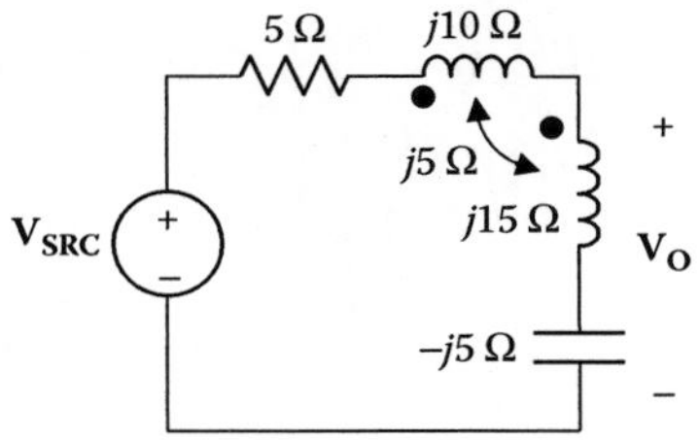

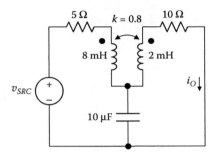

FIGURE P6.3.12

P6.3.13 Determine i_O in Figure P6.3.13 from: (a) the mesh-current equations; (b) the T-equivalent circuit of the linear transformer and the results of SE5.2. Assume $v_{SRC} = 50 \cos 500t$ V.

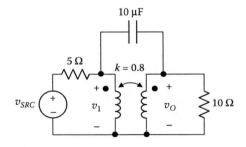

FIGURE P6.3.13

P6.3.14 Consider the same circuit as in Problem P6.3.13 but with the 10 μF capacitor connected as shown in Figure P6.3.14. Determine v_O from: (a) the node-voltage equations; (b) the π-equivalent circuit of the linear transformer and the results of SE5.2.

FIGURE P6.3.14

P6.3.15 Derive TEC between terminals ab in Figure P6.3.15.

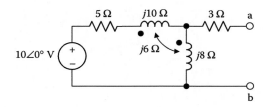

FIGURE P6.3.15

P6.3.16 (a) Show that when one of two coupled coils is short circuited,

$\dfrac{\mathbf{V}_1}{\mathbf{I}_1} = j\omega\left(L_1 - \dfrac{M^2}{L_2}\right)$ and $\mathbf{I}_{\text{SC}} = \dfrac{L_2 + M}{L_2}\mathbf{I}_1$, where $\mathbf{I}_{\text{SC}}$ is the current in the short cir-

cuit across coil 2, $\mathbf{I}_1$ is the current through coil 1, and $\mathbf{V}_1$ is the voltage across it. (b) Using the results in (a), derive TEC between terminals ab in Figure P6.3.16.

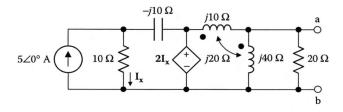

FIGURE P6.3.16

P6.3.17 Derive TEC between terminals ab in Figure P6.3.17.

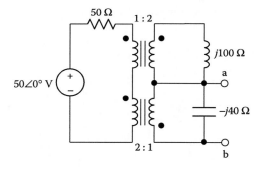

FIGURE P6.3.17

P6.3.18 Derive TEC between terminals ab in Figure P6.3.18.

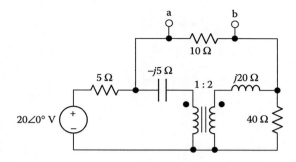

FIGURE P6.3.18

P6.3.19 Derive TEC between terminals ab in Figure P6.3.19.

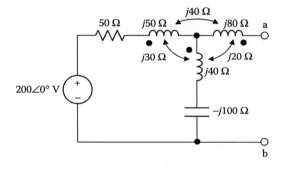

FIGURE P6.3.19

P6.3.20 Derive TEC between terminals ab in Figure P6.3.20.

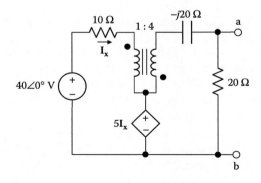

FIGURE P6.3.20

P6.3.21 Determine Z_{in} in Figure P6.3.21.

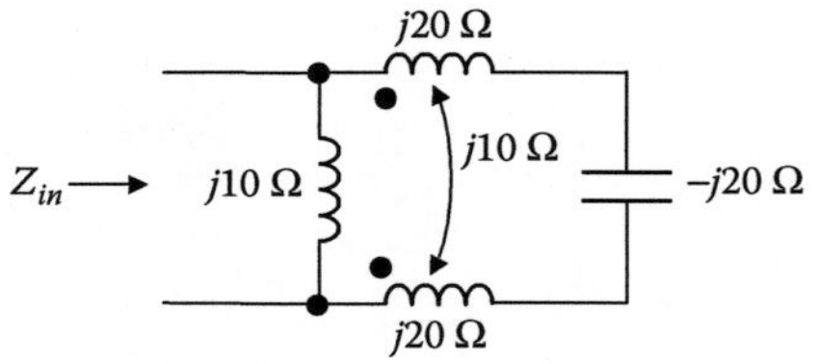

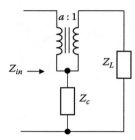

FIGURE P6.3.21

P6.3.22 Using the substitution theorem, show that for an ideal transformer with series coupling (Figure P6.3.22): $Z_{in} = a^2 Z_L + (1 \pm a)^2 Z_c$ where the plus or minus sign is used depending on the relative dot markings of the two windings. Verify the result by means of mesh-current analysis.

FIGURE P6.3.22

P6.4   Miscellaneous

P6.4.1 A 60-Hz transformer has a primary winding of 600 turns that is supplied from a sinusoidal source of 240 V rms. Assuming an ideal transformer, determine the peak flux in the core ϕ_m and the maximum rate of change of this flux. If the transformer is operated at 240 V rms, 50 Hz, what will ϕ_m be? By what factor will the core loss increase? If, because of nonlinearity of the B–H curve, the magnetizing current increases at twice the rate of increase in ϕ_m, and if at 60 Hz, this current is 30% of the total primary current, by how much does the copper loss in the primary winding increase? To what value must the primary voltage at 50 Hz be reduced to have the same ϕ_m?

P6.4.2 Consider an ideal step-up autotransformer. Derive the expressions for instantaneous power input, instantaneous power output, and input impedance for the cases where the assigned positive direction of output current and relative dots are reversed. Note that the effect of changing the relative dots on an autotransformer is not the same as in a two-winding transformer.

P6.4.3 A 10 kV/2.2 kV two-winding transformer is rated at 66 kVA. What is the current rating of each winding? What would be the magnitude of $\mathbf{V_O}$, $\mathbf{I_O}$, and the transformer rating if the transformer is connected as in: (a) Figure P6.4.3a; (b) Figure P6.4.3a but with the connections of the 2.2 kV winding reversed; (c) Figure P6.4.3b; (d) Figure P6.4.3b but with the connections of the 2.2 kV winding reversed. What should be the insulation level in kV of the 2.2 kilovolts winding in case (c)?

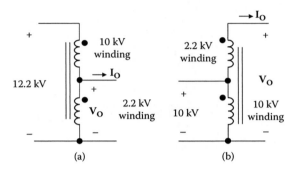

FIGURE P6.4.3

P6.4.4 Figure P6.4.4 shows the approximate equivalent circuit of a transformer at high frequencies at which the shunt inductance may be neglected. The leakage inductance L_{lk} is referred to the secondary. It is assumed that the primary:secondary voltage ratio is 1: a and that the secondary is open-circuited. Show that:

$$\frac{V_2}{V_1} = a\frac{1-\omega^2(L_{lk}C_i / a)}{1-\omega^2 L_{lk}(C_2 + C_i)}.$$ Plot $\dfrac{V_2}{V_1}$ as a function of frequency, identifying the pole ($\mathbf{V_2} \to \infty$) and zero ($\mathbf{V_2}=0$). Simulate with PSpice using appropriate numerical values.

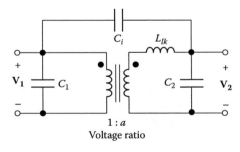

FIGURE P6.4.4

P6.4.5 Repeat Problem P6.4.4 with a reversed dot of the secondary of the ideal transformer.

P6.4.6 Show that the interwinding capacitance C_i in Figure P6.4.6a may be replaced by the capacitances shown in Figure P6.4.6b.

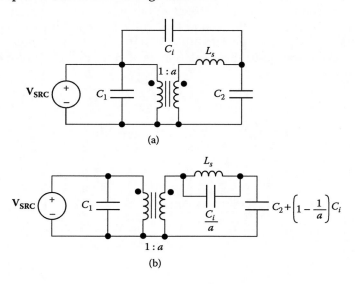

FIGURE P6.4.6

7

Power Relations and Circuit Measurements

Overview

In the sinusoidal steady state, power varies instantaneously over a cycle of the supply, at a frequency that is twice that of the supply. The instantaneous power can be averaged over a cycle to obtain the real power that is dissipated or otherwise expended. Energy storage elements alternately store energy and return this energy to the supply.

Power calculations are generally straightforward but tedious because they involve complex voltages and currents. They are considerably facilitated by utilizing the concept of complex power. The real part of complex power is real power, whereas its imaginary part is the reactive power associated with energy storage elements. The usefulness of complex power stems from its conservation, which implies that real and reactive power can be summed branch by branch in any given circuit.

The reactance of energy storage elements places added current and voltage burdens on the power system, which sometimes necessitates addition of capacitors that counteract the normally inductive reactances of power systems. Another important practical consideration is when a source inherently has a relatively large source impedance. It is of interest in such cases to determine the load impedance that results in maximum power being transferred to the load.

Finally, it is opportune at this point to discuss briefly how current, voltage, and power can be measured and the potential sources of error due to the finite impedance of measuring instruments.

Learning Objectives

- To be familiar with:
  - How current, voltage, and power can be measured
- To understand:
  - The concept of complex power and its application to circuit analysis
  - The implication of a low power factor and how power factor is corrected
  - The conditions for maximum power transfer
  - The errors that may be introduced by current- or voltage-measuring instruments because of their finite impedance

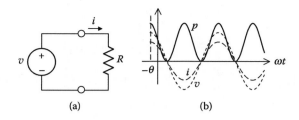

FIGURE 7.1.1
Instantaneous power in a resistor.

7.1   Instantaneous and Average Power

Resistor

Consider a voltage $v = V_m\cos(\omega t + \theta)$ applied to a resistor R (Figure 7.1.1a). The current through the resistor is $i = I_m \cos(\omega t + \theta)$, where $I_m = \dfrac{V_m}{R}$. The instantaneous power dissipated in the resistor at any time t is:

$$p = vi = V_m I_m \cos^2(\omega t + \theta)$$

$$= \frac{V_m I_m}{2}\left[1 + \cos 2(\omega t + \theta)\right] \tag{7.1.1}$$

as shown in Figure 7.1.1b. The instantaneous power varies at twice the supply frequency and is never negative because the resistor does not return power to the supply. Over a cycle, the cosine term averages to zero, so the average power dissipated over a cycle is

$$P = \frac{V_m I_m}{2} = \frac{V_m}{\sqrt{2}}\frac{I_m}{\sqrt{2}} = V_{\text{rms}} I_{\text{rms}} \tag{7.1.2}$$

Equation 7.1.2 agrees with what was derived earlier in Section 5.4 (Chapter 5).

Inductor

If the voltage $v = V_m\cos(\omega t + \theta)$ is applied across an inductor L (Figure 7.1.2a), the current through the inductor is $i = I_m\cos(\omega t + \theta - 90°) = I_m\sin(\omega t + \theta)$, where $I_m = \dfrac{V_m}{\omega L}$ (Section 5.4, Chapter 5). The instantaneous power delivered to the inductor at any time t is:

$$p = vi = V_m I_m \cos(\omega t + \theta)\sin(\omega t + \theta)$$

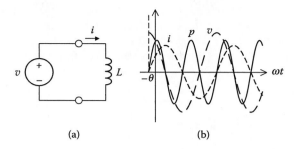

FIGURE 7.1.2
Instantaneous power in an inductor.

$$= \frac{V_m I_m}{2} \sin 2(\omega t + \theta) \qquad (7.1.3)$$

as shown in Figure 7.1.2b. Note that the average power is zero and that as much power flows in one direction as in the opposite direction.

Capacitor

Similarly, if the voltage $v = V_m\cos(\omega t + \theta)$ is applied across a capacitor C (Figure 7.1.3a), the current through the capacitor is $i = I_m\cos(\omega t + \theta + 90°) = -I_m\sin(\omega t + \theta)$, where $I_m = \omega C V_n$ (Section 5.4, Chapter 5). The instantaneous power delivered to the capacitor at any time t is:

$$p = vi = -V_m I_m\cos(\omega t + \theta)\sin(\omega t + \theta)$$

$$= -\frac{V_m I_m}{2} \sin 2(\omega t + \theta) \qquad (7.1.4)$$

as shown in Figure 7.1.3b. Again, the average power is zero and as much power flows in one direction as in the opposite direction.

The preceding results can be summarized as follows:

> *Concept: When v and i are sinusoidal functions of time of frequency ω, with v being a voltage drop in the direction of i, the instantaneous power p = vi is pulsating at a frequency 2ω. If v is in phase with i, as in the case of R, p ≥ 0 represents power dissipated. If v and i are in phase quadrature, as in the case of*

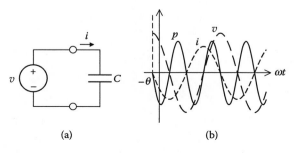

FIGURE 7.1.3
Instantaneous power in a capacitor.

L and C, p is purely alternating and of zero average, as no power is dissipated. In this case, when v and i have the same sign, p > 0 represents energy being stored in the energy-storage element. When v and i have opposite signs, p < 0 represents previously stored energy being returned to the rest of the circuit.

General Case

In the general case, the instantaneous power delivered to any given circuit N through a specified pair of terminals of N is:

$$p = vi \tag{7.1.5}$$

where i is the instantaneous current entering the terminals in the direction of the voltage drop v (Figure 7.1.4a). In general, the power p is partly dissipated in the resistors of the circuit and partly stored as electric and magnetic energy in the capacitors and inductors, respectively. If $v = V_m\cos(\omega t + \theta_v)$ and $i = I_m\cos(\omega t + \theta_i)$:

$$p = V_m I_m\cos(\omega t + \theta_v)\cos(\omega t + \theta_i) \tag{7.1.6}$$

as shown in Figure 7.1.4b.

Let us resolve v into two components: a component v_P in phase with i and a component v_Q in phase quadrature with i. The component of v in phase with i has a phase angle θ_i and a magnitude $V_m\cos(\theta_v - \theta_i)$, whereas the component in phase quadrature with i has a phase angle $(\theta_i + 90°)$ and a magnitude $V_m\sin(\theta_v - \theta_i)$. Thus:

$$v_P = [V_m\cos(\theta_v - \theta_i)]\cos(\omega t + \theta_i)] \tag{7.1.7}$$

$$v_Q = V_m\sin(\theta_v - \theta_i)\cos(\omega t + \theta_i + 90°) = -V_m\sin(\theta_v - \theta_i)\sin(\omega t + \theta_i) \tag{7.1.8}$$

These components can be verified readily with reference to a phasor diagram (Figure 7.1.4c), where it is assumed for the sake of argument that $\theta_v > \theta_i$. In the time domain, $v_P + v_Q = V_m[\cos(\theta_v - \theta_i)\cos(\omega t + \theta_i) - \sin(\theta_v - \theta_i)\sin(\omega t + \theta_i)] = V_m\cos(\omega t + \theta_v) = v.$

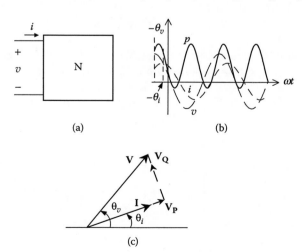

(a)

(b)

(c)

FIGURE 7.1.4
Instantaneous power in a circuit.

Multiplying each of the two components of v by i:

$$v_P i = V_m I_m \cos(\theta_v - \theta_i)\cos^2(\omega t + \theta_i)$$

$$= \frac{V_m I_m}{2}\cos(\theta_v - \theta_i)\left[1 + \cos 2(\omega t + \theta_i)\right]$$

$$= P[1 + \cos 2(\omega t + \theta_i)] \tag{7.1.9}$$

where,
$$P = \frac{V_m I_m}{2}\cos(\theta_v - \theta_i) = V_{rms}I_{rms}\cos(\theta_v - \theta_i) \tag{7.1.10}$$

and,
$$v_Q i = -V_m I_m \sin(\theta_v - \theta_i)\cos(\omega t + \theta_i)\sin(\omega t + \theta_i)$$

$$= -\frac{V_m I_m}{2}\sin(\theta_v - \theta_i)\sin 2(\omega t + \theta_i)$$

$$= \frac{V_m I_m}{2}\sin(\theta_v - \theta_i)\cos\left[2(\omega t + \theta_i) + 90°\right]$$

$$= Q\cos[2(\omega t + \theta_i) + 90°] \tag{7.1.11}$$

where,
$$Q = \frac{V_m I_m}{2}\sin(\theta_v - \theta_i) = V_{rms}I_{rms}\sin(\theta_v - \theta_i) \tag{7.1.12}$$

P is the **real** or **average power**. It appears in Equation 7.1.9 both as the average of $v_P i$, which is the power dissipated in the resistive elements of the circuit, and as the magnitude of the alternating component of $v_P i$. Q is the **reactive power** and is the power associated with the energy that is alternately stored and returned to the supply by the inductive and capacitive elements of the circuit. From Equation 7.1.12, Q is the magnitude of $v_Q i$, which is purely alternating. For a resistor $\theta_v = \theta_i$, so $Q = 0$ and $P = V_{rms}I_{rms}$ in accordance with Equation 7.1.2. For an inductor, $\theta_v - \theta_i = 90°$, so $Q = V_{rms}I_{rms}$ and $P = 0$. For a capacitor, $\theta_v - \theta_i = -90°$, so $Q = -V_{rms}I_{rms}$ and $P = 0$. Thus, Q is positive for an inductive reactance and is negative for a capacitive reactance. Whereas the unit of P is the watt (W), the unit of Q is the volt-ampere reactive (VAR).

EXERCISE 7.1.1
Show that Equation 7.1.11 reduces to Equation 7.1.3 for an inductor, and to Equation 7.1.4 for a capacitor.

Example 7.1.1: Real and Reactive Power
Consider a voltage $v_{SRC} = 100\cos(1000t + 30°)$ V applied to a 30 Ω resistor in series with a 40 mH inductor. It is required to determine the real and reactive power.

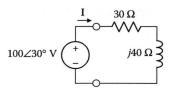

FIGURE 7.1.5
Figure for Example 7.1.1.

SOLUTION

$\omega L = 1000 \times 0.04 = 40$ Ω. The circuit in the frequency domain is shown in Figure 7.1.5. $\mathbf{I} =$

$\dfrac{100\angle 30°}{30 + j40} = 2\angle -23.13°$A, and $i = 2\cos(1000t - 23.13°)$A. From Equation 7.1.10, $P =$

$\dfrac{200}{2}\cos 53.13° = 100 \times \dfrac{30}{50} = 60$ W. From Equation 7.1.9, $v_P i = 60[1 + \cos 2(1000t - 23.13°)]$ W.

From Equation 7.1.12, $Q = \dfrac{200}{2}\sin(53.13°) = 100 \times \dfrac{40}{50} = 80$ VAR. From Equation 7.1.11,

$v_Q i = 80\cos[2(1000t - 23.13°) + 90°]$ VAR.

P is the average power dissipated in the resistor, which is also $RI_{rms}^2 = 30\left(\dfrac{2}{\sqrt{2}}\right)^2 = 60$ W.

The maximum instantaneous power dissipation is $2P = 120$ W and occurs

when $\cos 2(1000t - 23.13°) = 1$. The energy stored in the inductor at any instant is $w = \dfrac{1}{2}Li^2$

$= \dfrac{1}{2}LI_m^2 \cos^2(\omega t + \theta_i)$. The rate at which this energy is stored is $\dfrac{dw}{dt} = -\dfrac{1}{2}\omega LI_m^2 \sin 2(\omega t + \theta_i)$

$= \dfrac{1}{2}\omega LI_m^2 \cos[2(\omega t + \theta_i) + 90°]$, which is the same as Equation 7.1.11, where $Q = \dfrac{1}{2}\omega LI_m^2$

$= \dfrac{1}{2} \times 1000 \times 0.04 \times 4 = 80$ VAR.

EXERCISE 7.1.2

Repeat Example 7.1.1 with the inductor replaced by a 25 μF capacitor.
Answer: $P = 60$ W; $Q = -80$ VAR.

7.2   Complex Power

Complex Power Triangle

P and Q are the magnitudes of purely alternating components of $v_P i$, and $v_Q i$, namely, $P\cos 2$ $(\omega t + \theta_i)$ and $Q\cos[2(\omega t + \theta_i) + 90°]$, respectively (Equation 7.1.9 and Equation 7.1.11). As these

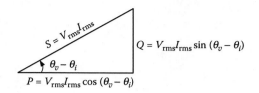

FIGURE 7.2.1
Complex power triangle.

components have the same frequency but differ in phase by 90°, they can be represented on an Argand diagram (Figure 7.2.1). The complex sum $S = P + jQ$ is the **complex power** having a magnitude $V_{rms}I_{rms}$ and a phase angle of $(\theta_v - \theta_i)$, assuming for the sake of argument that $\theta_v > \theta_i$. The magnitude of the complex power $|S| = \sqrt{P^2 + Q^2}$ is the **apparent power**. The unit of S is the volt-ampere (VA).

Although P, Q, and S are of the nature of phasors, they are not referred to as such, and are not drawn on the same diagram as voltage and current phasors because they have twice the frequency. S may be expressed as:

$$S = V_{rms}I_{rms}\angle(\theta_v - \theta_i) = V_{rms}\angle\theta_v \times I_{rms}\angle-\theta_i$$

$$= \mathbf{V}_{rms}\mathbf{I}_{rms}^* \tag{7.2.1}$$

where $\mathbf{I}_{rms}^*$ is the conjugate of $\mathbf{I}_{rms}$, having a magnitude I_{rms} and phase angle $-\theta_i$. Multiplication of $\mathbf{V}_{rms}$ by $\mathbf{I}_{rms}^*$ in Equation 7.2.1 is necessary to have the phase angle of S equal to $\theta_v - \theta_i$. Note that it is usually more convenient in power calculations to express magnitudes of voltages and currents as rms values.

Consider a circuit N that has resistances, reactances, and dependent sources but not independent sources. Let $\mathbf{V}_{rms} = V_{rms}\angle\theta_v$ and $\mathbf{I}_{rms} = I_{rms}\angle\theta_i$ be the voltage and current, respectively, at specified terminals of N as in Figure 7.2.2a. It follows that the impedance looking into these terminals is $\dfrac{\mathbf{V}_{rms}}{\mathbf{I}_{rms}} = Z = |Z|\angle(\theta_v - \theta_i) = R + jX$. Substituting $\mathbf{V}_{rms} = Z\mathbf{I}_{rms}$ in Equation 7.2.1:

$$S = Z\mathbf{I}_{rms}\mathbf{I}_{rms}^* = ZI_{rms}\angle\theta_i \times I_{rms}\angle-\theta_i = ZI_{rms}^2 \tag{7.2.2}$$

$$= RI_{rms}^2 + jXI_{rms}^2 \tag{7.2.3}$$

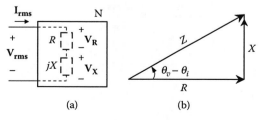

(a)   (b)

FIGURE 7.2.2
Complex power related to impedance.

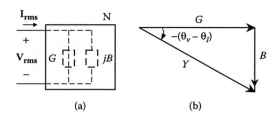

FIGURE 7.2.3
Complex power related to admittance.

or,
$$P = RI_{rms}^2 \text{ and } Q = XI_{rms}^2 \tag{7.2.4}$$

The impedance triangle is shown in Figure 7.2.2b. According to Equation 7.2.3, the complex power triangle of Figure 7.2.1 is simply a scaled version of the impedance triangle, the scaling factor being I_{rms}^2. Moreover, $\mathbf{V}_{rms} = \mathbf{V}_R + \mathbf{V}_X = R\mathbf{I}_{rms} + jX\mathbf{I}_{rms}$. It is seen that $R\mathbf{I}_{rms}$ in the time domain is v_P, the component of $\mathbf{V}_{rms}$ that is in phase with $\mathbf{I}_{rms}$ (Equation 7.1.7), whereas $X\mathbf{I}_{rms}$ in the time domain is v_Q.

In Figure 7.2.3a, $\mathbf{I}_{rms}$ and $\mathbf{V}_{rms}$ are related by the admittance: $\dfrac{\mathbf{I}_{rms}}{\mathbf{V}_{rms}} = Y = \dfrac{1}{|Z| \angle (\theta_v - \theta_i)} =$

$|Y| \angle -(\theta_v - \theta_i) = G + jB$ (Figure 7.2.3b), where B is negative for an inductive reactance $(\theta_v > \theta_i)$. S may be expressed as:

$$S = \mathbf{V}_{rms}\mathbf{I}_{rms}^* = \mathbf{V}_{rms}(Y\mathbf{V}_{rms})^* = \mathbf{V}_{rms}Y^*\mathbf{V}_{rms}^* = Y^*V_{rms}^2 = \frac{1}{Z^*}V_{rms}^2 \tag{7.2.5}$$

$$= (G - jB)V_{rms}^2 = G\,V_{rms}^2 - jB\,V_{rms}^2 \tag{7.2.6}$$

or
$$P = GV_{rms}^2 \text{ and } Q = -BV_{rms}^2 \tag{7.2.7}$$

The complex power triangle of Figure 7.2.1 is a scaled version of the admittance triangle, with B inverted, the scaling factor being V_{rms}^2. Complex power relations are summarized in Table 7.2.1.

TABLE 7.2.1

Complex Power Relations

	$S = P + jQ = \mathbf{V}_{rms}\,\mathbf{I}_{rms}^*$	
	Series Connection $Z = R + jX$	**Parallel Connection** $Y = G + jB$
S	ZI_{rms}^2	$Y^*V_{rms}^2 = V_{rms}^2 / Z^*$
P	RI_{rms}^2	GV_{rms}^2
Q	XI_{rms}^2	$-BV_{rms}^2$

Conservation of Complex Power

> Concept: *In any given circuit, complex power is conserved, which implies that real power and reactive power are also conserved.*

A rigorous justification is provided by Tellegen's theorem, which is discussed in Section ST7.1. In fact, according to Tellegen's theorem, not only is complex power conserved in any given circuit, but any fictitious power $v_k i_k$ that one cares to define is also conserved, as long as the v_k's satisfy KVL around every mesh, and the i_k's satisfy KCL at every node. Here v_k is an arbitrarily assigned voltage across the kth branch and i_k is the current in kth branch in the direction of the voltage drop v_k. Conservation of this fictitious power means that:

$$\sum_{b=1}^{B} v_k i_k = 0 \tag{7.2.8}$$

where the summation is over all the B branches of the circuit. The phasors $\mathbf{V_k}$ and $\mathbf{I_k}$ satisfy KVL and KCL, as explained in Section 5.6 (Chapter 5), so that Equation 7.2.8 takes the form $\sum_{k=1}^{B} \mathbf{V_k I_k} = 0$ in phasor notation. But because the $\mathbf{I_k}$'s satisfy KCL, the sums of their real parts and their imaginary parts are each separately equal to zero. It follows that the $\mathbf{I_k^*}$'s also satisfy KCL. Hence:

$$\sum_{k=1}^{B} \mathbf{V_k I_k^*} = \sum_{k=1}^{B} S_k = 0 \tag{7.2.9}$$

Equation 7.2.9 is an expression of the conservation of complex power. Moreover, as $S_k = P_k + jQ_k$, it follows that P_k and Q_k must each sum to zero.

$$\sum_{k=1}^{B} P_k = 0 \quad \text{and} \quad \sum_{k=1}^{B} Q_k = 0 \tag{7.2.10}$$

In other words, real and reactive powers are separately conserved. In making these summations, real power is considered positive if dissipated and negative if delivered. Inductive power is considered positive if absorbed and negative if delivered. Conversely, capacitive power is considered negative if absorbed and positive if delivered.

Both equalities in Equation 7.2.10 involve summation over branches. This has an important implication for analyzing circuits using complex power.

> Concept: *Real power and reactive power can each be summed branch by branch in a given circuit, with the total sum of each, over all branches of the circuit, equal to zero.*

Example 7.2.1: Application of Complex Power

Two loads L_1 and L_2 are connected across a 1000 $\angle 0°$ V rms supply (Figure 7.2.4a). L_1 absorbs real power of 40 kW and reactive, inductive power of 30 kVAR, whereas L_2 absorbs real power of 80 kW and reactive, inductive power of 60 kVAR. The loads are fed through

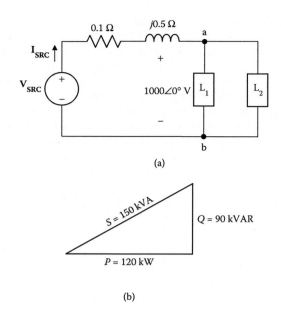

FIGURE 7.2.4
Figure for Example 7.2.1.

a power line having a resistance of 0.1 Ω and a reactance of 0.5 Ω. It is required to determine the voltage $\mathbf{V}_{SRC}$.

SOLUTION

The real powers of L_1 and L_2 are added together to give 120 kW, and their reactive powers are added together to give 90 kVAR inductive. The complex power triangle at terminals ab is shown in Figure 7.2.4b. Thus, $S = 120,000 + j90,000 = 1,000\angle0° \times \mathbf{I}_{SRC}^*$, which gives,

$$\mathbf{I}_{SRC}^* = \frac{120,000 + j90,000}{1,000} = 150\angle36.9° \text{ A. Hence, } \mathbf{I}_{SRC} = 150\angle-36.9° \text{A.}$$ The real power dissipated by the 0.1 Ω resistance is $0.1 \times (150)^2 = 2.25$kW, and the reactive power absorbed by the 0.5 Ω reactance is $0.5 \times (150)^2 = 11.25$ kVAR. The total real power at the inputs of the supply terminals is 122.25 kW and the total reactive power is 101.25 kVAR. The complex power at these terminals is therefore, $S = 122,250 + j101,250 = 158,734\angle39.6° = \mathbf{V}_{SRC}\angle\theta_v \times \mathbf{I}_{SRC}^* = V_{SRC}\angle\theta_v \times 150\angle36.9°$. Hence, $\mathbf{V}_{SRC} = 1058\angle2.7°$ V.

The real power delivered by the source is 122.25 kW. It is the sum of the real power dissipated in the line resistance and in loads L_1 and L_2. The reactive power delivered by the source is 101.25 kVAR. It is also the sum of the inductive reactive power in the line reactance and in the two loads. Real and reactive powers are conserved in the system as a whole.

7.3   Power Factor Correction

Concept: *Reactive power, although it averages to zero over a cycle, generally increases the voltage and current requirements of a load.*

In Figure 7.2.2a, the real power delivered to the load is $P = RI_{rms}^2$. To deliver a given I_{rms} in the presence of the reactance X requires a larger voltage across the series combination, and therefore a higher degree of insulation of the conductors subjected to the higher voltage. Similarly, in Figure 7.2.3a, the real power delivered to the load is $P = GV_{rms}^2$. To apply a given V_{rms} in the presence of the susceptance B requires a larger current through the parallel combination, and therefore a greater current-carrying capacity of the supply conductors.

The relative value of the reactive component of a load is indicated by the phase angle $(\theta_v - \theta_i)$; $\cos(\theta_v - \theta_i)$ is the **power factor** (abbreviated p.f.), and $\sin(\theta_v - \theta_i)$ is the **reactive factor**. For a purely resistive load, the p.f. is unity. Because the p.f. is the same for a positive $(\theta_v - \theta_i)$ as for a negative $(\theta_v - \theta_i)$, these two cases are distinguished by adding the attribute *lagging* or *leading*, respectively. For example, for a purely inductive load $(\theta_v - \theta_i = 90°)$, the p.f. is zero lagging; whereas for a purely capacitive load $(\theta_v - \theta_i = -90°)$, the p.f. is zero leading. In practice, ac loads generally have a lagging p.f., mainly because of ac motors as well as the inductances associated with power transformers and ballasts of fluorescent lights. The p.f. may be as low as 0.8 lagging or less, particularly during motor starting. A low p.f. is undesirable because of the additional current and voltage burdens placed on the supply, as explained earlier. In the case of large loads, the additional costs involved can be quite considerable, so that some measures are taken to improve the power factor. This **power factor correction** is achieved by adding capacitive reactance to counteract the inductive reactance of the load.

Example 7.3.1: Power Factor Correction

Assuming the supply frequency is 50 Hz, determine the capacitance that must be added in parallel with the loads of Figure 7.2.4 of Example 7.2.1, so as to make the p.f. unity at terminals ab. What is the effect of this capacitance on the supply current and voltage?

SOLUTION

The reactive power at terminals ab was found in Example 7.2.1 to be 90 kVAR. The reactive power of the added capacitor must be −90 kVAR. The value of capacitance follows from the relation: $-2\pi f C V^2 = Q$ (Equation 7.2.6). Thus, $2\pi f C \times (1000)^2 = 90 \times 10^3$ or $C = 286.5$ μF.

The total reactive power at terminals ab is now zero. Hence, $S = 120,000 = 1000\angle 0° \times I_{SRC}^*$. This gives $I_{SRC}^* = 120\angle 0°$ A. The real power dissipated by the 0.1 Ω resistance is $0.1 \times (120)^2 = 1.44$ kW, and the reactive power absorbed by the 0.5 Ω reactance is $0.5 \times (120)^2 = 7.2$ kVAR. The total real power at the inputs of the supply terminals is 121.44 kW and the total reactive power is 7.2 kVAR. The complex power at these terminals is therefore $S = 121,440 + j7,200 = 121,653\angle 3.4° = \mathbf{V}_{SRC}\mathbf{I}_{SRC}^* = V_{SRC}\angle\theta_v \times 120\angle 0°$. Hence, $\mathbf{V}_{SRC} = 1014\angle 3.4°$ V.

The p.f. at the load was initially $\dfrac{120}{150} = 0.8$. By correcting it to 1, the supply current was reduced from 150 A to 120 A, and the supply voltage was reduced by a relatively small amount in this case, from 1058 to 1014 V rms, because of the reduced voltage drop across the line impedance.

EXERCISE 7.3.1

Determine the capacitance that must be added in parallel with the loads of Example 7.2.1, so as to make the p.f. unity at the supply terminals. What are the new values of supply current and voltage?

Answer: 309.5 μF; 120.22∠3.45° A; 1010.2∠3.45° V.

EXERCISE 7.3.2

In Example 7.2.1, a wattmeter is used to measure the total real power delivered to the combined load. If the voltage terminal of the wattmeter is connected through a 20:1 voltage step-down transformer, and the current coil is connected through a 10:1 current step-down transformer, with normal polarities maintained, what would be the wattmeter reading? If the voltage and current at the load terminals are measured to be 1000 V and 150 A, respectively, what is the power factor of the combined load?

Answer: 600 W; 0.8.

7.4   Maximum Power Transfer

A matter of practical importance is to transfer maximum power to a load from a given source of specified open-circuit voltage and source impedance. Before considering the general case, we will start with the purely resistive case.

Purely Resistive Circuit

In Figure 7.4.1, a dc source having an open-circuit voltage V_{SRC} and a source resistance R_{src} supplies a resistive load R_L. The ideal voltage source and source resistance could represent, in general, TEC of a circuit connected to R_L.

The load current is $I_L = \dfrac{V_{SRC}}{R_{src} + R_L}$, and the power transferred to R_L is:

$$P_L = R_L \left(\frac{V_{SRC}}{R_{src} + R_L} \right)^2 = \frac{R_L}{\left(R_{src} + R_L \right)^2} V_{SRC}^2 \tag{7.4.1}$$

With V_{SRC} and R_{src} constant, we wish to find the value of R_L that maximizes P_L. If we derive $\dfrac{dP_L}{dR_L}$ and set it to zero, we find that P_L is maximum for R_L given by:

$$R_{Lm} = R_{src} \tag{7.4.2}$$

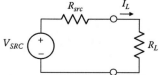

FIGURE 7.4.1
Purely resistive source and load.

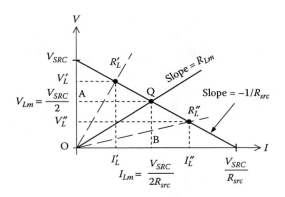

FIGURE 7.4.2
Graphical construction for maximum power transfer.

When Equation 7.4.2 is satisfied, the source and load resistances are *matched*. The voltage across R_L is $\dfrac{V_{SRC}}{2}$, and the power transferred to the load is $P_{Lm} = \dfrac{V_{SRC}^2}{4R_{Lm}}$. Figure 7.4.2 shows the source characteristic and load line, as previously derived in connection with Figure 4.1.13 in Chapter 4. The power transferred to the load is represented by the area of the rectangle OAQB. This area is a maximum when Equation 7.4.2 is satisfied. Any other load, such as R_L' or R_L'', results in a rectangle of smaller area.

Figure 7.4.3 illustrates the various power relations in the circuit under conditions of maximum power transfer. The power dissipated in R_{src} is $\dfrac{V_{SRC}^2}{4R_{Lm}}$, the same as that transferred to the load because the resistances and currents are equal. It is represented by the area of the rectangle ACDQ. The total power delivered by the ideal voltage source v_{SRC} is $\dfrac{V_{SRC}^2}{2R_{Lm}}$, represented by the area of the rectangle OCDB.

It should be emphasized that the condition of maximum power transfer applies to the case where V_{SRC} and R_{src} are kept constant and R_L is varied. If R_L is kept constant and R_{src} is varied, the results are quite different. It is seen from Figure 7.4.3 that if R_L is kept equal

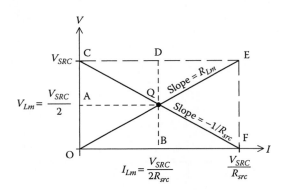

FIGURE 7.4.3
Power relations when maximum power is transferred.

to R_{Lm} while R_{src} is reduced, Q moves toward E and maximum power is transferred to R_{Lm} when $R_{src} = 0$. The source characteristic becomes horizontal, when all of V_{SRC} is applied to R_{Lm}, and the power transferred is $\dfrac{V_{SRC}^2}{R_{Lm}}$, corresponding to the area of the rectangle OCEF. Hence, given a source having a specified open-circuit voltage v_{SRC}, maximum power is transferred to the load when $R_{src} = 0$. However, in many practical cases, particularly in electronic circuits, the source inherently has a relatively large R_{src}, in which case maximum power is transferred to a load when $R_L = R_{src}$.

EXERCISE 7.4.1

Express Equation 7.4.1 as $P_L = \dfrac{V_{SRC}^2}{\left(R_{src}/\sqrt{R_L} + \sqrt{R_L} \right)^2}$. P_L is maximum when the term within parentheses is minimum. This term is of the general form $\dfrac{a}{x} + \dfrac{x}{b}$, where a and b are constants, and is encountered often in circuit analysis. Show that this expression has a minimum when $\dfrac{a}{x} = \dfrac{x}{b}$ or $x^2 = ab$. It follows that maximum P_L occurs when Equation 7.4.2 is satisfied.

Simulation Example 7.4.1: Maximum Power Transfer in Resistive Circuit

It is required to determine the value of R_L that should be connected between terminals ab in Figure 7.4.4a for maximum power transfer.

SOLUTION

We need only determine R_{Th} looking into terminals ab. R_{Lm} for maximum power transfer is then equal to R_{Th}.

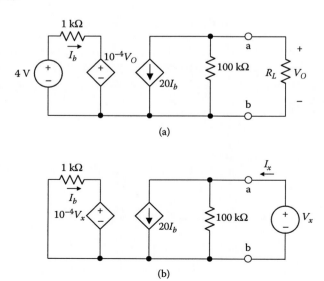

(a)

(b)

FIGURE 7.4.4
Figure for Example 7.4.1.

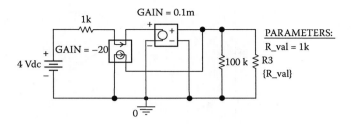

FIGURE 7.4.5
Figure for Example 7.4.1.

To determine R_{Th}, we apply a test source V_x between terminals ab with the 4 V source set to zero, and determine I_x. It is seen that $I_b \times 1 = -10^{-4}V_x$. On the output side, KCL gives $20 \times 10^{-4}V_x + I_x = \dfrac{V_x}{100}$. It follows that $\dfrac{V_x}{I_x} = R_{Th} = R_{Lm} = 125$ kΩ.

SIMULATION
We can determine R_{Th} using DC Sweep as was done in Example 4.1.2 in Chapter 4. Instead we will demonstrate the parametric sweep feature of PSpice that allows designating the value of a circuit component as a global parameter, which can then be varied continuously or for some discrete values. The schematic is shown in Figure 7.4.5. To use parametric sweep with R3, double click on the default resistance value displayed, which invokes the Display Properties window. In the Value field, enter a chosen designation enclosed in curly brackets, which tells PSpice that this is a parameter. In the present example, {R_val} is entered. The next step is to declare {R_val} a global parameter. Place the part PARAM from the SPECIAL library; this shows on the schematic as <u>PARAMETERS:</u>. When this word is double-clicked, the Property Editor spreadsheet is displayed. Press the New Column button to display the Add New Column dialog box. Enter R_val in the Name field and any value, say 1k, in the Value field. A new column R_val is added to the spreadsheet with the entry 1k. To have this displayed on the schematic, press the Display button and choose Name and Value in the Display Properties dialog box. R_val = 1k appears under <u>PARAMETERS:</u>, as shown in Figure 7.4.5.
To run the simulation, select DC Sweep in the Simulation Settings dialog box. Under Sweep variable, choose Global parameter and enter R_val for Parameter name. Under Sweep type choose Linear and enter 50k for Start value, 250k for End value, and 100 for Increment. It may be necessary, when one has no idea of the required value of R3, to try different ranges of values. When PSpice is run, a blank graph is displayed having the x-axis variable R_val. Add the trace W[R3] to display the power dissipated in R3. The plot of Figure 7.4.6 is displayed. Press the Toggle cursor button, then the Cursor Max button. The cursor reading gives 125.000K, 200.000. Thus, $R_{Lm} = 125$ kΩ, and the maximum power transferred is 200 W.

EXERCISE 7.4.2
Verify by analysis that the maximum power transferred in Example 7.4.1 is 200 W.

EXERCISE 7.4.3

The total power delivered by the source V_{SRC} varies with R_L in accordance with the

relation, $P_{SRC} = V_{SRC}I_L = \dfrac{V_{SRC}^2}{R_{src} + R_L}$. Because this power varies with R_L, maximum

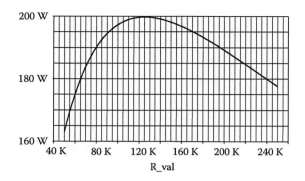

FIGURE 7.4.6
Figure for Example 7.4.1.

power dissipation in R_L does *not*, in general, coincide with minimum power dissipation in R_{src}, and conversely. Determine R_L and the power dissipated in R_L and R_{scr} when the power dissipated in R_{src} is: (a) maximum and (b) minimum.

Answer: (a) $R_L = 0$, $P_L = 0$, $P_{src} = V_{src}^2 / R_{src}$; (b) $R_L = \infty$, $P_L = 0$, $P_{src} = 0$.

Application Window 7.4.1: Maximum Power Output from a Solar Cell

Given a solar cell connected to a load R_L (Figure 7.4.7a), it is required to derive the condition for maximum power transfer to R_L.

SOLUTION

The solar cell generates a dc voltage and has a nonlinear source characteristic (Figure 7.4.7b). To obtain useful power, a solar panel consists of many such cells connected in series–parallel. The series connection builds up the load voltage, whereas the parallel connection supplies sufficient load current. For maximum efficiency, the load should be such that maximum power is delivered to it by the solar panel.

For a given value of the load resistance R_L, the operating point Q is the intersection of the source characteristic with the line of slope R_L drawn through the origin. At this point, the V–I relations of both the source and load are satisfied. Let the current be I and the voltage across R_L be V. The power delivered to the load is $P = VI$, and is equal to the area

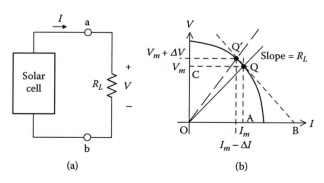

(a)   (b)

FIGURE 7.4.7
Application Window 7.4.1.

of the rectangle whose sides are V and I. To find the condition for maximum power transfer, we differentiate and set $dP = 0$: $dP = IdV + VdI = 0$. This gives

$$\frac{V_m}{I_m} = -\frac{dV}{dI}$$

(7.4.3)

where I_m and V_m are the coordinates of the point Q at which Equation 7.4.3 is satisfied (Figure 7.4.7b). To clarify the meaning of this equation, suppose that R_L is changed to $R_L + \Delta R_L$. The operating point moves from Q to Q'. The slope of the new resistance line is

$\dfrac{V_m + \Delta V}{I_m - \Delta I}$, and $-\dfrac{\Delta V}{\Delta I}$ represents the slope of the line Q'Q. In the limit, as $\Delta V \to 0$ and $\Delta I \to 0$,

$\dfrac{V_m + \Delta V}{I_m - \Delta I} \to \dfrac{V_m}{I_m} = R_L$ and $-\dfrac{\Delta V}{\Delta I} \to -\dfrac{dV}{dI}$, which is the slope of the source characteristic at Q. Equation 7.4.3 becomes:

$$R_{L\max} = -\frac{dV}{dI}$$

(7.4.4)

That is, maximum power is transferred to the load when the load resistance is equal in magnitude to the slope of source characteristic at the operating point. The tangent at this point and the line of slope R_L form an isosceles triangle OQB (Figure 7.4.7b). The area of the rectangle OAQC represents the maximum power transferred to the load.

How does a source with a nonlinear characteristic compare with a linear source having the same open-circuit voltage V_{oc} and short-circuit current I_{sc}? To simplify matters, let the characteristic of the nonlinear source be circular in shape (Figure 7.4.8). A load resistance at an angle of 45° results in maximum power transfer for both the nonlinear source and a linear source having the same V_{oc} and I_{sc}. The power transferred by the nonlinear source

is $\dfrac{V_{oc}}{\sqrt{2}} \dfrac{I_{sc}}{\sqrt{2}} = \dfrac{V_{oc}I_{sc}}{2}$ compared to $\dfrac{V_{oc}I_{sc}}{4}$ for the linear source. Considering the maximum

power that can be delivered by the source to be $V_{oc}I_{sc}$, the nonlinear source delivers 50% of this power, compared to 25% for the linear source. The performance of the nonlinear

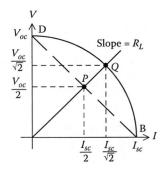

FIGURE 7.4.8
Application Window 7.4.1.

system is further improved if its characteristic is more square between V_{oc} and I_{sc} than the circular characteristic of Figure 7.4.8.

This example illustrates that, although linear systems are relatively easy to analyze and design, nonlinear systems can have a superior performance.

Source and Load Impedances

When the source or load is reactive, the condition for maximum power transfer should be generalized. Let $\mathbf{V_{SRC}}$ and $Z_{src} = R_{src} + jX_{src}$ represent, in general, TEC as seen from terminals ab of a given circuit connected to a load $Z_L = R_L + jX_L$. The current phasor $\mathbf{I_L}$ is given by:

$$\mathbf{I_L} = \frac{1}{Z_{src} + Z_L}\,\mathbf{V_{SRC}} = \frac{V_{src}\angle\theta_{src}}{(R_{src} + R_L) + j(X_{src} + X_L)} \qquad (7.4.5)$$

Assuming that $\mathbf{I_L}$ is expressed in terms of its rms value, the power transferred to R_L is

$$P_L = R_L|I|^2 = V_{src}^2\,\frac{R_L}{(R_{src} + R_L)^2 + (X_{src} + X_L)^2} \qquad (7.4.6)$$

We wish to determine the condition that maximizes P_L, assuming that V_{src}, R_{src}, and X_{src} are fixed, whereas R_L and X_L are variable. If R_L and X_L can be varied independently, it is clear that, with R_L fixed, P_L is maximum for X_L given by:

$$X_{Lm} = -X_{src} \qquad (7.4.7)$$

With this condition satisfied, Equation 7.4.6 reduces to Equation 7.4.1. P_L is maximum when Equation 7.4.2 is satisfied, that is, $R_{Lm} = R_{src}$. Combining this condition with that of Equation 7.4.7 gives the condition for maximum power transfer as:

$$Z_{Lm} = Z_{src}^* \qquad (7.4.8)$$

where Z_{src}^* is the conjugate of Z_{src}. When Equation 7.4.8 is satisfied, X_L is equal in magnitude but opposite in sign to X_{src} so that the two reactances cancel out. With the additional condition that $R_{Lm} = R_{src}$, maximum power is transferred to R_L.

What would be the condition for maximum power transfer if R_L and X_L can again be varied independently, but their range of variation is restricted so that Equation 7.4.8 cannot be satisfied? It is clear from Equation 7.4.6 that under these conditions, with R_L fixed, P_L is maximum when $(X_{src} + X_L)$ is as small as possible. With $(X_{src} + X_L)$ considered constant, the condition for maximum power transfer can be determined by deriving $\dfrac{dP_L}{dR_L}$ from Equation 7.4.6 and setting it to zero. This gives:

$$R_{Lm} = \sqrt{R_{src}^2 + (X_L + X_{src})^2} \qquad (7.4.9)$$

TABLE 7.4.1

Conditions for Maximum Power Transfer

Allowed Variation	Conditions
R_L and X_L can be varied independently over an arbitrary range	$R_{Lm} = R_{src}$ and $X_{Lm} = -X_{src}$
R_L is fixed, but X_L can be varied	$X_{Lm} = -X_{src}$
X_L is fixed, but R_L can be varied	$R_{Lm} = \sqrt{R_{src}^2 + (X_L + X_{src})^2}$
R_L and X_L can be varied independently over a restricted range	X_{Lm} as close to $-X_{src}$ as possible; R_{Lm} as close to $\sqrt{R_{src}^2 + (X_L + X_{src})^2}$ as possible
$|Z_L|$ can be varied, whereas $\angle\theta_L$ is constant	$|Z_{Lm}| = |Z_{src}|$

If R_L cannot be made equal to this value, then a value of R_L as close to it as possible will give maximum power transfer.

Another case of interest arises from the use of a transformer with Z_L. The load imped-ance $Z_L = |Z_L|\angle\theta_L$ reflected to the primary side is ideally $Z_{Lp} = |Z_{Lp}|\angle\theta_L = \left(\dfrac{N_p}{N_s}\right)^2 |Z_L|\angle\theta_L$,

where N_p and N_s are the number of turns of the primary and secondary windings, respec-tively. It follows that $|Z_{Lp}| = \left(\dfrac{N_p}{N_s}\right)^2 |Z_L|$. The effect of an ideal transformer is to vary $|Z_{Lp}|$

while the phase angle remains constant. To derive the condition for maximum power transfer under these conditions, we substitute in Equation 7.4.6 $R_L = |Z_L|\cos\theta_L$, $X_L = |Z_L|\sin\theta_L$,

$R_{src} = |Z_{src}|\cos\theta_{src}$, and $X_{src} = |Z_{src}|\sin\theta_{src}$ to obtain:

$$P_L = V_{src}^2 \frac{|Z_L|\cos\theta_L}{\left(|Z_L|\cos\theta_L + |Z_{src}|\cos\theta_{src}\right)^2 + \left(|Z_L|\sin\theta_L + |Z_{src}|\sin\theta_{src}\right)^2} \qquad (7.4.10)$$

where the only variable on the RHS is $|Z_L|$. Deriving $\dfrac{dP_L}{d|Z_L|}$ and setting it equal to zero gives the condition for maximum power transfer:

$$|Z_{Lm}| = |Z_{src}| \qquad (7.4.11)$$

The various conditions for maximum power transfer are summarized in Table 7.4.1.

EXERCISE 7.4.4
Verify Equation 7.4.9 and Equation 7.4.11.

Example 7.4.2: Maximum Power Transfer with Variable Load Impedance

Given the circuit of Figure 7.4.9, it is required to determine the condition for maximum power transfer and the power transferred to R_L, under the following conditions:

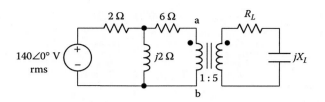

FIGURE 7.4.9
Figure for Example 7.4.2.

1. Both R_L and X_L are variable independently.
2. R_L is fixed at 35 Ω and X_L is variable.
3. X_L is fixed at –50 Ω and R_L is variable.
4. R_L is fixed at 35 Ω, X_L is fixed at –50 Ω, and the transformer can be selected to have any desired turns ratio.

SOLUTION
The first step is to derive TEC looking into terminals ab toward the source, and reflect TEC to the secondary side.

When terminals ab are open-circuited, $V_{Th} = 140\angle 0° \dfrac{j2}{2 + j2} = 70(1 + j)$ V rms. With the

source set to zero, the impedance looking into terminals ab is $Z_{Th} = 6 + \dfrac{j2 \times 2}{2 + j2} = 7 + j$ Ω.

Reflecting TEC to the secondary side, it becomes a voltage source of $70(1 + j) \times 5 =$
$350(1 + j) = 350\sqrt{2}\angle 45°$ V rms in series with $(7 + j) \times 25 = 175 + j25$ Ω (Figure 7.4.10).

1. If both R_L and X_L can be varied independently, the condition for maximum power transfer is given by Equation 7.4.8, $R_{Lm} = 175$ Ω and $X_{Lm} = -25$ Ω. The power

   transferred to R_{Lm} is $\dfrac{\left|350\sqrt{2}\right|^2}{4 \times 175} = 350$ W.

2. If $R_L = 35$ Ω and X_L is variable, maximum power transfer occurs when $X_{Lm} = -25$ Ω

   as in condition 1. The power transferred to R_L is $\dfrac{\left(350\sqrt{2}\right)^2 \times 35}{\left(175 + 35\right)^2} = 194.4$ W.

3. If $X_L = -50$ Ω and R_L is variable, the condition for maximum power transfer is

   given by Equation 7.4.9: $R_{Lm} = \sqrt{\left(175\right)^2 + \left(25 - 50\right)^2} = 176.8$ Ω. The magnitude of

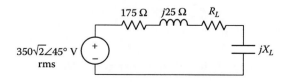

FIGURE 7.4.10
Figure for Example 7.4.2.

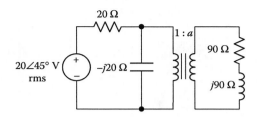

FIGURE 7.4.11
Figure for Exercise 7.4.5.

the current is $\dfrac{350\sqrt{2}}{\sqrt{(175+176.8)^2+(25)^2}} = 1.4$ A rms, and the power transferred to R_{Lm}

is $(1.4)^2 \times 176.8 = 348.2$ W.

4. If R_L is fixed at 35 Ω and X_L is fixed at –50 Ω, $|Z_L| = \sqrt{(35)^2+(50)^2} = 61.0$ Ω. The mag-

nitude of Z_{Th} is $\sqrt{49+1} = 7.1$ Ω. The turns ratio, instead of 1:5 should be $1:\sqrt{\dfrac{61.0}{7.1}}$ or

1:2.94, which is nearly 1:3. TEC reflected to the secondary side will be a source of

magnitude $70\sqrt{2} \times 3 \cong 300$ V rms in series with a resistance of $7 \times (3)^2 = 63$ Ω and a

reactance of $1 \times (3)^2 = 9$ Ω. The magnitude of the current is $\dfrac{300}{\sqrt{(35+63)^2+(9-50)^2}} =$

2.8 A rms, and the power transferred to R_L is $(2.8)^2 \times 35 = 274.4$ W.

EXERCISE 7.4.5

Determine a of the transformer in Figure 7.4.11, so that maximum power is transferred to the inductive load and calculate this power. Note that there are no dots on the transformer because these are irrelevant to the condition for maximum power transfer.

Answer: 3 and 5 W.

Admittance Relations

Let $\mathbf{I}_{SRC}$ and Y_{src} represent, in general, NEC as seen from terminals ab of a given circuit connected to a load $Y_L = G_L + jB_L$. The corresponding relations to Equation 7.4.8, Equation 7.4.9, and Equation 7.4.11 are, respectively (Exercise 7.4.5):

$$Y_{Lm} = Y_{src}^* \qquad (7.4.12)$$

$$G_{Lm} = \sqrt{G_{src}^2 + (B_L + B_{src})^2} \qquad (7.4.13)$$

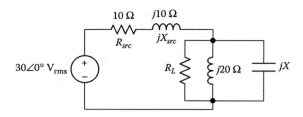

FIGURE 7.4.12
Figure for Example 7.4.3.

$$|Y_{Lm}| = |Y_{src}| \qquad\qquad (7.4.14)$$

EXERCISE 7.4.6
Verify Equation 7.4.12 to Equation 7.4.14.

Simulation Example 7.4.3: ac Power Relations with PSpice

Given the circuit of Figure 7.4.12, it is required to determine the condition for maximum power transfer and the power transferred to R_L assuming that:

1. Both R_L and X are variable independently.
2. Both R_L and X are variable independently, but only over the range 30 to 50 Ω for R_L and $-30\ \Omega$ to $-50\ \Omega$ for X_L.
3. If R_L is fixed at 30 Ω and X is variable, what is the condition for minimum power dissipated in R_{src}, and how much is this power?

SOLUTION

1. Let us convert the source and its impedance to its NEC. The Norton source current is: $I_N = \dfrac{30\angle 0°}{10 + j10} = \dfrac{30\angle 0°}{10\sqrt{2}\angle 45°} = 1.5\sqrt{2}\angle -45°$ A rms. The Norton admittance is:

$Y_N = \dfrac{1}{10 + j10} = \dfrac{10 - j10}{200} = \dfrac{1}{20} - \dfrac{j}{20}$ S. The circuit becomes as shown in Figure 7.4.13. According to Equation 7.4.12, maximum power is transferred to the load when $B_m = -B_N$ and $G_{Lm} = G_N$. The first condition gives: $-\dfrac{1}{20} - \dfrac{1}{X_m} = \dfrac{1}{20}$, from

which $X_m = -10\ \Omega$. The second condition gives: $G_L = G_s = \dfrac{1}{20}$ or $R_L = 20\ \Omega$. Under

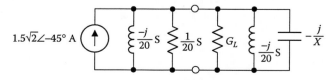

FIGURE 7.4.13
Figure for Example 7.4.3.

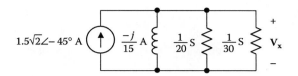

FIGURE 7.4.14
Figure for Example 7.4.3.

these conditions, the current in G_L is $\dfrac{I_N}{2} = \dfrac{1.5}{\sqrt{2}}$, and the power transferred to G_L is

$$\left(\frac{1.5}{\sqrt{2}}\right)^2 \frac{1}{G_{Lm}} = \frac{45}{2} = 22.5 \text{ W}.$$

2. If X is variable between -30 and -50 Ω, B is variable between $\dfrac{1}{30}$ and $\dfrac{1}{50}$ S. Maximum power transfer occurs when B has its largest positive value, which is $\dfrac{1}{30}$. Under these conditions, the total susceptance is $\left(-\dfrac{1}{20} - \dfrac{1}{20} + \dfrac{1}{30}\right) = -\dfrac{1}{15}$. G_{Lm} for maximum power transfer is then given by Equation 7.4.13 as, $G_{Lm} =$

$$\sqrt{\left(\frac{1}{20}\right)^2 + \left(\frac{1}{15}\right)^2} = \frac{1}{12} \text{ S. The value of } G_L \text{ nearest to this is } \frac{1}{30} \text{ S. Maximum power}$$

is transferred under these conditions when R_L and X have a magnitude of 30 Ω. To find the maximum power transferred, we note that the circuit reduces under these conditions to that shown in Figure 7.4.14. The admittance Y_x of the combi-

nation is $\left(-\dfrac{j}{15} + \dfrac{1}{20} + \dfrac{1}{30}\right) = \left(-\dfrac{j}{15} + \dfrac{1}{12}\right) = \dfrac{5 - j4}{60} = \dfrac{\sqrt{41}}{60} \angle -\tan^{-1} 0.8$. The voltage

$$\mathbf{V_x} = \frac{1}{Y_x}\,\mathbf{I_N}, \text{ so } |\mathbf{V_x}| = \frac{1}{|Y_x|}\,|\mathbf{I_N}| = 1.5\sqrt{2}\,\frac{60}{\sqrt{41}} \text{ and the power transferred to } G_L \text{ is:}$$

$G_L\,|\mathbf{V_x}|^2 = 13.2$ W.

3. If R_L is fixed and X is variable, minimum power is dissipated in R_{src} when the current in this resistor is minimum. With the source reactance fixed, the current in R_{src} is minimum when the impedance of the parallel combination of the load in Figure 7.4.12 is maximum; that is, when the admittance Y_L is a minimum.

$$Y_L = \frac{1}{30} - \frac{j}{20} - \frac{j}{X} = \frac{1}{30} + j\left(-\frac{1}{X} - \frac{1}{20}\right) \text{ or } |Y_L| = \sqrt{\left(\frac{1}{30}\right)^2 + \left(-\frac{1}{X} - \frac{1}{20}\right)^2} \text{ S. With } X \text{ as}$$

the only variable in this relation, $|Y_L|$ is minimum when the second term within parentheses vanishes, that is, when $X = -20$ Ω. The circuit reduces to that of

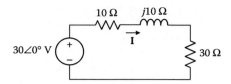

FIGURE 7.4.15
Figure for Example 7.4.3.

Figure 7.4.15. The current $\mathbf{I} = \dfrac{30\angle 0°}{10 + 30 + j10} = \dfrac{3}{4+j}$ and $|\mathbf{I}| = \dfrac{3}{\sqrt{17}}$ A. The power

dissipated in R_{src} is $R_{src}|\mathbf{I}|^2 = \dfrac{9}{17} \times 10 = 5.3$ W.

SIMULATION

The schematic is shown in Figure 7.4.16, where it has been assumed that $\omega = 1$ krad/s, $R_{Lm} = 20$ Ω and $X_m = -10$ Ω, so that the net reactance of the load is represented by a 50 μF capacitor. PSpice has a sinusoidal source whose waveform is described by:

$$v_{SRC} = \text{VOFF} + \left\{ \text{VAMPL}e^{-\text{DF}(t-\text{TD})} \sin[2\pi\text{FREQ}(t-\text{TD})] \right\} u(t - \text{TD})$$

That is, $v_{SRC} = \text{VOFF}$ for $0 \le t < \text{TD}$, where VOFF is a dc offset voltage and TD is a time delay for the start of the damped sinusoid, whose undamped amplitude is VAMPL and damping factor is DF. For $t > \text{TD}$, $u(t - \text{TD}) = 1$. The source is designated as VSIN in the SOURCE library. In the Property Editor Spreadsheet of the source enter 30 under VAMPL, 159.155 Hz (that is, $1000/2\pi$) under FREQ, and 0 under VOFF, TD, and DF. Nodes have been labeled VA, VB, and VC using net aliases (Appendix SD.2). The pin numbers of the passive circuit elements are shown for reference to current directions. Note that no printers are used for measuring voltages or currents; the Probe feature of PSpice can be used for this purpose.

To perform the simulation, select, from the Simulation Profile under PSpice, Time Domain (Transient) under Analysis type in the simulation Settings window. Enter 20m for Run to time, 0 for Start saving data after, and 10u for Maximum step size. After the simulation is run and the Schematic (active) window appears, select Add Trace from Trace

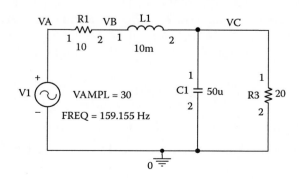

FIGURE 7.4.16
Figure for Example 7.4.3.

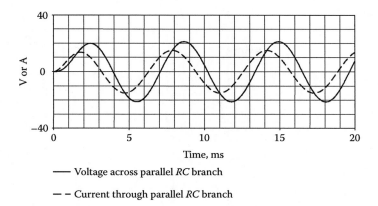

FIGURE 7.4.17
Figure for Example 7.4.3.

of the menu bar. To display the voltage across the parallel RC combination, choose V(VC) in the Add Traces window. To display the current through, and the voltage across, the parallel RC combination, choose I(L1), I(R1), or I(V2). To display the current through the parallel RC combination on the same plot, we have to multiply the current by 10 to have comparable amplitudes. Hence, we enter in the Trace Expression: V(VC), 10∗I(L1). The resulting plot is shown in Figure 7.4.17. Because this is a transient simulation, a steady state does not occur from the very beginning, at $t = 0$.

To derive quantitative information from the plot, press the Toggle cursor button. After the Probe Cursor is displayed, it can be moved between the two traces by double clicking on the label symbols below the plot, that is, the rectangle for the voltage trace and the diamond for the current trace. With the voltage display selected, place the cursor just before the third peak and press the Cursor Max button; the cursor displays 21.216, that is, very nearly the peak voltage ($15\sqrt{2}$) at that node. In a similar manner, we obtain 15.000 for the current trace which when divided by 10 gives 1.5 A for the peak current. From the difference between the times at which successive peaks occur, we can verify that the current leads the voltage by an eighth of a period or 45°. It should be noted that PSpice has some powerful tools for finding values at certain points, including search commands of appropriate syntax (Appendix SD.2).

To investigate power, we plot the instantaneous power p delivered to the parallel RC combination as V(VC)∗I(L1), the real power $P(1 + \cos 2\omega t)$ as V(VC)∗I(R3) or W(R3), and the reactive power $Q\sin 2\omega t$ as V(VC)∗I(C1). These plots are shown in Figure 7.4.18. The following measurements can be made from these plots using Probe cursor:

- Peak value of real power equals 22.5 W, so that the average P is half this value or

  11.25 W. This equals $\dfrac{15\sqrt{2} \times 1.5}{4} \cos 45°$ and is half the preceding value because the

  source is assumed here to have a peak value of 30 V, not rms value.

- Peak value of reactive power equals 11.25 W, which agrees with $\dfrac{15\sqrt{2} \times 1.5}{4} \sin 45°$.

- The instantaneous total power is the sum of the instantaneous real power and the instantaneous reactive power. The maximum value is 27.159 V, which

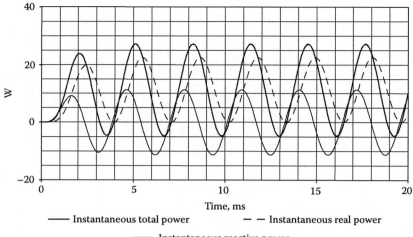

Time, ms

—— Instantaneous total power   — — Instantaneous real power

—— Instantaneous reactive power

FIGURE 7.4.18
Figure for Example 7.4.3.

equals $P + \sqrt{P^2 + Q^2}$ (P7.1.10) and the minimum value is –4.660, which equals $P - \sqrt{P^2 + Q^2}$.

7.5   Measurement of Current, Voltage, and Power

Current and Voltage Measurement

Instruments for measuring current are **ammeters**, whereas instruments for measuring voltage are **voltmeters**. The older instruments are analog, the deflection of the indicating pointer being proportional (or *analog*ous) to the current passing through the instrument. Modern instruments are digital, in which the quantity to be measured is sampled, digitized, and displayed as numerical digits (Section 17.6, Chapter 17). Compared to analog instruments, digital instruments: (1) are more accurate, (2) generally have negligible loading effects, (3) can measure very low values of current or voltage, and (4) can change ranges automatically. Nevertheless, analog instruments are useful for quick monitoring of the quantity being measured and for quick verification of normal or safe levels.

A basic type of analog movement, the moving coil or D'Arsonval movement, is useful for many applications. It consists essentially of a current-carrying coil that is free to rotate in the magnetic field of a permanent magnet (Figure 7.5.1). A pointer attached to the coil can move relative to a calibrated scale. When the coil is not carrying any current, helical springs hold the coil and pointer in a rest position that corresponds to a zero reading on the scale. When current passes through the coil, a torque is developed that rotates the coil against the restoring spring force. The pointer thus moves along the scale and indicates a certain reading. With proper design, the angle of rotation is directly proportional to the coil current, resulting in a linear scale. The D'Arsonval movement basically responds to dc, but can be made to respond to ac by adding rectifiers that convert an alternating current to a unidirectional one.

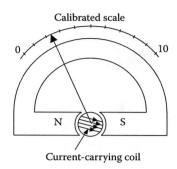

FIGURE 7.5.1
D'Arsonval movement.

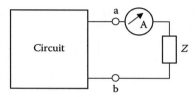

FIGURE 7.5.2
Current measurement.

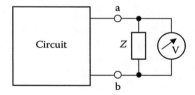

FIGURE 7.5.3
Voltage measurement.

   In order to measure the current passing through a given circuit element, an ammeter is connected in series with the element (Figure 7.5.2). An ideal ammeter has zero impedance so that it does not introduce additional impedance in series with Z, which would affect the current being measured. For example, let us assume that the circuit in Figure 7.5.2 has a TEC between terminals ab of 50 V in series with a 25 Ω resistance and that Z also equals 25 Ω. The current in the circuit is then 1 A. If an ammeter of 1 Ω resistance is used to measure the current, the total resistance in the circuit becomes 51 Ω, and the ammeter reading will be $\dfrac{50}{51} = 0.98$ A, which is in error by nearly –2%.

   In order to measure the voltage across a given circuit element, a voltmeter is connected in parallel with the element (Figure 7.5.3). An ideal voltmeter should have infinite impedance so that it does not introduce additional impedance in parallel with Z, which would affect the voltage being measured. For example, let us assume that the circuit in Figure 7.5.3 has an NEC between terminals ab of 1 A in parallel with a 50 Ω resistance and that Z also equals 50 Ω. It follows that $V_{ab} = 25$ V. If a voltmeter of 2.5 kΩ is used to measure the voltage, the total resistance becomes 25 ∥ 2,500 = 24.75 Ω. The voltmeter reading will be 24.75 V which is in error by –1%. It follows that:

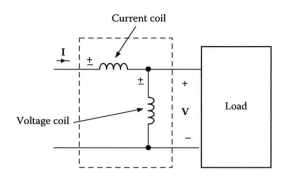

FIGURE 7.5.4
Power measurement.

> **Concept:** *For accurate current and voltage measurements, the impedance of an ammeter should be small compared with the impedance that appears in series with it, and the impedance of a voltmeter should be large compared with the impedance that appears in parallel with it.*

Power Measurement

As $|S| = |V_{rms}I^*_{rms}| = |V_{rms}||I_{rms}|$, the apparent power $|S|$ at any location is determined by measuring $|V_{rms}|$ and $|I_{rms}|$ at that location by means of an ac voltmeter and an ac ammeter, respectively. These instruments are usually calibrated to read rms values directly. To measure the real power, a **wattmeter** is used. The wattmeter has two coils: a current coil and a voltage coil with ± polarity markings. The current coil is connected like an ammeter so that the load current flows through it. The voltage coil is connected across the load terminals, like a voltmeter. When the polarities of the two coils with respect to I and V are as indicated in Figure 7.5.4, the wattmeter reads the power absorbed, $P = V_{rms}I_{rms}\cos(\theta_v - \theta_i)$. If the polarity of either coil is reversed, the wattmeter gives a negative indication. Knowing P and $|S|$, Q and the p.f. readily follow from the complex power triangle.

Application Window 7.5.1: Measurement of Large ac Currents and High ac Voltages

To measure large ac currents and to isolate the measuring and protection circuits from the high voltages that may be present, a current transformer is used with an ac ammeter. To consider the potential sources of error in such measurements, the transformer is represented by its equivalent circuit (Figure 6.5.3, Chapter 6), as shown in Figure 7.5.5. L_{sh} accounts for the finite inductance of the core, and L_{leak} accounts for the imperfect coupling between the primary and secondary windings. The resistances of the windings are assumed negligible. The effect of the total series impedance, including the impedance Z_A of the ammeter, is to produce a voltage drop across the primary winding. However, this effect is normally quite negligible because current transformers usually have a large secondary-to-primary turns ratio. In a 100 A/1 A transformer, for example, this turns ratio is 100. The series impedance reflected to the primary side is therefore divided by the square of the turns ratio. This impedance, including $j\omega L_{leak}$, is usually negligible compared to the circuit impedance in series with the primary of the current transformer. Of more significance is L_{sh}, as this has a shunting effect on the current through the ammeter. For high accuracy, therefore, the windings of a current transformer should have, at the lowest

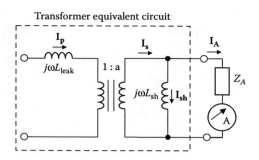

FIGURE 7.5.5
Figure for Application Window 7.5.1.

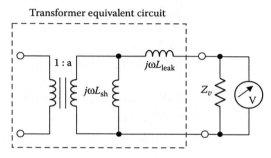

FIGURE 7.5.6
Figure for Application Window 7.5.1.

frequency of operation, a shunt reactance that is large compared to the impedance Z_A of the ammeter.

To measure high ac voltages, and to isolate the measuring and protection circuits from the high voltages that may be present, a voltage transformer is used with an ac voltmeter. Referring to the transformer equivalent circuit of Figure 7.5.6 (Figure ST6.2.5, Section ST6.2), the effect of L_{sh} is to draw additional current from the voltage source to which the primary winding is connected. This effect is normally quite negligible because voltage transformers usually have a large primary-to-secondary turns ratio. In a 10 kV/100 V transformer, for example, this turns ratio is 100. The shunt impedance reflected into the primary side is therefore multiplied by the square of the turns ratio. This impedance is usually much larger than the circuit impedance in parallel with the primary of the voltage transformer. Of more significance is the series impedance due to L_{leak}, as this has a voltage divider effect on the reading of the voltmeter. For high accuracy, therefore, a voltage transformer should have, at the highest frequency of operation, a leakage reactance that is small compared to the impedance of Z_v of the voltmeter.

Summary of Main Concepts and Results

- When v and i are sinusoidal functions of time of frequency ω, with v being a voltage drop in the direction of i, the instantaneous power $p = vi$ is pulsating at a frequency 2ω. If v is in phase with i, as in the case of R, $p \geq 0$ and represents power dissipated. If v and i are in phase quadrature, as in the case of L and C, p

is purely alternating, of zero average, because no power is dissipated. In this case, when v and i have the same sign, $p > 0$ represents energy being stored in the energy-storage element. When v and i have opposite signs, $p < 0$ represents previously stored energy being returned to the rest of the circuit.

- In any given circuit, complex power is conserved, which implies that real power and complex power are also conserved.
- Real power and reactive power can each be summed branch by branch in a given circuit, with the total algebraic sum of each, over all branches of the circuit, equal to zero.
- Reactive power, although it averages to zero over a cycle, generally increases the voltage and current requirements of a load.
- $\cos(\theta_v - \theta_i)$ is the p.f. It is the cosine of the angle between impedance and resistance in the impedance triangle, or between complex power and real power in the complex power triangle. The p.f. is unity for a pure resistance, is zero lagging for a purely inductive reactance, and is zero leading for a purely capacitive reactance.
- The p.f. can be brought to unity by adding capacitance to counteract the inductive reactance of the load.
- If a source of open-circuit voltage $\mathbf{V}_{\mathbf{SRC}}$ and source impedance Z_{src} is connected to a load $Z_L = R_L + jX_L$, and R_L and X_L can be varied independently without restriction, maximum power is transferred from the source to the load when $Z_{Lm} = Z_{src}^*$, where Z_{src}^* is the conjugate of Z_{src}. If R_L and X_L can be varied independently, but only over a restricted range, maximum power is transferred when $X_L + X_{src}$ is as small as possible and R_L is as close as possible to $\sqrt{R_{src}^2 + (X_L + X_{src})^2}$. If $|Z_L|$ can be varied, as when a transformer is used, maximum power is transferred when $|Z_{Lm}| = |Z_{src}|$. Similar relations apply if admittances are used instead of impedances.
- Current, voltage, and power are measured using ammeters, voltmeters, and wattmeters, respectively. An ammeter, as well as the current coil of a wattmeter, is connected so that the current of interest flows through these elements. Ideally, these elements should have zero impedance. A voltmeter, as well as the voltage coil of a wattmeter, is connected across the voltage of interest. Ideally, these measuring elements should have infinite impedance.

Learning Outcomes

- Apply complex power to analysis of power circuits.
- Derive the condition for maximum power transfer and determine the maximum power transferred.

Supplementary Topics and Examples on CD

ST7.1 Tellegen's theorem: Illustrates, discusses, and gives a proof of Tellegen's theorem.

ST7.2 Compensation theorem: Illustrates, discusses, and proves the theorem. The compensation theorem is useful in determining the effect of changing an impedance Z in a given circuit on the currents in the circuit. Only the current in Z and the circuit parameters need be known, and not the values of any excitations. The compensation theorem can also be applied to correct for the loading effects of nonideal ammeters and voltmeters.

SE7.1 Maximum power transfer when resistor other than load is varied: Investigates maximum power transfer when one resistor in a particular circuit is varied so as to maximize power dissipated in another resistor.

SE7.2 Maximum power transfer in transformer circuit with capacitor coupling: Examines maximum power transfer in a circuit having a capacitor connected between the primary and secondary windings of an ideal transformer.

SE7.3 Complex power in circuit with transformer having center-tapped secondary: Calculates complex power in three loads connected on the secondary side of an ideal transformer having a center-tapped secondary winding.

SE7.4 Complex power calculation using TEC: Determines power in a two-source *RLC* circuit using TEC.

Problems and Exercises

P7.1 Complex Power

Verify your results with PSpice simulation whenever appropriate.

P7.1.1 The conjugate of the complex power delivered by a current source is $400 - j400$ VA. If the source current is $10\angle45°$ A rms, determine the rms voltage across the source.

P7.1.2 Two inductive loads of 0.88 kW, 0.8 p.f. and 1.32 kW, 0.6 p.f. are connected across a 220 V rms, 50 Hz supply. (a) Calculate the total complex power of the loads and the supply current; (b) determine the capacitance that should be connected in parallel with the loads to bring the p.f. to 0.9 lagging.

P7.1.3 Determine the complex power delivered by the independent current source in Figure P7.1.3 and verify that it equals the complex power absorbed in the rest of the circuit.

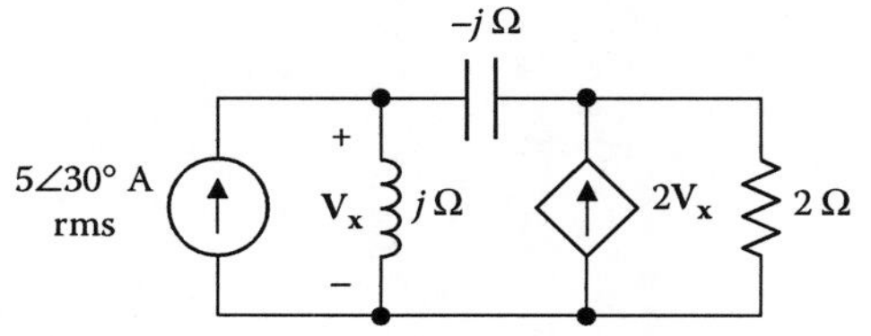

FIGURE P7.1.3

P7.1.4 An impedance $4 + j4$ Ω is connected in parallel with an impedance $12 + j16$ Ω. If the total reactive power absorbed is 2,500 VAR, what is the total real power absorbed? Verify your answer by working in terms of the overall impedance and admittance.

P7.1.5   Determine C in Figure P7.1.5, if the capacitor absorbs -5 VAR and the frequency is 50 Hz. Derive the power absorbed by C from conservation of power in the circuit.

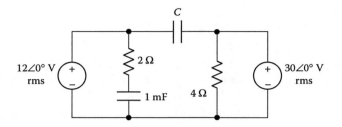

FIGURE P7.1.5

P7.1.6   Determine R and the rms magnitude of $\mathbf{V}_{SRC}$ in Figure P7.1.6, given that each resistor absorbs 2 W and $\omega = 1000$ rad/s.

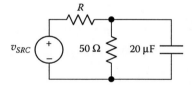

FIGURE P7.1.6

P7.1.7   Three loads are supplied in parallel at 500 V rms. L_1 absorbs 12 kW at 0.6 p.f. lagging, L_2 absorbs 15 kW at unity p.f., and L_2 absorbs 6 kVAR at 0.8 leading. The line has a resistance of 1 Ω and negligible reactance. Determine: (a) the rms magnitude of the source voltage, (b) the combined power factor of the load, and (c) the fraction of the real power delivered by the source that is absorbed by the loads.

P7.1.8   Given that the complex power absorbed by the inductive branch in Figure P7.1.8 is $12 + j16$ VA, determine C so that the power factor at terminals ab is unity, assuming that $\omega = 1$ rad/s.

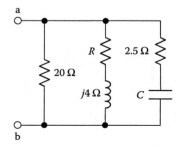

FIGURE P7.1.8

P7.1.9 Determine the complex power delivered by the two sources in Figure P7.1.9, given that L_1 absorbs 4 kW at a power factor of 0.6 lagging, L_2 absorbs 3 kW at a power factor of 0.6 leading, and the complex power absorbed by L_3 is $12 + j5$ kVA. Assume that $\mathbf{V}_{SRC1} = 400\angle0°$ V rms and $\mathbf{V}_{SRC2} = 400\angle90°$ V rms.

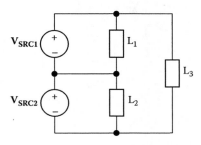

FIGURE P7.1.9

P7.1.10 Show that in Figure 7.1.4b the maximum and minimum instantaneous values of p are $P + \sqrt{P^2 + Q^2}$ and $P - \sqrt{P^2 + Q^2}$, respectively.

P7.1.11 In Figure P7.1.11, find the instantaneous power, the real power, and the reactive power delivered by the source, given that $v_{SRC} = 10\cos10^6 t$.

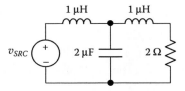

FIGURE P7.1.11

P7.1.12 Determine the reactance that must be placed in parallel with terminals ab in Figure P7.1.12 so that the power factor is unity at these terminals.

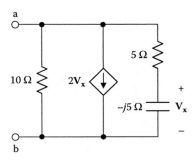

FIGURE P7.1.12

P7.1.13 Find the total complex power delivered by the two sources in Figure P7.1.13, if $v_{SRC1} = 5\cos(2t + 45°)$ V and $v_{SRC2} = 5\cos 2t$ V.

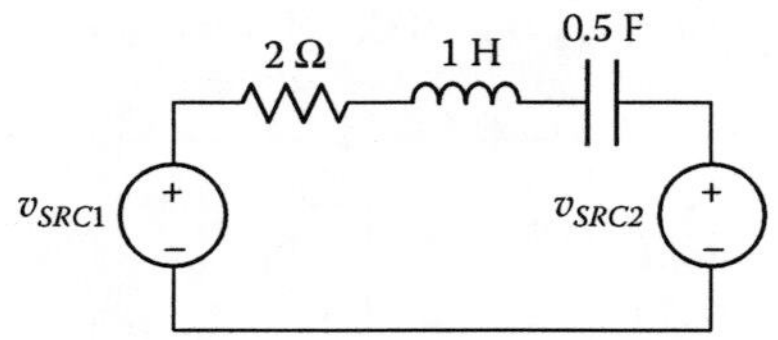

FIGURE P7.1.13

P7.2   Maximum Power Transfer

Verify your results with PSpice simulation.

P7.2.1 Determine R_L in Figure P7.2.1 that will absorb maximum power and calculate this power.

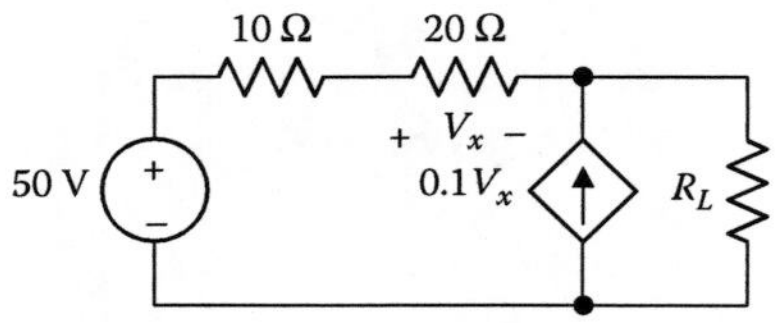

FIGURE P7.2.1

P7.2.2 Determine G_L in Figure P7.2.2 that will absorb maximum power and calculate this power.

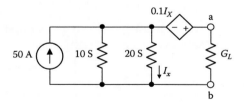

FIGURE P7.2.2

P7.2.3 Determine R_L in Figure P7.2.3 that will absorb maximum power and calculate this power.

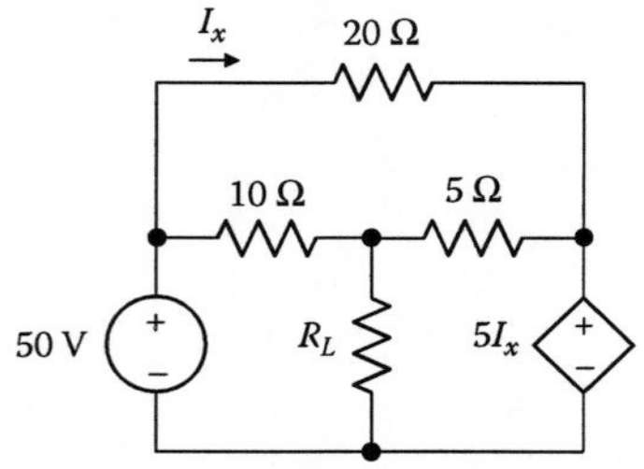

FIGURE P7.2.3

P7.2.4 Determine R_L that should be connected between terminals ab of Figure P5.3.37 (Chapter 5) for maximum power absorption in R_L and calculate this power.

P7.2.5 Determine R_L that should be connected between terminals ab of Figure P5.3.38 (Chapter 5) for maximum power absorption in R_L and calculate this power, assuming that a resistance of $20/17\ \Omega$ is added in series with terminal a of Figure P5.3.38 (Chapter 5).

P7.2.6 Determine ωM in Figure P7.2.6 so that maximum power is absorbed by the 20 Ω resistor and calculate this power.

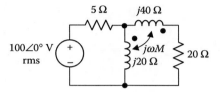

FIGURE P7.2.6

P7.2.7 Determine R_L in Figure P7.2.7 so that it absorbs maximum power and calculate this power.

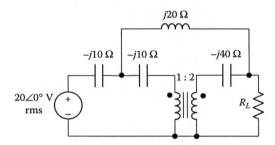

FIGURE P7.2.7

P7.2.8 Rework Problem P7.2.7, assuming that the positions of R_L and the $j20$ Ω inductor are interchanged.

P7.2.9 Determine Z_L in Figure P7.2.9 that makes it absorb maximum power and calculate this power.

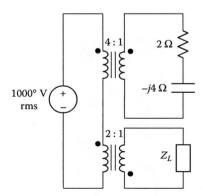

FIGURE P7.2.9

P7.2.10 Determine R_L that should be connected between terminals ab of Figure P6.3.15 (Chapter 6) for maximum power absorption in R_L and calculate this power.

P7.2.11 (a) Determine Z_L in Figure P7.2.11 that results in maximum power absorption in Z_L and calculate this power. (b) If Z_L consists of a resistor in parallel with a capacitor, what values of these elements will result in maximum power absorption in Z_L and how much is this power?

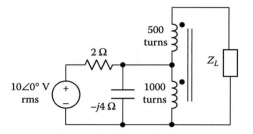

FIGURE P7.2.11

P7.2.12 (a) Determine Y_L in Figure P7.2.12 that results in maximum power absorption in Y_L and calculate this power. (b) If Y_L consists of a resistor in parallel with an inductor, what values of these elements will result in maximum power absorption in Z_L and how much is this power?

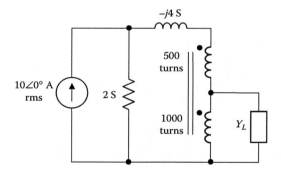

FIGURE P7.2.12

P7.2.13 Determine Z_L in Figure P7.2.13 that will absorb maximum power and calculate this power.

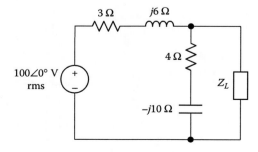

FIGURE P7.2.13

P7.2.14 Determine *a* in Figure P.7.2.14 so that maximum power is absorbed by the 20 Ω resistor and calculate this power.

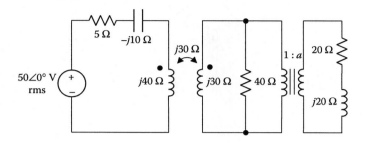

FIGURE P7.2.14

P7.2.15 Derive TEC between terminals cd in Figure SE7.4.1 (Example SE7.4) using superposition to determine the short-circuit current and the open-circuit voltage between these terminals.

P7.2.16 R_L in Figure P7.2.16 is restricted to the range 1 to 8 Ω. Determine the value of R_L that results in maximum power transfer to it and calculate this power.

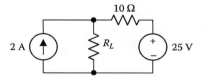

FIGURE P7.2.16

P7.2.17 Determine *a* of the transformer in Figure P7.2.17 so that maximum power is transferred to the 200 Ω load and calculate this power.

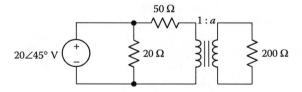

FIGURE P7.2.17

P7.2.18 A two-winding transformer having $N_1 = 1000$ turns and $N_2 = 200$ turns is used to match a 1 kΩ load to the primary side. If an autotransformer having $N_1 = 1000$ turns is to be used for this purpose, determine N_2 if (a) the windings are connected as in Figure P7.2.18 and (b) the connections of one of the windings are reversed.

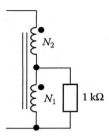

FIGURE P7.2.18

P7.2.19 (a) Determine the turns ratio a in Figure P7.2.19 so that maximum power is absorbed in the 10 Ω resistor and calculate this power. (b) Assuming $a = 2$ and $R_x = 10\ \Omega$, determine X that results in maximum power absorption in R_x and calculate this power. (c) Assuming $a = 2$ and $X = 15\ \Omega$, determine R_x that results in maximum power absorption in this resistor and calculate this power. (d) Assuming $a = 2$ and R_x and X are variable, determine R_x and X that will result in maximum power absorption in R_x and calculate this power.

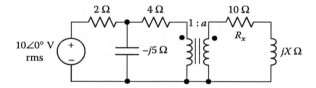

FIGURE P7.2.19

P7.2.20 (a) Determine the turns ratio a in Figure P7.2.20 so that maximum power is absorbed in the 10 S resistor and calculate this power. (b) Assuming $a = 2$ and $G_x = 10$ S, determine B that results in maximum power absorption in G_x and calculate this power. (c) Assuming $a = 2$ and $B = 15$ S, determine G_x that results in maximum power absorption in this resistor and calculate this power. (d) Assuming $a = 2$ and G_x and B are variable, determine G_x and B that will result in maximum power absorption in G_x and calculate this power.

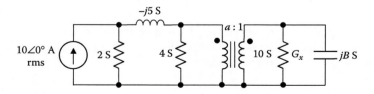

FIGURE P7.2.20

P7.3 Measurement of Current and Voltage

P7.3.1 A D'Arsonval movement has a resistance $R_A = 100 \ \Omega$ and requires 50 µA for full-scale deflection. Calculate R_{sh}, the resistance of a shunt that should be connected in parallel with the movement in Figure P7.3.1 so as to give an ammeter having a full-scale deflection of 1 mA.

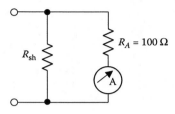

FIGURE P7.3.1

P7.3.2 Calculate R_v in Figure P7.3.2 so that the same D'Arsonval movement of Problem P7.3.1 can be used as a voltmeter having a full-scale deflection of 10 V.

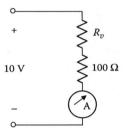

FIGURE P7.3.2

P7.3.3 In Figure P7.3.3, the voltmeter has a full-scale reading of 800 V and a 1 mA/ 100 mV D'Arsonval movement. Determine the percentage error in the reading of the voltmeter.

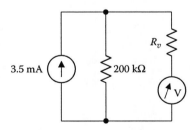

FIGURE P7.3.3

P7.3.4 A series resistance and a shunt are connected to two 1 mA/100 mV D'Arsonval movements to convert them to a voltmeter having a full-scale deflection of 10 V and to an ammeter having a full-scale deflection of 10 mA. The two meters are connected as shown in Figure P7.3.4 to measure the current through a resistor R and the voltage across R. If R is adjusted so that the ammeter reads full scale, what is the percentage error in the reading of each meter?

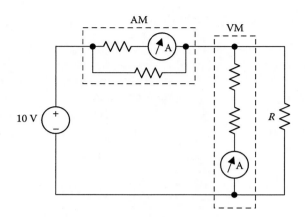

FIGURE P7.3.4

P7.3.5   A series resistance is used with a 100 μA /100 Ω D'Arsonval movement to make it read 1000 V full scale. The meter, including the series resistance, is to be converted to a voltmeter that reads 400 V full scale. Determine the resistance required.

P7.3.6   A shunt is used with a 100 μA /100 Ω D'Arsonval movement to make it read 1 mA full scale. The meter, including the shunt, is to be converted to a voltmeter that reads 500 V full scale. Determine the series resistance required.

P7.3.7   A 50 μA/100 Ω meter is converted to a voltmeter that reads 20 V full scale by adding a series resistance R_v. (a) Determine R_v. This same resistance is added in series with a 100 μA/100 Ω meter. (b) Determine the full-scale reading of the resulting voltmeter. The two voltmeters are connected in series across a voltage supply as shown in Figure P7.3.7. What should be the values of R_1 and R_2 if each voltmeter is to read full scale with the current from the supply being 200 μA?

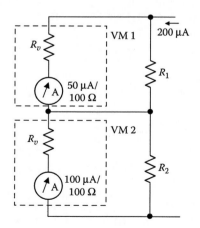

FIGURE P7.3.7

8

Balanced Three-Phase Systems

Overview

Modern electric power systems are gigantic networks that connect numerous power stations to a multitude of load centers dispersed over a wide geographical area that may span several neighboring countries. These networks incorporate sophisticated communications and control systems that are used to optimize the performance of the network. Three-phase systems are almost universally used for the generation, transmission, and distribution of electrical energy because of the many advantages they offer, compared to single-phase systems. Practical three-phase systems are nominally balanced under normal operating conditions; that is, their voltages and currents possess a certain symmetry.

Three-phase generators, transformers, and loads are connected in one of two basic connections: Y or Δ, each having its own characteristic features. In analyzing balanced three-phase systems, it is usually convenient to transform the Δ connection to Y and derive an equivalent single-phase system. Because three-phase systems are primarily power systems, it is important to examine power in these systems, both instantaneous and complex, as well as power measurement and power factor correction.

Some topics of practical importance are associated with power systems, such as the reasons for grounding these systems, protection of electrical installations against the hazards of fire and electric shock, and the physiological effects of electric current on the human body. These topics are discussed in the relevant sections of the Supplementary Topics.

Learning Objectives

- To be familiar with:
  - The terminology and general features of three-phase systems
  - Some general aspects of power generation, transmission, and distribution
- To understand:
  - Characteristics of the Y and Δ connections in balanced, three-phase systems
  - How balanced three-phase systems can be analyzed using a single-phase equivalent circuit based on a Y-Y representation
  - The interpretations of instantaneous power, real power, reactive power, and complex power in balanced three-phase systems

- The two-wattmeter method of measuring real power
- The advantages of three-phase systems compared to single-phase systems

8.1   Three-Phase Quantities

Consider the set y_a, y_b, and y_c representing either voltages or currents that vary sinusoidally with time, as follows:

$$y_a = Y_m \cos(\omega t + \theta) \tag{8.1.1}$$

$$y_b = Y_m \cos(\omega t + \theta - 120°) \tag{8.1.2}$$

$$y_c = Y_m \cos(\omega t + \theta + 120°) \tag{8.1.3}$$

y_a, y_b, and y_c are described as balanced three-phase quantities because of two distinguishing characteristics: they all have the same amplitude Y_m, and their phase angles differ by 120°, or 1/3 of the full angle of 360°. By extension, the set y_a, y_b, ..., y_n represents balanced n-phase quantities if all these quantities have the same amplitude and their phase angles differ by 360°/n. If either condition is not satisfied, the quantities are no longer balanced. Systems having $n > 2$ are described as **polyphase** systems.

Three-phase systems are by far the most important polyphase systems in practice. The simplest three-phase system consists of a three-phase generator (Section ST8.4) connected to a three-phase load. The generator and load may be connected in Y or Δ, which gives rise to the four possibilities illustrated in Figure 8.1.1a to Figure 8.1.1d.

The common node of the Y connection is the **neutral**. Each connection between the generator and the load (aA, bB, or cC in Figure 8.1.1a to Figure 8.1.1d) is a **line**. In power systems, the lines are often referred to as **feeders**. A line has a certain impedance, which in particular cases may be neglected. The line impedances are denoted by Z_{aA}, Z_{bB}, and Z_{cC} in Figure 8.1.1a to Figure 8.1.1d. The current that flows in a given line is a **line current**. The voltage between any two lines, whether at the generator end or the load end, is a **line voltage**.

The generator or load branches that are connected in Y or Δ are the **phases**. In Figure 8.1.1a to Figure 8.1.1d, each phase of the generator consists of an ideal voltage source in series with a source impedance. The phases of the load are the impedances Z_A, Z_B, and Z_C. These may be single-phase loads such as lamps or heaters, connected in Y or Δ, or they may be an inherently three-phase load, such as a three-phase motor. A **phase voltage** and a **phase current** are associated with each phase.

A three-phase system is balanced if all of the following conditions are satisfied:

1. The generator voltages are balanced in accordance with Equation 8.1.1 to Equation 8.1.3. That is:

$$v_{ga} = V_{gm} \cos(\omega t + \theta_v) \tag{8.1.4}$$

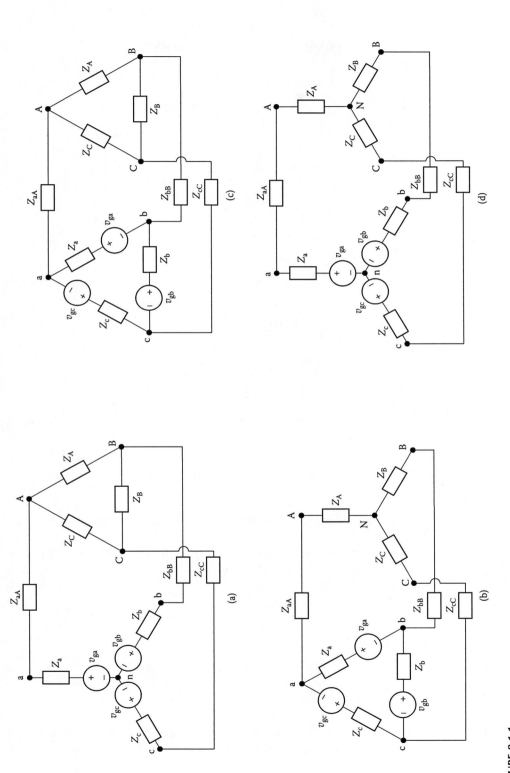

FIGURE 8.1.1

Δ and Y connections of three-phase systems.

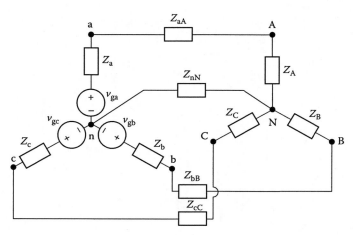

FIGURE 8.1.2
Four-wire three-phase system.

$$v_{gb} = V_{gm} \cos\left(\omega t + \theta_v - 120°\right) \tag{8.1.5}$$

$$v_{gc} = V_{gm} \cos\left(\omega t + \theta_v + 120°\right) \tag{8.1.6}$$

2. Phase impedances are equal, both for the generator and the load:

$$Z_a = Z_b = Z_c$$

$$Z_A = Z_B = Z_C \tag{8.1.7}$$

3. Line impedances are equal, that is:

$$Z_{aA} = Z_{bB} = Z_{cC} \tag{8.1.8}$$

Under these conditions, each of the sets of phase voltages, phase currents, line voltages, and line currents is a balanced set.

The three-phase systems illustrated in Figure 8.1.1a to Figure 8.1.1d are **three-wire** systems because three lines connect the generator to the load. With both the generator and load Y connected, the two neutral points may be connected together, resulting in a **four-wire** system (Figure 8.1.2).

Sum of Balanced Quantities

An important characteristic of balanced quantities is the following:

> **Concept:** *The sum of a set of balanced quantities is zero.*

Although this is true of any number of phases, we will illustrate it for the three-phase case. Thus, if y_a, y_b, and y_c are a balanced set given by Equation 8.1.1 to Equation 8.1.3, their sum is zero:

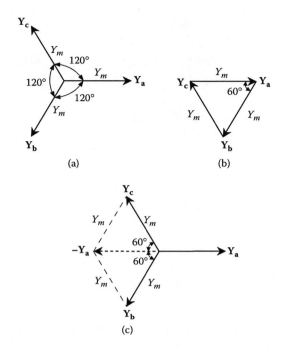

FIGURE 8.1.3
Sum of balanced three-phase quantities.

$$y_a + y_b + y_c = 0 \tag{8.1.9}$$

Equation 8.1.9 may be proved in a number of ways. As phasors (Figure 8.1.3a), $\mathbf{Y_a} = Y_m\angle 0°$, $\mathbf{Y_b} = Y_m\angle -120°$, and $\mathbf{Y_c} = Y_m\angle 120°$ may be added by laying them end to end. Because their magnitudes are equal and their phase angles differ by 120°, they form an equilateral triangle (Figure 8.1.3b). As this is a closed figure:

$$\mathbf{Y_a} + \mathbf{Y_b} + \mathbf{Y_c} = 0 \tag{8.1.10}$$

Alternatively, we may add any two phasors and show that the sum is the negative of the third phasor. For example, if we add $\mathbf{Y_b}$ and $\mathbf{Y_c}$ in Figure 8.1.3c using the parallelogram construction (Section 3.3, Chapter 3), the resultant is $-\mathbf{Y_a}$.

In the time domain, adding y_b and y_c in Equation 8.1.2 and Equation 8.1.3 gives $Y_m\cos(\omega t + \theta - 120°) + Y_m\cos(\omega t + \theta + 120°) = 2Y_m\cos(\omega t + \theta) \times \cos 120° = -Y_m\cos(\omega t + \theta) = -y_a$. Figure 8.1.4 illustrates the time variation of y_a, y_b, and y_c. At any instant of time, such as t_1 or t_2, the sum of y_a, y_b, and y_c is zero. At t_2, for example, if y_a represents current flowing toward a load, then the total current flowing away from the load, $(y_b + y_c)$, is equal and opposite. A useful way of looking at this is to consider that at any instant two line conductors are acting as the return line of the third conductor or, alternatively, one conductor is acting as the return line for the other two.

Phase Sequence

In Equation 8.1.1 to Equation 8.1.3, y_b lags y_a by 120° and y_c lags y_b by 120° because $\cos(\omega t + \theta + 120°) = \cos(\omega t + \theta - 240°)$. The **phase sequence** is abc and is a *positive* phase sequence. If the phase angles of y_b and y_c are interchanged:

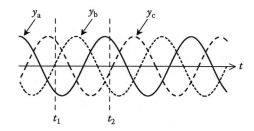

FIGURE 8.1.4
Time-variation of balanced three-phase quantities.

$$y_b = Y_m \cos(\omega t + \theta + 120°) \tag{8.1.11}$$

$$y_c = Y_m \cos(\omega t + \theta - 120°) \tag{8.1.12}$$

then y_c lags y_a by 120° and y_b lags y_c by 120°. The phase sequence is now acb and is a *negative* phase sequence.

The phasors in Figure 8.1.3a have a positive phase sequence. If the phasors are considered to rotate in the counterclockwise direction, they cross a fixed reference line, such as the real axis, in the order of the phase sequence. Alternatively, we may consider the phasors to be stationary, and we may move around the origin in the clockwise direction. The order in which the phasors are encountered is the phase sequence. In a negative phase sequence, phasors Y_b and Y_c are interchanged, but the same interpretations apply.

In practice, the phase sequence is reversed by interchanging any two line connections to a three-phase load. Assuming a positive phase sequence abc in Figure 8.1.5a, the load whose terminals are A, B, and C sees the same phase sequence abc. In Figure 8.1.5b, the same load connected to the same supply sees a negative phase sequence acb. If the load is a lighting or heating load, changing the phase sequence is generally of no consequence. However, if the load incorporates a three-phase motor, reversing the phase sequence reverses the direction of rotation of the motor (Section ST8.4). Although this provides a convenient way of reversing the direction of rotation of an ac motor, it may have disastrous mechanical consequences on loads already connected to the motor, such as compressors or machine tools. Moreover, it is common practice in medium and large three-phase power systems to have two or more three-phase generators connected in parallel, in which case they must have the same phase sequence.

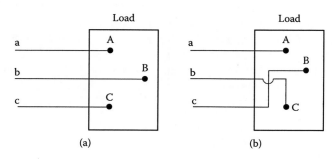

FIGURE 8.1.5
Reversal of phase sequence.

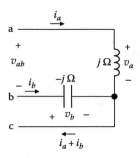

FIGURE 8.1.6
Figure for Exercise 8.1.2.

EXERCISE 8.1.1

A set of balanced six-phase voltages of 10 V amplitude is given. One of these voltages, say, that of phase 1, has a phase angle of 10°. Express the phase voltages in the time domain, with respect to phase 1, assuming the phase sequence is: (a) positive; (b) negative.

Answer: (a) $v_1 = 10\cos(\omega t + 10°)$ V, $v_2 = 10\cos(\omega t - 50°)$ V, $v_3 = 10\cos(\omega t - 110°)$ V, $v_4 = 10\cos(\omega t - 170°)$ V, $v_5 = 10\cos(\omega t + 130°)$ V, and $v_6 = 10\cos(\omega t + 70°)$ V. (b) $v_1 = 10\cos(\omega t + 10°)$ V, $v_2 = 10\cos(\omega t + 70°)$ V, $v_3 = 10\cos(\omega t + 130°)$ V, $v_4 = 10\cos(\omega t - 170°)$ V, $v_5 = 10\cos(\omega t - 110°)$ V, and $v_6 = 10\cos(\omega t - 50°)$ V.

EXERCISE 8.1.2

According to the definition of a polyphase system, a balanced two-phase system would be defined as $y_a = Y_m \cos(\omega t + \theta)$ and $y_b = Y_m \cos(\omega t + \theta - 180°)$. However, a two-phase system is usually defined as $y_a = Y_m \cos(\omega t + \theta)$ and $y_b = Y_m \cos(\omega t + \theta \pm 90°)$. Assume that in the two-phase system shown in Figure 8.1.6, $v_a = 10\cos\omega t$ V and $v_b = 10\cos(\omega t - 90°)$ V. Determine: (a) i_a; (b) i_b; (c) v_{ab}; and (d) $i_a + i_b$.

Answer: (a) $i_a = 10\cos(\omega t - 90°)$ A; (b) $i_b = 10\cos\omega t$ A; (c) $10\sqrt{2}\cos(\omega t + 45°)$ V; and (d) $10\sqrt{2}\cos(\omega t - 45°)$ A.

8.2 The Balanced Y Connection

Current Relations

Figure 8.2.1 shows a basic Y-Y configuration for a balanced three-phase system, where $\mathbf{V}_{ga} = V_g \angle 0°$, $\mathbf{V}_{gb} = V_g \angle -120°$, and $\mathbf{V}_{gc} = V_g \angle 120°$. It is clear from this figure that the phase currents are the same as the line currents:

$$\mathbf{I}_a = \mathbf{I}_{aA} = \mathbf{I}_A \ , \ \mathbf{I}_b = \mathbf{I}_{bB} = \mathbf{I}_B \ , \ \mathbf{I}_c = \mathbf{I}_{cC} = \mathbf{I}_C \qquad (8.2.1)$$

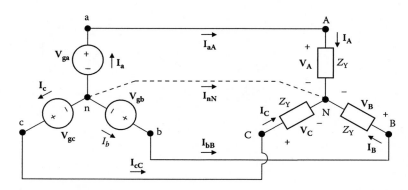

FIGURE 8.2.1
Balanced Y-Y connection.

An important feature of the balanced Y-Y connection follows from the fact that the currents form a balanced set as follows:

$$\mathbf{I}_{aA} + \mathbf{I}_{bB} + \mathbf{I}_{cC} = 0 \tag{8.2.2}$$

Concept: *In a balanced Y-Y connection, the neutral current is zero, $\mathbf{I}_{nN} = 0$.*

To show this, assume that the generators in Figure 8.2.1 are enclosed by a surface S. It follows from KCL, as discussed in connection with Figure 2.2.2, that $\mathbf{I}_{aA} + \mathbf{I}_{bB} + \mathbf{I}_{cC} + \mathbf{I}_{nN}$ = 0. Substituting Equation 8.2.2 gives $\mathbf{I}_{nN}$ = 0.

Voltage Relations

Because $\mathbf{I}_{nN}$ = 0, nodes n and N can be considered to be at the same voltage. If the line impedances are assumed to be zero, the phase voltages at the generator and load are equal:

$$\mathbf{V}_{ga} = \mathbf{V}_A, \ \ \mathbf{V}_{gb} = \mathbf{V}_B, \ \ \mathbf{V}_{gc} = \mathbf{V}_C \tag{8.2.3}$$

The line voltages at the load and generator are also equal in this case. From KVL:

$$\mathbf{V}_{ab} = \mathbf{V}_{ga} - \mathbf{V}_{gb}, \ \ \mathbf{V}_{bc} = \mathbf{V}_{gb} - \mathbf{V}_{gc}, \ \ \mathbf{V}_{ca} = \mathbf{V}_{gc} - \mathbf{V}_{ga} \tag{8.2.4}$$

Figure 8.2.2a shows the phasors of the three phase voltages, of phase sequence abc, and the three line voltages satisfying Equation 8.2.4. Triangle anb is isosceles having the two sides na and nb equal to V_g. Because α = 30°, then $\mathbf{V}_{ab} = \sqrt{3}V_g \angle 30°$. $\mathbf{V}_{bc}$ and $\mathbf{V}_{ca}$ are similarly determined from the phasor diagram. Thus,

$$\mathbf{V}_{ab} = \sqrt{3}V_g \angle 30°, \ \mathbf{V}_{bc} = \sqrt{3}V_g \angle -90°, \ \mathbf{V}_{ca} = \sqrt{3}V_g \angle 150° \tag{8.2.5}$$

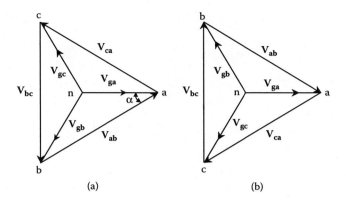

FIGURE 8.2.2
Phase and line voltages for positive and negative phase sequences.

For a negative phase sequence, Equation 8.2.4 still applies. From the phasor diagram (Figure 8.2.2b), the following relations may be derived:

$$\mathbf{V_{ab}} = \sqrt{3}V_g\angle - 30°, \ \mathbf{V_{ca}} = \sqrt{3}V_g\angle(-150°), \ \mathbf{V_{bc}} = \sqrt{3}V_g\angle 90° \tag{8.2.6}$$

> **Summary:** *In a balanced Y connection, the line currents are equal to the phase currents. The line voltages are balanced three-phase voltages whose magnitude is $\sqrt{3}$ times that of the phase voltages, the phase sequence being the same. For a positive phase sequence, the line voltage leads by 30° the voltage in the phase connected to the given line and having the same polarity of voltage at the connecting node. For a negative phase sequence, the phase angle is 30° lag.*

Thus, in Figure 8.2.1 $\mathbf{V_{ab}}$ leads $\mathbf{V_A}$ and $\mathbf{V_{ga}}$ by 30°.
Phase voltages are derived from line voltages in Section ST8.1.

Power Relations

Consider the load in Figure 8.2.1. If $Z_Y = |Z_Y|\angle\theta$, the phase currents are related to the corresponding phase voltages as follows:

$$\mathbf{I_A} = \mathbf{V_A}/Z_Y = \frac{V_g\angle 0°}{|Z_Y|\angle\theta} = \frac{V_g}{|Z_Y|}\angle - \theta$$

$$\mathbf{I_B} = \mathbf{V_B}/Z_Y = \frac{V_g\angle - 120°}{|Z_Y|\angle\theta} = \frac{V_g}{|Z_Y|}\angle(-120° - \theta)$$

$$\mathbf{I_C} = \mathbf{V_C}/Z_Y = \frac{V_g\angle + 120°}{|Z_Y|\angle\theta} = \frac{V_g}{|Z_Y|}\angle(120° - \theta) \tag{8.2.7}$$

Let us denote the rms value of the phase voltage by $V_{\phi rms}$ and the rms value of the phase current by $I_{\phi rms}$. The real power per phase is $P_\phi = V_{\phi rms} \times I_{\phi rms} \cos\theta$. The total real power in the three phases of the load is $P_T = 3P_\phi$. If V_{lrms} and I_{lrms} are the rms values of the line voltage and the line current, respectively, then,

$$P_T = 3V_{\phi rms} \times I_{\phi rms} \cos\theta = 3\frac{V_{lrms}}{\sqrt{3}} I_{lrms} \cos\theta = \sqrt{3}V_{lrms}I_{lrms}\cos\theta \qquad (8.2.8)$$

where the substitutions $I_{lrms} = I_{\phi rms}$ and $V_{lrms} = \sqrt{3}V_{\phi rms}$ were made.

The reactive power per phase is $Q_\phi = V_{\phi rms} \times I_{\phi rms} \sin\theta$. The total reactive power in the three phases of the load is $Q_T = 3Q_\phi$, which may be written as

$$Q_T = 3V_{\phi rms} \times I_{\phi rms} \sin\theta = 3\frac{V_{lrms}}{\sqrt{3}} I_{lrms} \sin\theta = \sqrt{3}V_{lrms}I_{lrms}\sin\theta \qquad (8.2.9)$$

The complex power is given by

$$S_T = P_T + jQ_T = \sqrt{3}V_{lrms}I_{lrms}\angle\theta \qquad (8.2.10)$$

The power factor angle of the load is $\tan^{-1}(Q_T / P_T) = \theta$. Hence, the power factor of the load is the same as that of the phase impedance.

Example 8.2.1: Phase Voltages and Currents in a Y-Connected Generator

A balanced three-phase load consumes 72 kW at 0.8 p.f. and is connected to a balanced, Y-connected, three-phase generator, the rms value of the line current being 100 A, line impedances being negligible. It is required to determine the phase voltages and phase currents of the generator, assuming a positive phase sequence.

SOLUTION

From Equation 8.2.8: $72,000 = \sqrt{3} \times V_{lrms} \times 100 \times 0.8$. This gives $V_{lrms} = \dfrac{900}{\sqrt{3}}$ V. The magnitude of the generator phase voltage is $V_{\phi rms} = \dfrac{1}{\sqrt{3}}V_{lrms} = 300$ V.

Because no information is given regarding phase angles, we can consider the phase voltage $\mathbf{V}_{ga}$ to have a zero phase angle. Then,

$$\mathbf{V}_{ga} = 300\angle 0° \text{ V}, \; \mathbf{V}_{gb} = 300\angle -120° \text{ V}, \; \mathbf{V}_{gc} = 300\angle 120° \text{ V}$$

all being rms values.

From Equation 8.2.5, the line voltages are:

$$\mathbf{V}_{ab} = \frac{900}{\sqrt{3}} \angle 30° \text{ V}, \ \mathbf{V}_{bc} = \frac{900}{\sqrt{3}} \angle -90° \text{ V}, \ \mathbf{V}_{ca} = \frac{900}{\sqrt{3}} \angle 150° \text{ V}$$

all being rms values.

The phase currents of the generator are the same as the corresponding line currents, whose magnitude is 100 A. To determine the phase angles, we note that the generator must deliver the same power as that consumed by the load, that is, 72 kW at 0.8 p.f., or 24 kW per phase at 0.8 p.f. Consider phase a of the generator; the power relation is: 24,000 = 300 × 100 × cosθ, so cosθ = 0.8, or θ = 36.9°. It follows that:

$$\mathbf{I}_{aA} = 100\angle -\theta \text{ A}, \mathbf{I}_{bB} = 100\angle(-120° - \theta) \text{ A}, \mathbf{I}_{cC} = 100\angle(120° - \theta) \text{ A}$$

Alternatively, we may assume that the load is Y connected, in which case the load phase voltages are the same as the generator phase voltages, the load phase currents are equal to the generator phase currents (Figure 8.2.1), and the load phase currents lag the corresponding phase voltages by θ. The preceding relations for the generator phase currents then follow.

EXERCISE 8.2.1

Assume that $\mathbf{V}_{an} = 240\angle 0°$ V for the generator in Example 8.2.1 and that the load is Y connected and of impedance $3 + j4$ Ω per phase. Determine: (a) the line voltages; and (b) the line currents, assuming a positive phase sequence.

Answer: (a) $\mathbf{V}_{ab} = 416\angle 30°$ V, $\mathbf{V}_{bc} = 416\angle -90°$ V, $\mathbf{V}_{ca} = 416\angle 150°$ V;
(b) $\mathbf{I}_{aA} = 48\angle -53.1°$ A, $\mathbf{I}_{bB} = 48\angle -173.1°$ A, $\mathbf{I}_{cC} = 48\angle 66.9°$ A.

EXERCISE 8.2.2

Repeat Exercise 8.2.1, assuming a negative phase sequence.

Answer: (a) $\mathbf{V}_{ab} = 416\angle -30°$ V, $\mathbf{V}_{ca} = 416\angle -150°$ V, $\mathbf{V}_{bc} = 416\angle 90°$ V;
(b) $\mathbf{I}_{aA} = 48\angle -53.2°$ A, $\mathbf{I}_{cC} = 48\angle -173.1°$ A, $\mathbf{I}_{bB} = 48\angle 66.9°$ A.

EXERCISE 8.2.3

Assume that $\mathbf{V}_{an} = 400\angle 0°$ V rms in Exercise 8.2.1 and that the complex power absorbed by the load is $4 + j3$ kVA. Determine: (a) the power factor of the load; (b) $|I_{aA}|$; and (c) the phase impedance of a Y-connected load.

Answer: (a) 0.8; (b) 4.17 A rms; (c) $96\angle 36.9°$ Ω.

8.3 The Balanced Δ Connection

Voltage Relations

Consider a balanced Δ-connected load, where $\mathbf{I}_A = I_m\angle 0°$, $\mathbf{I}_B = I_m\angle -120°$, and $\mathbf{I}_C = I_m\angle 120°$ (Figure 8.3.1). The phase voltages of the load are the same as the corresponding line voltages at the load and at the generator:

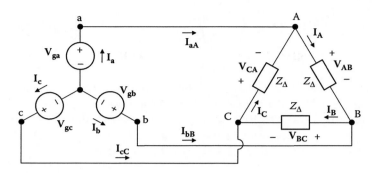

FIGURE 8.3.1
Balanced Y-Δ connection.

$$\mathbf{V_{AB}} = \mathbf{V_{ab}}, \ \mathbf{V_{BC}} = \mathbf{V_{bc}}, \ \mathbf{V_{CA}} = \mathbf{V_{ca}} \tag{8.3.1}$$

Current Relations

Given the phase currents, the line currents may be obtained from KCL at nodes A, B, and C:

$$\mathbf{I_{aA}} = \mathbf{I_A} - \mathbf{I_C}, \ \mathbf{I_{bB}} = \mathbf{I_B} - \mathbf{I_A}, \ \mathbf{I_{cC}} = \mathbf{I_C} - \mathbf{I_B} \tag{8.3.2}$$

Figure 8.3.2a shows the phasors of the three phase currents, of phase sequence abc, and the phasors of the three line currents satisfying Equations 8.3.2. Following an argument similar to that used in connection with Figure 8.2.2, it is seen that:

$$\mathbf{I_{aA}} = \sqrt{3}I_m\angle -30°, \ \mathbf{I_{bB}} = \sqrt{3}I_m\angle -150°, \ \mathbf{I_{cC}} = \sqrt{3}I_m\angle 90° \tag{8.3.3}$$

For a negative phase sequence, Equation 8.3.2 still applies. From the phasor diagram (Figure 8.3.2b), the following relations can be derived:

$$I_{aA} = \sqrt{3}I_m\angle 30°, \ \mathbf{I_{cC}} = \sqrt{3}I_m\angle -90°, \ \mathbf{I_{bB}} = \sqrt{3}I_m\angle 150° \tag{8.3.4}$$

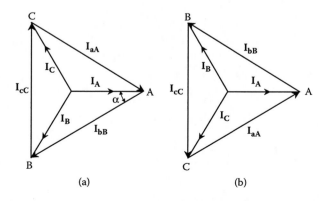

(a)   (b)

FIGURE 8.3.2
Phase and line currents for positive and negative phase sequences.

Summary: *In a balanced Δ connection, the line voltages are equal to the phase voltages. The line currents are balanced three-phase currents whose magnitude is $\sqrt{3}$ times that of the phase currents, the phase sequence being the same. For a positive phase sequence, the line current lags by 30° the current in the phase connected to the given line and whose direction is a continuation of that of the line current. For a negative phase sequence, the phase angle is 30° lead.*

Thus, in Figure 8.3.1, $\mathbf{I_{aA}}$ lags $\mathbf{I_A}$ by 30°.

Phase currents are derived from line currents in Section ST8.1.

Power Relations

Consider the load in Figure 8.3.1. If $Z_\Delta = |Z_\Delta| \angle\theta$, the phase currents are related to the corresponding phase voltages as follows:

$$\mathbf{I_A} = \mathbf{V_{AB}}/Z_\Delta = \frac{V_m \angle 0°}{|Z_\Delta| \angle \theta} = \frac{V_m}{|Z_\Delta|} \angle -\theta$$

$$\mathbf{I_B} = \mathbf{V_{BC}}/Z_\Delta = \frac{V_m \angle -120°}{|Z_\Delta| \angle \theta} = \frac{V_m}{|Z_\Delta|} \angle(-120° - \theta)$$

$$\mathbf{I_C} = \mathbf{V_{CA}}/Z_\Delta = \frac{V_m \angle +120°}{|Z_\Delta| \angle \theta} = \frac{V_m}{|Z_\Delta|} \angle(120° - \theta) \tag{8.3.5}$$

Let the rms value of the phase voltage be $V_{\phi rms}$, and the rms value of the phase current be $I_{\phi rms}$. The real power per phase is $P_\phi = V_{\phi rms} \times I_{\phi rms} \cos\theta$. The total real power in the three phases of the load is $P_T = 3P_\phi$, which may be written as:

$$P_T = 3V_{\phi rms} \times I_{\phi rms} \cos\theta = 3V_{lrms} \frac{I_{lrms}}{\sqrt{3}} \cos\theta = \sqrt{3} V_{lrms} I_{lrms} \cos\theta \tag{8.3.6}$$

where $V_{lrms} = V_{\phi rms}$ and $I_{lrms} = \sqrt{3} I_{\phi rms}$. Comparing Equation 8.2.8 and Equation 8.3.6, the expression for the total real power is the same in both cases. Following the same argument as in the case of the Y connection, the total reactive power and complex power are given by Equation 8.2.9 and Equation 8.2.10, respectively. The power factor of the load is again the same as that of the phase impedance.

Summary: *For both balanced* Y *and* Δ *connections:*

Real power $= \sqrt{3} \times$ *(rms line voltage)* $\times$ *(rms line current)* $\times$ *(p.f. of load phase)*

Reactive power = $\sqrt{3}$ × (rms line voltage) × (rms line current) × (reactive factor of load phase)

Apparent power = $\sqrt{3}$ × (rms line voltage) × (rms line current)

Example 8.3.1: Phase Currents in a Δ-Connected Load

Assume that the load in Example 8.2.1 is Δ connected (Figure 8.3.1). It is required to determine the load phase currents.

SOLUTION

The phase currents have a magnitude of $\dfrac{100}{\sqrt{3}}$ A rms. According to Equation 8.3.3, the line currents lag the corresponding phase currents by 30°, or the phase currents lead the corresponding line currents by 30°. Hence, using the same phase angles of the line currents as in Example 8.2.1:

$$\mathbf{I_{AB}} = \frac{100}{\sqrt{3}} \angle(30° - \theta) \text{ A}, \quad \mathbf{I_{BC}} = \frac{100}{\sqrt{3}} \angle(-90° - \theta) \text{ A}, \quad \mathbf{I_{CA}} = \frac{100}{\sqrt{3}} \angle(150° - \theta) \text{ A}$$

all being rms values.

It may also be noted that, from the relations between phase and line currents (Section ST8.1), the phase current $\mathbf{I_{CA}}$, for example, is given by: $\mathbf{I_{CA}} = \dfrac{1}{3}(\mathbf{I_{cC}} - \mathbf{I_{aA}}) =$

$\dfrac{1}{3}(100\angle(120° - \theta) - 100\angle - \theta) = \dfrac{100}{\sqrt{3}} \angle(150° - \theta)$, as determined earlier.

EXERCISE 8.3.1

Assume that $\mathbf{I_{aA}} = 17.3\angle0°$ A in Figure 8.3.1 and that the load is Δ connected, the impedance per phase being 3 + j4 Ω. Determine: (a) the phase currents; and (b) the line voltages, assuming a positive phase sequence.

Answer: (a) $\mathbf{I_A} = 10\angle30°$ A, $\mathbf{I_B} = 10\angle - 90°$ A, $\mathbf{I_C} = 10\angle150°$ A;

(b) $\mathbf{V_{AB}} = 50\angle83.1°$ V, $\mathbf{V_{BC}} = 50\angle - 36.9°$ V, $\mathbf{V_{CA}} = 50\angle - 156.9°$ V.

EXERCISE 8.3.2

Repeat Exercise 8.3.1, assuming a negative phase sequence.

Answer: (a) $\mathbf{I_A} = 10\angle - 30°$ A, $\mathbf{I_C} = 10\angle - 150°$ A, $\mathbf{I_B} = 10\angle90°$ A;

(b) $\mathbf{V_{AB}} = 50\angle23.1°$ V, $\mathbf{V_{CA}} = 50\angle - 96.9°$ V, $\mathbf{V_{BC}} = 50\angle143.1°$ V.

EXERCISE 8.3.3

Assume that in Figure 8.3.1, the load absorbs 9 kW at a lagging power factor of 0.8. If $\mathbf{V_{AB}} = 100\angle60°$ V rms, determine: (a) $\mathbf{I_{aA}}$ and (b) the complex power delivered by the source, assuming the line impedance is 0.1 + j0.5 Ω.

Answer: (a) $65.0\angle-6.9°$ A rms and (b) $10.3 + j13.1$ kVA.

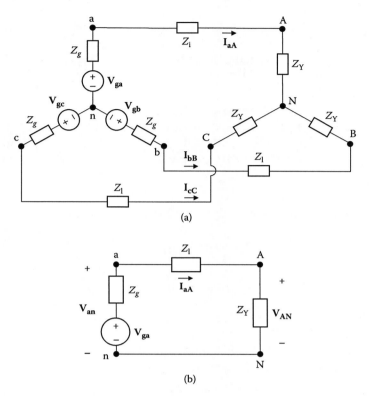

FIGURE 8.4.1
Single-phase equivalent circuit of Y-Y connection.

8.4 Analysis of Balanced Three-Phase Systems

Y-Y System

Three-phase systems, balanced or unbalanced, can be analyzed using the general methods discussed in previous chapters. The mesh-current method is advantageous because a three-wire Y-Y system has only two meshes, a Δ-Δ system has four meshes. A Y-Y unbalanced system can also be analyzed by the node-voltage method (Section ST8.2). However, it is generally simpler to analyze balanced, three-phase systems by reducing them to a single-phase equivalent circuit based on a Y-Y representation.

A balanced Y-Y system is shown in Figure 8.4.1a. Because nodes n and N are at the same voltage, KVL around loop naAN gives

$$(Z_g + Z_l + Z_Y) \mathbf{I}_{aA} = \mathbf{V}_{ga} \tag{8.4.1}$$

Equation 8.4.1 also applies to the circuit shown in Figure 8.4.1b, which is the single-phase equivalent circuit for phase a. Once $\mathbf{I}_{aA}$ is determined from Equation 8.4.1, the phase voltages $\mathbf{V}_{an}$ and $\mathbf{V}_{AN}$ are readily obtained as:

$$\mathbf{V}_{an} = \mathbf{V}_{ga} - Z_g \mathbf{I}_{aA}, \ \mathbf{V}_{AN} = Z_Y \mathbf{I}_{aA} \qquad (8.4.2)$$

Because all phase quantities in a balanced three-phase system are balanced sets, then once $\mathbf{I}_{aA}$, $\mathbf{V}_{an}$, and $\mathbf{V}_{AN}$ are determined, the corresponding quantities for the other two phases immediately follow; they have the same magnitude but are phase-shifted by 120°. The line voltages are obtained as discussed in Section 8.2.

Example 8.4.1: Analysis of Balanced Y-Y System

It is required to analyze the balanced Y-Y system shown in Figure 8.4.1a, given that $\mathbf{V}_{ga} = 120\angle0°$ V, $\mathbf{V}_{gb} = 120\angle-120°$ V, $\mathbf{V}_{gc} = 120\angle120°$ V, $Z_g = 1$ Ω, $Z_l = j2$ Ω, and the load phase impedance consists of a resistance of 10 Ω in parallel with a capacitive reactance of $-j10$ Ω.

SOLUTION

The single-phase equivalent circuit is shown in Figure 8.4.2. The load impedance Z_Y is the parallel combination of 10 Ω and $-j10$ Ω: $Z_Y = \dfrac{10(-j10)}{10-j10} = \dfrac{-j100(10+j10)}{200} = 5 - j5 =$

$5\sqrt{2}\angle-45°$Ω. Hence, $\mathbf{I}_{aA} = \dfrac{120\angle0°}{6-j3}$ A rms. It follows that:

$$\mathbf{I}_{aA} = 16 + j8 = 17.9\angle26.6° \text{A rms}$$
$$\mathbf{I}_{bB} = 17.9\angle(26.6° - 120°) = 17.9\angle-93.4° \text{ A rms}$$
$$\mathbf{I}_{cC} = 17.9\angle(26.6° + 120°) = 17.9\angle146.6° \text{ A rms}$$
$$\mathbf{V}_{an} = 120\angle0° - 1\times(16+j8) = 104.3\angle-4.40° \text{ V rms}$$
$$\mathbf{V}_{bn} = 104\angle(-4.40° - 120°) = 104.3\angle-124.4° \text{ V rms}$$
$$\mathbf{V}_{cn} = 104.3\angle(-4.40° + 120°) = 104.3\angle115.6° \text{ V rms}$$
$$\mathbf{V}_{ab} = \sqrt{3}\mathbf{V}_{an}e^{j30°} = 180.7\angle25.6° \text{ V rms}$$
$$\mathbf{V}_{bc} = 180.7\angle(25.6° - 120°) = 180.7\angle-94.4° \text{ V rms}$$
$$\mathbf{V}_{ca} = 180.7\angle(25.6° + 120°) = 180.7\angle145.6° \text{ V rms}$$
$$\mathbf{V}_{AN} = Z_Y\mathbf{I}_{aA} = 5\sqrt{2}\angle-45°\times17.9\angle26.6° = 126.5\angle-18.4° \text{ V rms}$$
$$\mathbf{V}_{BN} = 126.5\angle(-18.4° - 120°) = 126.5\angle-138.4° \text{ V rms}$$
$$\mathbf{V}_{CN} = 126.5\angle(-18.4° + 120°) = 126.5\angle101.6° \text{ V rms}$$
$$\mathbf{V}_{AB} = \sqrt{3}\mathbf{V}_{AN}e^{j30°} = 219.1\angle11.6° \text{ V rms}$$

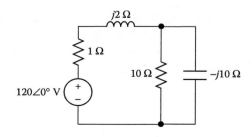

FIGURE 8.4.2
Figure for Example 8.4.1.

$$\mathbf{V}_{BC} = 219.1\angle(11.6° - 120°) = 219.1\angle-108.4° \text{ V rms}$$
$$\mathbf{V}_{CA} = 219.1\angle(+11.6° + 120°) = 219.1\angle131.6° \text{ V rms}$$

Δ-Δ System

When the generator, load, or both are Δ connected, the procedure is to transform the system to a Y-Y system, which is analyzed as discussed earlier. Any desired phase current of the Δ-connected load can then be derived.

A balanced Δ-Δ system is shown in Figure 8.4.3a. The Δ-connected load may be readily transformed to an equivalent Y simply by substituting $Z_Y = \frac{1}{3}Z_\Delta$. To transform the Δ-connected generator to its equivalent Y, we have to derive its TEC, which can be done from equivalence between every two of the three terminals abc, under two sets of conditions: (1) the same open-circuit voltage (Figure 8.4.3b), and (2) the same equivalent impedance with the ideal voltage sources set to zero (Figure 8.4.3c). The latter condition is the

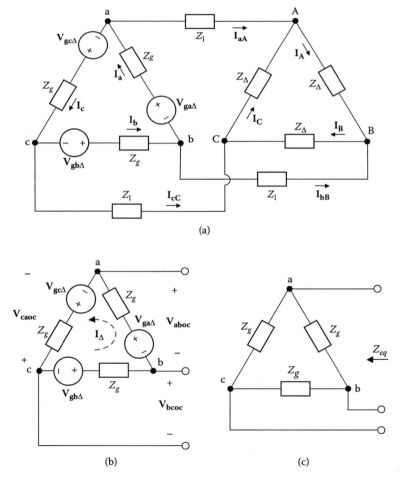

(a)

(b)   (c)

FIGURE 8.4.3
Single-phase equivalent circuit of Δ-Δ connection.

usual Y equivalent of Δ-connected impedances, which is a set of impedances $Z_{g\mathrm{Y}} = \dfrac{1}{3} Z_g$. To satisfy the first condition, we have to determine $\mathbf{I}_\Delta$, the current that circulates in the Δ under open-circuit conditions. From Figure 8.4.3b,

$$\mathbf{I}_\Delta = \frac{1}{3Z_g} \left(\mathbf{V}_{\mathrm{ga}\Delta} + \mathbf{V}_{\mathrm{gb}\Delta} + \mathbf{V}_{\mathrm{gc}\Delta} \right) \tag{8.4.3}$$

The sum of the balanced three-phase voltages is zero. Hence, $\mathbf{I}_\Delta = 0$. It follows that the line voltages under open-circuit conditions are equal to the phase voltages:

$$\mathbf{V}_{\mathrm{aboc}} = \mathbf{V}_{\mathrm{ga}\Delta}, \quad \mathbf{V}_{\mathrm{bcoc}} = \mathbf{V}_{\mathrm{gb}\Delta}, \quad \mathbf{V}_{\mathrm{caoc}} = \mathbf{V}_{\mathrm{gc}\Delta} \tag{8.4.4}$$

From the relations of Equation 8.2.5, the corresponding Y generator phase voltages are $\dfrac{1}{\sqrt{3}}$ times the magnitude, and lag the line voltages by 30°. Thus:

$$\mathbf{V}_{\mathrm{gaY}} = \frac{\mathbf{V}_{\mathrm{ga}}}{\sqrt{3}} e^{-j30°}, \quad \mathbf{V}_{\mathrm{gbY}} = \frac{\mathbf{V}_{\mathrm{gb}}}{\sqrt{3}} e^{-j30°}, \quad \mathbf{V}_{\mathrm{gcY}} = \frac{\mathbf{V}_{\mathrm{gc}}}{\sqrt{3}} e^{-j30°} \tag{8.4.5}$$

The equivalent Y connection will then have in each phase the corresponding voltage source in series with $\mathbf{Z}_{g\mathrm{Y}}$, as illustrated in Figure 8.4.4. Clearly, this Y connection satisfies the required conditions: under open circuit, the line voltages $\mathbf{V}_{\mathrm{ab}}$, $\mathbf{V}_{\mathrm{bc}}$, and $\mathbf{V}_{\mathrm{ca}}$ are the same as in Figure 8.4.3b; if the sources are set to zero in both cases, the resulting Y and Δ are equivalent, because $Z_{g\mathrm{Y}} = \dfrac{1}{3} Z_g$. The system of Figure 8.4.4 can now be analyzed as was done for the system of Figure 8.4.1a. Once the line currents are found, the phase currents can be determined, as illustrated by the following example.

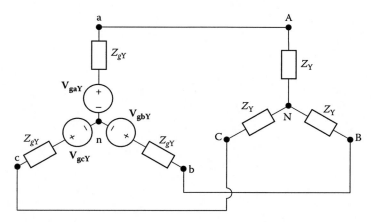

FIGURE 8.4.4
Equivalent Y-Y connection.

Example 8.4.2: Analysis of Balanced Δ-Δ System

It is required to analyze the balanced Δ-Δ system shown in Figure 8.4.3a, where $V_{ga\Delta}$ = 190∠120° V, $V_{gb\Delta}$ = 190∠0° V, $V_{gc\Delta}$ = 190∠−120° V, $Z_{g\Delta}$ = 1.5 + j1.5 Ω, Z_l = 0.5 + j0.5Ω, and the load phase impedance Z_Δ consists of a 12 Ω resistance in parallel with a capacitive reactance of −12 Ω.

SOLUTION

Considering the load first, the phase impedance is $Z_\Delta = \dfrac{12(-j12)}{12-j12} = 6 - j6 = 6\sqrt{2}\angle -45°\,\Omega.$

The equivalent Y has a phase impedance of $Z_Y = 2 - j2 = 2\sqrt{2}\angle -45°\ \Omega.$

As for the generator, the equivalent Y has a phase impedance Z_{gY} = 0.5 + j0.5 Ω. The ideal voltage sources are, from the relations of Equation 8.4.5:

$$V_{gaY} = \frac{190}{\sqrt{3}}\angle(120° - 30°) = 110\angle 90°\text{ V rms}$$

$$V_{gbY} = \frac{190}{\sqrt{3}}\angle(0° - 30°) = 110\angle -30°\text{ V rms}$$

$$V_{gcY} = \frac{190}{\sqrt{3}}\angle(-120° - 30°) = 110\angle -150°\text{ V rms}$$

The Y-Y equivalent system is shown in Figure 8.4.5a. The relationships between the voltage sources in the given Δ, and the equivalent Y are shown in Figure 8.4.5b. Note the correspondence between the phasors and the way the Δ and Y are drawn in Figure 8.4.3a and Figure 8.4.5a.

The equivalent single-phase diagram is shown in Figure 8.4.5c. The line currents are:

$$I_{aA} = \frac{110\angle 90°}{3 - j} = \frac{110\angle 90°}{\sqrt{10}\angle -18.4°} = 34.8\angle 108.4°\text{ A rms}$$

$$I_{bB} = 34.8\angle(108.4° - 120°) = 34.8\angle -11.6°\text{ A rms}$$

$$I_{cC} = 34.8\angle(-11.6° - 120°) = 34.8\angle -131.6°\text{ A rms}$$

From Equation 8.3.3, the phase currents of the load are:

$$I_A = \frac{I_{aA}}{\sqrt{3}}\angle(108.4° + 30°) = 20.1\angle 138.4°\text{ A rms}$$

$$I_B = \frac{I_{bB}}{\sqrt{3}}\angle(-11.6° + 30°) = 20.1\angle 18.4°\text{ A rms}$$

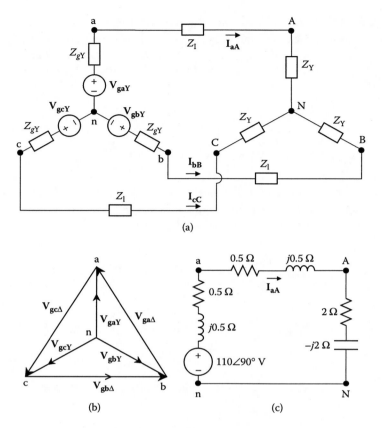

FIGURE 8.4.5
Figure for Example 8.4.2.

$$\mathbf{I_C} = \frac{I_{cC}}{\sqrt{3}} \angle(-131.6° + 30°) = 20.1\angle -101.6° \text{ A rms}$$

With the assignment of positive directions of phase currents in the generator and load as in Figure 8.4.3a, the corresponding phase currents are equivalent because for the load, $\mathbf{I_{aA}} = \mathbf{I_A} - \mathbf{I_C}$, for example, whereas for the generator, $\mathbf{I_{aA}} = \mathbf{I_a} - \mathbf{I_c}$. It follows that $\mathbf{I_a} = \mathbf{I_A}$, $\mathbf{I_b} = \mathbf{I_B}$, and $\mathbf{I_c} = \mathbf{I_C}$.
The line voltages at the load are:

$$\mathbf{V_{AB}} = Z_\Delta \mathbf{I_A} = 6\sqrt{2}\angle -45° \times 20.1\angle 138.4° = 170.5\angle 93.4° \text{ V rms}$$

$$\mathbf{V_{BC}} = 170.5\angle(93.4° - 120°) = 170.5\angle -26.6° \text{ V rms}$$

$$\mathbf{V_{CA}} = 170.5\angle(-26.6° - 120°) = 170.5\angle -146.6° \text{ V rms}$$

From Figure 8.4.5c, the phase voltage is: $\mathbf{V_{an}} = (2.5 - j1.5)\mathbf{I_{aA}} = 2.92\angle -31° \times 34.8\angle 108.4°$ $= 101.6\angle 77.4°$ V rms. It follows from the relations of Equation 8.2.5 that:

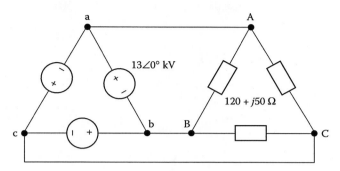

FIGURE 8.4.6
Figure for Exercise 8.4.1.

$$\mathbf{V}_{ab} = \sqrt{3}V_{an}e^{j30°} = 176\angle107.4° \text{ V rms}$$

$$\mathbf{V}_{bc} = 176\angle(107.4° - 120°) = 176\angle - 12.6° \text{ V rms}$$

$$\mathbf{V}_{ca} = 176\angle(-12.6° - 120°) = 176\angle - 132.6° \text{ V rms}$$

EXERCISE 8.4.1

Consider the balanced Δ-Δ system of Figure 8.4.6. (a) Determine the load phase currents $\mathbf{I}_{AB}$, $\mathbf{I}_{BC}$, and $\mathbf{I}_{CA}$ from the corresponding line voltages and the load impedance per phase; (b) determine the line currents $\mathbf{I}_{aA}$, $\mathbf{I}_{bB}$, $\mathbf{I}_{cC}$ from the phase currents and by using the single-phase equivalent circuit; and (c) using the results of Section ST8.1, determine the phase generator currents from the line currents to show that $\mathbf{I}_{ba} = \mathbf{I}_{AB}$, $\mathbf{I}_{cb} = \mathbf{I}_{BC}$, and $\mathbf{I}_{ac} = \mathbf{I}_{CA}$.

Answer: (a) $\mathbf{I}_{AB} = 100\angle - 22.6° \text{ A}$, $\mathbf{I}_{BC} = 100\angle - 142.6° \text{ A}$, $\mathbf{I}_{CA} = 100\angle97.4° \text{ A}$;
(b) $\mathbf{I}_{aA} = 100\sqrt{3}\angle - 52.6° \text{ A}$, $\mathbf{I}_{bB} = 100\sqrt{3}\angle - 172.6° \text{ A}$, $\mathbf{I}_{cC} = 100\sqrt{3}\angle67.4° \text{ A}$.

EXERCISE 8.4.2

Determine $\mathbf{I}_{aA}$ in Figure 8.4.6 if each line has an impedance of $j2 \, \Omega$.
Answer: $163.3\angle53.96° \text{ A}$.

Simulation Example 8.4.3: PSpice Simulation of Y-Δ Balanced System

It is required to determine: (a) $\mathbf{I}_{aA}$, (b) $\mathbf{I}_{CA}$, and (c) $\mathbf{V}_{AB}$ in the balanced three-phase system of Figure 8.4.7, assuming the frequency is 50 Hz, and to simulate it.

SOLUTION

$Z_Y = 10 + \dfrac{j\omega \times 0.12}{3} = 10 + j100\pi \times 0.04 = 10 + j12.57 \, \Omega$. The single-phase equivalent circuit is shown in Figure 8.4.8. It follows that:

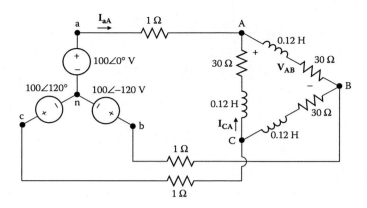

FIGURE 8.4.7
Figure for Example 8.4.3.

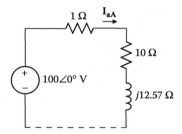

FIGURE 8.4.8
Figure for Example 8.4.3.

(a) $\mathbf{I_{aA}} = \dfrac{100\angle0°}{11+j12.57} = \dfrac{100\angle0°}{16.7\angle48.8°} = 5.988\angle-48.8°\ A$

(b) $\mathbf{I_{AB}} = \dfrac{1}{\sqrt{3}}\,\mathbf{I_{aA}}\angle30° = 3.457\angle-18.8°\ A;\ \mathbf{I_{CA}} = \mathbf{I_{AB}}\angle120° = 3.457\angle101.2°\ A$

(c) $\mathbf{Z_\Delta} = 3(10+j12.57)\ \Omega = 48.18\angle51.49°\ \Omega;\ \mathbf{V_{AB}} = \mathbf{I_{AB}Z_\Delta} = 166.6\angle32.69°\ V$

SIMULATION
The schematic is shown in Figure 8.4.9 and was prepared as described in the Simulation Examples of Chapter 5, with $f = 50$ Hz. Two IPRINT printers are used to print to the output file the line current $\mathbf{I_{aA}}$ and the phase current $\mathbf{I_{CA}}$. A VPRINT printer is used for the voltage $\mathbf{V_{AB}}$. The simulation gives:
   $\mathbf{I_{aA}} = 5.988\angle-48.8°\ A,\ \mathbf{I_{CA}} = 3.457\angle101.2°\ A,$ and $\mathbf{V_{AB}} = 166.6\angle32.69°\ V.$

EXERCISE 8.4.3

Determine the reading of the ideal voltmeter in Figure 8.4.10 if the line voltage is 220 V.
Answer: 190.5 V

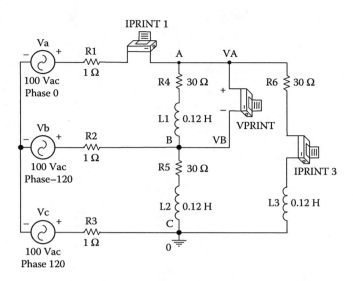

FIGURE 8.4.9
Figure for Example 8.4.3.

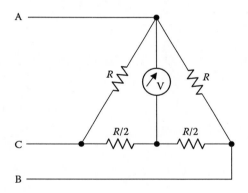

FIGURE 8.4.10
Figure for Exercise 8.4.3.

Application Window 8.4.1: Phase Sequence Indicator

A simple means of indicating phase sequence is to use two incandescent lamps and a capacitor connected in Y (Figure 8.4.11). One lamp will glow brighter than the other. The phase sequence is then in the order of the line connections to the (bright lamp)-(dim lamp)-(capacitor), that is, abc in Figure 8.4.11.

To see why this is so, let us assume that a 5 μF capacitor and two 60 W lamps are connected to a 220 V rms, 50 Hz, balanced three-phase supply (Figure 8.4.12). The reactance

of the capacitor is $-\dfrac{1}{2\pi \times 50 \times 5 \times 10^{-6}} = -637\ \Omega$. The resistance of a 60 W lamp at 220 V is

$\dfrac{(220)^2}{60} = 807\ \Omega$. It is assumed that the resistance of the lamps is constant at this value,

irrespective of the lamp current. This is not true, of course, because the lamp resistance increases with the current due to heating. But assuming a constant resistance allows the

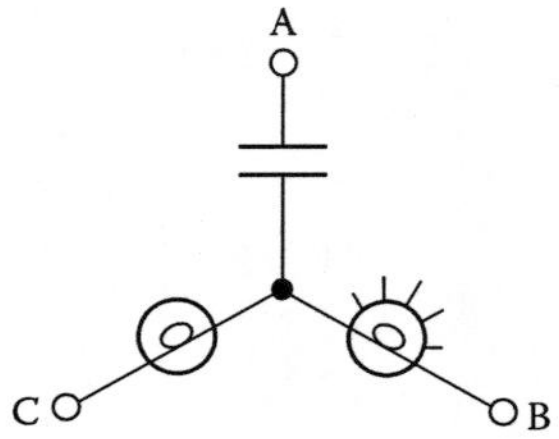

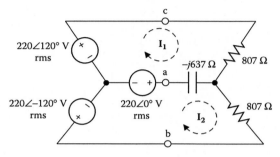

FIGURE 8.4.11
Figure for Application Window 8.4.1.

FIGURE 8.4.12
Figure for Application Window 8.4.1.

application of linear circuit theory and enables us to understand the principle of operation of the phase-sequence indicator.

The mesh-current equations are:

$$(807 - j637)\mathbf{I}_1 + j637\mathbf{I}_2 = 220\angle120° - 220\angle0° = 220\sqrt{3}\angle150°$$

$$j637\mathbf{I}_1 + (807 - j637)\mathbf{I}_2 = 220\angle0° - 220\angle -120° = 220\sqrt{3}\angle30°$$

Solution of these equations using MATLAB gives: $\mathbf{I}_1 = -0.12 + j0.051$ A rms, and $\mathbf{I}_2 = 0.12 + j0.421$ A rms. It is seen that $\mathbf{I}_2 > \mathbf{I}_1$. Although this means that the two lamp resistances will not be equal, it is still true that the lamp through which $\mathbf{I}_2$ flows will glow brighter. The phase sequence indication is therefore bcabca ..., or abc. It is seen from the mesh-current equations that if the phase sequence is reversed, $\mathbf{I}_1$ and $\mathbf{I}_2$ are interchanged, so the lamp through which $\mathbf{I}_1$ flows will glow brighter. The phase sequence indication is now cbacba ..., or acb.

Application Window 8.4.2: Star-Delta Starting of Induction Motors

A popular type of ac motor is the three-phase, squirrel-cage induction motor. In this type of motor, the magnetic field of the rotor is due to currents induced in the rotor winding by the rotating magnetic field of the stator. For this induction to take place, the rotor must rotate at an angular speed that is slightly less than that of the rotating magnetic field of the stator. The rotor winding need only be conducting bars embedded in slots in the rotor and connected to common conducting rings at each end of the rotor, leading to a very rugged construction.

A characteristic of this type of motor is that at starting, the motor resembles a transformer having a short-circuited secondary, which results in a starting current of low power factor and magnitude that is five to eight times the rated full-load current. This is objectionable as it causes an excessive, though momentary, voltage drop in the supply lines feeding the motor and other loads. A common method of avoiding this large starting current on medium-sized and small-sized motors is Y-Δ starting. The motor is designed to run with its stator windings Δ connected. At starting, however, the windings are Y connected. A short, preset interval after starting, the windings are connected in Δ for normal running.

Consider a 50 kW motor that is rated to run, Δ connected, at 0.8 p.f. from a three-phase supply having a line voltage of 400 V rms. If the starting current with full voltage across the motor windings is six times the rated full-load current, what would be the supply current at starting when the windings are Δ connected and when they are Y connected? The rated full-load line current is, from Equation 8.2.8 or Equation 8.3.6: $50 \times 1000 = \sqrt{3} \times 400 \times I_l \times 0.8$, which gives $I_l \cong 90$ A rms. If Δ connected at starting, the starting line current is: $I_{\text{lstΔ}} = 6 \times 90 = 540$ A rms, and the starting current per phase is

$$I_{\phi\text{stΔ}} = \frac{540}{\sqrt{3}} \cong 310 \text{ A rms.}$$

If Y connected at starting, the voltage per phase is $\frac{400}{\sqrt{3}} \cong 230$ V rms. The starting current per phase $I_{\phi\text{stY}}$ is reduced by the same factor as the voltage, so that $I_{\phi\text{stY}} = \frac{310}{\sqrt{3}} \cong 180$ A rms.

Because of the Y connection, the line current at starting is equal to the phase current and is 180 A. The Y connection has therefore reduced the line starting current by a factor of 3.

A limitation of Y-Δ starting is that the torque at starting is only about 50% of the full-load torque. Under these conditions, the motor will not start with full load applied. But many types of loads, such as fans, pumps, machine tools, and motor-generator sets have a reduced load at starting.

8.5   Power in Balanced Three-Phase Systems

Instantaneous Power

Consider, for the sake of argument, a balanced, Y-connected load (Figure 8.5.1). If the instantaneous phase voltages are denoted by $v_A = \sqrt{2}\, V_{\phi\text{rms}}\cos\omega t$, $v_B = \sqrt{2}\, V_{\phi\text{rms}}\cos(\omega t - 120°)$, and $v_C = \sqrt{2}\, V_{\phi\text{rms}}\cos(\omega t + 120°)$, the instantaneous phase, or line, currents lag the corresponding phase voltages by the phase angle θ of Z_Y, assuming the load is inductive. The instantaneous power in each of the three phases is:

$$p_A = 2V_{\phi\text{rms}}I_{\phi\text{rms}}\cos\omega t\cos(\omega t - \theta)$$

$$= V_{\phi\text{rms}}I_{\phi\text{rms}}[\cos(2\omega t - \theta) + \cos\theta]$$

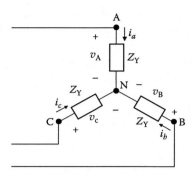

FIGURE 8.5.1
Instantaneous power.

$$p_B = 2V_{\phi rms}I_{\phi rms}\cos(\omega t - 120°)\cos(\omega t - 120° - \theta)$$

$$= V_{\phi rms}I_{\phi rms}\{\cos[(2\omega t - \theta) + 120°] + \cos\theta\}$$

$$p_C = 2V_{\phi rms}I_{\phi rms}\cos(\omega t + 120°)\cos(\omega t + 120° - \theta)$$

$$= V_{\phi rms}I_{\phi rms}\{\cos[(2\omega t - \theta) - 120°] + \cos\theta\} \tag{8.5.1}$$

In each phase, the instantaneous power has a steady component and an alternating component, just as in the single-phase case. The difference, however, is that the alternating components in the three phases form a balanced set. So, when p_A, p_B, and p_C are added to obtain the total power p_T delivered through the lines, the alternating components sum to zero, whereas the steady components add to give:

$$P_T = 3V_{\phi rms}I_{\phi rms}\cos\theta = \sqrt{3}\ V_{lrms}I_{lrms}\cos\theta \tag{8.5.2}$$

where $I_{lrms} = I_{\phi rms}$ is the rms value of the line current and V_{lrms} is the rms of the line voltage. The total instantaneous power is thus equal to the expression for the real power derived earlier for the Y and Δ connections.

> **Concept:** *The total instantaneous power delivered to a balanced three-phase load is steady with respect to time; it is not pulsating, as in the single-phase case.*

It is important to be clear about power relations in a three-phase system. Referring to Figure 8.5.1, the power relations per phase are exactly those for the single-phase case considered in Section 7.1 (Chapter 7). Thus, Equation 7.1.9, with $\theta_v = 0$ and $\theta_i = -\theta$ gives for the real power in terms of rms values per phase: $V_{\phi rms}I_{\phi rms}[\cos\theta + \cos\theta\cos 2(\omega t - \theta)]$. Equation 7.1.11 gives for the reactive power: $-V_{\phi rms}I_{\phi rms}\sin\theta\sin 2(\omega t - \theta)$. Adding these two components gives the instantaneous power per phase p_A in the relations of Equation 8.5.1. Both the real power per phase and the reactive power per phase in Figure 8.5.1 have time-varying alternating components that are also present in the corresponding lines. However, the alternating components of the three phases form a balanced set and cancel out for the three-phase load as a *whole*, leaving a steady component equal to the real power.

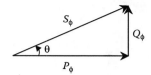

FIGURE 8.5.2
Complex power per phase.

An important practical consequence of the steadiness of instantaneous power is that, because the electrical power input to a three-phase motor is steady, the output torque is ideally steady at constant speed. In contrast, the output torque of a single-phase motor has a pulsating component that produces vibrations.

Complex Power

By analogy with the single-phase case, the alternating components of the relations of Equation 8.5.1 can be interpreted in terms of complex power in each phase. A power triangle may be constructed, as in the single-phase case, where the apparent power per phase $|S_\phi| = V_\phi I_\phi$ is the hypotenuse of a right triangle whose sides are the real power per phase, $P_\phi = V_\phi I_\phi \cos\theta$, and the reactive power per phase, $Q_\phi = V_\phi I_\phi \sin\theta$ (Figure 8.5.2). The load impedance angle θ is the angle between the hypotenuse and the side representing P_ϕ. Each of the total real, reactive, and complex power supplied to the load is three times the corresponding quantity in each phase, and is the same as that derived in Section 8.3.

As in the single-phase case, complex power can be very useful in solving three-phase problems because of its conservation in accordance with Tellegen's theorem. This is illustrated by Example 8.5.1, which also demonstrates power factor correction in a three-phase system.

Example 8.5.1: Power Factor Correction Using Complex Power

A balanced three-phase system is shown in Figure 8.5.3a, where the load absorbs 50 kW at 0.8 p.f. lagging. A capacitor bank of three capacitors C is connected across the load terminals so that the p.f. at terminals abc is unity. Given that the magnitude of the line current, with the capacitors connected, is 100 A rms, it is required to determine C, assuming the frequency is 50 Hz.

SOLUTION
From the complex power triangle for the load (Figure 8.5.3b), the reactive power of the load $Q_L = 50 \dfrac{0.6}{0.8} = 37.5$ kVAR. The total reactive power in the line impedances is $Q_l = 3 \times (100)^2 \times 0.1 = 3$ kVAR. If the reactive power of the capacitors is Q_C, then: $37.5 + 3 + Q_C = 0$ because the total reactive power at abc is zero. This gives $Q_C = -40.5$ kVAR.

To calculate C, we have to determine the line voltages at ABC. This can be done from the complex power. The total reactive power at ABC is $Q_{LC} = 37.5 - 40.5 = -3$ kVAR. Because the total real power at ABC is 50 kW, the complex power at ABC is

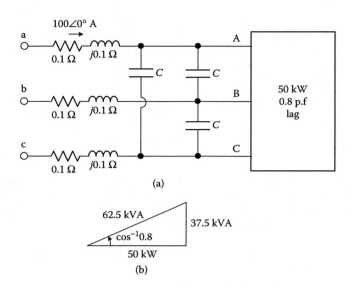

FIGURE 8.5.3
Figure for Example 8.5.1.

$S_{LC} = \sqrt{(50)^2 + (3)^2} = 50.1$ kVA. Let the line voltage at ABC be V_{LC} rms. Then, from Equation

8.2.10: $50.1 \times 10^3 = \sqrt{3} \times 100 \times V_{LC}$, which gives $V_{LC} = 289$ V rms.

   The magnitude of the reactive power per phase of the capacitor bank

is $\dfrac{40.5 \times 10^3}{3} = 13.5 \times 10^3$ VAR. This is equal to $(V_{LC})^2 \times \omega C$, where $\omega = 2\pi \times 50 = 100\pi$ rad/s.

It follows that $C = 0.5 \times 10^{-3} \equiv 510$ μF.

Two-Wattmeter Method of Power Measurement

In principle, the power in a three-phase system can be measured by using three wattmeters to measure the real power in each phase, and adding the three readings. However, in a three-wire system, only two wattmeters need be used. This is because, as mentioned in connection with Figure 8.1.4, one of the lines may be considered the return path for the currents in the two other lines. If, for the sake of argument, we consider lines aA and cC to be "input" lines and line bB to be the common return line, the two wattmeters can be connected as shown in Figure 8.5.4. The current coils are connected in accordance with the assigned positive directions of i_{aA} and i_{cC}, whereas the voltage coils are connected in accordance with the polarities of the line voltages v_{AB} and v_{CB}. The wattmeters are assumed ideal, so that the voltage drop across each current coil is zero, and the current through each of the voltage coils is zero. The sum of the readings of the two wattmeters gives the total real power consumed by the load, irrespective of whether the load is balanced or unbalanced.

   If the load is balanced, the readings of the two wattmeters can be readily related to the line voltages, the line currents, and the phase angle of the load. Assume, for the sake of argument, that the load is Y connected and that the phase sequence is abc. The phasor diagram of Figure 8.2.2a relates the phase and line voltages. This is reproduced in Figure 8.5.5, with only the relevant currents indicated. The currents are shown lagging the corresponding voltages by θ.

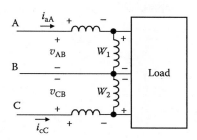

FIGURE 8.5.4
Two-wattmeter connection.

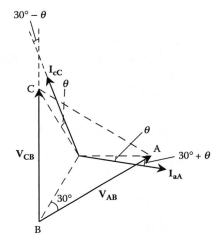

FIGURE 8.5.5
Phasor diagram for two-wattmeter method.

The reading of each wattmeter equals the product of the rms magnitude of the current through the current coil, the rms magnitude of the voltage across the voltage coil, and the cosine of the phase angle between the current and the voltage. From the phasor diagram of Figure 8.5.5, the phase angle between $\mathbf{V_{AB}}$ and $\mathbf{I_{aA}}$ is $30° + \theta$, whereas the phase angle between $\mathbf{V_{CB}}$ and $\mathbf{I_{cc}}$ is $30° - \theta$. The readings of the two wattmeters are therefore:

$$W_1 = V_{\text{lrms}} I_{\text{lrms}} \cos(30° + \theta)$$

$$W_2 = V_{\text{lrms}} I_{\text{lrms}} \cos(30° - \theta) \tag{8.5.3}$$

The sum of the two readings is

$$W_1 + W_2 = \sqrt{3}\, V_{\text{lrms}} I_{\text{lrms}} \cos\theta \tag{8.5.4}$$

which is the same as the total real power P_T.

In Figure 8.5.6, $\cos(30° + \theta)$ and $\cos(30° - \theta)$ are plotted for $90° \leq \theta \leq 90°$, where positive θ denotes a lagging p.f. and negative θ denotes a leading p.f. When $\theta = 0$, $W_1 = W_2$. If $|\theta| > 60°$, the reading of one of the wattmeters is negative, which means in practice that the connections of either the current coil or the voltage coil of that wattmeter are

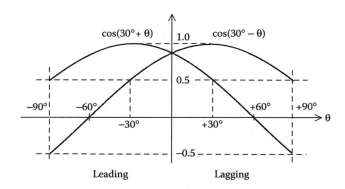

FIGURE 8.5.6
Variation of readings of the two wattmeters.

reversed, and its reading subtracted from that of the other wattmeter. If $|\theta| = 90°$, the sum of the readings of the two wattmeters is zero.

The difference between the readings of the two wattmeters is:

$$W_2 - W_1 = V_{\text{lrms}}I_{\text{lrms}}\sin\theta = \frac{Q_T}{\sqrt{3}} \tag{8.5.5}$$

It follows that:

$$\tan\theta = \sqrt{3}\,\frac{W_2 - W_1}{W_2 + W_1}\text{ , and the p.f. } = \frac{1}{\sqrt{1+\tan^2\theta}} \tag{8.5.6}$$

The following should be noted:

1. W_2, whose reading is proportional to $\cos(30° - \theta)$, is the wattmeter having its voltage coil connected to read a reverse line voltage, that is, $\mathbf{V_{CB}}$ instead of $\mathbf{V_{BC}}$ in Figure 8.5.4. This is true, irrespective of which line is the common, or "return," line.
2. If the phase sequence is reversed, the readings of W_1 and W_2 are interchanged (Problem P8.3.13).

Example 8.5.2: Two-Wattmeter Method of Measuring Power

In a balanced three-phase system, a load absorbs 30 kW at p.f. 0.75 lagging. If two wattmeters are connected to measure the real power in the load, as in Figure 8.5.4, determine the reading of each wattmeter.

SOLUTION
$\cos\theta = 0.75$, so $\theta = 41.4°$, and $\cos(30° + \theta) = 0.32$. The real power consumed by the load is 30 kW. Hence, $30 = \sqrt{3}(V_{\text{lrms}}I_{\text{lrms}})\cos\theta$, where V_{lrms} and I_{lrms} are expressed in appropriate units to give the power in kW. From Equation 8.5.6, $W_1 = (V_{\text{lrms}}I_{\text{ltms}})\cos(30° + \theta)$. Dividing these two equations gives $W_1 = \dfrac{30}{\sqrt{3}}\dfrac{\cos(30° + \theta)}{\cos\theta} = 7.4$ kW, so $W_2 = 22.6$ kW.

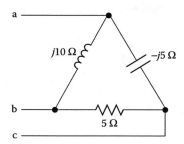

FIGURE 8.5.7
Figure for Exercise 8.5.2.

EXERCISE 8.5.1

The windings of a 10 kW three-phase motor are connected to a balanced, three-phase supply of 380 V rms line voltage. If the line current is 19 A rms, determine the p.f. of the motor.

Answer: 0.8.

EXERCISE 8.5.2

The load shown in Figure 8.5.7 is connected to a balanced, three-phase supply of 100 V rms line voltage. Determine the complex power absorbed by the load.

Answer: $2.24\angle-26.6°$ kVA.

EXERCISE 8.5.3

The apparent power in a balanced Y-connected load is 30 kVA at a line current of 50 A rms, and the real power is 15 kW. Calculate (a) the phase voltage and (b) the impedance per phase.

Answer: (a) 200 V rms and (b) $2 + j3.46$ Ω.

EXERCISE 8.5.4

A Δ-connected load is supplied from a balanced three-phase supply of 450 V rms line voltage. If the load absorbs 100 kVA at 0.65 p.f. lagging, what is the phase impedance?

Answer: $6.1\angle49.5°$ Ω.

8.6 Advantages of Three-Phase Systems

An important advantage of three-phase systems is in power transmission. Two main "figures of merit" may be used in comparing a three-phase system with a single-phase system that transmits the same power: the mass of line conductor used and the RI^2 loss in these conductors. The mass of conductor represents an installation cost of the system; the lighter the line conductors, whether in the form of underground cables or overhead transmission lines, the lower is their cost and, in the case of overhead transmission lines,

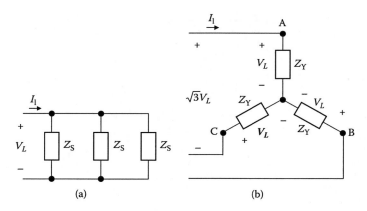

FIGURE 8.6.1
Comparison between single-phase and three-phase systems.

the lighter and less expensive are the transmission towers required. The RI^2 loss represents wasted power and is an operating cost of the system.

Consider a single-phase system supplying three loads Z_S (Figure 8.6.1a) and a three-phase system supplying three loads Z_Y connected in Y (Figure 8.6.1b). It is assumed that: (1) the magnitude of the load voltage is V_L in both cases, which means that the magnitude of the line voltage in the Y connection is $\sqrt{3}\ V_L$; (2) $Z_S = 3Z_Y$, which means Z_S and Z_Y have the same phase angle and $|Z_S| = 3|Z_Y|$; because $V_L = |Z_Y| I_1 = \dfrac{|Z_S|}{3} I_1$, this further implies that the magnitude of the line current is the same in both cases; and (3) the line conductors in both systems are of the same length, type, and material, which in practice is either copper or aluminum; because I_1 is the same, this implies that the line conductors have the same current-carrying capacity and hence the same cross-sectional area, resistance, mass, and RI^2 loss.

The power transmitted in the single phase is $P_S = V_L I_1 \cos\theta$, and the power transmitted per line conductor is $P_{S/c} = \dfrac{P_S}{2} = \dfrac{V_L I_1 \cos\theta}{2}$. The power transmitted in the Y connection is $P_Y = 3V_L I_1 \cos\theta$, and the power transmitted per line conductor is $P_{Y/c} = V_L I_1 \cos\theta = 2P_{S/c}$. The Y connection transmits twice as much power per line conductor, or per unit mass of line conductor, as the single-phase connection. The reason for this is easy to see from Figure 8.6.1. *A separate return conductor is required in a single-phase system but not in a balanced three-phase system* because, in the latter case, each line conductor also acts as the return for the other two conductors. If each line conductor in the Y connection had its own return conductor, the power transmitted per line conductor will be $\dfrac{3V_L I_1 \cos\theta}{6}$, which is the same as in single phase.

Because each line conductor in Figure 8.6.1 has the same RI^2 loss, the power transmitted per kilowatt of RI^2 loss varies between the two cases in the same manner as the power transmitted per line conductor. In other words, the three-phase system transmits twice as much power per kilowatt of RI^2 loss as the single-phase system. Moreover, the total line voltage drop in the single-phase case has a magnitude of $2|Z_1| I_1$, where $|Z_1|$ is the magnitude of the line impedance. In the three-phase case, the line-to-line voltage drop is the line impedance multiplied by the phasor difference of any two line currents. The magnitude of this difference is $\sqrt{3}\ I_l$, so that the magnitude of the line-to-line voltage drop in the three-phase case is $\sqrt{3}\ Z_l I_l$, which is less than that in the single-phase case. The three-phase system of Figure 8.6.1b is therefore more advantageous than the single-phase system of Figure 8.6.1a for transmitting power in all of the respects considered. Additional comparisons are made in Section ST8.5.

Other advantages of three-phase systems are: (1) as mentioned earlier, the output torque of a three-phase motor is steady and does not produce vibrations; (2) a three-phase motor is inherently self-starting because of the rotating magnetic field (Section ST8.4); and (3) three-phase transformers, motors, and generators have smaller frame sizes than their single-phase counterparts of the same power rating because of the elimination of a separate flux return path (Section ST8.3).

8.7 Power Generation, Transmission, and Distribution

A power system may be divided into the following functional parts:

1. *Power generation*: Electric power is generated in power stations from hydroelectric power, atomic energy, or burning fossil fuels such as oil, natural gas, or coal. Electric power is being increasingly generated from solar energy and wind power.
2. *Power transmission*: Because power stations are usually located a considerable distance from the load centers they serve, and to interconnect distant parts of the system, or different power systems, electric power is transmitted over long distances by means of overhead, high-voltage transmission lines.
3. *Medium-voltage distribution*: The high voltages are then stepped down to one or more medium voltages for distribution in the vicinity of load centers and within them.
4. *Low-voltage distribution*: The lowest medium voltage is eventually stepped down to a low voltage for utilization by commercial and domestic users.

In power generation, it is advantageous to have the highest practicable generator voltage because this reduces the current for a given power rating, and hence the cross-sectional area of the conductors in the generator windings. However, too high a generator voltage requires heavy, and costly, insulation. As a compromise, three-phase voltages are usually generated at few tens of kilovolts. Power is generated at 50 Hz throughout most of the world but at 60 Hz in the U.S.

For transmitting power over long distances, it is also advantageous to use the highest practicable voltages because this reduces the current for a given transmitted power and hence the RI^2 loss. Transmission voltages may be hundreds of thousands of volts or even a million volts or more. It is interesting to note that dc has some distinct advantages over ac for long-distance, high-voltage transmission, as is discussed in Application Window 8.7.1.

Medium-power distribution is normally at a few tens of kilovolts or less, using underground cables rather than the lower-cost overhead lines, in built-up areas. Low-voltage distribution is at standard voltages that differ between countries. In most countries, the low-voltage distribution transformer is a three-phase transformer having a Y-connected

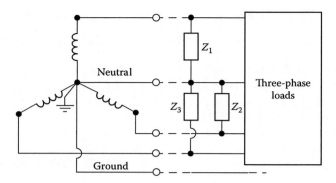

FIGURE 8.7.1
Three-phase, low-voltage distribution transformer.

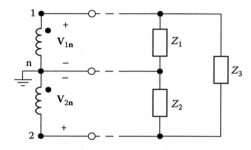

FIGURE 8.7.2
Single-phase, low-voltage distribution transformer.

secondary, the phase voltage being 220 to 240 V rms, and the line voltage being 380 to 415 V rms. The neutral point is grounded, and a ground wire connected to the neutral point is provided to consumers in some countries (Figure 8.7.1), in addition to the neutral line. Single-phase consumer loads are connected between line and neutral, and a three-phase supply is provided for relatively large loads.

In the U.S., the low-voltage distribution transformer is usually a single-phase transformer that steps down the line voltage to a three-wire single-phase system (Figure 8.7.2). The midpoint of the secondary winding is a grounded neutral point to which a neutral line is connected. The outer lines are at voltages of 110 to 120 V rms with respect to the neutral, and their phases are such that the voltage between lines is twice the line-to-neutral voltage. The higher voltage is used to supply relatively large loads, such as heating loads and air-conditioning units, whereas small loads, such as lamps, are connected between one line and neutral.

Application Window 8.7.1: Comparison between dc and ac Transmission

To minimize the level of insulation required, the dc system is grounded midway, so that the line voltage with respect to ground is $+V_{DC}$ V for one line and $-V_{DC}$ V for the other, which means that the insulation is required to withstand V_{DC}. For the ac case, a Y connection with grounded neutral is assumed, the line-to-neutral voltage being V_{ln} V rms. If the insulation is to withstand the same *maximum* voltage with respect to ground in both cases, $V_{DC} = \sqrt{2}\ V_{ln}$.

The dc power transmitted is $P_{DC} = 2V_{DC}I_{DC}$, whereas the ac power transmitted is $P_{AC} = 3V_{ln}I_1\cos\theta = 3\dfrac{V_{DC}}{\sqrt{2}}\,I_1\cos\theta = \dfrac{3}{2\sqrt{2}}\cos\theta P_{DC}$, assuming that $I_{DC} = I_1$. For the two transmitted powers to be equal, $\cos\theta = \dfrac{2\sqrt{2}}{3} = 0.9428$, or $\theta = 19.5°$. At unity power factor, $P_{AC} = 1.06 P_{DC}$.

For the same line current and line resistance in both systems, the conductor mass per line is the same, but only two lines are required with dc compared to three for ac. Hence, the ac case uses 50% more conductor and has 50% more RI^2 loss. Moreover, two line insulators per tower are required, compared to three for ac, and the dc towers are smaller because they have to support fewer conductors and insulators.

In the case of single conductors of relatively large cross-sectional area, of more than approximately 20 mm diameter, for example, the current-carrying capacity is less for ac than for dc because, unlike the dc current, the ac current is not uniformly distributed across the cross-section but tends to be more concentrated toward the surface. This "skin effect" increases with the cross-sectional area and the frequency and with a decrease of the resistivity of the material. It can be reduced by using conductors made of multiple strands that are insulated from one another, rather than a single core. This is often done at radio frequencies.

There is no reactance with dc, whereas the reactance of long ac transmission lines introduces problems of power system stability. The power systems connected by the dc line need not have exactly the same frequency. On the downside, because electric power is more conveniently generated and utilized as ac, ac must be converted to dc at high voltages for transmission, and is then converted back to ac for distribution. This introduces additional complexity and cost.

Summary of Main Concepts and Results

- The sum of a set of balanced quantities is zero.
- In a balanced Y-Y connection, the neutral current is zero.
- In a balanced Y connection, the line currents are equal to the phase currents. The line voltages are balanced three-phase voltages whose magnitude is $\sqrt{3}$ times that of the phase voltages, the phase sequence being the same. For a positive phase sequence, the line voltage leads by 30° the voltage in the phase connected to the given line and having the same polarity of voltage at the connecting node. For a negative phase sequence, the phase angle is 30° lag.
- In a balanced Δ connection, the line voltages are equal to the phase voltages. The line currents are balanced three-phase currents whose magnitude is $\sqrt{3}$ times that of the phase currents, the phase sequence being the same. For a positive phase sequence, the line current lags by 30° the current in the phase connected to the given line and whose direction is a continuation of that of the line current. For a negative phase sequence, the phase angle is 30° lead.
- For both balanced Y and Δ connections,

  Real power = $\sqrt{3} \times$ (rms line voltage) $\times$ (rms line current) $\times$ (p.f. of load phase).

Reactive power = $\sqrt{3}$ × (rms line voltage) × (rms line current) × (reactive factor of load phase).

Apparent power = $\sqrt{3}$ × (rms line voltage) × (rms line current).

- The total instantaneous power in a balanced three-phase load is steady with respect to time; it is not pulsating, as in the single-phase case.

Learning Outcomes

- Analyze balanced, three-phase systems connected in Y or in Δ.
- Understand how power in a three-phase system is measured using two wattmeters.

Supplementary Topics and Examples on CD

ST8.1 Phase quantities from line quantities: Derives phase voltages and currents graphically and analytically from corresponding line quantities.

ST8.2 General analysis of three-phase systems: Analyzes unbalanced systems using the mesh-current method, NEC, TEC, and Δ-Y transformation. An unbalanced Δ-Δ system is simulated to highlight an important feature of such a system.

ST8.3 Three-phase transformers: Discusses some practical aspects of three-phase transformers and their applications, including generation of polyphase systems using transformers, and open-Δ connection of three single-phase transformers.

ST8.4 Three-phase generators and motors: Explains the basic principles of generation of three-phase voltages, how a rotating magnetic field and torque are produced, and shows that the general transformer equation (Equation 6.3.7, Chapter 6) applies to the voltage generated in a coil rotating in a magnetic field.

ST8.5 Power transmission and distribution: Considers some practical aspects of power transmission and distribution, such as the reasons for grounding power systems, and different ways of comparing various power transmission systems.

ST8.6 Safety considerations in electrical installations: Explains the physiological effects of electric current, how electric shock and fire hazards arise, how leakage current is produced, the basic principles of protection against overcurrent and leakage currents, and how microshock can occur when devices connected to the ac mains are inserted inside the human body.

SE8.1 Wattmeter reading of reactive power: Analyzes the connection of a wattmeter to read $1/\sqrt{3}$ of the reactive power supplied to a balanced load.

SE8.2 Third harmonics in balanced three-phase systems: Discusses the effect of third harmonics in a four-wire connection of ballasts of fluorescent lights and in a Y-Y connection of three-phase transformers.

Problems and Exercises

In the following problems, all voltages and currents are rms, and the phase sequence is positive unless otherwise indicated.

P8.1 Basic Y and Δ Connections

P8.1.1 A balanced Y-connected load of $6\angle 20°$ Ω per phase is connected in parallel with a balanced Δ-connected load of $3\angle 40°$ Ω per phase. Determine the impedance per phase of the equivalent Δ-connected load.

P8.1.2 A balanced three-phase system of line voltage 240 V supplies a parallel combination of a balanced Y-connected load and a balanced Δ-connected load, having phase impedances of $8 + j8$ Ω and $12 - j24$ Ω, respectively. Determine the line current.

P8.1.3 A three-phase Δ-connected generator has an open-circuit line voltage of 380 V and a phase impedance of $0.1 + j0.5$ Ω. The generator supplies a balanced Y-connected resistive load of 50 Ω per phase. Determine the load phase voltage.

P8.1.4 A balanced Δ-connected load having $R_\Delta = 5$ Ω in parallel with $X_\Delta = -j5$ Ω is supplied from a balanced three-phase supply of negative phase sequence. If $\mathbf{V}_{AB} = 120\angle 0°$ V, determine $\mathbf{I}_{aA}$ and $\mathbf{I}_{AB}$.

P8.1.5 A three-phase, four-wire system is given. The load in phase A is a 360 W lamp, the current in phase B is 4 A at a lagging p.f. of 0.966, and the impedance in phase C is $60\angle -30°\Omega$. The supply is balanced and has a line voltage $\mathbf{V}_{ab} = 190\angle 0°$ V. Determine $\mathbf{I}_{nN}$.

P8.1.6 The impedances of a Δ-connected load are $Z_{AB} = 52\angle -30°\Omega$, $Z_{BC} = 52\angle 45°$ Ω, and $Z_{CA} = 104\angle 0°$ Ω. The load is supplied from a balanced three-phase supply of negative phase sequence. If $\mathbf{V}_{AB} = 208\angle 0°$ V, determine the magnitudes and phase angles of the three line currents.

P8.1.7 Given a three-phase generator having an open-circuit phase voltage of 400 V and a rated current of 50 A, determine the maximum magnitude of impedance per phase if the line voltage is not to drop by more than 2% when each phase is carrying its rated current, assuming that the generator phases are connected in (a) Y or (b) Δ. If the generator phases are connected in Δ, and there is a slight imbalance such that the open-circuit voltage of one of the phases is 395 V, what would be the magnitude of the circulating current in the windings with no load connected to the generator?

P8.1.8 Coils in the lines and phases of a balanced three-phase system are magnetically coupled as shown in Figure P8.1.8. If the current in line a is 10 ∠0° A, determine the current in the phase connected between lines ab.

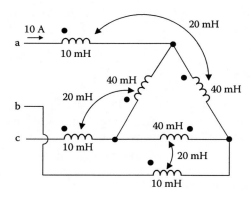

FIGURE P8.1.8

P8.2 Analysis of Three-Phase Systems

Use MATLAB to solve mesh-current equations and verify your analytical results with PSpice in the following problems.

P8.2.1 Verify the results of Example 8.4.1 using mesh-current analysis.

P8.2.2 Verify the results of Example 8.4.2 by solving the circuit of Figure 8.4.3a as a four-mesh circuit.

P8.2.3 A balanced Y-connected load consists of an inductive impedance of 50∠45° Ω in parallel with a 0.1 µF capacitor in each phase. The load is supplied from a balanced three-phase supply having a line voltage of 380 V, 50 Hz, through lines of 0.5 + j1.0 Ω impedance. Determine the magnitude of the line current.

P8.2.4 A balanced Δ-connected load consists of an inductive impedance of 30 + j45 Ω in each phase. The load is supplied from a balanced three-phase supply having a phase voltage of 220 V through lines of 0.1 + j0.2 Ω impedance. Determine the magnitude of the line current.

P8.2.5 The load of Figure P8.2.5 is supplied from a balanced three-phase system of 450 V line voltage, 50 Hz. Determine the current in the neutral before and after phase B is open-circuited at x.

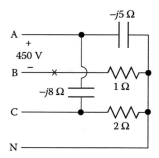

FIGURE P8.2.5

P8.2.6 Three resistors of 6, 10, and 15 Ω are Y connected to a balanced three-wire, three-phase supply of 300 V line voltage. Determine the magnitude of the voltage across each resistor.

P8.2.7 The load of Figure P8.2.7 is connected to a balanced three-wire, three-phase system of 400 V line voltage. Determine the current in each phase of the load.

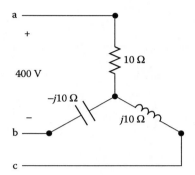

FIGURE P8.2.7

P8.2.8 Consider that the sources in P8.2.7 problem are Y connected. Determine the current in the 10 Ω resistor by superposition.

P8.2.9 A three-phase generator having an impedance of 0.9 + j0.9 Ω per phase is Δ connected. The open-circuit terminal voltage of the generator is 13.2 kV. The generator supplies a Δ-connected load of 650 + j170 Ω per phase through a transmission line of impedance 0.7 + j0.3 Ω per phase. Determine the magnitude of the line voltage at the load end.

P8.2.10 The load of Figure P8.2.10 is connected to a balanced three-phase supply. If $Z_Y = 5 + j10$ Ω and $V_{AB} = 240\angle 0°$ V, determine I_{aA}.

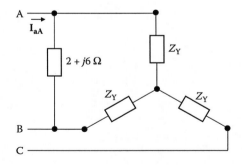

FIGURE P8.2.10

P8.2.11 The sources in Figure P8.2.11 form a balanced set, with $\mathbf{V}_{ga} = 100\angle 90°$ V. Determine $\mathbf{I}_{nN}$.

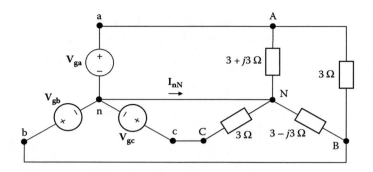

FIGURE P8.2.11

P8.2.12 In Figure 8.4.10 (Exercise 8.4.3), what would be the reading of the voltmeter if it had a resistance of $5R$?

P8.2.13 Given that in Figure P8.2.13, $\mathbf{I}_{aA} = 10\angle 20°$ A and $\mathbf{I}_{bB} = 12\angle -120°$ A, determine $\mathbf{V}_{AB}$.

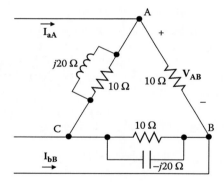

FIGURE P8.2.13

P8.2.14 Given that in Figure P8.2.14, $\mathbf{V}_{AB} = 100\angle 20°$ V and $\mathbf{V}_{BC} = 120\angle -120°$ V, determine $\mathbf{I}_C$.

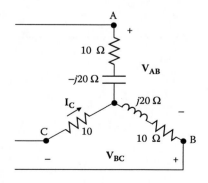

FIGURE P8.2.14

P8.2.15 Determine the single-phase equivalent impedance of the circuit of Figure P8.1.8, assuming a frequency of 50 Hz.

P8.2.16 In the balanced three-phase system shown in Figure P8.2.16, the three voltage sources form a balanced set, $\mathbf{V}_{ga} = 50\angle 0°$ V, $\mathbf{V}_{gb} = 50\angle -120°$ V, and $\mathbf{V}_{gc} = 50\angle 120°$ V; the current sources form a balanced set: $\mathbf{I}_A = 10\angle 30°$ A, $\mathbf{I}_B = 10\angle -90°$ A, $\mathbf{I}_C = 10\angle 150°$ A, and $R_\phi = 5$ Ω. Determine the single-phase equivalent circuit.

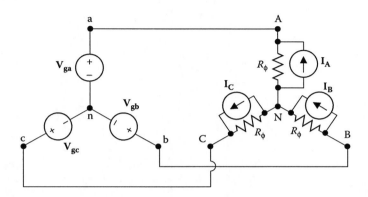

FIGURE P8.2.16

P8.2.17 In the balanced three-phase system shown in Figure P8.2.17, the three independent voltage sources form a balanced set: $\mathbf{V}_{ga} = 50\angle 0°$ V, $\mathbf{V}_{gb} = 50\angle -120°$ V, and $\mathbf{V}_{gc} = 50\angle 120°$ V; the dependent sources form a balanced set with $K = 1 - j$, and $R_\phi = 5$ Ω. Determine the single-phase equivalent circuit.

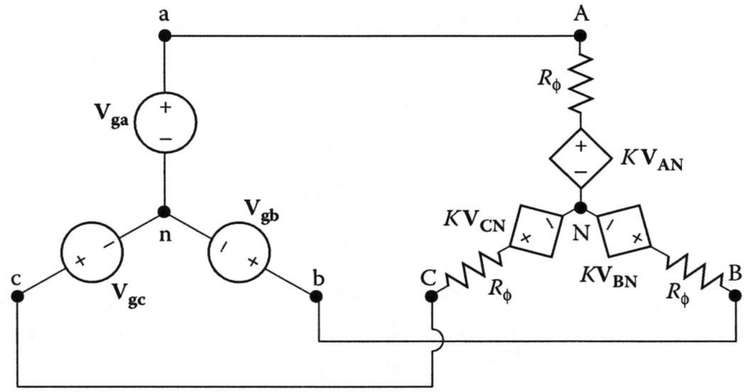

FIGURE P8.2.17

P8.3 Power in Three-Phase Systems

P8.3.1 A balanced Δ-connected, series-connected inductive load draws 10 A of line current and 3 kW at a line voltage of 220 V. Determine the load impedance per phase.

P8.3.2 A balanced three-phase system of 240 V line voltage, 50 Hz, supplies a 100 kW balanced load of 0.6 p.f. lagging. Determine the capacitance in each phase of a Y-connected capacitor bank that will give a power factor of 0.95 lagging.

P8.3.3 In the circuit of Figure P8.3.3, the Y-connected load and the Δ-connected load are both balanced with $R_Y = 4\ \Omega$, $X_Y = j4\ \Omega$, $R_\Delta = 6\ \Omega$, and $Y_\Delta = -j8\ \Omega$. Determine the real and reactive powers absorbed if the line voltage is 190 V.

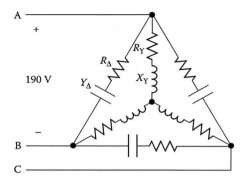

FIGURE P8.3.3

P8.3.4 A three-phase ac generator is rated at 50 MVA, 11 kV line voltage. If the generator is operated at its rated voltage and current, determine the percentage increase in the real power delivered if the power factor is increased from 0.7 to 0.95 lagging.

P8.3.5 Determine the real, reactive, and apparent powers absorbed by the load of Figure P8.3.5 when connected to a balanced three-phase supply of 220 V line voltage.

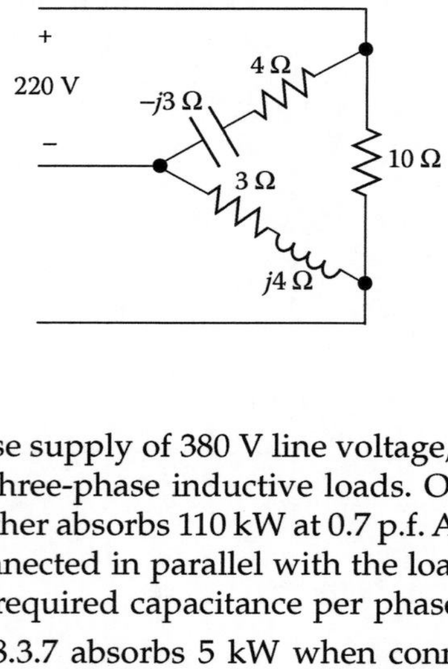

FIGURE P8.3.5

P8.3.6 A balanced three-phase supply of 380 V line voltage, 50 Hz, is connected to two paralleled, balanced three-phase inductive loads. One load absorbs 173 kW at 0.8 p.f., whereas the other absorbs 110 kW at 0.7 p.f. A bank of equal Δ-connected capacitors is to be connected in parallel with the loads so as to bring the p.f. to unity. Determine the required capacitance per phase.

P8.3.7 The load in Figure P8.3.7 absorbs 5 kW when connected to a balanced, four-wire, three-phase supply of 380 V rms line voltage. The power factors of phases B and C are 0.5 lagging and 0.5 leading, respectively, and the magnitudes of the currents in phases A and B are 5 A and 10 A, respectively. Determine: (a) the magnitude and phase angle of the current in phase C, assuming that the current in phase A has a phase angle of zero; (b) R_C and X_C; and (c) the reactive power of phases B and C.

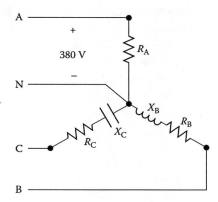

FIGURE P8.3.7

P8.3.8 Each phase of a Y-connected load consists of a resistance of 100 Ω in parallel with a capacitance of 31.8 µF. The load is connected to a balanced three-phase supply of 416 V rms line voltage, 50 Hz. Determine the real, reactive, and apparent powers absorbed by the load.

P8.3.9 A balanced bank of capacitors is connected in parallel with a balanced three-phase load in order to bring the p.f. at the supply end to unity. The apparent powers of the load and capacitor bank are 2 kVA and 500 VA, respectively. The load and capacitors are supplied from a balanced three-phase supply having a line impedance of $2 + j20$ Ω per line, the total real power dissipated in the lines being 24 W. Calculate the line voltage at the load end.

P8.3.10 A balanced three-phase system in which the load consumes 50 kW at 0.8 p.f. lagging is given, the line impedance being $0.5 + j0.5$ Ω. Capacitors are Δ connected so that the power factor at the supply terminals is 0.95. If the magnitude of the line current is 120 A, 50 Hz, determine the magnitude of the line voltage at the supply terminals.

P8.3.11 In the balanced three-phase system shown in Figure P8.3.11, $R = 0.5$ Ω and Z_Δ consists of a 50 Ω resistance in parallel with an inductive reactance of 50 Ω. Determine the total real power consumed by the load. What is the p.f. seen at the supply end?

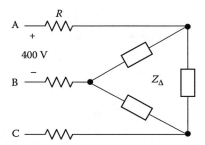

FIGURE P8.3.11

P8.3.12 A balanced source supplies 100 kVA at a line voltage of 450 V and 0.9 p.f. leading to a balanced load, the line impedance being $0.1 + j0.5$ Ω. Determine: (a) the magnitude of the line current; (b) the magnitude of the line voltage at the load; and (c) the total apparent power at the load.

P8.3.13 Verify that reversing the phase sequence interchanges the readings of the two wattmeters in the two-wattmeter method.

P8.3.14 Verify Equation 8.5.3 for a Δ connection.

P8.3.15 The power input to a three-phase synchronous motor is measured by the two-wattmeter method. When the p.f. of the motor is unity, each of the two wattmeters reads 50 kW. What would be the reading of each wattmeter if the power factor is changed to 0.866 leading, assuming the magnitudes of the line voltage and current stay the same? In a synchronous motor, the motor p.f. can be varied by varying the dc excitation to the rotor.

P8.3.16 A balanced load of 0.75 p.f. lagging is connected to a balanced three-phase, three-wire system. The sum of the readings of two wattmeters connected in the standard two-wattmeter connection is 26 kW. What is the reading of each wattmeter? This problem is to be solved using Equation 8.5.6 rather than the method of Example 8.5.2.

P8.3.17 A balanced three-phase load is connected to a balanced three-phase supply of 220 V line voltage. A wattmeter reads 600 W when its current coil is connected in line a and its voltage coil is connected between lines a and b, with the positive terminal of the coil connected to line a. When the voltage coil is connected between lines b and c, with the positive terminal of the coil connected to phase b, and the current coil connected in line a as before, the wattmeter again reads 600 W. What is the load p.f.?

P8.3.18 Two wattmeters W_1 and W_2 are connected to a balanced three-phase, inductive load of 0.8 p.f., as shown in Figure P8.3.18, the supply also being balanced. If W_1 reads 100 W, what is the reading of W_2?

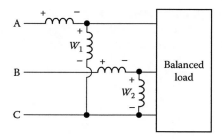

FIGURE P8.3.18

P8.3.19 In Figure P8.3.19, the source voltages V_1, V_2, and V_3 constitute a balanced set, with $V_1 = 300\angle -45°$ V, $V_{AB} = 381\angle 0°$ V, and $Z_\phi = 20\angle 30°$. The current coils C_1 and C_2 and voltage coils P_1 and P_2 of wattmeters W_1 and W_2 are connected as shown. Determine the readings of W_1 and W_2.

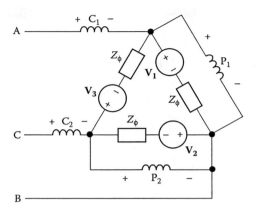

FIGURE P8.3.19

P8.3.20 Two wattmeters W_1 and W_2 are connected as shown in Figure P8.3.20 to measure the power in a balanced Δ-connected load, the line voltage being 380 V. If each wattmeter reads 1000 W, determine the load impedance per phase.

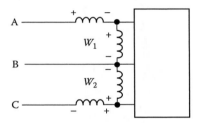

FIGURE P8.3.20

P8.3.21 Consider Figure SE8.2.2 (Example SE8.2). Show that if the phase voltages have a third-harmonic component, these components cancel out between lines but will cause the neutral to oscillate at the third harmonic frequency.

9

Responses to Periodic Inputs

Overview

Chapter 5 to Chapter 8 were concerned with the sinusoidal steady state in which all the sources applied to a given circuit are of the same frequency. This case can be readily generalized to inputs that are nonsinusoidal but periodic. The basis for this generalization is Fourier's theorem, according to which a periodic signal of a given frequency can be expressed, in general, as an infinite series of sine and cosine functions whose frequencies are integral multiples of the frequency of the periodic signal. The amplitudes of the higher-frequency sinusoids decrease fairly rapidly in most cases, so that the periodic signal can be represented in practice by a finite sum of sinusoids.

Once a periodic signal is represented as a series expansion of sinusoids of different frequencies, the response of an LTI circuit to the periodic signal is the sum of the responses to the individual frequency components. The different frequencies do not interact in an LTI circuit, so that each of these responses can be obtained using phasor analysis, as was done in Chapter 5.

Circuit responses to periodic signals are of considerable practical interest because these signals are very common. The steady state of any linear or nonlinear circuit, other than the dc steady state, is periodic. Thus, the outputs of free-running oscillators, the time bases of TV and computer displays, and continuous vibrations of all kinds are periodic.

The Fourier series expansion of periodic functions can be generalized to nonperiodic functions by means of the Fourier transform (Chapter 16).

Learning Objectives

- To be familiar with:
  - The Fourier series expansion of some commonly encountered periodic functions
- To understand:
  - The derivation and basic properties of Fourier series expansions, including symmetry considerations
  - The interpretation of amplitude and phase spectra
  - How the Fourier series expansion of a given function may be obtained from those of other functions through addition, subtraction, multiplication, differentiation, or integration

- How the response of an LTI circuit to a periodic signal can be derived
- Power relations of periodic inputs, including rms values

9.1   Fourier Series

The defining property of a periodic function $f(t)$ of period T is that $f(t) = f(t + nT)$, where n is any integer. That is, the function repeats every period T. The sinusoidal function discussed in Chapter 5 is a common example of a periodic function, so that the same definitions of cycle, frequency, and angular frequency for a sinusoidal function (Equation 5.1.2) apply to periodic functions in general. Strictly speaking, a periodic time function extends over all time, from $-\infty$ to $+\infty$. In practice, a function can be assumed periodic if it has been in a steady state for a time interval that is large compared to the period. Periodic functions need not be time functions. They can be functions of distance, in which case the period is the **wavelength**.

A remarkable theorem on periodic functions is Fourier's theorem:

> **Statement:** *A periodic function $f(t)$ can be expressed, in general, as an infinite series of cosine and sine functions*:

$$f(t) = a_0 + \sum_{k=1}^{\infty} (a_k \cos k\omega_0 t + b_k \sin k\omega_0 t) \tag{9.1.1}$$

where k is a positive integer, and a_0, a_k, and b_k are constants, known as **Fourier coefficients**, that depend on $f(t)$.

The component having $k = 1$ is the **fundamental**, whereas the component with k equal to a particular integer n is the **nth harmonic**. The series expression of Equation 9.1.1 is the **Fourier series expansion** (FSE) of $f(t)$.

Although periodic functions encountered in practice can be expressed as an FSE, it is of mathematical interest to determine if Fourier's theorem applies to any arbitrary periodic function. **Dirichlet's conditions**, which are sufficient to ensure that the FSE of a given periodic function $f(t)$ converges to $f(t)$ for any t, may be stated as follows:

1. The integral $\displaystyle\int_{t_0}^{t_0+T} |f(t)| dt$ exists, that is, is finite, for any arbitrary t_0

2. $f(t)$ is single valued over a period, with only a finite number of finite discontinuities, maxima, or minima.

Combining the sine and cosine terms, the FSE of Equation 9.1.1 becomes:

$$f(t) = c_0 + \sum_{k=1}^{\infty} c_k \cos(k\omega_0 t + \theta_k) \tag{9.1.2}$$

where
$$c_0 = a_0, \quad c_k = \sqrt{a_k^2 + b_k^2}, \quad \text{and} \quad \theta_k = -\tan^{-1}\frac{b_k}{a_k} \tag{9.1.3}$$

9.2 Fourier Analysis

The object of Fourier analysis is to derive the Fourier coefficients of a periodic function. Some integral trigonometric relations needed for this purpose can be summarized as follows:

Summary: *Given the four functions* $\cos m\omega_0 t$, $\sin m\omega_0 t$, $\cos n\omega_0 t$, *and* $\sin n\omega_0 t$, *where m and n are any integers, the integral of the product of any two of these functions over a period* $T = 2\pi/\omega_0$ *is zero, except the products* $\cos^2 n\omega_0 t$, *and* $\sin^2 n\omega_0 t$, *having m = n, in which case the integral is T/2.*

Thus:

$$\int_{t_0}^{t_0+T} \cos n\omega_0 t \sin m\omega_0 t\,dt = 0 = \int_{t_0}^{t_0+T} \sin n\omega_0 t \cos \omega_0 t\,dt \quad \text{for all } n \text{ and } m \tag{9.2.1}$$

$$\int_{t_0}^{t_0+T} \cos n\omega_0 t \cos m\omega_0 t\,dt = 0 = \int_{t_0}^{t_0+T} \sin n\omega_0 t \sin m\omega_0 t\,dt \quad \text{for } n \neq m \tag{9.2.2}$$

$$\int_{t_0}^{t_0+T} \cos^2 n\omega_0 t\,dt = \frac{T}{2} = \int_{t_0}^{t_0+T} \sin^2 n\omega_0 t\,dt \tag{9.2.3}$$

Equation 9.2.1 to Equation 9.2.3 are easily proved using basic trigonometric relations (Appendix) and are left as Exercise 9.2.1.

To determine a_0 in the FSE, we integrate both sides of Equation 9.1.1 over a period:

$$\int_{t_0}^{t_0+T} f(t)dt = \int_{t_0}^{t_0+T} a_0 dt + \sum_{k=1}^{\infty}\int_{t_0}^{t_0+T} a_k \cos k\omega_0 t\,dt + \sum_{k=1}^{\infty}\int_{t_0}^{t_0+T} b_k \sin k\omega_0 t\,dt$$

$$= a_0 T + 0 + 0$$

or,
$$a_0 = \frac{1}{T}\int_{t_0}^{t_0+T} f(t)dt \tag{9.2.4}$$

a_0 is therefore the average of $f(t)$ over a period. It is the *dc component of* $f(t)$, whereas the cosine and sine terms are the *ac component.*

To determine a_n, we multiply both sides of Equation 9.1.1 by $\cos n\omega_0 t$ and integrate over a whole period, invoking Equation 9.2.1 to Equation 9.2.3:

$$\int_{t_0}^{t_0+T} f(t)\cos n\omega_0 t dt = \int_{t_0}^{t_0+T} a_0 \cos n\omega_0 t dt + \sum_{k=1}^{\infty} \int_{t_0}^{t_0+T} a_k \cos k\omega_0 t \cos n\omega_0 t dt +$$

$$\sum_{k=1}^{\infty} \int_{t_0}^{t_0+T} b_k \sin k\omega_0 t \cos n\omega_0 t dt = 0 + \frac{T}{2} a_n + 0$$

This gives:

$$a_n = \frac{2}{T} \int_{t_0}^{t_0+T} f(t)\cos n\omega_0 t dt \qquad (9.2.5)$$

To determine b_n, we multiply both sides of Equation 9.1.1 by $\sin n\omega_0 t$ and integrate over a whole period, invoking Equation 9.2.1 to Equation 9.2.3:

$$\int_{t_0}^{t_0+T} f(t)\sin n\omega_0 t dt = \int_{t_0}^{t_0+T} a_0 \sin n\omega_0 t dt + \sum_{k=1}^{\infty} \int_{t_0}^{t_0+T} a_k \cos k\omega_0 t \sin n\omega_0 t dt +$$

$$\sum_{k=1}^{\infty} \int_{t_0}^{t_0+T} b_k \sin k\omega_0 t \sin n\omega_0 t dt = 0 + 0 + \frac{T}{2} b_n$$

This gives:

$$b_n = \frac{2}{T} \int_{t_0}^{t_0+T} f(t)\sin n\omega_0 t \, dt \qquad (9.2.6)$$

Summary: a_0 is the average of $f(t)$ over a period, a_n is twice the average of $f(t)\cos n\omega_0 t$ over a period, and b_n is twice the average of $f(t)\sin n\omega_0 t$ over a period.

Example 9.2.1: FSE of Sawtooth Waveform

It is required to derive the Fourier coefficients of the sawtooth waveform of Figure 9.2.1.

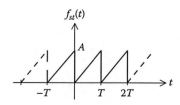

FIGURE 9.2.1
Figure for Example 9.2.1.

SOLUTION

During the interval $0 \le t < T$, $f_{st}(t) = \dfrac{A}{T} t$.

$$a_0 = \frac{1}{T} \int_0^T \frac{A}{T} t \, dt = \frac{A}{T^2} \left[\frac{t^2}{2} \right]_0^T = \frac{A}{2} \, ;$$

$a_n = \dfrac{2}{T} \int_0^T \dfrac{A}{T} t \cos n\omega_0 dt = \dfrac{2A}{T^2} \int_0^T t \cos n\omega_0 t \, dt$. Integrating by parts, or using the Table of

Integrals in the Appendix, $a_n = \dfrac{2A}{T^2} \left[\dfrac{1}{n^2 \omega_0^2} \cos n\omega_0 t + \dfrac{t}{n\omega_0} \sin n\omega_0 t \right]_0^{2\pi/\omega_0} = 0$. The FSE does

not have any cosine terms for reasons that will be explained in Section 9.3.

$$b_n = \frac{2}{T} \int_0^T \frac{A}{T} t \sin \omega_0 dt = \frac{2A}{T^2} \int_0^T t \sin n\omega_0 t \, dt = \frac{2A}{T^2} \left[\frac{1}{n^2 \omega_0^2} \sin n\omega_0 t - \frac{t}{n\omega_0} \cos n\omega_0 t \right]_0^{2\pi/\omega_0}$$

$$= \frac{2A}{T^2} \left[-\frac{2\pi}{n\omega_0^2} \right] = -\frac{A}{\pi n}$$

where $\omega_0 T = 2\pi$. The trigonometric form of $f_{st}(t)$ is therefore:

$$f_{st}(t) = \frac{A}{2} - \frac{A}{\pi} \sum_{n=1}^{\infty} \frac{\sin n\omega_0 t}{n} = \frac{A}{2} - \frac{A}{\pi} \left[\sin \omega_0 t + \frac{\sin 2\omega_0 t}{2} + \frac{\sin 3\omega_0 t}{3} + \dots \right] \quad (9.2.7)$$

At the points of discontinuity, $t = kT$, where k is an integer, all the sinusoidal terms vanish and $f(t) = A/2$, the average of the values of $f(kT^+)$ and $f(kT^-)$.

Exponential Form

The FSE can also be expressed in exponential form. It is convenient for this purpose to change the index k to n:

$$f(t) = a_0 + \sum_{n=1}^{\infty} \left[a_n \left(\frac{e^{jn\omega_0 t} + e^{-jn\omega_0 t}}{2} \right) + b_n \left(\frac{e^{jn\omega_0 t} - e^{-jn\omega_0 t}}{2j} \right) \right]$$

$$= a_0 + \sum_{n=1}^{\infty} \left[\left(\frac{a_n - jb_n}{2} \right) e^{jn\omega_0 t} + \left(\frac{a_n + jb_n}{2} \right) e^{-jn\omega_0 t} \right] \quad (9.2.8)$$

Let $C_n = \dfrac{1}{2}(a_n - jb_n)$. Substituting for a_n and b_n from Equation 9.2.5 and Equation 9.2.6:

$$C_n = \frac{1}{T} \int_{t_0}^{t_0+T} f(t)(\cos n\omega_0 t - j \sin n\omega_0 t)dt$$

$$= \frac{1}{T} \int_{t_0}^{t_0+T} f(t)e^{-jn\omega_0 t} dt \tag{9.2.9}$$

It follows that: $C_0 = \dfrac{1}{T} \int_{t_0}^{t_0+T} f(t)\, dt = a_0$, and:

$$C_n^* = \frac{1}{2}(a_n + jb_n) = \frac{1}{T} \int_{t_0}^{t_0+T} f(t)(\cos n\omega_0 t + j \sin n\omega_0 t)dt = \frac{1}{T} \int_{t_0}^{t_0+T} f(t)\, e^{jn\omega_0 t} \, dt = C_{-n} \tag{9.2.10}$$

where C_n^* is the complex conjugate of C_n. Equation 9.2.8 can be expressed as:

$$f(t) = C_0 + \sum_{n=1}^{\infty} C_n e^{jn\omega_0 t} + \sum_{n=1}^{\infty} C_{-n} e^{-jn\omega_0 t}$$

The last term on the RHS can be written in terms of negative values of n as:

$$f(t) = C_0 + \sum_{n=1}^{\infty} C_n e^{jn\omega_0 t} + \sum_{n=-1}^{-\infty} C_n e^{jn\omega_0 t} \tag{9.2.11}$$

Equation 9.2.11 can be expressed more compactly as:

$$f(t) = \sum_{n=-\infty}^{\infty} C_n e^{jn\omega_0 t} \tag{9.2.12}$$

It should be noted, however, that Equation 9.2.11 is the general exponential form because deriving C_n as in Equation 9.2.9 and then setting $n = 0$ does not always give a finite value for C_0 (Example 9.2.2). The relationships between C_n, a_n, and b_n readily follow from the definition of C_n:

$$a_n = 2\text{Re}(C_n), \quad \text{and} \quad b_n = -2\text{Im}(C_n) \tag{9.2.13}$$

$$|C_n| = \left| \frac{a_n - jb_n}{2} \right| = \frac{\sqrt{a_n^2 + b_n^2}}{2} = \frac{c_n}{2} \quad \text{and} \quad \angle C_n = -\tan^{-1} \frac{b_n}{a_n} = \theta_n \tag{9.2.14}$$

where c_n and θ_n are as in Equation 9.1.3.

Compared to the trigonometric form, the exponential form is advantageous where it is easier to apply Equation 9.2.9 rather than Equation 9.2.5 and Equation 9.2.6, particularly if the int command of MATLAB is used to evaluate C_n, as illustrated in the examples that follow. Moreover, deriving C_n gives both a_n and b_n simultaneously.

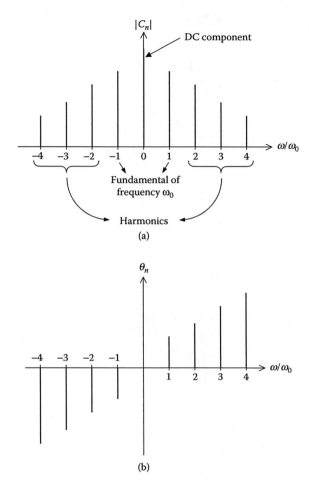

FIGURE 9.2.2
(a) Amplitude spectrum; (b) phase spectrum.

Frequency Spectrum

The plots of $|C_n|$ and θ_n against frequency are, respectively, the **amplitude spectrum** and the **phase** spectrum of $f(t)$. They both constitute the **frequency spectrum** of $f(t)$. Because frequencies in the FSE have discrete values only, the frequency spectrum of a periodic function is a line spectrum that consists of a series of lines at $\omega = n\omega_0$, where $n = 0,$ $\pm 1, \pm 2, \pm 3, \dots$ (Figure 9.2.2a). Because $C_{-n} = C_n^*$, it is seen that $|C_n| = |C_{-n}|$ and $\angle C_n = -\tan^{-1}\dfrac{b_n}{a_n} = -\angle C_{-n}$. The amplitude spectrum is therefore an even function, that is, it is symmetrical about the vertical axis, whereas the phase spectrum is an odd function (Figure 9.2.2b), except when C_n is real, so $b_n = 0$ and θ_n is either zero or 180°.

The reader may wonder about the physical meaning of negative frequencies. For present purposes, negative frequencies simply arise naturally from extending the integer index n to negative values (Equation 9.2.11). They are needed to combine with their positive counterparts in complex exponentials (Equation 9.2.12) so as to give real cosine and sine terms.

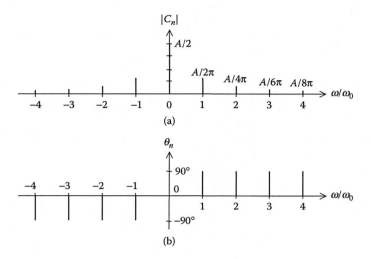

FIGURE 9.2.3
Figure for Example 9.2.2.

Example 9.2.2: Exponential Form of FSE of Sawtooth Waveform

It is required to derive the exponential Fourier coefficients of the sawtooth waveform of Figure 9.2.1 and plot its amplitude and phase spectra.

SOLUTION

$C_n = \dfrac{1}{T}\displaystyle\int_0^T \dfrac{A}{T}\, t e^{-jn\omega_0 t}\, dt$. Integrating by parts (Appendix), noting that $\omega_0 T = 2\pi$,

$$C_n = -\frac{A}{T^2 n^2 \omega_0^2}\left[-jn\omega_0 t e^{-jn\omega_0 t} - e^{-jn\omega_0 t}\right]_0^{2\pi/\omega_0} = -\frac{A}{4\pi^2 n^2}\left[-j2n\pi\right] = j\frac{A}{2\pi n} \qquad (9.2.15)$$

C_n is imaginary, which means that $a_n = 0$ (Equation 9.2.13). The average value of $f_{st}(t)$ is $A/2$, and cannot be obtained by setting $n = 0$ in Equation 9.2.15. The exponential form of $f_{st}(t)$ is:

$$f_{st}(t) = \frac{A}{2} + \frac{A}{2\pi}\sum_{\substack{n=-\infty \\ n\neq 0}}^{\infty}\frac{j}{n}e^{jn\omega_0 t} \qquad (9.2.16)$$

The amplitude spectrum consists of a line of height $A/2$ at $\omega = 0$ and lines of height $|C_n| = \left|\dfrac{A}{2\pi n}\right|$ at $\dfrac{\omega}{\omega_0} = \pm n$; the phase angle of C_n is $+90°$ for $n > 0$, and $-90°$ for $n < 0$ (Figure 9.2.3).

C_n can be obtained using MATLAB's int(E,t,a,b) command (Appendix SD.1). Ignoring for the moment $\dfrac{A}{T^2}$, the integral $\displaystyle\int_0^T t e^{-jn\omega_0 t}\, dt$ can be evaluated by entering the following code:

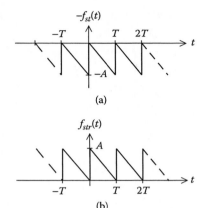

FIGURE 9.2.4
Figure for Example 9.2.2.

```
syms t n w
int(t*exp(-j*n*w*t),t,0,2*pi/w)
```

MATLAB returns a rather complicated expression. So enter: `simplify(ans)`. MATLAB returns:

```
1/n^2/w^2*(2*i*exp(-2*i*pi*n)*n*pi+exp(-2*i*pi*n)-1)
```

Recognizing that $e^{-j2\pi n} = 1$ for all n, this expression simplifies to $\dfrac{j2\pi n}{n^2 w^2}$. Multiplying by $\dfrac{A}{T^2}$ gives Equation 9.2.15.

If the function $f_{st}(t)$ of Figure 9.2.1 is negated (Figure 9.2.4a) and then shifted upward by A, it becomes the reversed sawtooth waveform $f_{str}(t)$ of Figure 9.2.4b. The FSE of $f_{str}(t)$ is obtained by adding A to the negation of the RHS of Equation 9.2.7:

$$f_{str}(t) = \frac{A}{2} + \frac{A}{\pi} \sum_{n=1}^{\infty} \frac{\sin n\omega_0 t}{n} \tag{9.2.17}$$

C_0' of $f_{str}(t)$ is $A/2$ and its C_n' is $-j\dfrac{A}{2\pi n}$. The amplitude spectrum is unchanged, but the phase spectrum is negated. The derivation of the FSE of $f_{str}(t)$ from that of $f_{st}(t)$ illustrates a useful technique of deriving the FSE of a function from that of another function.

EXERCISE 9.2.1
Prove Equation 9.2.1 to Equation 9.2.3.

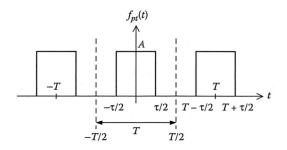

FIGURE 9.2.5
Figure for Example 9.2.3.

EXERCISE 9.2.2
Derive the Fourier coefficients of the reversed sawtooth waveform of Figure 9.2.4b by applying Equation 9.2.4, Equation 9.2.5, and Equation 9.2.6.

Example 9.2.3: FSE of Rectangular Pulse Train

It is required to derive the Fourier coefficients of the rectangular pulse train $f_{pt}(t)$ illustrated in Figure 9.2.5 and plot its amplitude and phase spectra.

SOLUTION
It is convenient to take a period that is symmetrical about the origin as shown. Hence, $t_0 = -T/2$ in Equation 9.2.9, so that:

$$C_n = \frac{1}{T}\int_{-T/2}^{T/2} f(t)e^{-jn\omega_0 t}\,dt = \frac{1}{T}\int_{-\tau/2}^{\tau/2} Ae^{-jn\omega_0 t}\,dt = \frac{A}{T}\left[\frac{e^{-jn\omega_0 t}}{-jn\omega_0}\right]_{-\tau/2}^{\tau/2} = \frac{A}{n\omega_0 T}\left(\frac{e^{jn\omega_0\tau/2}-e^{-jn\omega_0\tau/2}}{j}\right)$$

$$= \frac{2A}{n\omega_0 T}\sin(n\omega_0\tau/2) = A\frac{\tau}{T}\frac{\sin(n\omega_0\tau/2)}{(n\omega_0\tau/2)} = A\frac{\tau}{T}\operatorname{sinc}(n\omega_0\tau/2) \qquad (9.2.18)$$

where $\operatorname{sinc}(x) = \dfrac{\sin x}{x}$. When n is a nonzero integer, $\operatorname{sinc}(n\pi) = \dfrac{\sin n\pi}{n\pi} = 0$. However, according to L'Hopital's rule (Appendix), $\operatorname{sinc}(0) = 1$.

Because C_n is real, $b_n = 0$, and the FSE of $f_{pt}(t)$ function does not have any sine terms. The reason for this is that the function is even, as explained in Section 9.3. The average value of $f_{pt}(t)$ is $A\dfrac{\tau}{T}$. However, it can be obtained in this case by setting $n = 0$ in Equation 9.2.18.

If we set $\dfrac{\tau}{T} = \alpha$ and replace $\omega_0 T$ by 2π:

$$C_n = \alpha A\operatorname{sinc}(\alpha n\pi) = \frac{A}{n\pi}\sin(\alpha n\pi) \qquad (9.2.19)$$

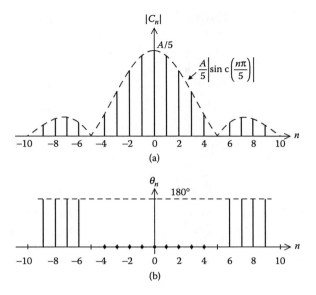

FIGURE 9.2.6
Figure for Example 9.2.3.

and

$$f_{pt}(t) = \alpha A \sum_{n=-\infty}^{\infty} \text{sinc}(\alpha n\pi)e^{jn\omega_0 t} \qquad (9.2.20)$$

To express the FSE in trigonometric form, we substitute $a_0 = C_0 = \alpha A$, $a_n = 2C_n$, and $b_n = 0$ in Equation 9.1.1:

$$f_{pt}(t) = \alpha A + \frac{2A}{\pi}\left[\sin\alpha\pi\cos\omega_0 t + \frac{\sin 2\alpha\pi}{2}\cos 2\omega_0 t + \frac{\sin 3\alpha\pi}{3}\cos 3\omega_0 t + ...\right] \qquad (9.2.21)$$

The amplitude spectrum is shown in Figure 9.2.6a for the case of $\alpha = 1/5$, so that $|C_n| = \frac{A}{5}\text{sinc}\left(\frac{n\pi}{5}\right)$ for all integer values of $n = \frac{\omega}{\omega_0}$. The amplitude is zero for n an integral multiple of 5. The lines are bounded by the dotted envelope representing the function $\frac{A}{5}\left|\text{sinc}\left(\frac{n\pi}{5}\right)\right|$ for continuous n. The phase spectrum is shown in Figure 9.2.6b. Because C_n is real, its phase angle is either zero ($C_n > 0$) or 180° ($C_n < 0$). The phase angle is zero when $n = 0$ because $C_n > 0$, and is not defined when $C_n = 0$, as when $n = \pm 5$, because the phase angle can be zero or 180° when the magnitude is zero. It can be readily verified that if $(m-1) < \left|\frac{n}{5}\right| < m$, where m is a positive (nonzero), odd integer, then $\frac{\sin(n\pi/5)}{n\pi/5} > 0$ and $\theta_n = 0$. On the other hand, if m is a positive, even integer, $\frac{\sin(n\pi/5)}{n\pi/5} < 0$ and $\theta_n = 180°$. Thus, if $m = 1$, $\theta_n = 0$ for $n = \pm 1, \pm 2, \pm 3$, and ± 4. If $m = 2$, $\theta_n = 180°$ for $n = \pm 6, \pm 7, \pm 8, \pm 9$, etc.

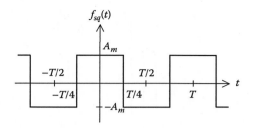

FIGURE 9.2.7
Figure for Example 9.2.3.

If α is small, it is seen from Equation 9.2.19 that all the harmonics have the same amplitude αA. This is an important result in signal analysis, according to which the narrower the pulses, the more significant are the higher harmonics (Section 16.6, Chapter 16).

To determine C_n using MATLAB's int(E,t,a,b) command, we enter the following:

```
syms t n w a
int(exp(-j*n*w*t),t,-a/2,a/2)
simplify(ans)
```

MATLAB returns:

```
2*sin(1/2*n*w*a)/n/w
```

Multiplying this by $\dfrac{A}{T}$ gives Equation 9.2.18.

We can deduce from Equation 9.2.21 the FSE $f(t)$ of a square wave of amplitude A_m and zero average value (Figure 9.2.7). To do so, we set $\alpha = \dfrac{1}{2}$, $A_m = \dfrac{A}{2}$, and remove the DC value by subtracting $\dfrac{A}{2}$ from $f(t)$. Noting that $\sin(n\pi/2) = 0$ for even n:

$$f_{sq}(t) = \frac{4A_m}{\pi}\left[\cos\omega_0 t - \frac{1}{3}\cos 3\omega_0 t + \frac{1}{5}\cos 5\omega_0 t - \frac{1}{7}\cos 7\omega_0 t + \ldots\right] \tag{9.2.22}$$

EXERCISE 9.2.3
Derive the Fourier coefficients of the rectangular waveform of Figure 9.2.5 by applying Equation 9.2.4, Equation 9.2.5, and Equation 9.2.6.

Translation in Time

The exponential form is convenient for determining the effect of a translation in time. If a periodic waveform $f(t)$ is delayed by t_d, it becomes $f(t - t_d)$ with respect to the same time origin. Replacing t by $(t - t_d)$ in Equation 9.2.12:

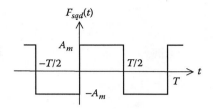

FIGURE 9.2.8
Figure for Example 9.2.4.

$$f(t - t_d) = \sum_{n=-\infty}^{\infty} C_n e^{jn\omega_0(t - t_d)} = \sum_{n=-\infty}^{\infty} \left[C_n e^{-jn\omega_0 t_d} \right] e^{jn\omega_0 t} \qquad (9.2.23)$$

The effect is to replace C_n by $C_n e^{-jn\omega_0 t_d}$. The magnitude of C_n, and hence the amplitude spectrum, remains unchanged. However, the new phase angle θ'_n is

$$\theta'_n = \theta_n - n\omega_0 t_d \qquad (9.2.24)$$

Conversely, if the function is advanced by t_d, C_n is replaced by $C_n e^{+jn\omega_0 t_d}$.

Example 9.2.4: Translation in Time of Square Wave
It is required to derive the FSE of the square wave (Figure 9.2.7) when delayed, or advanced, by $T/4$.

SOLUTION
Because $t_d = T/4$, $n\omega_0 t_d = n\omega_0 T/4 = n\pi/2$. If the function is delayed by $T/4$ (Figure 9.2.8), the phase angle of each of the terms in Equation 9.2.22 is decreased by $n\pi/2$:

$$f_{sqd}(t) = \frac{4A_m}{\pi} \left[\cos\left(\omega_0 t - \frac{\pi}{2} \right) - \frac{1}{3}\cos\left(3\omega_0 t - \frac{3\pi}{2} \right) + \frac{1}{5}\cos\left(5\omega_0 t - \frac{5\pi}{2} \right) - \frac{1}{7}\cos\left(7\omega_0 t - \frac{7\pi}{2} \right) + ... \right]$$

$$f_{sqd}(t) = \frac{4A_m}{\pi} \left[\sin\omega_0 t + \frac{1}{3}\sin 3\omega_0 t + \frac{1}{5}\sin 5\omega_0 t + ... \right] \qquad (9.2.25)$$

Intuitively, one would think of a sine function as a first rough approximation to the square wave of Figure 9.2.8.

If the square wave of Figure 9.2.7 is advanced by $T/4$, it becomes the negation of Figure 9.2.8, so that:

$$f_{sqa}(t) = -\frac{4A_m}{\pi} \left[\sin\omega_0 t + \frac{1}{3}\sin 3\omega_0 t + \frac{1}{5}\sin 5\omega_0 t + ... \right] \qquad (9.2.26)$$

EXERCISE 9.2.4

Derive: (a) the Fourier coefficients of the square waveforms of Equation 9.2.25 and Equation 9.2.26 by applying Equation 9.2.4, Equation 9.2.5, and Equation 9.2.6; (b) the exponential form of the square waves $f_{sq}(t)$, $f_{sqd}(t)$, and $f_{sqa}(t)$.

Answer: (b) $f_{sq}(t) = \dfrac{2A_m}{\pi} \displaystyle\sum_{\substack{n=-\infty \\ n\neq 0}}^{n=\infty} \dfrac{1}{n} \sin\left(\dfrac{n\pi}{2}\right) e^{jn\omega_0 t}$, $f_{sqd}(t) = -\dfrac{j2A_m}{\pi} \displaystyle\sum_{\substack{n=-\infty \\ n\neq 0}}^{n=\infty} \dfrac{1}{n} e^{jn\omega_0 t}$,

$f_{sqa}(t) = \dfrac{j2A_m}{\pi} \displaystyle\sum_{\substack{n=-\infty \\ n\neq 0}}^{n=\infty} \dfrac{1}{n} e^{jn\omega_0 t}$, where n is odd.

EXERCISE 9.2.5

Determine and compare the amplitude and phase spectra of the square waves of Equation 9.2.22, Equation 9.2.25, and Equation 9.2.26.

Answer: The amplitude spectrum is $|C_n| = \dfrac{2A_m}{|n|\pi}$ for the three cases. The phase spectrum of $f_{sq}(t)$ is zero for $n = 1, 5, 9, 13$, etc. and is 180° for $n = 3, 7, 11, 15$, etc. The phase spectrum of $f_{sqd}(t)$ is −90° for positive n, and 90° for negative n. The phase spectrum of $f_{sqa}(t)$ is 90° for positive n, and −90° for negative n.

9.3   Symmetry Properties of Fourier Series

Even-Function Symmetry

> **Concept:** *The FSE of an even periodic function does not contain any sine terms; its Fourier coefficients can be evaluated over half a period.*

The reason that the FSE of an even periodic function does not contain any sine terms is simply that, because the sine function is odd, the presence of sine terms introduces odd components in the function and destroys its even symmetry. An even function can have a dc component and still remain even because the dc component has even symmetry. Examples of even functions are the rectangular pulse train (Figure 9.2.5) and the square pulse train of Figure 9.2.7. The corresponding FSEs (Equation 9.2.21 and Equation 9.2.22) do not have any sine terms.

If the FSE of an even periodic function does not contain any sine terms, then $b_n = 0$ and C_n is real. Because a period of an even periodic function is centered about the vertical axis, C_n can be expressed as:

$$C_n = \frac{1}{T}\int_{-T/2}^{T/2} f(t)e^{-jn\omega_0 t}\,dt = \frac{1}{T}\left[\int_{-T/2}^{0} f(t)e^{-jn\omega_0 t}\,dt + \int_{0}^{T/2} f(t)e^{-jn\omega_0 t}\,dt\right] \qquad (9.3.1)$$

If we substitute $t = -t'$ in the first integral in brackets, this integral becomes $\int_{T/2}^{0} f(-t')e^{jn\omega_0 t'}(-dt') = \int_{0}^{T/2} f(-t')e^{jn\omega_0 t'}\,dt'$. Changing the dummy integration variable back to t and invoking the property of an even function that $f(t) = f(-t)$, the integral becomes

$\int_0^{T/2} f(t) e^{jn\omega_0 t} dt$. Substituting in Equation 9.3.1, combining with the second integral, and making use of the relation $e^{jn\omega_0 t} + e^{-jn\omega_0 t} = 2\cos n\omega_0$, we obtain:

$$C_n = \frac{2}{T}\int_0^{T/2} f(t)\cos n\omega_0 t\, dt = \frac{2}{T}\operatorname{Re}\left[\int_0^{T/2} f(t) e^{-jn\omega_0 t} dt\right] \qquad (9.3.2)$$

It follows that for an even function:

$$a_0 = \frac{2}{T}\int_0^{T/2} f(t)\, dt, \quad a_n = \frac{4}{T}\int_0^{T/2} f(t)\cos n\omega_0 t\, dt,\ \text{and}\ b_n = 0\ \text{ for all } n \qquad (9.3.3)$$

Odd-Function Symmetry

> **Concept:** *The FSE of an odd periodic function does not contain an average term nor any cosine terms; its Fourier coefficients can be evaluated over half a period.*

The reason that the FSE of an odd periodic function does not contain an average term nor any cosine terms is simply that these terms, being even, introduce even components and destroy the odd symmetry of the function. An example of an odd function is the square wave of Figure 9.2.8. The FSE (Equation 9.2.25) consists of sine terms only. However, it is possible that a function that appears to be neither odd nor even becomes odd when the dc component is removed. Examples are the sawtooth waveforms of Figure 9.2.1 and Figure 9.2.4. If the dc component $A/2$ is subtracted, the function becomes odd. Hence, a function can have an odd ac component but is not odd because of a dc component.

If the FSE of an odd periodic function does not contain any cosine terms, $a_n = 0$ and C_n is imaginary. Pursuing an argument similar to that mentioned earlier, for an even function (Exercise 9.3.1), it follows that for an odd periodic function:

$$C_0 = 0\ \text{ and }\ C_n = -\frac{2j}{T}\int_0^{T/2} f(t)\sin n\omega_0 t\, dt = \frac{2}{T}\operatorname{Im}\left[\int_0^{T/2} f(t) e^{-jn\omega_0 t} dt\right] \qquad (9.3.4)$$

or
$$a_0 = 0 = a_n\ \text{ for all } n,\ \text{and}\ b_n = \frac{4}{T}\int_0^{T/2} f(t)\sin n\omega_0 t\, dt \qquad (9.3.5)$$

EXERCISE 9.3.1
Verify Equation 9.3.4 and Equation 9.3.5 for an odd periodic function.

Half-Wave Symmetry

A periodic function $f(t)$ has half-wave symmetry if:

$$f(t) = -f(t + T/2)\ \text{ or }\ f(t) = -f(t - T/2) \qquad (9.3.6)$$

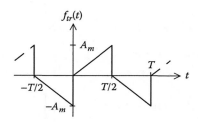

FIGURE 9.3.1
Half-wave symmetric function.

The two forms of the definition in Equation 9.3.6 are identical because changing the argument of a periodic function by nT does not change the function.

> **Concept:** *The FSE of a half-wave symmetric periodic function does not contain an average term nor any even harmonics; its Fourier coefficients can be evaluated over half a period. Thus:*

$$C_n = \frac{2}{T} \int_0^{T/2} f(t)\, e^{-jn\omega_0 t}\, dt \quad \text{for } n \text{ odd}$$

and
$$C_n = 0 \text{ for } n \text{ even or zero} \tag{9.3.7}$$

To prove this property, we express C_n as: $C_n = \frac{1}{T} \int_0^{T/2} f(t)\, e^{-jn\omega_0 t}\, dt + \frac{1}{T} \int_{T/2}^{T} f(t)\, e^{-jn\omega_0 t}\, dt$.

Substituting $t' = t - T/2$, the second integral becomes: $\frac{1}{T} \int_0^{T/2} f(t' + T/2)\, e^{-jn\omega_0(t' + T/2)}\, dt'$.

Changing the dummy variable t' back to t, invoking the half-symmetry property, and substituting $n\omega_0 \frac{T}{2} = n\pi$, the integral becomes: $\frac{1}{T} \int_0^{T/2} -f(t)\, e^{-jn\omega_0 t} e^{-jn\pi}\, dt$. But $e^{-jn\pi} = -1$ for odd n, and $e^{-jn\pi} = +1$ for even or zero n. Equation 9.3.7 then follows.

In terms of the coefficients a and b of the trigonometric form:

$$a_0 = 0 \,, \; a_n = 0 = b_n \quad \text{for } n \text{ even, and}$$

$$a_n = \frac{4}{T} \int_0^{T/2} f(t) \cos n\omega_0 t\, dt, \quad b_n = \frac{4}{T} \int_0^{T/2} f(t) \sin n\omega_0 t\, dt \quad \text{for } n \text{ odd} \tag{9.3.8}$$

The square wave of Figure 9.2.7 is both even and half-wave symmetric; its FSE (Equation 9.2.22) consists of odd cosine terms only. The square wave of Figure 9.2.8 is both odd and half-wave symmetric; its FSE (Equation 9.2.25) consists of odd sine terms only. Note that in this case, the same waveform can be made odd, or even, or neither odd nor even, by translation in time. The waveform of Figure 9.3.1 is half-wave symmetric, but is neither odd nor even, nor can it be made odd or even by translation in time. Its FSE consists of odd cosine and sine terms (Problem P9.1.10). If a dc component is added to a half-wave symmetric periodic waveform, the half-wave symmetric property is destroyed, but the ac component still does not contain any even harmonics.

Quarter-Wave Symmetry

A half-wave symmetric function that is also symmetrical about a vertical line through the middle of the positive or negative half-cycles is said to possess quarter-wave symmetry. Such a function can always be made either odd or even by translating it in time. The square waves of Figure 9.2.7 and Figure 9.2.8 are examples.

It is seen that the FSE of an odd, quarter-wave symmetric function consists of odd sine terms only, so that:

$$a_0 = 0, \ a_n = 0 \ \text{ for all } n, \ b_n = 0 \ \text{ for even } n$$

Moreover, b_n for odd n need only be evaluated over a quarter period, from $t = 0$ to $t = T/4$:

$$b_n = \frac{8}{T} \int_0^{T/4} f(t) \sin n\omega_0 t\, dt \quad \text{for odd } n \tag{9.3.9}$$

This is because both $f(t)$ and $\sin n\omega_0 t$, with n odd, are symmetrical about the middle of the half-cycle from $t = 0$ to $t = T/2$.

Similarly, the FSE of an even, quarter-wave symmetric function consists of odd cosine terms only, so that:

$$a_0 = 0, \ b_n = 0 \ \text{ for all } n, \ a_n = 0 \ \text{ for even } n$$

Moreover, a_n for odd n need be evaluated over a quarter period only:

$$a_n = \frac{8}{T} \int_0^{T/4} f(t) \cos n\omega_0 t\, dt \quad \text{for odd } n \tag{9.3.10}$$

Again, this is because both $f(t)$ and $\cos n\omega_0 t$, with n odd, are symmetrical about the middle of the half-cycle from $t = -T/4$ to $t = T/4$.

Quarter-wave symmetry is investigated in more detail in Section ST9.1. Symmetry properties are summarized in Table 9.3.1.

Example 9.3.1: FSE of Triangular Waveform

It is required to determine the FSE of the triangular waveform of Figure 9.3.2.

SOLUTION
The function has zero average, is even, and possesses half-wave symmetry. It is also quarter-wave symmetric. Its FSE must contain odd cosine terms only. Over the interval $0 \le t \le T/4$, $f_{tr}(t) = 4\dfrac{A_m}{T}(t - T/4)$. It follows from Equation 9.3.10 that:

$$a_n = \frac{8}{T} \int_0^{T/4} 4\frac{A_m}{T}(t - T/4) \cos n\omega_0 t\, dt = \frac{32 A_m}{T^2} \text{Re}\left[\int_0^{T/4}\left(t - \frac{T}{4}\right) e^{-jn\omega_0 t}\, dt \right] \tag{9.3.11}$$

TABLE 9.3.1

Summary of Symmetry Properties of Periodic Functions

Type of Symmetry	b_n	a_n	a_0
Neither odd nor even	$\dfrac{2}{T}\displaystyle\int_0^T f(t)\sin n\omega_0 t\,dt$	$\dfrac{2}{T}\displaystyle\int_0^T f(t)\cos n\omega_0 t\,dt$	$\dfrac{1}{T}\displaystyle\int_0^T f(t)dt$
Even	0	$\dfrac{4}{T}\displaystyle\int_0^{T/2} f(t)\cos n\omega_0 t\,dt$	$\dfrac{2}{T}\displaystyle\int_0^{T/2} f(t)dt$
Odd	$\dfrac{4}{T}\displaystyle\int_0^{T/2} f(t)\sin n\omega_0 t\,dt$	0	0
Half-wave symmetry			
Neither odd nor even	$\dfrac{4}{T}\displaystyle\int_0^{T/2} f(t)\sin n\omega_0 t\,dt$ n odd, 0 for n even	$\dfrac{4}{T}\displaystyle\int_0^{T/2} f(t)\cos n\omega_0 t\,dt$ n odd, 0 for n even	0
Quarter-wave symmetry			
Even	0	$\dfrac{8}{T}\displaystyle\int_0^{T/4} f(t)\cos n\omega_0 t\,dt$ n odd, 0 for n even	0
Odd	$\dfrac{8}{T}\displaystyle\int_0^{T/4} f(t)\ \sin n\omega_0 t\,dt$ n odd, 0 for n even	0	0

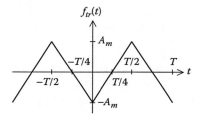

FIGURE 9.3.2
Figure for Example 9.3.1.

where using the exponential form makes the integration by parts somewhat simpler. It

follows that $a_n = \dfrac{32A_m}{T^2}\,\mathrm{Re}\left[\dfrac{te^{-jn\omega_0 t}}{-jn\omega_0} + \dfrac{e^{-jn\omega_0 t}}{(n\omega_0)^2} - \dfrac{T}{4}\dfrac{e^{-jn\omega_0 t}}{(-jn\omega_0)}\right]_0^{T/4} = -\dfrac{32A_m}{T^2(n\omega_0)^2} = -\dfrac{8A_m}{\pi^2 n^2}$. In eval-

uating this expression, even values of n *should not be used* because in applying Equation 9.3.11, we have already restricted n to be odd on account of half-wave symmetry. Hence,

$$a_n = -\frac{32A_m}{T^2(n\omega_0)^2} = -\frac{8A_m}{\pi^2 n^2} \text{ where } n \text{ is odd} \qquad (9.3.12)$$

The FSE of $f(t)$ is, therefore:

$$f_{tr}(t) = -\frac{8A_m}{\pi^2}\left(\cos\omega_0 t + \frac{1}{9}\cos 3\omega_0 t + \frac{1}{25}\cos 5\omega_0 t + ...\right) \qquad (9.3.13)$$

EXERCISE 9.3.2

Derive the FSE of the triangular waveform of Figure 9.3.2 when advanced by $T/4$, so that the origin is at the midpoint of the side of the positive slope.

Answer: $f_{tr}(t) = \dfrac{8A_m}{\pi^2}\sum\limits_{n}^{\infty}\left[\dfrac{1}{n^2}\sin\left(\dfrac{n\pi}{2}\right)\right]\sin n\omega_0 t$, where n is odd.

EXERCISE 9.3.3

Verify that Equation 9.3.13 for the triangular wave of Figure 9.3.2 can be obtained by evaluating over the first half period as in Equation 9.3.7 and assuming that n is odd.

EXERCISE 9.3.4

Give additional examples of each of the periodic functions included in Table 9.3.1. Note that an even or odd periodic function could be made neither even nor odd by shift of the time origin.

9.4 Derivation of FSEs from Those of Other Functions

Addition/Subtraction/Multiplication

> **Concept:** *The FSEs of some functions can be derived from the FSEs of other functions having the same period, through addition, subtraction, or multiplication.*

This is illustrated by the following example.

Example 9.4.1: FSE of Half-Wave and Full-Wave Rectified Waveforms

It is required to determine the FSE of: (a) the half-wave rectified waveform of Figure 9.4.1a; and (b) the full-wave rectified waveform of Figure 9.4.1b.

SOLUTION

(a) The given half-wave rectified waveform can be considered to be the product of a cosine function of amplitude A and a square pulse train of unity amplitude, both functions having the same period T (Figure 9.4.2). The FSE of the pulse train is that of Equation 9.2.21, with $A = 1$ and $\alpha = 1/2$. The FSE of the cosine function is the function itself. Hence:

$$f_{hw}(t) = A\cos\omega_0 t \left\{ \frac{1}{2} + \frac{2}{\pi} \left[\cos\omega_0 t - \frac{1}{3}\cos 3\omega_0 t + \frac{1}{5}\cos 5\omega_0 t - \frac{1}{7}\cos 7\omega_0 t + \ldots \right] \right\}$$

$$= \frac{A}{2}\cos\omega_0 t + \frac{A}{\pi}\left(\cos 2\omega_0 t + \cos 0 - \frac{1}{3}\cos 4\omega_0 t - \frac{1}{3}\cos 2\omega_0 t + \frac{1}{5}\cos 6\omega_0 t + \right.$$

$$\left. \frac{1}{5}\cos 4\omega_0 t - \frac{1}{7}\cos 8\omega_0 t - \frac{1}{7}\cos 6\omega_0 t + \ldots \right)$$

$$= \frac{A}{\pi} + \frac{A}{2}\cos\omega_0 t + \frac{2A}{\pi}\left(\frac{1}{3}\cos 2\omega_0 t - \frac{1}{15}\cos 4\omega_0 t + \frac{1}{35}\cos 6\omega_0 t + \ldots \right.$$

$$\left. + \frac{(-1)^{n+1}}{4n^2 - 1}\cos 2n\omega_0 t + \ldots \right), \quad n = 1, 2, 3, \ldots \tag{9.4.1}$$

The FSE contains a dc component of A/π, a fundamental component $A/2$, and even harmonics as cosine terms, as is to be expected of an even function.

(b) The FSE of the full-wave rectified waveform of Figure 9.4.1b may be derived by considering it as the sum of a half-wave rectified waveform of amplitude $2A$ and the function $-A\cos\omega_0 t$. When $-A\cos\omega_0 t$ is added to the RHS of Equation 9.4.1 multiplied by 2, the $\omega_0 t$ term cancels out, giving the FSE for a full-wave rectified waveform:

$$f_{fw}(t) = \frac{2A}{\pi} + \frac{4A}{\pi}\left(\frac{1}{3}\cos 2\omega_0 t - \frac{1}{15}\cos 4\omega_0 t + \frac{1}{35}\cos 6\omega_0 t - \ldots + \frac{(-1)^{n+1}}{4n^2 - 1}\cos 2n\omega_0 t + \ldots \right),$$

$$n = 1, 2, 3, \ldots \tag{9.4.2}$$

Note that the lowest angular frequency in Equation 9.4.2 is $2\pi \times \dfrac{2}{T} = 2\omega_0$. The FSE could be expressed in terms of $\omega_0' = 2\omega_0$. The ac component of the FSE then consists of a fundamental of frequency ω_0' and odd and even harmonics of this frequency.

The full-wave rectified waveform could also be considered as: (1) the sum of two half-wave rectified waveforms, with one waveform shifted by half a period with respect to the other waveform; and (2) the product of a square wave of zero average and $\cos\omega_0 t$ (Problem P9.1.17).

EXERCISE 9.4.1

Derive the FSE of the half-wave and full-wave rectified waveforms by applying Equation 9.2.4, Equation 9.2.5, and Equation 9.2.6.

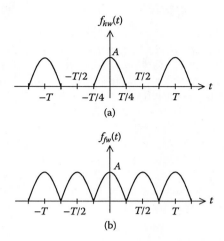

FIGURE 9.4.1
Figure for Example 9.4.1.

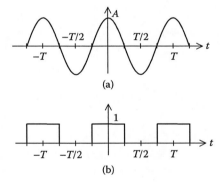

FIGURE 9.4.2
Figure for Example 9.4.1.

EXERCISE 9.4.2

Derive the FSE of the half-wave rectified and full-wave rectified waveforms when delayed by a quarter period of the supply so that they start as $\sin \omega_0 t$ functions.

Answer: Half-wave rectified waveform: $f(t) = \dfrac{A}{\pi} + \dfrac{A}{2}\sin \omega_0 t - \dfrac{2A}{\pi}\displaystyle\sum_{n=1}^{\infty}\dfrac{\cos 2n\omega_0 t}{4n^2 - 1}$

where ω_0 is the supply frequency; full-wave rectified waveform:

$$f(t) = \dfrac{2A}{\pi} - \dfrac{4A}{\pi}\displaystyle\sum_{n=1}^{\infty}\dfrac{\cos n\omega_0' t}{4n^2 - 1} \text{ where } \omega_0' = 2\omega_0.$$

Differentiation/Integration

> **Concept:** *The FSE of a given periodic function can be differentiated or integrated, term by term. The result is the FSE of a periodic function that is the derivative, or integral, of the given function, except that integrating the dc component destroys the periodicity of the function.*

This follows quite simply from differentiating, or integrating, both sides of the FSE. When a periodic function having a dc component is differentiated, the dc component vanishes and the resulting function is periodic with zero average. But when a periodic function having a dc component is integrated, the integral of the dc component increases linearly with time, which destroys the periodicity of the function, although the integral of the original ac component is still periodic.

Example 9.4.2: Integral of FSE of Square Waveform

It is desired to obtain the FSE of the triangular waveform as the integral of the square waveform.

SOLUTION

Consider the delayed square waveform of Figure 9.2.8, whose FSE is given by Equation 9.2.25. Integrating this FSE gives:

$$\int f_{sqd}(t)dt = -\frac{4A_m}{\pi\omega_0}\left[\cos\omega_0 t + \frac{1}{9}\cos 3\omega_0 t + \frac{1}{25}\cos 5\omega_0 t + ...\right] + C' \qquad (9.4.3)$$

where the constant of integration C' is the average value of the function. The RHS of Equation 9.4.3, with $C' = 0$, is identical with $f_{tr}(t)$ (Equation 9.3.13), bearing in mind that the peak-to-peak amplitude of the triangular wave equals the area under one half-cycle of the square wave. Thus, $A_{msq} \times \dfrac{T}{2} = 2A_{mtr}$. Hence, $\dfrac{4A_{msq}}{\pi\omega_0} = \dfrac{4}{\pi\omega_0} \times \dfrac{4A_{mtr}}{T} = \dfrac{8A_{mtr}}{\pi^2}$, as in Equation 9.3.13.

9.5   Concluding Remarks on FSEs

Rate of Attenuation of Harmonics

FSEs are theoretically infinite series. In practice, a periodic waveform is truncated, that is, approximated by a finite number of terms of the FSE, to any desired degree of accuracy. The more rapidly the magnitudes of the harmonics decrease with the order of the harmonic, the fewer are the number of terms of the FSE that have to be included to obtain a given degree of accuracy. The rate of attenuation of harmonics is related to the degree of smoothness of the function (Section 17.5, Chapter 17):

> **Concept:** *The smoother the function, the more rapidly the harmonics decrease in magnitude.*

More precisely:

> *If the mth derivative of a periodic function is discontinuous, with all the derivatives of lower order being continuous, the magnitudes of the harmonics decrease approximately as $\dfrac{1}{n^{m+1}}$, m = 0, 1, 2, ..., the derivative of order 0 being the function itself.*

Thus, in the case of the square wave, the function has a discontinuity at every half-cycle, so that $m = 0$, and the harmonics decrease as $1/n$. If the square waveform is integrated, a triangular waveform results, in which the function is continuous but the first derivative is discontinuous. Integration of the $\dfrac{1}{n}\cos n\omega_0 t$ or $\dfrac{1}{n}\sin n\omega_0 t$ terms of the square waveform makes the magnitude of the nth harmonic proportional to $1/n^2$, in accordance with the preceding statement, with $m = 1$. Whereas this statement holds exactly for the square and triangular waves, this is not the case with other functions. For example, the magnitudes of the harmonics of the pulse train decrease as $\dfrac{\sin n\pi\tau/2}{n\pi\tau/2}$ (Equation 9.2.21), so the $1/n$ is multiplied by a sine function. In the case of the half-wave and full-wave rectified waveforms (Equation 9.4.1) and (Equation 9.4.2), the magnitude decreases as $\dfrac{1}{4n^2-1}$. Nevertheless, this property is useful for checking purposes.

Application to Nonperiodic Functions

A nonperiodic function that satisfies the convergence criteria over a given interval can be represented over this interval by a Fourier series. It is convenient to choose this interval to be half a period, so that the FSE will consist of cosine terms only, if the periodic function is considered an even function, or of sine terms only, if the periodic function is considered an odd function.

9.6 Circuit Responses to Periodic Functions

> **Concept:** *The steady-state response of an LTI circuit to a periodic signal is the sum of the responses to each component acting alone.*

This follows readily from the principle of superposition (Section 16.6, Chapter 16), which applies to LTI systems. It does *not* apply to a nonlinear circuit (Section ST9.2).

Consider, for example, the circuit of Figure 9.6.1 with v_I a general periodic function of the form of Equation 9.1.2:

$$v_I = V_0 + \sum_{n=1}^{\infty} V_n \cos(n\omega_0 t + \theta_n) \tag{9.6.1}$$

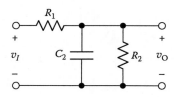

FIGURE 9.6.1
Response of circuit to a periodic input.

Because the ac components of the input are sinusoids, the steady-state output due to each of these components can be determined by phasor analysis. For the nth harmonic, we have, from voltage division:

$$\frac{\mathbf{V}_{On}}{\mathbf{V}_{In}} = \frac{R_2 \parallel (1/jn\omega_0 C_2)}{R_1 + R_2 \parallel (1/jn\omega_0 C_2)} = \frac{R_2}{R_1 + R_2} \frac{1}{1 + jn\omega_0 C_2(R_1 \parallel R_2)} \tag{9.6.2}$$

For $\omega_0 = 0$, the circuit is a simple voltage divider, and $V_O = \dfrac{R_2}{R_1 + R_2} V_I$. The nth harmonic in the output has a magnitude that is $\dfrac{R_2}{R_1 + R_2} \dfrac{1}{\sqrt{1 + [n\omega_0 C_2(R_1 \parallel R_2)]^2}}$ that of the corresponding input component and lags this component by $\tan^{-1} n\omega_0 C_2(R_1 \parallel R_2)$. The FSE of the output is therefore:

$$v_O(t) = \frac{R_2}{R_1 + R_2} V_0 + \sum_{n=1}^{\infty} \frac{R_2}{R_1 + R_2} \frac{1}{\sqrt{1 + \left[n\omega_0 C_2 (R_1 \parallel R_2)\right]^2}} V_n \cos\left[n\omega_0 t + \theta_n - \right.$$

$$\left. \tan^{-1} n\omega_0 C_2(R_1 \parallel R_2)\right] \tag{9.6.3}$$

Simulation Example 9.6.1: FSE of Response of *RC* Circuit to a Square Wave Input

The square wave of Figure 9.2.8 is applied to the *RC* circuit of Figure 9.6.2. It is required to determine the output v_O.

SOLUTION

The FSE of the input voltage is given by Equation 9.2.25, with A_m replaced by V_m. According to Equation 9.6.3, with $R_2 \to \infty$, $R_1 = R$, and $C_2 = C$, the nth harmonic in the output has a magnitude that is $\dfrac{1}{\sqrt{1 + (n\omega_0 CR)^2}}$ that of the corresponding input component and lags this component by $\tan^{-1} n\omega_0 CR$. The FSE of the output can therefore be expressed as:

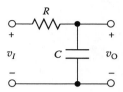

FIGURE 9.6.2
Figure for Example 9.6.1.

$$v_O = \frac{4}{\pi} V_m \sum_{\substack{n=1,\,3,\\5,\dots}}^{\infty} \frac{\sin\left(n\omega_0 t - \beta_n\right)}{n\sqrt{1 + \left(n\omega_0 CR\right)^2}}, \ \tan\beta_n = n\omega_0 CR \qquad (9.6.4)$$

SIMULATION

PSpice can be used to obtain a printout of the amplitude and phase of each frequency component of a periodic waveform and to display its amplitude spectrum. Both possibilities are illustrated in this example. Periodic sources are available in PSpice, such as VPULSE and VSIN from the SOURCE library, which can be modified using ABM parts. For example, a VSIN source followed by an ABS block can be used to generate a full-wave rectified waveform. VPULSE can be used to generate practically any periodic waveform of rectangular pulses. It can be used to provide a triangular waveform, if followed by an integrator block, and a trapezoidal waveform, if followed by an additional LIMIT block. Triangular and trapezoidal wave forms can also be generated by appropriate choices of TR, TF, and PW parameters of VPULSE.

In the present simulation, a VPULSE source is applied to the *RC* circuit with $R = 1 \text{ k}\Omega$ and $C = 1 \text{ μF}$. The initial value (IC) of the voltage across the capacitor is set in the Property Editor spreadsheet of the capacitor at –5 V. The following parameters of VPULSE are entered in the Property Editor spreadsheet of the source to produce the periodic square waveform of Figure 9.2.8, with $V_m = 5$ V and $T = 4$ ms:

V1 (initial voltage level at $t = 0$) = –5 V

V2 (pulsed voltage level, at which the pulse width, PW, is measured) = +5 V

TD (time delay, during which the voltage stays at V1) = 0

TR (time taken to go from V1 to V2) = 1u

TF (time taken to go from V2 to V1) = 1u

PW (pulse width, at V2 level, exclusive of rise and fall times) = 2m

PER (period) = 4m

The rise and fall times of 1 μs are very small compared to the time constant of 1 ms, so that the transitions between the two voltage levels are effectively instantaneous. For convenience, the source voltage and v_O have been labeled VIN and VOUT, respectively, using net aliases (Appendix SD.2).

The waveforms of the source voltage and the capacitor voltage v_O are shown in Figure 9.6.3 for the interval 0 to 12 ms. v_O starts at –5 V but quickly reaches a steady state that repeats indefinitely. Because the applied input is periodic, the steady-state output, after the initial transient, is also periodic.

To perform Fourier analysis, we could choose one period starting at $t = 12$ ms. In the Simulation Settings window, select Time Domain (Transient) analysis and enter 16m for "Run to time," 12m for "Start saving data after," and 0.1u for "Maximum step size."

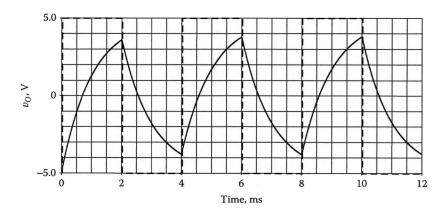

FIGURE 9.6.3
Figure for Example 9.6.1.

Then press the Output File Options button to display the Transient Output File Options window. Check the Perform Fourier Analysis box. For Center Frequency, enter the frequency in hertz, which is 250. Then enter the number of harmonics to be included in the Fourier analysis, say 9. Under Output Variables enter V(VIN), followed by a space, then V(VOUT) to instruct PSpice to perform Fourier analysis on the VIN and VOUT signals. After the simulation is run, the output file gives the following results of the Fourier analysis, for the input and output voltages, respectively:

```
FOURIER COMPONENTS OF TRANSIENT RESPONSE V(VIN)

DC COMPONENT = 2.546890E03
```

HARMONIC NO	FREQUENCY (HZ)	FOURIER COMPONENT	NORMALIZED COMPONENT	PHASE (DEG)	NORMALIZED PHASE (DEG)
1	2.500E+02	6.366E+00	1.000E+00	−9.399E−02	0.000E+00
2	5.000E+02	5.094E−03	8.001E−04	8.981E+01	9.000E+01
3	7.500E+02	2.122E+00	3.333E−01	−2.820E−01	−6.450E−09
4	1.000E+03	5.094E−03	8.001E−04	8.963E+01	9.000E+01
5	1.250E+03	1.273E+00	2.000E−01	−4.700E−01	−3.225E−08
6	1.500E+03	5.094E−03	8.001E−04	8.944E+01	9.000E+01
7	1.750E+03	9.094E−01	1.429E−01	−6.579E−01	−9.030E−08
8	2.000E+03	5.094E−03	8.001E−04	8.925E+01	9.000E+01
9	2.250E+03	7.073E−01	1.111E−01	−8.459E−01	−1.935E−07

```
   TOTAL HARMONIC DISTORTION = 4.287940E+01 PERCENT.

   FOURIER COMPONENTS OF TRANSIENT RESPONSE V(VOUT)

   DC COMPONENT = 2.498118E-03
```

HARMONIC NO	FREQUENCY (HZ)	FOURIER COMPONENT	NORMALIZED COMPONENT	PHASE (DEG)	NORMALIZED PHASE (DEG)
1	2.500E+02	3.419E+00	1.000E+00	−5.761E+01	0.000E+00
2	5.000E+02	1.515E−03	4.433E−04	1.746E+01	1.327E+02
3	7.500E+02	4.405E−01	1.288E−01	−7.830E+01	9.454E+01
4	1.000E+03	7.853E−04	2.297E−04	−8.656E+00	2.391E+02
5	1.250E+03	1.608E−01	4.704E−02	8.322E+01	2.048E+02

HARMONIC	FREQUENCY	FOURIER	NORMALIZED	PHASE	NORMALIZED
NO	(HZ)	COMPONENT	COMPONENT	(DEG)	PHASE (DEG)
6	1.500E+03	5.273E-04	1.542E-04	5.469E+00	3.511E+02
7	1.750E+03	8.239E-02	2.410E-02	-8.548E+01	3.178E+02
8	2.000E+03	3.963E-04	1.159E-04	3.748E+00	4.647E+02
9	2.250E+03	4.990E-02	1.460E-02	-8.681E+01	4.317E+02

```
TOTAL HARMONIC DISTORTION = 1.400322E+01 PERCENT
```

The magnitudes of the even harmonics are negligible because both waveforms are half-wave symmetric. The results for the input waveform are in accordance with Equation 9.2.25. The magnitude of the fundamental is $\dfrac{4 \times 5}{\pi} = 6.366\text{V}$. The harmonics decrease by a factor of $1/n$ with respect to the fundamental and have negligible phase difference. The results for the output waveform are in accordance with Equation 9.6.4. The magnitude of the fundamental is $\dfrac{1}{\sqrt{1 + (\omega_0 CR)^2}}$ times that of the fundamental component of the input, where $\omega_0 = 500\pi$ rad/s. The fundamental of the output lags the input component by $\tan^{-1}(\omega_0 CR)$, in agreement with the preceding results. PSpice gives the total harmonic distortion (THD), where $\text{THD} = \sum\limits_{n=2}^{\infty} \sqrt{D_n^2}$, where $D_n = \left|\dfrac{A_n}{A_1}\right| \times 100$, A_1 and A_n being the amplitudes of the fundamental and the nth harmonic, respectively. The total harmonic distortion of the output is less than that of the input because of the filtering effect of the lag RC circuit (Section 17.5, Chapter 17). Using the Probe cursor on the output voltage of Figure 9.6.3, we obtain ± 3.81 V for the peak and trough values. The output waveform is that of the charging and discharging of a capacitor and is derived in Example 13.3.2 (Chapter 13).

To obtain the amplitude spectrum using Probe, select Trace/Fourier, then Trace/Add Trace for V(VIN) and V(VOUT). Change the scale of the axis by selecting Plot/Axis Settings. In the Axis Settings window, select User Defined under XAxis Data range and enter 0 to 2.5k. The plot of Figure 9.6.4 is displayed. The peaks of the triangles are at the frequencies of the fundamental and the harmonics. The magnitudes of the peaks are the same as those obtained in the output file mentioned earlier.

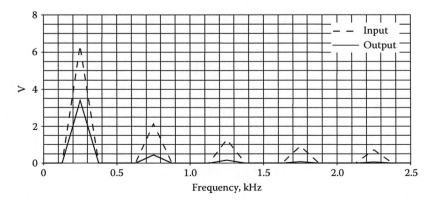

FIGURE 9.6.4
Figure for Example 9.6.1.

EXERCISE 9.6.1

Show that C_n in Example 9.6.1 evaluated over the interval $0 \le t \le T/2$ is $\dfrac{2V_m}{n\pi}\left(\dfrac{-j}{1+jn\omega_0 CR}\right)$

for n odd and zero for n even, in agreement with Equation 9.6.4.

9.7   Average Power and rms Values

> **Concept:** *In an LTI circuit, components of different frequencies do not interact, and the total average power is the sum of the average powers due to each component acting alone.*

To justify this, consider a periodic input voltage v_I of the form:

$$v_I = V_0 + \sum_{n=1}^{\infty} V_n \cos(n\omega_0 t + \theta_{vn}) \tag{9.7.1}$$

to be applied to two terminals of an LTI circuit. The input current i_I at these terminals is also periodic, of the same frequency, and can be expressed as:

$$i_I = I_0 + \sum_{n=1}^{\infty} I_n \cos(n\omega_0 t + \theta_{in}) \tag{9.7.2}$$

The average power input to the circuit is:

$$P = \frac{1}{T}\int_{t_0}^{t_0+T} v_I i_I \, dt \tag{9.7.3}$$

According to Equation 9.2.1 to Equation 9.2.3, the product terms involving trigonometric functions evaluate to zero over T, except those having the same n. Thus:

$$P = \frac{1}{T}V_0 I_0 t \Big|_{t_0}^{t_0+T} + \sum_{n=1}^{\infty}\frac{1}{T}\int_{t_0}^{t_0+T} V_n I_n \cos\left(n\omega_0 t + \theta_{vn}\right)\cos\left(n\omega_0 t + \theta_{in}\right)dt$$

$$= V_0 I_0 + \frac{1}{T}\sum_{n=1}^{\infty}\frac{V_n I_n}{2}\int_{t_0}^{t_0+T}\Big[\cos\left(\theta_{vn}-\theta_{in}\right)+\cos\left(2n\omega_0 t - \theta_{vn}-\theta_{in}\right)\Big]dt \tag{9.7.4}$$

The integral of the second term in the integrand vanishes over a period T, so that:

$$P = V_0 I_0 + \sum_{n=1}^{\infty}\frac{V_n I_n}{2}\cos\left(\theta_{vn}-\theta_{in}\right) = V_0 I_0 + \sum_{n=1}^{\infty} V_{nrms} I_{nrms}\cos\left(\theta_{vn}-\theta_{in}\right) \tag{9.7.5}$$

It is seen from Equation 9.7.5 that the average power is only due to components of voltage and current of the same frequency. The average power of each frequency component is given by the product of the rms voltage, the rms current, and the power factor $\cos(\theta_{vn} - \theta_{in})$, as in Equation 7.1.10 (Chapter 7). Components of different frequencies do not contribute to the average power. They do contribute to the instantaneous power, as does the component of frequency $2n\omega_0$ in Equation 9.7.4, causing a net power flow in or out of the circuit at any instant. However, this power averages to zero over a complete cycle, leaving only the contribution from components having the same frequency.

It should be noted that when more than one source of the *same* frequency is applied to an LTI circuit, the responses due to the individual sources are superposed, but the powers due to these sources do not superpose because the sum of squares of voltages or currents is not equal to the square of the sum, as discussed in Section 3.6, Chapter 3. When the various sources are of different frequencies, the responses due to the individual sources do not interact, that is, each response retains its individual identity, so that the powers due to the sources of different frequency can be superposed. In small-signal, linearized analysis of electronic circuits, for example, the power at signal frequencies is treated separately from that due to the dc bias supplies.

rms Value

It is of interest to determine the rms value of a periodic waveform $f(t)$. By definition, F_{rms}, the rms value of $f(t)$ is given by:

$$F_{\text{rms}} = \sqrt{\frac{1}{T} \int_{t_0}^{t_0+T} \left[f(t) \right]^2 dt} \tag{9.7.6}$$

In words, as its name implies, the rms value is obtained by squaring the given function, evaluating the mean of the square over a period, and then taking the square root of the mean. Assuming $f(t)$ to be given by Equation 9.1.2, and again using Equation 9.2.1 to Equation 9.2.3, it is clear that only squared terms do not vanish, so that:

$$F_{\text{rms}}^2 = \frac{1}{T} \int_{t_0}^{t_0+T} \left[c_0^2 + \sum_{n=1}^{\infty} c_n^2 \cos^2 \left(n\omega_0 t + \theta_n \right) \right] dt$$

$$= c_0^2 + \frac{1}{T} \int_{t_0}^{t_0+T} \sum_{n=1}^{\infty} \frac{c_n^2}{2} \left[1 - \sin 2 \left(n\omega_0 t + \theta_n \right) \right] dt$$

$$= c_0^2 + \sum_{n=1}^{\infty} \frac{c_n^2}{2}$$

or

$$F_{\text{rms}} = \sqrt{c_0^2 + \sum_{n=1}^{\infty} \frac{c_n^2}{2}} \tag{9.7.7}$$

Recall that c_n is the amplitude of the nth ac component, so $c_n^2/2$ is the square of the rms value of this component. By definition, the dc value is the same as the rms value. According to Equation 9.7.7, the rms value of a periodic function is the square root of the sum of the squares of the rms values of the individual components.

In terms of a and b coefficients (Equation 9.1.3), Equation 9.7.7 becomes:

$$F_{rms} = \sqrt{a_0^2 + \sum_{n=1}^{\infty} \frac{\left(a_n^2 + b_n^2\right)}{2}} \qquad (9.7.8)$$

If a periodic voltage v is applied across a resistor R, the rms current component I_{nrms} corresponding to a voltage component V_{nrms} is $I_{nrms} = \dfrac{V_{nrms}}{R}$, with $\theta_{in} = \theta_{vn}$. It follows from Equation 9.7.5 that:

$$P = \frac{1}{R}\left[V_0^2 + \sum_{n=1}^{\infty} V_{nrms}^2\right] = \frac{V_{rms}^2}{R} \qquad (9.7.9)$$

where V_{rms} is the rms value of the periodic voltage v. Similarly, in terms of current,

$$P = R\left[I_0^2 + \sum_{n=1}^{\infty} I_{nrms}^2\right] = I_{rms}^2 R \qquad (9.7.10)$$

where I_{rms} is the rms value of the periodic current through R.

If a periodic waveform $f(t)$ is expressed analytically, it is usually much simpler to determine its rms value from direct application of Equation 9.7.6 rather than from its Fourier coefficients (Equation 9.7.7). Consider, for example, the half-wave rectified waveform of Figure 9.4.1a given by $A\cos\omega_0 t$, $-\dfrac{\pi}{2} \le \omega_0 t \le \dfrac{\pi}{2}$, and zero over the rest of the period. The mean of its square over a period is $\dfrac{A^2}{2\pi}\displaystyle\int_{-\pi/2}^{\pi/2} \cos^2(\omega_0 t)\, d(\omega_0 t) = \dfrac{A^2}{4}$. Hence, the rms value is $\dfrac{A}{2}$. The dc component is $\dfrac{A}{\pi}$ (Equation 9.4.1). It follows from Equation 9.7.7 that:

$\dfrac{A}{2} = \sqrt{\left(\dfrac{A}{\pi}\right)^2 + A_{acrms}^2}$. This gives for the AC component of the half-wave rectified waveform,

$$A_{acrms} = \frac{A}{2}\sqrt{1 - \left(\frac{2}{\pi}\right)^2} = 0.39A.$$

Example 9.7.1: rms Value of Periodic Triangular Waveform

It is required to determine the rms value of the periodic triangular waveform $f(t)$ shown in Figure 9.7.1 and deduce the rms value of the ac component.

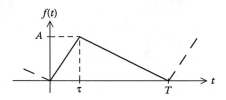

FIGURE 9.7.1
Figure for Example 9.7.1.

SOLUTION

For $0 \le t \le \tau$, $f(t) = \dfrac{A}{\tau} t$. The square of this waveform is $\left(\dfrac{A}{\tau}\right)^2 t^2$, and the area under the curve is:

$$\int_0^\tau \left(\frac{A}{\tau}\right)^2 t^2 \, dt = \left(\frac{A}{\tau}\right)^2 \frac{\tau^3}{3} = \frac{A^2\tau}{3} \tag{9.7.11}$$

For $\tau \le t \le T$, the area under the curve of the squared function is clearly the same as that for $f_1(t)$ in Figure 9.7.2. By analogy with Equation 9.7.11, the area under the square function is $\dfrac{A^2(T-\tau)}{3}$, $\tau \le t \le T$. The total squared area for one period of $f(t)$ is the sum $\dfrac{A^2T}{3}$. The mean is $\dfrac{A^2}{3}$ and the rms value is $\dfrac{A}{\sqrt{3}}$. Because it is independent of τ, this result applies to any triangular waveform that varies between 0 and A and repeats continuously without interruption, including a sawtooth waveform ($\tau = 0$ or $\tau = T$).

The dc, or average value of $f(t)$ is $\dfrac{1}{T}\left[\dfrac{A\tau}{2} + \dfrac{A(T-\tau)}{2}\right] = \dfrac{A}{2}$. If the rms value of the ac component is denoted by F_{acrms}, then $\dfrac{A}{\sqrt{3}} = \sqrt{\left(\dfrac{A}{2}\right)^2 + F_{acrms}^2}$. This gives $F_{acrms} = A\sqrt{\dfrac{1}{3} - \dfrac{1}{4}} = \dfrac{A}{\sqrt{12}} = \dfrac{A}{2\sqrt{3}}$.

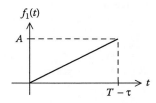

FIGURE 9.7.2
Figure for Example 9.7.1.

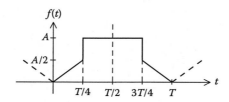

FIGURE 9.7.3
Figure for Exercise 9.7.1.

EXERCISE 9.7.1

(a) By inspection, what is the rms value of the full-wave rectified waveform of Figure 9.3.5? (b) Determine the rms value of the periodic waveform shown in Figure 9.7.3.

Answer: (a) $A/\sqrt{2}$, the same as a sinusoid of amplitude A because the squared function is the same; (b) $\dfrac{A}{2}\sqrt{\dfrac{13}{6}}$.

Summary of Main Concepts and Results

- A periodic function $f(t)$ can be expressed, in general, as an infinite series of cosine and sine functions: $f(t) = a_0 + \sum_{k=1}^{\infty}(a_k \cos k\omega_0 t + b_k \sin k\omega_0 t)$, where k is a positive integer, and a_0, a_k, and b_k are constants, known as Fourier coefficients, that depend on $f(t)$.

- Given the four functions $\cos m\omega_0 t$, $\sin m\omega_0 t$, $\cos n\omega_0 t$, and $\sin n\omega_0 t$, where m and n are any integers, the integral of the product of any two of these functions over a period $T = 2\pi/\omega_0$ is zero, except the products $\cos^2 n\omega_0 t$, and $\sin^2 n\omega_0 t$, having $m = n$, in which case the integral is $T/2$.

- The FSE of an even periodic function does not contain any sine terms.

- The FSE of an odd periodic function does not contain an average term nor any cosine terms.

- The FSE of a half-wave symmetric periodic function does not contain an average term nor any even harmonics; its Fourier coefficients can be evaluated over half a period.

- The FSE of some periodic waveforms may be derived from those of other periodic waveforms through addition, subtraction, multiplication, differentiation, or integration.

- The FSE of a given periodic function can be differentiated, or integrated, term by term. The result is the FSE of a periodic function that is the derivative or integral of the given function, except that integrating the dc component destroys the periodicity of the function.

- The steady-state response of an LTI circuit to a periodic signal is the sum of the responses to each component acting alone.
- In an LTI circuit, components of different frequencies do not interact, and the total average power is the sum of the average powers due to each component acting alone.
- The rms value of a periodic function is the square root of the sum of the squares of the rms values of the individual components.

Learning Outcomes

- Derive the FSE of a periodic function.
- Determine the response of a circuit to a periodic input.

Supplementary Topics and Examples on CD

ST9.1 Quarter-wave symmetry: Proves Equation 9.3.9, Equation 9.3.10, and that a quarter-wave symmetric function can always be made odd or even by a translation in time.

ST9.2 Nonlinear response to periodic input: Shows that when two signals of different frequencies are applied to a nonlinear element, the response contains harmonics of the individual frequencies, as well as sum and difference frequencies and a dc component.

SE9.1 Fourier series expansion of pulsed sinusoid: Derives the FSE of pulses of a given period and duty cycle, each pulse consisting of sinusoids of a given frequency.

SE9.2 Symmetrical nonlinear distortion: Analyzes two types of symmetrical distortion commonly encountered because of amplifier saturation and ferromagnetism. It is shown that when the input is half-wave symmetric, the output contains odd harmonics only.

SE9.3 Response of LC filter to a full-wave rectified input: Determines the dc and the first three harmonics of the output.

Problems and Exercises

P9.1 Fourier Series Expansion

P9.1.1 Find the period of each of the following signals: (a) $f(t) = 5 + 10\cos 100\pi t + 5\cos 200\pi t + 2\cos 400\pi t$, and (b) $g(t) = \dfrac{\cos 100\pi t \sin 300\pi t - \sin 100\pi t \cos 300\pi t}{\sin 200\pi t}$.

P9.1.2 Given the two functions, (a) $f(t) = \cos(100\pi t)\sin(200\pi t)$ and (b) $g(t) = \cos^2(100\pi t)\sin^2(200\pi t)$, show that $f(t)$ and $g(t)$ are periodic, determine their periods, and derive their FSEs.

P9.1.3 Two periodic functions of period 6 s are defined by: $f(t) = -t$, $-3 < t \le 0$, and $f(t) = t$, $0 \le t \le 3$, $g(t) = 1$, $0 < t < 3$, and $g(t) = -1$, $3 < t < 6$. What is the ratio of the amplitude of the third harmonic in $f(t)$ to that in $g(t)$?

P9.1.4 Verify the following properties of odd and even functions:

1. The product of two odd functions is an even function.

2. The product of an odd function and an even function is an odd function.

3. The product of even functions is an even function.

4. The sum, or difference, of two odd functions is an odd function.

5. The sum, or difference, of two even functions is an even function.

6. The sum, or difference, of an odd function and an even function is neither odd nor even.

P9.1.5 Express the function of Figure 9.7.1 as a Fourier series in both exponential and trigonometric forms, given that $A = 3$ units, $T = 4$ s, and $\tau = T/4$. Verify that the harmonics decrease at a rate of $1/n^2$ and that the function becomes odd if advanced by 0.5 s and the dc value is removed.

P9.1.6 Derive the FSE of the periodic triangular waveform of Figure P9.1.6. Show that it can be made that of an even function or an odd function by a shift in time.

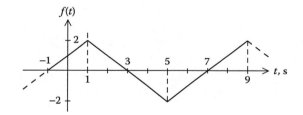

FIGURE P9.1.6

P9.1.7 Derive the FSE expansion of the function shown in Figure P9.1.7.

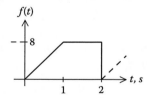

FIGURE P9.1.7

P9.1.8 Derive the FSE of a periodic triangular waveform defined over a period 0 to 2 as: $f(t) = t + 2, 0 < t < 1$, and $f(t) = 0$ elsewhere in the period.

P9.1.9 Given a full-wave rectified waveform of period T, as shown in Figure 9.4.1b, except that because of dissymmetry in the rectifier circuit, the half-sinusoids are not all of the same amplitude but alternate with amplitudes of 12 V and 10 V. Derive the FSE, assuming that the sinusoid centered at the origin has an amplitude of 12 V.

P9.1.10 Derive the FSE of the function shown in Figure 9.3.1.

P9.1.11 A function is defined over half a period by: e^t, $0 < t < 1$. Derive the FSE if the function is: (a) even, (b) odd.

P9.1.12 (a) Derive the FSE of the periodic waveform of Figure P9.1.12 by direct evaluation. (b) Show that if it is added to a delayed and negated version, the result agrees with Equation 9.3.13. (c) Indicate how the FSE can be obtained as the product of a rectangular pulse train of unit height (Figure 9.2.5) and a triangular waveform derived from that of Figure 9.3.2.

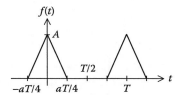

FIGURE P9.1.12

P9.1.13 (a) Derive the FSE of the waveform of Figure P9.1.13 by direct evaluation. (b) Indicate how the FSE can be obtained as the product of a rectangular pulse train of unit height (Figure 9.2.5) and a sawtooth waveform derived from that of Figure 9.2.1.

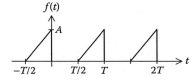

FIGURE P9.1.13

P9.1.14 Derive the FSE of the waveform of Figure P9.1.14 in two ways: (a) directly; (b) as the product of a rectangular pulse train of unit height (Figure 9.2.5) and a reversed sawtooth waveform derived from that of Figure 9.2.4.

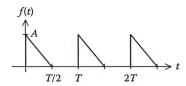

FIGURE P9.1.14

P9.1.15 (a) Show that if the FSE of Problem P9.1.13 is combined with the negated FSE of Problem P9.1.14, the FSE of a sawtooth waveform is obtained. (b) Derive the FSE of Problem P9.1.14 from that of Problem P9.1.13 and that of a rectangular pulse train of amplitude A and period T.

P9.1.16 Derive the FSE of the waveform of Figure P9.1.16 in three ways: (a) direct evaluation of coefficients; (b) shifting the waveform derived in Problem P9.1.5; and (c) from that of its derivative.

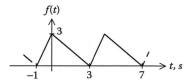

FIGURE P9.1.16

P9.1.17 Obtain the FSE of a full-wave rectified waveform in two ways: (a) as the sum of two half-wave rectified waveforms, with one waveform shifted by half a period with respect to the other waveform; (b) as the product of a square wave of zero average and $\cos \omega_0 t$.

P9.1.18 Derive the FSE of the waveform of Figure P9.1.18 in two ways: (a) direct evaluation of coefficients; (b) as the sum of two shifted rectangular pulse trains.

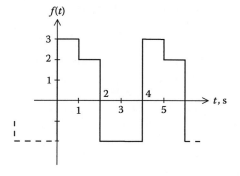

FIGURE P9.1.18

P9.1.19 A periodic function of period 1 s is defined as: $f(t) = 2t^3 - \dfrac{t}{2}, -\dfrac{1}{2} \le t \le \dfrac{1}{2}$. How do the magnitudes of the harmonics vary with the order n of the harmonic?

P9.1.20 Given a half-wave symmetric function $f(t)$, half a period of which is shown in Figure P9.1.20, determine the coefficients of the FSE up to and including the fifth harmonic, assuming that the function can be approximated by: $f(\pi/6) = 0.6, f(\pi/3) = 1.9, f(\pi/2) = 3.3, f(2\pi/3) = 4,$ and $f(5\pi/6) = 3.3$.

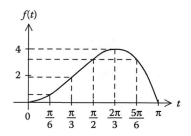

FIGURE P9.1.20

P9.1.21 Consider the square pulse train of Figure 9.2.8 whose FSE is given by Equation 9.2.25. Assume that $A_m = 1$ and $\omega_0 = 1$ rad/s. Using an appropriate computer program, plot the FSE expansion using an increasing number of harmonics. Observe that with the ninth harmonic, the waveform begins to look like a square wave. Note, however, that no matter how many harmonics you add (necessarily a finite number), there always remains a small overshoot and some damped oscillations following the discontinuities. This is referred to as *Gibb's phenomenon*.

P9.2   Responses to Periodic Inputs

P9.2.1 The triangular pulse train of Figure 9.3.2 is applied as the input v_I to the circuit of Figure P9.2.1. Determine v_O. Show that if $\omega CR \ll 1$, $v_O \to \dfrac{dv_I}{dt}$, and v_O becomes the square pulse train of Figure 9.2.8 with an appropriate amplitude.

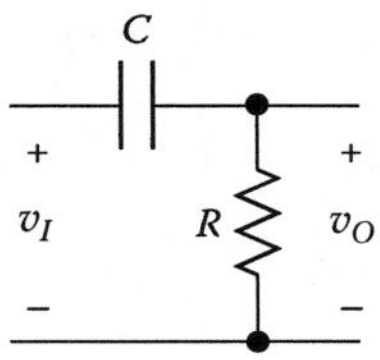

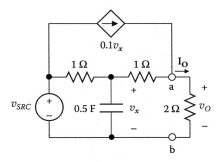

FIGURE P9.2.1

P9.2.2 v_{SRC} in Figure P9.2.2 is the triangular waveform of Figure 9.3.2. Determine the first three terms of v_O, assuming $\omega = 1$ rad/s and $A_m = 10$ V. Simulate with PSpice.

FIGURE P9.2.2

P9.2.3 In Figure P9.2.3 v_{SRC} is the full-wave rectified waveform of Figure 9.4.1b having $T = 1/50$ s and an amplitude of 50 V. Between v_{SRC} and the 4 kΩ load is the LC filter shown whose purpose is to attenuate the ac components, leaving a near-dc voltage across the load. Determine the first four nonzero terms in the FSE of v_O and compare with those of v_{SRC}. What is the rms value of the ac component in the first four nonzero terms in the FSE of v_O? Simulate with PSpice.

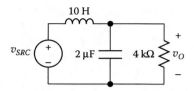

FIGURE P9.2.3

P9.2.4 Repeat Problem P9.2.3 assuming that v_{SRC} is a square wave 50 V peak-to-peak, 25 V average value, and $T = 1/50$ s. Compare with the results of Problem P9.2.3. Simulate with PSpice.

P9.2.5 The sawtooth waveform of Figure 9.2.1 is applied to the circuit of Figure P9.2.5, where C is a very large capacitor. Determine the first five terms in the FSE of v_O, assuming $A = 5$ V and $\omega = 10^6$ rad/s. What is the effect on the magnitude of the second harmonic? What is the purpose of C?

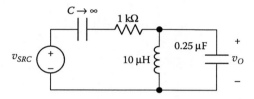

FIGURE P9.2.5

P9.2.6 A triangular current source having the waveform of Figure 9.3.2, with $A_m = 10$ mA and $\omega = 10^5$ rad/s, is applied to the circuit of Figure P9.2.6. Determine the first three terms in the FSE of v_O.

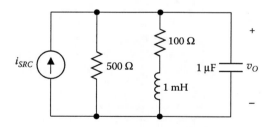

FIGURE P9.2.6

P9.2.7 A voltage $v_1 + v_2$ is applied to a series RLC circuit, where v_1 and v_2 are the sawtooth waveform of Figure 9.2.1 and the inverted sawtooth waveform of Figure 9.2.4b, respectively, having an amplitude of 10 V. Determine the voltage across the capacitor.

P9.2.8 A voltage $v = \cos 4000t + 2\sin 8000t$ V is applied across a 1 Ω resistor whose v–i relation is $i = v + 0.1v^2$. What are the frequency components present in the current?

P9.2.9 A voltage $v = \cos\omega_0 t + 2\sin 2\omega_0 t$ is applied to a circuit that gives an output $v_O = v + v^2$. Determine the FSE of the output.

P9.2.10 The output v_O of a nonlinear network is related to the input v_i by the power relation: $v_O = 5v_i + 0.5v_i^2 + 0.1v_i^3$. If $v_i = 4\sin 2000\pi t$, determine the magnitude of each term of the Fourier expansion of the output. Does the output posses half-wave symmetry? Why not?

P9.3 Power and RMS Values

P9.3.1 Consider $f(t)$ in Problem P9.1.1a. What is the rms value of the signal? If $f(t)$ is an approximation of a periodic voltage that dissipates 90 W in a 1 Ω resistor, what is the percentage error in real power involved in the approximation?

P9.3.2 Determine the rms value of i in Figure P9.3.2.

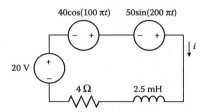

40cos(100 πt)   50sin(200 πt)

20 V

4 Ω   2.5 mH

FIGURE P9.3.2

P9.3.3 A voltage $v(t) = 400\sqrt{2}\sin(\omega t + \theta_1) + 180\sqrt{2}\sin(3\omega t + \theta_3)$ V is applied to a series RLC circuit having $R = 60$ Ω. At the frequency of the third harmonic, the reactances of L and C have equal magnitudes, and the ratio L/C is 900 Ω². Determine the rms current in the circuit.

P9.3.4 The voltage applied across a 1 Ω resistor is expressed as $e^{j100\pi t} + (2 + j4)e^{j200\pi t} +$
$(3 + j9)e^{j300\pi t}$ V rms. Determine the energy dissipated in the resistor in the interval 1/50 to 2/50 s.

P9.3.5 The current through a coil having a resistance of 1 Ω and an inductance of 10 mH consists of a fundamental and a third harmonic. The rms current through the coil is 5 A, and the rms voltage across the coil is 20 V. If the frequency of the fundamental is 300 rad/s, what are the rms values of the fundamental and third harmonic components of the coil current and voltage?

P9.3.6 Determine the rms value of the voltage waveform shown in Figure P9.3.6.

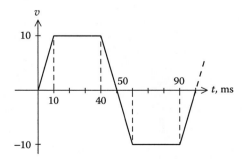

FIGURE P9.3.6

P9.3.7 Generalize the result of Example 9.7.1 to the triangular waveform of Figure P9.3.7a having two triangles per period of equal amplitude A but arbitrary values of τ_1, τ_2, and τ_3 within a period, where $\tau_1 \le \tau_2 \le \tau_3$. Show that the rms value is the same if one of the triangles is negative but of the same amplitude (Figure P9.3.7b). Deduce that the rms value of a function that alternates linearly between $+A$ and negative-going $-A$ a number of times is $A/\sqrt{3}$.

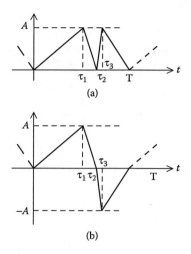

(a)

(b)

FIGURE P9.3.7

P9.3.8 A periodic signal consists of a fundamental of amplitude 20, frequency 1 krad/s and third, fifth, and seventh harmonics of relative amplitudes, with respect to the fundamental of 0.2, 0.05, and 0.01, respectively. The phase angle of the fundamental and fifth harmonics is $+90°$, whereas the phase angle of the third and seventh harmonics is $-90°$. (a) Determine the rms value of the signal. (b) Express the signal as an FSE. (c) Specify whether or not the signal is even or odd and whether or not it has half-wave symmetry. Repeat for the same signal but with the phase angles of all the components negated.

P9.3.9 The voltage at two given terminals of a circuit is: $v = 4\sin t + 5\cos 2t + 10\sin 4t$ V. The current input at these terminals, in the direction of the voltage drop, is: $i = 6\sin t + 8\cos 5t + 12\sin 8t$ A. Determine (a) the rms value of v, (b) the rms value of i, and (c) the average power delivered to the network at the given terminals.

10

Frequency Responses

Overview

The responses considered in the preceding chapter were concerned with the discrete frequencies of fundamental and harmonics. The next step is to consider circuit behavior in response to the steady-state sinusoidal excitation as the frequency is varied continuously over the range from dc to virtually infinite frequency. The study of this frequency response is important in two respects:

1. Energy-storage elements are present in all but the simplest physical systems, whether electrical or nonelectrical in nature. These energy-storage elements make the response frequency dependent. The responses of practically all physical systems are reduced at high frequencies because of these energy storage elements as well as inherent speed limitations on the maximum speed of physical processes (Section 16.6, Chapter 16). Moreover, the responses of some systems show pronounced peaks, or resonance, at particular frequencies, which can have a major impact on system performance. The frequency response is therefore relevant to understanding the behavior of physical systems, and for taking measures to improve their performance.

2. In the case of frequency selective circuits, or filters, the frequency response is deliberately shaped to modify system behavior in some desired manner. The design of frequency selective circuits is of critical importance in communications and control systems.

The frequency responses of first-order systems having a single, independent energy-storage element and that of second-order systems having two independent energy-storage elements are basic because the frequency response of higher order systems can be expressed in terms of the responses of these systems. The emphasis in this chapter is on understanding the fundamentals of the responses of first-order and second-order systems. Some aspects of filter design are discussed in the Supplementary Topics for this chapter and also in Chapter 18 of this book.

Learning Objectives

- To be familiar with:
  - The terminology used in connection with different types of filters
  - The general features of Bode magnitude and phase plots

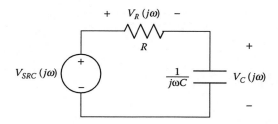

FIGURE 10.1.1
Series *RC* circuit.

- To understand:
  - The basic lowpass and highpass frequency responses of first-order circuits
  - The different types of responses of second-order circuits, including lowpass, highpass, bandpass, and bandstop
  - Peaking or resonance in bandpass, lowpass, and highpass responses

10.1 First-Order Frequency Responses

First-order circuits are those that contain a single energy-storage element, which could be either a capacitor or an inductor, or could contain several capacitors or several inductors, which can be combined into a single capacitor or a single inductor, respectively. They are referred to as first-order because they are described by a linear differential equation of the first order, as in the case of the *RL* circuit (Equation 5.2.1, Chapter 5).

A simple first-order circuit is the series *RC* circuit shown in Figure 10.1.1 in the frequency domain. It is convenient when discussing frequency response to represent phasors as functions of $j\omega$, as shown for the voltage phasors in Figure 10.1.1. From voltage division,

$$\frac{V_C(j\omega)}{V_{SRC}(j\omega)} = \frac{1/j\omega C}{R + 1/j\omega C} = \frac{1}{1 + j\omega CR} \tag{10.1.1}$$

$$\left|\frac{V_C(j\omega)}{V_{SRC}(j\omega)}\right| = \frac{|V_C(j\omega)|}{|V_{SRC}(j\omega)|} = \frac{1}{\sqrt{1 + \omega^2 C^2 R^2}} \text{ and } \angle[V_C(j\omega)/V_{SRC}(j\omega)] = -\tan^{-1}\omega CR \tag{10.1.2}$$

Assume that the magnitude of the source $|V_{SRC}(j\omega)|$ remains constant as its frequency is varied. When $\omega = 0$, *C* acts as a dc open circuit, no current flows, and $V_C = V_{SRC}$. As the frequency increases $|V_C(j\omega)|$ decreases continuously. When $\omega \to \infty$, the reactance of *C* approaches zero, that is, *C* acts as a short circuit, and $|V_C(j\omega)| \to 0$. Such a response is **lowpass**, because high-frequency signals are attenuated, whereas low-frequency signals are transmitted with little attenuation.

If the output is taken across *R*, then:

$$\frac{V_R(j\omega)}{V_{SRC}(j\omega)} = \frac{R}{R + 1/j\omega C} = \frac{j\omega CR}{1 + j\omega CR} \tag{10.1.3}$$

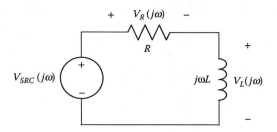

FIGURE 10.1.2
Series *RL* circuit.

$$\left|\frac{V_R(j\omega)}{V_{SRC}(j\omega)}\right| = \frac{|V_R(j\omega)|}{|V_{SRC}(j\omega)|} = \frac{\omega CR}{\sqrt{1+\omega^2 C^2 R^2}} \text{ and } \angle[V_R(j\omega)/V_{SRC}(j\omega)] = 90° - \tan^1\omega CR \quad (10.1.4)$$

When $\omega = 0$, no current flows and $V_R = 0$. As the frequency increases, $|V_R(j\omega)|$ increases continuously. When $\omega \to \infty$ the reactance of C approaches zero, and $|V_R(j\omega)| \to 1$. The response is now **highpass**, because low-frequency signals are attenuated, whereas high-frequency signals are transmitted with little attenuation.

Similar responses are obtained from an *RL* circuit (Figure 10.1.2). Following the same procedure as aforementioned, it is seen that

$$\frac{V_R(j\omega)}{V_{SRC}(j\omega)} = \frac{R}{R+j\omega L} = \frac{1}{1+j\omega L/R} \quad (10.1.5)$$

$$\left|\frac{V_R(j\omega)}{V_{SRC}(j\omega)}\right| = \frac{|V_R(j\omega)|}{|V_{SRC}(j\omega)|} = \frac{1}{\sqrt{1+\omega^2 L^2/R^2}} \text{ and } \angle[V_R(j\omega)/V_{SRC}(j\omega)] = -\tan^1\omega L/R \quad (10.1.6)$$

Moreover,
$$\frac{V_L(j\omega)}{V_{SRC}(j\omega)} = \frac{j\omega L}{R+j\omega L} \quad (10.1.7)$$

$$\left|\frac{V_L(j\omega)}{V_{SRC}(j\omega)}\right| = \frac{|V_L(j\omega)|}{|V_{SRC}(j\omega)|} = \frac{\omega L}{\sqrt{1+\omega^2 L^2/R^2}} \text{ and } \angle[V_L(j\omega)/V_{SRC}(j\omega)] = 90° - \tan^1\omega L/R \quad (10.1.8)$$

The response across R is now lowpass, whereas the response across L is highpass. The reason, of course, is that at $\omega = 0$, L acts as a dc short circuit so that $V_L = 0$ and $V_R = 1$. When $\omega \to \infty$, the reactance of L approaches infinity, that is, L acts as an open circuit and no current flows, so that $|V_L(j\omega)| \to 1$ and $|V_R(j\omega)| \to 0$.

Based on the aforementioned, it follows that:

> **Concept:** *The response of a first-order circuit is either lowpass or highpass, the variation of the response with frequency being due to the frequency-dependent reactance of energy storage elements.*

It should also be noted that in both Figure 10.1.1 and Figure 10.1.2, where a lowpass response is obtained across one element, a complementary highpass response is obtained across the other element. This is because the two elements are in series across the source, so that sum of the two responses is unity; that is, $\dfrac{V_C(j\omega)}{V_{SRC}(j\omega)} + \dfrac{V_R(j\omega)}{V_{SRC}(j\omega)} = 1$ in Figure 10.1.1.

Although it is straightforward enough to plot the magnitudes and phase angles of the preceding responses as a function of ω, it is much more useful and convenient, as will become apparent, to use logarithmic plots known as **Bode plots**. It is therefore necessary to digress a little to discuss logarithmic plots.

EXERCISE 10.1.1
Verify that RC and L/R have the dimensions of time, so that ωCR and $\omega L/R$ are dimensionless.

10.2 Bode Plots

Consider by way of example the function $y = ax^n$. Taking logarithms to base ten, $\log_{10} y = \log_{10} a + n \log_{10} x$. The plot of $\log_{10} y$ vs. $\log_{10} x$ is a line of slope n. The coefficient a shifts the plot vertically without affecting its slope. If $y = 0.4x^2$ and x assumes the values $x_1 = 5$, $x_2 = 20$, and $x_3 = 80$, the corresponding values of y are $y_1 = 10$, $y_2 = 160$, and $y_3 = 2{,}560$. The plot of $\log_{10} y$ vs. $\log_{10} x$ is shown in Figure 10.2.1, where the axes are scaled linearly

in terms of $\log_{10}$ of the variable. The slope of the line is $\dfrac{\log_{10} 2{,}560 - \log_{10} 10}{\log_{10} 80 - \log_{10} 5} =$

$\dfrac{\log_{10} 256}{\log_{10} 16} = \dfrac{\log_{10} 16^2}{\log_{10} 16} = 2$. Moreover, as $\dfrac{x_3}{x_2} = \dfrac{x_2}{x_1}$ and $\dfrac{y_3}{y_2} = \dfrac{y_2}{y_1}$, the points P_1, P_2, and P_3 are equally spaced along the two axes. Thus, *equal spacing along a given axis in a logarithmic plot implies equal ratios of the corresponding variable.* An alternative way of plotting $\log_{10}$ of

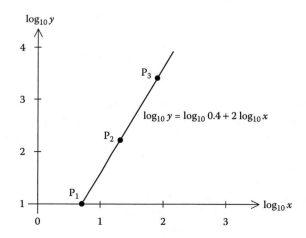

FIGURE 10.2.1
Plot of log values.

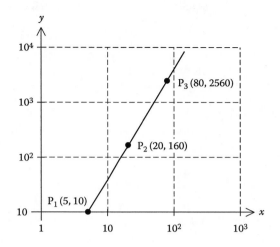

FIGURE 10.2.2
Plot using log axes.

variables is to have the axes scaled as $\log_{10}$ of the variable and labeled in terms of the argument of the log function, as shown in Figure 10.2.2. In this case, the rectangular coordinates x_1y_1, x_2y_2, and x_3y_3 are entered as usual using the corresponding values on the two axes, but what is actually plotted is $\log_{10} y = \log_{10} 0.4 + 2\log_{10} x$.

In Bode magnitude plots, the horizontal axis is scaled as $\log_{10} \omega$ and the vertical axis is scaled linearly in terms of $20\log_{10}(\text{magnitude})$, so that the plot is still log–log. The expression $20\log_{10} B$, where B is the magnitude of the ratio of two voltages or two currents, denotes the dB value of B, where dB is the abbreviation of **decibel**. This representation originated in the early days of telephony for specifying power loss in cascaded circuits, that is, a combination of circuits in which the output of one circuit, other than the last circuit, is the input to the next circuit in the cascade. The overall power loss is the product of the ratios of the power output to the power input of the individual circuits. Using logarithms was very convenient because: (1) it meant that the logarithms of the power ratios of the individual circuits could be added to obtain the logarithm of the overall power loss of the cascaded circuits, and (2) it allowed specifying a large range of power loss or gain, as a much smaller range of the logarithm of these values. The decibel was defined as $10\log_{10} A$, where A is a power ratio. As the power ratio for a given resistor is proportional to the square of the ratio of voltage or current, $20\log_{10}$ is used with the latter ratios. Thus, voltage ratios of 10^3, 10^4, and 10^5, are expressed as 60 dB, 80 dB, and 100 dB, respectively. Bode plots are discussed further in Section ST10.1.

Lowpass Response

Before deriving the Bode plots for the lowpass response, it is convenient to express frequency response in terms of a **transfer function**:

> **Definition:** *When a circuit is excited by a single input, the transfer function in the frequency domain, denoted as H(jω), is the ratio of the phasor of a designated response, or output, to the phasor of the input.*

In the earlier discussion, $\dfrac{V_C(j\omega)}{V_{SRC}(j\omega)}$, $\dfrac{V_R(j\omega)}{V_{SRC}(j\omega)}$, and $\dfrac{V_L(j\omega)}{V_{SRC}(j\omega)}$ are examples of transfer functions in the frequency domain. Strictly speaking, the transfer function is defined in terms of the Laplace transform (Chapter 15) and is a function of s, the complex frequency. However, when the input and response are steady-state sinusoidal functions of time, $H(j\omega)$ is the same as $H(s)$ with $s = j\omega$. We will therefore refer to $H(j\omega)$ as the transfer function in the frequency domain.

$H(j\omega)$ for the lowpass response may be expressed in normalized form as:

$$H(j\omega) = \frac{1}{1 + j\omega/\omega_{cl}} \tag{10.2.1}$$

where $\omega_{cl} = 1/RC$ in Equation 10.1.1 and $\omega_{cl} = R/L$ in Equation 10.1.5. The magnitude of the response is:

$$|H(j\omega)| = \frac{1}{\sqrt{1 + \left(\omega/\omega_{cl}\right)^2}} \tag{10.2.2}$$

A convenient feature of Bode plots is that they can be approximated by straight-line asymptotes. The low-frequency asymptote, as $\omega \to 0$, is $|H(j\omega)| = 1$. On the logarithmic plot, this represents the 0-dB line, or the horizontal axis because $20\log_{10}1 = 0$. The high-frequency asymptote, as $\omega \to \infty$, is $|H(j\omega)| = \dfrac{\omega_{cl}}{\omega}$. On the logarithmic plot, this asymptote is a line whose equation is:

$$20\log_{10}|H(j\omega)| = 20\log_{10}\omega_{cl} - 20\log_{10}\omega \tag{10.2.3}$$

When $\omega = \omega_{cl}$, $20\log_{10}|H(j\omega)| = 0$ so that the high-frequency asymptote intersects the horizontal axis at $\omega = \omega_{cl}$ (Figure 10.2.3a). The slope of the asymptote is expressed as so many decibels per frequency ratio. The ratios commonly used are two and ten, and are referred to as an *octave* and a *decade*, respectively. If $\omega = 2\omega_{cl}$, then from Equation 10.2.3, $20\log_{10}|H(j2\omega_{cl})| = -20\log_{10}2 = -20 \times 0.3010 \cong -6$ dB. Because $20\log_{10}|H(j\omega_{cl})| = 0$ in Equation 10.2.3, the slope of the line can be expressed as -6 dB/octave. Alternatively, if we consider $\omega = 10\omega_{cl}$, then $20\log_{10}|H(j10\omega_{cl})| = -20\log_{10}10 = -20$ dB. Hence, the slope can also be expressed as -20 dB/decade.

At $\omega = \omega_{cl}$, $|H(j\omega_{cl})| = \dfrac{1}{\sqrt{2}}$, and $20\log_{10}|H(j\omega_{cl})| = -3$ dB. ω_{cl} is therefore the **3-dB cutoff frequency**. A synonymous term is the **half-power frequency**. This is because if we imagine the input to be of constant amplitude but of variable frequency and the output to be across a fixed resistor R, then at very low frequencies the output voltage may be V_o rms and the average power dissipated in R is $P = V_o^2/R$. At $\omega = \omega_{cl}$, the output voltage drops to $V_o/\sqrt{2}$ rms and the power becomes $P/2$. Because ω_{cl} is the intersection of the two asymptotes, it is also the **corner frequency**. However, it should be noted that the half-power

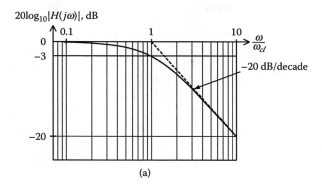

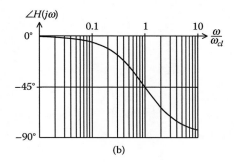

FIGURE 10.2.3
Bode plots of lowpass response.

frequency is, in general, not the same as the corner frequency, which is discussed further in connection with cascaded filters.

The lowpass Bode magnitude plot of Figure 10.2.3a is easily sketched, as it is -3 dB at $\omega = \omega_{cl}$, -1 dB at $\omega = \omega_{cl}/2$, and 1 dB below the high-frequency asymptote at $\omega = 2\omega_{cl}$ (Exercise 10.2.1).

EXERCISE 10.2.1
Show that in a lowpass Bode magnitude plot, the actual response an octave away from ω_{cl} is 1 dB below the closest asymptote.

From Equation 10.2.1, the phase angle $\angle H(j\omega)$ is given by:

$$\angle H(j\omega) = -\tan^{-1}(\omega / \omega_{cl}) \qquad (10.2.4)$$

It follows that when $\omega = 0$, $\angle H(j\omega) = 0$. At $\omega = \omega_{cl}$, $\angle H(j\omega) = -45°$, and as $\omega \to \infty$, $\angle H(j\omega) \to -90°$ (Figure 10.2.3b). At very low frequencies, $\angle H(j\omega) = -\tan^{-1}(\omega/\omega_{cl}) \cong -\omega/\omega_{cl}$.

Highpass Response

$H(j\omega)$ for the highpass response may be expressed in normalized form as:

$$H(j\omega) = \frac{j\omega/\omega_{ch}}{1 + j\omega/\omega_{ch}} \qquad (10.2.5)$$

where $\omega_{ch} = 1/RC$ in Equation 10.1.3 and $\omega_{ch} = R/L$ in Equation 10.1.7. Although ω_{cl} and ω_{ch} are given by the same expression, they refer to two different responses, lowpass and highpass. The magnitude and phase of the highpass response can be expressed as:

$$|H(j\omega)| = \frac{1}{\sqrt{1 + (\omega_{ch}/\omega)^2}} \qquad (10.2.6)$$

and
$$\angle H(j\omega) = 90° - \tan^{-1}(\omega/\omega_{ch}) \qquad (10.2.7)$$

The high-frequency asymptote is obtained by letting $\omega \to \infty$. This gives $|H(j\omega)| = 1$ and $20\log_{10} 1 = 0$ so that the high-frequency asymptote on the logarithmic plot is the horizontal axis. As $\omega \to 0$, $|H(j\omega)| \to \dfrac{\omega}{\omega_{ch}}$. On the logarithmic plot, the low-frequency asymptote is therefore a line whose equation is:

$$20\log_{10}|H(j\omega)| = 20\log_{10}\omega - 20\log_{10}\omega_{ch} \qquad (10.2.8)$$

Following an argument similar to that used for the lowpass response, it is seen that the low-frequency asymptote has a slope of +20 dB/decade or +6 dB/octave, and intersects the 0-dB line at the corner frequency ω_{ch}. This is also the 3-dB cutoff frequency or the half-power frequency in this case, as $|H(j\omega_{ch})| = \dfrac{1}{\sqrt{2}}$. The highpass Bode magnitude plot is shown in Figure 10.2.4a. Comparing Equation 10.2.4 and Equation 10.2.7, the Bode phase plot of the highpass response is that of the lowpass response shifted upward by 90° (Figure 10.2.4b).

The high-frequency asymptote of a lowpass response represents perfect integration, and the low-frequency asymptote of a highpass response represents perfect differentiation (Section ST10.2).

EXERCISE 10.2.2
Plot on the same graph a lowpass response and a highpass response having the same corner frequency $\omega_{cl} = \omega_{ch} = \omega_c$. Why are the two responses symmetrical on a logarithmic scale with respect to a line through $\omega = \omega_c$?

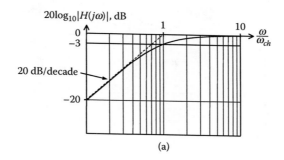

(a)

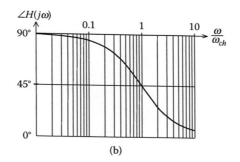

(b)

FIGURE 10.2.4
Bode plots of highpass response.

EXERCISE 10.2.3

Given $R = 1\,\text{k}\Omega$ and $C = 0.5\,\mu\text{F}$ in Figure 10.1.1, determine L in Figure 10.1.2 so as to obtain a highpass response, having ten times the corner frequency of the lowpass response of the RC circuit, with twice the value of R.

Answer: $L = 0.1\,\text{H}$.

Application Window 10.2.1: Crossover Circuit

In high-quality sound reproduction, different loudspeakers handle different ranges of audio frequencies. A woofer loudspeaker reproduces lower audio frequencies (up to approximately 3 kHz), whereas a tweeter loudspeaker reproduces higher audio frequencies (between 3 and 20 kHz). The output of the audio amplifier is applied to the loudspeakers through a *crossover circuit* that directs the appropriate range of frequencies to each loudspeaker. In its simplest form, this circuit consists of first-order lowpass and highpass filters that feed the woofer and tweeter, respectively (Figure 10.2.5). The responses of the two filters cross at the crossover frequency f_c, which is the -3-dB frequency of each filter. In more elaborate sound reproduction systems, a three-way crossover circuit is used in conjunction with a third loudspeaker for midrange audio frequencies.

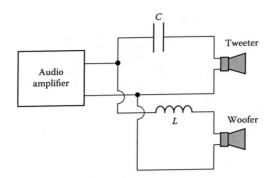

FIGURE 10.2.5
Figure for Application Window 10.2.1.

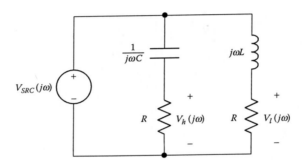

FIGURE 10.2.6
Figure for Application Window 10.2.1.

Because loudspeakers have a frequency-dependent impedance, it is customary to connect across the woofer and the tweeter a series RLC network that compensates for the variation of loudspeaker impedance with frequency. Moreover, a resistor is usually connected across the tweeter, or the tweeter is connected through a voltage divider, in order to reduce the audio power applied to the tweeter. If the loudspeaker is assumed to present a pure resistance R, the desired lowpass and highpass responses are obtained by connecting an inductor in series with the woofer and a capacitor in series with the

tweeter. From the equivalent circuit of Figure 10.2.6, $\left|\dfrac{V_l(s)}{V_{SRC}(s)}\right| = \dfrac{1}{\sqrt{1+(\omega/\omega_{cl})^2}}$, where ω_{cl}

$= R/L$, and $\left|\dfrac{V_h(s)}{V_{SRC}(s)}\right| = \dfrac{1}{\sqrt{1+(\omega_{ch}/\omega)^2}}$, where $\omega_{ch} = 1/RC$. Assume that $R = 8\ \Omega$, as is true

of most loudspeakers, and that a crossover frequency of 2.5 kHz is desired, that is, $\omega_{cl} = \omega_{ch}$

$= 2\pi \times 2.5$ rad/s. It follows that $L = \dfrac{R}{\omega_{cl}} = \dfrac{8}{2\pi \times 2.5 \times 10^3}$, which gives $L \cong 0.51$ mH. Moreover,

$C = \dfrac{1}{\omega_{ch}R} = \dfrac{1}{2\pi \times 2.5 \times 10^3 \times 8} \cong 8\ \mu F.$

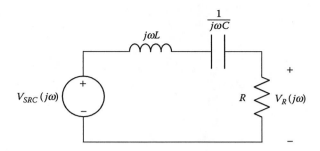

FIGURE 10.3.1
Series *RLC* circuit for deriving bandpass response.

10.3 Bandpass Response

Series Circuit

Consider the series *RLC* circuit of Figure 10.3.1. Because the circuit contains an inductor and a capacitor, it is a second-order circuit that is described by a linear differential equation of the second order (Section 12.2, Chapter 12). Let us determine how the response across the resistor varies with frequency. From voltage division,

$$H(j\omega) = \frac{V_R(j\omega)}{V_{SRC}(j\omega)} = \frac{R}{j\omega L + R + 1/j\omega C} = \frac{j\omega CR}{1 - \omega^2 LC + j\omega CR} \tag{10.3.1}$$

Compared to first-order responses, Equation 10.3.1 differs in one important respect, that is, the real part of the denominator $1 - \omega^2 LC$ is frequency-dependent and becomes zero at

$\omega = \omega_0 = \dfrac{1}{\sqrt{LC}}$. At this frequency, $V_R(j\omega_0) = V_{SRC}(j\omega_0)$, which means that the total voltage

drop across *L* and *C* is zero. This is because when $1 - \omega_0^2 LC = 0$, the two reactances $\omega_0 L$

and $-\dfrac{1}{\omega_0 C}$ are equal in magnitude but opposite in sign, so their sum is zero. At $\omega = 0$, no

current flows because *C* acts as an open circuit, so $V_R = 0$. Similarly, as $\omega \to \infty$, *L* acts as an open circuit, so $V_R \to 0$. The response shows a maximum at the intermediate frequency ω_0 and falls off on either side of this maximum. The resulting response is described as **bandpass**.

To investigate the frequency response in detail, it is convenient to define a quantity *Q* for the series *RLC* circuit as:

$$Q = \frac{\omega_0 L}{R} = \frac{1}{\omega_0 CR} \tag{10.3.2}$$

where *Q* is the **quality factor**. It fundamentally depends on the ratio of the peak energy stored in *L* or *C* to the power dissipated in *R* during a cycle (Section ST10.8). For given *L* and *C*, the smaller *R*, the larger is *Q*.

Substituting for Q and ω_0 and simplifying:

$$H(j\omega) = \frac{1}{1 + jQ\left(\dfrac{\omega}{\omega_0} - \dfrac{\omega_0}{\omega}\right)} \qquad (10.3.3)$$

It follows that

$$|H(j\omega)| = \frac{1}{\sqrt{1 + Q^2\left(\dfrac{\omega}{\omega_0} - \dfrac{\omega_0}{\omega}\right)^2}} \qquad (10.3.4)$$

and
$$\angle H(j\omega) = -\tan^{-1} Q\left(\frac{\omega}{\omega_0} - \frac{\omega_0}{\omega}\right) \qquad (10.3.5)$$

It is seen from Equation 10.3.4 that when $\omega = \omega_0$, $|H(j\omega)|$ has its maximum value of unity. $|H(j\omega)|$ decreases on either side of ω_0 and is the same at $\omega = k\omega_0$ and at $\omega = \omega_0/k$, where k is a positive constant, which means that on a logarithmic plot $|H(j\omega)|$ is symmetrical about the line $\omega = \omega_0$. As $\omega \to 0$, $|H(j\omega)| \to \dfrac{1}{Q}\dfrac{\omega}{\omega_0}$, so that the low-frequency asymptote has a slope of +20 dB/decade, as for a first-order highpass response. As $\omega \to \infty$, $|H(j\omega)| \to \dfrac{1}{Q}\dfrac{\omega_0}{\omega}$ and the high-frequency asymptote has a slope of –20 dB/decade, as for a first-order lowpass response.

The Bode magnitude and phase plots are shown in Figure 10.3.2a and Figure 10.3.2b, respectively, for two values of Q, namely, 10 and 5. The phase angle is zero at $\omega = \omega_0$ because the reactances cancel out and $V_R(j\omega_0) = V_{SRC}(j\omega_0)$. When $\omega < \omega_0$, the capacitive reactance predominates so that the current, and hence $V_R(j\omega)$, leads $V_{SRC}(j\omega)$. As $\omega \to 0$, $\tan \angle H(j\omega) \to +\infty$, and $\angle H(j\omega) \to +90°$, as for an ideal capacitor. When $\omega > \omega_0$, the inductive reactance predominates so that the current, and hence $V_R(j\omega)$, lags $V_{SRC}(j\omega)$. As $\omega \to \infty$, $\tan \angle H(j\omega) \to -\infty$, and $\angle H(j\omega) \to -90°$, as for an ideal inductor.

At the half-power frequencies $|H(j\omega_c)| = 1/\sqrt{2}$. This gives, from Equation 10.3.4:

$\dfrac{\omega_c}{\omega_0} - \dfrac{\omega_0}{\omega_c} = \pm\dfrac{1}{Q}$. Solving for ω_c and retaining only positive roots:

$$\omega_{c1} = \omega_0\left[\sqrt{1 + \left(\frac{1}{2Q}\right)^2} - \frac{1}{2Q}\right], \quad \omega_{c2} = \omega_0\left[\sqrt{1 + \left(\frac{1}{2Q}\right)^2} + \frac{1}{2Q}\right] \qquad (10.3.6)$$

The **bandwidth** (BW) is defined as $\omega_{c2} - \omega_{c1}$. It follows from Equation 10.3.6 that:

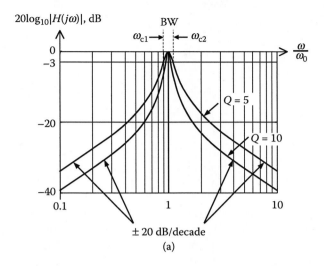

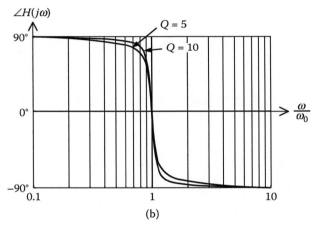

FIGURE 10.3.2
Bandpass response.

$$\text{BW} = \omega_{c2} - \omega_{c1} = \frac{\omega_0}{Q} \quad \text{and} \quad \sqrt{\omega_{c1}\omega_{c2}} = \omega_0 \qquad (10.3.7)$$

For a given ω_0, the larger the Q, the narrower is the BW, that is, the sharper is the peak of the bandpass response. If the BW does not exceed 10% of the center frequency, corresponding to a Q of at least 10, the bandpass response is described as **narrowband**.

The pronounced peaking of the frequency response at some frequency ω_0 is described as **resonance**, ω_0 being the **resonant frequency**. As illustrated in the preceding paragraph, resonance results in a bandpass response that allows selection of a relatively narrow band of frequencies out of a wide range of frequencies that may be present. A resonant circuit is also referred to as a **tuned circuit**. Dynamic systems, in general, exhibit resonance, which could be quite pronounced if Q is high. A structure, such as a bridge, may resonate at some frequency. If excited at this frequency, the amplitude of vibration may be large enough to cause failure of the bridge. An example of resonance in a liquid-filled tube is considered in Section 19.2 (Chapter 19).

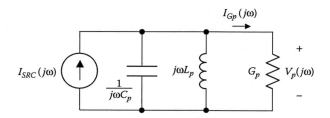

FIGURE 10.3.3
Parallel *GCL* circuit.

Parallel Circuit

The case of a parallel *GCL* circuit excited by a current source (Figure 10.3.3) is quite analogous to that of the series circuit excited by a voltage source, as considered earlier. The circuit of Figure 10.3.3 is of special interest because it is the small-signal representation of a parallel-tuned circuit driven by a transistor amplifier.

From current division,

$$H(j\omega) = \frac{I_{Gp}(j\omega)}{I_{SRC}(j\omega)} = \frac{G_p}{j\omega C_p + G_p + 1/j\omega L_p} = \frac{j\omega L_p G_p}{1 - \omega^2 L_p C_p + j\omega L_p G_p} \quad (10.3.8)$$

It should be noted that Equation 10.3.8 can, in fact, be derived from Equation 10.3.1 by making the following substitutions:

$$V_{SRC}(j\omega) \to I_{SRC}(j\omega) \, , \, V_R(j\omega) \to I_{Gp}(j\omega) \, , \, R \to G_p \, , \, L \to C_p \, , \text{ and } C \to L_p \quad (10.3.9)$$

These substitutions follow from the concept of duality, which will be developed in detail in the following chapter. With the preceding substitutions in Equation 10.3.2,

$$Q_p = \frac{\omega_0 C_p}{G_p} = \omega_0 C_p R_p = \frac{R_p}{\omega_0 L_p} = \frac{\omega_0}{\text{BW}} \quad (10.3.10)$$

where $\omega_0 = 1/\sqrt{L_p C_p}$ and ω_{c1} and ω_{c2} are given by Equation 10.3.6. It should be noted that for a parallel resonant circuit, the larger R_p, the larger is Q_p.

Design Example 10.3.1: Frequency Response of Parallel *GCL* Circuit

Given a parallel *GCL* circuit excited by a current source, as in Figuare 10.3.3, if a current input of 10 mA peak is to produce a maximum response of 1 V peak at 100 krad/s, it is required to determine L and $R_p = 1/G$ to give a BW of 2 krad/s, and to calculate ω_{c1} and ω_{c2} .

SOLUTION

At resonance, the susceptances of L and C are equal in magnitude but opposite in sign. They add to zero, so that only the resistance $R_p = 1/G$ is seen by the source. For a 10 mA

current to produce a 1 V response at this frequency, $R_p = \dfrac{1}{0.01} = 100 \; \Omega$.

If BW = 2 krad/s and $\omega_0 = 100$ krad/s, it follows from Equation 10.3.10 that

$Q_p = \dfrac{\omega_0}{\text{BW}} = 50$. With $Q_p = \omega_0 C_p R_p$, $C_p = \dfrac{Q_p}{\omega_0 R_p} \equiv 5$ μF. Then $L_p = \dfrac{1}{\omega_0^2 C_p} = 20$ μH.

From Equation 10.3.6, $\omega_{c1} = 100(\sqrt{1+10^{-4}} - 0.01) = 99.005$ krad/s and $\omega_{c2} = 100(\sqrt{1+10^{-4}} +$

$0.01) = 101.005$ krad/s. The geometric mean $\sqrt{\omega_{c1}\omega_{c2}}$ is 100 krad/s (Equation 10.3.7) and

is very nearly equal to the arithmetic mean $(\omega_{c1} + \omega_{c2})/2$.

Example 10.3.2 illustrates the following concepts:

> **Concept:** *A second-order response can be obtained from two independent energy storage elements of the same type, that is, elements that cannot be combined into a single element. However, the Q in these circuits does not exceed 0.5 (Section ST10.3).*

> **Concept:** *A bandpass response can result from cascading a lowpass response and a high pass response (Section 17.5, Chapter 17).*

The lowpass response produces high overall attenuation at high frequencies, whereas the highpass response produces high overall attenuation at low frequencies. Hence, the overall response would have a maximum at some intermediate frequency, thereby producing a bandpass response. However, when Q is very low, the maximum is quite flat and cannot be described as resonance.

Simulation Example 10.3.2: Second-Order RC Response

A lowpass RC circuit is cascaded with a highpass RC circuit having the same value of RC product (Figure 10.3.4). It is required to determine the overall response and to simulate the circuit with PSpice assuming $C = 1$ μF and R chosen to give a maximum response at 1 kHz.

SOLUTION

The cascaded filters are shown in Figure 10.3.4. It follows that $\dfrac{V_O(j\omega)}{V_1(j\omega)} = \dfrac{j\omega CR}{j\omega CR + 1}$ and

$$\frac{V_1(j\omega)}{V_{SRC}(j\omega)} = \frac{\dfrac{(R+1/j\omega C)/j\omega C}{R+2/j\omega C}}{R+\dfrac{(R+1/j\omega C)/j\omega C}{R+2/j\omega C}} = \frac{j\omega CR+1}{1-\omega^2 C^2 R^2 + 3j\omega CR} \text{ . Multiplying the two expressions,}$$

$$\frac{V_O(j\omega)}{V_{SRC}(j\omega)} = \frac{j\omega CR}{1-\omega^2 C^2 R^2 + 3j\omega CR} \tag{10.3.11}$$

Equation 10.3.11 is of exactly the same form as Equation 10.3.1 and is therefore a bandpass response. To determine Q, we have to compare coefficients of the terms in the

denominator. In Equation 10.3.1 the coefficient of ω^2 is $\dfrac{1}{\omega_0^2}$. Hence, $\omega_0^2 = \dfrac{1}{C^2 R^2}$ in this

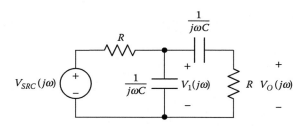

FIGURE 10.3.4
Figure for Example 10.3.2.

example, or $\omega_0 = \dfrac{1}{CR}$. In Equation 10.3.1, the coefficient of $j\omega$ is CR, and is $\dfrac{1}{\omega_0 Q}$. In

Equation 10.3.11, the coefficient of $j\omega$ in the denominator is $3CR$, and is $\dfrac{3}{\omega_0}$. It follows

that the Q of the circuit is $1/3$.

SIMULATION

For $\omega_0 = 2\pi \times 10^3$ rad/s, $R = \dfrac{1}{\omega_0 C} = \dfrac{500}{\pi} = 159.1549\ \Omega$. The schematic is that of Figure 10.3.4.

Using the steady-state sinusoidal source VAC with the default amplitude of 1 V, name the output node VO. In the simulation settings, choose AC Sweep/Noise for Analysis type. Under AC Sweep Type choose Logarithmic/Decade. Enter 100 for Start Frequency, 10k for End Frequency, and 100 Points/Decade to obtain a smooth plot over the range 100 Hz to 10 kHz. After the simulation is run, select Trace/Add Traces. From the Functions or Macros window, choose DB[]. When this appears in the Trace Expression field, select V[Vo] from the Simulation Output Variables window. The plot of Figure 10.3.5 is displayed.

To obtain quantitave information, press the Toggle cursor button and then the Cursor Max button. The cursor reads the max as –9.542 dB at 1 kHz. At $\omega = \omega_0$, the response from

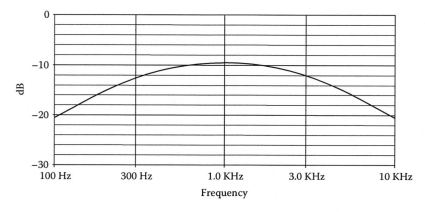

FIGURE 10.3.5
Figure for Example 10.3.2.

Equation 10.3.11 is 1/3, and $20\log_{10}(1/3) = -9.542$. To determine the half-power points, position the cursor towards the low-frequency end, select/trace/cursor/search/commands, and enter the cursor search command `sfle(max - 3)`. The cursor moves to the lower half-power point and reads a frequency of 303.4 Hz. Repeat the command and the cursor moves to the upper half-power point and reads a frequency of 3.2963 kHz. The BW is therefore 2.9929 kHz. The calculated value is $3f_0 = 3$ kHz.

To obtain the maximum possible Q of 0.5, the second RC circuit should be isolated from the first; that is, the first circuit should not be loaded by the second circuit drawing current from it. This may be achieved in one of two ways: (1) inserting a unity-gain, ideal amplifier between the two circuits (Section 18.1, Chapter 18). Such an amplifier has infinite input impedance, so it draws no current from the output of the first circuit, and behaves as an ideal voltage source at its output; (2) increasing the impedance level of the second circuit without affecting the RC product. If R of the second circuit is multiplied by k, and C is divided by k, where k is a large positive constant, RC is not altered, but the current drawn by the second circuit is negligibly small. In other words, the load impedance presented by the second circuit is large compared to the source impedance of the first circuit. It then follows from TEC at the output terminals of the first circuit that

$$\frac{V_1(j\omega)}{V_{SRC}(j\omega)} = \frac{1}{j\omega CR+1}$$. With $$\frac{V_O(j\omega)}{V_1(j\omega)} = \frac{j\omega CR}{j\omega CR+1}$$, the overall transfer function becomes,

$$\frac{V_O(j\omega)}{V_{SRC}(j\omega)} = \frac{j\omega CR}{(j\omega CR+1)^2} = \frac{j\omega CR}{1-\omega^2 C^2 R^2 + 2j\omega CR} \qquad (10.3.12)$$

Comparing with Equation 10.3.11, it is seen that $Q = 0.5$.

10.4 Bandstop Response

Consider the response $V_{LC}(j\omega) = V_C(j\omega) + V_L(j\omega)$ in Figure 10.3.1. It follows from voltage division that

$$H(j\omega) = \frac{V_{LC}(j\omega)}{V_{SRC}(j\omega)} = \frac{j\omega L + 1/j\omega C}{j\omega L + R + 1/j\omega C} = \frac{1-\omega^2 LC}{1-\omega^2 LC + j\omega CR} \qquad (10.4.1)$$

The denominator is the same as that of Equation 10.3.1. When $\omega = 0$, the capacitor behaves as an open circuit, no current flows, $V_{LC} = V_{SRC}$ and $H(j\omega) = 1$. As $\omega \to \infty$, the inductor behaves as an open circuit, so again no current flows, $V_{LC}(j\omega) = V_{SRC}(j\omega)$ and $H(j\omega) = 1$. However, at $\omega = \omega_0 = \dfrac{1}{\sqrt{LC}}$, $H(j\omega) = 0$. This response is described as **bandstop**, and is the complement, with respect to the source, of the bandpass response, because the sum of the two responses is unity.

To investigate the frequency response in detail, we again express the response in terms of Q and ω_0. This gives:

$$H(j\omega) = \frac{jQ\left(\dfrac{\omega}{\omega_0} - \dfrac{\omega_0}{\omega}\right)}{1 + jQ\left(\dfrac{\omega}{\omega_0} - \dfrac{\omega_0}{\omega}\right)} \tag{10.4.2}$$

It follows that:

$$\left|H(j\omega)\right| = \frac{\left|Q\left(\dfrac{\omega}{\omega_0} - \dfrac{\omega_0}{\omega}\right)\right|}{\sqrt{1 + Q^2\left(\dfrac{\omega}{\omega_0} - \dfrac{\omega_0}{\omega}\right)^2}} \tag{10.4.3}$$

To obtain the phase angle (Exercise 10.4.1) it is convenient to divide the numerator and denominator of Equation 10.4.2 by $jQ\left(\dfrac{\omega}{\omega_0} - \dfrac{\omega_0}{\omega}\right)$ to give:

$$\angle H(j\omega) = \tan^{-1}\left[1 \bigg/ Q\left(\dfrac{\omega}{\omega_0} - \dfrac{\omega_0}{\omega}\right)\right] \tag{10.4.4}$$

The magnitude and phase plots are shown in Figure 10.4.1a and Figure 10.4.1b for two values of Q, namely, 5 and 10. $\left|H(j\omega)\right|$ is plotted and not its logarithm because $\left|H(j\omega)\right| = 0$ at $\omega = \omega_0$. At low frequencies, the reactance of the capacitor is very large and that of the inductor negligibly small. The response is like that of a first-order lowpass RC circuit. $V_{LC}(j\omega) \cong V_{SRC}(j\omega)$ and the phase angle is small and negative, as in Figure 10.2.3b. At high frequencies, the reactance of the inductor is very large and that of the capacitor negligibly small. The response is like that of a first-order highpass RL circuit. $V_{LC}(j\omega) \cong V_{SRC}(j\omega)$ and the phase angle is small and positive, as in Figure 10.2.4b.

At $\omega = \omega_0$, $V_{LC}(j\omega) = 0$ as the reactances of the capacitor and the inductor add to zero. Just below ω_0, the net reactance in the circuit is small and capacitive, as in the first-order lowpass RC circuit at high frequencies, and the phase angle approaches $-90°$. Just above ω_0, the net reactance in the circuit is small and inductive, as in the first-order highpass RL circuit at low frequencies, and the phase angle approaches $+90°$. At the half-power frequencies $\left|H(j\omega)\right| = 1/\sqrt{2}$. This gives, from Equation 10.4.3, $\dfrac{\omega_c}{\omega_0} - \dfrac{\omega_0}{\omega_c} = \pm\dfrac{1}{Q}$ as for the bandpass case. The half-power frequencies are given by Equation 10.3.6 and the BW by Equation 10.3.7.

EXERCISE 10.4.1
Deduce Equation 10.4.4 by rationalizing Equation 10.4.2 and determining the phase angle of the resulting expression.

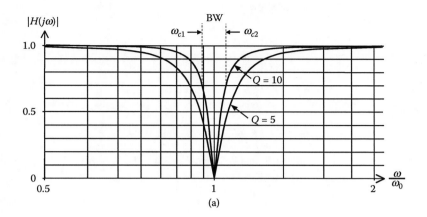

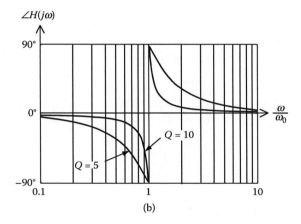

FIGURE 10.4.1
Bandstop response.

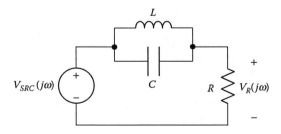

FIGURE 10.4.2
Figure for Example 10.4.1.

Design Example 10.4.1: Bandstop Response of *RLC* Circuit

1. Verify that the response of the circuit of Figure 10.4.2 is bandstop and interpret the result.

2. If $R = 1\ k\Omega$, determine L and C, so that the half-power frequencies are 1000 rad/s and 1200 rad/s.

SOLUTION

1. $H(j\omega) = \dfrac{V_R(j\omega)}{V_{SRC}(j\omega)} = \dfrac{R}{R + \dfrac{L/C}{j\omega L + 1/j\omega C}}$. For changes in current in R, L, or C, the ideal

    voltage source behaves as a short circuit so that the circuit reduces to a parallel

    GCL circuit (Example 10.3.1). Substituting $Q = \omega_0 CR = \dfrac{R}{\omega_0 L}$ gives Equation 10.4.2.

    The interpretation of circuit behavior is that at very low frequencies the inductor
    provides a low-impedance path between input and output so that
    $V_R(j\omega) \cong V_{SRC}(j\omega)$. At very high frequencies, the capacitor provides a low-imped-

    ance path between input and output, and again $V_R(j\omega) \cong V_{SRC}(j\omega)$. At $\omega = \omega_0$, the

    total susceptance of L and C is zero, so $V_R(j\omega) = 0$.

2. $\omega_0 = \sqrt{1000 \times 1200} = 1095.4$ rad/s. $C = \dfrac{Q}{\omega_0 R} = \dfrac{1}{BW \times R} \equiv 5$ µF. $L = \dfrac{1}{\omega_0^2 C} = 0.17$ H.

10.5   Second-Order Lowpass and Highpass Responses

Lowpass Response

If the response in Figure 10.3.1 is taken across the capacitor,

$$H(j\omega) = \frac{V_C(j\omega)}{V_{SRC}(j\omega)} = \frac{1/j\omega C}{j\omega L + R + 1/j\omega C} = \frac{\omega_0^2}{\omega_0^2 - \omega^2 + \dfrac{\omega_0}{Q} j\omega} \tag{10.5.1}$$

This is clearly a lowpass response. $|H(j\omega)| = 1$ when $\omega = 0$, and $|H(j\omega)| \to 0$ as $\omega \to \infty$. In
terms of circuit behavior, the reactance of the capacitor is high at low frequencies, whereas
the reactance of the inductor is small, and $V_C(j\omega) \cong V_{SRC}(j\omega)$. At high frequencies, the
reactance of the inductor is high, whereas the reactance of the capacitor is small, so
that $V_C(j\omega) \cong 0$.

To derive the Bode magnitude and phase plots, it is convenient to express $H(j\omega)$ in terms
of a dimensionless **damping parameter** ξ defined as:

$$\xi = \frac{1}{2Q} = \frac{R}{2\omega_0 L} = \frac{\omega_0 CR}{2} \tag{10.5.2}$$

Substituting for ξ and setting $\dfrac{\omega}{\omega_0} = u$, Equation 10.5.1 becomes

$$H(j\omega) = \frac{1}{1 - u^2 + j2u\xi} \tag{10.5.3}$$

Hence,

$$|H(j\omega)| = \frac{1}{\sqrt{(1-u^2)^2 + (2\xi u)^2}} = \frac{1}{\sqrt{y}} \tag{10.5.4}$$

where

$$y = (1-u^2)^2 + (2\xi u)^2 = u^4 + 2u^2(2\xi^2 - 1) + 1 \tag{10.5.5}$$

and

$$\angle H(j\omega) = \tan^{-1}\frac{2\xi u}{u^2 - 1} \tag{10.5.6}$$

We wish to determine if $|H(j\omega)|$ has a maximum at some frequency, which means that y has a minimum. To do so, we set $\frac{dy}{du} = 0$ and retain only positive nonzero values of u. This gives:

$$u = \sqrt{1-2\xi^2} \tag{10.5.7}$$

Hence, $|H(j\omega)|$ will have a maximum at some nonzero values of u if $\xi < \frac{1}{\sqrt{2}}$, and the maximum value is $|H(j\omega)|_{\max} = \frac{1}{2\xi\sqrt{1-\xi^2}}$. For very low damping, $R \to 0$, $\xi \to 0$, and $|H(j\omega)|_{\max} \to \infty$. The response becomes highly peaked at low damping. At high frequencies, $|H(j\omega)| \to \frac{1}{u^2}$, and $20\log_{10}|H(j\omega)| \mapsto -40\log_{10}\omega$. The high-frequency asymptote has a slope of -40 dB/decade, which is twice that of a first-order highpass response. The Bode magnitude and phase plots are derived in Simulation Example 10.5.1 and illustrated by Figure 10.5.1.

The general shape of the phase plot may be readily argued from Equation 10.5.6. If $u = 0$, $\angle H(j\omega) = 0$. As u increases from zero, $\tan \angle H(j\omega)$ increases in magnitude but is negative in sign. $\angle H(j\omega)$ is therefore a negative angle whose magnitude increases with frequency. As $u \to 1$, $\tan \angle H(j\omega)$ is large and negative and $\angle H(j\omega)$ approaches $-90°$. When u becomes just larger than 1, $\tan \angle H(j\omega)$ is large and positive, $\angle H(j\omega)$ is a negative angle in the third quadrant whose magnitude exceeds $90°$. If u is very large, $\tan \angle H(j\omega)$ is small and positive. $\angle H(j\omega)$ approaches $-180°$ from the third quadrant. That the phase angle is just less negative than $-180°$ at high frequencies may be ascertained from the circuit. At these frequencies, the current lags the applied voltage by almost $90°$ because of the dominance of the reactance of the inductor. The voltage across the capacitor lags the current by $90°$. Hence the voltage across the capacitor lags the applied voltage by almost $180°$. The smaller ξ, the smaller is $\tan \angle H(j\omega)$ for a given ω, and the greater is the magnitude of the slope at $\angle H(j\omega) = -90°$.

Simulation Example 10.5.1: Second-Order Lowpass Response

It is required to obtain the Bode magnitude and phase plots of the lowpass response of the series RLC circuit assuming that $L = 0.5$ H, $C = 12.5$ μF, and for $R = 10, 100, 283,$ and $400\ \Omega$.

SIMULATION

It is seen that $\omega_0 = \dfrac{1}{\sqrt{0.5 \times 12.5 \times 10^{-6}}} = 400$ rad/s, $2\omega_0 L = 400$ Ω, and the chosen values of

R correspond to $\xi = 0.025$ ($R = 10$ Ω), $\xi = 0.25$ ($R = 100$ Ω), $\xi = 1/\sqrt{2}$ ($R = 283$ Ω), and $\xi = 1$ ($R = 400$ Ω). R is declared a global parameter, as described in Example 7.4.1 (Chapter 7). Select AC Sweep/Noise analysis, then check the Parametric Sweep box. Choose Global parameter under Sweep variable and enter R_val in the Parameter name field. Under Sweep type, enter in the Value list: 100,200,400,700. When the simulation is run, press the All and OK buttons in the Available Sections window. After selecting DB(V(VC)) in the Add Traces window, assuming VC is the output voltage, the magnitude plot for the four resistance values is displayed, as in Figure 10.5.1a. We can add the high frequency asymptote by entering its line equation. Select Trace/Add traces, then enter the following equation in the Trace Expression field: 40*LOG10[63.662] −40*LOG10[Frequency]. The LOG10[]

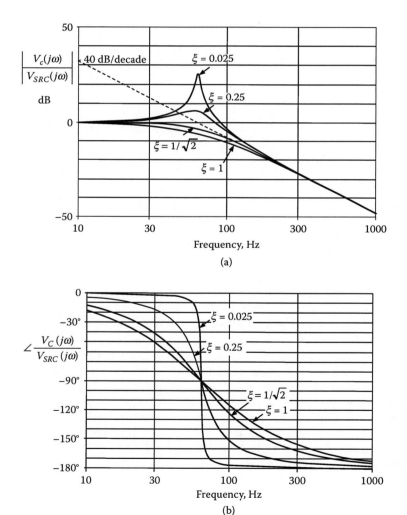

(a)

(b)

FIGURE 10.5.1
Figure for Example 10.5.1.

function is selected from the Functions or Macros window and Frequency is selected from the Simulation Output Variables window. The value of ω_0 is 63.662 Hz. The cursor can be used to obtain quantitative information. To select a particular trace, such as that for $\xi = 0.025$ for example, click on its colored marker just below the horizontal axis on the LHS. Press the Toggle cursor button and then the Cursor Max button. The cursor reads the max

as 26.015 dB at 63.551 Hz. The calculated values are $20\log_{10}\left(\dfrac{1}{2\xi\sqrt{1-\xi^2}}\right) = 26.0$ dB at

$63.662\sqrt{1-(0.025)^2} = 63.64$ Hz. If the lowest trace ($R = 400\ \Omega$) is selected for cursor movement, and the cursor search command `sfle(-6)` is entered, with the cursor positioned at the beginning of the trace, we obtain 63.511 Hz as the frequency for -6 dB. The calculated value is $f_0 = 63.66$ Hz.

To derive the phase plot, select Trace/Add Traces. From the Functions or Macros window, choose P[]. When this appears in the Trace Expression field, select V[VC] from the Simulation Output Variables window. The phase plot is shown in Figure 10.5.1b.

Highpass Response

If the response in Figure 10.3.1 is taken across the inductor,

$$H(j\omega) = \frac{V_L(j\omega)}{V_{SRC}(j\omega)} = \frac{j\omega L}{j\omega L + R + 1/j\omega C} = \frac{-\omega^2}{\omega_0^2 - \omega^2 + \dfrac{\omega_0}{Q}j\omega} \qquad (10.5.8)$$

This is a highpass response. $|H(j\omega)| = 0$ when $\omega = 0$, and $|H(j\omega)| \rightarrow 1$ as $\omega \rightarrow \infty$. In terms of circuit behavior, the reactance of the inductor is high at high frequencies, whereas the reactance of the capacitor is small, and $V_L(j\omega) \cong V_{SRC}(j\omega)$. At low frequencies, the reactance of the inductor is small, whereas the reactance of the capacitor is high, so that $V_L(j\omega) \cong 0$. Equation 10.5.8 can be written in terms of ξ and $r = \omega_0 / \omega$ as:

$$H(j\omega) = \frac{-\omega^2}{\omega_0^2 - \omega^2 + j2\xi\omega_0\omega} = \frac{1}{1 - r^2 - j2\xi r} \qquad (10.5.9)$$

This is of the same form as Equation 10.5.3 with r replacing $u = \omega / \omega_0$ and with the sign of the imaginary part negated. Hence,

$$|H(j\omega)| = \frac{1}{\sqrt{\left(1-r^2\right)^2 + \left(2\xi r\right)^2}} \qquad (10.5.10)$$

It follows that the lowpass and highpass Bode magnitude plots are symmetrical on a log ω scale with respect to the line $\omega = \omega_0$, because $r = 1/u$. Figure 10.5.2a shows the Bode magnitude plot of the high pass response as a function of ω / ω_0 and for various values of ξ. Note that replacing ω / ω_0 by ω_0 / ω in the magnitude response is tantamount to interchanging capacitors and inductors.

From Equation 10.5.9 the phase angle is:

$$\tan\left(\angle H(j\omega)\right) = \frac{2\xi r}{1-r^2} = \frac{2\xi u}{u^2 - 1} \qquad (10.5.11)$$

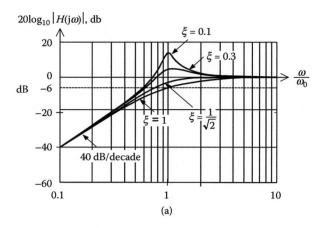

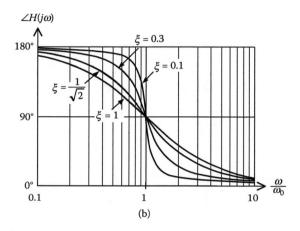

FIGURE 10.5.2
Bode plots for highpass response.

This is the same as Equation 10.5.6, when expressed as a function of $u = \omega / \omega_0$. However, $\tan(\alpha + \pi) = \tan\alpha$ and the phase angle of the highpass response is 180° plus that of the lowpass response (Figure 10.5.2b), as can be readily interpreted in the same manner as the lowpass response. At very low frequencies, the capacitive reactance predominates so that the current leads the applied voltage by almost 90°. The voltage across the inductor leads the current by 90°. Hence, the voltage across the inductor leads the applied voltage by almost 180°.

EXERCISE 10.5.1
Show that for the highpass magnitude response: (1) the peak occurs at a frequency $\omega_p = \omega_0/\sqrt{1-2\xi^2}$, (2) the response crosses the 0-dB axis ($y = 1$) at a frequency $\omega_0/\sqrt{2(1-2\xi^2)}$.

TABLE 10.5.1

Prototypical, Second-Order Responses

Types of Respons	Transfer Functions
Lowpass	$\dfrac{\omega_n^2}{s^2 + \dfrac{\omega_n}{Q}s + \omega_n^2}$
Bandpass	$\dfrac{\dfrac{\omega_n}{Q}s}{s^2 + \dfrac{\omega_n}{Q}s + \omega_n^2}$
Highpass	$\dfrac{s^2}{s^2 + \dfrac{\omega_n}{Q}s + \omega_n^2}$
Bandstop	$\dfrac{s^2 + \omega_n^2}{s^2 + \dfrac{\omega_n}{Q}s + \omega_n^2}$
Allpass	$\dfrac{s^2 - \dfrac{\omega_n}{Q}s + \omega_n^2}{s^2 + \dfrac{\omega_n}{Q}s + \omega_n^2}$

Summary of Second-Order Responses

Prototypical second-order responses are summarized in Table 10.5.1, rationalized to a maximum response magnitude of 1, and expressed in terms of s. Substituting $s = j\omega$ gives the relations derived earlier. Also, the frequency ω_n is used instead of ω_0 because the latter is usually reserved for $1/\sqrt{LC}$ in circuits containing inductance and capacitance, whereas second-order, frequency-selective circuits may be RC or RL circuits. The allpass response has a magnitude of unity at all frequencies but produces a phase shift as a function of frequency (Problem 10.2.11).

In conclusion, it should be noted that in all the first-order and second-order responses discussed earlier, the phase angle $\angle H(j\omega)$ becomes more lagging as the frequency increases. The maximum change in the phase angle is 90° in first-order responses and 180° in the second-order responses of Table 10.5.1.

Scaling

It is often convenient in filter design to assume that $L = 1$ H and $C = 1$ F, so that the cutoff frequency, or center frequency, is 1 rad/s. R is specified to give the desired Q. After the design is completed, the results are scaled to desired practical values. Scaling generally involves changing impedance magnitudes and frequency.

In magnitude scaling, impedances are multiplied by a positive, real scale factor k_m, which could be less than or greater than unity. This means that all resistances and inductances are multiplied by k_m and all capacitances are divided by k_m. Thus, if unprimed parameters represent the initial assumed values, the primed parameters represent the final scaled values:

$$R' = k_m R , \quad L' = k_m L , \text{ and } C' = C / k_m \qquad (10.5.12)$$

Magnitude scaling neither affects cutoff frequencies, such as R/L or $1/RC$, nor center frequencies, such as $1/\sqrt{LC}$.

In frequency scaling, frequencies are changed without affecting impedances. As resistance is not a function of frequency, it is not affected by frequency scaling. Impedances of inductors and capacitors are $j\omega L$ and $1/j\omega C$, respectively. If $\omega' = k_f \omega$, that is, frequency is scaled by a positive real scale factor k_f, then in order to keep the same impedances, inductances and capacitances must be divided by k_f:

$$R' = R , \quad \omega' = k_f \omega , \quad L' = L / k_f , \text{ and } C' = C / k_f \qquad (10.5.13)$$

If both magnitude and frequency scaling are used, then combining Equation 10.5.12 and Equation 10.5.13:

$$R' = k_m R , \quad \omega' = k_f \omega , \quad L' = \frac{k_m}{k_f} L , \text{ and } C' = \frac{1}{k_m k_f} C \qquad (10.5.14)$$

BW, being a frequency, scales like ω. $Q = \omega_0/\text{BW}$ is unchanged. Thus,

$$(\text{BW})' = k_f \times \text{BW} \text{ and } Q' = Q \qquad (10.5.15)$$

An example of scaling is considered later in connection with the Butterworth response.

Application Window 10.5.1: Touch-Tone Telephone

Tone dialing has practically replaced pulse dialing for selecting a telephone number, as it is much faster to process. The touch-tone keypad typically has 12 buttons arranged in 3 columns and 4 rows (Figure 10.5.3). The four rows are assigned four frequencies in a low-frequency group, whereas the three columns are assigned three frequencies in a high-frequency group. Pressing the number 5, for example, generates the tones 770 Hz and 1336 Hz. Thus, only 7 tones are required for the 12 buttons.

The tones are produced by connecting a capacitor to different taps on a coil, as schematically illustrated in Figure 10.5.4, for the high-frequency group. The normally open switches SC1, SC2, and SC3 are closed when any switch in columns 1, 2, or 3, respectively, is pressed. When any of these switches is halfway through its travel, the normally open switch S makes contact with node D, which establishes an initial dc current through the coil. As S travels further, contact D moves with it, so that at full travel of S, contacts S, D, and E are connected together. This sets the LC circuit into oscillation and connects it to the transistor oscillator circuit, which compensates for the power loss in the LC circuit, thereby allowing sustained oscillations. If SC2 is pressed, for example, the frequency of the tone is determined by C in conjunction with L_2 and L_3. At the central office, an RLC bandpass filter is used for each frequency group, followed by digital filters that determine the tone frequency from the outputs of the two bandpass filters for the frequency groups.

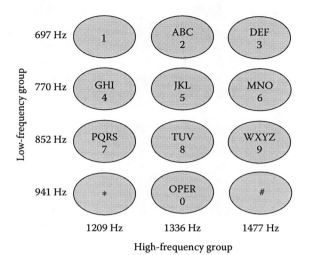

FIGURE 10.5.3
Figure for Application Window 10.5.1.

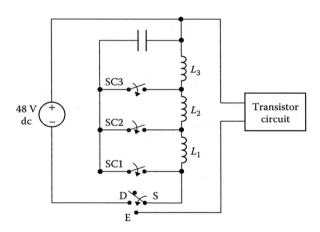

FIGURE 10.5.4
Figure for Application Window 10.5.1.

Let us investigate, by way of illustration, the performance of the bandpass filter for the low-frequency group, considering that R in standard telephone circuits is 600 Ω. The two outermost frequencies of the group are located at the edges of the pass band, that is, they are the 3-dB cutoff frequencies ω_{c1} and ω_{c2}. Thus, $\omega_{c1} = 2\pi \times 697 = 4379$ rad/s and $\omega_{c2} = 2\pi \times 941 = 5912$ rad/s. The BW is, therefore, BW $= \dfrac{R}{L} = 5912 - 4379 = 1533$ rad/s, which gives $L = \dfrac{600}{1533} = 0.39$ H. The center frequency is $\sqrt{4379 \times 5912} = 5088$ rad/s, and

$$C = \frac{1}{0.39 \times 4379 \times 5912} = 0.1 \ \mu F.$$

The other two frequencies of the low-frequency group, 770 Hz and 852 Hz, are symmetrically located with respect to the center frequency of the passband because

$2\pi\sqrt{770\times852} = 5089$ rad/s . Moreover, the four tones were chosen to be harmonically unrelated. That is, none of the frequencies is a linear combination of any of the others. This minimizes the risk of a false identification of a tone due to any circuit nonlinearities.

That there is sufficient selectivity to discriminate against tones in the high-frequency group may be ascertained by calculating the relative magnitude of a signal at the lowest frequency in this group, that is, 1209 Hz. From Equation 10.3.4 and 10.3.7, which apply to a parallel

circuit, $Q^2\left(\dfrac{\omega}{\omega_0}-\dfrac{\omega_0}{\omega}\right)^2 = \left[\dfrac{\omega_0}{\beta}\left(\dfrac{\omega}{\omega_0}-\dfrac{\omega_0}{\omega}\right)\right]^2$. Substituting $\omega_0 = 5088$ rad/s , BW = 1533 rad/s

and $\omega = 2\pi \times 1029 = 7596$ rad/s, gives: $\dfrac{\omega_0}{\text{BW}}\left(\dfrac{\omega}{\omega_0}-\dfrac{\omega_0}{\omega}\right)=2.732$. Hence, the ratio of the

response at 1209 Hz to that at the center of the passband is $\dfrac{1}{\sqrt{1+(2.732)^2}} = 0.344$, which is

sufficiently small for discrimination on the basis of amplitude.

10.6 Improved Responses of Filters

Cascaded Filters

In an ideal frequency-selective circuit, or filter, the response should be flat in the passband and zero in the stopband with a sharp transition in-between (Section 17.5, Chapter 17). The first-order lowpass and highpass responses depart significantly from this ideal. Trying to improve performance by cascading such filters does not help. Let us suppose, for example, that two lowpass filters having the same corner frequency are cascaded, without the second filter having any loading effect on the first. If $H(j\omega)$ of one filter is

$H(j\omega)=\dfrac{1}{1+j\omega/\omega_{cl}}$ (Equation 10.2.1), the transfer function of the two identical, isolated, and cascaded filters is:

$$H(j\omega) = \frac{1}{\left(1+j\omega/\omega_{cl}\right)^2} \qquad (10.6.1)$$

$|H(j\omega)|$ is the square of a single filter and is given by:

$$|H(j\omega)| = \frac{1}{1+\left(\omega/\omega_{cl}\right)^2} \qquad (10.6.2)$$

At the corner frequency ω_{cl} , the response is $-20\log_{10} 2 = -6$ dB. The corner frequency is no longer the same as the half-power frequency. To find this frequency, we substitute $|H(j\omega_{1/2})|=1/\sqrt{2}$ in Equation 10.6.2 to obtain: $\omega_{1/2} = \omega_{cl}\sqrt{\sqrt{2}-1}=0.64\omega_{cl}$. The 3-dB BW, that is, the range of frequencies that are attenuated by not more than 3 dB, is therefore reduced by approximately 36%. For large ω, $|H(j\omega)| \cong (\omega_{cl}/\omega)^2$, so that the high-frequency

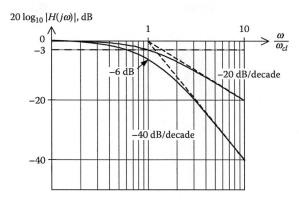

FIGURE 10.6.1
Cascading of two first-order lowpass responses.

asymptote of the response is -40 dB/decade. Figure 10.6.1 compares the responses of a single lowpass filter and a cascade of two of these filters, with isolation. It follows that:

> **Concept:** *Cascading of first-order filters to increase attenuation in the stopband reduces the BW.*

EXERCISE 10.6.1

Show that for n identical, isolated, and cascaded lowpass filters $\omega_{1/2} = \omega_{cl}\sqrt{2^{1/n}-1}$ and derive the corresponding expression for n identical cascaded highpass filters.

Answer: $\omega_{1/2} = \omega_{ch}/\sqrt{2^{1/n}-1}$.

Butterworth Response

If we consider second-order responses, it is seen from Equation 10.5.4 that if $\xi = 1$ the response is also given by Equation 10.6.2. Thus, the response for $\xi = 1$ in Figure 10.5.1 is the same as the two-filter response in Figure 10.6.1. However, it follows from Figure 10.5.1 that having $\xi = 1/\sqrt{2}$ gives a wider BW. The magnitude response is

$$|H(j\omega)| = \frac{1}{\sqrt{1+\left(\omega/\omega_{cl}\right)^4}} \qquad (10.6.3)$$

where for a lowpass response, $\omega_{cl} = \omega_0$. Equation 10.6.3 is an example of a maximally flat, or Butterworth, response of the second-order. A lowpass **Butterworth response** of the nth order is defined as:

$$|H(j\omega)| = \frac{1}{\sqrt{1+\left(\omega/\omega_{cl}\right)^{2n}}} \qquad (10.6.4)$$

Such a response has two important features: (1) as the *only* power in ω is ω^{2n}, then as ω increases from zero, the response stays closer to unity than if lower powers of ω were present, which means that the Butterworth response has the widest 3-dB BW for a response that is flat within the passband. (2) The response is 3 dB at $\omega = \omega_{cl}$ irrespective of n. This is

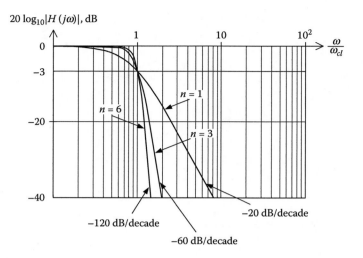

FIGURE 10.6.2
Butterworth responses.

TABLE 10.6.1

Normalized Butterworth Polynomials of Order n

n	Factors of Polynomial $B_n(s)$
1	$(s + 1)$
2	$(s^2 + 1.414s + 1)$
3	$(s + 1)(s^2 + s + 1)$
4	$(s^2 + 0.765s + 1)(s^2 + 1.848s + 1)$
5	$(s + 1)(s^2 + 0.618s + 1)(s^2 + 1.618s + 1)$
6	$(s^2 + 0.518s + 1)(s^2 + 1.414s + 1)(s^2 + 1.932s + 1)$
7	$(s + 1)(s^2 + 0.445s + 1)(s^2 + 1.247s + 1)(s^2 + 1.802s + 1)$
8	$(s^2 + 0.390s + 1)(s^2 + 1.111s + 1)(s^2 + 1.663s + 1)(s^2 + 1.962s + 1)$

in contrast to n cascaded filters, where the 3-dB BW decreases with n (Exercise 10.6.1). Figure 10.6.2 illustrates Butterworth responses of order one, which is the same as a simple first-order filter, and of orders three and six. Normalized Butterworth polynomials are given in Table 10.6.1. Their derivation and other aspects of Butterworth lowpass and highpass filters are discussed in Section ST10.4.

Design Example 10.6.1: Scaling of Second-Order Butterworth Response

Given a series RLC circuit with $L = 1$ H and $C = 1$ F, it is desired to have a second-order, lowpass, normalized Butterworth response when the output is taken across C. Scale the parameters so as to have $\omega_0 = 100$ krad/s and $L = 1$ mH.

SOLUTION

The transfer function for a second-order, lowpass normalized Butterworth response is from Table 10.6.1 with $s = j\omega$:

$$H(j\omega) = \frac{1}{(j\omega)^2 + j\sqrt{2}\omega + 1} \tag{10.6.5}$$

The response is normalized to a dc gain of 1, as setting $\omega = 0$ gives $H(0) = 1$. That this is a second-order Butterworth response with $\omega_{cl} = \omega_0 = \dfrac{1}{\sqrt{LC}} = 1\,\text{rad/s}$, can be readily verified from Equation 10.6.5:

$$|H(j\omega)| = \frac{1}{\sqrt{\left(1-\omega^2\right)^2 + 2\omega^2}} = \frac{1}{\sqrt{1+\omega^4}} \tag{10.6.6}$$

which is of the form of Equation 10.6.3 with $\omega_{cl} = 1\,\text{rad/s}$. We proceed by assuming that $L = 1\,\text{H}$ and $C = 1\,\text{F}$, so as to obtain $\omega_{cl} = \omega_0 = \dfrac{1}{\sqrt{LC}} = 1\,\text{rad/s}$, and determine R, so as to obtain $H(j\omega)$ of Equation 10.6.5. Comparing the coefficient of $j\omega$ in the denominators of the transfer function of Equation 10.6.5 with that in Table 10.5.1 with $\omega_n = 1\,\text{rad/s}$, we should have $\dfrac{1}{Q} = \sqrt{2}$

or $Q = \dfrac{1}{\sqrt{2}} = \dfrac{\omega_0 L}{R} = \dfrac{1}{R}$, or $R = \sqrt{2}$. To scale the frequency to 100 krad/s requires $k_f = 10^5$

(Equation 10.5.13). From Equation 10.5.14, $L' = \dfrac{k_m}{k_f}L$, which gives $k_m = k_f \dfrac{L'}{L} = 10^5 \dfrac{10^{-3}}{1} = 100$.

This makes $R = 100\sqrt{2}\ \Omega$. From Equation 10.5.14, $C' = \dfrac{1}{k_m k_f}C = \dfrac{1}{100 \times 10^5} \equiv 0.1\ \mu\text{F}$. It is seen

that $\omega_0 = \dfrac{1}{\sqrt{10^{-3} \times 10^{-7}}} = 10^5\,\text{rad/s}$ and $Q = \dfrac{10^5 \times 10^{-3}}{100\sqrt{2}} = \dfrac{1}{\sqrt{2}}$ as required.

Improved Lowpass and Highpass Responses

It is generally true that:

> Concept: *The bandwidth of a filter can be increased if some ripple in the response is allowed.*

In Figure 10.5.1, for example, if $\xi = 0.25$, the 3-dB frequency is about $1.48\omega_0$ compared to ω_0 for the Butterworth response (Problem 10.2.14). However, the response now has a maximum of 6.3 dB in the passband which is considered a ripple in the passband above the response at $\omega = 0$. A filter based on Chebyshev polynomials (Section ST10.5) has a greater roll-off than a Butterworth filter of the same order, that is, a faster increase in attenuation near the edge of the passband, but at the expense of allowing some ripple in the response in the passband. Elliptic filters have a greater roll-off still, but have ripple in the responses in both the passband and the stopband.

Improved Bandpass and Bandstop Responses

Narrowband Butterworth bandpass or bandstop filters can be derived from Butterworth lowpass and highpass filters, respectively, using an appropriate transformation of the frequency response, as illustrated in Section ST10.6. An improved bandpass response can be obtained if the coils of two tuned *RLC* circuits are magnetically coupled, as discussed in Section ST10.7.

Active Filters

Operational amplifiers, or op amps (Chapter 18), may be used with *RC* circuits to construct active filters. *RC*, rather than *RL*, circuits are used for this purpose because capacitors are less bulky and expensive than inductors, and are closer to an ideal circuit element. Op amps in active filters provide gain, isolation, load drive, and a higher Q for improved frequency selectivity, as well as for overcoming the limitation of a maximum Q of 0.5 of second-order *RC* circuits. Active filters are extensively used for frequencies up to a few megahertz, limited by the frequency response of op amps. For higher frequencies, passive filters are generally preferred.

Summary of Main Concepts and Results

- The response of a first-order circuit is either lowpass or highpass, the variation of the response with frequency being due to the frequency-dependent reactance of energy storage elements.

- The responses of second-order systems are more varied and can be lowpass, highpass, bandpass, bandstop, or allpass. The first three types of response show pronounced peaking, or resonance, when the damping is small.

- When a circuit is excited by a single input, the transfer function in the frequency domain, denoted as $H(j\omega)$, is the ratio of the phasor of a designated response, or output, to the phasor of the input.

- A second-order response can be obtained from two independent energy storage elements of the same type, that is, elements that cannot be combined into a single element. However, the Q in these circuits does not exceed 0.5.

- A bandpass response can result from cascading a lowpass response and a highpass response.

- Cascading of first-order filters to increase attenuation in the stopband reduces the BW.

- The 3-dB BW of bandpass and bandstop circuits equals to ω_0/Q. The higher the Q, the narrower is the BW for a given ω_0, where Q depends on the power dissipated in the circuit relative to the maximum energy stored.

- A lowpass or highpass Butterworth response has the widest 3-dB BW without attenuation ripple in the passband.

- The BW of a filter can be increased if some ripple in the response is allowed.

Learning Outcomes

- Derive the frequency responses of the first-order and second-order systems and sketch their Bode plots.

Supplementary Topics and Examples on CD

ST10.1 Straight-line Bode plots: Explains how to approximate a frequency response by means of straight-line Bode magnitude and phase plots when the corner frequencies are sufficiently far apart.

ST10.2 First-order responses related to differentiation and integration: Shows that a lowpass response approximates perfect integration at high frequencies, whereas a highpass response approximates perfect differentiation at low frequencies, and interprets this in both frequency and time domains.

ST10.3 Maximum Q of RC or RL second-order responses: Argues that the maximum Q of a second-order RC or RL circuit does not exceed 0.5.

ST10.4 Butterworth response: Derives Butterworth polynomials in normalized form and shows how to determine the order of a Butterworth low pass filter from a *brickwall specification*.

ST10.5 Chebyshev filters: Illustrates Chebyshev lowpass filters and compares them to Butterworth filters of the same specification.

ST10.6 Lowpass to narrowband bandpass transformation: Derives a narrowband band pass response from a Butterworth lowpass response and applies it to the design of three stagger-tuned circuits.

ST10.7 Magnetically coupled tuned circuits: Derives the frequency response of two magnetically coupled tuned circuits and illustrates the effects of the degree of coupling.

ST10.8 Quality factor: Identifies Q in series or parallel tuned circuits, as well as for nonideal inductors and capacitors, as the ratio of the peak instantaneous energy stored in a reactive element to the energy dissipated per radian of a cycle.

SE10.1 Transfer function of Wien bridge: Derives the transfer function of a Wien bridge in the frequency domain, which is that of a bandpass response of $Q = 1/3$.

SE10.2 High-Q approximation: Discusses the approximation often made in electronic circuits, according to which a high-Q lossy inductor is represented by an ideal inductor in parallel with a resistor.

SE10.3 Resonance in parallel LC circuit with coil resistance: Explores two criteria for resonance in a circuit consisting of a capacitor in series with a coil, represented by an inductance in series with a resistance. One criterion is that the voltage response is in phase with the applied current. The other criterion is that of maximum impedance seen by the current source.

Problems and Exercises

P10.1   Responses of First-Order Circuits

P10.1.1   Given the transfer function $H(s) = \dfrac{10^7}{s + 10^6}$, where $s = j\omega$, determine the output voltage when the input voltage is 0.1sin ωt, where: (a) $\omega = 0.3 \times 10^6$ rad/s, (b) $\omega = 10^6$ rad/s, and (c) $\omega = 3 \times 10^6$ rad/s.

P10.1.2   Determine the response $V_O(j\omega)/I_{SRC}(j\omega)$, both in magnitude and phase, the passband gain, and the corner frequency in Figure P10.1.2.

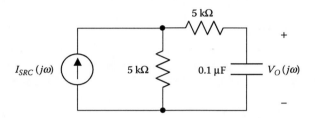

FIGURE P10.1.2

P10.1.3   Repeat Problem P10.1.2 (in Chapter 6) with $V_O(j\omega)$ taken across the 5 kΩ resistor in series with the capacitor.

P10.1.4   A first-order lowpass *RC* filter having $R = 50$ kΩ and $C = 0.1$ µF is loaded with a 200-kΩ resistance. Derive the modified transfer function and determine the effect of the load on: (a) the magnitude of the low-frequency response and (b) the corner frequency.

P10.1.5   Determine the response $V_O(j\omega)/V_{SRC}(j\omega)$, both in magnitude and phase, the passband gain, and the corner frequency in Figure P10.1.5.

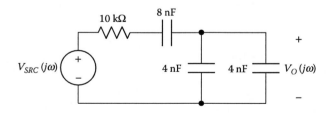

FIGURE P10.1.5

P10.1.6   Repeat Problem P10.1.5 with $V_O(j\omega)$ taken across the 10 kΩ resistor.

P10.1.7 Determine the response $V_O(j\omega)/I_{SRC}(j\omega)$, both in magnitude and phase, the passband gain, and the corner frequency in Figure P10.1.7.

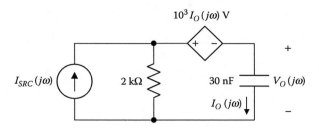

FIGURE P10.1.7

P10.1.8 Determine the response $V_O(j\omega)/V_{SRC}(j\omega)$, both in magnitude and phase, the passband gain, and the corner frequency in Figure P10.1.8.

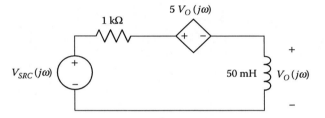

FIGURE P10.1.8

P10.1.9 Repeat Problem P10.1.8 with the inductor replaced by a 50 nF capacitor.

P10.1.10 Determine the response $V_O(j\omega)/V_{SRC}(j\omega)$, both in magnitude and phase, the passband gain, and the corner frequency in Figure P10.1.10.

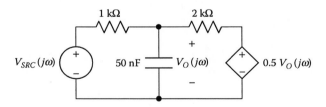

FIGURE P10.1.10

P10.1.11 Repeat Problem P10.1.10 with the capacitor replaced by a 50 mH inductor.

P10.1.12 Determine the response $V_O(j\omega)/V_{SRC}(j\omega)$, both in magnitude and phase, the passband gain, and the corner frequency in Figure P10.1.12.

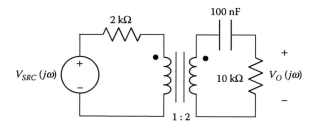

FIGURE P10.1.12

P10.1.13 Repeat Problem P10.1.12 with the capacitor replaced by a 0.1 H inductor.

P10.1.14 Express Equation 10.2.6 as $|H(j\omega)| = \dfrac{\omega/\omega_{ch}}{\sqrt{1+\left(\omega/\omega_{ch}\right)^2}}$ and plot the highpass response as the sum, on a logarithmic scale, of a lowpass response and response ω/ω_{ch} of a perfect differentiator.

P10.2   Responses of Second-Order Circuits

P10.2.1   Is it possible to have first-order bandpass and bandstop filters? Why?

P10.2.2   Determine R, L, and C in Figure P10.2.2 such that: (a) the maximum response of $V_O(s)$ is 1 V, (b) the BW is 4 krad/s, and (c) the center frequency is 100 krad/s.

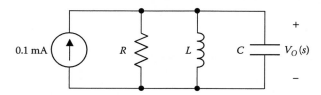

FIGURE P10.2.2

P10.2.3   Given the transfer function $H(s) = \dfrac{4s^2+1000s+100}{5s^2+500s+125}$, determine the maximum response in dB.

P10.2.4 In the circuit of Figure P10.2.4, the magnitude of the transfer function is unity at 1 Mrad/s and zero at 0.5 Mrad/s. Determine L_1 and L_2 .

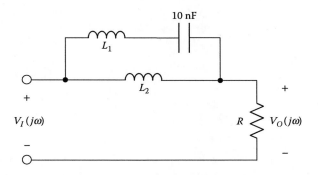

FIGURE P10.2.4

P10.2.5 Consider the phase angle of the bandpass response (Equation 10.3.5). Let $\omega' = \omega - \omega_0$ be the frequency deviation from resonance. Show that in the vicinity of resonance, the phase angle is proportional to ω' .

P10.2.6 Determine the frequency at which the response $V_O(j\omega)/V_I(j\omega)$ in Figure P10.2.6 is maximum. Verify with PSpice.

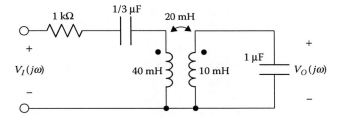

FIGURE P10.2.6

P10.2.7 Determine: (a) the transfer function of the circuit of Figure P.10.2.7, (b) the resonant frequency, (c) the Q, and (d) the bandwidth.

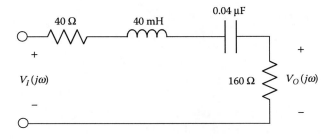

FIGURE P10.2.7

P10.2.8   In the circuit of Figure P10.2.8, determine the effect of the 5 MΩ load resistor on the Q.

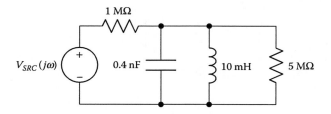

FIGURE P10.2.8

P10.2.9   Given the circuit of Figure P10.2.9, where k is a positive constant, determine Q when: (a) $k = 1$ and (b) k is very large. What is the interpretation of the value of Q when k is large?

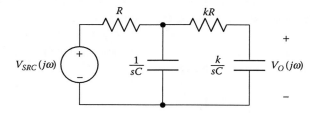

FIGURE P10.2.9

P10.2.10 In Figure P10.2.10, the response $I_C(j\omega)/I_{SRC}(j\omega)$ is to remain −3 dB at the same frequency, whether the switch is opened or closed. (a) Determine R and the corner frequency, assuming $L = 20$ mH, and (b) calculate the response in dB at one tenth the corner frequency for the two cases when the switch is opened or closed.

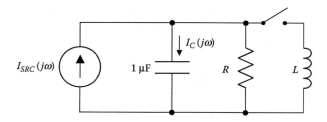

FIGURE P10.2.10

P10.2.11 For the allpass response in Table 10.5.1, determine: (a) the magnitude of the response at any frequency ω and (b) the phase shift as ω varies between zero and infinity.

P10.2.12 Show that for the lowpass response having $\xi = 1/\sqrt{2}$: (a) the peak value is given by $20\log_{10}|H(j\omega_p)| = -10\log_{10}[4\xi^2(1-\xi^2)]$ at a frequency $\omega_p = \omega_0\sqrt{1-2\xi^2}$ and (b) the response crosses the 0-dB axis ($y = 1$) at a frequency $\omega_0\sqrt{2(1-2\xi^2)} = \sqrt{2}\omega_p$.

P10.2.13 Show that for the lowpass response at the frequency ω_0 , $20\log_{10}|H(j\omega_0)| = -20\log_{10}2\xi$. Note that if $\xi = 1$, $-20\log_{10}2\xi = -6$ dB.

P10.2.14 Show that for the lowpass response, the half-power frequency is given by $\omega_{1/2} = \omega_0\sqrt{\sqrt{(2\xi^2-1)^2+1}-((2\xi^2-1))}$. Verify this result for the cases of critical damping ($\xi = 1$) and Butterworth response.

P10.2.15 Show that if the lowpass response is peaked, the 3-dB frequencies, with respect to the peak, are given by: $\omega_{c1} = \omega_0\sqrt{(1-2\xi^2)-2\xi\sqrt{1-\xi^2}}$ and $\omega_{c2} = \omega_0\sqrt{(1-2\xi^2)+2\xi\sqrt{1-\xi^2}}$. Verify that if ξ is small, the BW is approximately $2\xi\omega_0$.

P10.2.16 Consider a plot of the phase angle of Equation 10.5.6 as a function of $\log_{10}u$. Show that the slope at $\omega = \omega_0$ is $-\dfrac{2.3}{\xi}$ rad/decade. Deduce that a line of this slope intersects the 0° line at $u_1 = 10^{-\frac{\pi}{4.6}\xi} \equiv (4.81)^{-\xi}$ and intersects the −180° line at $u_2 = 10^{\frac{\pi}{4.6}\xi} \equiv (4.81)^{\xi}$. Note that such a line may be used to approximate the Bode phase plot in the vicinity of $-90°$.

P10.2.17 The response of a second-order lowpass filter is −5 dB at the corner frequency ω_0 at which the high-frequency asymptote intersects the 0 dB line. Determine Q.

P10.2.18 Consider the transfer function $H(s) = \dfrac{s^2}{s^2+10s+10^6}$. (a) How do you classify this response? (b) Determine the magnitude of the peak response in dB. (c) Compute the half-power frequencies with respect to the peak. How do these values compare with those for a bandpass filter?

P10.3   Butterworth Response and Scaling

Obtain the Bode plots for Problem P6.3.1 to Problem P6.3.5 (in Chapter 6) using PSpice.

P10.3.1 Given a series RLC circuit with $L = 1$ H and $C = 1$ F, select R so as to have a second-order, lowpass, normalized Butterworth response when the output is taken across C. Scale the parameters so as to have $\omega_0 = 10$ krad/s and $C = 100$ nF.

P10.3.2   Given a series RLC circuit with $L = 1$ H and $C = 1$ F, select R so as to have a second-order, highpass, normalized Butterworth response when the output is taken across L. Scale the parameters so as to have $\omega_0 = 10$ krad/s and $C = 100$ nF.

P10.3.3   Derive the transfer function of the circuit shown in Figure P10.3.3, and determine L and C so as to have a second-order, lowpass, normalized, Butterworth response. Scale the parameters so as to have $R_L = 10$ kΩ and the 3-dB frequency $f_0 = 10$ kHz. Draw the Bode magnitude plot.

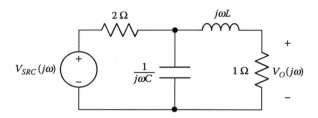

FIGURE P10.3.3

P10.3.4   Derive the transfer function of the circuit shown in Figure P10.3.3 with the inductor and capacitor interchanged. Determine L and C so as to have a second-order, highpass, normalized Butterworwth response. Scale the parameters so as to have $R_L = 10$ kΩ and the 3-dB frequency $\omega_0 = 10$ kHz.

P10.3.5   Derive the transfer function of the circuit shown in Figure P10.3.5 and determine L and C so as to have a third-order, highpass, normalized Butterworth response. Scale the parameters so as to have resistance values of 2 kΩ and a cutoff frequency $f_0 = 10$ kHz.

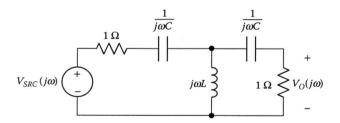

FIGURE P10.3.5

P10.4   Miscellaneous

P10.4.1   Show that a capacitance C_p in parallel with a resistance R_p is equivalent to a capacitance $C_p\left(1+\dfrac{1}{Q_C^2}\right)$ in series with a resistance $\dfrac{R_p}{1+Q_C^2}$, where $Q_C = \omega C_p R_p$.

P10.4.2   Show that the Q of a capacitor having effectively a resistance R_p in parallel with it is $\omega C R_p$.

P10.4.3 Given the circuit of Figure P10.4.3, (a) determine the transfer function $V_0(j\omega)/V_{SRC}(j\omega)$, (b) show that in the case of high Q of the two branches, the response becomes of the same form as the prototypical band pass response, and (c) calculate the resonant frequency and the overall Q of the circuit if $R_{src} = 1$ kΩ, $L = 10$ mH, $R_L = 4$ Ω, $C = 1$ μF, and $R_C = 2$ Ω. Obtain the Bode plots using PSpice and compare ω_0 and Q.

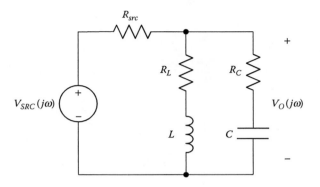

FIGURE P10.4.3

P10.4.4 Given the straight-line, Bode plot approximation of Figure P10.4.4, determine: (a) the X dB level and (b) the transfer function approximated by this Bode plot.

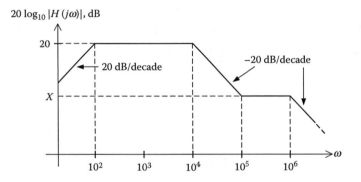

FIGURE P10.4.4

P10.4.5 Verify Equation SE10.3.8.

11

Duality and Energy-Storage Elements

Overview

The circuits considered so far were in a steady state, in response to dc or sinusoidal excitations. We are now ready to consider the transient behavior of electric circuits in response to different types of input. An added complication introduced in these cases is the initial energy stored in capacitors and inductors, because the initial voltages across capacitors and the initial currents in inductors affect the response to a given input. These initial conditions are ignored in steady-state analysis because their effects would have died out when a steady-state is reached.

Before embarking on the study of electric circuits under transient conditions in the following chapters, we will consider in this chapter some basic concepts that govern the behavior of energy storage elements. Of fundamental importance in analyzing circuits containing these elements is the behavior of capacitors and inductors in response to sudden changes in voltages, currents, or circuit configuration. In this regard, the concept of duality in electric circuits provides a helpful and instructive analogy between certain quantities and relations, which can be usefully applied to capacitive and inductive circuits.

When capacitors or inductors are connected in series or in parallel, with or without initial energy storage, an equivalent capacitor or inductor is defined, whose properties are related in a well-defined manner to those of the individual elements. These equivalent circuit elements play a basic role in analyzing some types of circuits involving more than one capacitor or inductor. Charge or flux linkage may be conserved in these cases under certain conditions.

Learning Objectives

- To be familiar with:
  - The nature and significance of the principle of duality in electric circuits
- To understand:
  - Principles underlying the response of capacitors and inductors to sudden changes in voltages, currents, or circuit configuration
  - How capacitances and inductances combine in series or in parallel
  - How charge is distributed in capacitors connected in series or in parallel, including charge due to initial energy storage

- How flux linkage is distributed in inductors connected in series or in parallel, including flux linkage due to initial energy storage
- Conservation of charge and of flux linkage in electric circuits

11.1 Duality

Concept: *Two v–i relations are duals if one relation reduces to the other when the following are interchanged: (1) v and i, (2) R and G, and (3) L and C.*

The quantities involved in these interchanges are *dual quantities*. Thus, v and i are dual circuit variables, R and G are dual circuit parameters, as are L and C. In making the interchanges, all known quantities are replaced by duals having the *same numerical value*.

To illustrate this concept, consider a 10 Ω resistor carrying a current of 2 A. According to Ohm's law: v (V) = (10 Ω) × (2 A). The dual relation is: i (A) = (10 S) × (2 V), and represents a conductance of 10 S across which 2 V are applied. As another example, consider a 2 H inductor across which 10 V are applied. The v–i relation is: 10 (V) = (2 H) × $\dfrac{di}{dt}$ (A/s). The dual relation is: 10 (A) = (2 F) × $\dfrac{dv}{dt}$ (V/s), and represents a 2 F capacitor to which a 10 A current source is applied.

Concept: *Two circuits are duals if the node-voltage equations of one circuit are the dual relations of the mesh-current equations of the other circuit.*

To illustrate this concept, consider the series RLC circuit of Figure 11.1.1a with numerical values assigned. The mesh-current equation, or Kirchhoff's voltage law (KVL), in phasor notation gives:

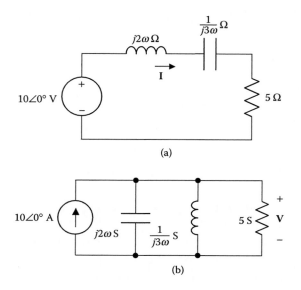

(a)

(b)

FIGURE 11.1.1
Dual circuits.

$$5\mathbf{I} + j2\omega\mathbf{I} + \frac{1}{j3\omega}\mathbf{I} = 10 \tag{11.1.1}$$

To derive the dual circuit, we have to determine the dual of the series connection. Because in a series connection the current is common to all the elements, then in the dual connection the voltage is common to all the elements, that is, the connection is parallel. The dual circuit will therefore have the dual elements of those of the series circuit connected in parallel (Figure 11.1.1b). The node-voltage equation, or Kirchhoff's current law (KCL), gives:

$$5\mathbf{V} + j2\omega\mathbf{V} + \frac{1}{j3\omega}\mathbf{V} = 10 \tag{11.1.2}$$

which is the dual relation of Equation 11.1.1.

This example illustrates one advantage of duality, namely, that when a circuit is analyzed, its dual circuit is automatically analyzed at the same time. Making use of this can save considerable time and effort. An added advantage that will be demonstrated shortly is that duality forms a basis for a common understanding of the behavior of capacitive and inductive circuits and aids in remembering the relations for these circuits.

Table 11.1.1 lists the duality relations that apply to the circuit quantities, connections, and concepts discussed in previous chapters. A general procedure for deriving the dual of a planar circuit is discussed in Section ST11.1. For our purposes, duals of circuits follow from the duality between circuit connections listed in Table 11.1.1.

Of immediate interest and worthy of emphasis is the close correspondence between capacitors and inductors exemplified by duality. For a capacitor, $i = dq/dt$ irrespective of v and C; q and v are related by the capacitance: $q = Cv$. For an inductor, $v = d\lambda/dt$, irrespective of i and L; λ and i are related by the inductance: $\lambda = Li$.

Example 11.1.1: Dual of Voltage Source

It is required to derive the dual of the voltage source of Example 2.6.1 (Chapter 2), reproduced in Figure 11.1.2a.

SOLUTION
From KVL,

$$0.5I_L + 5.5I_L = 12 \tag{11.1.3}$$

The dual circuit will consist of the dual elements connected in parallel, as shown in Figure 11.1.2b. From KCL,

$$0.5V_L + 5.5V_L = 12 \tag{11.1.4}$$

which is the dual relation of Equation 11.1.3.

Note that the dual circuit is of the same form as the equivalent current source (Figure 2.6.4b, Chapter 2), but the values of the circuit elements are different. The V_L–I_L relation for the voltage source is:

$$V_L = 12 - 0.5I_L \tag{11.1.5}$$

with $I_L = 2$ A and $V_L = 11$ V.

TABLE 11.1.1

Dual Circuit Entities

Dual Circuit Variables and Quantities

v	i
q	λ

Dual Circuit Parameters

R	G
C	L

Dual Relations

Resistor: $v = Ri$ Resistor: $i = Gv$

Capacitor: $i = \dfrac{dq}{dt}$ Inductor: $v = \dfrac{d\lambda}{dt}$

$\quad\quad\quad\quad q = Cv$ $\quad\quad\quad\quad \lambda = Li$

$\quad\quad\quad\quad i = C\dfrac{dv}{dt}$ $\quad\quad\quad\quad v = L\dfrac{di}{dt}$

$\quad\quad\quad V(s) = \dfrac{1}{sC}I(s) + \dfrac{v(0^-)}{s}$ $\quad\quad\quad I(s) = \dfrac{1}{sL}V(s) + \dfrac{i(0^-)}{s}$

KVL: $\displaystyle\sum v = 0$ KCL: $\displaystyle\sum i = 0$

algebraically around a loop   algebraically at a node

Impedance and Admittance

Admittance: $j\omega C$ Impedance: $j\omega L$
Susceptance ωC Reactance: ωL

Dual Sources

Ideal voltage source: Ideal current source:
 v_{SRC} is specified for all i i_{SRC} is specified for all v
Voltage-controlled voltage source: Current-controlled current source:
 $v_{SRC} = \alpha v_\phi$ $i_{SRC} = \alpha i_\phi$
Current-controlled voltage source: Voltage-controlled current source:
 $v_{SRC} = \beta i_\phi$ $i_{SRC} = \beta v_\phi$

Dual Circuit Connections

Open circuit or open switch: Short circuit or closed switch:
 $i = 0$ for all v $v = 0$ for all i
Series connection: Parallel connection:
 Same current flows in circuit elements   Same voltage appears across circuit elements
Mesh: Node:
 Voltages add algebraically   Currents add algebraically

Circuit Duals

Y or T Δ or π
Ideal transformer of turns ratio 1: a Ideal transformer of turns ratio a:1

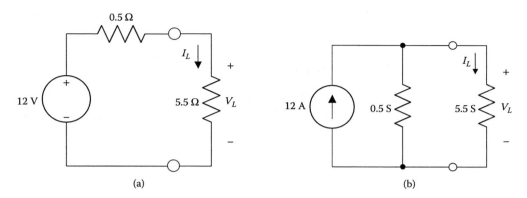

FIGURE 11.1.2
Figure for Example 11.1.1.

By definition, the equivalent current source has the same V_L–I_L relation as Equation 11.1.5 and the same values of I_L and V_L. In contrast, the dual circuit has the dual V_L–I_L relation: $I_L = 12 - 0.5V_L$, and the values of I_L and V_L are interchanged, so that $I_L = 11$ A and $V_L = 2$ V. The power delivered to the load is the same in the three cases, however, because the $V_L I_L$ product is the same.

11.2 Instantaneous Changes

Current Source Applied to a Capacitor

In Figure 11.2.1a, the current source applies a step function of time such that $i_{SRC} = 0$, $t < 0$, and $i_{SRC} = I_{SRC}$, $t > 0$; i_{SRC} is discontinuous at the origin (Figure 11.2.1b) and is therefore undefined at $t = 0$. The variation of v with time is obtained from the v–i relation for a capacitor: $v(t) = \dfrac{1}{C}\displaystyle\int_0^t i_{SRC}dt + v(0)$. Assuming that the capacitor is initially uncharged, $v(0) = 0$, so that:

$$v(t) = \frac{I_{SRC}}{C}t \ , \ t \geq 0 \tag{11.2.1}$$

The capacitor voltage increases linearly with time (a ramp function) beginning at $t = 0$ (Figure 11.2.1c). At $t = 0$, v is still zero, despite the stepwise jump in current. The reason is simply that it takes a finite time for I_{SRC} to build charge on the capacitor and hence for the capacitor voltage to change. This is analogous to suddenly turning on a water supply to an empty vessel. At the instant water is turned on, it has not yet accumulated in the vessel, so the volume of water in the vessel (analogous to charge on the capacitor) is zero at that instant.

The question that may be asked is: what would it then take to suddenly change the capacitor voltage? To answer this question, let i_{SRC} in Figure 11.2.1a be a pulse of current of amplitude I_{SRC} that starts at $t = 0$ and ends at $t = a$ (Figure 11.2.2a). $v = 0$ for $t < 0$,

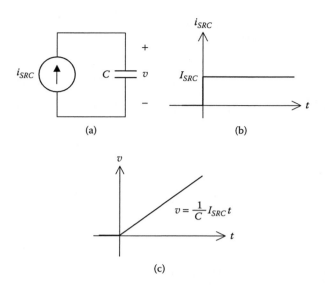

FIGURE 11.2.1
Current step applied to a capacitor.

assuming that the capacitor was initially uncharged. During the time interval $0 \le t \le a$, v is given by Equation 11.2.1. The voltage increases linearly during this interval, as in the case of the current step (Figure 11.2.2b). At $t = a$, $v(a) = \dfrac{aI_{SRC}}{C}$. The charge transferred by the current pulse is aI_{SRC}, and the voltage at the end of the pulse is therefore $1/C$ times this charge. Because the capacitor is assumed ideal, the charge is maintained for $t \ge a$, and v remains at its value $v(a)$. The ideal current source does not allow the capacitor to discharge because $i_{SRC} = 0$ for $t > a$. In other words, the ideal current source behaves as an open circuit for $t > a$.

Let the amplitude of the pulse be increased to I'_{SRC} and its duration reduced to a' (Figure 11.2.2a), while keeping the area of the pulse the same, that is, $a'I'_{SRC} = aI_{SRC}$. Now $\dfrac{dv(t)}{dt} = \dfrac{I'_{SRC}}{C} > \dfrac{I_{SRC}}{C}$. The change in v is steeper, whereas the final value of v remains the same because the charge transferred is the same. Thus, from Equation 11.2.2, the final

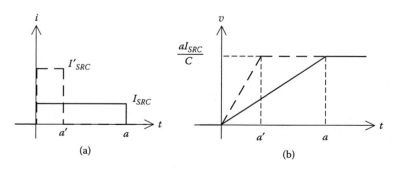

FIGURE 11.2.2
Current pulse applied to a capacitor.

value of $v(t)$ is $\dfrac{a'I'_{SRC}}{C} = \dfrac{aI_{SRC}}{C}$. As the pulse width is reduced, v approaches a step function at the origin. The current pulse now becomes an **impulse function**, or a **Dirac delta function**. It is a pulse of infinitesimal duration and infinite amplitude but having a finite area, or strength aI_{SRC} that is current × time, or charge. The impulse function will be considered in more detail in the following section.

The voltage response to a sudden change in capacitor current may be argued in general terms from the v–i relation: $i = C\dfrac{dv}{dt}$. If i is changing between t_1 and t_2, then, integrating both sides of the equation from t_1 and t_2,

$$v(t_2) - v(t_1) = \frac{1}{C}\int_{t_1}^{t_2} i\, dt \qquad (11.2.2)$$

As $t_2 \rightarrow t_1$ the integral approaches zero as long as i remains finite during the interval t_1 to t_2, which implies that $v(t_2) = v(t_1)$. In other words, the capacitor voltage does not change instantaneously because of a finite change in capacitor current. Only if i goes infinite as $t_2 \rightarrow t_1$ will the integral have a finite value, so that $v(t_2) \neq v(t_1)$.

Voltage Source Applied to an Inductor

In accordance with duality, the case of a voltage source applied to an inductor is exactly analogous to that of the current source applied to a capacitor. As will be shown, exactly the same expressions apply with v and i interchanged, q replaced by λ, and C replaced by L.

Consider a step of voltage applied to an inductor (Figure 11.2.3a and Figure 11.2.3b). The variation of i with time is obtained from the v–i relation for an inductor $i(t) = \dfrac{1}{L}\int_0^t v_{SRC}dt + i(0)$. Assuming that the inductor current is initially zero, $i(0) = 0$, so that

$$i(t) = \frac{V_{SRC}}{L}t \ , \ t \geq 0 \qquad (11.2.3)$$

The inductor current increases linearly with time, starting with $i = 0$ at $t = 0$.

When a voltage pulse of amplitude V_{SRC} and duration a is applied to the inductor (Figure 11.2.4a), the current increases linearly during the interval $0 \leq t \leq a$, to a value $\dfrac{aV_{SRC}}{L}$ (Figure 11.2.4b). The voltage pulse increases the flux linkage in the inductor by aV_{SRC}. The increase in current is the increase in flux linkage divided by L. This current is maintained for $t > a$ because the voltage source presents a short circuit that keeps the current flowing through the ideal inductor.

If the pulse amplitude is increased to V'_{SRC} and its duration reduced to a' so that the area of the pulse remains the same, $aV_{SRC} = a'V'_{SRC}$, the inductor current increases more rapidly, but the final value is the same. As the pulse width is reduced, i approaches a step function at the origin. The voltage pulse now becomes an impulse of infinite amplitude, infinitesimal duration, but finite area equal to aV_{SRC}. This is summarized in the following concept.

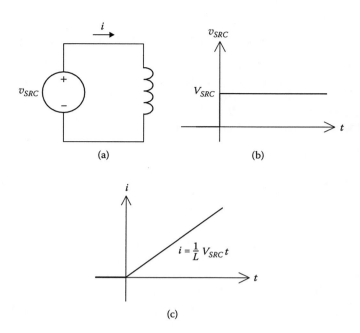

FIGURE 11.2.3
Voltage step applied to an inductor.

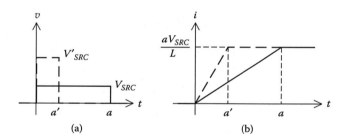

FIGURE 11.2.4
Voltage pulse applied to an inductor.

> **Concept:** *The voltage across a capacitor cannot be changed instantaneously except by a current impulse, in which case the change in the charge of the capacitor is that represented by the area of the impulse. Similarly, the current through an inductor cannot be changed instantaneously except by a voltage impulse, in which case the change in the flux linkage of the inductor is that represented by the area of the impulse.*

EXERCISE 11.2.1

A square wave of 0 to 2 V and 1 s period is applied to a 5 μF capacitor that is initially uncharged. Determine the capacitor current.

Answer: A series of positive impulses of strength 10 A s at $t = 0$ s, 1 s, 2 s, 3 s, ... and a series of negative impulses of the same strength at $t = 0.5$ s, 1.5 s, 2.5 s, 3.5 s, ...

A square wave of 0 to 2 A and 1 s period is applied to a 5 µH inductor that has zero initial current. Determine the inductor voltage.

Answer: A series of positive impulses of strength 10 V s at $t = 0$ s, 1 s, 2 s, 3 s, ... and a series of negative impulses of the same strength at $t = 0.5$ s, 1.5 s, 2.5 s, 3.5 s, ...

11.3 The Impulse Function

The impulse function is of fundamental importance in circuit theory, as will become apparent from future discussions. Mathematically, an impulse function at the origin of strength K is denoted as $K\delta(t)$ and defined as:

$$\int_{-\infty}^{+\infty} K\delta(t)dt = K \text{ , with } \delta(t) = 0 \quad \text{for } t \neq 0 \qquad (11.3.1)$$

where $\delta(t)$ is the unit impulse at the origin. $K\delta(t)$ is represented as shown in Figure 11.3.1. The relationship between an impulse function and a step function is illustrated in Figure 11.3.2. The integral of the current pulse of Figure 11.3.2a is $\frac{K}{\Delta}t$, $0 \leq t \leq \Delta$, and is K for $t \geq \Delta$, as shown in Figure 11.3.2b. Conversely, the derivative of the function shown in Figure 11.3.2b is $\frac{K}{\Delta}$, $0 \leq t \leq \Delta$, and is zero $t > \Delta$. As $\Delta \to 0$, $f(t)$ and $f^{(1)}(t)$ become a step function and an impulse function, respectively. We may therefore write:

$$\int_{-\infty}^{+\infty} K\delta(t)dt = K\int_{0^-}^{0^+} \delta(t)dt = Ku(t) \quad \text{and} \quad \frac{d(Ku(t))}{dt} = K\delta(t) \qquad (11.3.2)$$

where $u(t)$ is the unit step at the origin. The second integral is from $t = 0^-$ to $t = 0^+$, which ensures that the impulse is included. $t = 0$ is ambiguous in the case of an impulse at the

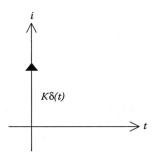

FIGURE 11.3.1
Impulse function.

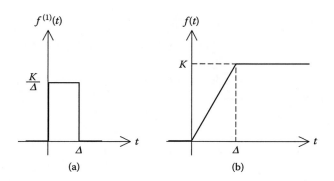

FIGURE 11.3.2
Relation between step and impulse functions.

TABLE 11.3.1

Some Useful Properties of the Impulse Function

Evenness:	$\delta(t) = \delta(-t)$
Time scaling:	$\displaystyle\int_{-\infty}^{\infty} \delta(at)dt = \frac{1}{|a|}$
Sampling of continuous function:	$f(t)\delta(t-a) = f(a)\delta(t-a)$
	$\qquad\qquad\qquad = 0$ if $f(a) = 0$
Sampling of derivative of continuous function:	$\displaystyle\int_{-\infty}^{\infty} f(t)\delta^{(1)}(t-a)dt = -f^{(1)}(a)$

origin. At $t < 0^-$, the impulse has not yet occurred, and at $t > 0^+$ the impulse is over, although $t = 0^-$ and $t = 0^+$ are infinitesimally separated. Similarly, $u(t) = 0$ at $t = 0^-$ and $u(t) = 1$ at $t = 0^+$. The impulse function has some peculiar properties, some of which are encountered in problems at the end of this chapter and in future chapters. In fact, the impulse function is not a function in the ordinary sense but is a *singularity function*. Singularity functions are used to extend the concept of ordinary functions to what are known as *generalized functions*.

Table 11.3.1 summarizes some useful properties of the impulse function. The sampling property is discussed in Example 11.3.1. The remaining properties are left to exercises and problems at the end of the chapter.

EXERCISE 11.3.1
Consider that the pulse in Figure 11.3.2a extends from $t = -\Delta/2$ to $t = +\Delta/2$. Show that in the limit, $f(t) = u(t)$ and $f^{(1)}(t) = \delta(t)$, as mentioned earlier.

Example 11.3.1: Sampling of Continuous Function by an Impulse
A useful relation for the impulse function is:

$$f(t)\delta(t-a) = f(a)\delta(t-a) \qquad\qquad (11.3.3)$$

where $\delta(t-a)$ is a unit impulse function that occurs at $t = a$ and $f(t)$ is assumed to be continuous at $t = a$. In words, Equation 11.3.3 states that multiplying a function by a unit impulse that occurs at $t = a$ gives an impulse at $t = a$ of strength equal to the value of the function at $t = a$. In effect, the impulse samples the function at $t = a$.

PROOF If the LHS of Equation 11.3.3 is integrated, and the integral is split over three intervals — before, during, and after the occurrence of the impulse — we obtain:

$$\int_{-\infty}^{+\infty} f(t)\delta(t-a)dt = \int_{-\infty}^{a^-} f(t)\delta(t-a)dt + \int_{a^-}^{a^+} f(t)\delta(t-a)dt + \int_{a^+}^{\infty} f(t)\delta(t-a)dt \qquad (11.3.4)$$

The first and third integrals on the RHS vanish because the impulse function is zero in the interval of integration. The interval of integration of the second integral is from $t = a^-$ to $t = a^+$. As long as $f(t)$ is continuous at $t = a$, $f(t)$ over this interval equals $f(a)$. Hence,

$$\int_{-\infty}^{+\infty} f(t)\delta(t-a)dt = f(a)\int_{a^-}^{a^+} \delta(t-a)dt = f(a)u(t-a) \qquad (11.3.5)$$

$$= 0,\ t < a$$

$$= f(a),\ t > a$$

Differentiating both sides of Equation 11.3.5 gives Equation 11.3.3. If $f(a) = 0$, the RHS of Equation 11.3.5 is zero and $f(a)\delta(t-a) = 0$.

EXERCISE 11.3.2

By changing variables in the integral definition of the impulse (Equation 11.3.1) show that: (a) $\delta(-t) = \delta(t)$; (b) $\int_{-\infty}^{\infty} \delta(at)dt = \dfrac{1}{|a|}$.

11.4 Series and Parallel Connections with Zero Initial Energy

We will derive from first principles the cases of a series connection of capacitors and a series connection of inductors, and deduce the other two cases from duality, leaving their derivation from first principles to Section ST11.2. The relations for the four cases, with and without energy storage, are summarized in Table 11.4.1.

Capacitors in Series

Consider a series connection of n capacitors through which a current i_s flows (Figure 11.4.1a). From KVL,

$$v_s = v_1 + v_2 + \ldots + v_n \qquad (11.4.1)$$

TABLE 11.4.1

Energy Storage Elements in Series and in Parallel

	Parallel Connection	**Series Connection**
Capacitors	$\{C_{eqp} = C_1 + C_2 + \ldots + C_n\}$ $\{q_{eqp} = q_1 + q_2 + \ldots + q_n\}$	$\left\{\dfrac{1}{C_{eqs}} = \dfrac{1}{C_1} + \dfrac{1}{C_2} + \ldots + \dfrac{1}{C_n}\right\}$ $\left\{q_m = \displaystyle\int_0^t i_s dt + q_m(0)\right\}$ $q_1 = q_2 = \ldots = q_n$, with no initial energy storage
Inductors	$\left\{\dfrac{1}{L_{eqs}} = \dfrac{1}{L_1} + \dfrac{1}{L_2} + \ldots + \dfrac{1}{L_n}\right\}$ $\left\{\lambda_m = \displaystyle\int_0^t v_p dt + \lambda_m(0)\right\}$ $\lambda_1 = \lambda_2 = \ldots = \lambda_n$, with no initial energy storage	$\{L_{eqs} = L_1 + L_2 + \ldots + L_n\}$ $\{\lambda_{eqs} = \lambda_1 + \lambda_2 + \ldots + \lambda_n\}$

Note: Relations in curly brackets apply irrespective of initial stored energy.

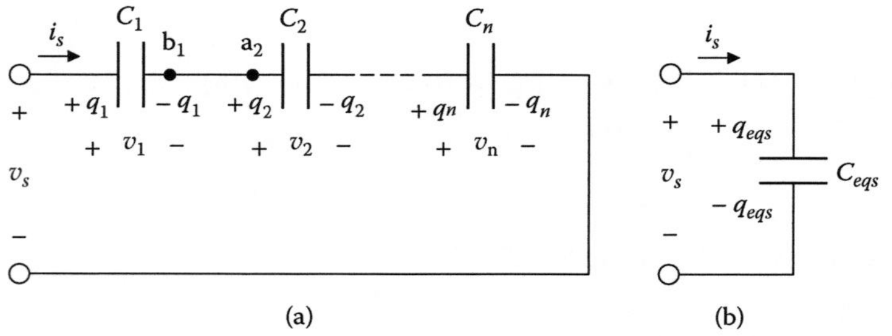

(a)   (b)

FIGURE 11.4.1
Capacitors in series.

The voltage across any capacitor is of the form $v = \dfrac{1}{C}\displaystyle\int_0^t i_s dt + v(0)$. Assuming zero initial energy storage, $v(0) = 0$. Substituting for each current in Equation 11.4.1 and factoring out the current integral,

$$v_s = \left(\frac{1}{C_1} + \frac{1}{C_2} + \ldots + \frac{1}{C_n}\right)\int_0^t i_s dt \qquad (11.4.2)$$

The bracketed sum defines an equivalent series capacitor that would have the same v_s and $\displaystyle\int_0^t i_s dt$ (Figure 11.4.2b). Thus,

$$\frac{1}{C_{eqs}} = \frac{1}{C_1} + \frac{1}{C_2} + \ldots + \frac{1}{C_n} \qquad (11.4.3)$$

The reciprocals of capacitances in series add, just like conductances. This is to be expected because the corresponding v–i relations are $v = \frac{1}{G}i$ and $v = \frac{1}{C}\int i\,dt$. In both cases, $1/G$ and $1/C$ are multipliers of the common circuit variable i or a function of i. Fundamentally, this results in Equation 11.4.3 and Equation 2.3.8 (Chapter 2) being of the same form.

It follows from Equation 11.4.1 that voltage divides between capacitors in series as in the case of conductances. The voltage across C_m, where $1 \le m \le n$, is $v_m = \frac{1}{C_m}\int_0^t i_s\,dt$. Dividing by Equation 11.4.2,

$$\frac{v_m}{v_s} = \frac{1/C_m}{1/C_1 + 1/C_2 + ... + 1/C_n} , \quad m = 1, 2, ..., n \tag{11.4.4}$$

Applying Equation 11.4.4 to any two voltages, v_k and v_r, and dividing,

$$\frac{v_k}{v_r} = \frac{C_r}{C_k}, \quad k, r = 1, 2, ..., n \tag{11.4.5}$$

The charge on any capacitor C_m is $q_m = \int_0^t i_s\,dt + q_m(0)$. Because the initial charge is zero and i_s is common to all the capacitors, it follows that the charge is the same on all the capacitors and on the equivalent series capacitor:

$$q_1 = q_2 = ... = q_n = q_{eqs} \tag{11.4.6}$$

This equality of charge may be physically justified in terms of the perfect insulation in an ideal capacitor. Consider any two adjacent capacitors such as C_1 and C_2 in Figure 11.4.1a. Because plates b_1 and a_2 of these capacitors are assumed not to have any initial charge, and because they have no external connections, there can be no net charge on them after the passage of current; charge can only be displaced from one plate to the other. If the charge on b_1 is $-q_1$, that on a_2 must be $+q_2 = +q_1$, and C_1 and C_2 must therefore have the same charge.

Inductors in Parallel

This is shown in Figure 11.4.2 and is the dual of capacitors in series. It follows that the duals of Equation 11.4.1 to Equation 11.4.6 are, respectively,

$$i_p = i_1 + i_2 + ... + i_n \tag{11.4.7}$$

$$i_p = \left(\frac{1}{L_1} + \frac{1}{L_2} + ... + \frac{1}{L_n} \right) \int_0^t v_p\,dt \tag{11.4.8}$$

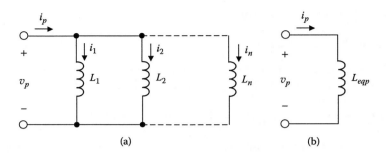

FIGURE 11.4.2
Inductors in parallel.

$$\frac{1}{L_{eqp}} = \frac{1}{L_1} + \frac{1}{L_2} + ... + \frac{1}{L_n}$$

(11.4.9)

$$\frac{i_m}{i_p} = \frac{1/L_m}{1/L_1 + 1/L_2 + ... + 1/L_n} \, , \, m = 1, 2, ..., n$$

(11.4.10)

$$\frac{i_k}{i_r} = \frac{L_r}{L_k} \, , \quad k,r = 1, 2, ..., n$$

(11.4.11)

$$\lambda_1 = \lambda_2 = ... = \lambda_n = \lambda_{eqp}$$

(11.4.12)

Equation 11.4.12 follows from Faraday's law in the form of $\lambda = \int_0^t v dt$, with no initial current in the inductors. Since v is the same for all inductors, λ is the same.

EXERCISE 11.4.1
The dual of the conservation of charge at any node between capacitors connected in series is the conservation of flux linkage around any mesh or loop involving two inductors in parallel. Use Equation 11.4.11 to justify this (see Section 11.6).

Inductors in Series

Consider a series connection of n inductors through which a current i_s flows (Figure 11.4.3a). From KVL,

$$v_s = v_1 + v_2 + ... + v_n$$

(11.4.13)

The voltage across any inductor is of the form: $v = L\frac{di_s}{dt}$. Substituting in Equation 11.4.13 for the voltage across each inductor and factoring out the $\frac{di_s}{dt}$ term,

$$v_s = \left(L_1 + L_2 + ... + L_n\right)\frac{di_s}{dt}$$

(11.4.14)

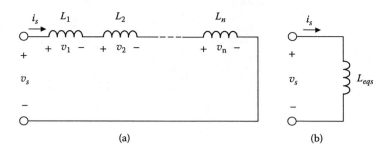

FIGURE 11.4.3
Inductors in series.

The bracketed sum defines an equivalent series inductor that would have the same v_s and $\dfrac{di_s}{dt}$ (Figure 11.4.3b). Thus,

$$L_{eqs} = L_1 + L_2 + \ldots + L_n \tag{11.4.15}$$

Inductances in series add, just like resistances. This is to be expected because the corresponding v–i relations are $v = Ri$ and $v = L\dfrac{di}{dt}$. In both cases, R and L are multipliers of the common circuit variable i, or a function of i, so that, fundamentally, Equation 2.3.7 (Chapter 2) and Equation 11.4.15 are of the same form.

It follows from Equation 11.4.13 that voltage divides between the inductors in series, as in the case of resistors. The voltage across L_m, where $1 \le m \le n$, is $v_m = L_m \dfrac{di_s}{dt}$. Dividing by Equation 11.4.14,

$$\frac{v_m}{v_s} = \frac{L_m}{L_1 + L_2 + \ldots + L_n} \ , m = 1, 2, \ldots, n \tag{11.4.16}$$

Applying Equation 11.4.16 to any two voltages v_k and v_r and dividing,

$$\frac{v_k}{v_r} = \frac{L_k}{L_r} \ , k, r = 1, 2, \ldots, n \tag{11.4.17}$$

If we multiply both sides of Equation 11.4.15 by i and substitute the relation $\lambda = Li$ for each element, the resulting equation is

$$\lambda_{eqs} = \lambda_1 + \lambda_2 + \ldots + \lambda_n \tag{11.4.18}$$

The flux linkage in the equivalent series inductor is the sum of the flux linkages in the individual inductors.

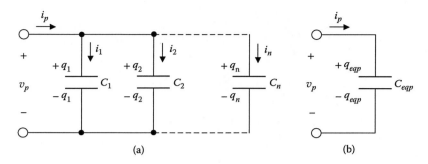

FIGURE 11.4.4
Capacitors in parallel.

Capacitors in Parallel

This is shown in Figure 11.4.4 and is the dual of inductors in series. It follows that the duals of Equation 11.4.13 to Equation 11.4.18 are, respectively,

$$i_p = i_1 + i_2 + \dots + i_n \tag{11.4.19}$$

$$i_p = \left(C_1 + C_2 + \dots + C_n \right) \frac{dv_p}{dt} \tag{11.4.20}$$

$$C_{eqp} = C_1 + C_2 + \dots + C_n \tag{11.4.21}$$

$$\frac{i_m}{i_p} = \frac{C_m}{C_1 + C_2 + \dots + C_n} , \, m = 1, 2, \dots, n \tag{11.4.22}$$

$$\frac{i_k}{i_r} = \frac{C_k}{C_r} , \quad k, r = 1, 2, \dots, n \tag{11.4.23}$$

$$q_{eqp} = q_1 + q_2 + \dots + q_n \tag{11.4.24}$$

Application Window 11.4.1: Generation of High dc Voltages

High dc voltages, such as those used for testing the breakdown of insulating materials, may be generated by charging a bank of capacitors in parallel at a relatively low voltage, then connecting them in series to obtain a much higher voltage. For example, if 20 capacitors of 1 µF each are charged to 5 kV in parallel, and then connected in series, the resulting voltage is 20×5 kV = 100 kV. The energy delivered is that initially stored in the 20 capacitors. This energy is: $20 \times \frac{1}{2} \times (10^{-6})(5 \times 10^3)^2 = 250$ J; $C_{eqs} = \frac{1 \, \mu F}{20} = 0.05$ µF. The energy stored in this capacitor is $\frac{1}{2} \times (0.05 \times 10^{-6})(100 \times 10^3)^2 = 250$ J, as it should be.

The electric eel has been using this principle for millions of years. In order to stun its prey or ward off a predator, the eel delivers between its head and tail an electric discharge of more than 600 V at a peak current of about 2 A. The average power of the discharge is about 200 W, enough to disable a frog at a distance of 1 m. The discharge consists of a burst of two to eight 2 ms pulses separated by about 10 ms. The electric organ that produces the discharge is composed of some 300,000 specialized units, or **electroplaxes**, that are charged in parallel at a fraction of a volt. They are then discharged in series–parallel; the series connection builds up the voltage and the parallel connection provides the necessary current. The nervous system and musculature of the eel are embedded in a thick, insulating layer of fat so that the eel is not affected by the discharge.

11.5 Series and Parallel Connections with Initial Stored Energy

To extend the results of the preceding section to the case of initial energy storage, the following questions should be answered:

1. Are the relations for the equivalent series and parallel inductance and capacitance derived in the preceding section valid in the case of initial energy storage? The answer is yes, as can be justified by considering changes in applied voltage or current (Section ST11.3). Here we note quite simply that L and C are constant, independent of any initial energy storage, in accordance with the definition of an LTI circuit (Section 1.10, Chapter 1). In Figure 11.4.1, for example, if any capacitor has an initial voltage, due to a charge, the capacitance of that capacitor does not depend on its charge. C_{eqs} is therefore the same as without any energy storage. The voltage across C_{eqs} is the sum of the voltages across the individual series-connected capacitors. Similar considerations apply in the case of initial energy storage in Figure 11.4.2 to Figure 11.4.4. Hence,

    Concept: *The constancy of L and C in LTI circuits implies that the expressions for the equivalent series or parallel capacitance or inductance (Table 11.4.1) are independent of initial energy storage.*

2. What happens if two capacitors of different voltages are connected in parallel, or two inductors of different currents are connected in series? If the voltages on the two capacitors were to remain unchanged, as in the case of voltage sources, the connection would violate KVL and would be invalid. What happens, under idealized conditions, is that the voltages of the two capacitors are forced to become equal instantaneously by a current impulse, and charge is conserved, as discussed in detail in Example 11.5.1. Similarly, if the currents in the series inductors were to remain unchanged, as in the case of current sources, the connection would violate KCL and would be invalid. What happens, under idealized conditions, is that the currents in the two inductors are forced to become equal instantaneously by a voltage impulse, and flux linkage is conserved, as discussed in detail in Example 11.5.2. It follows that:

    Concept: *When capacitors of different voltages are paralleled, the voltage across them is equalized by current impulses that instantaneously transfer charge between the capacitors, charge being conserved in the process. Similarly, when inductors carrying different currents are connected in series, the current through them is equalized by voltage impulses that instantaneously transfer flux linkage between the inductors, flux linkage being conserved in the process.*

After equalization of voltage across capacitors in parallel, any change in this voltage is accompanied by changes in the charge across each capacitor. Multiplying Equation 11.4.21

by the common voltage v_p and substituting the relation $q = Cv$ for each capacitor, including C_{eqp}, gives Equation 11.4.24. Similarly, after equalization of current in each of the inductors connected in series, any change in this current is accompanied by changes in the flux linkage in each inductor. Multiplying Equation 11.4.15 by the common current i_s and substituting the relation $\lambda = Li$ for each inductor, including L_{eqs}, gives Equation 11.4.18.

When capacitors of different charges are connected in series, each capacitor retains its charge. The voltage across C_{eqs} is the sum of the voltages across the individual capacitors. Passing a current through the combination would simply add to, or subtract from, the charge on each capacitor, as illustrated by Example 11.5.3. Similarly, when inductors carrying different currents are connected in parallel, each inductor retains its current. The current through L_{eqp} is the sum of the currents in the individual inductors. Changing the voltage across the combination would simply add to, or subtract from, the flux linkage in each conductor, as illustrated by Example 11.5.4.

It will be noted from the examples and exercises that follow in this section and Section 11.6 that the concept of an equivalent series or parallel capacitor or inductor provides a powerful and convenient tool for analyzing many types of problems. Table 11.4.1 summarizes the relations for capacitors or inductors connected in series or in parallel with or without initial energy storage.

Example 11.5.1: Charge Distribution in Two Paralleled Capacitors

A 2 µF capacitor C_1 having an initial charge $q_1(0) = 16$ µC is connected at $t = 0$ to a 6 µF capacitor C_2 having an initial charge $q_2(0) = 24$ µC. It is required to determine the final charge on each capacitor and on the equivalent capacitor, and to compare the initial and final energies stored in C_1 and C_2.

SOLUTION

The initial voltages of C_1 and C_2 are $\dfrac{16}{2} = 8$ V and $\dfrac{24}{6} = 4$ V, respectively. When connected together, the voltages across C_1 and C_2 are forced to change instantaneously to a new value. A current impulse therefore occurs that instantaneously transfers charge between the two capacitors so that they will have the same voltage across them. Because C_1 has a higher voltage than C_2, charge flows from C_1 to C_2, although C_2 has more charge. Note that because C_1 and C_2 are considered to have a common voltage across them, they are effectively in parallel. The equivalent parallel capacitance is $2 + 6 = 8$ µF.

Let the charge transferred by the current impulse be Δq. At $t = 0^+$ when the impulse is over, q_1 and q_2 assume their final values q_{1f} and q_{2f}, respectively, corresponding to a final voltage v_f across the capacitors. It follows that $v_f = \dfrac{q_{1f}}{C_1} = \dfrac{q_1(0) - \Delta q}{C_1} = \dfrac{q_{2f}}{C_2} = \dfrac{q_2(0) + \Delta q}{C_2}$. These equations give $\Delta q = \dfrac{C_2 q_1(0) - C_1 q_2(0)}{C_1 + C_2} = 6$ µC, $q_{1f} = 16 - 6 = 10$ µC $= \dfrac{C_1}{C_1 + C_2}[q_1(0) + q_2(0)]$, and $q_{2f} = 24 + 6 = 30$ µC $= \dfrac{C_2}{C_1 + C_2}[q_1(0) + q_2(0)]$. It is seen that $q_{eqp} = q_{1f} + q_{2f} = q_1(0) + q_2(0) = 40$ µC. Charge is therefore conserved during charge transfer because Δq subtracts from the charge on C_1 and adds to the charge on C_2.

The final voltage is $v_f = \dfrac{q_{1f}}{C_1} = \dfrac{q_{2f}}{C_2} = \dfrac{q_{eqp}}{C_{eqp}} = \dfrac{q_{1f} + q_{2f}}{C_1 + C_2} = \dfrac{q_1(0) + q_2(0)}{C_1 + C_2} = 5$ V.

The initial energy stored in the two capacitors is $\dfrac{1}{2}\dfrac{(q_1(0))^2}{C_1}+\dfrac{1}{2}\dfrac{(q_2(0))^2}{C_2}=112$ μJ, and the

final energy is $\dfrac{1}{2}(C_1+C_2)v_f^2=100$ μJ. 12 μJ are apparently lost as a result of redistribution of charge.

That the final energy is less than the initial energy does not violate conservation of energy. Mathematically, this discrepancy is manifested as the product of zero resistance, the square of an infinite current, and an infinitesimal duration. This product is indeterminate, but not zero. Physically, energy is dissipated in some residual resistance of the connection, which is inevitably present, and in electromagnetic radiation due to charge acceleration. If the capacitors are connected through a resistor, it can be readily shown that the power dissipated in the resistor accounts for the difference in the initial and final energies of the capacitors (Example 13.3.5, Chapter 13).

EXERCISE 11.5.1

Three capacitors of 2.5, 1, and 0.5 μF are connected in parallel. The initial charges on these capacitors are 0.5, 0.3, and 0 μC, respectively. Determine the final charge on each capacitor and the voltage across the capacitors. (Hint: determine the voltage from the equivalent parallel capacitance and conservation of charge.)

Answer: 0.5, 0.2, and 0.1 μC, respectively; 0.2 V

Example 11.5.2: Distribution of Flux Linkage in Two Series-Connected Inductors

A 2 μH inductor L_1 having a current $i_1(0) = 8$ A is connected at $t = 0$ to a 6 μH inductor L_2 having a current $i_2(0) = 4$ A by simultaneously moving the two switches, as shown in Figure 11.5.1. It is required to determine the final current in each inductor and in the equivalent inductor and to compare the initial and final energies stored in L_1 and L_2. The numerical values are chosen so as to make this example the dual of Example 11.5.1.

The switches are of the make-before-break type that is, they make contact with the final terminal before breaking with the initial terminal. This ensures that the inductor current is not interrupted during switching. Otherwise, interrupting the inductor current would

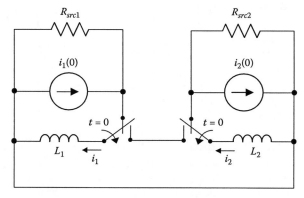

FIGURE 11.5.1
Figure for Example 11.5.2.

give rise to a large Ldi/dt voltage across the switch contacts as they interrupt the current. The large voltage causes breakdown of the air dielectric between the contacts, leading to an arc discharge between the contacts. The high current density of the arc can damage the contacts because of localized burning. The resistances across the current sources ensure that these sources are not left open-circuited after the switches move to their final positions.

SOLUTION

Moving the two switches to their final positions forces the currents in the two inductors to change instantaneously to a new value. A voltage impulse therefore occurs that instantaneously equalizes the currents in the two inductors. Because the final current in L_1 and L_2 is the same, the inductors are effectively in series. The equivalent series inductance is $2 + 6 = 8$ μH.

The voltage impulse reduces the current in L_1 and increase that in L_2. But it can only do so by affecting the flux linkages because voltage and flux linkage are directly related

through Faraday's law, $v = \dfrac{d\lambda}{dt}$. The initial flux linkages in L_1 and L_2 are $(2\ \mu H) \times (8\ A) =$ 16 μWb-turns and $(6\ \mu H) \times (4\ A) = 24$ μWb-turns, respectively. The voltage impulse results

in a change of flux linkage $\Delta\lambda = \displaystyle\int_{0^-}^{0^+} v dt$ in each inductor, which reduces the flux linkage,

and hence the current, in L_1, and increases the flux linkage in L_2. At $t = 0^+$, λ_1 and λ_2 assume their final values λ_{1f} and λ_{2f}, respectively, corresponding to the final values of currents i_{1f}

and i_{2f}, where $i_{1f} = i_{2f} = i_{eqs}$. It follows that $i_{1f} = \dfrac{\lambda_1(0) - \Delta\lambda}{L_1} = i_{2f} = \dfrac{\lambda_2(0) + \Delta\lambda}{L_2}$. This

gives $\Delta\lambda = \dfrac{L_2\lambda_1(0) - L_1\lambda_2(0)}{L_1 + L_2} = 6\mu$ Wb-turns. The final values λ_{1f} and λ_{2f} are: $\lambda_{1f} = 16 - 6 =$

10 μWb-turns $= \dfrac{L_1}{L_1 + L_2}(\lambda_1(0) + \lambda_2(0))$, and $\lambda_{2f} = 24 + 6 = 30$ μWb-turns $= \dfrac{L_2}{L_1 + L_2}(\lambda_1(0) + \lambda_2(0))$.

It is seen that $\lambda_{eqs} = \lambda_{1f} + \lambda_{2f} = \lambda_1(0) + \lambda_2(0) = 40$ μWb-turns. In other words, flux linkage is conserved during current readjustment because $\Delta\lambda$ subtracts from the flux linkage in L_1 and adds to the flux linkage in L_2.

The final value of the current is $i_f = \dfrac{\lambda_{1f}}{L_1} = \dfrac{\lambda_{2f}}{L_2} = \dfrac{\lambda_{eqs}}{L_{eps}} = \dfrac{\lambda_{1f} + \lambda_{2f}}{L_1 + L_2} = \dfrac{\lambda_1(0) + \lambda_2(0)}{L_1 + L_2} = 5$ A. The

initial energy stored in the two inductors is $\dfrac{1}{2}L_1(i_1(0))^2 + \dfrac{1}{2}L_2(i_2(0))^2 = 112$ μJ. The final

energy stored in the two inductors is $\dfrac{1}{2}(L_1 + L_2)i_f^2 = 100$ μJ. 12 μJ are apparently lost as a

result of readjustment of current.

As explained in connection with Example 11.5.1, the discrepancy between the initial and final values of energy does not violate conservation of energy. The discrepancy is attributed to an impulse of voltage that appears across the two inductors, without a parallel or shunt path. In the presence of such a path, it can be readily shown that the power dissipated in the shunt resistor accounts for the difference in the initial and final energies of the inductors (Example SE13.1, Chapter 13).

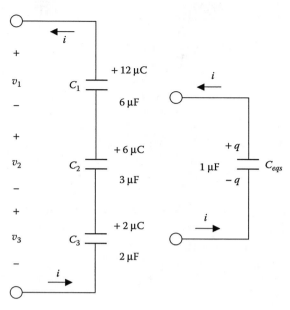

FIGURE 11.5.2
Figure for Example 11.5.3.

EXERCISE 11.5.2

Three inductors of 2.5, 1, and 0.5 μH are connected in series. The initial flux linkages in these inductors are 0.5, 0.3, and 0 μWb-turns, respectively. It is required to determine the final flux linkage in each inductor and the current through the inductors. Note that this exercise is the dual of Exercise 11.5.1. (Hint: determine the current from the equivalent series inductance and conservation of flux linkage.)

Answer: 0.5, 0.2, and 0.1 μWb-turns, respectively; 0.2 A.

Example 11.5.3: Series-Connected Capacitors

Three capacitors $C_1 = 6$ μF, $C_2 = 3$ μF, and $C_3 = 2$ μF are connected in series (Figure 11.5.2). The initial charges on these capacitors are, respectively: 12 μC, 6 μC, and 2 μC. Assume that a discharging current flows in these capacitors that transfers a total of 4 μC of charge. It is required to calculate the initial voltages across the capacitors and across the equivalent series capacitor, and the final charges on all these capacitors and their voltages.

SOLUTION

$C_{eqs} = \dfrac{6 \times 3 \times 2}{6 \times 3 + 3 \times 2 + 2 \times 6} = 1$ μF. The initial voltages across the capacitors are

$v_1(0) = \dfrac{12\ \mu C}{6\ \mu F} = 2$ V, $v_2(0) = \dfrac{6\ \mu C}{3\ \mu F} = 2$ V, and $v_3(0) = \dfrac{2\ \mu C}{2\ \mu F} = 1$ V. The initial voltage across C_{eqs} is therefore 5 V, and the initial charge on this capacitor is 5 V × 1 μF = 5 μC.

After the current i transfers 4 μC, this charge subtracts from that on each capacitor, including C_{eqs}. The final charges on C_1, C_2, C_3, and C_{eqs}, are 8, 2, −2, and 1 μC, respectively.

The final voltages on C_1, C_2, and C_3, are, respectively, 8/6, 2/3, and −2/2 V, which add up to 1 V. The final voltage across C_{eqs} will also be $\dfrac{1\ \mu C}{1\ \mu F} = 1$ V, as it should be.

EXERCISE 11.5.3

A 1 A pulse of 2 µs duration is passed through three capacitors of 0.2, 0.8, and 0.04 µF connected in series. The capacitors were initially uncharged. It is required to determine, at the end of the pulse: (a) the charge of each capacitor and the voltage across each of them; (b) the capacitance of the equivalent capacitor, its charge, and the voltage across the capacitor.

Answer: (a) 2 µC; 10, 2.5, and 50 V, respectively; (b) 0.032 μF, 2 µC; 62.5 V.

Example 11.5.4: Paralleled Inductors

Three inductors $L_1 = 6$ µH, $L_2 = 3$ µH, and $L_3 = 2$ µH are connected in parallel (Figure 11.5.3). The initial currents in these inductors are, respectively, 2, 2, and 1 A. Assume that

$$v \text{ changes negatively such that } \lambda = \int_0^t v\, dt = -4 \ \mu\text{Wb-turns. It is required to calculate the}$$

initial flux linkages in the inductors and in the equivalent parallel inductor, and the final flux linkages and currents in all these inductors. The numerical values are chosen so as to make this example the dual of Example 11.5.3.

SOLUTION

$L_{eqp} = \dfrac{6 \times 3 \times 2}{6 \times 3 + 3 \times 2 + 2 \times 6} = 1$ µH. Because $\lambda = Li$, the initial flux linkages in L_1, L_2, and L_3 are, respectively, 12, 6, and 2 µWb-turns. The initial current through L_{eqp} is 5 A, and the initial flux linkage in this inductor is 5 A × 1 µH = 5 µWb-turns.

After the voltage v transfers 4 µWb-turns, this flux linkage subtracts from that on each inductor, including L_{eqp}. The final flux linkages on L_1, L_2, L_3, and L_{eqp} are 8, 2, −2, and 1 µWb-turns, respectively. The final currents in L_1, L_2, and L_3 are, respectively, 8/6, 2/3, and −2/2 A, which add up to 1 A. This is the final current in L_{eqp}, which also agrees with the fact that L_{eqp} is 1 µH and has a final flux linkage of 1 µWb-turn.

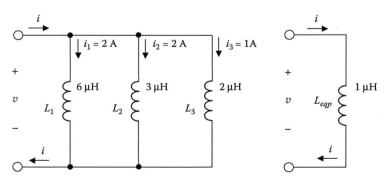

FIGURE 11.5.3
Figure for Example 11.5.4.

EXERCISE 11.5.4

A 1 V pulse of 2 μs duration is applied to three inductors of 0.2, 0.8, and 0.04 μH connected in parallel. The inductors did not have any initial energy storage. It is required to determine, at the end of the pulse: (a) the flux linkage in each inductor and the current through it; (b) the inductance of the equivalent inductor, its flux linkage, and the current through the inductor. Note that this exercise is the dual of Exercise 11.5.3.

Answer: (a) 2 μWb-turns, 10, 2.5, and 50 A, respectively; (b) 0.032 μH, 2 μWb-turns, 62.5 A.

EXERCISE 11.5.5

Two capacitors $C_1 = 200$ nF and $C_2 = 400$ nF have initial charges of 8 and 4 μC, respectively. At $t = 0$, they are connected in parallel and to a dc current source of 6 mA. (a) Determine the current in each capacitor for $t > 0^+$ (b) Express the voltage across the combination as a function of time for $t > 0^+$. (c) How much charge is delivered by the source during the first 2 msec? (d) What is the charge on each capacitor at the end of 2 ms?

Answer: (a) $i_1 = 2$ mA, $i_2 = 4$ mA. (b) $(20 + 10t)$ V, where t is in ms. (c) 12 μC. (d) $q_1 = 8$ μC, $q_2 = 16$ μC.

EXERCISE 11.5.6

Two inductors $L_1 = 2$ H and $L_2 = 4$ H connected in parallel have initial currents of 4 and −1 A, respectively. What is the equivalent inductance and the current through the inductance?

Answer: 4/3 H, 3 A.

EXERCISE 11.5.7

Assume that two inductors L_1 and L_2 are supplied from a time-varying current source i_t, and that a dc current I circulates between the two inductors. If i_{t1} and i_{t2} are the time-varying components of current in L_1 and L_2, respectively, such that $i_t = i_{t1} + i_{t2}$, show that i_t, i_{t1}, and i_{t2} satisfy the current division equations.

11.6 Conservation of Charge and of Flux Linkage

After having considered several examples of switching operations in circuits that include capacitors or inductors, that is, operations involving a sudden change in the circuit, this section consolidates the important concept of conservation of charge and of flux linkage.

Conservation of charge is a universal, inviolate principle that is always satisfied at the atomic level. However, in circuit terms, charge is said to be conserved if it is not neutralized, partially or completely, by a charge of opposite sign. When a capacitor discharges fully, complete charge neutralization takes place. The opposite charges that had been separated during charging of the capacitor neutralize each other at the macroscopic level but are not destroyed at the atomic level. In circuit terms, however, the charge has not been conserved.

On the other hand, conservation of flux linkage is not a universal, inviolate principle. When an inductor is charged, flux linkage is created, and when the inductor is discharged, flux linkage is destroyed. But in circuit terms, the creation of flux linkage is the dual of charge separation, and destruction of flux linkage is the dual of charge neutralization.

In the circuit context, charge conservation at a given node is exemplified by KCL and the basic relation $i = dq/dt$. Thus, from KCL at the given node, $\sum_{k=1}^{k=B} i_k = 0$, where B is the number of branches connected to the node, and the currents being summed are those entering or leaving the node. It is assumed that all the B branches have capacitors connected to the given node. If switching takes place at t_0, then we can integrate $\sum_{k=1}^{k=B} i_k$ between t_{0^-} and t_{0^+} to obtain:

$$\sum_{k=1}^{k=B} \int_{t_{0^-}}^{t_{0^+}} i_k dt = \sum_{k=1}^{k=B} \int_{t_{0^-}}^{t_{0^+}} \frac{dq_k}{dt} dt = \sum_{k=1}^{k=B} \int_{t_{0^-}}^{t_{0^+}} dq_k = \sum_{k=1}^{k=B} q_k(0^+) - \sum_{k=1}^{k=B} q_k(0^-) \qquad (11.6.1)$$

If no external current impulse is applied to the given node at t_0, the first sum on the LHS is zero, so that

$$\sum_{k=1}^{k=B} q_k(0^+) = \sum_{k=1}^{k=B} q_k(0^-) \qquad (11.6.2)$$

In other words, the total charge at the node is conserved at the instant of switching. We can conclude, therefore,

> **Concept:** *The total charge at a given node is conserved at the instant of switching as long as it is not affected by external current impulses that will add, or subtract, charge at the node.*

Thus, when capacitors of unequal initial voltages are connected in parallel, q, which in this case is the net charge on the capacitor plates that are connected together, is redistributed by current impulses. However, these impulses are *not* external to the set of charges q as a whole; they cancel out in the KCL relation and q is conserved. If an external current impulse were applied in parallel with the capacitors during switching, the total initial and final charges will differ by the amount of charge added or removed by the external impulse. When capacitors are connected in series, q is redistributed through charge displacement. The charge q on connected plates is conserved at the instant of switching, in the absence of external current impulses, as illustrated by Example 11.6.1.

Analogous considerations apply to the dual case of flux linkage, whose conservation around a given loop is exemplified by KVL and Faraday's law $v = d\lambda/dt$. Thus, from KVL around the given loop, $\sum_{k=1}^{k=B} v_k = 0$, where B is the number of branches in the loop, and the voltages being summed are the voltage drops, or rises, in the loop. It is assumed that all the B branches contain inductors. If switching takes place at t_0, then we can integrate $\sum_{k=1}^{k=B} v_k$ between t_{0^-} and t_{0^+} to obtain:

$$\sum_{k=1}^{k=B}\int_{t_{0^-}}^{t_{0^+}} v_k dt = \sum_{k=1}^{k=B}\int_{t_{0^-}}^{t_{0^+}} \frac{d\lambda_k}{dt} dt = \sum_{k=1}^{k=B}\int_{t_{0^-}}^{t_{0^+}} d\lambda_k = \sum_{k=1}^{k=B}\lambda_k(0^+) - \sum_{k=1}^{k=B}\lambda_k(0^-) \qquad (11.6.3)$$

If no external voltage impulse is applied in the given loop at t_0, the first sum on the LHS is zero, so that:

$$\sum_{k=1}^{k=B}\lambda_k(0^+) = \sum_{k=1}^{k=B}\lambda_k(0^-) \qquad (11.6.4)$$

In other words, the total flux linkage around the loop is conserved at the instant of switching. We can conclude, therefore,

> **Concept:** *The total flux linkage around a given loop is conserved at the instant of switching as long as it is not affected by external voltage impulses that will add, or subtract, flux linkage in the loop.*

These concepts are illustrated by Example 11.6.1 and by examples in following chapters. A more rigorous treatment of conservation of charge and of flux linkage is given in Section ST 11.4.

Example 11.6.1: Conservation of Charge and Flux Linkage

A capacitor $C_1 = 4$ F having an initial charge of 7 C is connected in series with uncharged capacitors C_2 and C_3 having capacitances of 2 and 1 F, respectively (Figure 11.6.1). It is required: (a) to determine the final capacitor charges and voltages; (b) to analyze the dual circuit.

SOLUTION

(a) An efficient and general method of solving this type of problem is by using the equivalent series capacitor C_{eqs}. Assume that the three capacitors are connected

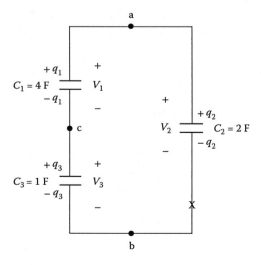

FIGURE 11.6.1
Figure for Example 11.6.1.

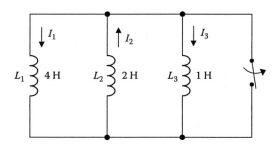

FIGURE 11.6.2
Figure for Example 11.6.1.

in a string, as with an open connection at X in Figure 11.6.1. Then C_{eqs}

$= \dfrac{4 \times 1 \times 2}{4 + 2 + 8} = \dfrac{4}{7}$ F. The initial voltage across C_1 is $\dfrac{7}{4}$ V, which is also the voltage across the string because the initial voltages of C_2 and C_3 are zero. The charge on

C_{eqs} is therefore $\dfrac{4}{7} \times \dfrac{7}{4} = 1$ C. When the two ends of the string are joined together so as to connect the three capacitors in series, C_{eqs} is discharged. 1 C flows clockwise around the loop in Figure 11.6.1, and the capacitor charges are modified accordingly. q_{1f}, the final charge on C_1, becomes 6 C, q_{2f} becomes 1 C, and q_{3f} becomes -1 C. The final voltages are $V_{1f} = 1.5$ V, $V_{2f} = 0.5$ V, and $V_{3f} = -1$ V. KVL around the mesh gives $1.5 - 0.5 - 1 = 0$, which is correct.

Let us look at conservation of charge. The sets of charges that are conserved are those at nodes a, b, and c, which are formed after the capacitors are connected in series and which have no external connections. Initially, the total positive charge on the two upper plates of C_1 and C_2, which when connected form node a, is $7 + 0 = 7$ C. The final charge at a is $6 + 1 = 7$, so charge is conserved at this node. Similarly, the initial and final charges at node b are 0, and $1 - 1 = 0$, whereas the initial and final charges on node c are $-7 + 0 = -7$, and $-6 - 1 = -7$. The total initial charge on the capacitors is 7 C, but the total final charge is 8 C in magnitude, so it is meaningless to talk about conservation of total charge on the capacitors.

(b) In the dual circuit, an inductor L_1 of 4 H, having an initial flux linkage of 7 Wb-turns, is paralleled with $L_2 = 2$ H and $L_3 = 1$ H, both having no initial current, by opening the switch (Figure 11.6.2). The equivalent parallel inductor L_{eqp} will be used to determine the final flux linkages and currents in the three inductors. To preserve duality, we note that $V_1 + V_3 = V_2$ in Figure 11.6.1. Hence, the assigned positive directions of currents in Figure 11.6.2 are such that $I_1 + I_3 = I_2$. L_{eqp}

$= \dfrac{4 \times 1 \times 2}{4 + 2 + 8} = \dfrac{4}{7}$ H. The current in this inductor is the $\dfrac{7}{4}$ A in L_1 and its flux linkage is 1 Wb-turn. When the switch is opened, the current in L_{eqp} is zero, and the flux linkage of 1 Wb-turn disappears. Because the voltage across the inductors is common, the flux linkage in each inductor is reduced by 1 Wb-turn. Hence, $\lambda_{1f} = 7 - 1 = 6$ Wb-turns, $\lambda_{2f} = 0 + 1 = 1$ Wb-turn, and $\lambda_{3f} = 0 - 1 = -1$ Wb-turn. The assigned positive direction of λ_{2f} is the same as that of i_2, so that a decrease of flux linkage for a current flowing away from the common upper node is equivalent to an increase in flux linkage for a current in the opposite direction. It follows that $i_{1f} = 1.5$ A, $i_{2f} = 0.5$ A, and $i_{3f} = -1$ A, in accordance with KCL.

Flux linkage is conserved around the three closed paths formed by two inductors at a time in Figure 11.6.2. Thus, if we move counterclockwise around the mesh formed by L_1 and L_2, the initial flux linkage is $7 + 0 = 7$ Wb-turns, and the final flux linkage is $6 + 1 = 7$ Wb-turns. The initial flux linkage around the mesh formed by L_2 and L_3 is zero, and the final flux linkage is $1 - 1 = 0$. The initial flux linkage around the loop formed by L_1 and L_3 is $7 + 0 = 7$ Wb-turns, and the final flux linkage is $6 - (-1) = 7$ Wb-turns.

Note that because impulses are involved in establishing the final state, the final energy in both cases (5.25 J) is less than the initial energy (6.125 J).

Summary of Main Concepts and Results

- Two v–i relations are duals if one relation reduces to the other when the following are interchanged: (1) v and i, (2) R and G, and (3) L and C.

- Two circuits are duals if the node-voltage equations of one circuit are the dual relations of the mesh-current equations of the other circuit.

- The voltage across a capacitor cannot be changed instantaneously except by a current impulse, in which case the change in the charge of the capacitor is that represented by the area of the impulse.

- The current through an inductor cannot be changed instantaneously except by a voltage impulse, in which case the change in the flux linkage of the inductor is that represented by the area of the impulse.

- When capacitors are connected in parallel, $C_{eqp} = C_1 + C_2 + \dots + C_n$, and when connected in series, $\dfrac{1}{C_{eqs}} = \dfrac{1}{C_1} + \dfrac{1}{C_2} + \dots + \dfrac{1}{C_n}$, irrespective of any initial energy storage.

- When inductors are connected in series, $L_{eqs} = L_1 + L_2 + \dots + L_n$, and when connected in parallel, $\dfrac{1}{L_{eqs}} = \dfrac{1}{L_1} + \dfrac{1}{L_2} + \dots + \dfrac{1}{L_n}$, irrespective of any initial energy storage.

- When capacitors of different voltages are paralleled, the voltage across them is equalized by current impulses that instantaneously transfer charge between the capacitors, charge being conserved in the process.

- When inductors carrying different currents are connected in series, the current through them is equalized by voltage impulses that instantaneously transfer flux linkage between the inductor, flux linkage being conserved in the process.

- When capacitors are connected in parallel, the charge on the equivalent parallel capacitor is the sum of the charges on the individual capacitors. Similarly, when inductors are connected in series, the flux linkage in the equivalent series inductor is the sum of the flux linkages in the individual inductors.

- The total charge at a given node is conserved at the instant of switching as long as it is not affected by external current impulses that will add, or subtract, charge at the node.

- The total flux linkage around a given loop is conserved at the instant of switching as long as it is not affected by external voltage impulses that will add, or subtract, flux linkage in the loop.

Learning Outcomes

- Apply duality to basic circuit relations and simple circuits.
- Analyze basic circuits of capacitors or inductors in series or in parallel.

Supplementary Topics and Examples on CD

ST11.1 Derivation of duals of planar circuits: Describes a general procedure for deriving the dual of a planar circuit.

ST11.2 Parallel connection of capacitors and inductors with zero initial energy: Derives from first principles the relations for the parallel connection of capacitors and the parallel connection of inductors.

ST11.3 Series and parallel relations of circuit parameters: Justifies the relations for the equivalent series and parallel parameters by considering changes in applied voltage or current.

ST11.4 Conservation of charge and of flux linkage: Provides a more rigorous proof of conservation of these entities and gives additional examples.

SE11.1 Excitation of capacitor from current source: Determines the voltage, charge, and energy of a capacitor in response to a step plus a ramp of current.

SE11.2 Excitation of capacitor from voltage source: Determines the current, charge, and energy of a capacitor in response to a step plus a ramp of voltage.

SE11.3 Excitation of inductor from voltage source: Determines the current, flux linkage, and energy of an inductor in response to a step plus a ramp of voltage. This is the dual of Example SE 11.1.

SE11.4 Excitation of inductor from current source: Determines the voltage, flux linkage, and energy of an inductor in response to a step plus a ramp of voltage. This is the dual of Example SE11.2.

Problems and Exercises

P11.1 Duality

P11.1.1 Of the circuits of Figure P1.2.7 to Figure P1.2.10 (Chapter 1), which are duals?

P11.1.2 Deduce that the dual of an ideal transformer of turns ratio N_1:N_2 is an ideal transformer of turns ratio N_2:N_1. Verify by considering the reflected impedance of a load Z_L. What is the dual of a step-up autotransformer having N_1 turns on the input winding and N_2 in the other winding?

P11.1.3 Given a T-equivalent circuit supplied from a voltage source and connected to a load $Z_L = (25\ \Omega\,||\,j50\ \Omega)$, where the series impedances are each $8 - j12\ \Omega$ and the shunt impedance is $(20\ \Omega\,||\,j30\ \Omega)$, show that the dual is a π-equivalent circuit and determine the value of each element in this circuit.

P11.1.4 Show that the dual of a bridge circuit is also a bridge circuit.

P11.1.5 Show that the dual of the bridged-T circuit of Figure P5.3.7 (Chapter 5) is a bridged-T circuit in which L and C are interchanged.

P11.1.6 Derive the dual of the circuit of Figure P7.2.1 (Chapter 7), and determine R_L that will absorb maximum power. Calculate this power.

P11.1.7 Derive the dual of the circuit of Figure P7.2.11 (Chapter 7) and rework Problem P7.2.11 using admittances.

P11.1.8 Derive the dual of the circuit of Figure P7.2.19 (Chapter 7) and rework Problem P7.2.19 (Chapter 7) using admittances.

P11.1.9 Determine the dual circuit to that of Figure P11.1.9 and the dual transfer function. Is the response of the dual circuit low pass or high pass?

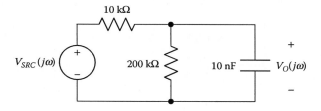

FIGURE P11.1.9

P11.2 Impulse Function

P11.2.1 The voltage drop across a device is $u(t)$ V and the current through it, in the direction of voltage drop, is $\delta(t)$ As. Determine the instantaneous power and the total energy delivered to the device.

P11.2.2 The voltage drop across a device is $\delta(t-2)$ Vs and the current through it, in the direction of voltage drop, is $5t$ A. Determine the instantaneous power and energy delivered to the device.

P11.2.3 The voltage drop across a device is $\delta(t-1)$ Vs and the current through it, in the direction of voltage drop, is $10(1-e^{0.5t})$ A. Determine the instantaneous power and energy delivered to the device.

P11.2.4 Evaluate the following integrals involving impulse functions: (a) $\int_{-\infty}^{\infty} 10e^t \sin 2\pi t\ \delta(t-0.75)dt$; (b) $\int_{-\infty}^{\infty} [4\delta(t) + \cos 2\pi t\delta(t-1) + 2t^2\delta(t-2)]\ dt$.

P11.2.5 Using integration by parts, show that $\int_{-\infty}^{\infty} f(t)\delta^{(1)}(t-a)dt = -f^{(1)}(a)$, where the (1) superscript denotes the first derivative.

P11.2.6 Evaluate $\int_{-\infty}^{\infty} 24\delta(1-12t)\cos 4\pi t\, dt$.

P11.2.7 Verify the integrals of Problem 11.2.4 to Problem 11.2.6 using the symbolic processing feature of MATLAB. Note that in MATLAB infinity is entered as Inf, π as pi, $\delta(t-a)$ as dirac(t −a), the derivative of a function f as diff(f), and that f is declared a function of t using the command f = sym('f(t)').

P11.2.8 Derive the FSE of a train of positive impulses of strength $2A_m$ and period T. Deduce the FSE of the delayed square wave as the sum of a train of positive impulses and a train of negative impulses delayed by $T/2$, both having the same period and magnitude of impulse strength. Sketch the frequency spectra in the three cases.

P11.2.9 Obtain the FSE of an impulse train of period T and impulse strength B as a limiting case of the rectangular pulse train as $\alpha \to 0$ and $A \to \infty$.

P11.3 Capacitors and Inductors

P11.3.1 Determine the equivalent capacitance between terminals ab in Figure P11.3.1 if all the capacitances are 1 F.

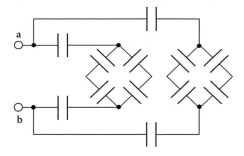

FIGURE P11.3.1

P11.3.2 Determine the equivalent inductance between terminals ab in Figure P11.3.2.

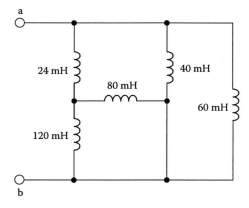

FIGURE P11.3.2

P11.3.3   Determine the equivalent inductance between terminals ab in Figure P11.3.3 if all inductances are 1 H.

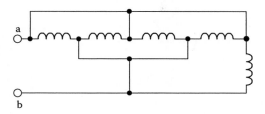

FIGURE P11.3.3

P11.3.4   Determine the equivalent capacitance between terminals ab in the ladder circuit of Figure P11.3.4 if all the capacitances are 1 F.

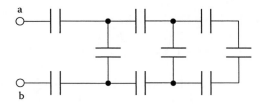

FIGURE P11.3.4

P11.3.5   Determine the equivalent capacitance between terminals ab in Figure P11.3.5 if all the capacitances are 1 F.

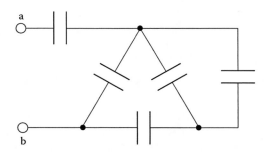

FIGURE P11.3.5

P11.3.6   Determine the equivalent inductance between terminals ab in Figure P11.3.6.

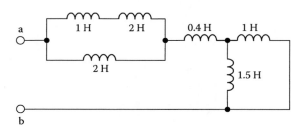

FIGURE P11.3.6

P11.3.7   If a time-varying voltage v is applied across two capacitors C_1 and C_2 in series, show that the voltages across the capacitors are: $v_{C_1} = \dfrac{C_2}{C_1 + C_2} v$ and $v_{C_2} = \dfrac{C_1}{C_1 + C_2} v$. Similarly, show that if v is applied across two inductors L_1 and L_2 in series, the voltages across the inductors are $v_{L_1} = \dfrac{L_1}{L_1 + L_2} v$ and $v_{L_2} = \dfrac{L_2}{L_1 + L_2} v$. Assume no initial energy storage in both cases.

P11.3.8   Capacitors $C_1 = 4$ F, $C_2 = 6$ F, and $C_3 = 10$ F, having initial charges of 12, 15, and 20 C, respectively, are connected in parallel, the corresponding positive and negative terminals being connected together. Determine: (a) the final voltage and charge of each capacitor; (b) the total initial and final energies.

P11.3.9   The capacitors of Problem P11.3.8, with the same initial charges, are connected end to end in a closed loop, the polarities being in the same sense. Determine: (a) the final voltage and charge of each capacitor; (b) the total initial and final energies.

P11.3.10 Inductors $L_1 = 4$ H, $L_2 = 6$ H, and $L_3 = 12$ H, having initial currents of 4, 2, and 3 A, respectively, are connected in parallel between nodes a and b, with all the initial currents directed from a to b. Determine: (a) the final current and flux linkage of each inductor; (b) the total initial and final energies. Verify that the same result is obtained if L_1 is considered to be in series with the parallel combination of L_2 and L_3. Also verify that the initial and final flux linkages are the same when taken two at a time in the same sense.

P11.3.11 The inductors of Problem 11.3.10 (Chapter 2), with the same initial currents, are connected in series, with the initial current in L_2 opposing the initial currents in L_1 and L_3. Determine: (a) the final current and flux linkage of each inductor; (b) the total initial and final energies.

P11.3.12 Derive and solve the duals of the circuits of Problem P11.3.8 to Problem P11.3.11.

P11.3.13 Capacitors C_1 to C_4 having the values and assigned positive polarities of charge indicated in Figure P11.3.13 are connected as shown. Determine the final voltages and charges of the capacitors, assuming that $Q_{1i} = 2$ C, $Q_{2i} = -3$ C, $Q_{3i} = -6$ C, and $Q_{4i} = 9$ C.

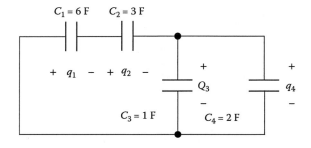

FIGURE P11.3.13

P11.3.14 Derive and solve the dual of the circuit problem of Problem P11.3.13.

P11.3.15 The capacitors in Figure P11.3.15 are initially charged to the voltages shown. The switch is closed at $t = 0$. Determine v_O and the voltage across each capacitor for $t \geq 0^+$.

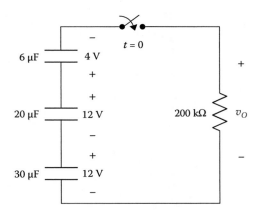

FIGURE P11.3.15

P11.3.16 Verify that when two capacitors C_1 and C_2, having initial charges $q_1(0)$ and $q_2(0)$, respectively, are connected in parallel, the final energy is

$$\frac{1}{2}\frac{[q_1(0)+q_2(0)]^2}{C_1+C_2}.$$ Show that this is always less than the initial

energy $\dfrac{1}{2}\dfrac{[q_1(0)]^2}{C_1}+\dfrac{1}{2}\dfrac{[q_2(0)]^2}{C_2}$. (Hint: manipulate the inequality to show that the

final energy is less than the initial energy as long as

$$\left[\sqrt{\frac{C_2}{C_1}}q_1(0)-\sqrt{\frac{C_1}{C_2}}q_2(0)\right]^2 > 0.$$ $C_2 q_1(0) = C_1 q_2(0)$ is of course the case when the

voltages across C_1 and C_2, are equal.

12

Natural Responses and Convolution

Overview

In preceding chapters we considered the steady-state response of circuits to dc, sinusoidal, and periodic excitations. These responses depend on both the applied excitation and the circuit. When an excitation is applied to a given circuit, and before a steady state is reached, the response depends, particularly, on the initial conditions in the circuit, which are due to the energy stored in capacitors or inductors at the time the excitation is applied. Therefore, to determine the complete response, we must determine not only the steady-state response of the circuit but also its response to initial conditions. The circuit response resulting from some initial conditions in the circuit is the **natural response** of the circuit.

One way of establishing initial conditions is by applying a voltage or current impulse to a circuit in a quiescent or relaxed state, that is, a state in which no energy is stored in capacitors or inductors. In the case of first-order circuits, which essentially contain only a single energy-storage element, the impulse response and the natural response are one and the same because the effect of the impulse is to establish an initial condition in the given energy-storage element. Second-order circuits contain more than one independent energy-storage element. A single applied impulse establishes, in general, initial conditions in one of the energy-storage elements, so that the impulse response is a special case of the natural response. The impulse response of a given circuit is characteristic of the circuit, just like the frequency response discussed in Chapter 10. The impulse and frequency responses are therefore related, as will be highlighted in subsequent discussions.

It may be thought that the impulse is an abstract concept that is of no practical importance. In fact, the impulse response is of great theoretical interest and can be approximated in practice by a sufficiently narrow pulse of large amplitude, as subsequently demonstrated. If the impulse response of a circuit is known, it will be shown that the response to an arbitrary input can be determined by applying a convolution operation.

Learning Objectives

- To be familiar with:
  - The definition of terms used in describing the responses of first-order and second-order circuits, such as time constant, damping factor, damped natural frequency, etc.

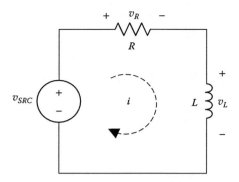

FIGURE 12.1.1
RL circuit excited by voltage impulse.

- To understand:
  - How a voltage or current impulse applied to a circuit establishes an initial condition in the circuit
  - The nature and interpretation of the natural response of first-order systems
  - The three types of natural response of second-order systems and how they arise
  - The nature of convolution, its graphical interpretation, and the derivation of the convolution integral

12.1 Natural Responses of First-Order Circuits

RL Circuit

Consider the circuit of Figure 12.1.1, where v_{SRC} is an impulse of strength K. From Kirchhoff's Voltage Law (KVL),

$$L\frac{di}{dt} + Ri = K\delta(t) \tag{12.1.1}$$

Integrating both sides of Equation 12.1.1 over the duration of the impulse, from $t = 0^-$ to $t = 0^+$:

$$L\int_{0^-}^{0^+} di + R\int_{0^-}^{0^+} idt = K\int_{0^-}^{0^+} \delta(t)dt \tag{12.1.2}$$

We have seen in Section 11.2 (Chapter 11) that a voltage impulse of strength K applied to an inductor causes a step change of K in flux linkage, and hence a finite change in current of K/L. Adding a resistor in series with the inductor cannot cause an infinite change in i, so that the second integral on the LHS of Equation 12.1.2 is zero. It follows that $i(0^+) - i(0^-) = \dfrac{K}{L}$. Assuming no initial energy storage, $i(0^-) = 0$, so that:

$$i(0^+) = \frac{K}{L} \tag{12.1.3}$$

The following should be noted:

1. $i(0^+)$ does not depend on R. The interpretation is that during the impulse, $v_{SRC} \to \infty$, while i, and hence Ri, remain finite. This implies that the voltage impulse essentially appears across L, so that the presence of R is of no consequence as far as $i(0^+)$ is concerned. In other words, the inductor behaves as an open circuit to the impulse, except that an open circuit does not pass any current, whereas the current through the inductor increases to K/L at the end of the impulse.
2. Once the impulse is over, the voltage source behaves as a short circuit and allows current to flow. The circuit for $t \geq 0^+$ reduces to a resistor connected across an inductor having an initial current K/L.

How does i vary for $t \geq 0^+$? From energy considerations, it must decrease with time. This is because i flowing in R dissipates energy, which can only come from the energy stored in the inductor. This energy, and hence the current, must therefore decrease con-tinuously, so that eventually all the energy $\frac{1}{2}L\left(\frac{K}{L}\right)^2$ initially stored in L is dissipated in R (Exercise 12.1.1).

To determine the time course of i, KVL gives, after the impulse is over, $L\frac{di}{dt} + Ri = 0$. Dividing by L,

$$\frac{di}{dt} + \frac{i}{\tau} = 0 \tag{12.1.4}$$

where $\tau = L/R$ is the time constant. Note that $\tau = 1/\omega_c$, where ω_c is the 3-dB cutoff frequency considered in Section 10.2 (Chapter 10).

To solve Equation 12.1.4, which is a first-order, linear differential equation with zero forcing function, we rearrange the variables i and t: $\frac{di}{i} = -\frac{dt}{\tau}$. Integrating both sides, $\ln i = -\frac{t}{\tau} + A$, where A is a constant of integration. Expressing this relation in expo-nential form, $i = Be^{-\frac{t}{\tau}}$, where $B = e^A$ is a new constant to be determined from the initial conditions. At $t = 0^+$, $i = K/L = B$. Hence,

$$i = \frac{K}{L}e^{-t/\tau}, \quad t \geq 0^+ \tag{12.1.5}$$

Equation 12.1.5 is plotted in normalized form in Figure 12.1.2, where the x-axis represents t/τ and the y-axis represents the ratio of a voltage or a current y to its initial value.

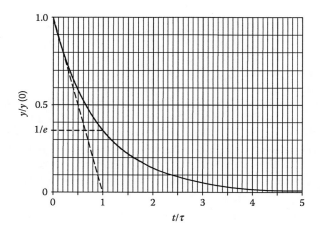

FIGURE 12.1.2
Normalized decaying exponential.

The normalized plot is therefore that of e^{-x} vs. x. At $t = \tau$, the response decreases to $1/e$ of its initial value. Another interpretation of τ is that the magnitude of the slope at $t = 0$ is $1/\tau$. In other words, if the response decreases at the initial high rate, instead of exponentially, the response would be reduced to zero at $t = \tau$. Theoretically, it takes an infinitely long time for the response to decay to zero. In practice, the response decays to $e^{-5} = 0.0067$ of its initial value, or 0.67%, after five time constants. The following concept is observed:

> **Concept:** *The time constant of a first-order circuit is a measure of its speed of response. The larger the time constant, the slower, or more sluggish, is the circuit, and the lower is the 3-dB cutoff frequency.*

A lower 3-dB cutoff frequency means a reduced bandwidth of the lowpass response and an increased bandwidth of the highpass response (Section 10.3, Chapter 10).

The voltage across the inductor is $v_L = +L\dfrac{di}{dt}$. From Equation 12.1.5,

$$v_L = -K\frac{R}{L}e^{-t/\tau}, \quad t \geq 0^+ \tag{12.1.6}$$

v_L is again an exponential function but is negative because after the impulse is over, $v_R = Ri$ and $v_L = -Ri$ (Figure 12.1.3).

The interpretation of the behavior of the circuit is that the polarity of the impulse is that of positive v_L (Figure 12.1.1) and establishes an initial flux linkage K Vs or Wb-turns. When i flows, v_L becomes negative (Equation 12.1.6), and $\int v_L dt$ increases negatively with time, which decreases the flux linkage in the inductor, and hence, the current. At any time t, $\lambda = K - \int_0^t v_L dt = K\left[1 - \dfrac{1}{\tau}\int_0^t e^{-t/\tau}\right] = Ke^{-t/\tau} = Li$. The magnitude of the rate at which λ decreases is $\left|\dfrac{d\lambda}{dt}\right| = |v_L|$, so that as v_L becomes smaller in magnitude, the rate of decrease of

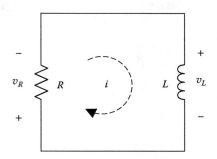

FIGURE 12.1.3
RL circuit for $t \geq 0^+$.

λ, and hence i, becomes progressively smaller in magnitude, which causes flattening of the response that is characteristic of an exponential.

The case of a current impulse applied to a parallel *RL* circuit or to a series *RL* circuit is discussed in Section ST12.1.

EXERCISE 12.1.1
Show that the total energy dissipated in *R* equals that initially stored in *L*.

Simulation Example 12.1.1: Impulse Response of *RL* Circuit with Initial Current

An impulse of 1 Vs is applied to an *RL* circuit consisting of 10 Ω in series with 1 H. An initial current of 1 A flows in the inductor in the same direction as would flow due to the impulse. It is required to determine how the current and voltage across the inductor vary with time, and to simulate the circuit.

SOLUTION
Equation 12.1.1 and Equation 12.1.2 apply. Integrating Equation 12.1.2 gives as before $i(0^+) - i(0^-) = \dfrac{K}{L}$. The only difference is that $i(0^-) = 1$ A instead of zero.

Hence, $i(0^+) = \dfrac{1}{1} + 1 = 2$ A. Note that the voltage impulse causes a jump of 1 A in the inductor current irrespective of the initial current. Thus, zero initial energy storage may be assumed and the strength of the applied impulse doubled to account for the initial current of 1 A.

$\tau = \dfrac{1}{10} = 0.1$ s. The solution to Equation 12.1.4 is Be^{-10t}, where $B = i(0^+) = 2$ A. It follows that $i = 2e^{-10t}$ A and $v_L = 1 \times \dfrac{di}{dt} = -20e^{-20t}$ V.

SIMULATION
The schematic is entered as in Figure 12.1.4. To simulate a 1 Vs impulse, a VPULSE source is used and a single pulse is applied of 0.1 ms duration, which is small compared to the time constant of 0.1 s. To have a pulse area of 1 Vs, the amplitude of the pulse must be 10 kV. The parameters of VPULSE are set as shown. The initial voltage level V1 = 0. To

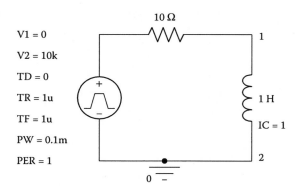

FIGURE 12.1.4
Figure for Example 12.1.1.

have a pulse amplitude of 10 kV, V2 = 10k. The time delay of the pulse is TD = 0. The rise time (TR) of the pulse and its fall time (TF) are chosen to be 1 μs, which is small compared to the pulse duration (PW) of 0.1 ms. Because the simulation is to run for five time constants, that is, 0.5 s, the period of the pulses (PER) is made larger than this, say 1 s, so that only a single pulse is applied during the simulation. The initial current is entered in the Property Editor spreadsheet of the inductor as IC = 1. PSpice considers this current to flow from terminal 1 of the inductor to terminal 2.

In the Simulation Settings window, choose Time Domain (Transient) for Analysis Type, enter 0.5s for Run to Time, 0 for Start Saving Data after, and 10u for Maximum Step Size. After the simulation is run, select Plot/Axis Settings/*x*-axis and enter a User Defined Data Range of 0.12 to 500ms so as to display data after the end of the pulse. Choose a User Defined Data Range of −2 to +2 for the *y*-axis. Select Trace/Add Traces and enter first I(L1) then V(L1:1)/10 so as to display inductor current and voltage to appropriate scales on the same plot. To display the line tangent enter: 2 − 20*Time. The graph of Figure 12.1.5 is displayed.

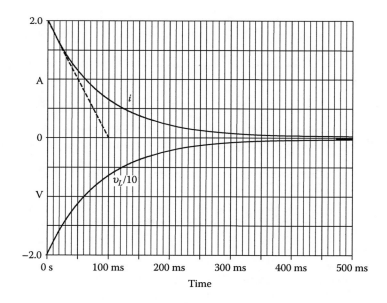

FIGURE 12.1.5
Figure for 12.1.1.

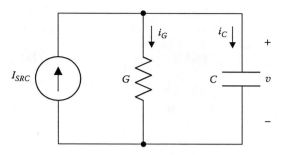

FIGURE 12.1.6
GC circuit excited by current impulse.

TABLE 12.1.1

Dual Circuits

RL Circuit	GC Circuit
$v_{SRC} = K\delta(t)$ Vs	$i_{SRC} = K\delta(t)$ As
i	v
R	G
L	C
v_R	i_G
v_L	i_C
L/R	C/G

GC Circuit

The dual of the *RL* circuit of Figure 12.1.1 is the *GC* circuit of Figure 12.1.6, where the dual quantities are listed in Table 12.1.1. We can immediately write the dual relations of Equation 12.1.5 and Equation 12.1.6 as

$$v = \frac{K}{C} e^{-\frac{t}{\tau}}, \quad t \geq 0^+ \tag{12.1.7}$$

and

$$i_C = -K\frac{G}{C} e^{-\frac{t}{\tau}}, \quad t \geq 0^+ \tag{12.1.8}$$

where the time constant is $\tau = C/G = 1/\omega_c$, and ω_c is the 3-dB frequency considered in Section 10.2 (Chapter 10). The time course of v and i_C is exponential, as in Figure 12.1.2.

To gain some insight into the behavior of the circuit, we can write the differential equation from KCL in terms of v as:

$$C\frac{dv}{dt} + Gv = K\delta(t) \tag{12.1.9}$$

which is the dual relation of Equation 12.1.1. Integrating Equation 12.1.1 over the duration of the impulse, from $t = 0^-$ to $t = 0^+$:

$$C\int_{0^-}^{0^+} dv + G\int_{0^-}^{0^+} vdt = K\int_{0^-}^{0^+} \delta(t)dt \tag{12.1.10}$$

We have seen in Section 11.2 (Chapter 11) that a current impulse applied to a capacitor causes a step change of K in the charge, and hence a finite change in voltage of K/C. Adding a resistor in parallel with the capacitor cannot cause an infinite change in v, so that the second integral on the LHS of Equation 12.1.10 is zero. It follows that $v(0^+) - v(0^-) = \dfrac{K}{L}$. Assuming no initial energy storage, $v(0^-) = 0$, so that

$$v(0^+) = \frac{K}{C} \qquad\qquad (12.1.11)$$

The following comments can be made, analogous to those for the inductor:

1. $v(0^+)$ does not depend on G. The interpretation is that during the impulse, $i_{SRC} \to \infty$, while v, and hence Gv, remain finite. This implies that the current impulse essentially flows through C, so that the presence of G is of no consequence as far as $v(0^+)$ is concerned. In other words, the capacitor behaves as a short circuit as far as the impulse is concerned, except that the voltage across a short circuit is zero, whereas the voltage across the capacitor increases to K/C at the end of the impulse.

2. Once the impulse is over, the current source behaves as an open circuit. The circuit for $t \ge 0^+$ reduces to a resistor connected across a capacitor having an initial voltage K/C; i_C goes negative, as C discharges through R.

The time course of v for $t \ge 0^+$ is governed by the differential equation: $C\dfrac{dv}{dt} + Gv = 0$. Dividing by C gives Equation 12.1.4 but with $\tau = C/G$. Solving this equation, as was done earlier, gives Equation 12.1.7. i_C is then $+C\dfrac{dv}{dt}$.

The interpretation of the behavior of the circuit is that the current impulse establishes an initial charge K coulombs and an initial voltage $v = K/C$. For $t \ge 0^+$, the capacitor discharges, the current in the circuit being $i = -i_C$ (Figure 12.1.7). When i flows, the charge q on C decreases. At any time t, $q = K - \displaystyle\int_0^t i_C dt = K\left[1 - \frac{1}{\tau}\int_0^t e^{-t/\tau}\right] = Ke^{-t/\tau} = Cv$. The magnitude

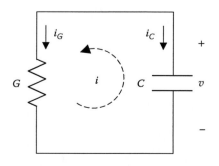

FIGURE 12.1.7
GC circuit for $t \ge 0+$.

of the rate at which q decreases is $|i_C|$, so that as i_C becomes smaller in magnitude, the rate of decrease of q, and hence v, becomes progressively smaller in magnitude.

The case of a voltage impulse applied to a series GC circuit or to a parallel GC circuit is discussed in Section ST12.2.

It can therefore be concluded that:

> **Concept:** *The natural response of a first-order circuit consisting of R across L, or G across C, is a decaying exponential function of time having a time constant L/R or C/G, respectively.*

EXERCISE 12.1.2

Formulate, analyze, and simulate the dual of the circuit of Example 12.1.1.

12.2 Natural Responses of Second-Order Circuits

We will illustrate a second-order response by considering the series RLC circuit of Figure 12.2.1 with $v_{SRC} = K\delta(t)$ and without any initial energy storage. From KVL,

$$L\frac{di}{dt} + Ri + \frac{1}{C}\int idt = K\delta(t) \tag{12.2.1}$$

Integrating both sides of Equation 12.2.1 over the duration of the impulse, from $t = 0^-$ to $t = 0^+$,

$$L\int_{0^-}^{0^+} di + R\int_{0^-}^{0^+} idt + \frac{1}{C}\int_{0^-}^{0^+} qdt = K\int_{0^-}^{0^+} \delta(t)dt \tag{12.2.2}$$

We will assume, as in the case of the RL circuit, that there will be a stepwise change in the current due to the impulse. The presence of C cannot make i infinite. Nevertheless, it should be noted that *in solving a problem one can always make certain assumptions and then check the consistency of the results based on these assumptions. If the results are consistent, the*

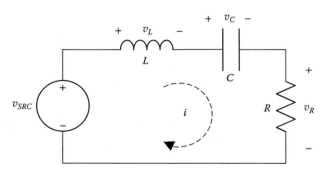

FIGURE 12.2.1
RLC circuit excited by voltage impulse.

assumptions are justified. When the change in i is finite, $q = \int_{0^-}^{0^+} i \, dt = 0$, so that the second

and third integrals on the LHS of Equation 12.2.2 are zero. It follows that $i(0^+) = \dfrac{K}{L}$, as in Equation 12.1.3. In other words, because the change in i is finite, the change in v_R is finite, the change in v_C is zero, and the infinite voltage of the impulse appears across L. This causes a stepwise change in λ of K Wb-turns and hence a change in current of K/L.

To determine how i varies for $t \geq 0^+$, we have to solve Equation 12.2.1 with the RHS equal to zero. Differentiating the LHS of the resulting equation so as to remove the integral of i:

$$L\frac{d^2i}{dt^2} + R\frac{di}{dt} + \frac{i}{C} = 0 \; ; \; t \geq 0^+ \tag{12.2.3}$$

Equation 12.2.3 is a second-order linear differential equation whose solution is subject to the following considerations: (1) i is an exponential-type function of time, which includes sinusoidal functions, because the derivative of an exponential is also an exponential, which means that such a function can satisfy Equation 12.2.3; (2) the equation is linear, so that sums of exponential functions would also satisfy the equation; and (3) because the equation contains a second-order derivative, its solution basically involves two integrations with respect to time, which means that the solution should have two arbitrary constants. It follows that the general solution is of the form:

$$i = Ae^{s_1t} + Be^{s_2t} \; ; \; t \geq 0^+ \tag{12.2.4}$$

where A and B are the arbitrary constants determined by initial conditions, and s_1 and s_2 depend on the coefficients of the terms in Equation 12.2.3. To determine s_1 and s_2, we substitute $i = e^{st}$ in Equation 12.2.3, which gives: $e^{st}\left(s^2L + sR + \dfrac{1}{C}\right) = 0$. To satisfy this equation for any s, the bracketed sum must be zero. Dividing by L, this result can be expressed as:

$$s^2 + 2\alpha s + \omega_0^2 = 0 \tag{12.2.5}$$

where

$$\alpha = \frac{R}{2L} \text{ and } \omega_0 = \frac{1}{\sqrt{LC}} \tag{12.2.6}$$

Equation 12.2.5 is the **characteristic equation** of the differential equation (Equation 12.2.3). ω_0 is the same **resonant frequency** of the bandpass response (Section 10.3, Chapter 10). α is the **damping factor, damping coefficient**, or **neper frequency**. α and ω_0 are related to Q and ξ, defined by Equation 10.3.2 and Equation 10.5.2 of Chapter 10, respectively. Thus,

$$Q = \frac{\omega_0}{2\alpha} \text{ and } \xi = \frac{\alpha}{\omega_0} \tag{12.2.7}$$

The solutions s_1 and s_2 of Equation 12.2.5 are:

$$s_1 = -\alpha + \sqrt{\alpha^2 - \omega_0^2} \tag{12.2.8}$$

$$s_2 = -\alpha - \sqrt{\alpha^2 - \omega_0^2} \tag{12.2.9}$$

It is seen from Equation 12.2.8 and Equation 12.2.9 that there are three cases to consider, corresponding to (1) $\alpha > \omega_0$, (2) $\alpha = \omega_0$, and (3) $\alpha < \omega_0$. In all cases, however,

$$s_1 + s_2 = -2\alpha, \; s_1 - s_2 = 2\sqrt{\alpha^2 - \omega_0^2}, \text{ and } s_1 s_2 = \omega_0^2 \tag{12.2.10}$$

Moreover, the initial conditions at $t = 0^+$ are $i(0^+) = K/L$ and $q(0^+) = 0$. The functions $e^{s_1 t}$ and $e^{s_2 t}$ in Equation 12.2.4 are continuous at $t = 0$, and therefore have the same value at $t = 0$ as at $t = 0^+$. Substituting $t = 0$ in Equation 12.2.4 gives:

$$A + B = \frac{K}{L} \tag{12.2.11}$$

To obtain $q(t)$ at any t, we integrate Equation 12.2.4 to obtain:

$$q(t) = \frac{A}{s_1} e^{s_1 t} + \frac{B}{s_2} e^{s_2 t} \tag{12.2.12}$$

where the constant of integration, which sets the initial value of q, is ignored because A and B have already been included for this purpose. Equation 12.2.12 gives at $t = 0^+$:

$$\frac{A}{s_1} + \frac{B}{s_2} = 0 \tag{12.2.13}$$

Solving Equation 12.2.11 and Equation 12.2.13 for A and B, $A = \frac{s_1}{s_1 - s_2} \frac{K}{L}$ and $B = -\frac{s_2}{s_1 - s_2} \frac{K}{L}$. Hence,

$$i = \frac{K}{L(s_1 - s_2)} \left[s_1 e^{s_1 t} - s_2 e^{s_2 t} \right], \; t \geq 0^+ \tag{12.2.14}$$

An alternative, and somewhat less direct, method for determining A and B is to recognize that at $t = 0^+$, with $v_C(0^+) = 0$, KVL gives $v_L(0^+) = -Ri(0^+)$, or $\left.\dfrac{di}{dt}\right|_{t=0^+} = -\dfrac{R}{L}i(0^+) = -2\alpha i(0^+)$. Substituting from Equation 12.2.4, we obtain

$As_1 + Bs_2 = -2\alpha\dfrac{K}{L}$. Solving with Equation 12.2.11 gives the same values of A and B.

Before considering the three cases referred to earlier, we wish to clarify the relation between the impulse response and the natural response of a second-order circuit. Both of these responses are the solutions to Equation 12.2.3 with zero on the RHS. The impulse response of the circuit of Figure 12.2.1 is that for an initial current in the circuit but no initial voltage across the capacitor. On the other hand, the natural response may include an initial voltage across the capacitor as well.

Overdamped Response

> **Statement:** *In an overdamped circuit, R is large enough so that $\alpha > \omega_0$, s_1 and s_2 are negative real and distinct, $Q < 1/2$, and $\xi > 1$.*

It follows from Equation 12.2.14 that the response is the difference of two real exponentials with negative exponents. Current $i = K/L$ when $t = 0^+$, $i = 0$ when the expression in square brackets is zero, and $i \to 0$ as $t \to \infty$, because s_1 and s_2 are negative real. Moreover,

$$v_L = L\frac{di}{dt} = \frac{K}{(s_1 - s_2)}\left[s_1^2 e^{s_1 t} - s_2^2 e^{s_2 t}\right], \quad t \ge 0^+ \tag{12.2.15}$$

$$v_C = \frac{1}{C}\int_0^t i\,dt = \frac{K\omega_0^2}{(s_1 - s_2)}\left[e^{s_1 t} - e^{s_2 t}\right], \quad t \ge 0^+ \tag{12.2.16}$$

It is seen that $v_R + v_L + v_C = 0, t \ge 0^+$, in accordance with KVL. All the results obtained are consistent, which validates the assumption made concerning the jump in i. The variation of i, v_L, and v_C with time is discussed in Simulation Example 12.2.1.

Simulation Example 12.2.1: Overdamped Response

The circuit to be simulated is that of Example 10.5.1 (Chapter 10) having $L = 0.5$ H, $C = 12.5$ μF, but with $R = 500$ Ω and $K = 0.5$ Vs. This gives $\omega_0 = \dfrac{1}{\sqrt{LC}} = 400$ rad/s,

$\alpha = \dfrac{R}{2L} = 500$ rad/s. Hence, $s_1 = -500 + \sqrt{(500)^2 - (400)^2} = -200$ rad/sec, and $s_2 = -500 - \sqrt{(500)^2 - (400)^2} = -800$ rad/s.

SIMULATION

The schematic is shown in Figure 12.2.2. The parameters of the VPULSE source are as explained in Simulation Example 12.1.1. After the simulation is run, the plots of Figure 12.2.3 are obtained showing the time variation of i, v_L, and v_C.

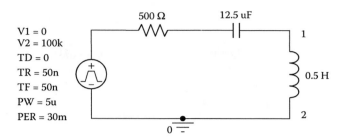

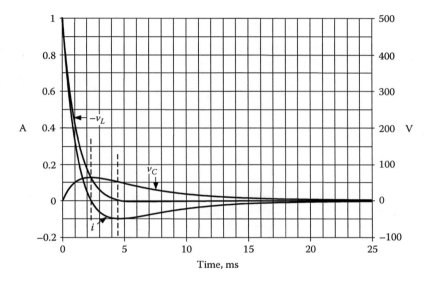

FIGURE 12.2.2
Figure for Example 12.2.1.

FIGURE 12.2.3
Figure for Example 12.2.1.

At $t = 0^+$, $v_{SRC} = 0$, $i = \dfrac{0.5 \text{ Vs}}{0.5 \text{ H}} = 1$ A, and $v_C = 0$. It follows from KVL that $v_R + v_L = 0$, or

$-v_L = Ri = 500$ V. The polarity of v_L is such that $\int v_L dt$ decreases the flux linkage established by the impulse. As t increases, L discharges, as described in Section 12.1. Simultaneously, C is charged by i, and v_C increases. i continues to drop, passing through zero. When $i = 0$, v_C equals $-v_L$, and v_C is a maximum because $i = C\dfrac{dv_C}{dt}$. Beyond this time, C discharges and drives i in the negative direction. Note that because $v_L = Ldi/dt$, then $-v_L > 0$ when i is decreasing, $-v_L < 0$ when i is increasing, and $v_L = 0$ when i is minimum. Eventually, all the energy initially stored in L by the impulse is dissipated in R, and all the circuit responses decay to zero.

EXERCISE 12.2.1

Show that $i = 0$ at $t_0 = \dfrac{\ln(s_2/s_1)}{s_1 - s_2}$, and that $v_L = 0$ at $2t_0$. Calculate these quantities for the values of Example 12.2.1.

Answer: 2.31 ms and 4.62 ms.

EXERCISE 12.2.2

Verify that Equation 12.2.14 to Equation 12.2.16 give: $v_R + v_L + v_C = 0, t \geq 0^+$.

Critically Damped Response

Statement: *In a critically damped circuit, R is such that $\alpha = \omega_0$, which makes $s_1 = s_2 = \omega_0$, $Q = 1/2$, and $\xi = 1$.*

Equation 12.2.14 is not valid in this case because Equation 12.2.4 is not the most general form of the response. With $s_1 = s_2 = \omega_0$, Equation 12.2.4 becomes $i = (A+B)e^{\omega_0 t}$ and contains only one arbitrary constant $(A + B)$. The general solution is of the form:

$$i = Ae^{-\omega_0 t} + Bte^{-\omega_0 t} \tag{12.2.17}$$

as can be verified by substituting Equation 12.2.17 in Equation 12.2.3. The initial condition that $i(0^+) = K/L$ at $t = 0^+$ gives $A = K/L$. Integrating Equation 12.2.17 by parts, or referring to the Table of Integrals in the Appendix,

$$q(t) = -\frac{A}{\omega_0}e^{-\omega_0 t} - \frac{Be^{-\omega_0 t}}{\omega_0^2}(\omega_0 t + 1) \tag{12.2.18}$$

where the constant of integration has been ignored, as explained previously. At $t = 0^+$, $q(0^+) = -\dfrac{A}{\omega_0} - \dfrac{B}{\omega_0^2} = 0$, or $B = -\omega_0 A$. It follows that:

$$i = \frac{K}{L}e^{-\omega_0 t}\left(1 - \omega_0 t\right), \; t \geq 0^+ \tag{12.2.19}$$

$$v_L = L\frac{di}{dt} = -K\omega_0 e^{-\omega_0 t}\left[2 - \omega_0 t, \; t \geq 0^+\right. \tag{12.2.20}$$

$$v_C = \frac{1}{C}\int_0^t idt = K\omega_0^2 te^{-\omega_0 t}, \; t \geq 0^+ \tag{12.2.21}$$

Although Equation 12.2.14 to Equation 12.2.16 look quite different from Equation 12.2.19 to Equation 12.2.21, the general time course of the variation of i, v_L, and v_C is the same as in Figure 12.2.3, as can be readily verified by repeating the simulation of Example 12.2.1 with $R = 2\omega_0 L = 400\ \Omega$. This gives: $i = 0$ at $t_0 = 1/\omega_0 = 2.5$ ms and $v_L = 0$ at $2t_0 = 1/\omega_0 = 5$ ms. These values are slightly larger than for the overdamped case with $R = 500\ \Omega$. For the same initial current, the larger R the greater, the power dissipation. and the faster is the decrease in current.

EXERCISE 12.2.3

Verify that Equation 12.2.17 satisfies Equation 12.2.3.

EXERCISE 12.2.4

Verify that Equation 12.2.19 to Equation 12.2.21 give $v_R + v_L + v_C = 0, t \geq 0^+$.

Underdamped Response

Statement: *In an underdamped circuit, R is small enough so that $\alpha < \omega_0$, s_1 and s_2 are complex, $Q > 1/2$, and $\xi < 1$.*

In this case,

$$s_1 = -\alpha + j\omega_d \tag{12.2.22}$$

$$s_2 = -\alpha - j\omega_d \tag{12.2.23}$$

where

$$\omega_d = \sqrt{\omega_0^2 - \alpha^2} \tag{12.2.24}$$

is the **damped natural frequency.**
Substituting in Equation 12.2.14 and simplifying,

$$i = \frac{K}{L}e^{-\alpha t}\left(\cos \omega_d t - \frac{\alpha}{\omega_d}\sin \omega_d t\right), \quad t \geq 0^+ \tag{12.2.25}$$

It follows that

$$v_L = L\frac{di}{dt} = -Ke^{-\alpha t}\left[2\alpha \cos \omega_d t + \left(\omega_d - \frac{\alpha^2}{\omega_d}\right)\sin \omega_d t\right], \quad t \geq 0^+ \tag{12.2.26}$$

$$v_C = \frac{1}{C}\int_0^t i\,dt = \left(K\frac{\omega_0^2}{\omega_d}e^{-\alpha t}\sin \omega_d t\right), \quad t \geq 0^+ \tag{12.2.27}$$

The variation of i, v_L, and v_C is now oscillatory, as illustrated by Example 12.2.2.

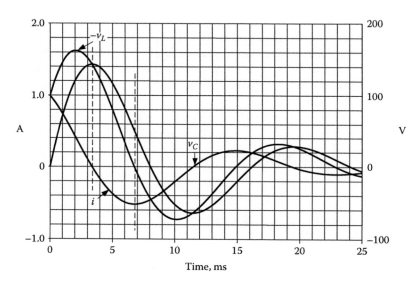

FIGURE 12.2.4
Figure for Example 12.2.2.

Simulation Example 12.2.2: Underdamped Response

The same circuit of Example 12.2.1 is be simulated but with $R = 100\ \Omega$. This makes $\alpha = 100$ rad/s, $\omega_d = 100\sqrt{15} = 387.3$ rad/s, $Q = 2$, and $\xi = 0.25$.

SIMULATION

The simulated response is shown in Figure 12.2.4.

At $t = 0^+$, $v_{SRC} = 0$, $i = 1$ A, and $v_C = 0$. It follows from KVL that $v_R + v_L = 0$, or $-v_L = Ri = 100$ V. However, because of the sinusoidal variation of i, the curvature of the variation of i for small values of t is such that $\dfrac{di}{dt}$ becomes more negative, that is, $\dfrac{d^2 i}{dt^2} < 0$, $\dfrac{dv_L}{dt} < 0$, or $\dfrac{d(-v_L)}{dt} > 0$, so that v_L increases for small values of t before reaching a maximum and then decreasing. Note the marked oscillatory nature of the response. The relationship between i, v_L, and v_c can be interpreted in the same manner as in Simulation Example 12.2.1.

EXERCISE 12.2.5

Show that $i = 0$ at $t_0 = \dfrac{1}{\omega_d} \tan^{-1}\left(\dfrac{\omega_d}{\alpha}\right)$, and that $v_L = 0$ at $2t_0$. Calculate these quantities

for the values of Example 12.2.2.

Answer: 3.40 ms and 6.81 ms.

EXERCISE 12.2.6

Verify that Equation 12.2.25 to Equation 12.2.27 give $v_R + v_L + v_C = 0$, $t \geq 0^+$.

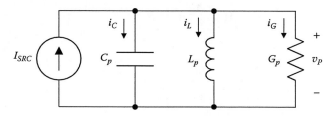

FIGURE 12.2.5
GCL circuit excited by current impulse.

EXERCISE 12.2.7

Repeat the simulation of Example 12.2.2 with $R = 283\ \Omega$ for a Butterworth response. Note that although the Butterworth response is flat in the frequency domain, it is slightly oscillatory in the time domain.

Parallel *GCL* Circuit

The dual of the series *RLC* circuit considered previously is the parallel *GCL* circuit to which a current impulse of strength K is applied (Figure 12.2.5). The resonant frequency ω_0 remains the same when L and C are interchanged. The dual damping factor is

$$\alpha_p = \frac{G_p}{2C_p} = \frac{1}{2C_p R_p} \qquad (12.2.28)$$

Q_p and ξ_p are related as in Equation 12.2.7:

$$Q_p = \frac{\omega_0}{2\alpha_p} \text{ and } \xi_p = \frac{\alpha_p}{\omega_0} \qquad (12.2.29)$$

s_{1p}, s_{2p}, and ω_{dp} are given by

$$s_{1p} = -\alpha_p + \sqrt{\alpha_p^2 - \omega_0^2} \qquad (12.2.30)$$

$$s_{2p} = -\alpha_p - \sqrt{\alpha_p^2 - \omega_0^2} \qquad (12.2.31)$$

$$\omega_{dp} = \sqrt{\omega_0^2 - \alpha_p^2} \qquad (12.2.32)$$

v, i_G, i_C, and i_L are the duals of i, v_R, v_L, and v_C, respectively, and are given by the corresponding dual relations. Table 12.2.1 summarizes the impulse responses of the two circuits.

It may be concluded that:

TABLE 12.2.1

Impulse Responses of Dual Circuits

Series *RLC* Circuit	Parallel *GCL* Circuit

Overdamped Response

$$i = \frac{K}{L(s_1 - s_2)}\left[s_1 e^{s_1 t} - s_2 e^{s_2 t} \right] \qquad v = \frac{K}{C_p(s_{1p} - s_{2p})}\left[s_{1p} e^{s_{1p} t} - s_{2p} e^{s_{2p} t} \right]$$

$$v_L = \frac{K}{(s_1 - s_2)}\left[s_1^2 e^{s_1 t} - s_2^2 e^{s_2 t} \right] \qquad i_C = \frac{K}{(s_{1p} - s_{2p})}\left[s_{1p}^2 e^{s_{1p} t} - s_{2p}^2 e^{s_{2p} t} \right]$$

$$v_C = \frac{K\omega_0^2}{(s_1 - s_2)}\left[e^{s_1 t} - e^{s_2 t} \right] \qquad i_L = \frac{K\omega_0^2}{(s_{1p} - s_{2p})}\left[e^{s_{1p} t} - e^{s_{2p} t} \right]$$

Critically Damped Response

$$i = \frac{K}{L} e^{-\omega_0 t}\left(1 - \omega_0 t \right) \qquad v = \frac{K}{C_p} e^{-\omega_0 t}\left(1 - \omega_0 t \right)$$

$$v_L = -K\omega_0 e^{-\omega_0 t}\left[2 - \omega_0 t \right] \qquad i_C = -K\omega_0 e^{-\omega_0 t}\left[2 - \omega_0 t \right]$$

$$v_C = K\omega_0^2 t e^{-\omega_0 t} \qquad i_L = K\omega_0^2 t e^{-\omega_0 t}$$

Underdamped Response

$$i = \frac{K}{L} e^{-\alpha t}\left(\cos\omega_d t - \frac{\alpha}{\omega_d}\sin\omega_d t \right) \qquad v = \frac{K}{C_p} e^{-\alpha_p t}\left(\cos\omega_{dp} t - \frac{\alpha_p}{\omega_{dp}}\sin\omega_{dp} t \right)$$

$$v_L = -K e^{-\alpha t}\left[2\alpha\cos\omega_d t + \left(\omega_d - \frac{\alpha^2}{\omega_d} \right)\sin\omega_d t \right] \qquad i_C = -K e^{-\alpha_p t}\left[2\alpha_p \cos\omega_{dp} t + \left(\omega_{dp} - \frac{\alpha_p^2}{\omega_d} \right)\sin\omega_{dp} t \right]$$

$$v_C = \left(K\frac{\omega_0^2}{\omega_d} e^{-\alpha t}\sin\omega_d t \right) \qquad i_L = \left(K\frac{\omega_0^2}{\omega_{dp}} e^{-\alpha_p t}\sin\omega_{dp} t \right)$$

Conclusion: *A voltage impulse instantly establishes in a series RLC circuit an initial energy storage in the inductor, whereas a current impulse instantly establishes in a parallel GCL circuit an initial energy storage in the capacitor. The stored energy is manifested as an initial condition in the circuit, that is, an initial current in an inductor, or an initial voltage across a capacitor.*

Sustained Oscillations

The case of zero R or G_p deserves special consideration. It might seem like this is an ideal case that is of no practical interest. Whereas the first part of the statement is true, the second part is not. Active devices, such as transistors, can be used to introduce a negative resistance that effectively reduces R or G_p to zero (Example SE2.5, Chapter 2). In other words, the active device introduces sufficient energy to compensate for the power dissipated in the circuit. Considering the series circuit, when $R = 0$, $\alpha = 0$. The current and voltages become purely sinusoidal functions, so that the oscillations at the frequency ω_0 can, in principle, be sustained indefinitely (Example SE12.2). This is an example of an ***LC* oscillator**.

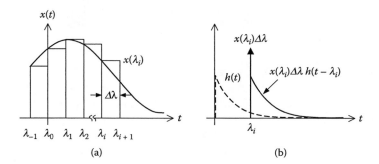

FIGURE 12.3.1
Approximation by means of impulses.

12.3 Convolution

Convolution is basically an integration operation applied to two time functions in a particular way, producing a new time function. To motivate our study of convolution, consider a complicated circuit or system whose response $h(t)$ to a unit impulse can be determined experimentally by applying a large, brief input of a duration that is much shorter than the smallest time constant of the system, as was done in the simulation examples earlier. The question is: knowing $h(t)$, how can one determine the circuit response $y(t)$ to an arbitrary excitation $x(t)$? We will show that for a linear time-invariant (LTI) circuit,

$$y(t) = \int_{-\infty}^{\infty} x(\lambda) h(t - \lambda) d\lambda \tag{12.3.1}$$

Equation 12.3.1 defines a convolution operation, represented as

$$y(t) = x(t) * h(t) \tag{12.3.2}$$

In words, $x(t)$ is said to be convolved with $h(t)$.

To derive Equation 12.3.1, we first approximate $x(t)$ by a series of pulses of equal width, as illustrated in Figure 12.3.1a. The ith pulse starts at $t = \lambda_i$, is of duration $\Delta\lambda$, and height $x(\lambda_i)$. As $\Delta\lambda$ is made very small, the response $y(\lambda_i)$ to the ith pulse is the same as the response to an impulse at $t = \lambda_i$ whose strength is the same as the area of the pulse, which is $x(\lambda_i)\Delta\lambda$. In other words, $y(\lambda_i) = x(\lambda_i)\Delta\lambda \times h(t - \lambda_i), t \geq \lambda_i$, where $h(t - \lambda_i)$ is the response of the circuit to an impulse at $t = \lambda_i$ (Figure 12.3.1b). Because the system is LTI, the total response is the sum of the responses to all the individual pulses representing the input:

$$y(t) = \sum_{i=-\infty}^{\infty} x(\lambda_i) h(t - \lambda_i) \Delta\lambda \tag{12.3.3}$$

In the limit, $\Delta\lambda \to d\lambda$, and the summation becomes an integration, which yields Equation 12.3.1. It follows that:

Concept: *The response of an LTI circuit to an arbitrary input can be considered as the superposition of responses to sufficiently narrow pulses, each having an amplitude determined by the input.*

Mathematically, the integral of Equation 13.3.1 represents an infinite sum, over a continuum of time, of responses to impulses, each having an infinitesimal strength. Practically, the summation of Equation 12.3.3 is the sum of responses to pulses whose duration is small compared with the smallest time constant of the circuit or the reciprocal of the highest frequency in the circuit's natural response.

It is convenient, though by no means essential, to assume that $x(t) = 0$ for $t < 0$, which means that the lower limit of integration in Equation 12.3.1 can be taken as 0 instead of $-\infty$. This is usually done because of the link between convolution and the one-sided Laplace transform. Moreover, physical systems operating in real time are **causal**, or nonanticipatory (Section 16.6, Chapter 16); that is, they do not respond to an input before the input is applied. This means that if an impulse is applied at time λ, there cannot be a response to the impulse at $t < \lambda$, before the impulse is applied. Thus, $h(t - \lambda) = 0$ for $t < \lambda$, or $\lambda > t$. The upper limit of integration becomes t instead of ∞. Equation 12.3.1 reduces to:

$$y(t) = \int_0^t x(\lambda)h(t - \lambda)d\lambda \qquad (12.3.4)$$

It should be appreciated that in Equation 12.3.4 t is a constant that has a particular value. The convolution integral evaluates $y(t)$ at a single value of time. However, in a given problem, t can assume any value within a specified range, in which case $y(t)$ must be obtained over a given range of t.

The convolution integral Equation 12.3.4 can be expressed as

$$y(t) = \int_0^t x(\lambda)h(t - \lambda)d\lambda = \int_0^t h(\lambda)x(t - \lambda)d\lambda \qquad (12.3.5)$$

The second integral can be derived from the first by substituting $u = t - \lambda$, which gives

$$= \int_t^0 x(t - u)h(u)[-du].$$ Interchanging the limits of integration and replacing the dummy integration variable u by λ gives the second integral. According to Equation 12.3.5, convolution is *commutative*. That is,

$$y(t) = x(t) * h(t) = h(t) * x(t) \qquad (12.3.6)$$

EXERCISE 12.3.2

Determine the convolution of $(1 - e^{-t})$ and e^{-2t}.

Answer: $\dfrac{1}{2} - e^{-t} + \dfrac{e^{-2t}}{2}$.

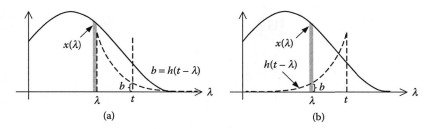

FIGURE 12.3.2
Graphical interpretation of convolution.

Graphical Interpretation

Figure 12.3.2a shows the excitation $x(\lambda)$ with respect to the time variable λ. Consider a narrow pulse of height $x(\lambda)$ and width $d\lambda$ at an arbitrary value of λ. As discussed earlier, the response to this pulse is, in the limit, the same as that to an impulse at λ of strength $x(\lambda)d\lambda$. The response to this impulse is determined by drawing $h(t)$ starting at λ as shown in Figure 12.3.2a. At any time $t > \lambda$, the response to the pulse $x(\lambda)d\lambda$ is $x(\lambda)d\lambda h(t-\lambda)$, where $h(t-\lambda) = b$ is the magnitude of the impulse response at a time $(t-\lambda)$ after its start. The circuit response at t is then the summation of all responses, such $x(\lambda)d\lambda h(t-\lambda)$, from $\lambda = 0$ to $\lambda = t$, as expressed by the first integral in Equation 12.3.5. For example, the pulse $x(0)d\lambda$ at $\lambda = 0$ is multiplied by $h(t)$, whereas the pulse at $x(t)d\lambda$ at $\lambda = t$ is multiplied by $h(0)$.

Although the graphical construction of Figure 12.3.2a is easy to interpret, the integral $\int_0^t x(\lambda)h(t-\lambda)d\lambda$ should represent an area. But this area is difficult to visualize in Figure 12.3.2a because the ordinate at λ is multiplied by the ordinate at a different location on the time axis, namely t. This difficulty can be overcome by drawing the impulse response differently. $h(-\lambda)$ is the folded impulse response drawn backwards, that is, in the negative λ direction, starting at $\lambda = 0$. $h(t-\lambda)$ is then $h(-\lambda)$ shifted by t in the positive λ direction, as shown in Figure 12.3.2b. The vertical intercept with the pulse $x(\lambda)d\lambda$ at λ is the same b as in Figure 12.3.2a, namely $h(t-\lambda)$. The product $x(\lambda)d\lambda \times h(t-\lambda)$ now represents the product of two quantities both occurring at time λ: $x(\lambda)d\lambda$ from the applied excitation and $h(t-\lambda)$ from the impulse response. The integral now represents the area subtended by the product of the two functions from $\lambda = 0$ to $\lambda = t$.

Note that when a function $x(t)$ is shifted by a in the positive direction of t, it becomes $x(t-a)$ with respect to the same time origin, where a could be positive or negative. If x is a function $x(-t)$ of $-t$, then replacing t by $(t-a)$, $-(t-a) = -t+a$, so that a is added to $-t$ for a shift in the positive direction of t.

Figure 12.3.2b forms the basis for the graphical evaluation of the convolution integral for a particular value of t, according to the following steps:

Procedure

1. Fold the impulse response, that is, draw it backwards as $h(-\lambda)$ at $\lambda = 0$.
2. Shift $h(-\lambda)$ to the right by t to obtain $h(t-\lambda)$.
3. Calculate the area corresponding to the product $x(-\lambda)h(t-\lambda)$ over the range $0 \le \lambda \le t$. The result is $y(t)$ at the chosen value of t.
4. Steps 2 and 3 are repeated for various values of t to obtain $y(t)$ over the desired range of t.

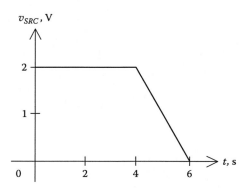

FIGURE 12.3.3
Figure for Example 12.3.1.

The aforementioned procedure is illustrated by Example 12.3.1. From the commutative property of convolution, the preceding graphical procedure could just as well be applied with the roles of $x(t)$ and $h(t)$ interchanged (Exercise 12.3.2).

Simulation Example 12.3.1: Response of *RL* Circuit Using Convolution

Consider a series *RL* circuit with $R = 1\ \Omega$, $L = 1$ H, and v_{SRC} as shown in Figure 12.3.3. It is required to determine i using convolution, assuming zero initial conditions.

SOLUTION
The first step is to derive the impulse response. From Equation 12.1.5, with $K = 1$, $L = 1$, $\tau = 1$, and t replaced by λ:

$$h(\lambda) = e^{-\lambda}u(\lambda)\,\text{A/Vs} \tag{12.3.7}$$

Because of the shape of the excitation, the integration involved in step 3 has to be carried out over three time intervals corresponding to the three distinct regions of time: $0 \le t \le 4$ s, $4\ \text{s} \le t \le 6$ s, and $t \ge 6$ sec. For integration over the first interval, the folded impulse response is displaced t to the right, $0 \le t \le 4$ s, as shown in Figure 12.3.4a. Its analytical expression is obtained by substituting $t - \lambda$ for λ in Equation 12.3.7:

$$h(t - \lambda) = e^{-(t-\lambda)}u(t - \lambda)\text{A/Vs} \tag{12.3.8}$$

Over the interval $0 \le t \le 4$ s, $v_{SRC}(\lambda) = 2$ V. The contribution to the current response over the first interval of integration is:

$$i' = \int_0^t 2e^{-(t-\lambda)}d\lambda = 2\left(1 - e^{-t}\right)\text{A},\ \ 0 \le t \le 4\ \text{s} \tag{12.3.9}$$

i' corresponds to the shaded area in Figure 12.3.4a, which is the area, from 0 to t, under the curve representing the product of $e^{-(t-\lambda)}$ and $v_{SRC} = 2$ V. Thus, i' is zero for $t' = 0$ and is maximum at $t = 4$ s. Note how the result of the integration in Equation 12.3.9 has the dimensions of current.

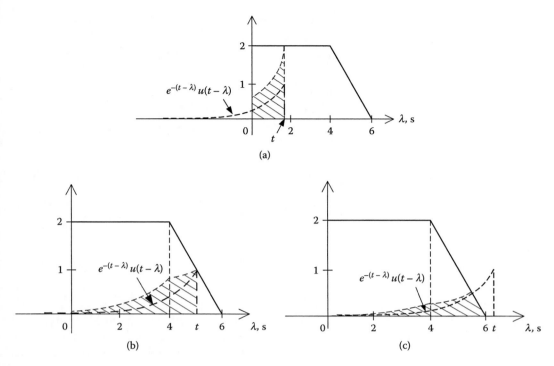

FIGURE 12.3.4
Figure for Simulation Example 12.3.1.

For $4 \text{ s} \leq t \leq 6 \text{ s}$, the impulse response is placed as shown in Figure 12.3.4b. Over this interval, $v_{SRC}(\lambda) = 6 - \lambda$, and

$$i'' = \int_0^4 2e^{-(t-\lambda)}d\lambda + \int_4^t (6-\lambda)e^{-(t-\lambda)}d\lambda = 7 - t - 2e^{-t} - e^{-(t-4)} \text{ A}, \; 4 \text{ s} \leq t \leq 6 \text{ s} \quad (12.3.10)$$

For $t \geq 6 \text{ s}$ (Figure 12.3.4c),

$$i''' = \int_0^4 2e^{-(t-\lambda)}d\lambda + \int_4^6 (6-\lambda)e^{-(t-\lambda)}d\lambda = -2e^{-t} - e^{-(t-4)} + e^{-(t-6)} \text{ A}, \; t \geq 6 \text{ s} \quad (12.3.11)$$

The total response is the combination of the responses of Equation 12.3.9 to Equation 12.3.11 over the respective time intervals and is shown in the simulation plot of Figure 12.3.5. Current i and its first derivative are continuous at the breakpoints $t = 4$ and $t = 6$ s. The continuity of i follows from the fact that the current through the inductor is not being forced to change by any voltage impulse. The continuity of $\dfrac{di}{dt}$ follows from KVL, $v_{SRC} = Ri + L\dfrac{di}{dt}$, and the continuity of v_{SRC} and i. Thus for $t = 4$ s, $v_{SRC}\big|_{t=4^-} = v_{SRC}\big|_{t=4^+}$ and $i\big|_{t=4^-} = i\big|_{t=4^+}$, so that $\dfrac{di}{dt}\big|_{t=4^-} = \dfrac{di}{dt}\big|_{t=4^+}$, which means that $\dfrac{di}{dt}$ is continuous at $t = 4$ s. Similar considerations apply for $t = 6$ s.

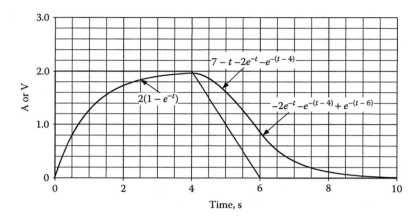

FIGURE 12.3.5
Figure for Example 12.3.1.

SIMULATION
The source used in the present example is VPWL, a piecewise linear voltage source from the SOURCE library that allows the application of an excitation having a waveform consisting of straight-line segments up to eight breakpoints. To apply the waveform of Figure 12.3.3, we enter in the Property Editor spreadsheet of the source the following breakpoints and the corresponding voltage levels:

$$T1 = 0, V1 = 0; T2 = 1u, V2 = 2; T3 = 4, V3 = 2; T4 = 6, V4 = 0$$

The waveform starts at zero, at $t = 0$, then jumps to 2 V, at $t = 1$ μs. As this is much smaller than the time constant of the circuit, a step of 2 V is effectively applied to the circuit at $t = 0$. The voltage is maintained at 2 V until $t = 4$ s, then decreases linearly to zero at $t = 6$ s and remains zero thereafter.

The schematic is entered as in Figure 12.3.3. The simulation is run using Time Domain (Transient) analysis for a duration of 10 s, with a maximum step size of 0.01 s. The plots of the voltage applied and the resulting current are shown in Figure 12.3.5.

EXERCISE 12.3.2
Consider that the excitation v_{SRC} is folded and displaced instead of the impulse response. Verify that the integration over the three time intervals is given by:

$$i' = \int_0^t 2e^{-\lambda}d\lambda, \ 0 \le t \le 4 \text{ s}; \ i'' = \int_0^{t-4}\left(\lambda - t + 6\right)e^{-\lambda}d\lambda + \int_{t-4}^t 2e^{-\lambda}d\lambda, \ 4 \text{ s} \le t \le 6 \text{ s};$$

$$i''' = \int_{t-6}^{t-4}\left(\lambda - t + 6\right)e^{-\lambda}d\lambda + \int_{t-4}^t 2e^{-\lambda}d\lambda, \ t \ge 6 \text{ s}.$$

Show that these integrals evaluate to the same expressions of Equation 12.3.9 to Equation 12.3.11. (Note the limits of integration in the preceding integrals, and that these integrals are identical to Equation 12.3.9 to Equation 12.3.11 when a change of variable $u = t - \lambda$ is made.)

A more thorough study of convolution involves the Laplace transform and the convolution theorem (Section 15.4, Chapter 15). We have already proved the commutative property and will prove, in the following paragraph, the distributive and associative properties.

Distributive property

$$x(t) * \left[f(t) + g(t) \right] = x(t) * f(t) + x(t) * g(t) \tag{12.3.12}$$

The proof of this property readily follows from the distributive property of integration. Thus, $x(t) * \left[f(t) + g(t) \right] = \int_0^t x(\lambda) \left[f(t - \lambda) + g(t - \lambda) \right] d\lambda = \int_0^t x(\lambda) f(t - \lambda) d\lambda +$

$\int_0^t x(\lambda) g(t - \lambda) d\lambda = x(t) * f(t) + x(t) * g(t)$.

Associative property

$$x(t) * \left[f(t) * g(t) \right] = \left[x(t) * f(t) \right] * g(t) \tag{12.3.13}$$

To prove this property, we note that $f(t) * g(t) = \int_0^t f(\lambda) g(t - \lambda) d\lambda$. Hence, using the commutative property, $x(t) * \left[f(t) * g(t) \right] = \int_0^t \left[\int_0^t f(\lambda) g(t - \lambda) d\lambda \right] x(t - \sigma) d\sigma$, which can be expressed

as $\int_0^t \int_0^t f(\lambda) g(t - \lambda) x(t - \sigma) d\lambda d\sigma$. Similarly, $x(t) * f(t) = \int_0^t f(\sigma) x(t - \sigma) d\sigma$, and $\left[x(t) * f(t) \right] *$

$g(t) = \int_0^t \left[\int_0^t f(\sigma) x(t - \sigma) d\sigma \right] g(t - \lambda) d\lambda = \int_0^t \int_0^t f(\sigma) g(t - \lambda) x(t - \sigma) d\lambda d\sigma$, as before, because $f(\sigma)$ and $f(\lambda)$ are the same functions.

12.4 Special Cases of Convolution

Convolution of three types of functions is discussed in this section, namely staircase, step, and impulse functions. Convolution with step and impulse functions is particularly simple, and functions consisting of linear segments and step discontinuities, when differentiated, yield a combination of impulse and step functions, as illustrated by Example 12.4.2.

Staircase Functions

A function of finite magnitude and duration can, in general, be approximated by a series of steps at regular intervals (Figure 12.4.1). The smaller the interval, the better the

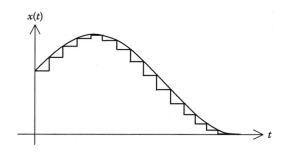

FIGURE 12.4.1
Staircase approximation.

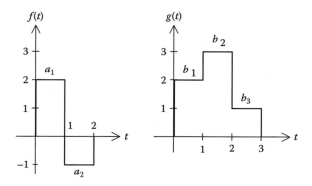

FIGURE 12.4.2
Figure for Example 12.4.1.

approximation. Such a stepped function is a staircase function. The convolution of two such functions reduces to the multiplication of two polynomials and can be readily performed using MATLAB's conv command, as illustrated by Example 12.4.1.

Example 12.4.1: Convolution of Staircase Functions

It is required to convolve $f(t)$ and $g(t)$ of Figure 12.4.2.

SOLUTION
In order to perform the convolution graphically, $f(\lambda)$ is folded and shifted by t to the right. There are five cases to consider, corresponding to the various breakpoints:

1. $0 \le t \le 1$, during which the area varies linearly with t, from zero to $a_1 b_1 = 4$ (Figure 12.4.3 a).
2. $1 \le t \le 2$, during which the expression for the area changes linearly from $a_1 b_1$ to $(a_1 b_2 + a_2 b_1) = (6 - 2) = 4$ (Figure 12.4.3b).
3. $2 \le t \le 3$, during which the area varies linearly from $(a_1 b_2 + a_2 b_1)$ to $(a_1 b_3 + a_2 b_2) = (2 - 3) = -1$ (Figure 12.4.3c).
4. $3 \le t \le 4$, during which the expression for the area changes linearly from $(a_1 b_3 + a_2 b_2)$ to $a_2 b_3 = -1$ (Figure 12.4.3d).
5. $4 \le t \le 5$, during which the area varies linearly from $a_2 b_3$ to zero (Figure 12.4.3e).

$y(t)$ consists of a series of line segments, as shown in Figure 12.4.3f.

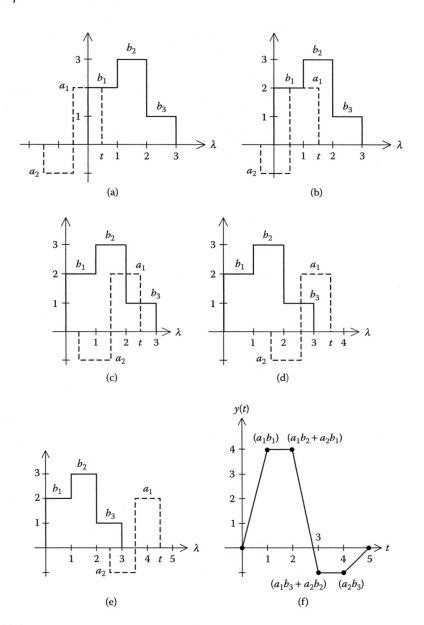

FIGURE 12.4.3
Figure for Example 12.4.1.

If we form two polynomials corresponding to $f(t)$ and $g(t)$ having coefficients equal to the various levels, with the last level as the constant term, we obtain:

$$F(x) = a_1 x + a_2$$

$$G(x) = b_1 x^2 + b_2 x + b_3$$

When $F(x)$ and $G(x)$ are multiplied together, the product polynomial is:

$$Y(x) = a_1b_1x^3 + \left(a_1b_2 + a_2b_1\right)x^2 + \left(a_1b_3 + a_2b_2\right)x + a_2b_3$$

It is seen that the coefficients of $Y(x)$ are the values of $y(t)$ at the successive breakpoints.

The multiplication of large polynomials is facilitated by MATLAB's conv command. If we enter the coefficients of $F(x)$ and $G(x)$ as arrays,

```
>> F = [2 -1]
>> G = [2 3 1]
```

followed by the command,

```
>> Y = conv(F, G)
```

MATLAB returns

```
Y=
     4    4    -1    -1
```

corresponding to the values of $y(t)$ at the successive breakpoints other than the zero at the $t = 0$ and $t = 5$. Note that $t = 5$ is the sum of the largest abscissas of $f(t)$ and of (t). (See Problem P12.3.10.)

Convolution with Step Function

It is required to convolve a function $x(t)$, which is assumed to be zero for $t < 0$, with a delayed unit step function $u(t - a)$, where a is a positive constant or zero. Replacing t by λ and reflecting the step function with respect to the vertical axis, it becomes $u(-\lambda - a)$, as shown in Figure 12.4.4. To avoid any possible confusion about the interpretation of this function, recall that by definition, $u(x) = 0$, if $x < 0$, and $u(x) = 1$, if $x > 0$. This means that when $\lambda < -a$, $(-\lambda - a) > 0$ and $u(-\lambda - a) = 1$. Conversely, when $\lambda > -a$, $(-\lambda - a) < 0$ and $u(-\lambda - a) = 0$, as it should be. For example, if $a = 2$ and $\lambda = -1$, $-\lambda - a = -1$, whereas if $\lambda = -3$, $-\lambda - a = 1$.

$u(t - \lambda - a) = 0$ for $\lambda > (t - a)$. For $0 < \lambda < (t - a)$, $u(t - \lambda - a) = 1$, and the convolution integral is $\int x(\lambda)u(t - \lambda - a)d\lambda = \int x(\lambda)d\lambda$. The lower limit of integration is $\lambda = 0$, and the upper limit for any t is $t - a$, as shown for the shaded area in Figure 12.4.4. Hence,

$$x(t) * u(t - a) = 0, \ t < a, \text{ and } \ x(t) * u(t - a) = \int_0^{t-a} x(\lambda)d\lambda, \ t > a,$$

or

$$x(t) * u(t - a) = \left[\int_0^{t-a} x(\sigma)d\sigma\right]u(t - a) \tag{12.4.1}$$

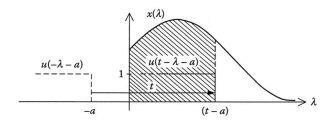

FIGURE 12.4.4
Convolution with step function.

Multiplying the integral by $u(t-a)$ is equivalent to assuming that the convolution function is zero for $t < a$ and is equal to the integral for $t > a$. For $a = 0$, Equation 12.4.1 becomes:

$$x(t) * u(t) = \left[\int_0^t x(\sigma)d\sigma \right] u(t) \tag{12.4.2}$$

In words, convolving a function with $u(t)$ is equivalent to integrating the function from 0 to t.

EXERCISE 12.4.1
Show that if the function $x(t)$ is delayed by a nonnegative constant b, so that it becomes $x(t - b)u(t - b)$, its convolution with $u(t - a)$ is:

$$x(t-b) * u(t-a) = \left[\int_b^{t-a} x(\sigma-b)d\sigma \right] u(t-b-a) \tag{12.4.3}$$

Convolution with Impulse Function

It is required to convolve a function $x(t)$, which is assumed to be zero for $t < 0$, with a delayed unit impulse function $\delta(t - a)$, where a is a positive constant or zero. Replacing t by λ and reflecting the impulse function with respect to the vertical axis, it becomes $\delta(-\lambda - a)$, as shown in Figure 12.4.5. The impulse function is nonzero when its argument is zero, that is, when $\lambda = -a$. After shifting by t, the function becomes $\delta(t - \lambda - a)$. The convolution integral $\int x(\lambda)\delta(t-\lambda-a)d\lambda$ is nonzero over the interval $\lambda = (t-a)^-$ to $\lambda = (t-a)^+$ and evaluates to $x(t - a)$, assuming that $x(\lambda)$ is continuous at $\lambda = t - a$. Thus,

$$x(t) * \delta(t-a) = x(t-a)u(t-a) \tag{12.4.4}$$

where $u(t - a)$ has been included because $x(t)$ is assumed to be zero for $t < 0$, so that $x(t - a)$ is zero for $t < a$ and is nonzero for $t > a$. When $a = 0$, Equation 12.4.4 reduces to:

$$x(t) * \delta(t) = x(t)u(t) \tag{12.4.5}$$

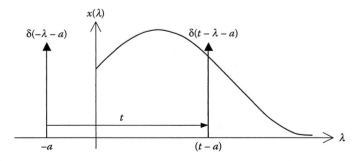

FIGURE 12.4.5
Convolution with impulse function.

Equation 12.4.5 has an instructive interpretation. Recall that the convolution of $x(t)$ and $\delta(t)$ is the response to $x(t)$ of a circuit whose impulse response is $\delta(t)$. Because convolution is commutative, the circuit response is the same if the excitation is $\delta(t)$ and the impulse response is $x(t)$. Evidently, the impulse response is $x(t)$ in this case.

EXERCISE 12.4.2

Integrate by parts $\displaystyle\int_{0^-}^{t-a} x(t)\delta^{(1)}(t-\lambda-a)d\lambda$, $t > a$, to show that:

$$x(t) * \delta^{(1)}(t-a) = \frac{dx(t)}{dt}\bigg|_{t=t-a} u(t-a) \tag{12.4.6}$$

Note that $\displaystyle\int \delta^{(1)}(-u)du = -\delta(-u)$, as readily follows by a change of variable $\sigma = -u$.

To make use of convolution with step functions and impulse functions, we invoke a relation that follows quite readily from the convolution theorem for two functions $f(t)$ and $g(t)$ that are zero for $t < 0$, namely (Section 15.4, Chapter 15):

$$f^{(n)}(t) * g^{(-n)}(t) = f^{(-n)}(t) * g^{(n)}(t) = f(t) * g(t) \tag{12.4.7}$$

where n is a positive integer, (n) is the nth derivative, and $(-n)$ is the nth integral. Equation 12.4.7 is applied in Example 12.4.2.

Example 12.4.2: Application of Properties of Convolution

It is required to evaluate the convolution integral of Example 12.3.1 using Equation 12.4.7 with $n = 1$ and $n = 2$.

SOLUTION

We will first apply Equation 12.4.7 identifying $f(t)$ with v_{SRC} and $g(t)$ with $h(t) = e^{-t}u(t)$; $v_{SRC}^{(1)}(t)$ is the function shown in Figure 12.4.6 and $h^{(-1)}(t) = \displaystyle\int_0^t e^{-t}dt = (1-e^{-t})u(t)$. It follows that

$$i' = h^{(-1)}(t) * 2\delta(t) = 2\left(1-e^{-t}\right) \; ; \; 0 \le t \le 4 \text{ s} \tag{12.4.8}$$

which is identical with Equation 12.3.9 over the interval $0 \le t \le 4$ s.

At $t = 4$, a unit step $-u(t-4)$ is applied. From Equation 12.4.1,

$$\left[\left(1-e^{-t}\right)u(t)\right] * \left(-u(t-4)\right) = -\int_0^{t-4}\left(1-e^{-\lambda}\right)d\lambda = -t+5-e^{-(t-4)}, \; 4 \le t \le 6 \text{ s} \tag{12.4.9}$$

Adding Equation 12.4.8 and Equation 12.4.9 gives Equation 12.3.10.

At $t = 6$, a unit step $u(t-6)$ is applied. Proceeding as for Equation 12.4.9,

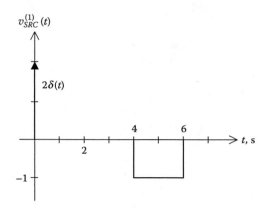

FIGURE 12.4.6
Figure for Example 12.4.2.

$$\left[\left(1-e^{-t}\right)u(t)\right]*\left(u(t-6)\right)=\int_{0}^{t-6}\left(1-e^{-\lambda}\right)d\lambda=t-7+e^{-(t-6)}, \quad t\geq6 \text{ s} \qquad (12.4.10)$$

Adding Equation 12.4.8, Equation 12.4.9, and Equation 12.4.10 gives Equation 12.3.11.

To apply Equation 12.4.6 with $n = 2$, we note that $f^{(2)}(t) = 2\delta^{(1)}(t) - \delta(t-4) + \delta(t-6)$. For $0 \leq t \leq 4$ s, it follows from Equation 12.4.6 that

$$h^{(-2)}(t) * 2\delta^{(1)}(t) = 2h^{(-1)}(t) = 2(1 - e^t)$$

as in Equation 12.3.9, where the derivative of $h^{(-2)}(t)$ is $h^{(-1)}(t)$.

To apply Equation 12.4.7 with $n = 2$, we note that $f^{(2)}(t) = 2\delta^{(1)}(t) - \delta(t-4) + \delta(t-6)$. For $0 \leq t \leq 4$s, it follows from Equation 12.4.6 that:

$$h^{(-2)}(t) * 2\delta^{(1)}(t) = 2h^{(-1)}(t) = 2(1 - e^{-t})$$

as in Equation 12.3.9, where the derivative of $h^{(-2)}(t)$ is $h^{(-1)}(t)$, and

$h^{(-2)}(t) = \int_{0}^{t}(1-e^{-t})dt = [1-e^{-t}]_{0}^{t} = t + e^t - 1$. Convolving this with the impulse $-\delta(t-4)$:

$$-(t + e^t - 1) * \delta(t-4) = -t + 5 - e^{(t-4)}$$

as in Equation 12.4.9. Convolving $h^{(-2)}(t)$ with the impulse $\delta(t-6)$:

$$(t + e^t - 1) * \delta(t-6) = t - 7 - e^{(t-6)}$$

as in Equation 12.4.10.

Summary of Main Concepts and Results

- The time constant of a first-order circuit is a measure of its speed of response. The larger the time constant, the slower, or more sluggish, is the circuit, and the lower is the 3-dB cutoff frequency.

- The natural response of a first-order circuit consisting of R in series with L, or G across C, is a decaying exponential function of time having a time constant L/R or C/G, respectively.

- The natural response of a second-order circuit could be overdamped, critically damped, or underdamped. In an overdamped circuit, R is large enough so that $\alpha > \omega_0$, s_1 and s_2 are negative real and distinct, $Q < 1/2$, and $\xi > 1$. In a critically damped circuit, R is such that $\alpha = \omega_0$, which makes $s_1 = s_2 = \omega_0$, $Q = 1/2$, and $\xi = 1$. In an underdamped circuit, R is small enough so that $\alpha < \omega_0$, s_1 and s_2 are complex, $Q > 1/2$, and $\xi < 1$.

- A voltage impulse instantly establishes an initial energy storage in an inductor, whereas a current impulse instantly establishes an initial energy storage in a capacitor. The stored energy is manifested as an initial condition in the circuit, that is, an initial current in an inductor, or an initial voltage across a capacitor.

- The response of an LTI circuit to an arbitrary input can be considered as the superposition of responses to sufficiently narrow pulses, each having an amplitude determined by the input. This naturally leads to a convolution integral involving the input function and a folded and shifted impulse response.

- Practical functions may be approximated as staircase functions. Convolution of staircase functions reduces to multiplication of two polynomials.

- Given a function $x(t)$ that is zero for $t < 0$, the convolution of $x(t)$ with $\delta(t)$ is $x(t)u(t)$, and the convolution of $x(t)$ with $u(t)$ is the integral of $x(t)$ from $t = 0$ to t.

Learning Outcomes

- Derive the natural response of a first-order circuit.
- Derive the different types of natural responses of second-order circuits.
- Convolve relatively simple functions.

Supplementary Topics and Examples on CD

ST12.1 Current impulse applied to an RL circuit: Analyzes the case where a current impulse is applied to a parallel *RL* circuit or to a series *RL* circuit.

ST12.2 Voltage impulse applied to a GC circuit: Analyzes the case where a voltage impulse is applied to a series *GC* circuit or to a parallel *GC* circuit.

SE12.1 Transient response of series RLC to a current impulse across C: Analyzes the response of an *RLC* circuit following a current impulse applied across *C*.

SE12.2 Response of LC circuit to current inputs: Analyzes the response of a parallel *LC* circuit to a current impulse and to a sinusoidal excitation. Shows that the impulse response is continuous oscillation, and that the response to sinusoidal excitation is unbounded if the frequency of the excitation equals the natural frequency of the circuit.

SE12.3 Convolution of function starting at negative time: Derives the convolution integral by different methods.

SE12.4 Convolution with step and impulse functions: Derives the convolution of two functions by three methods: direct integration, convolving with step functions, and convolving with impulse functions.

SE12.5 Response of RL circuit to exponential input: Derives, by means of convolution, the response of an *RL* circuit to an exponential input $10e^{-3|t|}$ V.

Problems and Exercises

Verify the solutions of Problem P12.1.1 to Problem P12.2.15 by simulating with PSpice.

P12.1 Responses of First-Order Circuits

P12.1.1 $i_{SRC} = 0.1\delta(t)$ As in Figure P12.1.1. Determine i_L and v_1, $t \geq 0^+$, by (a) transforming the current source to a voltage source, and (b) considering the effect of the current impulse on the circuit as is.

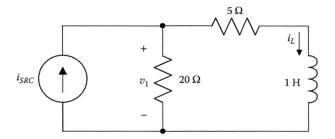

FIGURE P12.1.1

P12.1.2   Derive and solve the dual of the circuit in Problem P12.1.1.

P12.1.3 $i_{SRC} = \delta(t)$ μAs in Figure P12.1.3 with zero initial currents in the inductors. Determine v, $t \geq 0^+$.

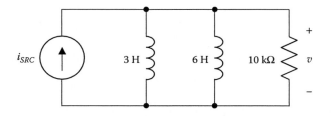

FIGURE P12.1.3

P12.1.4   Derive and solve the dual of the circuit in Problem P12.1.3.

P12.1.5   Repeat Problem P12.1.3 assuming initial currents of 2 mA flowing downward in each of the two inductors.

P12.1.6   Derive and solve the dual of the circuit in Problem P12.1.5.

P12.1.7 $v_{SRC} = 4\delta(t)$ Vs in Figure P12.1.7 with $i_1 = 1$ A, $i_2 = 1$A, and $i = 2$ A at $t = 0$. Determine i_1 and i_2 for $t \geq 0^+$.

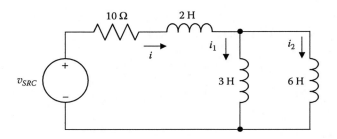

FIGURE P12.1.7

P12.1.8   Derive and solve the dual of the circuit in Problem P12.1.7.

P12.1.9 $v_{SRC} = 10\delta(t)$ Vs in Figure P12.1.9 with zero initial conditions. Determine v_L for $t \geq 0^+$.

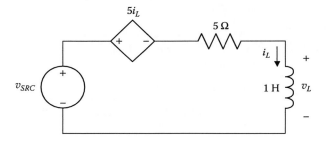

FIGURE P12.1.9

P12.1.10 Derive and solve the dual of the circuit in Problem P12.1.9.

P12.1.11 $v_{SRC} = 10\delta(t)$ Vs in Figure P12.1.11 with zero initial conditions. Determine v_L for $t \geq 0^+$.

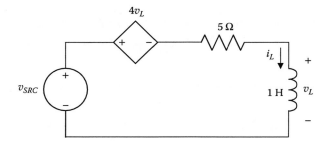

FIGURE P12.1.11

P12.1.12 Derive and solve the dual of the circuit in Problem P12.1.11.

P12.2 Responses of Second-Order Circuits

P12.2.1 $i_{SRC} = 20\delta(t)$ µAs in Figure P12.2.1, with zero energy storage. (a) Choose R so that the response is critically damped; (b) deduce that at $t = 0^+$, $v_C(0^+) = 10$ V and $i_L(0^+) = 0$; (c) show that with these initial values, $A = 0$ and $B = -20$ A/s

in Equation 12.2.17, which gives $i_L = -\dfrac{dq}{dt} = 20te^{-t}$ mA, where t is in ms; (d)

determine $v_L = L\dfrac{di_L}{dt}$, v_C, and verify that $v_C = v_L + Ri_L$ for $t \geq 0^+$.

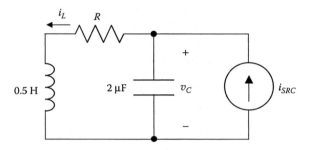

FIGURE P12.2.1

P12.2.2 (a) Show from the circuit equations that the circuit of Figure P12.2.2, with $v_{SRC} = 20\delta(t)$ µVs is the dual of that of Figure P12.2.1; (b) deduce from duality: R_p

for critical damping, v_{Cp}, $i_{Cp} = C_p\dfrac{dv_{Cp}}{dt}$, and i_{Lp}.

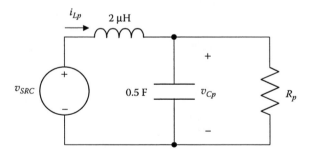

FIGURE P12.2.2

P12.2.3 In the circuit of Figure P12.2.1, with R having the value for critical damping, the current impulse of $20\delta(t)$ µAs is applied across L, as shown in Figure P12.2.3. (a) Deduce that at $t = 0^+$, $v_C(0^+) = 10$ V and $i_L(0^+) = 40$ mA; (b) show that with these initial values, $A = 0.04$ A and $B = -20$ A/s in Equation 12.2.17,

which gives $i_L = 20e^{-t}(2-t)$ mA, where t is in ms; (c) determine $v_L = L\dfrac{di_L}{dt}$, v_C,

and verify that $v_L = v_C - Ri_L$ for $t \geq 0^+$.

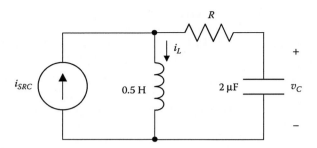

FIGURE P12.2.3

P12.2.4 Assume that R in Problem P12.2.3 has twice the value required for critical damping. (a) Determine s_1, s_2, Q, $v_C(0^+)$, $i_L(0^+)$, A, and B; (b) show that i_L may

be expressed as: $20e^{-2t}\left(4\cosh\sqrt{3}t - \dfrac{7}{\sqrt{3}}\sinh\sqrt{3}t\right)$ mA, where t is in ms; (c)

determine $v_L = L\dfrac{di_L}{dt}$, v_C, and verify that $v_L = v_C - Ri_L$ for $t \geq 0^+$.

P12.2.5 Assume that R in Problem P12.2.3 has one fifth of the value required for critical

damping. Determine: (a) s_1, s_2, Q, $v_C(0^+)$, $i_L(0^+)$, A, and B; (b) i_L, $v_L = L\dfrac{di_L}{dt}$, v_C,

and verify that $v_L = v_C - Ri_L$ for $t \geq 0^+$.

P12.2.6 A voltage impulse of 10 μVs is applied in the circuit of Figure P12.2.6. (a) Choose R so that the circuit is critically damped; (b) deduce that at $t = 0^+$, $v_C(0^+) = 10$ V and $i_L(0^+) = 10$ mA; (c) considering that for $t \geq 0^+$, the duals of Equation 12.2.17 and Equation 12.2.18 are: $v_L = v_C = Ae^{-\omega_0 t} + Bte^{-\omega_0}$

and $\lambda = Li_L = -\dfrac{A}{\omega_0}e^{-\omega_0 t} - \dfrac{Be^{-\omega_0 t}}{\omega_0^2}(\omega_0 t + 1)$, show that $A = 0.04$ V and $B = 2.5 \times 10^6$

V/s, which gives $v_L = 2.5e^{-t/2}(t-4)$ V, where t is in μs; (d) determine i_L,

$i_C = C\dfrac{dv_C}{dt}$ and verify that $i_L + i_C + i_R = 0$, for $t \geq 0^+$.

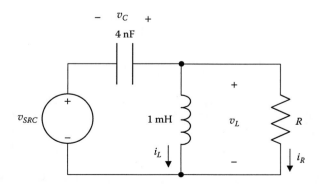

FIGURE P12.2.6

P12.2.7 Assume that R in Problem P12.2.6 is $1250/13\ \Omega$. Determine: (a) s_1, s_2, Q, $v_C(0^+)$, $i_L(0^+)$, A, and B; (b) v_L, i_L, and $i_C = C\dfrac{dv_C}{dt}$, v_C, and verify that $i_L + i_C + i_R = 0$ for $t \geq 0^+$.

P12.2.8 Assume that R in Problem P12.2.6 is $312.5\ \Omega$. Determine: (a) s_1, s_2, Q, $v_C(0^+)$, $i_L(0^+)$, A, and B; (b) i_L, $v_L = L\dfrac{di_L}{dt}$, i_C, and verify that $i_L + i_C + i_R = 0$ for $t \geq 0^+$.

P12.2.9 Determine β in Figure P12.2.9 so that the response is: (a) critically damped; (b) a Butterworth response ($Q = \xi = 1/\sqrt{2}$).

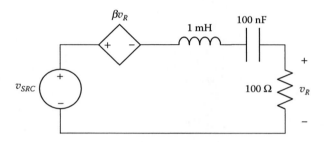

FIGURE P12.2.9

P12.2.10 Derive and solve the dual of the circuit in Problem P12.2.9.

P12.2.11 Determine β and σ in Figure P12.2.11 so that $\omega_0 = 10^4\ \text{rad/s}$ and the response is Butterworth.

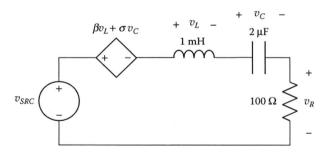

FIGURE P12.2.11

P12.2.12 Derive and solve the dual of the circuit in Problem P12.2.11.

P12.2.13 In the circuit of Figure P12.2.13, $i = 20e^{-400t}(1-400t)$ mA. Determine L, R, and v_C, assuming zero initial charge on the capacitor.

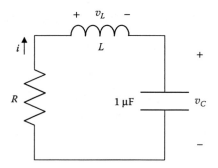

FIGURE P12.2.13

P12.2.14 Consider a critically damped series RLC circuit. Show that: (a) if half of R is associated with L and the other half with C, the two time constants $2L/R$ and $RC/2$ are equal; (b) if the initial inductor current is I_0 and the initial capacitor voltage is V_0, with $\dfrac{V_0}{LI_0} = \omega_0$, and the polarities of I_0 and V_0 are such that I_0 tends to neutralize the charge on C, circuit responses reduce to that of a single time constant $1/\omega_0$; (c) under these conditions, the initial stored energies in the inductor and capacitor are equal; and (d) I_0 just neutralizes the charge on C.

P12.2.15 Formulate the case of a parallel circuit that is the dual of the series circuit of P12.2.14 and verify the analogous conclusions.

P12.2.16 In the circuit of Figure P12.2.16, $v = 0$, $I_{10} = 1$ A, and $I_{20} = 3$ A at $t = 0$. What is the total instantaneous energy in the inductors as a function of time?

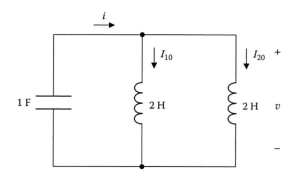

FIGURE P12.2.16

P12.3 Convolution

P12.3.1 Convolve e^{-t} with $(1-e^{-2t})$.

P12.3.2 Evaluate the following convolution operations: (a) $\sin\omega t * \cos\omega t$; (b) $\cos\omega t * \cos\omega t$; (c) $\sinh at * \sin\omega t$. (Note that $\sinh x = -j\sin jx$).

P12.3.3 Evaluate the convolution of the two functions shown in Figure P12.3.3. Determine the value at $t = 5$ s.

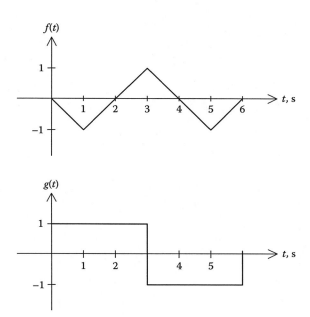

FIGURE P12.3.3

P12.3.4 Repeat Problem P12.3.3 after differentiating one function and integrating the other.

P12.3.5 Evaluate the convolution of the two functions $f(t)$ and $g(t)$ shown in Figure P12.3.5.

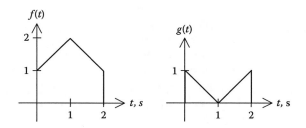

FIGURE P12.3.5

P12.3.6 Repeat Problem P12.3.5 after differentiating one function and integrating the other.

P12.3.7   Evaluate the convolution of the two staircase functions shown in Figure P12.3.7. Verify using MATLAB's conv command.

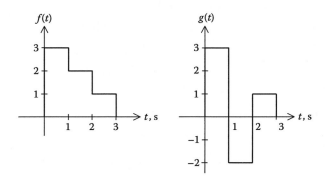

FIGURE P12.3.7

P12.3.8   Evaluate the convolution of the two staircase functions shown in Figure P12.3.8. Verify using the MATLAB's conv command.

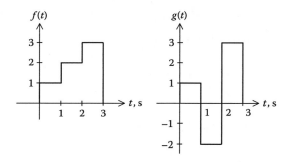

FIGURE P12.3.8

P12.3.9   An input voltage e^{-at} is applied from $t = -\infty$ to R and C in series. Obtain an expression for the output voltage across C in terms of a convolution integral taking the lower limit as $t = -\infty$. Show that the response at $t = 0$ is: (a) finite if $a < 1/CR$; and (b) infinite if $a \geq 1/CR$.

P12.3.10 Given two functions $f(t)$ and $g(t)$ that have the following properties: (a) they are both zero for $t < 0$; (b) $f(0)$ and $g(0)$ are finite; and (c) $f(t) = 0$ for $t \geq a$ and $g(t) = 0$ for $t \geq b$, where a and b are some positive constant. Verify that their convolution is zero at $t = 0$ and for $t \geq a + b$.

P12.3.11 The impulse response of a circuit is $h(t) = 4u(t) - 2u(t - 5)$. Determine the circuit output when an excitation $x(t) = 2\delta(t - 1) - 3\delta(t - 3)$ is applied with zero initial conditions.

P12.3.12 An input defined as: $v_I = t+1$, $-1 \leq t \leq 0$, $v_I = -t+1$, $0 \leq t \leq 1$, and $v_I = 0$, elsewhere, is applied to a circuit having the impulse response $h(t) = e^{-t}u(t)$. Determine the output for all t.

P12.3.13 Evaluate the convolution of the functions $f(t)$ and $g(t)$ shown in Figure P12.3.13 (a) graphically and (b) analytically by differentiating one function and integrating the other.

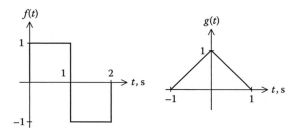

FIGURE P12.3.13

P12.3.14 Express the functions of Example 12.4.1 as sums of step functions and evaluate the convolution function analytically.

13

Switched Circuits

Overview

After considering the impulse responses of first-order and second-order circuits in the preceding chapter, we turn our attention in this chapter to switched circuits; that is, circuits in which the movement of a switch applies an input to the circuit or changes the circuit configuration. A practically important case is the application of a step input to a circuit. The responses of prototypical first-order and second-order circuits to a step input are analyzed in this chapter and related to the impulse response.

We begin by answering a fundamental question as to the need for having a transient component in the response of a circuit. Considering first-order circuits, it turns out that a general procedure can be readily developed that expresses the time variation of any variable in the circuit in terms of the initial value of the variable, its final value, and the effective time constant.

When it comes to second-order circuits, the situation is more complicated. The general case of switched circuits that have relatively complicated connections of resistors, inductors, and capacitors, and which may include linear transformers and dependent sources, is best handled using the Laplace transform, as will be discussed in Chapter 15. The discussion is therefore restricted to prototypical second-order circuits that can be reduced to series or parallel connections of resistors, inductors, and capacitors. This serves to highlight some basic circuit principles, without introducing unnecessary complications at this stage.

Learning Objectives

- To be familiar with:
  - The general features of the responses of switched first-order and second-order circuits
- To understand:
  - The need for a transient component in the response of a circuit
  - The step responses of first-order and second-order circuits and their relation to the natural responses

- How a general first-order system can be analyzed
- The analysis of prototypical, switched second-order circuits

13.1 Purpose of Transient

The response of a circuit to a given input consists, in general, of two components: (1) a transient component that dies out with time, and (2) a steady-state component that persists as long as the input is applied. It may be wondered why there should be a transient component. In other words, why does the circuit not assume a steady state immediately when the input is applied, at say $t = 0$? The answer is simply that the steady state may entail a certain value of the response at $t = 0$ that is different from the value that is compatible with the initial conditions at this time. Something has to happen, therefore, to accommodate these conflicting requirements.

> **Concept:** *The purpose of a transient response is to match the initial conditions in a given circuit to the steady-state response in the absence of a transient. The time course of the transient is that of the natural response of the circuit.*

This concept may be illustrated by considering a voltage $V_m\cos(\omega t + \theta)$ applied at $t = 0$ to a series RL circuit as in Section 5.2 (Chapter 5). KVL gives the differential equation:

$$L\frac{di}{dt} + Ri = V_m\cos\left(\omega t + \theta\right)$$

(13.1.1)

The complete solution of this differential equation is the sum of two components:

1. A steady-state component i_{SS} that satisfies this equation, and which was determined as Equation 5.2.6, (Chapter 5): $i_{SS} = \dfrac{V_m}{\sqrt{R^2 + \omega^2 L^2}}\cos(\omega t + \theta - \alpha)$, where $\tan\alpha = \dfrac{\omega L}{R}$.

2. A transient component that is the solution of the differential equation $L\dfrac{di}{dt} + Ri = 0$, and which was determined in Section 12.1 (Chapter 12) as the natural circuit response of the general form: $i = Be^{-\frac{t}{\tau}}$, where B is an arbitrary constant.

Because the circuit is linear, the complete, general response is the sum of these two components:

$$i = \frac{V_m}{\sqrt{R^2 + \omega^2 L^2}}\cos\left(\omega t + \theta - \alpha\right) + Be^{-\frac{t}{\tau}}, \quad t \geq 0$$

(13.1.2)

i, as given by Equation 13.1.2, satisfies Equation 13.1.1. The arbitrary constant *B* in this equation is determined by the initial conditions. If $i = 0$ at $t = 0$, then $B = -\dfrac{V_m \cos(\theta - \alpha)}{\sqrt{R^2 + \omega^2 L^2}}$, and Equation 13.1.2 becomes:

$$i = \frac{V_m}{\sqrt{R^2 + \omega^2 L^2}} \cos\left(\omega t + \theta - \alpha\right) - \frac{V_m \cos\left(\theta - \alpha\right)}{\sqrt{R^2 + \omega^2 L^2}} e^{-\frac{R}{L}t}, \quad t \geq 0 \tag{13.1.3}$$

The following should be noted concerning Equation 13.1.3:

1. The magnitude of the transient, $-\dfrac{V_m \cos(\theta - \alpha)}{\sqrt{R^2 + \omega^2 L^2}}$, at $t = 0$ is equal and opposite to the steady-state value at $t = 0$, so as to make $i = 0$ at $t = 0$, as specified.
2. The time course of the transient is dictated by the natural response of the circuit.
3. As *t* becomes large, the transient component dies out in a stable circuit (Section 16.6, Chapter 16), and the total response merges smoothly with the steady-state response.

The steady-state and transient components are illustrated in Figure 13.1.1 for a hypothetical circuit, together with the total response that is the sum of these components.

This interpretation of the transient suggests that if the voltage is applied to the *RL* circuit having $i = 0$ at $t = 0$ when the would-be steady-state current is zero, the transient is not required at all. If $\alpha = 30°$, for example, then if $\theta = -60°$, $\cos(\theta - \alpha) = \cos{-90°} = 0$. Thus, if we select $t = 0$ to be the instant when the input is $V_m \cos(-60°)$, the transient term is zero. The circuit will go into the steady-state right from $t = 0$. Conversely, if θ is such that the steady-state current i_{SS} is at its peak at $t = 0$, $\cos(\theta - \alpha) = \pm 1$, and the transient term will have its maximum magnitude.

EXERCISE 13.1.1
Sketch the waveforms corresponding to those of Figure 13.1.1 for the cases of (a) zero transient and (b) maximum transient.

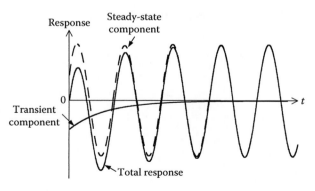

FIGURE 13.1.1
Steady-state, transient, and total responses.

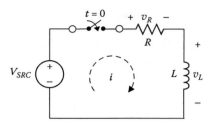

FIGURE 13.2.1
Charging of an inductor.

13.2 Charging of Energy Storage Elements

Charging of an Inductor

Consider an inductor L, with zero initial current, to be connected at $t = 0$ to a dc voltage source V_{SRC} through a resistance R (Figure 13.2.1). It is required to determine how the current i varies for $t \geq 0$.

From KVL,

$$L\frac{di}{dt} + Ri = V_{SRC}, \quad t \geq 0 \tag{13.2.1}$$

As explained earlier, the complete solution is the sum of two components: a transient component $i = Be^{-\frac{t}{\tau}}$ that is the solution to the equation $L\frac{di}{dt} + Ri = 0$, and a steady-state component I_{SS} that is a solution of Equation 13.2.1. Because V_{SRC} is dc, I_{SS} would also be dc, which means that $\frac{dI_{SS}}{dt} = 0$. It follows from Equation 13.2.1 that $I_{SS} = \frac{V_{SRC}}{R}$, which is to be expected because under dc conditions, the inductor behaves as a short circuit so that $V_{SRC} = RI_{SS}$. The complete solution is, therefore, $i = Be^{-\frac{t}{\tau}} + \frac{V_{SRC}}{R}$, where B is such that the initial condition $i = 0$ at $t = 0$ is satisfied, which gives $B = -\frac{V_{SRC}}{R}$. The complete response is:

$$i = \frac{V_{SRC}}{R}\left(1 - e^{-t/\tau}\right), \quad t \geq 0 \tag{13.2.2}$$

where $\tau = \frac{L}{R}$ is the time constant.

Equation 13.2.2 is plotted in Figure 13.2.2 in normalized form, where the horizontal axis is t/τ and the vertical axis is i/I_{SS}. The plot is a *saturating exponential*. A tangent to the

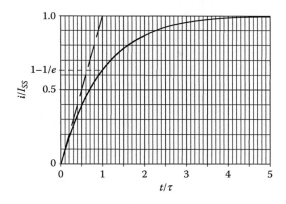

FIGURE 13.2.2
Normalized saturating exponential.

curve at the origin intersects the line $\dfrac{i}{I_{SS}} = 1$ at $\dfrac{t}{\tau} = 1$. At this value of time, $i = I_{SS}\left(1 - \dfrac{1}{e}\right) =$ $0.63 I_{SS}$.

The time course of i can be interpreted as follows. Current i is zero for $t < 0$, by assumption. When the switch is closed at $t = 0$, i remains zero because there is no voltage impulse to force it to change instantly. With $i = 0$, $v_L = V_{SRC}$ at $t = 0$, and the inductor behaves as an *open circuit* at this instant. As t increases, $\displaystyle\int_0^t v_L dt = \lambda$ increases, as does $i = \lambda/L$. The increase in i reduces $v_L = \dfrac{d\lambda}{dt}$, because of the voltage drop across R, so that the rate of increase of λ, and hence i, becomes progressively smaller with time, and the plot becomes flatter. This process continues until as $t \to \infty$, $i_L \to I_{SS}$, $v_L \to 0$, and the inductor behaves as a *short-circuit*.

v_L can be determined in one of two ways, either as $L\dfrac{di_L}{dt}$, or as $V_{SRC} - Ri$. Either method gives:

$$v_L = V_{SRC}e^{-\frac{t}{\tau}}, \quad t \geq 0 \tag{13.2.3}$$

This is a decaying exponential, as in Figure 12.1.2 (Chapter 12).

An alternative derivation of Equation 13.2.2 is instructive. The application of V_{SRC} by closing a switch at $t = 0$ (Figure 13.2.1), with no initial energy stored in the inductor, is equivalent to applying V_{SRC} as a step function at $t = 0$. We may thus write,

$$L\dfrac{di}{dt} + Ri = V_{SRC}u(t), \text{ for all } t \tag{13.2.4}$$

Comparing Equation 13.2.4 with Equation 12.1.1 (Chapter 12), it is seen that the forcing function is a step function in Equation 13.2.4 and an impulse function in Equation 12.1.1 (Chapter 12). However, $u(t)$ is the integral of $\delta(t)$, and because the circuit is LTI, the response

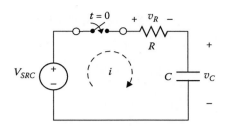

FIGURE 13.2.3
Charging of a capacitor.

to $u(t)$ is the integral of the response to $\delta(t)$. Replacing K in Equation 12.1.5 (Chapter 12), by V_{SRC} and integrating from 0 to t gives Equation 13.2.2.

> **Conclusion:** *When an inductor is charged from a dc source, with zero initial current, the inductor current increases with time as a saturating exponential, and the inductor voltage decreases as a decaying exponential.*

EXERCISE 13.2.1
Consider an inductor in parallel with a resistance R_p being charged from a dc current source I_{SRC}, with zero initial current. (a) Using source transformation in Figure 13.2.1, show that the inductor current is $i_L = I_{SRC}(1 - e^{-t/\tau})$, where $\tau = L/R_p$, and interpret this result; (b) show that the energy dissipated in R_p over the period 0 to t is: $\frac{1}{2}LI_{SRC}^2(1 - e^{-2t/\tau})$, the energy stored in L is $\frac{1}{2}LI_{SRC}^2(1 - e^{-t/\tau})^2$, and the energy supplied by the source is $LI_{SRC}^2(1 - e^{-t/\tau})$. Verify the latter by evaluating $\int_0^t I_{SRC}v_L dt$.

EXERCISE 13.2.2

> A 2 mH inductor that is initially uncharged is charged at $t = 0$ from a 10 V dc source through a 1 kΩ resistor. Determine: (a) the inductor current and voltage at $t = 4 \ \mu s$, and (b) the final flux linkage in the inductor and the final energy stored in the inductor.

Answer: (a) 8.65 mA; 1.35 V, (b) 20 μWb-turns; 0.1 μJ.

Charging of a Capacitor

Consider a capacitor C, with zero initial voltage, to be connected at $t = 0$ to a dc voltage source V_{SRC} through a resistance R (Figure 13.2.3). It is required to determine how the current i varies for $t \geq 0$.

The dual of a series connection of a voltage source V_{SRC}, R, and L is a parallel connection of I_{SRC}, G_p, and C. The dual of Equation 13.2.2 is:

$$v_C = \frac{I_{SRC}}{G_p}\left(1 - e^{-t/\tau}\right), \quad t \geq 0 \tag{13.2.5}$$

where $\tau = C/G_p$ and G_p is numerically equal to R in Figure 13.2.1. The current source is transformed to a voltage source I_{SRC}/G_p in series with G_p. Identifying I_{SRC}/G_p with V_{SRC} and G_p with $1/R$ in Figure 13.2.3 gives:

$$v_C = V_{SRC}\left(1 - e^{-t/\tau}\right), \ t \geq 0 \tag{13.2.6}$$

where $\tau = RC$. Alternatively, KVL applied to the circuit of Figure 13.2.3 gives $Ri + v_C = V_{SRC}$. Substituting $i = C\dfrac{dv_C}{dt}$,

$$RC\frac{dv_C}{dt} + v_C = V_{SRC}, \ t \geq 0 \tag{13.2.7}$$

Solving this equation in the same manner as Equation 13.2.1 gives Equation 13.2.6. The current i may be obtained as $i = C\dfrac{dv_C}{dt}$ or from $Ri = V_{SRC} - v_C$:

$$i = \frac{V_{SRC}}{R}e^{-\frac{t}{\tau}}, \ t \geq 0 \tag{13.2.8}$$

The interpretation of these equations is that at $t = 0^+$ just after the switch is closed, $v_C = 0$ because the capacitor voltage, initially at zero, remains at zero in the absence of a current impulse. The voltage V_{SRC} appears across R, so $i = \dfrac{V_{SRC}}{R}$ at $t = 0$. As t increases, the current builds up charge on the capacitor, and v_C increases. This means that the voltage across the resistor, and hence i, must also decrease, but at a progressively smaller rate, as the capacitor voltage builds up. This process continues until as $t \to \infty$, the charge on the capacitor is such that $v_C \to V_{SRC}$ and $i \to 0$. As $t \to \infty$, the circuit approaches a dc steady state, in which none of the circuit variables changes with time. At $t = 0$, $v_C = 0$, and the capacitor behaves as a *short-circuit*. As $t \to \infty$, $v_C = V_{SRC}$, $i = 0$, and the capacitor behaves as an *open circuit*.

> **Conclusion:** *When a capacitor is charged from a dc source, with zero initial voltage, the capacitor voltage increases with time as a saturating exponential, and the capacitor current decreases as a decaying exponential.*

It is interesting to note that when a capacitor is charged from a voltage source through a resistance R, the total energy dissipated in R is equal to that stored in C. This can be argued quite simply by noting that the total energy delivered by the source is $V_{SRC}\displaystyle\int_0^\infty i\,dt = qV_{SRC}$, where q is the final charge on C. The final energy stored in C is $\dfrac{1}{2}qV_{SRC}$, so the energy dissipated in R must also be $\dfrac{1}{2}qV_{SRC}$. This is true irrespective of the nature of R, that is,

R could be nonlinear, time-varying, or any power absorbing device whose current is zero when the voltage across it is zero.

EXERCISE 13.2.3
Derive the equivalent case of an inductor whose total energy stored after charging equals that dissipated in a resistance.

EXERCISE 13.2.4
Show that the energy dissipated in R in Figure 13.2.3 over the period 0 to t is

$\frac{1}{2}CV_{SRC}^2(1-e^{-2t/\tau})$ and that stored in C is $\frac{1}{2}CV_{SRC}^2(1-e^{-t/\tau})^2$. Deduce that the total energy

delivered by the voltage source is $CV_{SRC}^2(1-e^{-t/\tau})$ and verify this by direct evaluation

of $\int_0^t V_{SRC}i\,dt$.

EXERCISE 13.2.5

A 0.5 µF capacitor that is initially uncharged is charged at $t = 0$ from a 10 V dc source through a 1 kΩ resistor. Determine: (a) the capacitor voltage and current at $t = 1.5$ ms; (b) the final charge on the capacitor and the final energy stored in the capacitor; (c) the energy delivered by the source as $t \to \infty$

Answer: (a) 9.50 V, 0.50 mA; (b) 5 µC, 25 µJ; (c) 50 µJ.

EXERCISE 13.2.6

Repeat Exercise 13.2.5 assuming that the capacitor is charged from the equivalent current source.

Answer: (a) 9.50 V, 0.50 mA; (b) 5 µC, 25 µJ; (c) ∞ because the current source continues to dissipate energy in R.

13.3 General Solution for First-Order Circuits

All the linear, first-order differential equations derived so far for the charging or discharging of an energy storage element are of the general form:

$$\frac{dy}{dt} + \frac{y}{\tau} = A \tag{13.3.1}$$

where y could be either v or i, and A is a constant, which is zero in the case of the discharging of an energy storage element or in the case of the impulse response, for $t \ge 0^+$.

As $t \to \infty$, a steady state is reached, so that $\dfrac{dy}{dt} \to 0$ and y assumes a final, steady value y_f, which from Equation 13.3.1 equals $A\tau$.

Equation 13.3.1 may be solved by separating the variables: $\dfrac{dy}{y - A\tau} = -\dfrac{dt}{\tau}$. Integrating both sides, $\ln(y - A\tau) = -\dfrac{t}{\tau} + K'$ where K' is a constant of integration. Expressing this relation in exponential form, $y = A\tau + Ke^{-\frac{t}{\tau}}$, where $K = e^{K'}$ is a new constant to be determined from initial conditions. Suppose that at $t = t_0$, $y = y(t_0)$. It follows that $K = \left[y(t_0) - A\tau\right]e^{\frac{t_0}{\tau}}$. Substituting for K and setting $y_f = A\tau$, the general solution of Equation 13.3.1 is:

$$y = y_f + \left[y(t_0) - y_f\right]e^{-\frac{t-t_0}{\tau}}, \quad t \geq 0 \tag{13.3.2}$$

Equation 13.3.2 is a useful relation that applies to any system having effectively a single energy storage and subject to a constant, or zero, excitation. It enables writing the time variation of a circuit variable directly from knowledge of the initial and final values of the circuit variable and the time constant that is *effective* in the circuit.

The term effective refers to cases where there is more than one resistance, capacitance, or inductance in the circuit, but the circuit can be reduced to a single energy storage element in series or in parallel with a resistor, as illustrated by Example 13.3.1, Example 13.3.2, and problems at the end of the chapter. As for sources, it is important to note the following:

> **Concept:** *Independent sources affect initial or final values of circuit responses but not the time constant. Dependent sources affect initial or final values of circuit responses as well as the time constant.*

The reason is that during charging or discharging of energy storage elements, circuit variables change with time. If the current through an ideal, independent voltage source varies by Δi_{SRC}, the voltage across the source does not vary, by definition of an ideal, independent voltage source. Alternatively, it could be argued that, because an ideal voltage source has zero source resistance, $\Delta v_{SRC} = 0 \times \Delta i_{SRC} = 0$. Because Δi_{SRC} produces no change in voltage across the voltage source, the source appears as a short-circuit as far as the changing current is concerned. Similarly, if the voltage across an ideal, independent current source changes by Δv_{SRC}, the source current does not change. Hence, the source appears as an open circuit as far as the changing voltage is concerned. The effective time constant is therefore determined with ideal, independent voltage sources replaced by short-circuits and ideal, independent current sources by open circuits. On the other hand, dependent sources must not be replaced in this manner, because the source values change as the circuit variables change with time during the charging or discharging of energy storage elements. The effective time constant under these conditions is R_{Th} of the circuit appearing across the energy storage element, and is determined in the same manner.

The following should be noted concerning Equation 13.3.2:

1. The same time constant applies to all the voltages and currents in the circuit.

2. The initial and final values of y must apply to the circuit *after* switching has occurred. If y is a capacitor voltage or an inductor current, then the initial values of these variables just after switching are the same as the final values just before switching, assuming that they are not being forced to change by any impulses at the instant of switching. However, if y is, say, the voltage across a resistor, then its value generally changes upon switching, so that $y(t_0)$ is the value after switching. This is illustrated by Example 13.3.3 and Example 13.3.4.

3. Equation 13.3.2 can be interpreted in terms of what was said about transients in Section 13.1. y_f is the final, steady-state value of y. The second term is a transient term that dies out as $t \to \infty$. Its magnitude depends on $(y(t_0) - y_f)$, which is the difference between the initial and final values. The transient term thus acts as a transition between the initial and the final values of the variable. If the initial and final values are equal, there is no transient.

EXERCISE 13.3.1

A 1 μF capacitor is initially charged to 1 V and disconnected from the charging source at $t = 0$. If the capacitor voltage drops to 0.9 V after 100 h, determine: (a) the time constant, and (b) the insulation resistance R_p that is effectively in parallel with the capacitor.

Answer: (a) 949 h, (b) 3.42×10^{12} Ω.

Simulation Example 13.3.1: Charging of Two Series-Connected Capacitors

A 3 μF and a 6 μF capacitor having initial voltages of 6 V each are connected at $t = 0$ to an 18 V supply in series with a 2 kΩ resistor. It is required to analyze and simulate the charging process.

ANALYSIS

The time variation of i, v_1, and v_2 (Figure 13.3.1) may be obtained from the initial and final values of these quantities and the effective time constant.

At $t = 0$, v_1 and v_2 do not change instantaneously. The voltage drop across the resistor will therefore be: $18 - (6 + 6) = 6$ V, and $i(0) = \dfrac{6}{2} = 3$ mA. As the capacitors charge, i decreases, eventually becoming zero. The effective time constant is that of R, C_1, and C_2 in series, $\tau = RC_{eqs} = 2\dfrac{3 \times 6}{3 + 6} = 4$ ms. Hence, $i = 0 + (3 - 0)e^{-\frac{t}{4}} = 3e^{-\frac{t}{4}}$ mA, $t \geq 0$, where t is in ms.

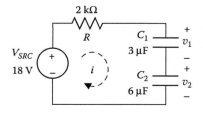

FIGURE 13.3.1
Figure for Simulation Example 13.3.1.

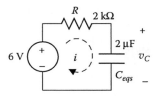

FIGURE 13.3.2
Figure for Example 13.3.1.

To find v_1 and v_2, we note that:

$$v_1 = \frac{1}{C_1} \int_0^t i\,dt + v_1(0) = \frac{1}{3} \int_0^t 3e^{-\frac{t}{4}}\,dt + 6 = 10 - 4e^{-\frac{t}{4}} \text{ V.}$$

$$v_2 = \frac{1}{C_2} \int_0^t i\,dt + v_2(0) = \frac{1}{6} \int_0^t 3e^{-\frac{t}{4}}\,dt + 6 = 8 - 2e^{-\frac{t}{4}} \text{ V.}$$

It is instructive to determine v_{1f} and v_{2f} by an alternative method and apply Equation 13.3.2. Initially, $v_1(0) = 6$ V, $q_1(0) = (3\ \mu\text{F}) \times (6\ \text{V}) = 18\ \mu\text{C}$, $v_2(0) = 6$ V, and $q_2(0) = (6\ \mu\text{F}) \times (6\ \text{V}) = 36\ \mu\text{C}$. $C_{eqs} = 2\ \mu\text{F}$; its initial voltage is $v_C(0) = 12$ V, and its initial charge is $q_{eqs} = (2\ \mu\text{F}) \times (12\ \text{V}) = 24\ \mu\text{C}$. The charge, and hence voltage, acquired by C_1, C_2, and C_{eqs} during charging adds to the initial values of charge and voltage on these capacitors. These initial values may be set to zero, the problem solved assuming C_1 and C_2 are initially uncharged, and the initial values added afterwards.

If the initial voltage on C_{eqs} is subtracted from V_{SRC}, the problem reduces to a capacitor C_{eqs} of 2 μF capacitance and zero initial charge connected at $t = 0$ to a supply of $18 - 12 = 6$ V through a resistance of 2 kΩ (Figure 13.3.2). The final voltage on C_{eqs} is 6 V. Because initial charges are assumed to be zero, the final charge q on C_{eqs} will be the same as on C_1 and C_2 (Section 11.4, Chapter 11), where $q = (2\ \mu\text{F}) \times (6\ \text{V}) = 12\ \mu\text{C}$. The voltages acquired by C_1 and C_2 through charging are $\dfrac{12\ \mu\text{C}}{3\ \mu\text{F}} = 4$ V and $\dfrac{12\ \mu\text{C}}{6\ \mu\text{F}} = 2$ V, respectively. The final voltages of C_1 and C_2 are, therefore, $v_{1f} = 10$ V and $v_{2f} = 8$ V. It follows that

$$v_1 = 10 + (6 - 10)e^{-\frac{t}{4}} = 10 - 4e^{-\frac{t}{4}} \text{ V, and } v_2 = 8 + (6 - 8)e^{-\frac{t}{4}} = 8 - 2e^{-\frac{t}{4}} \text{ V, as before.}$$

SIMULATION
The schematic is shown in Figure 13.3.3 using a normally open switch, Sw_tClose, from the EVAL library. Although there is no need to change the switch parameters in this simulation, the default values in the Property Editor spreadsheet should be noted. These are: closure time (TCLOSE = 0), time it takes to close (TTRAN = 1u), resistance of the switch when closed (RCLOSED = 0.01 Ω), and resistance of the switch when open (ROPEN = 1 MΩ). The switch is included for illustration; the simulation could be performed without it because PSpice effectively applies the 18 V at $t = 0$.

Because PSpice does not allow nodes having no dc continuity to ground, a large 10 MΩ resistor is added in parallel with C_2, without significantly affecting the response. The initial voltages on the capacitors are entered as IC values in the Property Editor spreadsheet,

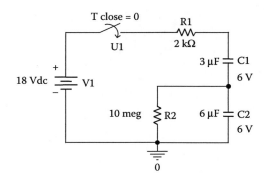

FIGURE 13.3.3
Figure for Simulation Example 13.3.1.

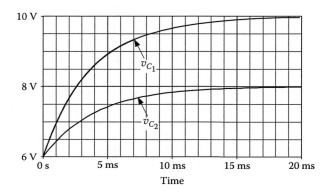

FIGURE 13.3.4
Figure for Simulation Example 13.3.1.

with proper polarities. PSpice assigns the initial voltage as a voltage drop from pin 1 to pin 2 of the capacitor. C having marked terminals can be placed from the ANALOG_P Library. The plots of the current and the capacitor voltages are shown in Figure 13.3.4 and Figure 13.3.5.

EXERCISE 13.3.2

A 1 µF capacitor is charged to 10 V and connected at $t = 0$ to an uncharged 4 µF capacitor in series with a 1.25 kΩ resistor. Determine the current for $t \geq 0$.

Answer: $8e^{-t}$ mA, where t is in ms.

Example 13.3.2: Neon Lamp Flasher

A flashing circuit may be constructed using a neon lamp, a resistor, a capacitor, and a battery connected as shown in Figure 13.3.6. When the lamp is not conducting, it behaves as an open circuit. The capacitor charges, through R, toward the battery voltage V_{SRC}. But when the capacitor voltage reaches a voltage V_{max}, the lamp is turned on and presents a low resistance R_L across the capacitor. The capacitor discharges and the voltage across it

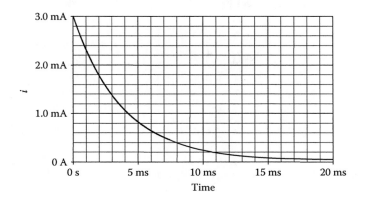

FIGURE 13.3.5
Figure for Simulation Example 13.3.1.

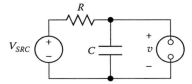

FIGURE 13.3.6
Figure for Example 13.3.2.

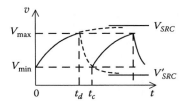

FIGURE 13.3.7
Figure for Example 13.3.2.

falls. When the capacitor voltage reaches a voltage V_{min}, the lamp turns off and the cycle repeats. It is required to determine the frequency of flashing of the neon lamp.

SOLUTION

Let the charging of C from V_{min} begin at $t = 0$ and the discharging from V_{max} begin at $t = t_d$ and continue to $t = t_c$, when charging begins again (Figure 13.3.7).

During $0 \leq t \leq t_d$, the initial value for v_C is V_{min}, the final value is V_{SRC}, and not V_{max}, because if the lamp does not turn on, C continues to charge to V_{SRC}. The time constant is RC. It follows from Equation 13.3.2, with $t_0 = 0$ that:

$$v_C = V_{SRC} + \left[V_{min} - V_{SRC} \right] e^{-\frac{t}{RC}}, \ 0 \leq t \leq t_d \tag{13.3.3}$$

At $t = t_d$, $v_C = V_{max}$. Substituting in Equation 13.3.3,

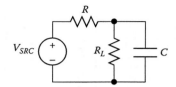

FIGURE 13.3.8
Figure for Example 13.3.2.

$$V_{max} = V_{SRC} + \left[V_{min} - V_{SRC}\right]e^{-\frac{t_d}{RC}} \qquad (13.3.4)$$

Solving for t_d,

$$t_d = RC\ln\frac{V_{SRC} - V_{min}}{V_{SRC} - V_{max}} \qquad (13.3.5)$$

During capacitor discharge, the lamp presents a resistance R_L across the capacitor. The circuit becomes that of Figure 13.3.8. Thevenin's voltage is $V_{Th} = V'_{SRC} = V_{SRC}\dfrac{R_L}{R + R_L}$ and $R_{Th} = R' = (R \parallel R_L)$. Hence, during the discharge period, the initial value of v_C is V_{max} at $t = t_d$, the final value is V'_{SRC}, and the time constant is $\tau' = R'C$. It follows from Equation 13.3.2, with $t_o = t_d$,

$$v_C = V'_{SRC} + \left[V_{max} - V'_{src}\right]e^{-\frac{t - t_d}{R'C}}, \ \ t_d \leq t \leq t_c \qquad (13.3.6)$$

At $t = t_c$, $v_C = V_{min}$. Substituting in Equation 13.3.6,

$$V_{min} = V'_{SRC} + \left[V_{max} - V'_{SRC}\right]e^{-\frac{t_c - t_d}{R'C}} \qquad (13.3.7)$$

Solving for $t_c - t_d$,

$$t_c - t_d = R'C\ln\frac{V_{max} - V'_{SRC}}{V_{min} - V'_{SRC}} \qquad (13.3.8)$$

The time t_d represents the period that the lamp is off; that is, the time between flashes. The time $t_c - t_d$ represents the flash duration. When t_c is in seconds, the flashing frequency is $\dfrac{1}{t_c}$ flashes per second. This circuit is an example of a **relaxation oscillator.**

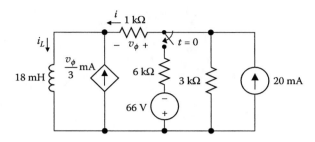

FIGURE 13.3.9
Figure for Example 13.3.3.

Example 13.3.3: Inductor Current Determined from Initial Value, Final Value, and Effective Time Constant

In Figure 13.3.9, the switch is closed at $t = 0$ after being open for a long time, which allows the circuit to reach a steady state. It is required to determine i_L and v_ϕ for $t \geq 0$.

SOLUTION

To apply Equation 13.3.2, we should determine the initial values of i_L and v_ϕ, their final values, and the effective time constant that appears in parallel with the inductor.

Considering i_L first, its initial value i_{L0}, just after the switch is closed, is the same as its value, just before the switch is closed. To determine this value, the 20 mA current source in parallel with 3 kΩ is first transformed to an equivalent 60 V source in series with 3 kΩ. With the switch open, the circuit becomes as shown in Figure 13.3.10. In the steady state, the voltage across the inductor is zero. Applying KVL to the outer loop gives $(1 + 3)i_0 = 60$, or $i_0 = 15$ mA. The dependent source current is 5 mA, so that $i_{L0} = 15 + 5 = 20$ mA.

After the switch is closed, the 66 V source in series with 6 kΩ is transformed to an 11 mA current source of polarity opposite that of the 20 mA source in parallel with 6 kΩ. The two current sources are then combined in a single current source of 9 mA of the same polarity as the 20 mA source in parallel with a resistance of 3 || 6 = 2 kΩ. This source is transformed back to a voltage source of 18 V in series with 2 kΩ (Figure 13.3.11). Proceeding in a similar fashion gives $i_f = \dfrac{18}{3} = 6$ mA and $i_{Lf} = 6 + 2 = 8$ mA.

To find the resistance seen by the inductor, a test voltage v_T is applied in place of the inductor in Figure 13.3.11, with the independent voltage source replaced by a short-circuit,

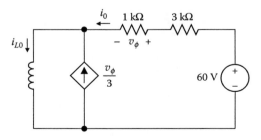

FIGURE 13.3.10
Figure for Example 13.3.3.

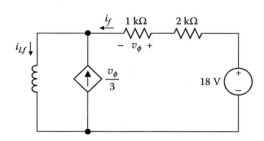

FIGURE 13.3.11
Figure for Example 13.3.3.

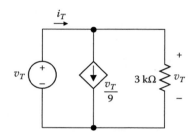

FIGURE 13.3.12
Figure for Example 13.3.3.

and v_T/i_T determined (Figure 13.3.12). Now the 1 kΩ and the 2 kΩ resistors form a voltage divider across v_T, so that $v_\phi = -v_T/3$, and the current source becomes $v_T/9$, with its polarity reversed. From KCL, $i_T = \dfrac{v_T}{3} + \dfrac{v_T}{9}$, which gives $\dfrac{v_T}{i_T} = \dfrac{9}{4}$ kΩ. Alternatively, we note that the source current is $v_T/9$ in the direction of a voltage drop v_T. It may therefore be replaced by a resistance $\dfrac{v_T}{v_T/9} = 9$ kΩ. This, in parallel with the 3 kΩ, gives an effective resistance of 9/4 kΩ, as before. The effective time constant is therefore $\tau = (18 \times 4)/9 = 8$ µs. Note that the effective resistance in this case is R_{Th} as seen from the inductor terminals and could be obtained as $V_{Th} = V_{oc} = 18$ V divided by $I_{sc} = i_{Lf} = 8$ mA.

From Equation 13.3.2: $i_L = 8 + (20 - 8)e^{-t/8} = 8 + 12e^{-t/8}$ mA, $t \geq 0$, where t is in µsec.

Considering v_ϕ, it is seen from Figure 13.3.9 that $v_\phi = 1 \times i$ and $i_L = i + v_\phi/3$, so that $v_\phi = \dfrac{3}{4}i_L$,

irrespective of the position of the switch. Hence, $v_\phi = 6 + 9e^{-t/8}$ V, $t \geq 0$, where t is in µs. Note that, in this case, v_ϕ is directly related to i_L, so that v_ϕ does not change when the switch is moved. Example 13.3.4 illustrates a different case.

Example 13.3.4: Time Variation of Voltage across a Resistor

In Figure 13.3.13, the switch is opened at $t = 0$ after being closed for a long time. It is required to determine i_L and v_1 for $t \geq 0$.

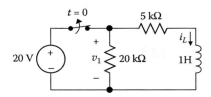

FIGURE 13.3.13
Figure for Example 13.3.4.

SOLUTION
After the switch has been closed for a long time, the voltage across the inductor is zero, so that $i_L = \dfrac{20}{5} = 4$ mA. This is also i_{L0} at $t = 0^+$ because the inductor current does not change at the instant the switch is opened. The final value of i_L, after the switch is opened is zero. The resistance seen by L after the switch is opened is 25 kΩ, so that $\tau = 1$ H/25 kΩ = 1/25 ms. Hence, $i_L = 4e^{-2.5t}$ mA, $t \geq 0$, where t is in ms.

As for v_1, it follows from Figure 13.3.13 that $v_1 = -80e^{-25t}$ V, $t \geq 0$, where t is in ms. If Equation 13.3.2 is to be applied to v_1, its initial value at $t = 0^+$ is that after the switch is opened, which is $-20i_{L0}$, or -80 V. Its final value is zero. Hence Equation 13.3.2 gives the same expression for v_1.

EXERCISE 13.3.3

Two inductors, one of 3 mH having an initial current of 20 mA, the other of 6 mH and having an initial current of 30 mA, are paralleled together and with a 500 Ω resistor at $t = 0$ (Figure 13.3.14). Determine the voltage across the combination after two time constants.

Answer: 3.83 V.

Example 13.3.5: Discharge of Two Series-Connected Capacitors through a Resistor

Capacitors C_1 and C_2 of 3 F and 2 F, respectively, are charged to 10 V each and connected at $t = 0$ in series with a resistor R of 2 Ω resistance (Figure 13.3.15). It is required to determine the charges q_1 and q_2 on C_1 and C_2 in the steady state.

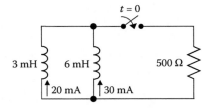

FIGURE 13.3.14
Figure for Exercise 13.3.3.

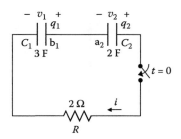

FIGURE 13.3.15
Figure for Example 13.3.5.

SOLUTION
Method 1: Immediately the switch is closed, 20 V is impressed across R, causing a current
i in the direction that discharges both capacitors. i continues to flow until the voltage
across R is zero. This happens when i transfers enough positive charge from the positive
plate of C_2 to the negative plate of C_1, so that the voltage of one of the capacitors reverses
and becomes equal and opposite to that on the other capacitor.

The initial charge on C_1 is $q_1(0) = 3 \times 10 = 30$ C, whereas the initial charge on C_2 is
$q_2(0) = 2 \times 10$ V $= 20$ C. The charge on C_2, being the smaller of the two, will therefore be
reversed first. Let the final values of v_1 and v_2 be v_{1f} and v_{2f}, respectively, corresponding
to final values of charge q_{1f} and q_{2f}. From KVL, with zero voltage across R,

$$v_{1f} + v_{2f} = 0 \quad \text{or} \quad \frac{q_{1f}}{3} + \frac{q_{2f}}{2} = 0 \tag{13.3.9}$$

Let the total charge transferred from $t = 0$ and until the steady state is reached ($t \rightarrow \infty$)
be Δq. Because of the direction of i, Δq subtracts from both q_1 and q_2. The final value of
the charge on either capacitor is the initial charge minus Δq.

$$q_{1f} = 30 - \Delta q \quad \text{and} \quad q_{2f} = 20 - \Delta q \tag{13.3.10}$$

Substituting for q_{1f} and q_{2f} in the second part of Equation 13.3.9 gives $\Delta q = 24$ C. Thus,
$q_{1f} = 6$ C and $v_{1f} = 2$ V, whereas $q_{2f} = -4$ C and $v_{2f} = -2$ V.

A variation of this method is to recognize that because the same current flows in C_1 and
C_2, the initial charge difference of 10 C between C_1 and C_2 will always be maintained.
Thus: $q_{1f} - q_{2f} = 10$ C. This equation, together with the second part of Equation 13.3.9, gives
$q_{1f} = 6$ C and $q_{2f} = 4$ C. An alternative interpretation of the relation $q_{1f} - q_{2f} = 10$ C is that
the charge on the two plates of C_1 and C_2 that are connected together must be conserved.
Initially, $q_1(0) - q_2(0) = 30 - 20 = 10$ C. Hence, $q_{1f} - q_{2f} = 10$ C also. The initial 30 C on plate
b_1, for example, completely neutralizes the -20 C on plate a_2, and the net charge of 10 C
is redistributed as 6 C on b_1 and 4 C on a_2.

Method 2: Consider the equivalent series capacitor $C_{eqs} = \dfrac{3 \times 2}{3 + 2} = 1.2$ F. The initial voltage
of C_{eqs} is $v_1(0) + v_2(0) = 20$ V. The initial charge on C_{eqs} is $q_{eqs} = (1.2$ F$) \times (20$ V$) = 24$ C. If
C_{eqs} with this initial charge were connected across R, then it will discharge completely,
with no residual charge. The current that discharges C_{eqs} is the same current that flows

through the series combination of C_1 and C_2. This current would therefore decrease the charge on each capacitor by 24 C, leaving 6 C on C_1 and −4 C on C_2.

A variation of this method is to consider the total initial energy in C_1 and C_2, $w_{12} =$

$\frac{1}{2} \times 3 \times (10)^2 + \frac{1}{2} \times 2 \times (10)^2 = 25$ J. The initial energy in C_{eqs} is $w_{eqs} = \frac{1}{2} \times 1.2 \times (20)^2 = 240$ J. The initial energy in C_{eqs} is completely dissipated in R, so that the difference in the two

energies is because of the residual charges. Thus, $w_{12f} = \frac{1}{2} \frac{q_{1f}^2}{3} + \frac{1}{2} \frac{q_{2f}^2}{2} = 10$ J. This equation,

together with the second part of Equation 13.3.9, gives $q_{1f} = \pm 6$ C and $q_{2f} = \mp 4$ C. The solutions $q_{1f} = -6$ C and $q_{2f} = 4$ C must be rejected because v_1 does not reverse polarity with $q_1(0) > q_2(0)$.

Method 3: We can start with the variation of i with time. The initial value of i is $\frac{20\ V}{2\ \Omega} = 10$ A, and its final value is zero. The capacitance seen from the terminals of R is C_{eqs}. Hence, the effective time constant is $RC_{eqs} = (2\ \Omega) \times (1.2\ F) = 2.4$ s. The time variation of i is:

$$i = 10e^{-\frac{t}{2.4}} \text{ A} \tag{13.3.11}$$

It follows that

$$q_1 = -\int_0^t i\, dt + q_1(0) = (10\ \text{A})(2.4\ \text{s})\left[e^{-\frac{t}{2.4}}\right]_0^t + 30 = 6 + 24e^{-\frac{t}{2.4}} \text{ C} \tag{13.3.12}$$

The minus sign of the integral in Equation 13.3.12 accounts for the fact that i is a discharge current that decreases q_1. According to Equation 13.3.12, $q_1 = 30$ C at $t = 0$, and $q_{1f} = 6$ C when t is infinite. Similarly,

$$q_2 = -\int_0^t i\, dt + q_2(0) = (10\ \text{A})(2.4\ \text{s})\left[e^{-\frac{t}{2.4}}\right]_0^t + 20 = -4 + 24e^{-\frac{t}{2.4}} \text{ C} \tag{13.3.13}$$

where $q_2 = 20$ C at $t = 0$, and $q_{2f} = -4$ C when t is infinite.

The total energy dissipated in R from Equation 13.3.11 is,

$$w_R = \int_0^\infty \left(10e^{-\frac{t}{2.4}}\right)^2 2\, dt \text{ J} = 200(-1.2)\left[e^{-\frac{t}{1.2}}\right]_0^\infty = 240 \text{ J}.$$

The energy dissipated in R is that stored in C_{eqs} and is equal to the difference between the initial and final energies of C_1 and C_2.

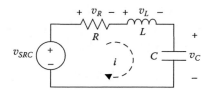

FIGURE 13.4.1
RLC circuit excited by voltage step.

13.4 Step Response of Second-Order Circuits

To analyze the step response of a second-order circuit, we consider a series *RLC* circuit excited by a source $v_{SRC} = Ku(t)$, where $u(t)$ is the unit step at the origin (Figure 13.4.1).

From KVL, $L\dfrac{di}{dt} + Ri + v_C = Ku(t)$. It is convenient to write the circuit equation in terms of

v_C. Substituting $i = C\dfrac{dv_C}{dt}$ and $\dfrac{di}{dt} = C\dfrac{d^2v_C}{dt^2}$, $LC\dfrac{d^2v_C}{dt^2} + RC\dfrac{dv_C}{dt} + v_C = Ku(t)$. Dividing by

LC and setting $\omega_0 = \dfrac{1}{\sqrt{LC}}$ and $\alpha = \dfrac{R}{2L}$, as in Section 12.2 (Chapter 12),

$$\frac{d^2v_C}{dt^2} + 2\alpha\frac{dv_C}{dt} + \omega_0^2 v_C = K\omega_0^2 u(t) \tag{13.4.1}$$

As usual, the general solution to this equation is the sum of the following two components:

1. A transient component that is the solution to the differential equation with zero forcing function on the RHS. As argued in Section 12.2 (Chapter 12), this solution is of the form $Ae^{s_1 t} + Be^{s_2 t}$, where A and B are arbitrary constants.
2. A steady-state component that satisfies the equation. Because the input is constant
   for $t > 0$, the steady state is dc, in which $\dfrac{dv_C}{dt}$ and $\dfrac{d^2v_C}{dt^2}$ are zero. Equation 13.4.1
   then gives $v_C = K$. This makes sense, of course, because under dc conditions, C acts as an open circuit, $i = 0$, and $v_C = v_{SRC} = K$.

The complete solution is therefore:

$$v_C = Ae^{s_1 t} + Be^{s_2 t} + K, \ t \geq 0 \tag{13.4.2}$$

Equation 13.4.2 applies for $t \geq 0$ because v_C and i are initially zero at $t = 0^-$, before the step is applied, and remain zero at $t = 0^+$, just after the step is applied because no voltage or current impulses are applied at $t = 0$ that will change these initial values. Substituting $t = 0$ in Equation 13.4.2 gives:

$$A + B + K = 0 \tag{13.4.3}$$

To apply the initial condition $i = C\dfrac{dv_C}{dt} = 0$ at $t = 0$, we differentiate Equation 13.4.2 and set the RHS equal to zero:

$$s_1 A + s_2 B = 0 \tag{13.4.4}$$

Solving Equation 13.4.3 and Equation 13.4.4 for A and B, we obtain $A = \dfrac{Ks_2}{s_1 - s_2}$

and $B = -\dfrac{Ks_1}{s_1 - s_2}$. Substituting in Equation 13.4.2,

$$v_C = K\left[\frac{s_2}{s_1 - s_2}e^{s_1 t} - \frac{s_1}{s_1 - s_2}e^{s_2 t} + 1\right], \ t \geq 0 \tag{13.4.5}$$

where, as derived in Section 12.2 (Chapter 12),

$$s_1 = -\alpha + \sqrt{\alpha^2 - \omega_0^2} \tag{13.4.6}$$

$$s_2 = -\alpha - \sqrt{\alpha^2 - \omega_0^2} \tag{13.4.7}$$

$$s_1 + s_2 = -2\alpha, \ \ s_1 - s_2 = 2\sqrt{\alpha^2 - \omega_0^2}, \text{ and } \ s_1 s_2 = \omega_0^2 \tag{13.4.8}$$

It follows that:

$$i = C\frac{dv_C}{dt} = \frac{K}{L(s_1 - s_2)}\left[e^{s_1 t} - e^{s_2 t}\right], \ t \geq 0 \tag{13.4.9}$$

and

$$v_L = L\frac{di}{dt} = \frac{K}{(s_1 - s_2)}\left[s_1 e^{s_1 t} - s_2 e^{s_2 t}\right], \ t \geq 0 \tag{13.4.10}$$

Whereas A and B generally depend on the excitation, initial conditions, and circuit parameters, α and ω_0 depend on circuit parameters only. Hence, the circuit that has an overdamped, critically damped, or underdamped natural response also has the same type of response when subjected to a step of excitation. We will therefore consider each of these responses in turn.

Recall that in an overdamped circuit, R is large enough so that $\alpha > \omega_0$, which makes s_1 and s_2 real and distinct. v_C, i, and v_L are given by Equation 13.4.5, Equation 13.4.9, and Equation 13.4.10, respectively. In a critically damped circuit, $\alpha = \omega_0$, so that $s_1 = s_2 = \omega_0$. As in the case of the natural response, the transient component is no longer given by

Equation 13.4.2 but by Equation 12.2.17 (Chapter 12). The general response is therefore of the form

$$v_C = Ae^{-\omega_0 t} + Bte^{-\omega_0 t} + K \tag{13.4.11}$$

Substituting the initial conditions $i = 0$ and $v_C = 0$ at $t = 0$, as obtained earlier, gives $A = K$ and $B = K\omega_0$. It follows that

$$v_C = K\left[1 - e^{-\omega_0 t} - \omega_0 t e^{-\omega_0 t}\right], \ t \geq 0 \tag{13.4.12}$$

$$i = C\frac{dv_C}{dt} = \frac{K}{L}te^{-\omega_0 t}u(t), \ t \geq 0 \tag{13.4.13}$$

and

$$v_L = L\frac{di}{dt} = Ke^{-\omega_0 t}\left(1 - \omega_0 t\right), \ t \geq 0 \tag{13.4.14}$$

For the underdamped response, $\alpha < \omega_0$. v_C, i, and v_L are given by Equation 13.4.5, Equation 13.4.9, and Equation 13.4.10, respectively, but with s_1 and s_2 complex, as given by Equation 12.2.22 and Equation 12.2.23 (Chapter 12). Substituting $s_1 = -\alpha + j\omega_d t$, $s_2 = -\alpha - j\omega_d t$, $s_1 - s_2 = j2\omega_d t$, $\omega_0^2 = \omega_d^2 + \alpha^2$, and simplifying gives:

$$v_C = K\left[1 - e^{-\alpha t}\left(\cos\omega_d t + \frac{\alpha}{\omega_d}\sin\omega_d t\right)\right], \ t \geq 0 \tag{13.4.15}$$

$$i = C\frac{dv_C}{dt} = \frac{K}{\omega_d L}e^{-\alpha t}\sin\omega_d t, \ t \geq 0 \tag{13.4.16}$$

and

$$v_L = L\frac{di}{dt} = Ke^{-\alpha t}\left(\cos\omega_d t - \frac{\alpha}{\omega_d}\sin\omega_d t\right), \ t \geq 0 \tag{13.4.17}$$

The overdamped, critically damped, and underdamped responses are compared in Example 13.4.1.

EXERCISE 13.4.1

Given a critically damped *RLC* circuit having $L = 1$ mH and $R = 2$ kΩ, how much is C? Answer: 1 nF.

EXERCISE 13.4.2

An underdamped circuit has $\omega_d = 30$ krad/s and $\alpha = 40$ krad/s. If $C = 1$ μF, how much is R? Answer: 32 Ω.

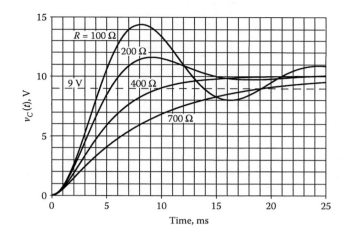

FIGURE 13.4.2
Figure for Simulation Example 13.4.1.

Simulation Example 13.4.1: Series *RLC* Circuit Excited by Voltage Step

It is required to simulate a series *RLC* circuit in response to a voltage step of 10 V, with $L = 0.5$ H, $C = 12.5$ μF, and to plot the voltage across the capacitor for $R = 100$, 200, 400, and 700 Ω. The circuit is the same as that of Example 10.5.1 (Chapter 10).

SIMULATION

The schematic is entered as usual with a 10 V VDC source connected. Zero initial conditions are entered in the IC columns of the Property Editor spreadsheets of L and C. R is declared as a global parameter, as described in Example 7.4.1 (Chapter 7). Select Time domain (Transient) analysis, then check the Parametric Sweep box. Choose Global parameter under Sweep variable and enter R_val in the Parameter name field. Under Sweep type, enter in the Value list: 100,200,400,700. When the simulation is run, press the All and OK buttons in the Available Sections window. After selecting V[VO] from the Simulation Output Variables window, assuming VO is the output voltage. When the simulation is run, the resulting plots are shown in Figure 13.4.2. $R = 700$ Ω gives an overdamped response, $R = 400$ Ω gives a critically damped response, whereas $R = 200$ Ω and $R = 100$ Ω give underdamped responses.

The following conclusions may be drawn from Simulation Example 13.4.1:

CONCLUSIONS

1. The overdamped response is the slowest of the responses to approach the steady-state value.

2. The critically damped response is the fastest of the responses to reach the steady-state value without overshoot.

3. The underdamped response is the fastest to reach the steady-state value, but overshoots this value and oscillates about it at a frequency ω_d before settling to the final value.

If we consider the 9 V level, for example, the overdamped response (700 Ω) reaches this level at 19.17 ms, the critically damped response at 9.72 ms, the underdamped response (200 Ω) at 5.31 ms, and the underdamped response (100 Ω) at 4.32 ms. However, the underdamped response can have unacceptable overshoot and undershoot. The underdamped response is analyzed in more detail in Section ST13.1.

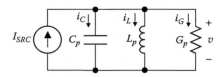

FIGURE 13.4.3
GCL circuit excited by current step.

Settling Time

An important consideration in systems that are required to change the value of an output in response to a sudden change of input is the **settling time**, or time taken to reach a specified output. In the case of the circuit of Example 13.4.1, for example, the settling time may be specified as the time it takes v_C to reach 9.5 V, that is, to within 5% of the final value. If no overshoot is allowed, the shortest settling time is achieved with the critically damped response. If some overshoot is allowed, the settling time can be reduced (Problem P13.2.1).

Parallel GCL Circuit

The dual of the series RLC circuit excited by a voltage step is the parallel GCL circuit excited by a current step, as shown in Figure 13.4.3, where v, i_G, i_C, and i_L are the duals of i, v_R, v_L, and v_C, respectively, as in Section 12.2 (Chapter 12), and are given by the corresponding dual relations. Table 13.4.1 summarizes the step responses of the two circuits.

Step Response from Impulse Response

As discussed in connection with Equation 13.2.4, the step response of an LTI circuit is the integral of its impulse response. It can be readily verified that the entries in Table 13.4.1 are the time integrals of the corresponding entries in Table 12.4.1 (Chapter 12).

EXERCISE 13.4.3

A parallel GCL circuit has $C_p = 10$ nF and $L_p = 100$ μH. Determine R_p if the maximum percentage overshoot is not to exceed 5%, given that the maximum percentage

overshoot is $100\, e^{-\frac{\alpha_p}{\omega_{dp}}\pi}$ (Section ST13.1).

Answer: 72.5 Ω.

13.5 Switched Second-Order Circuits

As mentioned in the overview, discussion of switched second-order circuits in this section is restricted to prototypical second-order circuits, that is, circuits that can be reduced to simple series RLC or parallel GCL circuits. This serves to emphasize the following points:

Governing Princples

1. A capacitor voltage (v_C) does not change at the instant of switching, that is, between $t = 0^-$ and $t = 0^+$, unless forced to by a current impulse. Similarly, an

TABLE 13.4.1

Step Responses of Dual Circuits, $t \geq 0^+$

Series *RLC* Circuit	Parallel *GCL* Circuit

Overdamped Response

$$i = \frac{K}{L(s_1 - s_2)}\left[e^{s_1 t} - e^{s_2 t}\right]$$

$$v = \frac{K}{C_p(s_{1p} - s_{2p})}\left[e^{s_{1p} t} - e^{s_{2p} t}\right]$$

$$v_L = \frac{K}{(s_1 - s_2)}\left[s_1 e^{s_1 t} - s_2 e^{s_2 t}\right]$$

$$i_C = \frac{K}{(s_{1p} - s_{2p})}\left[s_{1p} e^{s_{1p} t} - s_{2p} e^{s_{2p} t}\right]$$

$$v_C = K\left[\frac{s_2}{s_1 - s_2}e^{s_1 t} - \frac{s_1}{s_1 - s_2}e^{s_2 t} + 1\right]$$

$$i_L = K\left[\frac{s_{2p}}{s_{1p} - s_{2p}}e^{s_{1p} t} - \frac{s_{1p}}{s_{1p} - s_{2p}}e^{s_{2p} t} + 1\right]$$

Critically Damped Response

$$i = \frac{K}{L}te^{-\omega_0 t}$$

$$v = \frac{K}{C_p}te^{-\omega_0 t}$$

$$v_L = Ke^{-\omega_0 t}\left(1 - \omega_0 t\right)$$

$$i_C = Ke^{-\omega_0 t}\left(1 - \omega_0 t\right)$$

$$v_C = K\left[1 - e^{-\omega_0 t} - \omega_0 t e^{-\omega_0 t}\right]$$

$$i_L = K\left[1 - e^{-\omega_0 t} - \omega_0 t e^{-\omega_0 t}\right]$$

Underdamped Response

$$i = \frac{K}{\omega_d L}e^{-\alpha t}\sin\omega_d t$$

$$i = \frac{K}{\omega_{dp}C_p}e^{-\alpha_p t}\sin\omega_{dp} t$$

$$v_L = Ke^{-\alpha t}\left(\cos\omega_d t - \frac{\alpha}{\omega_d}\sin\omega_d t\right)$$

$$i_C = Ke^{-\alpha_p t}\left(\cos\omega_{dp} t - \frac{\alpha_p}{\omega_{dp}}\sin\omega_{dp} t\right)$$

$$v_C = K\left[1 - e^{-\alpha t}\left(\cos\omega_d t + \frac{\alpha}{\omega_d}\sin\omega_d t\right)\right]$$

$$i_L = K\left[1 - e^{-\alpha_p t}\left(\cos\omega_{dp} t + \frac{\alpha_p}{\omega_{dp}}\sin\omega_{dp} t\right)\right]$$

inductor current (i_L) does not change at the instant of switching unless forced to by a voltage impulse.

2. KCL and KVL must be satisfied at all times, before switching, just after switching, and in the steady state.

3. From preceding statements 1 and 2, it is possible to determine dv_C/dt and di_L/dt at $t = 0^+$. These values, together with those of statement 1, determine the arbitrary constants in the response.

4. The type of response, whether overdamped, critically damped, or underdamped, is determined by comparing α and ω_0 in the circuit after switching has occurred, with all independent sources set to zero (Section 13.3).

The aforementioned points are illustrated by Example 13.5.1.

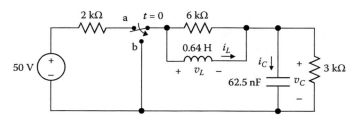

FIGURE 13.5.1
Figure for Example 13.5.1.

Example 13.5.1: Switched *GCL* Circuit

The switch in Figure 13.5.1 is moved at $t = 0$ to position b after having been in position a for a long time. It is required to determine how i_L and v_C change with time for $t \geq 0$.

SOLUTION

At $t = 0^-$, just before the switch is moved, $v_L(0^-) = 0$, $i_L(0^-) = \dfrac{50}{5} = 10$ mA. This 10 mA current flows through the 3 kΩ resistor, so that $v_C(0^-) = 30$ V. At $t = 0^+$, i_L and v_C remain unchanged because there are no impulses to cause them to change instantaneously. However, although i_L and v_C do not change at $t = 0^+$, $\dfrac{di_L}{dt}$ and $\dfrac{dv_C}{dt}$ do change at $t = 0^+$ so that

$v_L = L\dfrac{di_L}{dt}$ satisfies KVL and $i_C = C\dfrac{dv_C}{dt}$ satisfies KCL. The voltages and currents at $t = 0^+$ are shown in Figure 13.5.2. To satisfy KVL, $v_L(0^+) = -30$ V. This implies that a 5 mA current flows through the 6 kΩ resistor, as shown. The 10 mA current continues to flow through the 3 kΩ resistor because $v_C(0^+) = 30$ V. To satisfy KCL, $i_C(0^+) = -5$ mA.

For $t \geq 0^+$, the circuit reduces to the parallel *GCL* circuit shown in Figure 13.5.3, with initial values: $v_C(0^+) = 30$ V, $i_L(0^+) = 10$ mA, $\dfrac{dv_C}{dt}\Big|_{t=0^+} = \dfrac{i_C(0^+)}{C}$, and $\dfrac{di_L}{dt}\Big|_{t=0^+} = -\dfrac{v_C(0^+)}{L}$ because $v_L = -v_C$ at $t = 0^+$.

After having determined the initial values, the next step is to decide whether the circuit is overdamped, critically damped, or underdamped. $\omega_0 = \dfrac{1}{\sqrt{L_p C_p}} =$

$: \dfrac{1}{\sqrt{0.64 \times 62.5 \times 10^{-9}}} \equiv 5 \text{ krad/s}; \alpha_p = \dfrac{1}{2R_p C_p} = \dfrac{1}{4 \times 10^3 \times 62.5 \times 10^{-9}} \equiv 4 \text{ krad/s}.$ Because

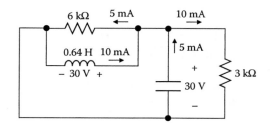

FIGURE 13.5.2
Figure for Example 13.5.1.

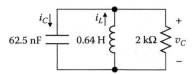

FIGURE 13.5.3
Figure for Example 13.5.1.

$\alpha_p < \omega_0$, the response is underdamped, with $\omega_{dp} = \sqrt{\omega_0^2 - \alpha_p^2} = 3$ krad/s. To determine the response, we could begin with either v_C or i_L. Suppose we begin with v_C. We express the general solution in terms of v_C as:

$$v_C = Ae^{s_1 t} + Be^{s_2 t} + 0 \text{ V} \tag{13.5.1}$$

where $s_{1p} = -4 + j3$ krad/s, $s_{2p} = -4 - j3$ krad/s, and the zero has been included to emphasize that the steady-state value of v_C is zero. Because s_{1p} and s_{2p} are in krad/s, t is in ms in Equation 13.5.1 and in all the expressions that follow. Note that v_C in Equation 13.5.1 is expressed in volts, so that A and B are also in volts. At $t = 0^+$, $v_C(0^+) = 30$ V. Equation 13.5.1 gives:

$$A + B = 30 \tag{13.5.2}$$

A second equation relating A and B is derived from $\dfrac{dv_C}{dt}\Big|_{t=0^+} = \dfrac{i_C(0^+)}{C}$. Differentiating

Equation 13.5.1: $\dfrac{dv_C}{dt} = As_{1p}e^{s_{1p}t} + Bs_{2p}e^{s_{2p}t}$, which at $t = 0^+$ gives $(As_{1p} + Bs_{2p})$ V/ms, or

$10^3 \times (As_{1p} + Bs_{2p})$ V/s. Equating this to $\dfrac{i_C(0^+)}{C} = \dfrac{-5 \times 10^{-3}}{62.5 \times 10^{-9}}$ gives $As_{1p} + Bs_{2p} = -80$ V/ms or:

$$(4 - j3)A + (4 + j3)B = 80 \tag{13.5.3}$$

Solving Equation 13.5.2 and Equation 13.5.3 gives, $A = 15 - j20/3$ and $B = 15 + j20/3$. Substituting in Equation 13.5.1, $v_C = (15 - j20/3)e^{(-4+j3)t} + (15 + j20/3)e^{(-4-j3)t} = $

$e^{-4t}\left[15(e^{j3t} + e^{-j3t}) - \dfrac{j20}{3}(e^{j3t} + e^{-j3t})\right]$ V. It follows that:

$$v_C = 10e^{-4t}\left(3\cos 3t + \frac{4}{3}\sin 3t\right) \text{ V}, \; t \geq 0 \text{ ms} \tag{13.5.4}$$

It is seen that $v_C = 30$ V at $t = 0$, and $v_C \to 0$ as $t \to \infty$, as it should. i_L could be obtained

from v_C in several ways. Because $i_L = \dfrac{1}{L}\int v_L dt = -\dfrac{1}{L}\int v_C dt$, the integration could be performed on Equation 13.5.1 or on Equation 13.5.4. Integrating Equation 13.5.1,

$\dfrac{A}{s_{1p}}e^{s_{1p}t} + \dfrac{B}{s_{2p}}e^{s_{2p}t} = -Li_L$ Vms. Dividing by L in henries gives i_L in mA. Thus, $i_L =$

$\dfrac{e^{-4t}}{0.64}\left[\dfrac{(15 - j20/3)}{4 - j3}e^{j3t} + \dfrac{(15 + j20/3)}{4 + j3}e^{-j3t}\right]$. After simplification, this gives

$$i_L = 5e^{-4t}\left(2\cos 3t - \dfrac{11}{24}\sin 3t\right) \quad \text{mA}, \; t \geq 0 \text{ ms} \tag{13.5.5}$$

It is seen that $i_L = 10$ mA at $t = 0$, and $i_L \to 0$ as $t \to \infty$, as it should be. Note that no constant of integration is needed when integrating Equation 13.5.1 because the values of A and B already give the correct value of $i_L(0^+)$. If a constant of integration is included, it will evaluate to zero.

If Equation 13.5.4 is to be integrated using the Table of Integrals in the Appendix,

$$i_L = -\dfrac{30}{0.64}\int e^{-4t}\cos 3t\,dt \; -\dfrac{55}{24\times 0.64}\int e^{-4t}\sin 3t\,dt = -\dfrac{30e^{-4t}}{0.64\times 25}\left(-4\cos 3t + 3\sin 3t\right) -$$

$$\dfrac{40e^{-4t}}{3\times 0.64\times 25}\left(-4\sin 3t - 3\cos 3t\right) + K = 10e^{-4t}\cos 3t - \dfrac{55e^{-4t}}{24}\sin 3t + K$$

Because $i_L = 10$ mA at $t = 0^+$, $K = 0$, which gives Equation 13.5.5.

Alternatively, from KCL in Figure 13.5.3, $i_L = i_C + v_C/2$ mA, where $i_C = C\dfrac{dv_C}{dt}$. From Equation

13.5.4, $\dfrac{dv_C}{dt} = -e^{-4t}\left[80\cos 3t + \dfrac{430}{3}\sin 3t\right]$ V/ms. Multiplying by 10^3 makes $\dfrac{dv_C}{dt}$ V/s. Multi-

plying by C in farads gives i_C in amperes, so it has to be multiplied by 10^3 to express i_C in mA.

This gives: $i_C = -e^{-4t}\left[5\cos 3t + \dfrac{215}{24}\sin 3t\right]$ mA. Adding $v_C/2$ mA gives Equation 13.5.5.

EXERCISE 13.5.1
Rework Example 13.5.1 starting with i_L expressed as $Be^{s_{1p}t} + De^{s_{2p}t}$ mA.

Summary of Main Concepts and Results

- The purpose of a transient response is to match the initial conditions in a given circuit to the steady-state response in the absence of a transient. The time course of the transient is that of the natural response of the circuit.
- When an inductor is charged from a dc source, with zero initial current, the inductor current increases with time as a saturating exponential, and the inductor voltage decreases as a decaying exponential.

- When a capacitor is charged from a dc source, with zero initial voltage, the capacitor voltage increases with time as a saturating exponential, and the capacitor current decreases as a decaying exponential.

- Any voltage or current response in a first-order circuit can be expressed in terms of the initial value of the response, its final value, and the effective time constant.

- Any voltage or current response in a first-order circuit can be expressed in terms of the initial value of the response, its final value, and the effective time constant.

- Independent sources affect initial or final values of circuit responses but not the time constant. Dependent sources affect initial or final values of circuit responses as well as the time constant.

- The overdamped response is the slowest of the responses to approach the steady-state value. The critically damped response is the fastest of the responses to reach the steady-state value without overshoot. The underdamped response is the fastest to reach the steady-state value, but overshoots this value and oscillates about it at a frequency ω_d before settling to the final value.

- A capacitor voltage (v_C) does not change at the instant of switching, that is, between $t = 0^-$ and $t = 0^+$, unless forced to by a current impulse. Similarly, an inductor current (i_L) does not change at the instant of switching unless forced to by a voltage impulse.

- KCL and KVL must be satisfied at all times, before switching, just after switching, and in the steady state, which allows determining dv_C/dt and di_L/dt at $t = 0^+$.

Learning Outcomes

- Be able to analyze the step responses of first-order and second-order circuits
- Be able to derive responses of first-order circuits from their initial values, final values, and the effective time constant
- Be able to analyze a switched second-order circuit that is reducible to a series RLC circuit or a parallel GCL circuit

Supplementary Topics and Examples on CD

ST13.1 Underdamped second-order response to a step input: Derives expressions for the maxima, minima, their times of occurrence, and other characteristics of the underdamped response.

SE13.1 Discharging of two parallel-connected inductors through a resistor: Analyzes the dual case of Example 13.3.5.

SE13.2 Discharging of two parallel-connected capacitors through a resistor: Two capacitors, initially charged to different voltages, are paralleled together and with a resistor at $t = 0$. It is argued that the presence of the resistor has no effect on the current impulse that equalizes the voltage across the two capacitors.

Problems and Exercises

P13.1 First-Order Systems

P13.1.1 The switch in Figure P13.1.1 is moved to position b at $t = 0$ after having been in position a for a long time. Determine v_O.

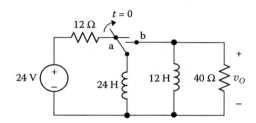

FIGURE P13.1.1

P13.1.2 In Figure P13.1.2, the switch is moved at $t = 0$ to position b after being in position a for a long time. Determine v_C and the time at which it becomes zero.

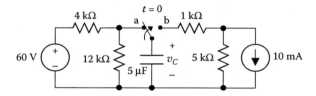

FIGURE P13.1.2

P13.1.3 In Figure P13.1.3, the switch is closed at $t = 0$ after being open for a long time. Determine i_x, $t \geq 0$.

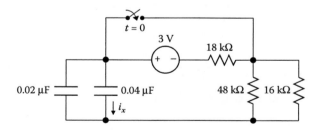

FIGURE P13.1.3

P13.1.4 Calculate the total energy stored in Figure P13.1.4 after the switch has been closed for a long time.

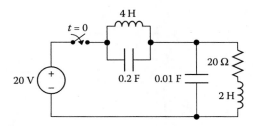

FIGURE P13.1.4

P13.1.5 In Figure P13.1.5, R may be considered infinite while $0 < v_C < 3$ V. When $v_C = 3$ V, R becomes zero, instantly discharging the capacitor, and immediately becomes infinite again. Determine the frequency of oscillation.

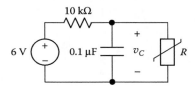

FIGURE P13.1.5

P13.1.6 In Figure P13.1.6, the capacitor is initially uncharged and the switch is closed at $t = 0$. Determine v_C and i_x for $t \geq 0$.

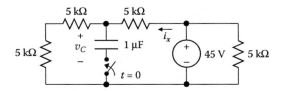

FIGURE P13.1.6

P13.1.7 In Figure P13.1.7, the switch is closed at $t = 0$ after being open for a long time. Determine v_O for $t \geq 0$

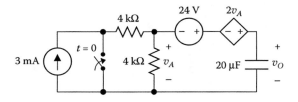

FIGURE P13.1.7

P13.1.8   Both switches in Figure P13.1.8 have been open for a long time. S_1 is closed
at $t = 0$ and S_2 at $t = 1$ s Determine v_C for $t \geq 1$ s.

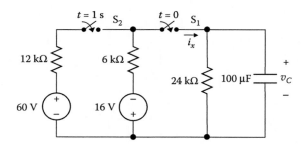

FIGURE P13.1.8

P13.1.9   In Figure P13.1.9, switch S_1 has been in position a for a long time, with switch
S_2 closed. At $t = 0$, S_1 is moved to position b and remains in this position. S_2
is opened at $t = 50$ μs and closed again at $t = 100$ μs. Derive v_x for: (a) $0 \leq t \leq$
50 μs, (b) $50 \leq t \leq 100$ μs; (c) $t \geq 100$ μs.

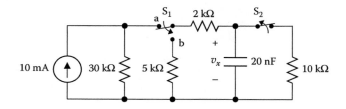

FIGURE P13.1.9

P13.1.10   Determine v_C for $t \geq 0$ in Figure P13.1.10 if $C = 1.25$ μF and $v_C(0) = 10$ V.

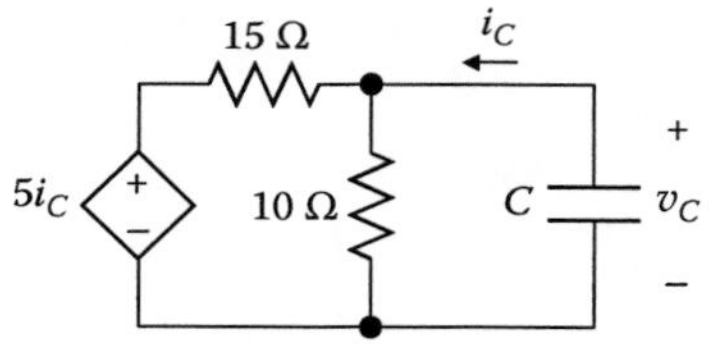

FIGURE P13.1.10

P13.1.11 In Figure P13.1.11, the switch is moved to position b at $t = 0$ after being in position a for a long time. Determine i_x, $t \geq 0$.

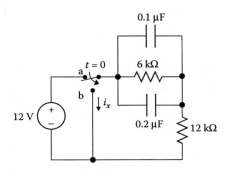

FIGURE P13.1.11

P13.1.12 In Figure P13.1.12, the switch is moved to position b at $t = 0$ after being in position a for a long time. Determine v_C for $t \geq 0$.

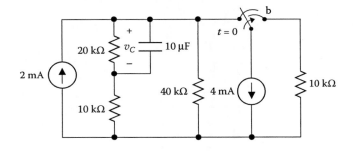

FIGURE P13.1.12

P13.1.13 In Figure P13.1.13, the switch is initially in position a, with the capacitors uncharged. At $t = 0$, the switch is moved to position b. When $v_C = 10$ V, the switch is moved to position c. Determine v_C for the time when the switch is in position b and after it was moved to position c.

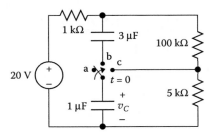

FIGURE P13.1.13

P13.1.14 In Figure P13.1.14, the switch is closed at $t = 0$ after being open for a long time. Determine v_x for $t \geq 0$.

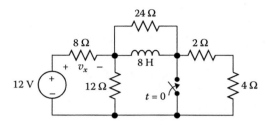

FIGURE P13.1.14

P13.1.15 In Figure P13.1.15, the capacitor was initially uncharged and the switch was in position a. At $t = 0$, the switch is moved to position b. Determine for $t \geq 0$: (a) v_C and i_C ; (b) the energy delivered by the supply and that absorbed by the battery; (c) the energy dissipated in the resistor; verify this by integrating the power dissipated by i_C from $t = 0$ to $t = \infty$; (d) If after a long time $t' = 0$, the switch is moved to position c; repeat part a to part c for $t' \geq 0$.

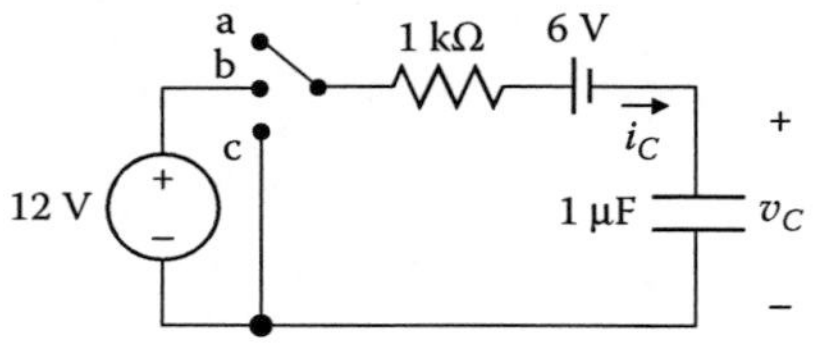

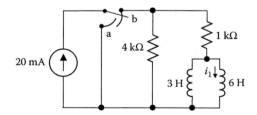

FIGURE P13.1.15

P13.1.16 In Figure P13.1.16, the inductors do not have any initial stored energy, with the switch in position a. The switch is moved to position b at $t = 0$ and returned to position a at $t = 1$ ms. Determine i_1 for $t \geq 0$. Simulate with PSpice using a pulse input of 20 mA amplitude and 1 ms duration.

FIGURE P13.1.16

P13.1.17 The capacitors in Figure P13.1.17 were initially charged as shown. When the switch is closed at $t = 0$, the current i_ϕ is found to be $0.18e^{-t}$ mA where t is in msec. Determine v_ϕ for $t \geq 0$. What are the final voltages across the capacitors? What is the value of R?

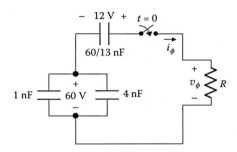

FIGURE P13.1.17

P13.1.18 The inductors in Figure P13.1.18 had initial currents as shown. When the switch is opened at $t = 0$, the voltage v_ϕ is found to be $50e^{-2.5t}$ V, where t is in seconds. Determine i_ϕ for $t \geq 0$. What is inside the black box? What are the final currents in the inductors? What is the value of R.

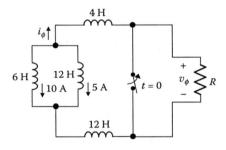

FIGURE P13.1.18

P13.1.19 The three inductors in Figure P13.1.19 have the initial currents shown. The switch is opened at $t = 0$. Determine v_O for $t \geq 0$ and the final currents in the inductors.

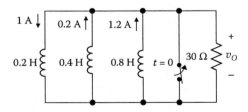

FIGURE P13.1.19

P13.1.20 Determine v_L for $t \geq 0$ in Figure P13.1.20, if $v_L(0) = 10$ V.

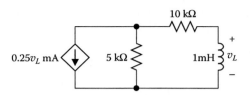

FIGURE P13.1.20

P13.1.21 In Figure P13.1.21, the switch is moved to position b after being in position a for a long time. Determine v_O for $t \geq 0$.

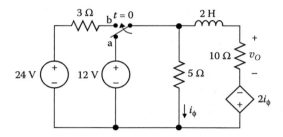

FIGURE P13.1.21

P13.1.22 In Figure P13.1.22, the switch is closed at $t = 0$ after being open for a long time. Determine i_1 for $t \geq 0$.

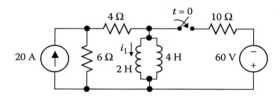

FIGURE P13.1.22

P13.1.23 In Figure P13.1.23, the capacitors are initially uncharged, with the switch in position a. The switch is moved to position b at $t = 0$ and to position c at $t = 3$ ms. Determine v_1 for $t \geq 0$. Simulate with PSpice using a pulse input of 12 V amplitude and 3 ms duration.

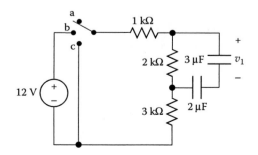

FIGURE P13.1.23

P13.2 Second-Order Systems

P13.2.1 In the circuit of Example 13.4.1, choose R for a maximum overshoot of (a) 2% and (b) 5%. Simulate with PSpice to determine the settling time in each case.

(Note: the maximum percentage overshoot is $100\,e^{-\frac{\alpha}{\omega_d}\pi}$ (Section ST13.1)).

P13.2.2 (a) Determine the initial values of i, i_L, v_C, and v_L at $t = 0^+$ in the circuit of Figure P13.2.2, assuming zero initial conditions, and interpret the result.

(b) Is the circuit overdamped, critically damped, or underdamped? (c) Determine i_L, v_L, v_C, and i for $t \geq 0^+$.

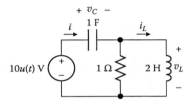

FIGURE P13.2.2

P13.2.3 Derive and solve the dual of the circuit in Problem P13.2.2.

P13.2.4 In Figure P13.2.4, the switch is moved to position b at $t = 0$, after being in position a for a long time. (a) Choose R so that the response is critically damped. (b) Determine the initial values of v_C, i_L, and v_L at $t = 0^+$. (c) Determine v_C and i_L for $t \geq 0^+$.

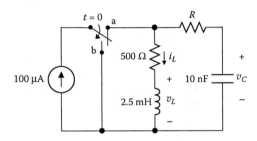

FIGURE P13.2.4

P13.2.5   In Figure P13.2.5, the switch is moved to position b at $t = 0$, after being in position a for a long time. (a) Choose G so that the response is critically damped. (b) Determine the initial values of i_L, v_C, and i_C at $t = 0^+$. (c) Determine i_L and v_C for $t \geq 0$. Note that this problem is the dual of the circuit in Problem 13.2.4.

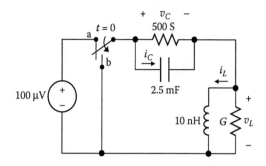

FIGURE P13.2.5

P13.2.6   In Figure P13.2.6 determine: (a) ρ for critical damping and (b) v, i_C, and i_L for $t \geq 0$, assuming zero initial conditions.

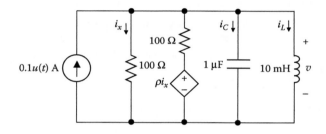

FIGURE P13.2.6

P13.2.7   In Figure P13.2.7 determine: (a) ρ for critical damping and (b) i, v_C, and v_L for $t \geq 0$, assuming zero initial conditions. Note that the circuit configuration is that of the dual of the circuit in Problem P13.2.6, but the numerical values are not.

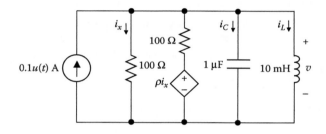

FIGURE P13.2.7

P13.2.8   Repeat Problem P13.2.7 with $\rho = 8 \times 10^{-3}\,\text{A}/\text{V}$.

P13.2.9   In Figure P13.2.9, the switch is moved to position b at $t = 0$ after having been in position a for a long time. Determine i and v_O.

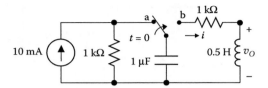

FIGURE P13.2.9

P13.2.10 In Figure P13.2.10, the switch is moved to position b at $t = 0$ after having been in position a for a long time. Determine v_O.

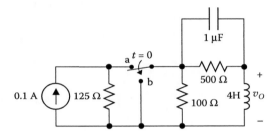

FIGURE P13.2.10

P13.2.11 In Figure P13.2.11, the switch is opened at $t = 0$ after having been in position a for a long time. Determine v_O. Note that the circuit configuration is that of the dual circuit of Problem 13.2.10, but the numerical values are not.

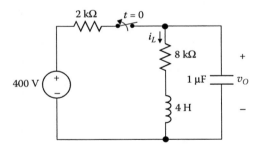

FIGURE P13.2.11

P13.2.12 Determine v_O in Figure P13.2.12 if the initial current in the inductor is 0.1 A and the initial voltage across the capacitor is zero.

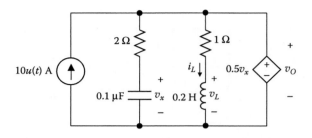

FIGURE P13.2.12

P13.2.13 Determine v_O in Figure P13.2.12 if the initial current in the inductor is zero and the initial voltage across the capacitor is 2 V and of the same polarity as v_x.

14

Two-Port Circuits

Overview

Any circuit that is excited by a single, independent source applied to one port, and having a load connected to another port, can be considered as a two-port circuit. Active, electronic devices and their associated circuits are a prime example of two-port circuits. A two-port circuit is characterized, in general, by sets of four parameters defined in terms of open-circuit or short-circuit terminations at each port. The analysis based on this characterization provides a useful and powerful alternative to analysis based on the node-voltage or mesh-current methods.

Terminal voltages and currents of two-port circuits are usually considered to be in the s domain, that is, as Laplace transforms of the corresponding functions in the time domain. To make the discussion of two-port circuits independent of the Laplace transform, we will consider the terminal voltages and currents as phasors, which detracts only slightly from the generality of the discussion. Where it is more natural to express transfer functions in terms of s, we will do so, with the understanding that s simply denotes $j\omega$ in the sinusoidal steady state. Similarly, where it is desired to express a transfer function in terms of s, this can readily be done by replacing $j\omega$ by s in the frequency-domain representation of the circuit.

Moreover, to utilize the full power of two-port analysis, matrix methods are needed. Examples of these methods, as well as an appendix on matrices, are included in the supplementary material on the CD. We will introduce some of these methods in the following sections, mainly as a convenient, compact notation that follows certain rules that are particularly simple for 2×2 matrices.

Learning Objectives

- To be familiar with:
  - The terminology and description of two-port circuits
  - The application of some matrix operations to the analysis of two-port circuits
- To understand:
  - The interpretation of the parameters of two-port circuits and their inter-relations, including those of reciprocal and symmetric circuits
  - The equivalent circuits of two-port circuits

- The derivation of the composite parameters when two-port circuits are connected in cascade, in parallel, in series, in series–parallel, or in parallel–series, and their application to some special cases of interest
- The methods of analysis of terminated two-port circuits

14.1 Circuit Description

As its name implies, a two-port circuit has a pair of input terminals, or input port, and a pair of output terminals, or output port (Figure 14.1.1), the assigned positive directions of input and output voltages and currents being as indicated. As explained in the overview, terminal voltages and currents are considered to be phasors. By convention, the circuit may contain passive, linear circuit elements and dependent sources, but no independent sources.

Clearly, if a voltage V_1 is applied to a given two-port circuit of known parameters, such as a simple resistive T-circuit, I_1, V_2, and I_2 are not determined unless another constraint is imposed, such as $I_2 = 0$, or a relation between V_2 and I_2. In general, two of the four terminal variables may be specified independently, in which case the other two variables are determined by the parameters of the given circuit. The two-port circuit may therefore be described in terms of two simultaneous equations. However, because four variables are involved, there are six ways of choosing two of these variables as independent, which gives rise to six sets of simultaneous equations, listed in Table 14.1.1. Each of these equations is written in terms of four coefficients, or parameters, that characterize the equation. Thus, one speaks of the z-parameter equations, the y-parameter equations, etc. The two equations in the same row in Table 14.1.1 are inversely related, in the sense that the independent variables in one set are the dependent variables in the other. As discussed below, this means that the matrices of parameters in the two cases are inversely related.

It is seen that a two-port circuit is specified, in general, by four nonzero parameters in one of the set of six equations. The z- and y-parameters, being impedances and admittances, respectively, are **immittance parameters** (Chapter 5, Section 5.5). They express

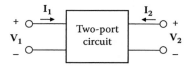

FIGURE 14.1.1
Terminal voltages and currents of a two-port circuit.

TABLE 14.1.1

Two-Port Circuit Equations

$V_1 = z_{11}I_1 + z_{12}I_2$	$I_1 = y_{11}V_1 + y_{12}V_2$
$V_2 = z_{21}I_1 + z_{22}I_2$	$I_2 = y_{21}V_1 + y_{22}V_2$
$V_1 = a_{11}V_2 - a_{12}I_2$	$V_2 = b_{11}V_1 - b_{12}I_1$
$I_1 = a_{21}V_2 - a_{22}I_2$	$I_2 = b_{21}V_1 - b_{22}I_1$
$V_1 = h_{11}I_1 + h_{12}V_2$	$I_1 = g_{11}V_1 + g_{12}I_2$
$I_2 = h_{21}I_1 + h_{22}V_2$	$V_2 = g_{21}V_1 + g_{22}I_2$

terminal voltages (or currents) in terms of terminal currents (or voltages). The *a*- and *b*-parameters are **transmission parameters** because they express voltage and current at one port in terms of voltage and current at the other port. The *a*- and *b*-parameters are also referred to as *ABCD* or *abcd* parameters, respectively. The *h*- and *g*-parameters are **hybrid parameters** because they express an input voltage (or current) and an output current (or voltage) in terms of an input current (or voltage) and an output voltage (or current). Finally, it should be noted that a two-port circuit may be describable by some, but not all, of the six equations. An ideal transformer is an example (Example SE14.1).

EXERCISE 14.1.1

Of the six sets of equations in Table 14.1.1, which sets are dual relations?

Answer: *z*-parameter and *y*-parameter equations are dual relations, as are the *h*-parameter and *g*-parameter equations, provided the corresponding coefficients have equal numerical values.

14.2 Parameter Interpretation and Relations

Interpretation of Parameters

The parameters of two-port circuits can be interpreted in terms of voltage and current ratios under specified open-circuit and short-circuit terminations, which provides a convenient means of evaluation or measurement of these parameters. Starting with the *z*-parameter equations, it is seen that the *z*-parameters can be interpreted as follows:

$$z_{11} = \frac{\mathbf{V}_1}{\mathbf{I}_1}\Big|_{I_2=0} \ \Omega: \text{impedance looking into port 1 with port 2 open-circuited}$$

$$z_{12} = \frac{\mathbf{V}_1}{\mathbf{I}_2}\Big|_{I_1=0} \ \Omega: \text{ratio of voltage at port 1 to current at port 2 with port 1 open-circuited}$$

$$z_{21} = \frac{\mathbf{V}_2}{\mathbf{I}_1}\Big|_{I_2=0} \ \Omega: \text{ratio of voltage at port 2 to current at port 1 with port 2 open-circuited}$$

$$z_{22} = \frac{\mathbf{V}_2}{\mathbf{I}_2}\Big|_{I_1=0} \ \Omega: \text{impedance looking into port 2 with port 1 open-circuited}$$

Whereas z_{11} and z_{22} are input impedances at ports 1 and 2, respectively, z_{12} and z_{21} are transfer impedances, being the ratio of voltage at one port to current at the other port. The other parameters may be similarly interpreted, as summarized in Table 14.2.1. Circuit parameters are, in general, frequency dependent.

As the same circuit variables are involved in the six equations, the parameters in these equations must be related. To find the relation between any two sets of parameters, we simply express one set of equations in the same form as the other set and compare coefficients. For example, to express the *z*-parameters in terms of the *a*-parameters, we eliminate $\mathbf{V}_2$ between the two equations of the *a*-parameters to obtain:

TABLE 14.2.1

Interpretation of Circuit Parameters

$z_{11} = \dfrac{V_1}{I_1}\Big|_{I_2=0}\ \Omega$	$z_{12} = \dfrac{V_1}{I_2}\Big|_{I_1=0}\ \Omega$	$y_{11} = \dfrac{I_1}{V_1}\Big|_{V_2=0}\ S$	$y_{12} = \dfrac{I_1}{V_2}\Big|_{V_1=0}\ S$
$z_{21} = \dfrac{V_2}{I_1}\Big|_{I_2=0}\ \Omega$	$z_{22} = \dfrac{V_2}{I_2}\Big|_{I_1=0}\ \Omega$	$y_{21} = \dfrac{I_2}{V_1}\Big|_{V_2=0}\ S$	$y_{22} = \dfrac{I_2}{V_2}\Big|_{V_1=0}\ S$
$a_{11} = \dfrac{V_1}{V_2}\Big|_{I_2=0}$	$a_{12} = -\dfrac{V_1}{I_2}\Big|_{V_2=0}\ \Omega$	$b_{11} = \dfrac{V_2}{V_1}\Big|_{I_1=0}$	$b_{12} = -\dfrac{V_2}{I_1}\Big|_{V_1=0}\ \Omega$
$a_{21} = \dfrac{I_1}{V_2}\Big|_{I_2=0}\ S$	$a_{22} = -\dfrac{I_1}{I_2}\Big|_{V_2=0}$	$b_{21} = \dfrac{I_2}{V_1}\Big|_{I_1=0}\ S$	$b_{22} = -\dfrac{I_2}{I_1}\Big|_{V_1=0}$
$h_{11} = \dfrac{V_1}{I_1}\Big|_{V_2=0}\ \Omega$	$h_{12} = \dfrac{V_1}{V_2}\Big|_{I_1=0}$	$g_{11} = \dfrac{I_1}{V_1}\Big|_{I_2=0}\ S$	$g_{12} = \dfrac{I_1}{I_2}\Big|_{V_1=0}$
$h_{21} = \dfrac{I_2}{I_1}\Big|_{V_2=0}$	$h_{22} = \dfrac{I_2}{V_2}\Big|_{I_1=0}\ S$	$g_{21} = \dfrac{V_2}{V_1}\Big|_{I_2=0}$	$g_{22} = \dfrac{V_2}{I_2}\Big|_{V_1=0}\ \Omega$

$$V_1 = \frac{a_{11}}{a_{21}}I_1 + \frac{\Delta a}{a_{21}}I_2 \tag{14.2.1}$$

where $\Delta a = a_{11}a_{22} - a_{12}a_{21}$. Rearranging the second a-parameter equation,

$$V_2 = \frac{1}{a_{21}}I_1 + \frac{a_{22}}{a_{21}}I_2 \tag{14.2.2}$$

Comparing Equation 14.2.1 and Equation 14.2.2 with the corresponding z-parameter equations gives: $z_{11} = \dfrac{a_{11}}{a_{21}}$, $z_{12} = \dfrac{\Delta a}{a_{21}}$, $z_{21} = \dfrac{1}{a_{21}}$, and $z_{22} = \dfrac{a_{22}}{a_{21}}$. Table 14.2.2 gives the relations between the different sets of parameters.

Inverse Relations

Any set of equations in Table 14.1.1, say the z-parameter equations, may be expressed in matrix form (Appendix SC) as follows:

$$\begin{bmatrix} V_1 \\ V_2 \end{bmatrix} = \begin{bmatrix} z_{11} & z_{12} \\ z_{21} & z_{22} \end{bmatrix}\begin{bmatrix} I_1 \\ I_2 \end{bmatrix} \tag{14.2.3}$$

Applying the rules of matrix multiplication to Equation 14.2.3 gives the z-parameter equations. Equation 14.2.3 is inverted by multiplying both sides by the inverse of the z-parameter matrix as follows:

TABLE 14.2.2

Relations between Sets of Parameters

$$z_{11} = \frac{y_{22}}{\Delta y} = \frac{a_{11}}{a_{21}} = \frac{b_{22}}{b_{21}} = \frac{\Delta h}{h_{22}} = \frac{1}{g_{11}} \qquad y_{11} = \frac{z_{22}}{\Delta z} = \frac{a_{22}}{a_{12}} = \frac{b_{11}}{b_{12}} = \frac{1}{h_{11}} = \frac{\Delta g}{g_{22}}$$

$$z_{12} = -\frac{y_{12}}{\Delta y} = \frac{\Delta a}{a_{21}} = \frac{1}{b_{21}} = \frac{h_{12}}{h_{22}} = -\frac{g_{12}}{g_{11}} \qquad y_{12} = -\frac{z_{12}}{\Delta z} = -\frac{\Delta a}{a_{12}} = -\frac{1}{b_{12}} = -\frac{h_{12}}{h_{11}} = \frac{g_{12}}{g_{22}}$$

$$z_{21} = -\frac{y_{21}}{\Delta y} = \frac{1}{a_{21}} = -\frac{\Delta b}{b_{21}} = -\frac{h_{21}}{h_{22}} = \frac{g_{21}}{g_{11}} \qquad y_{21} = -\frac{z_{21}}{\Delta z} = -\frac{1}{a_{12}} = \frac{\Delta b}{b_{12}} = \frac{h_{21}}{h_{11}} = -\frac{g_{21}}{g_{22}}$$

$$z_{22} = \frac{y_{11}}{\Delta y} = \frac{a_{22}}{a_{21}} = \frac{b_{11}}{b_{21}} = \frac{1}{h_{22}} = \frac{\Delta g}{g_{11}} \qquad y_{22} = \frac{z_{11}}{\Delta z} = \frac{a_{11}}{a_{12}} = \frac{b_{22}}{b_{12}} = \frac{\Delta h}{h_{11}} = \frac{1}{g_{22}}$$

$$a_{11} = \frac{z_{11}}{z_{21}} = -\frac{y_{22}}{y_{21}} = \frac{b_{22}}{\Delta b} = -\frac{\Delta h}{h_{21}} = \frac{1}{g_{21}} \qquad b_{11} = \frac{z_{22}}{z_{12}} = -\frac{y_{11}}{y_{12}} = \frac{a_{22}}{\Delta a} = \frac{1}{h_{12}} = -\frac{\Delta g}{g_{12}}$$

$$a_{12} = \frac{\Delta z}{z_{21}} = -\frac{1}{y_{21}} = \frac{b_{12}}{\Delta b} = -\frac{h_{11}}{h_{21}} = \frac{g_{22}}{g_{21}} \qquad b_{12} = \frac{\Delta z}{z_{12}} = -\frac{1}{y_{12}} = \frac{a_{12}}{\Delta a} = \frac{h_{11}}{h_{12}} = -\frac{g_{22}}{g_{12}}$$

$$a_{21} = \frac{1}{z_{21}} = -\frac{\Delta y}{y_{21}} = \frac{b_{21}}{\Delta b} = -\frac{h_{22}}{h_{21}} = \frac{g_{11}}{g_{21}} \qquad b_{21} = \frac{1}{z_{12}} = -\frac{\Delta y}{y_{12}} = \frac{a_{21}}{\Delta a} = \frac{h_{22}}{h_{12}} = -\frac{g_{11}}{g_{12}}$$

$$a_{22} = \frac{z_{22}}{z_{21}} = -\frac{y_{11}}{y_{21}} = \frac{b_{11}}{\Delta b} = -\frac{1}{h_{21}} = \frac{\Delta g}{g_{21}} \qquad b_{22} = \frac{z_{11}}{z_{12}} = -\frac{y_{22}}{y_{12}} = \frac{a_{11}}{\Delta a} = \frac{\Delta h}{h_{12}} = -\frac{1}{g_{12}}$$

$$h_{11} = \frac{\Delta z}{z_{22}} = \frac{1}{y_{11}} = \frac{a_{12}}{a_{22}} = \frac{b_{12}}{b_{11}} = \frac{g_{22}}{\Delta g} \qquad g_{11} = \frac{1}{z_{11}} = \frac{\Delta y}{y_{22}} = \frac{a_{21}}{a_{11}} = \frac{b_{21}}{b_{22}} = \frac{h_{22}}{\Delta h}$$

$$h_{12} = \frac{z_{12}}{z_{22}} = -\frac{y_{12}}{y_{11}} = \frac{\Delta a}{a_{22}} = \frac{1}{b_{11}} = -\frac{g_{12}}{\Delta g} \qquad g_{12} = -\frac{z_{12}}{z_{11}} = \frac{y_{12}}{y_{22}} = -\frac{\Delta a}{a_{11}} = -\frac{1}{b_{22}} = -\frac{h_{12}}{\Delta h}$$

$$h_{21} = -\frac{z_{21}}{z_{22}} = \frac{y_{21}}{y_{11}} = -\frac{1}{a_{22}} = -\frac{\Delta b}{b_{11}} = -\frac{g_{21}}{\Delta g} \qquad g_{21} = \frac{z_{21}}{z_{11}} = -\frac{y_{12}}{y_{22}} = \frac{1}{a_{11}} = \frac{\Delta b}{b_{22}} = \frac{h_{21}}{\Delta h}$$

$$h_{22} = \frac{1}{z_{22}} = \frac{\Delta y}{y_{11}} = \frac{a_{21}}{a_{22}} = \frac{b_{21}}{b_{11}} = \frac{g_{11}}{\Delta g} \qquad g_{22} = \frac{\Delta z}{z_{11}} = \frac{1}{y_{22}} = \frac{a_{12}}{a_{11}} = \frac{b_{12}}{b_{22}} = \frac{h_{11}}{\Delta h}$$

$$\Delta z = z_{11}z_{22} - z_{12}z_{21} \qquad\qquad \Delta y = y_{11}y_{22} - y_{12}y_{21}$$

$$\Delta a = a_{11}a_{22} - a_{12}a_{21} \qquad\qquad \Delta b = b_{11}b_{22} - b_{12}b_{21}$$

$$\Delta h = h_{11}h_{22} - h_{12}h_{21} \qquad\qquad \Delta g = g_{11}g_{22} - g_{12}g_{21}$$

$$\begin{bmatrix} I_1 \\ I_2 \end{bmatrix} = \begin{bmatrix} z_{11} & z_{12} \\ z_{21} & z_{22} \end{bmatrix}^{-1} \begin{bmatrix} V_1 \\ V_2 \end{bmatrix} \tag{14.2.4}$$

Comparing this with the *y*-parameter equations, it is seen that

$$\begin{bmatrix} y_{11} & y_{12} \\ y_{21} & y_{22} \end{bmatrix} = \begin{bmatrix} z_{11} & z_{12} \\ z_{21} & z_{22} \end{bmatrix}^{-1} \tag{14.2.5}$$

Similar inverse relations hold between the parameter matrices of any two sets of equations in the same row of Table 14.1.1. Thus,

$$\begin{bmatrix} z_{11} & z_{12} \\ z_{21} & z_{22} \end{bmatrix} = \begin{bmatrix} y_{11} & y_{12} \\ y_{21} & y_{22} \end{bmatrix}^{-1}, \quad \begin{bmatrix} y_{11} & y_{12} \\ y_{21} & y_{22} \end{bmatrix} = \begin{bmatrix} z_{11} & z_{12} \\ z_{21} & z_{22} \end{bmatrix}^{-1} \tag{14.2.6}$$

$$\begin{bmatrix} a_{11} & -a_{12} \\ a_{21} & -a_{22} \end{bmatrix} = \begin{bmatrix} b_{11} & -b_{12} \\ b_{21} & -b_{22} \end{bmatrix}^{-1}, \quad \begin{bmatrix} b_{11} & -b_{12} \\ b_{21} & -b_{22} \end{bmatrix} = \begin{bmatrix} a_{11} & -a_{12} \\ a_{21} & -a_{22} \end{bmatrix}^{-1} \tag{14.2.7}$$

$$\begin{bmatrix} h_{11} & h_{12} \\ h_{21} & h_{22} \end{bmatrix} = \begin{bmatrix} g_{11} & g_{12} \\ g_{21} & g_{22} \end{bmatrix}^{-1}, \quad \begin{bmatrix} g_{11} & g_{12} \\ g_{21} & g_{22} \end{bmatrix} = \begin{bmatrix} h_{11} & h_{12} \\ h_{21} & h_{22} \end{bmatrix}^{-1} \tag{14.2.8}$$

Note the negative signs in the matrices of the *a*- and *b*-parameters, which arise because of the negative signs in the equations of these parameters.

The inverse of a 2×2 parameter matrix is expressed as (Equation SC2.7, Appendix SC):

$$\begin{bmatrix} m_{11} & m_{12} \\ m_{21} & m_{22} \end{bmatrix}^{-1} = \frac{1}{\Delta m} \begin{bmatrix} m_{22} & -m_{12} \\ -m_{21} & m_{11} \end{bmatrix} \tag{14.2.9}$$

where $\Delta m = m_{11}m_{22} - m_{12}m_{21}$ is the determinant of the matrix. The relations in Table 14.2.2 are seen to conform to Equation 14.2.9.

EXERCISE 14.2.1

Derive the expressions for the *z*-parameters in terms of the *y*-, *a*-, *b*-, *h*-, and *g*-parameters.

Example 14.2.1: Determination of *a*-Parameters

It is required to determine the *a*-parameters of the circuit of Figure 14.2.1 at $\omega = 1$ rad/s.

SOLUTION

With $\mathbf{I}_2 = 0$, $\mathbf{V}_2 = \dfrac{1/5j}{2 + 1/5j} \times 10\mathbf{I}_1$, which gives $a_{21} = \dfrac{\mathbf{I}_1}{\mathbf{V}_2}\Big|_{\mathbf{I}_2=0} = 0.1 + j$ S. On the input side,

$\mathbf{V}_1 = (20 + j10)\mathbf{I}_1 + 10\mathbf{I}_1$. Hence, $\dfrac{\mathbf{V}_1}{\mathbf{I}_1} = 30 + j10$ Ω. But, $\dfrac{\mathbf{V}_1}{\mathbf{V}_2}\Big|_{\mathbf{I}_2=0} = \dfrac{\mathbf{V}_1}{\mathbf{I}_1}\Big|_{\mathbf{I}_2=0} \times \dfrac{\mathbf{I}_1}{\mathbf{V}_2}\Big|_{\mathbf{I}_2=0}$. It follows

that $a_{11} = \dfrac{\mathbf{V}_1}{\mathbf{V}_2}\Big|_{\mathbf{I}_2=0} = (30 + j10)(0.1 + j) = -7 + 31j$.

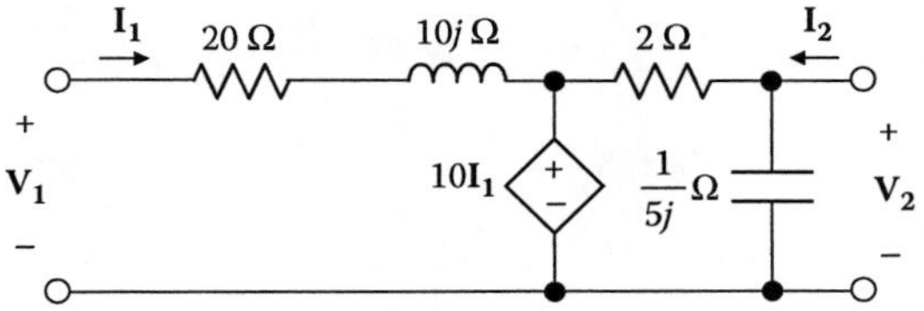

FIGURE 14.2.1
Figure for Example 14.2.1.

With $\mathbf{V}_2 = 0$, $\mathbf{I}_2 = -\dfrac{10}{2}\mathbf{I}_1$. Hence, $a_{22} = -\dfrac{\mathbf{I}_1}{\mathbf{I}_2}\Big|_{V_2=0} = 0.2$. On the input side, we have $\dfrac{\mathbf{V}_1}{\mathbf{I}_1} =$

$30 + j10$ as before. Because $\dfrac{\mathbf{V}_1}{\mathbf{I}_2}\Big|_{V_2=0} = \dfrac{\mathbf{V}_1}{\mathbf{I}_1}\Big|_{V_2=0} \times \dfrac{\mathbf{I}_1}{\mathbf{I}_2}\Big|_{V_2=0}$, it follows that $a_{12} = -\dfrac{\mathbf{V}_1}{\mathbf{I}_2}\Big|_{V_2=0}$·

$-(30 + j10)(-0.2) = 6 + j2\Omega.$

Example 14.2.2: Determination of *h*-Parameters

The following dc measurements were made on a two-port resistive circuit:

Port 2 open-circuited: $V_1 = 10$ mV, $I_1 = 50$ μA, $V_2 = 20$ V.
Port 2 short-circuited: $V_1 = 40$ mV, $I_1 = 100$ μA, $I_2 = -1$ mA.

It is required to find the *h*-parameters of the circuit.

SOLUTION

With port 2 short-circuited, $h_{11} = \dfrac{V_1}{I_1}\Big|_{V_2=0} = \dfrac{40\text{ mV}}{100\text{ μA}} \equiv 400\ \Omega$ and $h_{21} = \dfrac{I_2}{I_1}\Big|_{V_2=0} = -10$;

$h_{12} = \dfrac{V_1}{V_2}\Big|_{I_1=0}$ and $h_{22} = \dfrac{I_2}{V_2}\Big|_{I_1=0}$ cannot be obtained from the given measurements because these do not include the case of port 1 open-circuited. However, we can obtain the *a*-parameters from the given measurements. Thus,

$$a_{11} = \frac{V_1}{V_2}\Big|_{I_2=0} = \frac{10 \times 10^{-3}}{20} = 5 \times 10^{-4}$$

$$a_{12} = -\frac{V_1}{I_2}\Big|_{V_2=0} = -\frac{40}{-1} = 40\ \Omega$$

$$a_{21} = \frac{I_1}{V_2}\Big|_{I_2=0} = \frac{50 \times 10^{-6}}{20} = 2.5 \times 10^{-6}\text{ S}$$

$$a_{22} = -\frac{I_1}{I_2}\bigg|_{V_2=0} = -\frac{100\times10^{-6}}{-1\times10^{-3}} = 0.1$$

$$\Delta a = a_{11}a_{22} - a_{12}a_{21} = 5\times10^{-4}\times0.1 - 40\times2.5\times10^{-6} = -5\times10^{-5}$$

As a check, we find $h_{11} = \dfrac{a_{12}}{a_{22}} = \dfrac{40}{0.1} = 400\ \Omega$ and $h_{21} = -\dfrac{1}{a_{22}} = -\dfrac{1}{0.1} = -10$, as shown earlier.

It follows that $h_{12} = \dfrac{\Delta a}{a_{22}} = \dfrac{-5\times10^{-5}}{0.1} = -5\times10^{-4}$ and $h_{22} = \dfrac{a_{21}}{a_{22}} = \dfrac{2.5\times10^{-6}}{0.1} \equiv 25\ \mu S$.

Reciprocal Circuits

The z-parameter equations are of the form of mesh-current equations, where $\mathbf{I}_1$ is a current flowing clockwise in an input mesh, and $\mathbf{I}_2$ is a current flowing counterclockwise in an output mesh. z_{11} and z_{22} are thus self-impedances of the meshes, whereas z_{12} and z_{21} are mutual impedances (Section 3.4, Chapter 3). As discussed in Section 3.4, if an LTI circuit does not contain dependent sources, the array, or matrix, of coefficients is symmetrical about the diagonal, that is $z_{12} = z_{21}$. Such a two-port circuit is said to be reciprocal and obeys the reciprocity theorem (Section ST14.1). Similarly, the y-parameter equations are of the form of node-voltage equations. If the circuit does not contain dependent sources, it is reciprocal, and $y_{12} = y_{21}$. It follows that in a reciprocal circuit, only three of the four circuit parameters are independent. It is of interest to determine the dependency relations for the a-, b-, h-, and g-parameters in a reciprocal circuit.

The required dependency relations may be derived from the expressions for z_{12} and z_{21}, or y_{12} and y_{21}, in Table 14.2.2. If $z_{12} = z_{21}$, it follows that $\dfrac{\Delta a}{a_{21}} = \dfrac{1}{a_{21}}$, $\dfrac{1}{b_{21}} = \dfrac{\Delta b}{b_{21}}$, $\dfrac{h_{12}}{h_{22}} = -\dfrac{h_{21}}{h_{22}}$,

and $-\dfrac{g_{12}}{g_{11}} = \dfrac{g_{21}}{g_{11}}$. These relations give $\Delta a = 1 = \Delta b$, $h_{12} = -h_{21}$, and $g_{12} = -g_{21}$. Parameter relations in reciprocal circuits are summarized in Table 14.2.3.

It should be noted that having no dependent sources in a two-port circuit is a sufficient condition for reciprocity but not a necessary one. In other words, a circuit that does not have dependent sources is reciprocal. But a circuit can have dependent sources and still be reciprocal, as illustrated by Example 14.2.3. Hence, we define a reciprocal circuit as follows:

Definition: *A reciprocal circuit is one whose two-port parameters satisfy the reciprocity relations (Table 14.2.3).*

TABLE 14.2.3

Parameter Relations in Reciprocal Circuits

$z_{12} = z_{21}$	$y_{12} = y_{21}$
$\Delta a = a_{11}a_{22} - a_{12}a_{21} = 1$	$\Delta b = b_{11}b_{22} - b_{12}b_{21} = 1$
$h_{12} = -h_{21}$	$g_{12} = -g_{21}$

TABLE 14.2.4

Parameter Relations in Symmetric Circuits

$z_{11} = z_{22}$	$y_{11} = y_{22}$
$z_{12} = z_{21}$	$y_{12} = y_{21}$
$a_{11} = a_{22}$	$b_{11} = b_{22}$
$\Delta a = a_{11}a_{22} - a_{12}a_{21} = 1$	$\Delta b = b_{11}b_{22} - b_{12}b_{21} = 1$
$\Delta h = h_{11}h_{22} - h_{12}h_{21} = 1$	$\Delta g = g_{11}g_{22} - g_{12}g_{21} = 1$
$h_{12} = -h_{21}$	$g_{12} = -g_{21}$

Symmetric Circuits

> **Definition:** *A reciprocal circuit is symmetric if terminal voltages and currents remain the same when the two ports are interchanged.*

To determine the relations between parameters in such a circuit, consider, to begin with, the z-parameter equations. If $\mathbf{V}_1$ is interchanged with $\mathbf{V}_2$ in these equations, and $\mathbf{I}_1$ is interchanged with $\mathbf{I}_2$, the equations become

$$\mathbf{V}_2 = z_{11}\mathbf{I}_2 + z_{12}\mathbf{I}_1 \tag{14.2.13}$$

and
$$\mathbf{V}_1 = z_{21}\mathbf{I}_2 + z_{22}\mathbf{I}_1$$

With $z_{12} = z_{21}$, as the circuit is reciprocal, these equations are identical to the original equations if $z_{11} = z_{22}$. The same considerations apply to the y-parameters, so that the circuit is symmetric if $y_{11} = y_{22}$. In a symmetric circuit, therefore, only two of the four circuit parameters are independent.

The relations between the other two-port parameters in a symmetric circuit may be derived by equating z_{11} to z_{22} or y_{11} to y_{22} in Table 14.2.2. These relations are summarized in Table 14.2.4.

EXERCISE 14.2.2
Verify the relations for the a-, b-, h-, and g-parameters in Table 14.2.4.

Simulation Example 14.2.3: Symmetric Two-Port Circuit
It is required to determine: (a) R and α in Figure 14.2.2 so that the two-port circuit is symmetric, and (b) the a-parameters by simulation, with $R = 3\ \Omega$ and $\alpha = 0.5$.

SOLUTION
The procedure is to write the Kirchhoff's voltage law (KVL) equations for the three meshes, then eliminate I_3 so as to have two equations of the form of the z-parameter equations. It follows from the figure that

$$\text{Mesh 1: } V_1 + \alpha V_2 = RI_1 - RI_3 \tag{14.2.14}$$

$$\text{Mesh 2: } 0 = -(R+1)I_1 + I_2 + (R+2)I_3 \tag{14.2.15}$$

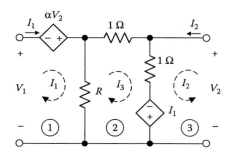

FIGURE 14.2.2
Figure for Example 14.2.3.

$$\text{Mesh 3: } V_2 = -I_1 + I_2 + I_3 \tag{14.2.16}$$

Eliminating I_3 between the last two equations gives:

$$V_2 = -\frac{1}{R+2}I_1 + \frac{R+1}{R+2}I_2 \tag{14.2.17}$$

Eliminating I_3 by adding the first two equations, multiplying the third equation by 2, and subtracting it follows that:

$$V_1 + (\alpha - 2)V_2 = I_1 - I_2 \tag{14.2.18}$$

Substituting for V_2 from Equation 14.2.17 in Equation 14.2.18 and rearranging:

$$V_1 = \frac{\alpha + R}{R+2}I_1 - \frac{\alpha R + \alpha - R}{R+2}I_2 \tag{14.2.19}$$

Equation 14.2.17 and Equation 14.2.19 are of the form of the z-parameter equations, from which it follows that:

$$z_{11} = \frac{\alpha + R}{R+2}, \quad z_{12} = -\frac{\alpha R + \alpha - R}{R+2}, \quad z_{21} = -\frac{1}{R+2}, \quad z_{22} = \frac{R+1}{R+2}$$

For a circuit to be symmetric, $z_{11} = z_{22}$ and $z_{12} = z_{21}$. Equating z_{11} and z_{22} gives $\alpha = 1$. With this value of α any value of R makes $z_{12} = z_{21}$. If $R = 5\ \Omega$ and $\alpha = 0.5$, then $z_{11} = 0.7\ \Omega$, $z_{12} = 0.2\ \Omega$, $z_{21} = -0.2\ \Omega$, and $z_{22} = 0.8\ \Omega$.

SIMULATION
The simulation is straightforward and is performed with the output port open-circuited and then short-circuited. The schematic for the former case is shown in Figure 14.2.3. Although a 1 V source may be applied, it is convenient to use a 7 V source in this case so as to obtain round figures for the voltages and currents. When the simulation is run, PSpice gives $V_2 = -2$ V and $I_1 = 10$ A. It follows that $a_{11} = -\frac{7}{2} = -3.5$ and $a_{21} = -\frac{10}{2} = -5$ S.

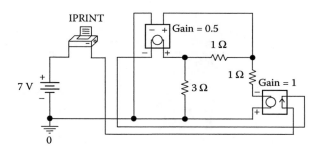

FIGURE 14.2.3
Figure for Example 14.2.3.

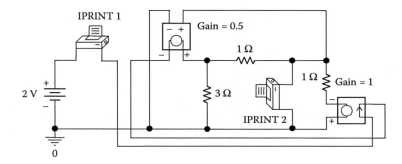

FIGURE 14.2.4
Figure for Example 14.2.3.

Figure 14.2.4 shows the schematic with the output port short-circuited by IPRINT2. It is convenient in this case to use a source voltage of 21 V so as to obtain round figures for the voltages and currents. When the simulation is run, PSpice gives $I_1 = 28$ A and $I_2 = 7$ A.

It follows that $a_{12} = -\dfrac{21}{7} = -3\ \Omega$ and $a_{22} = -\dfrac{28}{7} = -4$. It is seen that $z_{11} = \dfrac{-3.5}{-5} = 0.7\ \Omega$,

$z_{12} = \dfrac{(-3.5)(-4)-(-3)(-5)}{-5} = 0.2\ \Omega$, $z_{21} = \dfrac{1}{-5} = -0.2\ \Omega$, and $z_{22} = \dfrac{-4}{-5} = 0.8\ \Omega$, in agreement

with the earlier results.

EXERCISE 14.2.3
Verify that the two-port circuit of Figure 14.2.2 is symmetric if $\alpha = 1$ and $R = 0$ or $R = \infty$. Express Equation 14.2.14 to 14.2.16 in standard three-mesh form for $\alpha = 1$ and for an arbitrary R, and note that the matrix of coefficients is not symmetrical about the diagonal, although $z_{12} = z_{21}$.

14.3 Equivalent Circuits

In the z-parameter, y-parameter, h-parameter, and g-parameter equations, the first equation expresses an input variable in terms of the other input variable and an output variable, whereas the second equation expresses an output variable in terms of the other output

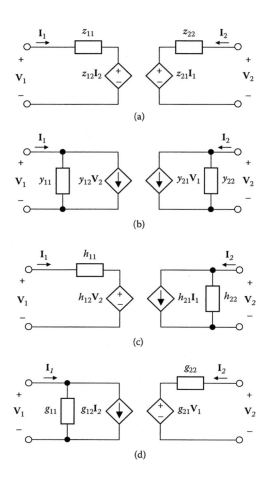

FIGURE 14.3.1
Equivalent circuits.

variable and an input variable. These two-port equations may therefore be represented by equivalent circuits with appropriate dependent sources on the input and output sides, as illustrated in Figure 14.3.1a to Figure 14.3.1d.

The effect of the input on the output side, that is, the *forward transmission*, is expressed by the dependent source on the output side and its associated "21" subscript parameter. On the other hand, the effect of the output on the input side, that is, the *reverse transmission*, is expressed by the dependent source on the input side and its associated "12" subscript parameter.

The parameters may be readily interpreted in terms of these circuits. For example, in the case of the z-parameter circuit, if port 2 is open-circuited, the dependent voltage source $z_{12}I_2 = 0$, so that the input impedance is z_{11}, and V_2 equals the source voltage $z_{21}I_1$. In general, the "11" subscript parameter describes an input immittance when the dependent source on the input side is set to zero, which imposes a constraint on the output port, such as an open circuit or a short circuit. Similarly, the "22" subscript parameter describes an output immittance when the dependent source on the output side is set to zero, which imposes a constraint on the input port.

Because the z-parameter equations are of the form of mesh-current equations, it follows that if the circuit is reciprocal, and if the lower terminal is common between the two ports,

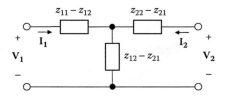

FIGURE 14.3.2
T-equivalent circuit.

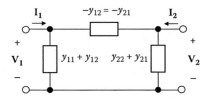

FIGURE 14.3.3
π-equivalent circuit.

these equations may be simply represented by a T-equivalent circuit, as shown in Figure 14.3.2. Similarly, the y-parameter equations of a reciprocal circuit having a common terminal between the two ports may be represented by a π-equivalent circuit (Figure 14.3.3). It is evident from these representations that a reciprocal circuit can be completely described by three independent parameters, which are the elements of the T- or π-equivalent circuits. The z-parameter and y-parameter equations of a nonreciprocal circuit may be represented by the addition of a dependent source to the T- or π-equivalent circuits, respectively, as illustrated by Example 14.3.1.

Example 14.3.1: T- and π-Equivalent Circuits of Nonreciprocal Circuits

It is required to derive the z-parameter and y-parameter equations, respectively, of the circuits shown in Figure 14.3.4a and Figure 14.3.4b.

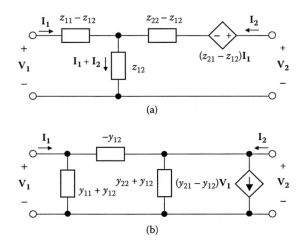

FIGURE 14.3.4
Figure for Example 14.3.1.

SOLUTION

Considering the circuit of Figure 14.3.4a, the KVL equations are:

$$\text{Mesh 1: } \mathbf{V}_1 = (z_{11} - z_{12})\mathbf{I}_1 + z_{12}(\mathbf{I}_1 + \mathbf{I}_2) = z_{11}\mathbf{I}_1 + z_{12}\mathbf{I}_2 \tag{14.3.1}$$

$$\text{Mesh 2: } \mathbf{V}_2 = (z_{22} - z_{12})\mathbf{I}_2 + z_{12}(\mathbf{I}_1 + \mathbf{I}_2) + (z_{21} - z_{12})\mathbf{I}_1 = z_{21}\mathbf{I}_1 + z_{22}\mathbf{I}_2 \tag{14.3.2}$$

Equation 14.3.1 and Equation 14.3.2 are the z-parameter equations for a two-port circuit in which z_{12} and z_{21} need not be equal.

Considering the circuit of Figure 14.3.4b, the node equations are:

$$\text{Node 1: } \mathbf{I}_1 = (y_{11} + y_{12})\mathbf{V}_1 - y_{12}(\mathbf{V}_1 - \mathbf{V}_2) = y_{11}\mathbf{V}_1 + y_{12}\mathbf{V}_2 \tag{14.3.3}$$

$$\text{Node 2: } \mathbf{I}_2 = (y_{22} + y_{12})\mathbf{V}_2 - y_{12}(\mathbf{V}_2 - \mathbf{V}_1) + (y_{21} - y_{12})\mathbf{V}_1 = y_{21}\mathbf{V}_1 + y_{22}\mathbf{V}_2 \tag{14.3.4}$$

Equation 14.3.3 and Equation 14.3.4 are the y-parameter equations for a two-port circuit in which y_{12} and y_{21} need not be equal.

EXERCISE 14.3.1

Modify the circuits of Figure 14.3.4 so that the dependent sources appear at the input instead of the output.

14.4 Composite Two-Port Circuits

> **Concept:** *An important feature of two-port circuit analysis is the analysis in terms of the parameters of a composite two-port circuit that is a combination of simple two-port circuits.*

The simple two-port circuits may be combined in one of five types of connection: cascade, parallel, series, series–parallel, or parallel–series.

Cascade Connection

When two or more circuits are cascaded, the output of any circuit other than the last is applied as the input of the following circuit, as illustrated in Figure 14.4.1 for two cascaded

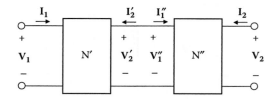

FIGURE 14.4.1
Cascaded two-port circuits.

two-port circuits N' and N''. A composite circuit is formed whose input variables are V_1 and I_1 and whose output variables are V_2 and I_2. It is desired to express the parameters of the composite circuit in terms of the parameters of the individual circuits.

Although, in principle, one could choose to work in terms of any of the six sets of parameters, the a-parameters are particularly convenient for this purpose because they have the input variables on one side of the equation and the output variables on the other. To derive the parameters of the composite circuit, we begin by writing the a-parameter equations for the two circuits:

$$V_1 = a'_{11}V'_2 - a'_{12}I'_2 \qquad V''_1 = a''_{11}V_2 - a''_{12}I_2$$

$$I_1 = a'_{21}V'_2 - a'_{22}I'_2 \qquad I''_1 = a''_{21}V_2 - a''_{22}I_2 \qquad (14.4.1)$$

Because of the cascade connection, $V'_2 = V''_1$ and $I'_2 = -I''_1$. Substituting for V'_2 and I'_2 from the equations for N'' in the equations for N' gives:

$$V_1 = (a'_{11}a''_{11} + a'_{12}a''_{21})\, V_2 - (a'_{11}a''_{12} + a'_{12}a''_{22})\, I_2$$

$$I_1 = (a'_{21}a''_{11} + a'_{22}a''_{21})\, V_2 - (a'_{21}a''_{12} + a'_{22}a''_{22})\, I_2 \qquad (14.4.2)$$

The terms in parentheses represent the a-parameters of the composite circuit. Thus,

$$a_{11} = a'_{11}a''_{11} + a'_{12}a''_{21} \qquad a_{12} = a'_{11}a''_{12} + a'_{12}a''_{22}$$

$$a_{21} = a'_{21}a''_{11} + a'_{22}a''_{21} \qquad a_{22} = a'_{21}a''_{12} + a'_{22}a''_{22} \qquad (14.4.3)$$

Matrix notation allows a more direct and compact derivation of Equation 14.4.3. Using this notation, the a-parameter equations of N' and N'' may be written as:

$$\begin{bmatrix} V_1 \\ I_1 \end{bmatrix} = \begin{bmatrix} a'_{11} & a'_{12} \\ a'_{21} & a'_{22} \end{bmatrix} \begin{bmatrix} V'_2 \\ -I'_2 \end{bmatrix} \quad \text{and} \quad \begin{bmatrix} V''_1 \\ I''_1 \end{bmatrix} = \begin{bmatrix} a''_{11} & a''_{12} \\ a''_{21} & a''_{22} \end{bmatrix} \begin{bmatrix} V_2 \\ -I_2 \end{bmatrix} \qquad (14.4.4)$$

Because of the cascade connection,

$$\begin{bmatrix} V'_2 \\ -I'_2 \end{bmatrix} = \begin{bmatrix} V''_1 \\ I''_1 \end{bmatrix}$$

Substituting from the second equation in the first,

$$\begin{bmatrix} \mathbf{V_1} \\ \mathbf{I_1} \end{bmatrix} = \begin{bmatrix} a'_{11} & a'_{12} \\ a'_{21} & a'_{22} \end{bmatrix} \begin{bmatrix} a''_{11} & a''_{12} \\ a''_{21} & a''_{22} \end{bmatrix} \begin{bmatrix} \mathbf{V_2} \\ -\mathbf{I_2} \end{bmatrix} \tag{14.4.5}$$

But the *a*-parameter equation of the composite circuit in matrix form is

$$\begin{bmatrix} \mathbf{V_1} \\ \mathbf{I_1} \end{bmatrix} = \begin{bmatrix} a_{11} & a_{12} \\ a_{21} & a_{22} \end{bmatrix} \begin{bmatrix} \mathbf{V_2} \\ -\mathbf{I_2} \end{bmatrix} \tag{14.4.6}$$

Comparing Equation 14.4.5 and Equation 14.4.6:

$$\begin{bmatrix} a_{11} & a_{12} \\ a_{21} & a_{22} \end{bmatrix} = \begin{bmatrix} a'_{11} & a'_{12} \\ a'_{21} & a'_{22} \end{bmatrix} \begin{bmatrix} a''_{11} & a''_{12} \\ a''_{21} & a''_{22} \end{bmatrix} \tag{14.4.7}$$

Equation 14.4.3 follows from Equation 14.4.7 by applying the rules of matrix multiplication (Appendix SC.2).

The preceding results may be generalized to the cascading of *m* two-port circuits. Thus,

$$[a] = [a_1][a_2]...[a_m] \tag{14.4.8}$$

where [*a*] is the matrix of *a*-parameters of the composite circuit and [a_k] is the matrix of *a*-parameters of the *k*th circuit, *k* = 1, 2, …, *m*. Note that matrix multiplication is not commutative so that the order of multiplication must correspond to that of the cascaded circuits.

Example 14.4.1: *a*-Parameters of Two Cascaded Circuits

In Figure 14.4.2, $v_{SRC} = 10\cos t$ V. It is required to determine v_O with $i_O = 0$, given that circuit N is symmetric and that it gave the following measurements: with port 2 open-circuited,

$$\frac{\mathbf{V_2}}{\mathbf{V_1}} = 0.2 \text{ and } \frac{\mathbf{V_1}}{\mathbf{I_1}} = 5\,\Omega\,.$$

SOLUTION

The procedure is to determine $a_{11} = \left.\dfrac{\mathbf{V_{src}}}{\mathbf{V_o}}\right|_{I_O=0}$ for the composite circuit from the *a*-parameters of the two circuits.

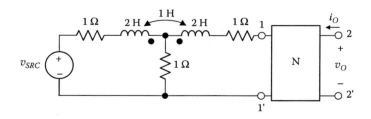

FIGURE 14.4.2
Figure for Example 14.4.1.

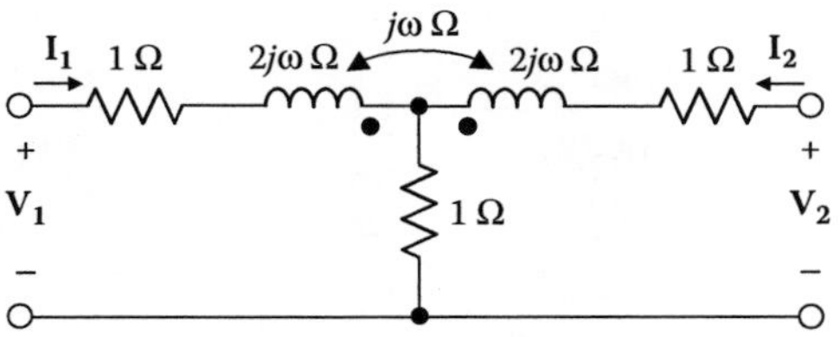

FIGURE 14.4.3
Figure for Example 14.4.1.

The first circuit, redrawn in the frequency domains Figure 14.4.3, is symmetric also.

With $I_2 = 0$, $V_1 = (2 + 2j\omega)I_1$ and $V_2 = (1 + j\omega)I_1$. It follows that $a'_{21} = \left.\dfrac{I_1}{V_2}\right|_{I_2=0} = \dfrac{1}{1 + j\omega}$ S and

that $a'_{11} = \left.\dfrac{V_1}{V_2}\right|_{I_2=0} = \dfrac{2(1 + j\omega)}{1 + j\omega} = 2$. From symmetry, $a'_{22} = a'_{11} = 2$ and $\Delta a' = a'_{11}a'_{22} - a'_{12}a'_{21} =$

$4 - \dfrac{a'_{12}}{jw + 1} = 1$, which gives $a'_{12} = 3(1 + j\omega)$ Ω.

For circuit N, $a''_{11} = \left.\dfrac{V_1}{V_2}\right|_{I_2=0} = \dfrac{1}{0.2} = 5$, and $z''_{11} = \left.\dfrac{V_1}{I_1}\right|_{I_2=0} = 5$ Ω. From Table 14.2.2, $z''_{11} = \dfrac{a''_{11}}{a''_{21}}$.

Hence, $a''_{21} = \dfrac{a''_{11}}{z''_{11}} = \dfrac{5}{5} = 1$ S. Because the circuit is symmetric, $a''_{22} = a''_{11} = 5$ and

$\Delta a'' = a''_{11}a''_{22} - a''_{12}a''_{21} = 25 - a''_{12} \times 1 = 1$, which gives $a''_{12} = 24$ Ω. Note that the loading effect of circuit N due to its finite input impedance is automatically taken care of.

From Equation 14.4.3, $a_{11} = a'_{11}a''_{11} + a'_{12}a''_{21} = 2 \times 5 + 3(1 + j\omega) \times 1 = 13 + 3j\omega$.

$$\left.\dfrac{V_O}{V_{src}}\right|_{I_O=0} = \dfrac{1}{a_{11}} = \dfrac{1}{13 + j3} = \dfrac{1}{\sqrt{178}\angle 13°}.$$ It follows that $v_O = 0.75 \cos(t - 13°)$ V.

EXERCISE 14.4.1

Consider the cascade of a lowpass filter and a highpass filter shown in Figure 10.3.4 (Chapter 10). Determine the *a*-parameters of each filter separately, then derive the overall transfer function (Equation 10.3.11, Chapter 10) from the cascade of two-port circuits.

Answer: Lowpass filter: $a'_{11} = (1 + j\omega CR)$, $a'_{12} = R$, $a'_{21} = j\omega C$, $a'_{22} = 1$;

Highpass filter: $a''_{11} = 1 + 1/j\omega CR$, $a''_{12} = 1/j\omega C$, $a''_{21} = 1/R$, $a''_{22} = 1$;

$$a_{11} = 3 + j\omega CR + \dfrac{1}{j\omega CR}; \quad \dfrac{1}{a_{11}} = \dfrac{V_O}{V_{SRC}}$$ given by Equation 10.3.11.

A useful special application of Equation 14.4.8 is in deriving the transfer function of ladder circuits of the general form shown in Figure 14.4.4. The idea is to consider each series impedance and each shunt admittance as a two-port circuit, derive the *a*-parameter matrix of each element, then determine the transfer function from the a_{11} parameter of the

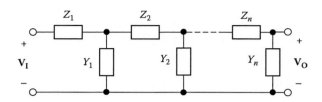

FIGURE 14.4.4
Ladder circuit.

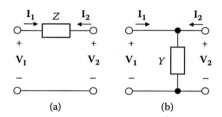

FIGURE 14.4.5
Figure for Example 14.4.2.

cascaded combination derived from the product of the matrices of the individual elements. Example 14.4.2 illustrates the procedure.

Example 14.4.2: Ladder Circuit as a Cascade of Two-Port Circuits

It is required to derive the transfer function of a ladder circuit as a cascade of two-port circuits.

SOLUTION

Consider a series impedance element Z in Figure 14.4.5a. The a-parameters follow readily.

With port 2 open-circuited, $a_{11} = \left.\dfrac{V_1}{V_2}\right|_{I_2=0} = 1$ and $a_{21} = \left.\dfrac{I_1}{V_2}\right|_{I_2=0} = 0$. With port 2 short-circuited,

$a_{12} = -\left.\dfrac{V_1}{I_2}\right|_{V_2=0} = Z$ and $a_{22} = -\left.\dfrac{I_1}{I_2}\right|_{V_2=0} = 1$. The a-parameter matrix is, therefore,

$$[a] = \begin{bmatrix} 1 & Z \\ 0 & 1 \end{bmatrix}$$
(14.4.9)

For a shunt admittance element Y (Figure 14.4.5b), when port 2 is open-circuited: $a_{11} = \left.\dfrac{V_1}{V_2}\right|_{I_2=0} = 1$, $a_{21} = \left.\dfrac{I_1}{V_2}\right|_{I_2=0} = Y$. When port 2 is short-circuited, $V_1 = V_2 = 0$ and $I_1 = -I_2$. It follows that $a_{12} = 0$ and $a_{22} = 1$. The a-parameter matrix is therefore,

$$[a] = \begin{bmatrix} 1 & 0 \\ Y & 1 \end{bmatrix}$$
(14.4.10)

Both two-port circuits of Figure 14.4.5 are symmetric, so that $a_{11} = a_{22}$ and $\Delta a = 1$. When elements Z and Y are cascaded, with Y following Z, the a-parameter matrix becomes:

$$[a] = \begin{bmatrix} 1 & Z \\ 0 & 1 \end{bmatrix} \begin{bmatrix} 1 & 0 \\ Y & 1 \end{bmatrix} = \begin{bmatrix} 1 + ZY & Z \\ Y & 1 \end{bmatrix} \tag{14.4.11}$$

according to the rules of matrix multiplication (Appendix SC.2). Note that the resulting circuit is reciprocal but not symmetric.

As an example, consider the ladder circuit of Figure 14.4.6. The a-parameter matrix of the cascaded combination is:

$$[a] = \begin{bmatrix} 1 & 1 \\ 0 & 1 \end{bmatrix} \begin{bmatrix} 1 & 0 \\ 2j\omega & 1 \end{bmatrix} \begin{bmatrix} 1 & j\omega \\ 0 & 1 \end{bmatrix} \begin{bmatrix} 1 & 0 \\ 2 & 1 \end{bmatrix} \tag{14.4.12}$$

To evaluate this product, we can multiply together the first two matrices, the last two matrices, and finally the two product matrices. Thus,

$$\begin{bmatrix} 1 & 1 \\ 0 & 1 \end{bmatrix} \begin{bmatrix} 1 & 0 \\ 2j & 1 \end{bmatrix} = \begin{bmatrix} 2j\omega + 1 & 1 \\ 2j\omega & 1 \end{bmatrix} \tag{14.4.13}$$

$$\begin{bmatrix} 1 & j\omega \\ 0 & 1 \end{bmatrix} \begin{bmatrix} 1 & 0 \\ 2 & 1 \end{bmatrix} = \begin{bmatrix} 2j\omega + 1 & j\omega \\ 2 & 1 \end{bmatrix} \tag{14.4.14}$$

$$\begin{bmatrix} 2j\omega + 1 & 1 \\ 2j\omega & 1 \end{bmatrix} \begin{bmatrix} 2j\omega + 1 & j\omega \\ 2 & 1 \end{bmatrix} = \begin{bmatrix} (3 - 4\omega^2) + 4j\omega & (1 - 2\omega^2) + j\omega \\ (2 - 4\omega^2) + 2j\omega & 1 - 2\omega^2 \end{bmatrix} \tag{14.4.15}$$

It follows that $\dfrac{V_2}{V_1} = \dfrac{1}{a_{11}} = \dfrac{1}{3 - 4\omega^2 + 4j\omega}$. MATLAB can of course be used to multiply the matrices. Enter: syms w, then each of the matrices, and obtain their product in the correct order.

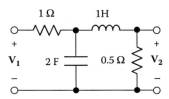

FIGURE 14.4.6
Figure for Example 14.4.2.

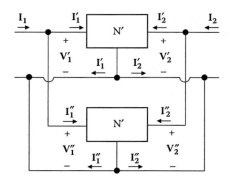

FIGURE 14.4.7
Parallel-connected three-terminal circuits.

EXERCISE 14.4.2
Rework Exercise 14.4.1 based on the procedure of Example 14.4.2.

Parallel Connection

Consider two parallel three-terminal circuits (Figure 14.4.7). It is desired to express the parameters of the composite circuit in terms of the parameters of the individual circuits. Because the input and output currents of the composite circuit are, respectively, the sums of the input and output currents of the individual circuits, the y-parameter representation is the most appropriate. The y-parameter equations for the two circuits are:

$$\mathbf{I}_1' = y_{11}'\mathbf{V}_1' + y_{12}'\mathbf{V}_2' \quad \mathbf{I}_1'' = y_{11}''\mathbf{V}_1'' + y_{12}''\mathbf{V}_2''$$

$$\mathbf{I}_2' = y_{21}'\mathbf{V}_1' + y_{22}'\mathbf{V}_2' \quad \mathbf{I}_2'' = y_{21}''\mathbf{V}_1'' + y_{22}''\mathbf{V}_2'' \tag{14.4.16}$$

For the parallel connection, $\mathbf{V}_1 = \mathbf{V}_1' = \mathbf{V}_1''$, $\mathbf{V}_2 = \mathbf{V}_2' = \mathbf{V}_2''$, $\mathbf{I}_1 = \mathbf{I}_1' + \mathbf{I}_1''$, and $\mathbf{I}_2 = \mathbf{I}_2' + \mathbf{I}_2''$. Substituting in Equation 14.4.16,

$$\mathbf{I}_1 = (y_{11}' + y_{11}'')\mathbf{V}_1 + (y_{12}' + y_{12}'')\,\mathbf{V}_2$$

$$\mathbf{I}_2 = (y_{21}' + y_{21}'')\mathbf{V}_1 + (y_{22}' + y_{22}'')\,\mathbf{V}_2 \tag{14.4.17}$$

The corresponding y-parameters of the individual circuits simply add to give the y-parameters of the composite circuit. In matrix notation,

$$\begin{bmatrix} y_{11} & y_{12} \\ y_{21} & y_{22} \end{bmatrix} = \begin{bmatrix} y_{11}' & y_{12}' \\ y_{21}' & y_{22}' \end{bmatrix} + \begin{bmatrix} y_{11}'' & y_{12}'' \\ y_{21}'' & y_{22}'' \end{bmatrix} = \begin{bmatrix} y_{11}' + y_{11}'' & y_{12}' + y_{12}'' \\ y_{12}' + y_{21}'' & y_{22}' + y_{22}'' \end{bmatrix} \tag{14.4.18}$$

according to the rules of matrix addition (Appendix SC.2). The preceding results may be generalized to the paralleling of m three-terminal circuits. Thus,

$$[y] = [y_1] + [y_2] + \ldots + [y_m] \qquad (14.4.19)$$

where $[y]$ is the matrix of y-parameters of the composite circuit and $[y_k]$ is the matrix of y-parameters of the kth circuit, $k = 1, 2, \ldots, m$.

Example 14.4.3: y-Parameters of Notch Filter

An application of the paralleling of two three-terminal circuits that is of practical importance is the notch filter, shown in Figure 14.4.8. It is required to derive the y-parameters of the filter from those of the two individual T-circuits.

SOLUTION

Consider the T-circuit of Figure 14.4.8a. From the definition of the y-parameters,

$$y'_{11} = \left.\frac{I'_1}{V'_1}\right|_{V'_2=0} = \frac{1+2sCR}{2R(1+sCR)} \text{ S and } y'_{21} = \left.\frac{I'_2}{V'_1}\right|_{V'_2=0} = -\frac{1}{2R(1+sCR)} \text{ S, where } s = j\omega. \text{ Because the}$$

circuit is symmetric, $y'_{11} = y'_{22}$ and $y'_{12} = y'_{21}$. The T-circuit of Figure 14.4.8b may be derived from that of Figure 14.4.8a by replacing R by $1/sC$ and sC by $1/R$, which gives:

$$y''_{11} = \frac{sCR(2+sCR)}{2R(1+sCR)} \text{ S and } y''_{21} = -\frac{s^2C^2R^2}{2R(1+sCR)} \text{ S}$$

It follows that the y-parameters of the notch filter (Figure 14.4.8c) are:

$$y_{11} = y_{22} = \frac{s^2C^2R^2 + 4sCR + 1}{2R(1+sCR)} = \frac{C}{2}\frac{s^2 + 4\omega_n s + \omega_n^2}{s + \omega_n} \text{ S} \qquad (14.4.20)$$

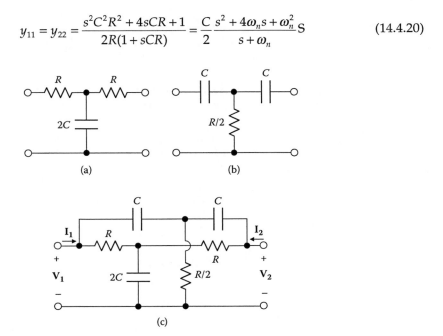

(a)

(b)

(c)

FIGURE 14.4.8
Figure for Example 14.4.3.

$$y_{12} = y_{21} = -\frac{s^2 C^2 R^2 + 1}{2R(1 + sCR)} = -\frac{C}{2}\frac{s^2 + \omega_n^2}{s + \omega_n} \text{ S} \qquad (14.4.21)$$

where $\omega_n = 1/RC$.

Let us determine the transfer function $H(s) = \left.\frac{V_2}{V_1}\right|_{I_2=0}$. It follows from the second y-

parameter equation (Table 14.1.1) that $\left.\frac{V_2}{V_1}\right|_{I_2=0} = -\frac{y_{21}}{y_{22}}$. Hence,

$$H(s) = \frac{s^2 + \omega_n^2}{s^2 + 4\omega_n s + \omega_n^2} \qquad (14.4.22)$$

The following should be noted concerning the preceding results:

1. The transfer function (Equation 14.4.22) is that of a bandstop filter (Table 10.5.1, Chapter 10). It has a zero at $s = j\omega_n$. At this frequency, $V_2 = 0$. This means that the output terminals may be shorted together and no current flows. $y_{21} = \left.\frac{I_2}{V_1}\right|_{V_2=0}$ (Equation 14.4.21) confirms this.

2. From the coefficient of S in Equation 14.4.22, $Q = 0.25$.

EXERCISE 14.4.3
Derive the y-parameters of the notch filter of Example 14.4.3 by first determining the a-parameters of each T-circuit using Equation 14.4.8. Then determine the y-parameters of each T-circuit using Table 14.2.2.

It should be noted that Equation 14.4.19 does not, in general, apply to two-port circuits that are not three-terminal. The reason is that when two-port circuits are paralleled, currents are, in general, redistributed so that the current entering one terminal of a given port is no longer equal to the current leaving the other terminal of the same port, and the two-port equations no longer apply. This is illustrated in Figure 14.4.9, where, for N',

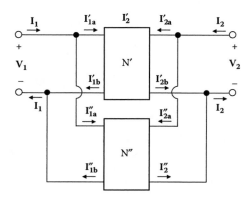

FIGURE 14.4.9
Parallel-connected two-port circuits.

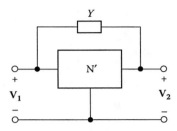

FIGURE 14.4.10
Bridged-T circuit.

$I'_{1a} \neq I'_{1b}$ and $I'_{2a} \neq I'_{2b}$. The same applies for N". When N' and N" are both three-terminal, with the common terminals of the two circuits connected together, this situation does not arise, and the two-port equations apply. If either N' or N" is coupled through an ideal 1:1 transformer having no direct connection between primary and secondary, then the currents entering and leaving at any port are forced to be equal and the two-port equations apply. This is discussed in more detail in Section ST14.2.

A useful application of paralleling two-port circuits is the derivation of the responses of a three-terminal circuit that is bridged by an admittance between the input and output terminals, as shown in Figure 14.4.10. The admittance Y may itself be considered a three-terminal circuit, as in Figure 14.4.5a. With $\mathbf{V}_2 = 0$, $\mathbf{I}_1 = -\mathbf{I}_2 = Y\mathbf{V}_1$ the y-parameter matrix is, therefore,

$$\begin{bmatrix} Y & -Y \\ -Y & Y \end{bmatrix} \tag{14.4.23}$$

Hence, the y-parameter matrix of the bridged circuit is:

$$\begin{bmatrix} y_{11} & y_{12} \\ y_{21} & y_{22} \end{bmatrix} = \begin{bmatrix} y'_{11} + Y & y'_{12} - Y \\ y'_{21} - Y & y'_{22} + Y \end{bmatrix} \tag{14.4.24}$$

where the primed parameters are those of the circuit N'.

Example 14.4.4: Transfer Function of Bridged-T Circuit

It is required to determine the transfer function $\dfrac{\mathbf{V}_2}{\mathbf{V}_1}$ in Figure 14.4.11.

SOLUTION

Consider the T-circuit consisting of the two resistors and inductor. $y'_{11} = \dfrac{\mathbf{I}'_1}{\mathbf{V}'_1}\Big|_{V'_2=0} = \dfrac{s+1}{2s+1}$ S,

$y'_{21} = \dfrac{\mathbf{I}'_2}{\mathbf{V}'_1}\Big|_{V'_2=0} = -\dfrac{s}{2s+1}$ S, where $s = j\omega$. Because the circuit is symmetric, $y'_{11} = y'_{22}$ and

$y'_{12} = y'_{21}$. From Equation 14.4.23, the y-parameter matrix is:

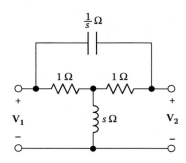

FIGURE 14.4.11
Figure for Example 14.4.4.

$$[y] = \begin{bmatrix} s & -s \\ -s & s \end{bmatrix}$$

It follows that:

$$y_{11} = y_{22} = \frac{s+1}{2s+1} + s = \frac{2s^2 + 2s + 1}{2s+1} \tag{14.4.25}$$

$$y_{12} = y_{21} = -\frac{s}{2s+1} - s = -\frac{2s(s+1)}{2s+1} \tag{14.4.26}$$

The transfer function is

$$\frac{V_2}{V_1} = -\frac{y_{21}}{y_{22}} = \frac{2s(s+1)}{2s^2 + 2s + 1} \tag{14.4.27}$$

EXERCISE 14.4.4

Assume that in Figure 14.4.10, circuit N' has the following z-parameters: $z_{11} = 8 \; \Omega$, $z_{12} = 3 \; \Omega$, $z_{21} = 5 \; \Omega$, and $z_{22} = 2 \; \Omega$. Determine the y-parameters of N' when bridged by a 1 H inductor.

Answer: $y_{11} = 2 + 1/j\omega$ S, $y_{12} = -(3 + 1/j\omega)$ S, $y_{21} = -(5 + 1/j\omega)$ S, and $y_{22} = 8 + 1/j\omega$ S.

Series Connection

Consider two two-port circuits connected as in Figure 14.4.12. This is a series connection because the current of port 1 of circuit N' is the same as the current of port 1 of circuit N'', and the voltages at these two ports add to give port 1 voltage of the composite network; the same is true for port 2.

It is desired to express the parameters of the composite circuit in terms of the parameters of the individual circuits. Because the port voltages add, the z-parameter representation is the most appropriate. The z-parameter equations for the two circuits are:

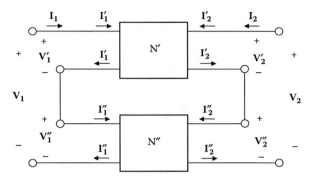

FIGURE 14.4.12
Series-connected two-port circuits.

$$V_1' = z_{11}'I_1' + z_{12}'I_2' \quad V_1'' = z_{11}''I_1'' + z_{12}''I_2''$$

$$V_2' = z_{21}'I_1' + z_{22}'I_2' \quad V_2'' = z_{21}''I_1'' + z_{22}''I_2'' \tag{14.4.28}$$

For the series connection, $I_1 = I_1' = I_1''$, $I_2 = I_2' = I_2''$, $V_1 = V_1' + V_1''$, and $V_2 = V_2' + V_2''$. Substituting in Equation 14.4.28:

$$V_1 = (z_{11}' + z_{11}'')I_1 + (z_{12}' + z_{12}'')I_2$$

$$V_2 = (z_{21}' + z_{21}'')I_1 + (z_{22}' + z_{22}'')I_2 \tag{14.4.29}$$

The corresponding z-parameters of the individual circuits simply add to give the z-parameters of the composite circuit. In matrix notation,

$$\begin{bmatrix} z_{11} & z_{12} \\ z_{21} & z_{22} \end{bmatrix} = \begin{bmatrix} z_{11}' & z_{12}' \\ z_{21}' & z_{22}' \end{bmatrix} + \begin{bmatrix} z_{11}'' & z_{12}'' \\ z_{21}'' & z_{22}'' \end{bmatrix} = \begin{bmatrix} z_{11}' + z_{11}'' & z_{12}' + z_{12}'' \\ z_{12}' + z_{21}'' & z_{22}' + z_{22}''' \end{bmatrix} \tag{14.4.30}$$

according to the rules of matrix addition. The preceding results may be generalized to the series connection of m three-terminal circuits. Thus,

$$\begin{bmatrix} z \end{bmatrix} = \begin{bmatrix} z_1 \end{bmatrix} + \begin{bmatrix} z_2 \end{bmatrix} + \dots + \begin{bmatrix} z_m \end{bmatrix} \tag{14.4.31}$$

where $[z]$ is the matrix of z-parameters of the composite circuit and $[z_k]$ is the matrix of z-parameters of the kth circuit, $k = 1, 2, \dots, m$.

As in the case of the parallel connection and for the same reasons, Equation 14.4.31 does not apply, in general, to two-port circuits that are not three-terminal and have their common terminals connected together, unless one of the circuits is coupled through an ideal 1:1 transformer having no direct connection between primary and secondary. This is discussed in more detail in Section ST14.2.

A useful application of the series connection of two-port circuits is the derivation of the responses of a three-terminal circuit having a coupling impedance as shown in Figure

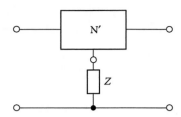

FIGURE 14.4.13
Two-port circuit with a series coupling impedance.

14.4.13. The impedance Z may itself be considered a three-terminal circuit, like the shunt admittance in Figure 14.4.5b. The z-parameter matrix is directly derived as:

$$\begin{bmatrix} Z & Z \\ Z & Z \end{bmatrix} \tag{14.4.32}$$

Hence, the z-parameter matrix of the composite circuit is

$$\begin{bmatrix} z_{11} & z_{12} \\ z_{21} & z_{22} \end{bmatrix} = \begin{bmatrix} z'_{11} + Z & z'_{12} + Z \\ z'_{21} + Z & z'_{22} + Z \end{bmatrix} \tag{14.4.33}$$

where the primed parameters are those of the circuit N′.

EXERCISE 14.4.5
Verify that the z-parameter matrix of the impedance Z is the same as that of Equation 14.4.10, using the conversion of a- to z-parameters of Table 14.2.2.

Example 14.4.5: z-Parameters of Coupled Coils with Series Coupling Capacitor

It is required to derive the z-parameters of the circuit of Figure 14.4.14 using Equation 14.4.33.

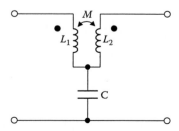

FIGURE 14.4.14
Figure for Example 14.4.5.

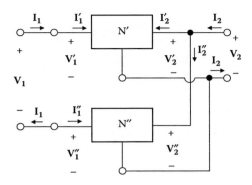

FIGURE 14.4.15
Series–parallel connection of two-port circuits.

SOLUTION

For the linear transformer, $z_{11} = sL_1$, $z_{12} = z_{21} = sM$, $z_{22} = sL_2$ (Example SE14.1). It follows that the z-parameters of the composite circuit are $z_{11} = j\omega L_1 + 1/j\omega C$, $z_{12} = z_{21} = j\omega M + 1/j\omega C$, $z_{22} = j\omega L_2 + 1/j\omega C$.

Series–Parallel Connection

Two other connections of two-port circuits are possible, series–parallel and parallel–series. As discussed for the parallel and series connections earlier, the series–parallel and parallel–series connections will, in general, lead to violation of the condition that each individual circuit should have the same current entering a given port as leaving it. We will therefore restrict the discussion to three-terminal networks that are connected in such a way so as not to cause the aforementioned violation.

A series–parallel connection of two three-terminal circuits is shown in Figure 14.4.15. The input ports of the two circuits are series connected, with $I_1 = I_1' = -I_1''$ and $V_1 = V_1' - V_1''$. The output ports are paralleled because $V_2 = V_2' = V_2''$ and $I_2 = I_2' + I_2''$. With port 1 voltages and port 2 currents additive, the *h*-parameter representation is the most appropriate. The *h*-parameter equations for the two circuits are:

$$V_1' = h_{11}'I_1' + h_{12}'V_2' \quad V_1'' = h_{11}''I_1'' + h_{12}''V_2''$$

$$I_1' = h_{21}'I_1' + h_{22}'V_2' \quad I_2'' = h_{21}''I_1'' + h_{22}''V_2'' \tag{14.4.34}$$

Substituting,

$$V_1 = V_1' - V_1'', \; I_1 = I_1' = -I_1'', \; V_2 = V_2' = V_2'', \text{ and } I_2 = I_2' + I_2'' :$$

$$V_1 = (h_{11}' + h_{11}'')I_1 + (h_{12}' - h_{12}'')V_2$$

$$I_2 = (h_{21}' - h_{21}'')I_1 + (h_{22}' + h_{22}'')V_2 \tag{14.4.35}$$

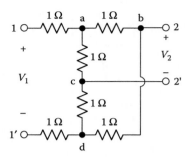

FIGURE 14.4.16
Figure for Example 14.4.6.

It follows that:

$$h_{11} = h'_{11} + h''_{11} \qquad h_{12} = h'_{12} - h''_{12}$$

$$h_{21} = h'_{21} - h''_{21} \qquad h_{22} = h'_{22} + h''_{22} \qquad (14.4.36)$$

Example 14.4.6: *h*-Parameters of Series–Parallel Three-Terminal Circuits

Consider two T-circuits connected as shown in Figure 14.4.16. It is required to verify Equation 14.4.36.

SOLUTION

The *h*-parameters of the T-circuit are readily found to be:

$$h'_{11} = \frac{V'_1}{I'_1}\bigg|_{V'_2=0} = \frac{3}{2}\ \Omega \qquad h'_{12} = \frac{V'_1}{V'_2}\bigg|_{I'_1=0} = \frac{1}{2}$$

$$h'_{21} = \frac{I'_2}{I'_1}\bigg|_{V'_2=0} = -\frac{1}{2} \qquad h'_{22} = \frac{I'_2}{V'_2}\bigg|_{I'_1=0} = \frac{1}{2}\ \text{S} \qquad (14.4.37)$$

It follows from Equation 14.4.36 that the *h*-parameters of the composite circuit are:

$$h_{11} = 3\ \Omega,\ h_{12} = 0,\ h_{21} = 0,\ h_{22} = 1\ \text{S} \qquad (14.4.38)$$

EXERCISE 14.4.6

Verify Equation 14.4.38 by redrawing the circuit of Figure 14.4.16 as a bridge.

Parallel–Series Connection

A parallel–series connection of two three-terminal circuits is shown in Figure 14.4.17. Port 1 currents and port 2 voltages are additive, so that the *g*-parameter representation is the most appropriate. The *g*-parameter equations for the two circuits are:

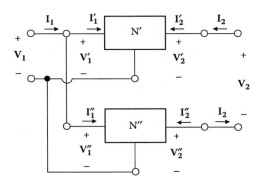

FIGURE 14.4.17
Parallel–series connection of two-port circuits.

$$I_1' = g_{11}'V_1' + g_{12}'I_2' \quad I_1'' = g_{11}''V_1'' + g_{12}''I_2''$$

$$V_2' = g_{21}'V_1' + g_{22}'I_2' \quad V_2'' = g_{21}''V_1'' + g_{22}''I_2'' \tag{14.4.39}$$

For the connections in Figure 14.4.17, $V_1 = V_1' = V_1''$, $I_1 = I_1' + I_1''$, $V_2 = V_2' - V_2''$, and $I_2 = I_2' = -I_2''$. Substituting in Equation 14.4.39,

$$I_1 = (g_{11}' + g_{11}'')I_1 + (g_{12}' - g_{12}'')I_2$$

$$V_2 = (g_{21}' + g_{21}'')I_1 + (g_{22}' - g_{22}'')I_2 \tag{14.4.40}$$

It follows that:

$$g_{11} = g_{11}' + g_{11}'' \quad g_{12} = g_{12}' - g_{12}''$$

$$g_{21} = g_{21}' - g_{21}'' \quad g_{22} = g_{22}' - g_{22}'' \tag{14.4.41}$$

Example 14.4.7: g-Parameters of Parallel–Series Three-Terminal Circuits

Consider two T-circuits connected as shown in Figure 14.4.18. It is required to verify Equation 14.4.41.

SOLUTION

The g-parameters of the T-circuit are readily found to be

$$g_{11}' = \left.\frac{I_1'}{V_1'}\right|_{I_2'=0} = \frac{1}{2}\,\text{S} \quad g_{12}' = \left.\frac{I_1'}{I_2'}\right|_{V_1'=0} = -\frac{1}{2}$$

$$g_{21}' = \left.\frac{V_2'}{V_1'}\right|_{I_2'=0} = \frac{1}{2} \quad g_{22}' = \left.\frac{V_2'}{I_2'}\right|_{V_1'=0} = \frac{3}{2}\,\Omega \tag{14.4.42}$$

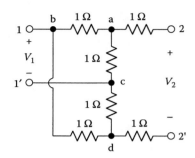

FIGURE 14.4.18
Figure for Exercise 14.4.7.

It follows from Equation 14.4.42 that the *g*-parameters of the composite circuit are:

$$g_{11} = 1\,\text{S}, \quad g_{12} = 0, \quad g_{21} = 0, \quad g_{22} = 3\,\Omega \tag{14.4.43}$$

EXERCISE 14.4.7
Verify Equation 14.4.43 by redrawing the circuit of Figure 14.4.18 as a bridge.

14.5 Analysis of Terminated Two-Port Circuits

A terminated two-port circuit has a source of impedance Z_{src} connected to one port, say port 1, and an impedance Z_L connected to the other port, as illustrated in Figure 14.5.1. It is required to analyze this circuit and derive the following expressions that are generally of interest in describing amplifier circuits:

1. The input impedance $Z_{in} = V_1/I_1$, with Z_L connected to port 2.
2. The current gain I_2/I_1.
3. TEC looking into port 2, where Z_{Th} is the output impedance looking into this port.
4. The ratio I_2/V_{src}.
6. The voltage gain V_2/V_1.
7. The voltage gain V_2/V_{src}.

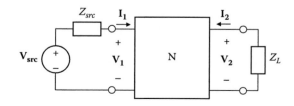

FIGURE 14.5.1
Terminated two-port circuit.

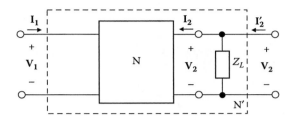

FIGURE 14.5.2
Two-port circuit cascaded with load impedance.

It is straightforward enough to replace the circuit N in Figure 14.5.1 by the equivalent z-parameter circuit of Figure 14.3.1a and to derive the preceding expressions from conventional mesh-current analysis (Exercise 14.5.1). What we will do instead in this section is derive these expressions using two-port relations.

An interesting approach based on the z-parameters is considered in Section ST14.3. Another approach, used here, is to consider N and Z_L as two cascaded two-port circuits constituting a composite two-port circuit N' (Figure 14.5.2). N and N' have the same V_1, I_1, and V_2, but different port 2 currents. When $I_2' = 0$, the same conditions prevail at port 2 of N' as at the output in Figure 14.5.1.

The a-parameter matrix of Z_L is, from Equation 14.4.10:

$$\begin{bmatrix} 1 & 0 \\ 1/Z_L & 1 \end{bmatrix}$$

It follows from Equation 14.4.7 that:

$$\begin{bmatrix} a_{11}' & a_{12}' \\ a_{21}' & a_{22}' \end{bmatrix} = \begin{bmatrix} a_{11} & a_{12} \\ a_{21} & a_{22} \end{bmatrix} \begin{bmatrix} 1 & 0 \\ 1/Z_L & 1 \end{bmatrix} = \begin{bmatrix} a_{11} + a_{12}/Z_L & a_{12} \\ a_{21} + a_{22}/Z_L & a_{22} \end{bmatrix} \tag{14.5.1}$$

To find Z_{in}, we note that $Z_{in} = \dfrac{V_1}{I_1}\Big|_{I_2=0} = z_{11}' = \dfrac{a_{11}'}{a_{21}'}$ (Table 14.2.2). Substituting from Equation 14.5.1 gives:

$$Z_{in} = \frac{a_{11} + a_{12}/Z_L}{a_{21} + a_{22}/Z_L} = \frac{a_{11}Z_L + a_{12}}{a_{21}Z_L + a_{22}} \tag{14.5.2}$$

To find V_2/V_1, we note that $a_{11}' = \dfrac{V_1}{V_2}\Big|_{I_2=0}$. Hence,

$$\frac{V_2}{V_1} = \frac{1}{a_{11}'} = \frac{1}{a_{11} + a_{12}/Z_L} = \frac{Z_L}{a_{11}Z_L + a_{12}} \tag{14.5.3}$$

To determine I_2/I_1, we note that $a_{21}' = \dfrac{I_1}{V_2}\Big|_{I_2=0}$. But with $I_2' = 0$, $V_2 = -I_2 Z_L$. Substituting, we obtain:

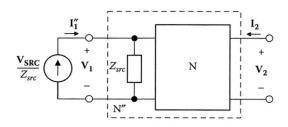

FIGURE 14.5.3
Two-port circuit cascaded with source impedance.

$$\frac{I_2}{I_1} = -\frac{1}{a'_{21}Z_L} = -\frac{1}{a_{21}Z_L + a_{22}} \tag{14.5.4}$$

To derive TEC looking into port 2 in Figure 14.5.1, we transform V_{SRC} in series with Z_{src} to its equivalent current source and consider Z_{src} cascaded with N to be a new two-port circuit N'' (Figure 14.5.3). N and N'' have the same V_1, V_2, and I_2. However, $I''_1 = V_{SRC}/Z_{src}$. The a-parameter matrix of N'' is given by:

$$\begin{bmatrix} a''_{11} & a''_{12} \\ a''_{21} & a''_{22} \end{bmatrix} = \begin{bmatrix} 1 & 0 \\ 1/Z_L & 1 \end{bmatrix} \begin{bmatrix} a_{11} & a_{12} \\ a_{21} & a_{22} \end{bmatrix} = \begin{bmatrix} a_{11} & a_{12} \\ a_{11}/Z_{src} + a_{21} & a_{12}/Z_{src} + a_{22} \end{bmatrix} \tag{14.5.5}$$

Now $a''_{21} = \dfrac{I''_1}{V_2}\Big|_{I_2=0}$, where $I''_1 = V_{SRC}/Z_{src}$ and $V_2 = V_{Th}$ when $I_2 = 0$. Substituting,

$$V_{Th} = \frac{1}{a''_{21}Z_{src}}V_{SRC} = \frac{1}{a_{11} + a_{21}Z_{src}}V_{SRC} \tag{14.5.6}$$

To find Z_{Th}, we note that $Z_{Th} = z''_{22} = \dfrac{V_2}{I_2}\Big|_{I''_1=0} = \dfrac{a''_{22}}{a''_{21}}$ (Table 14.2.2). Hence,

$$Z_{Th} = \frac{a''_{22}}{a''_{21}} = \frac{a_{12}/Z_{src} + a_{22}}{a_{11}/Z_{src} + a_{21}} = \frac{a_{12} + a_{22}Z_{src}}{a_{11} + a_{21}Z_{src}} \tag{14.5.7}$$

Finally, to derive I_2/V_{SRC} and V_1/V_{SRC}, we make use of the fact that the source sees an input impedance Z_{in} at the input of the two-port circuit N in Figure 14.5.1. This gives $\dfrac{I_1}{V_{SRC}}$

$= \dfrac{1}{Z_{src} + Z_{in}}$. Substituting for I_2/I_1 from Equation 14.5.4 and for Z_{in} from Equation 14.5.2,

$$\frac{I_2}{V_{SRC}} = -\frac{1}{(Z_g + Z_{in})(a_{21}Z_L + a_{22})} = -\frac{1}{Z_{src}(a_{21}Z_L + a_{22}) + a_{11}Z_L + a_{12}} \tag{14.5.8}$$

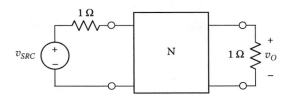

FIGURE 14.5.4
Figure for Example 14.5.1.

From Figure 14.5.1, $\dfrac{V_1}{V_{SRC}} = \dfrac{1}{1 + Z_{src}/Z_{in}}$. Substituting for V_2/V_1 from Equation 14.5.3 and for Z_{in} from Equation 14.5.2:

$$\frac{V_2}{V_{SRC}} = \frac{1}{(1 + Z_{src}/Z_{in})} \frac{Z_L}{(a_{11}Z_L + a_{12})} = \frac{Z_L}{Z_{src}(a_{21}Z_L + a_{22}) + a_{11}Z_L + a_{12}} \quad (14.5.9)$$

Alternatively, because $V_2 = -Z_L I_2$, Equation 14.5.9 follows directly from Equation 14.5.8.

To express items 1 to 6 described earlier in terms of parameters other than a-parameters, we substitute in the expressions derived earlier for the a-parameters in terms of any other set of parameters. The resulting expressions are listed in Table 14.5.1.

EXERCISE 14.5.1

Derive the expressions for items 1 to 6 listed at the beginning of the section directly in terms of the z-parameters by using the equivalent circuit of Figure 14.3.1a for N in Figure 14.5.1 and solving the two mesh-current equations. Verify your results against Table 14.5.1.

Example 14.5.1: Analysis of Terminated Two-Port Circuit

In the circuit of Figure 14.5.4, N has the following h-parameters: $h_{11} = \dfrac{-\omega^2 + 2j\omega}{1 + j\omega}$, $h_{12} =$

$\dfrac{1}{1 + j\omega}$, $h_{21} = -\dfrac{1}{1 + j\omega}$, $h_{22} = \dfrac{1}{1 + j\omega}$. It is required to find v_O if $v_{SRC} = 10\cos t$ V.

SOLUTION
From Table 14.5.1,

$$\frac{V_O}{V_2} = \frac{-h_{21}Z_L}{(h_{11} + Z_{src})(1 + h_{22}Z_L) - h_{12}h_{21}Z_L} = \frac{1 + j\omega}{3 + 7j\omega - 5\omega^2 - j\omega^3} =$$

$$\frac{1 + j\omega}{(1 + j\omega)(3 - \omega^2 + j4\omega)} = \frac{1}{3 - \omega^2 + j4\omega}$$

Substituting $\omega = 1\,\text{rad/s}$,

TABLE 14.5.1

Circuit Relations of Terminated Two-Port Circuits

	z-Parameters		*y*-Parameters
$Z_{in} =$	$z_{11} - \dfrac{z_{12}z_{21}}{z_{22} + Z_L}$	$Y_{in} =$	$y_{11} - \dfrac{y_{12}y_{21}Z_L}{1 + y_{22}Z_L}$
$\dfrac{V_{Th}}{V_{SCR}} =$	$\dfrac{z_{21}}{z_{11} + Z_{src}}$		$-\dfrac{y_{21}}{y_{22} + \Delta y Z_{src}}$
$Z_{Th} =$	$\dfrac{\Delta z + z_{22}Z_{src}}{z_{11} + Z_{src}} = z_{22} - \dfrac{z_{12}z_{21}}{z_{11} + Z_{src}}$		$\dfrac{1 + y_{11}Z_{src}}{y_{22} + \Delta y Z_{src}} \; ; \; Y_{Th} = y_{22} - \dfrac{y_{12}y_{21}Z_{src}}{1 + y_{11}Z_{src}}$
$\dfrac{I_2}{V_{SRC}} =$	$-\dfrac{z_{21}}{(z_{11} + Z_{src})(z_{22} + Z_L) - z_{12}z_{21}}$		$\dfrac{y_{21}}{1 + y_{11}Z_{src} + y_{22}Z_L + \Delta y Z_{src}Z_L}$
$\dfrac{I_2}{I_1} =$	$\dfrac{-z_{21}}{z_{22} + Z_L}$		$\dfrac{y_{21}}{y_{11} + \Delta y Z_L}$
$\dfrac{V_2}{V_1} =$	$\dfrac{z_{21}Z_L}{z_{11}Z_L + \Delta z}$		$-\dfrac{y_{21}Z_L}{1 + y_{22}Z_L}$
$\dfrac{V_2}{V_{SRC}} =$	$\dfrac{z_{21}Z_L}{(z_{11} + Z_{src})(z_{22} + Z_L) - z_{12}z_{21}}$		$\dfrac{y_{21}Z_L}{y_{12}y_{21}Z_{src}Z_L - (1 + y_{11}Z_{src})(1 + y_{22}Z_L)}$

	a-Parameters	*b*-Parameters
$Z_{in} =$	$\dfrac{a_{11}Z_L + a_{12}}{a_{21}Z_L + a_{22}}$	$\dfrac{b_{22}Z_L + b_{12}}{b_{21}Z_L + b_{11}}$
$\dfrac{V_{Th}}{V_{SRC}} =$	$\dfrac{1}{a_{11} + a_{21}Z_{src}}$	$\dfrac{\Delta b}{b_{22} + b_{21}Z_{src}}$
$Z_{Th} =$	$\dfrac{a_{12} + a_{22}Z_{src}}{a_{11} + a_{21}Z_{src}}$	$\dfrac{b_{12} + b_{11}Z_{src}}{b_{22} + b_{21}Z_{src}}$
$\dfrac{I_2}{V_{SRC}} =$	$-\dfrac{1}{Z_{src}(a_{21}Z_L + a_{22}) + a_{11}Z_L + a_{12}}$	$-\dfrac{\Delta b}{Z_{src}(b_{21}Z_L + b_{11}) + b_{22}Z_L + b_{12}}$
$\dfrac{I_2}{I_1} =$	$-\dfrac{1}{a_{21}Z_L + a_{22}}$	$-\dfrac{\Delta b}{b_{21}Z_L + b_{11}}$
$\dfrac{V_2}{V_1} =$	$\dfrac{Z_L}{a_{11}Z_L + a_{12}}$	$\dfrac{\Delta b Z_L}{b_{22}Z_L + b_{12}}$
$\dfrac{V_2}{V_{SRC}} =$	$\dfrac{Z_L}{Z_{src}(a_{21}Z_L + a_{22}) + a_{11}Z_L + a_{12}}$	$\dfrac{\Delta b Z_L}{Z_{src}(b_{21}Z_L + b_{11}) + b_{22}Z_L + b_{12}}$

	h-parameters		*g*-Parameters
$Z_{in} =$	$h_{11} - \dfrac{h_{12}h_{21}Z_L}{1 + h_{22}Z_L}$	$Y_{in} =$	$g_{11} - \dfrac{g_{12}g_{21}}{g_{22} + Z_L}$
$\dfrac{V_{Th}}{V_{SRC}} =$	$-\dfrac{h_{21}}{h_{22}Z_{src} + \Delta h}$		$\dfrac{g_{21}}{1 + g_{11}Z_{src}}$
$Z_{Th} =$	$\dfrac{Z_{src} + h_{11}}{h_{22}Z_{src} + \Delta h} \; ; \; Y_{Th} = h_{22} - \dfrac{h_{12}h_{21}}{h_{11} + Z_{src}}$		$\dfrac{g_{22} + \Delta g Z_{src}}{1 + g_{11}Z_{src}} = g_{22} - \dfrac{g_{12}g_{21}Z_{src}}{1 + g_{11}Z_{src}}$

TABLE 14.5.1 *(Continued)*

Circuit Relations of Terminated Two-Port Circuits

	h-Parameters	*g*-Parameters
$\dfrac{I_2}{V_{SRC}}$	$\dfrac{h_{21}}{(h_{11}+Z_{src})(1+h_{22}Z_L)-h_{12}h_{21}Z_L}$	$-\dfrac{g_{21}}{(1+g_{11}Z_{src})(g_{22}+Z_L)-g_{12}g_{21}Z_{src}}$
$\dfrac{I_2}{I_1}$	$\dfrac{h_{21}}{1+h_{22}Z_L}$	$-\dfrac{g_{21}}{g_{11}Z_L+\Delta g}$
$\dfrac{V_2}{V_1}$	$-\dfrac{h_{21}Z_L}{h_{11}+\Delta h Z_L}$	$\dfrac{g_{21}Z_L}{g_{22}+Z_L}$
$\dfrac{V_2}{V_{SRC}}$	$-\dfrac{h_{21}Z_L}{(h_{11}+Z_{src})(1+h_{22}Z_L)-h_{12}h_{21}Z_l}$	$\dfrac{g_{21}Z_L}{(1+g_{11}Z_{src})(g_{22}+Z_L)-g_{12}g_{21}Z_{src}}$

$$\frac{V_O}{V_{SRC}} = \frac{1}{2+j4} = \frac{1}{2\sqrt{5}}\angle - \tan^{-1}2 \; .$$

Hence,

$$v_O = \sqrt{5}\cos(t-63.4°)\;\text{V}.$$

Summary of Main Concepts and Results

- A two-port circuit may be specified in terms of one of six sets of two simultaneous equations, each of which involves four parameters. In general, these parameters are nonzero and independent.
- A reciprocal two-port circuit is specified in general by three independent, nonzero parameters. A symmetric circuit is specified in general by two independent, non-zero parameters.
- The *z*-parameter, *y*-parameter, *h*-parameter, and *g*-parameter equations may be represented in terms of equivalent circuits in which forward and reverse transmission are described by dependent sources on the output and input sides.
- Two-port circuits may be connected in cascade, in parallel, in series, in series–parallel, or in parallel–series. The two-port parameters of the composite circuit can be derived from those of the individual circuits, as long as each individual circuit has the same current entering a given port as leaving it. This condition is assured in the cascade connection. It is assured in the other connections either by restricting the two-port circuits to be three-terminal circuits with their common terminals connected together, or by coupling one of the two-port circuits using an ideal 1:1 transformer having no direct connection between primary and secondary.
- A two-port circuit that is terminated by source and load impedances may be analyzed by considering the two-port circuit to be cascaded with the load impedance, when looking into port 1, and to be cascaded with the source impedance when looking into port 2. An alternative approach is to combine the source immittance with the "11" immittance parameter and to combine the load immittance with the "22" immittance parameter.

Learning Outcomes

- Derive the two-port parameters of a given circuit.
- Analyze a circuit using the methods of two-port analysis.

Supplementary Topics and Examples on CD

ST14.1 Reciprocity theorem: The theorem is stated, proved, and illustrated by examples.

ST14.2 Invalid connections of two-port circuits: Illustrates how the two-port equations do not apply in general to two-port circuits connected in other than the cascade connection, unless one of the circuits is coupled through an ideal 1:1 transformer.

ST14.3 Analysis of terminated two-port circuit using z-parameters: Analyzes a two-port circuit using the z-parameter equivalent circuit, with the addition of Z_{src} to z_{11} and Z_L to z_{22}.

ST14.4 Indefinite admittance matrix: Given a three-terminal circuit specified by its y-parameters, the indefinite admittance matrix can be used to derive the y-parameters when any of the other two terminals is made the common terminal.

ST14.5 Ideal transformer and gyrator: Shows how consideration of lossless, two-port circuits that do not store energy, naturally leads to the circuit characterization of an ideal transformer as well as another important device of theoretical and practical interest, namely, the **gyrator**, which converts capacitance to inductance, and conversely.

SE14.1 Two-port equations of linear and ideal transformers: Derives the two-port parameters for a linear transformer, with or without perfect coupling, and for an ideal transformer. It is shown that a linear transformer with perfect coupling does not have y-parameters, whereas an ideal transformer does not have z- or y-parameters.

SE14.2 Matrix partitioning: The y-parameters of a three-node circuit are determined using the method of matrix partitioning.

SE14.3 y-Parameters of cascaded circuits: This an instructive problem in which the y-parameters of two cascaded circuits are derived. One circuit has a resistive voltage divider connected at its input terminals, whereas the other circuit is bridged by a resistor.

Problems and Exercises

P14.1   Parameters and Equivalent Circuits

Verify the results of P14.1.1 to P14.1.19 with PSpice simulation.

P14.1.1 Determine the *z*-parameters and *y*-parameters of the circuit shown in Figure P14.1.1 from the direct definition of these parameters or from the node-voltage equations in the case of *y*-parameters. Verify the results using matrix inversion.

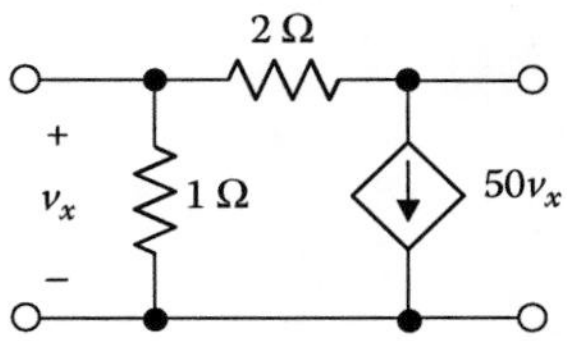

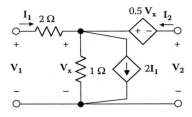

FIGURE P14.1.1

P14.1.2 Repeat P14.1.1 for the *h*- and *g*-parameters.

P14.1.3 Repeat P14.1.1 for the *a*- and *b*-parameters.

P14.1.4 Determine the *z*-parameters and *y*-parameters of the circuit shown in Figure P14.1.4 from the direct definition of these parameters. Verify the results using matrix inversion.

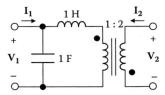

FIGURE P14.1.4

P14.1.5 Repeat P14.1.4 for the *h*- and *g*-parameters.

P14.1.6 Repeat P14.1.4 for the *a*- and *b*-parameters.

P14.1.7 Determine the *z*-parameters and *y*-parameters of the circuit shown in Figure P14.1.7 from the direct definition of these parameters, assuming $\omega = 1$ rad/s. Verify the results using matrix inversion.

FIGURE P14.1.7

P14.1.8 Repeat P14.1.7 for the *h*- and *g*-parameters.

P14.1.9 Repeat P14.1.7 for the *a*- and *b*-parameters.

P14.1.10 Determine the z-parameters and y-parameters of the circuit shown in Figure P14.1.10 from the direct definition of these parameters. Verify the results using matrix inversion.

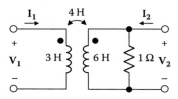

FIGURE P14.1.10

P14.1.11 Repeat P14.1.10 for the *h*- and *g*-parameters.

P14.1.12 Repeat P14.1.10 for the *a*- and *b*-parameters.

P14.1.13 Determine the z-parameters of the circuit shown in Figure P14.1.13. Verify by considering the circuit as a series connection of two three-terminal circuits across the dotted line.

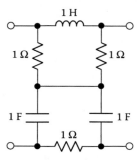

FIGURE P14.1.13

P14.1.14 Determine the y-parameters of the circuit shown in Figure P14.1.14. Verify by considering the circuit as a series connection of two three-terminal circuits.

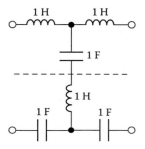

FIGURE P14.1.14

P14.1.15 Determine the z-parameters of the circuit shown in Figure P14.1.15. Verify by considering the circuit as a cascade connection of two three-terminal circuits at the dotted line.

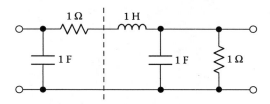

FIGURE P14.1.15

P14.1.16 Determine the z-parameters of the circuit shown in Figure P14.1.16.

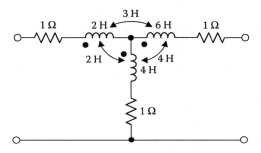

FIGURE P14.1.16

P14.1.17 Determine the z-parameters of the circuit shown in Figure P14.1.17. Verify by considering it as a cascade of three-terminal circuits.

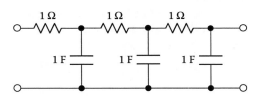

FIGURE P14.1.17

P14.1.18 Determine the y-parameters of the circuit shown in Figure P14.1.18. Verify by considering the circuit as a parallel connection of two three-terminal circuits.

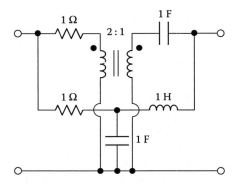

FIGURE P14.1.18

P14.1.19 Determine the *h*-parameters of the high-frequency equivalent circuit of a bipolar junction transistor shown in Figure P14.1.19.

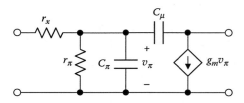

FIGURE P14.1.19

P14.2 Two-Port Circuit Analysis

P14.2.1 Perform the analysis of Section 14.5 using *h*-parameters. Consider Z_{src} to add to h_{11} and Y_L to add to h_{22}.

P14.2.2 Perform the analysis of Section 14.5 using *y*-parameters. Consider Y_L to add to h_{22} and transform the voltage source of impedance Z_{src} to an equivalent current source.

P14.2.3 Perform the analysis of Section 14.5 using *g*-parameters. Consider Z_L to add to g_{22} and transform the voltage source of impedance Z_{src} to an equivalent current source.

P14.2.4 A linear transformer is connected to a 100 Ω load and is driven by a voltage source having an open-circuit voltage $v_{SRC} = 10\cos 1000t$ V and a source impedance of 5 Ω (Figure P14.2.4). Derive TEC seen by the load and determine the steady-state load voltage. Simulate with PSpice.

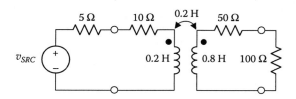

FIGURE P14.2.4

P14.2.5 A two-port circuit has the following *g*-parameters: $g_{11} = 1 - j$ S, $g_{12} = -2 + j2$, $g_{21} = 2 - j2$, and $g_{22} = 4 + j4$ Ω. It is connected at port 2 to a resistive load of 50 Ω and at port 1 to a source having an open-circuit voltage $v_{SRC} = 20\cos t$ V and zero source resistance. Determine the average power delivered by the source to the two-port circuit and to the load.

P14.2.6 Determine the load impedance in P14.2.5 for maximum power transfer to the load and calculate this power.

P14.2.7 An amplifier has the following *y*-parameters: $y_{11} = 1$ μS, $y_{12} = -2$ μS, $y_{21} = 100$ μS, and $y_{22} = 50$ μS. The amplifier is connected to a source of 10 mV rms open-circuit voltage and 10 kΩ source impedance. Determine the load resistance for maximum power transfer to the load and calculate this power.

P14.2.8   A two-port resistive, symmetrical circuit connected between a source and an 8 Ω load gave the measurements indicated in Figure P14.2.8. Determine the *a*-parameters of the circuit.

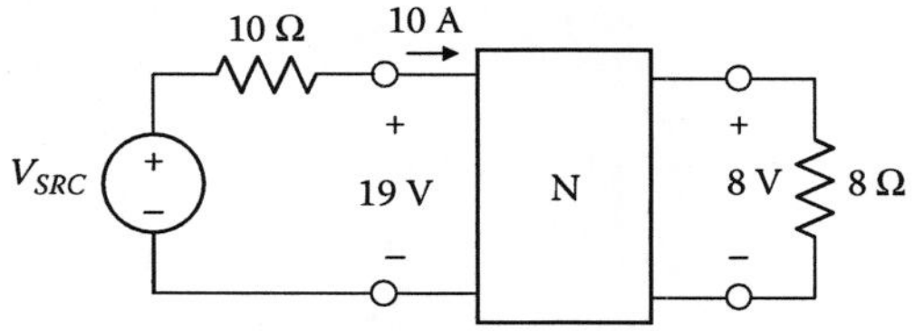

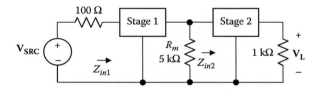

FIGURE P14.2.8

P14.2.9   A two-stage amplifier is shown in Figure P14.2.9. The *y*-parameters of the two stages are as follows:

Stage 1: $y_{11} = 1$ μS, $y_{12} = -2$ μS, $y_{21} = 500$ μS $y_{22} = 0.1$ μS

Stage 2: $y_{11} = 5$ μS, $y_{12} = -1$ μS, $y_{21} = 100$ μS $y_{22} = 0.4$ μS

Determine the input impedances Z_{in1} and Z_{in2} and the overall voltage gain $\mathbf{V_L}/\mathbf{V}_{SRC}$.

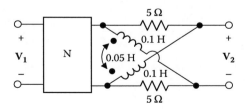

FIGURE P14.2.9

P14.2.10 In Problem P14.2.9, determine R_m so that maximum power is transferred to a 2 kΩ load.

P14.2.11 Circuit N in Figure P14.2.11 is symmetric and described by its *b*-parameters: $b_{11} = 1/2$ and $b_{12} = 2$ Ω. Determine $\mathbf{V_2}/\mathbf{V_1}$, assuming $\omega = 100$ rad/s.

FIGURE P14.2.11

P14.2.12 If the two-port circuit N in Figure P14.2.12 is described by its *y*-parameters,

show that $\dfrac{\mathbf{V_2}}{\mathbf{V_1}} = -\dfrac{y_{21}}{y_{22} + sC}$. If the circuit is described by its *h*-parameters, show that $\dfrac{\mathbf{V}_2}{\mathbf{I}_1} = -\dfrac{h_{21}}{h_{22} + sC}$. Obtain the input impedance by dividing the latter expression by the former and express the result in terms of: (a) the y-parameters and (b) the h-parameters. Verify by comparing with Table 14.14.1.

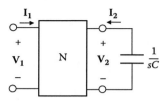

FIGURE P14.2.12

P14.2.13 If the two-port circuit N in Figure P14.2.13 is described by its g-parameters,

show that $\dfrac{\mathbf{V}_2}{\mathbf{V}_1} = \dfrac{sL}{g_{22} + sL}\,g_{21}$. If the circuit is described by its z-parameters,

show that $\dfrac{\mathbf{V}_2}{\mathbf{I}_1} = \dfrac{sL}{z_{22} + sL}\,z_{21}$. Obtain the input impedance by dividing the latter expression by the former and express the result in terms of: (a) the g-parameters; and (b) the z-parameters. Verify by comparing with Table 14.14.1.

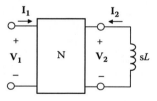

FIGURE P14.2.13

P14.2.14 In Figure P14.2.14, $v_{SRC} = 20\cos 1000t$ V. Determine Z_{in}, Z_{out}, and the steady-state value of v_2. Simulate with PSpice.

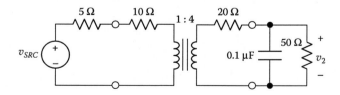

FIGURE P14.2.14

P14.2.15 Derive TEC seen by the load at terminals 2 and 3 in Figure P14.2.15 using the *b*-parameters of the three-terminal circuit between the voltage source and the load, and determine v_2, assuming $v_{SRC} = 20 \cos 1000t$ V.

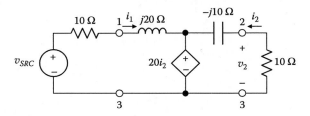

FIGURE P14.2.15

P14.2.16 Derive TEC seen by the load at terminals 2 and 3 in Figure P14.2.16 using the *g*-parameters of the three-terminal circuit between the current source and the load, assuming $v_{SRC} = 10 \cos 1000t$ V. Determine v_2.

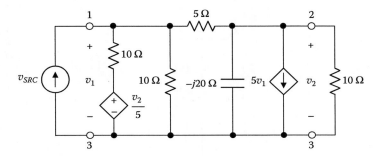

FIGURE P14.2.16

P14.2.17 Determine Z_{in}, Z_{out}, and V_2/V_1 in the circuit of Figure P14.2.17, assuming $\omega = 10^3$ rad/s and that circuit N is symmetric having $z_{11} = 1 + j$ kΩ and $z_{12} = 1 + j$ kΩ.

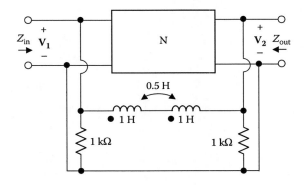

FIGURE P14.2.17

P14.2.18 Determine $\mathbf{V_O}/\mathbf{V_{SRC}}$ in the circuit of Figure P14.2.18, where $s = j\omega$ and $\omega = 1$ rad/s.

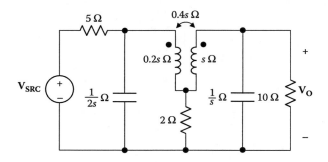

FIGURE P14.2.18

P14.2.19 Determine the 3-dB frequency of the response $\mathbf{V_O}/\mathbf{V_{SRC}}$ in the circuit of Figure P14.2.19. Simulate with PSpice.

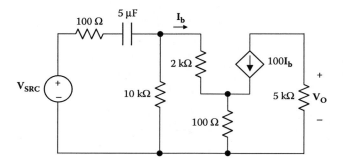

FIGURE P14.2.19

P14.2.20 Determine the 3-dB frequency of the response $\mathbf{V_O}/\mathbf{V_{SRC}}$ in the circuit of Figure P14.2.20. Simulate with PSpice.

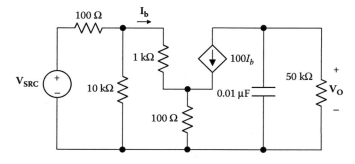

FIGURE P14.2.20

P14.2.21 Two three-terminal circuits are connected in series–parallel as in Figure 14.4.15, to a load of 5 Ω and to a voltage source having a source resistance 2 Ω and an open-circuit voltage $\mathbf{V_{SRC}}$. The *h*-parameters of the two circuits are:

Circuit 1: $h_{11} = 2\ \Omega$, $h_{12} = -(j + 2)$, $h_{21} = -2$, $h_{22} = j$ S

Circuit 2: $h_{11} = 4 \ \Omega$, $h_{12} = -j$, $h_{21} = (j+1)$, $h_{22} = j2 \ \text{S}$

Determine the input impedance of the combination, the output impedance, and the voltage gain $\mathbf{V_O}/\mathbf{V_{SRC}}$, where $\mathbf{V_O}$ is the load voltage.

P14.2.22 Two three-terminal circuits connected in parallel–series as in Figure 14.4.17 to a load of 5 Ω, and to a current source having a source resistance 2 Ω and a short-circuit current $\mathbf{I_{SRC}}$. The *g*-parameters of the two circuits are:

Circuit 1: $g_{11} = 1 \ \text{S}$, $g_{12} = j$, $g_{21} = j2$, $g_{22} = 2 \ \Omega$

Circuit 2: $g_{11} = j \ \text{S}$, $g_{12} = (j+2)$, $g_{21} = j$, $g_{22} = j2 \ \Omega$

Determine the input impedance of the combination, the output impedance, and the gain $\mathbf{V_O}/\mathbf{V_{SRC}}$, where $\mathbf{V_O}$ is the load voltage.

P14.2.23 Show that the input and output impedances of a terminated circuit may be

expressed as follows: $IM_{in} = \lambda_{11} - \dfrac{\lambda_{12}\lambda_{21}}{\lambda_{22} + IM_i}$, $IM_{out} = \lambda_{22} - \dfrac{\lambda_{12}\lambda_{21}}{\lambda_{11} + IM_{sn}}$, where IM_{in}, IM_{out}, IM_L, and IM_{src} are, respectively, the input, output, load, and source immittances, and the λ's may be *z*-, *y*-, *h*-, or *g*-parameters. The units of λ_{11} and λ_{22} just after the equality sign determine whether the expression is an impedance or admittance, and the units of λ_{11} and λ_{22} in the denominators determine whether a load, source impedance, or admittance is to be used.

P14.2.24 Derive the T-π transformation by considering three impedances connected as a T-circuit, then determining the *y*-parameters. Deduce the values of the π-equivalent circuit from Figure 14.3.3.

P14.2.25 Derive the π-T transformation by considering three impedances connected as a π-circuit, then determining the *z*-parameters. Deduce the values of the T-equivalent circuit from Figure 14.3.2.

P14.2.26 Consider the circuit shown in Figure P14.2.26. When the input impedance equals the source resistance and the output impedance equals the load resistance, the circuit is said to be *image-terminated*. Note that under these conditions, maximum power is transferred to the load. From the expressions for Z_{in}

and Z_{Th} (Table 14.5.1), show that in an image-terminated circuit $\dfrac{R_S}{R_L} = \dfrac{z_{11}}{z_{22}}$.

Substituting back in the expressions for Z_{in} and Z_{Th}, show that $R_S = $

$\sqrt{z_{11}/y_{11}} = \sqrt{a_{11}a_{12}/a_{21}a_{22}}$ and that $R_L = \sqrt{z_{22}/y_{22}} = \sqrt{a_{12}a_{22}/a_{11}a_{21}}$.

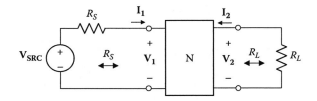

FIGURE P14.2.26

P14.2.27 Consider an image-terminated circuit (Problem P14.2.26). The *image transmis-sion constant* is defined as: $\gamma = \alpha + j\beta = \dfrac{1}{2}\ln\dfrac{V_1 I_1}{V_2 I_2'}$, where $I_2' = -I_2$ is the current flowing out of port 2. Show that if the circuit is reciprocal, $\tanh\gamma = \dfrac{1}{\sqrt{z_{11} y_{11}}} = \dfrac{1}{\sqrt{z_{22} y_{22}}} = \sqrt{\dfrac{a_{12} a_{21}}{a_{11} a_{22}}}$, $\sinh\gamma = \sqrt{a_{12} a_{21}}$, and $\cosh\gamma = \sqrt{a_{11} a_{22}}$. Note that $\alpha = \dfrac{1}{2}\ln\dfrac{P_i}{P_L}$, where P_i is the power delivered to the two-port circuit and P_L is the power delivered to the load. Also $\beta = \dfrac{1}{2}\big[\angle(V_1 I_1) - \angle(V_2 I_2')\big]$.

P14.2.28 Show that if n circuits are cascaded, the overall γ, defined in Problem P14.2.27, is the sum of the individual γ's, and that if the image-terminated circuit is reciprocal, γ is the same if the input and output are interchanged.

P14.2.29 Show that if an image-terminated circuit is symmetric, $R_s = \sqrt{\dfrac{a_{12}}{a_{21}}} = R_L$, $\gamma = \alpha + j\beta = \ln\dfrac{V_1}{V_{2s}} = \ln\dfrac{I_1}{I_{2s}'}$, so that $\alpha = \ln\left|\dfrac{V_1}{V_{2s}}\right| = \ln\left|\dfrac{I_1}{I_{2s}'}\right|$ and $\beta = \angle V_1 - \angle V_{2s}$, where the s subscript refers to the output of a symmetric circuit. A nonsymmetric circuit N can be made symmetric by inserting a transformer of turns ratio $\sqrt{R_L / R_S}$ between the output and load, as shown in Figure P14.2.29. Deduce that even in a nonsymmetric circuit N, $\beta = \angle V_1 - \angle V_2 = \angle I_1 - \angle I_2'$.

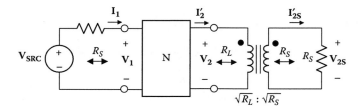

FIGURE P14.2.29

15

Laplace Transform

Overview

The Laplace transform (LT) provides an extremely powerful tool for analyzing linear systems. It will be recalled that the power and usefulness of the phasor approach is in transforming ordinary, linear differential equations to algebraic equations for analyzing the sinusoidal steady state. The LT extends this approach by transforming ordinary, linear differential equations to algebraic equations for deriving the complete response, that is, steady-state plus transient, to any arbitrary excitation that has an LT. To fully utilize the power of this approach, the circuit problem should be solved entirely in the s domain, where s is the complex frequency. To achieve this, it is necessary to be able to represent circuit elements in the s domain, including any initial conditions. Once this is done, conventional circuit analysis techniques can be applied.

The power of the LT method is extended by some useful theorems. The initial value theorem allows determining the circuit variables just after switching that may involve impulsive readjustment. The convolution theorem provides an alternative method of deriving the convolution function and leads to some interesting deductions. Moreover, the concept of transfer function allows some important interpretations of stability, sinusoidal steady-state response, and responses of first-order and second-order circuits.

Learning Objectives

- To be familiar with:
  - LTs of functions commonly encountered in circuit analysis
- To understand:
  - The basic properties of the LT
  - The application of the LT to the solution of ordinary, linear differential equations
  - The final value theorem, the initial value theorem, and the convolution theorem
  - The representation of circuit elements and magnetically coupled coils in the s domain, including source representations of initial values in energy storage elements
  - The general procedure for solving circuit problems, including switching problems, using the LT method

- The concept of the transfer function
- The interpretation of stability and the sinusoidal steady-state response, in terms of the LT
- The natural responses of first-order and second-order systems in the s domain

15.1   Definition of the Laplace Transform

The LT can be derived from the Fourier transform, which is derived in turn as a limiting case of the Fourier series expansion as the period becomes infinitely large. This is done in Section ST15.1. Here, we will simply state the definition of the LT without derivation.

Given a function of time $f(t)$, its LT, denoted as $\mathcal{L}\{f(t)\}$, represents a transformation from the time domain to the s domain, where $s = \sigma + j\omega$ is a complex frequency. The function $f(t)$ becomes $F(s)$, a function of s, according to the following integral relation:

$$\mathcal{L}\{f(t)\} = F(s) = \int_0^\infty f(t)e^{-st}dt \tag{15.1.1}$$

The following should be noted concerning this definition:

1. s assumes values in the complex s-plane whose real, horizontal axis is σ, and whose vertical, imaginary axis is $j\omega$. Theoretically, σ is chosen greater than some constant c so that multiplying $f(t)$ by $e^{-\sigma t}$ ensures convergence of the integral, that is, gives a finite value of the integral. The region in the s-plane where $\sigma > c$ is therefore known as the **region of convergence**. This convergence property ensures the existence of the LT for many, but not all, functions. For example, functions like t^t or e^{t^n} increase so rapidly with t that $t^t e^{-\sigma t}$ and $e^{t^n}e^{-\sigma t}$ do not tend to zero as $t \to \infty$, no matter how large σ is. Such functions do not have LTs.

2. Sufficient conditions for the existence of the LT are: (i) convergence of the integral from zero to infinity, and (ii) that the function is piecewise continuous on every finite interval of time for $t \geq 0$, which means that $f(t)e^{-st}$ is integrable over any such interval of time. Functions of interest in circuit analysis normally possess LTs.

3. Because of the restriction of the range of integration in Equation 15.1.1 to $t \geq 0$, the LT is *one-sided* or *unilateral*. For a function $f(t)$ defined for $t < 0$, what is being derived, in effect, is the LT of $f(t)u(t)$. This is not a limitation in circuit applications because it allows inclusion of initial conditions, as discussed later.

4. The inverse LT (ILT) transforms the function $F(s)$ back to $f(t)$ according to the following integral relation:

$$\mathcal{L}^{-1}\{F(s)\} = f(t) = \frac{1}{2\pi j}\int_{\sigma - j\infty}^{\sigma + j\infty} F(s)e^{st}ds \tag{15.1.2}$$

5. The LT of a given function, as defined by Equation 15.1.1, is unique. Conversely, if two functions have the same LT, these functions cannot differ over any time interval of finite length, although they may differ at isolated points. Functions having isolated points in time are not of practical interest.

6. An important question in connection with Equation 15.1.1 is whether the lower limit of integration is 0^- or 0^+. The distinction is important, for example, when $f(t)$ has an impulse at the origin. The impulse is included in the integration if the lower limit is 0^- but not if it is 0^+. In order to include impulse functions at the origin, we will consider the lower limit of integration to be 0^-. The LT of $\delta(t)$ can then be derived quite simply as:

$$\mathcal{L}\{\delta(t)\} = \int_{0^-}^{\infty} \delta(t)e^{-st}dt = \int_{0^-}^{0^+} \delta(t).1.dt = 1 \qquad (15.1.3)$$

The LT is, in general, derived in the following ways:

1. Direct evaluation of the integral, as in the case of the impulse function (Equation 15.1.3). Another useful example is the LT of the exponential function $e^{-at}u(t)$, where a is a real or imaginary constant:

$$\mathcal{L}\left\{e^{-at}u(t)\right\} = \int_{0^-}^{\infty} e^{-at}e^{-st}dt = \frac{1}{s+a} \qquad (15.1.4)$$

2. From known LTs of functions, using the properties of the LT discussed in Section 15.2.
3. Look-up in tables of LT pairs. Extensive tables of the LTs of many types of functions and their inverses are available.
4. Use of MATLAB's `laplace` command. For example, if we enter:

    ```
    >>syms a t
    >>laplace(exp(-a*t))
    ```

    MATLAB returns: `1/(s+a)`

15.2 Properties of the Laplace Transform

Multiplication by a Constant

If $\mathcal{L}\left\{f(t)\right\} = F(s)$, then:

$$\mathcal{L}\left\{Kf(t)\right\} = KF(s) \qquad (15.2.1)$$

PROOF

$$\mathcal{L}\left\{Kf(t)\right\} = \int_{0^-}^{\infty} Kf(t)e^{-st}dt = K\int_{0^-}^{\infty} f(t)e^{-st}dt = KF(s)$$

Thus, the LT of $K\delta(t)$ is K, and the LT of $Ke^{-at}u(t)$ is $\dfrac{K}{s+a}$.

Addition/Subtraction

If $\mathcal{L}\left\{f_1(t)\right\} = F_1(s)$, and $\mathcal{L}\left\{f_2(t)\right\} = F_2(s)$, then:

$$\mathcal{L}\left\{f_1(t) \pm f_2(t)\right\} = F_1(s) \pm F_2(s) \qquad (15.2.2)$$

PROOF

$$\mathcal{L}\left\{f_1(t) \pm f_2(t)\right\} = \int_{0^-}^{\infty}\left[f_1(t) \pm f_2(t)\right]e^{-st}dt = \int_{0^-}^{\infty} f_1(t)e^{-st}dt \pm \int_{0^-}^{\infty} f_2(t)e^{-st}dt = F_1(s) \pm F_2(s)$$

We can use this property to derive the LTs of $\cos\omega t$ and $\sin\omega t$, using Equation 15.1.4:

$$\mathcal{L}\left\{\cos\omega t\right\} = \mathcal{L}\left\{\frac{e^{j\omega t} + e^{-j\omega t}}{2}\right\} = \frac{1}{2}\left[\frac{1}{s - j\omega} + \frac{1}{s + j\omega}\right] = \frac{s}{s^2 + \omega^2} \qquad (15.2.3)$$

$$\mathcal{L}\left\{\sin\omega t\right\} = \mathcal{L}\left\{\frac{e^{j\omega t} - e^{-j\omega t}}{2j}\right\} = \frac{1}{2j}\left[\frac{1}{s - j\omega} - \frac{1}{s + j\omega}\right] = \frac{\omega}{s^2 + \omega^2} \qquad (15.2.4)$$

EXERCISE 15.2.1
Determine the LT of $\cos\omega t$ and $\sin\omega t$ from Equation 15.1.1 using integration by parts.

Time Scaling

If $\mathcal{L}\{f(t)\} = F(s)$, then:

$$\mathcal{L}\left\{f(at)\right\} = \frac{1}{a}F\left(\frac{s}{a}\right), \text{ where } a > 0 \qquad (15.2.5)$$

PROOF

$$\mathcal{L}\left\{f(at)\right\} = \int_{0^-}^{\infty} f(at)e^{-st}dt$$

Changing the variable of integration to $x = at$,

$$\mathcal{L}\left\{f(at)\right\} = \frac{1}{a}\int_{0^-}^{\infty} f(x)e^{-\frac{s}{a}x}dx = \frac{1}{a}F\left(\frac{s}{a}\right)$$

The restriction $a > 0$ ensures that $f(at)$ is defined only for $t > 0$.

Integration

If $\mathcal{L}\left\{f(t)\right\} = F(s)$, then:

$$\mathcal{L}\left\{\int_{0^-}^t f(x)dx\right\} = \frac{1}{s}F(s) \qquad (15.2.6)$$

PROOF

$$\mathcal{L}\left\{\int_{0^-}^t f(x)dx\right\} = \int_{0^-}^\infty \left[\int_{0^-}^t f(x)dx\right] e^{-st}\ dt$$

Let

$$\int_{0^-}^t f(x)dx = u \quad \text{and} \quad e^{-st}\ dt = dv$$

Integrating by parts:

$$\int_{0^-}^\infty \left[\int_{0^-}^t f(x)dx\right] e^{-st}\ dt = \left[-\frac{e^{-st}}{s}\int_{0^-}^t f(x)dx\right]_{t=0^-}^{t=\infty} + \frac{1}{s}\int_{0^-}^t f(t)e^{-st}dt$$

The first term vanishes at both limits, and the second term is $\dfrac{1}{s}F(s)$.

To evaluate the LT of an indefinite integral, let $g(t) = \int f(t)\ dt = \int_{0^-}^t f(x)\ dx + g(0^-)$, then using Equation 15.2.6 and Equation 15.2.9:

$$\mathcal{L}\left\{\int f(t)dt\right\} = \frac{1}{s}F(s) + \frac{1}{s}g(0^-) \qquad (15.2.7)$$

Example 15.2.1: LT of Powers of t

It is required to prove that:

$$\mathcal{L}\left\{t^n u(t)\right\} = \frac{n\ !}{s^{n+1}} \qquad (15.2.8)$$

SOLUTION

We will successively apply the integration property, starting with $\mathcal{L}\left\{\delta(t)\right\} = 1$ (Equation 15.1.3)

$$\mathcal{L}\left\{u(t)\right\} = \mathcal{L}\left\{\int_{0^-}^t \delta(x)dx\right\} = \frac{1}{s} \qquad (15.2.9)$$

and

$$\mathcal{L}\left\{ tu(t) \right\} = \mathcal{L}\left\{ \int_{0^-}^{t} u(x)dx \right\} = \frac{1}{s}\cdot\frac{1}{s} = \frac{1}{s^2}, \quad \mathcal{L}\left\{ t^2 u(t) \right\} = 2\mathcal{L}\left\{ \int_{0^-}^{t} xdx \right\} = 2\cdot\frac{1}{s}\cdot\frac{1}{s^2} = \frac{2}{s^3},$$

$$\mathcal{L}\left\{ t^3 u(t) \right\} = 3\mathcal{L}\left\{ \int_{0^-}^{t} x^2 dx \right\} = 3\cdot\frac{1}{s}\cdot\frac{2}{s^3} = \frac{6}{s^4}, \text{ and so on.}$$

EXERCISE 15.2.2

Determine $\mathcal{L}\left\{ \int_{0^-}^{t} xdx \right\}$ by first integrating and then taking the LT.

Differentiation

If $\mathcal{L}\left\{ f(t) \right\} = F(s)$, then:

$$\mathcal{L}\left\{ \frac{df(t)}{dt} \right\} = sF(s) - f(0^-) \tag{15.2.10}$$

and in general:

$$\mathcal{L}\left\{ \frac{df^n(t)}{dt^n} \right\} = s^n F(s) - s^{n-1} f(0^-) - s^{n-2} f^{(1)}(0^-) - \ldots - f^{(n-1)}(0^-) \tag{15.2.11}$$

where $f^{(m)}(0^-)$ is the *m*th derivative of $f(t)$ evaluated at $t = 0^-$.

PROOF

$$\mathcal{L}\left\{ \frac{df(t)}{dt} \right\} = \int_{0^-}^{\infty} \frac{df(t)}{dt} e^{-st}\, dt = \left[f(t)e^{-st} \right]_{0^-}^{\infty} - \int_{0^-}^{\infty} (-s)f(t)e^{-st}dt = sF(s) - f(0^-)$$

Note that $f(t)e^{-st}$ evaluates to zero as $t \to \infty$ because $f(t)$ has an LT of $F(s)$.

Similarly, if $g(t) = \dfrac{df(t)}{dt}$, then using Equation 15.2.10, $\mathcal{L}\left\{ \dfrac{dg(t)}{dt} \right\} = s\mathcal{L}\left\{ g(t) \right\} - g(0^-)$. But

$\dfrac{dg(t)}{dt} = \dfrac{d^2 f(t)}{dt^2}$, $\mathcal{L}\left\{ g(t) \right\} = \mathcal{L}\left\{ \dfrac{df(t)}{dt} \right\} = sF(s) - f(0^-)$, and $g(0^-) = f^{(1)}(0^-)$. Substituting term by

term: $\mathcal{L}\left\{ \dfrac{d^2 f(t)}{dt^2} \right\} = s^2 F(s) - sf(0^-) - f^{(1)}(0^-)$. Successive applications of the differentiation property lead to Equation 15.2.11.

According to the differentiation property, $\mathcal{L}\left\{ \dfrac{d\delta(t)}{dt} \right\} = s$, $\mathcal{L}\left\{ \dfrac{d^2\delta(t)}{dt^2} \right\} = s^2$, and:

$$\mathcal{L}\left\{ \frac{d\delta^n(t)}{dt^n} \right\} = s^n \tag{15.2.12}$$

Example 15.2.2: Application of Differentiation and Integration Properties

It is required to verify that the LTs of $\sin\omega t$ and $\cos\omega t$, as given by Equation 15.2.3 and Equation 15.2.4, satisfy the integration and differentiation properties.

SOLUTION

From the differentiation property:

$$\mathcal{L}\left\{\frac{d}{dt}(\cos\omega t)\right\} = s\mathcal{L}\left\{(\cos\omega t)u(t)\right\} - \cos\omega t\big|_{t=0^-} = s\mathcal{L}\left\{(\cos\omega t)u(t)\right\} - 1 \qquad (15.2.13)$$

But $\dfrac{d}{dt}(\cos\omega t) = -\omega\sin\omega t$. Equation 15.2.13 gives:

$$-\omega\mathcal{L}\left\{(\sin\omega t)u(t)\right\} = s\mathcal{L}\left\{(\cos\omega t)u(t)\right\} - 1 \qquad (15.2.14)$$

or $-\dfrac{\omega^2}{s^2+\omega^2} = \dfrac{s^2}{s^2+\omega^2} - 1 = -\dfrac{\omega^2}{s^2+\omega^2}$. Similarly:

$$\mathcal{L}\left\{\frac{d}{dt}(\sin\omega t)\right\} = s\mathcal{L}\left\{(\sin\omega t)u(t)\right\} - \sin\omega t\big|_{t=0^-} = s\mathcal{L}\left\{(\sin\omega t)u(t)\right\} \qquad (15.2.15)$$

But $\dfrac{d}{dt}(\sin\omega t) = \omega\cos\omega t$. Substituting in Equation 15.2.15:

$$\omega\mathcal{L}\left\{(\cos\omega t)u(t)\right\} = s\mathcal{L}\left\{(\sin\omega t)u(t)\right\} \qquad (15.2.16)$$

or $\dfrac{s\omega}{s^2+\omega^2} = \dfrac{s\omega}{s^2+\omega^2}$.

From the integration property: $\mathcal{L}\left\{\left[\displaystyle\int_{0^-}^{t}\cos\omega t\,dt\right]u(t)\right\} = \dfrac{1}{s}\mathcal{L}\left\{(\cos\omega t)u(t)\right\}$. But $\displaystyle\int_{0^-}^{t}\cos\omega t\,dt$

$\dfrac{1}{\omega}\sin\omega t$. Hence:

$$\frac{1}{\omega}\mathcal{L}\left\{(\sin\omega t)u(t)\right\} = \frac{1}{s}\mathcal{L}\left\{(\cos\omega t)u(t)\right\} \qquad (15.2.17)$$

or $\dfrac{1}{s^2+\omega^2} = \dfrac{1}{s^2+\omega^2}$. Similarly, $\mathcal{L}\left\{\left[\displaystyle\int_{0^-}^{t}\sin\omega t\,dt\right]u(t)\right\} = \dfrac{1}{s}\mathcal{L}\left\{(\sin\omega t)u(t)\right\}$. But $\displaystyle\int_{0^-}^{t}\sin\omega t\,dt =$

$\dfrac{1}{\omega}(1-\cos\omega t)$. It follows that:

$$\mathscr{L}\left\{\frac{1}{\omega}\left[1-\cos\omega t\right]u(t)\right\} = \frac{1}{\omega s} - \frac{1}{\omega}\mathscr{L}\left\{(\cos\omega t)u(t)\right\} = \frac{1}{s}\mathscr{L}\left\{(\sin\omega t)u(t)\right\}$$

or

$$1 - s\,\mathscr{L}\left\{(\cos\omega t)u(t)\right\} = \omega\,\mathscr{L}\left\{(\sin\omega t)u(t)\right\} \tag{15.2.18}$$

which gives: $1 - \dfrac{s^2}{s^2 + \omega^2} = \dfrac{\omega^2}{s^2 + \omega^2}$. It is seen that the LTs of $\sin\omega t$ and $\cos\omega t$, as given by Equation 15.2.3 and Equation 15.2.4, satisfy Equation 15.2.14 and Equation 15.2.16 through Equation 15.2.18.

This example emphasizes that whereas the LT of $f(t)$ is *ipso facto* the LT of $f(t)u(t)$, the initial value of $f(t)$ at $t = 0^-$ is used when substituting for $f(0^-)$, in accordance with Equation 15.2.10. Thus, in Equation 15.2.13, we substitute the value of $\cos\omega t$ at $t = 0^-$, which is 1, and not the value of $\cos\omega t u(t)$, which is 0.

Translation in s-Domain

If $\mathscr{L}\left\{f(t)\right\} = F(s)$, then:

$$\mathscr{L}\left\{e^{-at}f(t)\right\} = F(s + a) \tag{15.2.19}$$

PROOF

$$\mathscr{L}\left\{e^{-at}f(t)\right\} = \int_{0^-}^{\infty} e^{-at}f(t)e^{-st}dt = K\int_{0^-}^{\infty} f(t)e^{-(s+a)t}dt = F(s + a)\ .$$

Equation 15.2.19 allows easy derivation of LTs of functions multiplied by e^{-at}. For example, $\mathscr{L}\left\{e^{-at}u(t)\right\} = \dfrac{1}{s+a}$ (Equation 15.1.4), given that $\mathscr{L}\left\{u(t)\right\} = \dfrac{1}{s}$ (Equation 15.2.7). It also follows that:

$$\mathscr{L}\left\{e^{-at}\cos\omega t\right\} = \frac{s+a}{\left(s+a\right)^2 + \omega^2} \tag{15.2.20}$$

and

$$\mathscr{L}\left\{e^{-at}\sin\omega t\right\} = \frac{\omega}{\left(s+a\right)^2 + \omega^2} \tag{15.2.21}$$

Translation in Time

If $\mathscr{L}\{f(t)\} = F(s)$, then:

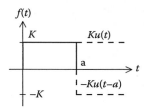

FIGURE 15.2.1
Rectangular pulse.

$$\mathscr{L}\left\{f(t-a)u(t-a)\right\} = e^{-as}F(s), \text{ where } a \text{ is a positive constant} \tag{15.2.22}$$

PROOF

$$\mathscr{L}\left\{f(t-a)u(t-a)\right\} = \int_{0^-}^{\infty} f(t-a)u(t-a)e^{-st}dt = \int_{a^-}^{\infty} f(t-a)e^{-st}dt$$

Substituting $x = t - a$:

$$\mathscr{L}\left\{f(t-a)u(t-a)\right\} = \int_{0^-}^{\infty} f(x)e^{-as}e^{-xt}dx = e^{-as}\int_{0^-}^{\infty} f(t)e^{-st}dt = e^{-as}F(s)$$

where the dummy variable of integration was changed back from x to t. Note the similarity to the Fourier series, where a delay t_d multiplied C_n by $e^{-jn\omega_0 t_d}$ (Equation 9.2.23, Chapter 9).
As a simple example, consider the rectangular pulse of Figure 15.2.1. The pulse can be expressed as: $f(t) = Ku(t) - Ku(t-a)$. As $\mathscr{L}\left\{Ku(t)\right\} = \dfrac{K}{s}$ and $\mathscr{L}\left\{Ku(t-a)\right\} = \dfrac{K}{s}e^{-as}$:

$$\mathscr{L}\left\{f(t)\right\} = \frac{K}{s} - \frac{K}{s}e^{-as} = \frac{K}{s}\left(1 - e^{-as}\right) \tag{15.2.23}$$

The translation-in-time property can be used to determine the LT of a periodic function $f(t)$ of period T that starts at $t = 0$, given that the LT of the first period is $G(s)$. As periods after the first are successively translated by T:

$$\mathscr{L}\left\{f(t)\right\} = G(s)\left[1 + e^{-sT} + e^{-2sT} + \ldots\right] = \frac{G(s)}{1 - e^{-sT}} \tag{15.2.24}$$

We also have:

$$\mathscr{L}\left\{\delta(t-a)\right\} = e^{-as} \tag{15.2.25}$$

Note that $f(t-a)u(t-a)$ is the function $f(t)u(t)$ shifted by a and not the function $f(t)$ shifted by a. This distinction is illustrated, for example, by the LT of periodic functions or of a single half sinusoid (Exercise 15.2.5). The translation-in-time property can be generalized to:

$$\mathcal{L}\left\{f(t-b)u(t-a)\right\} = e^{-as}\,\mathcal{L}\left\{f(t-b+a)\right\} \tag{15.2.26}$$

(Problem P15.1.16). It follows that $\mathcal{L}\left\{f(t)u(t-a)\right\} = e^{-as}\,\mathcal{L}\left\{f(t+a)\right\}$.

Multiplication by t

If $\mathcal{L}\left\{f(t)\right\} = F(s)$, then:

$$\mathcal{L}\left\{tf(t)\right\} = -\frac{dF(s)}{ds} \tag{15.2.27}$$

and

$$\mathcal{L}\left\{t^n f(t)\right\} = (-1)^n \frac{dF^n(s)}{ds^n} \tag{15.2.28}$$

PROOF

$$\mathcal{L}\left\{tf(t)\right\} = \int_{0^-}^{\infty} tf(t)e^{-st}dt = -\int_{0^-}^{\infty} f(t)\frac{d}{ds}(e^{-st})dt = -\frac{d}{ds}\int_{0^-}^{\infty} f(t)e^{-st}dt = -\frac{dF(s)}{ds}$$

Hence,

$$\mathcal{L}\left\{t.tf(t)\right\} = -\frac{d}{ds}[\mathcal{L}\left\{tf(t)\right\}] = \frac{d^2F(s)}{ds^2}, \text{ and so on.}$$

We can deduce that:

$$\mathcal{L}\left\{t.\cos\omega t\right\} = -\frac{d}{ds}\left[\frac{s}{s^2+\omega^2}\right] = \frac{s^2-\omega^2}{\left(s^2+\omega^2\right)^2} \tag{15.2.29}$$

$$\mathcal{L}\left\{t.\sin\omega t\right\} = -\frac{d}{ds}\left[\frac{\omega}{s^2+\omega^2}\right] = \frac{2\omega s}{\left(s^2+\omega^2\right)^2} \tag{15.2.30}$$

Division by t

If $\mathcal{L}\left\{f(t)\right\} = F(s)$, then, assuming that $\dfrac{f(t)}{t}$ has an LT:

$$\mathcal{L}\left\{\frac{f(t)}{t}\right\} = \int_{s}^{\infty} F(x)dx \tag{15.2.31}$$

PROOF

$$\int_s^\infty F(u)du = \int_s^\infty \left[\int_{0^-}^\infty f(t)e^{-ut}dt \right] du$$

Reversing the order of integration:

$$\int_s^\infty F(u)du = \int_{0^-}^\infty \left[\int_s^\infty f(t)e^{-ut}du \right] dt = \int_{0^-}^\infty \left[-\frac{f(t)}{t} e^{-ut} \right]_s^\infty dt = \int_{0^-}^\infty \left[\frac{f(t)}{t} e^{-st} \right] dt$$

For example, $\mathcal{L}\left\{u(t)\right\} = \mathcal{L}\left\{ \dfrac{tu(t)}{t} \right\} = \int_s^\infty \dfrac{1}{x^2} dx = \dfrac{1}{s}.$

EXERCISE 15.2.3
Verify Equation 15.2.29 and Equation 15.2.30 by direct evaluation of the LT (Equation 15.1.1).

EXERCISE 15.2.4

From the LT of the square pulse of Figure 15.2.2a, deduce the LT of the single sawtooth pulse of Figure 15.2.2b using: (a) the integration property; (b) the multiplication-by-t property. (Note: do not forget $u(t-1)$ that must be introduced.)

Answer: $A\left[-\dfrac{e^{-sT}}{s} + \dfrac{1-e^{-sT}}{Ts^2} \right].$

EXERCISE 15.2.5

Determine the LT of a half-sinusoid defined by $f(t) = \sin t,\ 0 \le t \le \pi,$ and $f(t) = 0,$ $t \ge \pi,$ using: (a) translation in the time domain; (b) direct evaluation, considering $\sin t = \text{Im}(e^{jt})$. Note that in (a) $\sin t$ is added to $\sin(t-\pi)u(t-\pi)$ and not to $\sin(t-\pi)$.

Answer: $\dfrac{1+e^{-\pi s}}{s^2+1}$

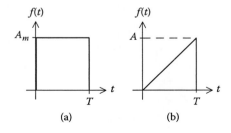

(a)   (b)

FIGURE 15.2.2
Figure for Exercise 15.2.4.

The reader may have noted that in Section 15.1 we referred to the region of convergence in the s-plane, which assured convergence of the integral (Equation 15.1.1), yet no reference to this region of convergence was made in subsequent discussions of the LT. In fact, the validity of the unilateral LT is not restricted to the region of convergence but extends to the whole of the s-plane except at the poles (Section 15.3), where the LT becomes infinite.

Table 15.2.1 summarizes the basic properties of the LT.

TABLE 15.2.1

Basic Properties of the Laplace Transform $\mathcal{L}\{f(t)\} = F(s)$

$\mathcal{L}\{Kf(t)\}$	$KF(s)$
$\mathcal{L}\{f_1(t) \pm f_2(t)\}$	$F_1(s) \pm F_2(s)$
$\mathcal{L}\{f(at)\},\quad a > 0$	$\dfrac{1}{a}F\left(\dfrac{s}{a}\right)$
$\mathcal{L}\left\{\displaystyle\int_{0^-}^{t} f(x)dx\right\}$	$\dfrac{1}{s}F(s)$
$\mathcal{L}\left\{\displaystyle\int_{0^-}^{t} f(x)dx + f(0^-)\right\}$	$\dfrac{1}{s}F(s) + \dfrac{1}{s}f(0^-)$
$\mathcal{L}\left\{\dfrac{df(t)}{dt}\right\}$	$sF(s) - f(0^-)$
$\mathcal{L}\left\{\dfrac{df^n(t)}{dt^n}\right\}$	$s^n F(s) - s^{n-1}f(0^-) - s^{n-2}f^{(1)}(0^-) - \ldots - f^{(n-1)}(0^-)$
$\mathcal{L}\{e^{-at}f(t)\}$	$F(s+a)$
$\mathcal{L}\{tf(t)\}$	$-\dfrac{dF(s)}{ds}$
$\mathcal{L}\{t^n f(t)\}$	$(-1)^n \dfrac{dF^n(s)}{ds^n}$
$\mathcal{L}\left\{\dfrac{f(t)}{t}\right\}$	$\displaystyle\int_{s}^{\infty} F(u)du$
$\mathcal{L}\{f(t-a)u(t-a)\}$	$e^{-as}F(s)$
$\mathcal{L}\{f(t-b)u(t-a)\}$	$e^{-as}\,\mathcal{L}\{f(t-b+a)\}$
$\mathcal{L}\{f(t)\}$, where $f(t)$ is periodic of period T	$\dfrac{G(s)}{1-e^{-sT}}$, where $G(s)$ is LT of first period
$\mathcal{L}\{f(t) * g(t)\}$	$F(s)G(s)$

15.3 Solution of Ordinary, Linear Differential Equations

To illustrate the application of the LT to the solution of ordinary, linear differential equations, we consider, for simplicity and without loss of generality, a second-order differential equation of the form:

$$a\frac{d^2y(t)}{dt^2} + b\frac{dy(t)}{dt} + cy(t) = f(t) \tag{15.3.1}$$

where $f(t)$ is some arbitrary function that has an LT. Taking the LT of both sides, making use of the differentiation property:

$$as^2Y(s) - asy(0^-) - ay^{(1)}(0^-) + bsY(s) - by(0^-) + cY(s) = F(s) \tag{15.3.2}$$

where $Y(s)$ and $F(s)$ are the LTs of $y(t)$ and $f(t)$, respectively. Solving for $Y(s)$:

$$Y(s) = \frac{(as+b)y(0^-) + ay^{(1)}(0^-)}{as^2 + bs + c} + \frac{F(s)}{as^2 + bs + c} \tag{15.3.3}$$

The following should be noted in connection with Equation 15.3.3, these being important features of the LT method:

> **Concept:** *A linear differential equation is transformed by the LT to an algebraic equation in powers of s that can be solved for the LT of the variable of the equation Y(s), as in any algebraic equation. When this is done, the initial conditions y(0⁻) and y⁽¹⁾(0⁻) appear like applied inputs.*

Formally, the solution to the differential equation, $y(t)$, can be expressed as:

$$y(t) = \mathcal{L}^{-1}\left\{ \frac{(as+b)y(0^-) + ay^{(1)}(0^-)}{as^2 + bs + c} \right\} + \mathcal{L}^{-1}\left\{ \frac{F(s)}{as^2 + bs + c} \right\} \tag{15.3.4}$$

where $\mathcal{L}^{-1}$ denotes the ILT.

The first term on the RHS of Equation 15.3.4 is the response $y(t)$ for nonzero initial conditions, with no forcing function. It is the **natural response,** or **zero-input response.** The second term is the response $y(t)$ to the forcing function alone. It is the **forced response,** or **zero-state response.** The steady-state response is the forced response after the transient dies out. Because the system is linear, the total response is the sum of the natural and forced responses, the initial conditions being treated as inputs at $t = 0$ (Section 15.5). The denominator $as^2 + bs + c$ is analogous, term by term, to the LHS of Equation 15.3.1 with the first and second derivatives replaced by s and s^2, respectively. The equation $as^2 + bs + c = 0$ is the **characteristic equation** previously encountered in Equation 12.2.5.

> **Concept:** *The characteristic equation of a linear differential equation is a polynomial in s obtained by taking the LT of the equation with zero forcing function, zero initial conditions, and assuming a nonzero value of the variable of the equation. The LHS of the characteristic equation appears in all the responses derived from the differential equation.*

Thus, if we take the LT of Equation 15.3.1, with $f(t) = 0$ and zero initial conditions, we obtain: $(as^2 + bs + c)Y(s) = 0$. If $Y(s) \neq 0$, then $as^2 + bs + c = 0$ is the characteristic equation. If $Y(s)$ is, say, the current $I(s)$ in a series RLC circuit, then $V_R(s) = RI(s)$, $V_L(s) = sLI(s)$, and $V_C(s) = I(s)/sC$. It follows that the expression $as^2 + bs + c$ will also appear in the denominators of $V_R(s)$, $V_L(s)$, and $V_C(s)$.

Inverse Laplace Transform

The ILT can be obtained in one of the following ways:

1. Look-up in tables of LT pairs, as previously mentioned.
2. Numerical inversion of the LT.
3. The LTs of responses of lumped-parameter LTI circuits are usually rational functions of s. These can be expressed, at least in part, as proper rational functions. Such functions can be expanded as partial fractions, which are then inverted term by term.
4. Use of MATLAB's ilaplace command. For example, if we enter:

   ```
   >> syms s
   >> ilaplace (1/s^5)
   ```
   MATLAB returns: 1/24*t^4

For convenience in determining ILTs, Table 15.3.1 summarizes the LTs of some commonly encountered functions in circuit analysis, based on the properties of the LT discussed in the previous section.

Partial Fraction Expansion

Consider an $F(s)$ that is a rational function of s, that is, the ratio of two polynomials in s:

$$F(s) = \frac{a_m s^m + a_{m-1} s^{m-1} + \ldots + a_1 s + a_0}{b_n s^n + b_{n-1} s^{n-1} + \ldots + b_1 s + b_0} \tag{15.3.5}$$

where the a's and b's are real coefficients, and m and n are integers. $F(s)$ can be expressed in terms of factors of the numerator and denominator as:

$$F(s) = K \frac{(s + z_1)(s + z_2)\ldots(s + z_m)}{(s + p_1)(s + p_2)\ldots(s + p_n)} \tag{15.3.6}$$

where $K = a_m / b_n$. The values $-z_1, -z_2, \ldots, -z_m$ are **zeros** of $F(s)$, because $F(s) = 0$ when s assumes any of these values. The values $-p_1, -p_2, \ldots, -p_n$ are **poles** of $F(s)$, because $F(s) \to \infty$ when s assumes any of these values. In general, zeros and poles may be complex, in which case they must occur in complex conjugate pairs, because the a's and b's are real in the LTs of physical systems.

In general, $m > n$ in Equation 15.3.5, so by dividing the numerator by the denominator, $F(s)$ can be expressed as:

TABLE 15.3.1

Laplace Transform Pairs

$f(t)$	$F(s)$	Reference Equation
$\delta(t)$	1	15.1.3
$\delta(t-a)$	e^{-as}	15.2.25
$\delta^{(n)}(t)$	s^n	15.2.12
$u(t)$	$\dfrac{1}{s}$	15.2.9
$tu(t)$	$\dfrac{1}{s^2}$	15.2.8
$t^n u(t)$	$\dfrac{n!}{s^{(n+1)}}$	15.2.8
$e^{-at}u(t)$	$\dfrac{1}{s+a}$	15.1.4
$te^{-at}u(t)$	$\dfrac{1}{\left(s+a\right)^2}$	15.1.4 and 15.2.19
$t^n e^{-at}u(t)$	$\dfrac{n!}{\left(s+a\right)^{(n+1)}}$	15.1.4 and 15.2.19
$\sin\omega t\, u(t)$	$\dfrac{\omega}{s^2+\omega^2}$	15.2.4
$\cos\omega t\, u(t)$	$\dfrac{s}{s^2+\omega^2}$	15.2.3
$\sin\left(\omega t+\theta\right)u(t)$	$\dfrac{s\sin\theta+\omega\cos\theta}{s^2+\omega^2}$	15.2.3 and 15.2.4
$\cos\left(\omega t+\theta\right)u(t)$	$\dfrac{s\cos\theta-\omega\sin\theta}{s^2+\omega^2}$	15.2.3 and 15.2.4
$e^{-at}\sin\omega t\, u(t)$	$\dfrac{\omega}{\left(s+a\right)^2+\omega^2}$	15.2.21
$e^{-at}\cos\omega t\, u(t)$	$\dfrac{s+a}{\left(s+a\right)^2+\omega^2}$	15.2.20
$t\sin\omega t\, u(t)$	$\dfrac{2\omega s}{\left(s^2+\omega^2\right)^2}$	15.2.30
$t\cos\omega t\, u(t)$	$\dfrac{s^2-\omega^2}{\left(s^2+\omega^2\right)^2}$	15.2.29
$te^{-at}\sin\omega t\, u(t)$	$\dfrac{2\omega\left(s+a\right)}{\left(\left(s+a\right)^2+\omega^2\right)^2}$	15.2.19 and 15.2.30

Continued.

TABLE 15.3.1 *(Continued)*

Laplace Transform Pairs

$f(t)$	$F(s)$	Reference Equation
$te^{-at}\cos\omega t u(t)$	$\dfrac{(s+a)^2 - \omega^2}{\left((s+a)^2 + \omega^2\right)^2}$	15.2.19 and 15.2.29
$\dfrac{2\lvert K\rvert t^{n-1}}{(n-1)!}e^{-\alpha t}\cos(\omega t + \theta),\ K=\lvert K\rvert e^{j\theta}$	$\dfrac{K}{(s+\alpha-j\omega)^n} + \dfrac{K^*}{(s+\alpha+j\omega)^n}$	15.3.13

$$F(s) = k_{m-n}s^{m-n} + k_{m-n-1}s^{m-n-1} + \ldots + k_1 s + k_0 + \frac{c_r s^r + c_{r-1}s^{r-1} + \ldots + c_1 s + c_0}{b_n s^n + b_{n-1}s^{n-1} + \ldots + b_1 s + b_0} \quad (15.3.7)$$

where $r < n$, so that the ratio of polynomials on the RHS is a *proper* rational function. The inverse transform of k_0 is $k_0\delta(t)$, that of $k_1 s$ is $k_1 \delta^{(1)}(t)$, and that of $k_{m-n}s^{m-n}$ is $k_{m-n}\delta^{(m-n)}(t)$ (Table 15.3.1). It remains to determine the ILT of the proper rational function, which we will denote as $\dfrac{N(s)}{D(s)}$.

Most textbooks discuss methods for deriving the partial fraction expansion (PFE) of proper rational functions. These methods are explained in Section ST15.2 and are based on examples in which the coefficients of s in $N(s)$ and $D(s)$ are relatively small whole numbers, and the power of s in $D(s)$ is usually 2 or 3. Such examples, however, are rather artificial, and the aforementioned methods become impractical in more realistic cases where the coefficients of s are not small whole numbers and the power of s in the denominator exceeds 3. In practice, the ILT is obtained in these cases using MATLAB's ilaplace command, without deriving the PFE, as illustrated by Example 15.3.2. Moreover, this command can be used irrespective of whether or not the rational function is proper. Nevertheless, the PFE is of fundamental importance. We will therefore consider in the following paragraphs some aspects of the PFE that are important for the discussion that follows.

Consider the proper rational function:

$$F(s) = \frac{N(s)}{(s+p_1)(s+p_2)\ldots(s+p_n)} \quad (15.3.8)$$

where $N(s)$ is a numerator of power less than n and all the poles have different, real values. $F(s)$ can be expressed in partial fraction form as:

$$F(s) = \frac{K_1}{s+p_1} + \frac{K_2}{s+p_2} + \ldots + \frac{K_n}{s+p_n} \quad (15.3.9)$$

where the K's are constant coefficients known as the **residues** of $F(s)$. An easy way of determining any of these coefficients, say K_1, is to multiply both sides of Equation 15.3.9 by the corresponding denominator, $(s+p_1)$ in this case, and set $s=-p_1$. The RHS of Equation 15.3.9 reduces simply to K_1. The LHS becomes:

$$\frac{N(s)}{(s+p_2)...(s+p_n)}\Big|_{s=-p_1}$$

which can be readily evaluated to give K_1. This is the residue method for determining the PFE.

EXERCISE 15.3.1

Determine the PFE of $F(s)=\dfrac{2s^2+8s+7}{(s+1)(2s+3)(s+2)}$ and derive its ILT.

Answer: $\dfrac{1}{s+1}+\dfrac{2}{2s+3}-\dfrac{1}{s+3}$; $f(t)=[e^{-t}+e^{-1.5t}-e^{-2t}]\,u(t)$.

Next, consider the proper rational function $F(s)=\dfrac{3s^2+8s+4}{(s+1)^3}$. To see how $F(s)$ can be represented by a PFE, we express the numerator in terms of powers of $(s + 1)$. Thus, $3s^2 + 8s + 4 = 3(s + 1)^2 + 2(s + 1) - 1$. Hence:

$$F(s)=\frac{3s^2+8s+4}{(s+1)^3}=\frac{3}{s+1}+\frac{2}{(s+1)^2}-\frac{1}{(s+1)^3} \tag{15.3.10}$$

Similar reasoning leads to the conclusion that if the denominator of $F(s)$ consists of the repeated root $(s + p)^n$, the PFE of $F(s)$ is of the general form:

$$F(s)=\frac{N(s)}{(s+p)^n}=\frac{K_1}{s+p}+\frac{K_2}{(s+p)^2}+...+\frac{K_{n-1}}{(s+p)^{n-1}}+\frac{K_n}{(s+p)^n} \tag{15.3.11}$$

The residue method used with Equation 15.3.11 can only determine K_n if both sides of Equation 15.3.11 are multiplied by $(s + p)^n$ and s is set equal to $-p$. Multiplying by $(s + p)$ raised to a power less than n and setting s equal to $-p$ makes the RHS infinite. The residues K_1 to K_{n-1} can be determined analytically either through differentiation, or by equating the coefficients of the same power of s on both sides of the equation (Section ST15.2). Alternatively, the residues, if required, can be obtained in more complicated cases using MATLAB's residue command, as illustrated in Example 15.3.1. Whereas the poles in Equation 15.3.9 are simple poles, $-p$ in Equation 15.3.11 is a multiple pole of order n.

The preceding roots considered are real, but the same considerations apply if the roots are complex, except that, as the coefficients in $N(s)$ and $D(s)$ are real, complex poles must occur in conjugate pairs and their residues must also be conjugate pairs. We will illustrate working with complex poles by deriving the ILT for a pair of complex conjugate poles of order n:

$$F(s)=\frac{K}{(s+\alpha-j\omega)^n}+\frac{K^*}{(s+\alpha+j\omega)^n} \tag{15.3.12}$$

Recall that $\mathcal{L}\{t^n e^{-at}u(t)\} = \dfrac{n!}{(s+a)^{(n+1)}}$ (Table 15.3.1). Hence, $\mathcal{L}^{-1}\left\{\dfrac{1}{(s+a)^n}\right\} = \dfrac{t^{n-1}e^{-at}}{(n-1)!}u(t)$,

where a could be real or complex. Taking the ILT of both sides of Equation (15.3.12) and substituting $a = \alpha - j\omega$ and $K = |K|e^{j\theta}$, in the first term, and $a = \alpha + jw$ and $K^* = |K|e^{-j\theta}$ in the second term, it follows that:

$$f(t) = \mathcal{L}^{-1}\{F(s)\} = \frac{|K|t^{n-1}e^{-\alpha t}}{(n-1)!}\left[e^{j(\omega t + \theta)} + e^{-j(\omega t + \theta)}\right]u(t)$$

$$= \frac{2|K|t^{n-1}}{(n-1)!}e^{-\alpha t}\cos(\omega t + \theta)u(t) \tag{15.3.13}$$

Example 15.3.1: ilaplace and Residue Commands of MATLAB

It is required to determine the ILT of $F(s) = \dfrac{25}{s(s^2 + 2s + 5)^2}$ and its residues using MATLAB.

Solution

We invoke MATLAB's ilaplace command by entering:

```
>> syms s
>> ilaplace (25/(s*(s^2+2*s+5)^2))
```

MATLAB returns: 1+(5/8*t-1)*exp(-t)*cos(2*t)+(-13/16-5/4*t)*sin(2*t)*exp(-t)

This can be rearranged as:

$$1 - e^{-t}\left[\cos 2t + \frac{13}{16}\sin 2t\right] + \frac{5}{8}te^{-t}\left[\cos 2t - 2\sin 2t\right]$$

MATLAB's residue command can be used to determine the residues of a rational function. The procedure is to enter, as appropriately named arrays, the coefficients of the polynomials in s of the numerator and denominator of the given rational function, including the constant terms. The residue command is then entered, with the named arrays of the numerator and denominator as arguments. The command returns an array r of the residues, a matched array p for the corresponding poles, and an array k for the quotient terms if the order of the polynomial in the numerator is not less than that of the denominator.

In the case of $F(s) = \dfrac{25}{s(s^2 + 2s + 5)^2}$, for example, the following code is entered:

```
>> num = [25]
>> den = [1 4 14 20 25 0]
```

```
>> [r,p,k] = residue(num, den)
```
MATLAB returns:

r = −0.5000 + 0.4063i, 0.3125 + 0.6250i, −0.5000 − 0.4063i, 0.3125 − 0.6250i, 1.0000

p = −1.0000 + 2.0000i, −1.0000 + 2.0000i, −1.0000 − 2.0000i, −1.0000 − 2.0000i, 0

k = []

The first element of p is the simple pole $s = -1 + 2j$ and the first element of r is the corresponding residue $K_1 = -\dfrac{1}{2} + j\dfrac{13}{32}$. The second element of p is the double pole $s = -1 + 2j$ and the second element of r is the corresponding residue $K_2 = \dfrac{5}{16}(1 + j2)$. Similarly for the third and fourth elements, which correspond to the conjugate poles. The last pole is that at the origin, and its residue is $K_0 = 1$. The array k is empty because the order of the numerator is less than that of the denominator. The PFE is:

$$\frac{25}{s(s^2 + 2s + 5)^2} = \frac{1}{s} - \frac{1/2 - j13/32}{s + 1 - j2} - \frac{1/2 + j13/32}{s + 1 + j2} + \frac{5}{16}\frac{1 + j2}{(s + 1 - j2)^2} + \frac{5}{16}\frac{1 - j2}{(s + 1 + j2)^2}$$

The residue command can be used to combine complex poles and their residues to obtain real terms. If, after obtaining the preceding result, we enter:

```
>> [num1, den1] = residue([r(1) r(3)], [p(1) p(3)], [])
```
MATLAB returns:

```
num1 = −1.0000 − 2.6250
den1 = 1.0000 2.0000 5.0000
```
The two simple poles combine to give:

$$-\frac{s + 21/8}{s^2 + 2s + 5}$$

To combine the double poles we enter, after obtaining the preceding result:

```
>> [num2, den2] = residue([0 r(2) 0 r(4)], [p(2) p(2) p(4) p(4)], [])
```
MATLAB returns:

```
num2 = 0 0.6250 −3.750 − 6.8750
den2 = 1.0000 4.0000 14.0000 20.0000 25.0000
```
The two double poles combine to give:

$$\frac{5}{8} \frac{s^2 - 6s - 11}{(s^2 + 2s + 5)^2}$$

Note that when a pole is entered as [p(2) p(2)], MATLAB interprets the first entry as a simple pole, the second entry as a double pole, etc. As we are interested in combining the double poles only, we enter the residue of the simple pole as zero.

EXERCISE 15.3.2

This is an exercise in working with s multiplied by integer powers of 10. Consider the function $F(s) = \dfrac{s + 10^3}{10^{-3} s^2 + 5s + 6 \times 10^3}$. If we substitute $s = 10^3 s'$, we obtain a more convenient function $F'(s') = \dfrac{s' + 1}{(s')^2 + 5s' + 6} = \dfrac{2}{s' + 3} - \dfrac{1}{s' + 2}$, where s' is in krad/s. The inverse of this function is $f'(t') = 2e^{-3t'} - e^{-2t'}$, where t' is in ms. But what we are really interested in is $f(t)$, the inverse of the original $F(s)$. We note that the inverse transform is

$$f(t) = \frac{1}{2\pi j} \int_{\sigma - j\infty}^{\sigma + j\infty} F(s) e^{st} ds \text{ (Equation 15.1.2). If we substitute } s = 10^3 s', \text{ ignoring the change}$$

in the limits at infinity, we obtain $f(t) = \dfrac{10^3}{2\pi j} \displaystyle\int_{\sigma - j\infty}^{\sigma + j\infty} F'(s') e^{10^3 s' t} ds' = \dfrac{10^3}{2\pi j} \displaystyle\int_{\sigma - j\infty}^{\sigma + j\infty} F'(s') e^{s' t'} ds' =$

$10^3 f'(t')$, where $f'(t')$ is the inverse transform of $F'(s')$ and t' is in ms. Hence, $f(t) =$

$2 \times 10^3 e^{-3t'} - 10^3 e^{-2t'}$. Verify that this result agrees with inverting the original $F(s)$.

If $F'(s')$ is multiplied by s' and the limit obtained as s' tends to infinity or zero, the result must also be multiplied by 10^3 in order to obtain the correct limiting value of $f(t)$.

15.4 Theorems on the Laplace Transform

Final Value Theorem

This is a useful theorem that gives the final value of a function from its LT without having to invert the transform.

If $\mathcal{L}\{f(t)\} = F(s)$, where the poles of $F(s)$ have negative real parts, except for a simple pole at the origin, if such a pole exists, then:

$$\lim_{t \to \infty} f(t) = \lim_{s \to 0} sF(s) \tag{15.4.1}$$

PROOF If $F(s)$ has only a simple pole at the origin, its PFE can be expressed as:

$$F(s) = k(s) + \frac{K_1}{s} + \dots + \frac{K_r}{\left(s + a_r\right)^m} + \dots \tag{15.4.2}$$

It follows from this equation that $\lim\limits_{s \to 0} sF(s) = K_1$. The ILT is:

$$f(t) = (\text{Impulse function and its derivatives}) + K_1 + \dots + \frac{K_r t^{m-1}}{(m-1)\,!} e^{-a_r t} + \dots \tag{15.4.3}$$

The impulse function and its derivatives are over at $t = 0^+$. As the poles have negative real parts, all terms having $e^{-a_r t}$ vanish as $t \to \infty$, and $f(t) \to K_1$.

Initial Value Theorem

If $\mathcal{L}\left\{f(t)\right\} = F(s)$ where $F(s)$ is a proper rational function, then:

$$\lim_{t \to 0^+} f(t) = \lim_{s \to \infty} sF(s) \tag{15.4.4}$$

PROOF

$$\mathcal{L}\left\{\frac{df(t)}{dt}\right\} = sF(s) - f(0^-) = \int_{0^-}^{\infty} \frac{df(t)}{dt} e^{-st} dt = \int_{0^-}^{0^+} \frac{df(t)}{dt} 1\, dt + \int_{0^+}^{\infty} \frac{df(t)}{dt} e^{-st} dt$$

If we take the limit as $s \to \infty$, the last integral vanishes. As $F(s)$ is a proper rational function, $f(t)$ does not have any impulses or their derivatives at the origin, and

$$\int_{0^-}^{0^+} \frac{df(t)}{dt} 1\, dt = f(0^+) - f(0^-).$$

It follows that:

$$\lim_{s \to \infty} sF(s) = f(0^-) + f(0^+) - f(0^-) = f(0^+).$$

If $F(s)$ is not a proper rational function, $f(t)$ has impulses or their derivatives at the origin. But these are over at $t = 0^+$. The correct value of $f(0^+)$ is obtained by applying the initial value theorem to the proper rational fraction part of $F(s)$.

The initial and final value theorems are useful for checking the LT for a given variable, because it is often obvious from the nature of the problem what the initial and final values of the variable are. The initial value theorem is particularly useful in the case of impulsive readjustments at $t = 0$, when the voltage across a capacitor, or the current through an inductor, is forced to change instantaneously. The initial value theorem gives the correct values at $t = 0^+$, after the impulsive readjustments are over, as illustrated by Example 15.4.1.

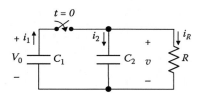

FIGURE 15.4.1
Figure for Example 15.4.1.

Example 15.4.1: Impulsive Response of Switched Capacitive Circuit

In the circuit of Figure 15.4.1, C_1 is initially charged to V_0 and C_2 is uncharged. The switch is closed at $t = 0$. It is required to determine the charge $q_1(t)$ at $t = 0^+$.

SOLUTION

Just after the switch is closed, the voltages across C_1 and C_2 are forced to change instantaneously from their initial values to a common voltage. This voltage can be determined from conservation of charge between $t = 0^-$ and $t = 0^+$, the charge due to a finite i_R being vanishingly small during this infinitesimal interval. The initial charge $q_1(0^-)$ is C_1V_0, where from conservation of charge, $q_1(0^-) = q_1(0^+) + q_2(0^+)$. The voltage $v(0^+)$ is such that:

$$v(0^+) = \frac{q_1(0^+)}{C_1} = \frac{q_2(0^+)}{C_2}$$

Solving for $q_1(0^+)$ gives:

$$q_1(0^+) = \frac{C_1^2}{C_1 + C_2} V_0 \qquad (15.4.5)$$

We wish to verify Equation 15.4.5 by deriving the LT of $q_1(t)$ and applying the initial value theorem. After the switch is closed ($t > 0$):

$$v = \frac{q_1}{C_1} = \frac{q_2}{C_2} \qquad (15.4.6)$$

$$i_1 = i_2 + i_R \quad \text{or} \quad -\frac{dq_1}{dt} = \frac{dq_2}{dt} + \frac{q_2}{C_2R} \qquad (15.4.7)$$

where $i_R = \dfrac{v}{R} = \dfrac{q_2}{C_2R}$, and the minus sign accounts for the fact that i_1 decreases q_1. Taking the LTs of Equation 15.4.6 and Equation 15.4.7 and substituting $q_1(0^-) = C_1V_0$ and $q_2(0^-) = 0$ in the expressions for the derivative:

$$\frac{Q_1(s)}{C_1} = \frac{Q_2(s)}{C_2} \quad \text{and} \quad -sQ_1(s) + C_1V_0 = \left(s + \frac{1}{CR_2}\right)Q_2(s) \tag{15.4.8}$$

where $Q_1(s)$ and $Q_2(s)$ are the LTs of $q_1(t)$ and $q_2(t)$, respectively. Solving for $Q_1(s)$:

$$Q_1(s) = \frac{C_1^2 V_0}{C_1 + C_2} \frac{1}{s + \dfrac{1}{(C_1 + C_2)R}} \tag{15.4.9}$$

If we multiply the RHS of Equation 15.4.9 by s and let $s \to \infty$, we obtain $q_1(0^+)$ as in Equation 15.4.5. If after multiplying by s we set $s = 0$, $q_1(\infty) = 0$, as the capacitors completely discharge through R. The initial value theorem could be applied to $Q_1(s)$ because although $q_1(t)$ is discontinuous at $t = 0$, it does not have an impulse at the origin.

If we derive $I_1(s) = -sQ_1(s) + C_1V_0$, we obtain:

$$I_1(s) = C_1V_0 \frac{sC_2R + 1}{s(C_1 + C_2)R + 1} \tag{15.4.10}$$

If the RHS is multiplied by s and we set $s = 0$, $i_1(\infty) = 0$, which is correct. But if we multiply s and let $s \to \infty$, then $I_1(s) \to \infty$. The initial value theorem cannot be applied to $I_1(s)$ as an improper rational function because of the impulse at the origin. If we divide the numerator in Equation 15.4.10 by the denominator, we obtain:

$$I_1(s) = \frac{C_1 C_2 V_0}{C_1 + C_2} + \frac{C_1^2 V_0}{C_1 + C_2} \frac{1}{s(C_1 + C_2)R + 1} \tag{15.4.11}$$

The first term on the RHS represents an impulse of strength $\dfrac{C_1 C_2 V_0}{C_1 + C_2}$. We can readily

check this value because the impulse transfers an amount of charge equal to $q_1(0^-) - q_1(0^+) =$

$C_1V_0 - \dfrac{C_1^2}{C_1 + C_2}V_0 = \dfrac{C_1 C_2 V_0}{C_1 + C_2}$. We can obtain $i_1(0^+)$, after the impulse is over, by applying the initial value theorem to the second term on the RHS of Equation 15.4.11:

$$i_1(0^+) = \frac{C_1^2}{(C_1 + C_2)^2} \frac{V_0}{R} \tag{15.4.12}$$

To check this, we note from the ILT of $Q_1(s)$ in Equation 15.4.9 that:

$$q_1(t) = \frac{C_1^2 V_0}{C_1 + C_2} e^{-\frac{t}{(C_1 + C_2)R}} \tag{15.4.13}$$

But $i_1(0^+) = -\dfrac{dq_1(t)}{dt}\Big|_{t=0^+}$, which gives the same result as Equation 15.4.12.

Equation 15.4.11 illustrates a remarkable feature of the LT in that Equation 15.4.7 applies to the circuit for $t \geq 0^+$, yet Equation 15.4.11 includes the impulse, which is applied between $t = 0^-$ and $t = 0^+$. This is because the differentiation property (Equation 15.2.10) introduces the value of the function at $t = 0^-$.

Convolution Theorem

If $F(s) = \mathcal{L}\{f(t)\}$ and $G(s) = \mathcal{L}\{g(t)\}$, then:

$$F(s)G(s) = \mathcal{L}\left\{f(t) * g(t)\right\} \tag{15.4.14}$$

PROOF We start with $F(s) = \int_{0^-}^{\infty} f(\lambda)\, e^{-s\lambda} d\lambda$. As $G(s)$, being a function of s, is a constant as far as integration with respect to λ is concerned, $F(s)G(s)$ can be written as:

$$F(s)G(s) = \int_{0^-}^{\infty} G(s)e^{-s\lambda} f(\lambda)\, d\lambda \tag{15.4.15}$$

From the time-shift property of the LT:

$$G(s)e^{-s\lambda} = \int_{0^-}^{\infty} g(t-\lambda)u(t-\lambda)e^{-s\lambda}\, dt \tag{15.4.16}$$

Substituting in Equation 15.4.15:

$$F(s)G(s) = \int_{0^-}^{\infty}\left[\int_{0^-}^{\infty} g(t-\lambda)u(t-\lambda)e^{-s\lambda}\, dt\right]f(\lambda)\, d\lambda \tag{15.4.17}$$

$f(\lambda)$ and $e^{-s\lambda}$ are constants as far as integration with respect to t is concerned. Hence, $f(\lambda)$ could be moved inside the inner integral and $e^{-s\lambda}$ could be moved outside it:

$$F(s)G(s) = \int_{0^-}^{\infty}\left[\int_{0^-}^{\infty} f(\lambda)g(t-\lambda)u(t-\lambda)\, dt\right]e^{-s\lambda}d\lambda \tag{15.4.18}$$

$u(t-\lambda)$ ensures that $g(t-\lambda)u(t-\lambda) = 0$ for $\lambda < t$, or $\lambda > t$. Hence, it can be omitted if the upper limit of integration is made t instead of infinity. Interchanging the order of integration, Equation 15.4.18 becomes:

$$F(s)G(s) = \int_{0^-}^{\infty}\left[\int_{0^-}^{t} f(\lambda)g(t-\lambda)\, d\lambda\right]e^{-s\lambda}dt \tag{15.4.19}$$

The inner integral is $f(t) * g(t)$, which proves Equation 15.4.14.

In addition to direct evaluation and the graphical method, the convolution theorem provides an alternative means for deriving the convolution integral, analytically or

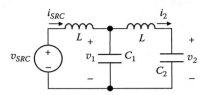

FIGURE 15.4.2
Figure for Example 15.4.2.

numerically. Equation 12.4.7 (Chapter 12) readily follows from the convolution theorem.

Thus, $[s^n F(s)]\left[\dfrac{G(s)}{s^n}\right] = F(s)G(s)$. If $f(t)$ and $g(t)$ are zero for $t < 0$, $s^n F(s)$ is the LT of $f^{(n)}(t)$,

the nth derivative of $f(t)$, and $\dfrac{G(s)}{s^n}$ is the LT of $g^{(-n)}(t)$, the nth integral of $g(t)$. Hence:

$$\left[s^n F(s)\right]\left[\dfrac{G(s)}{s^n}\right] = \mathcal{L}\left\{f^{(n)}(t) * g^{(-n)}(t)\right\} \tag{15.4.20}$$

As the LHS of Equation 15.4.14 and Equation 15.4.20 are equal, the RHS must be equal.

This gives: $f(t) * g(t) = f^{(n)}(t) * g^{(-n)}(t)$. In an analogous manner, as $F(s)G(s) = \left[\dfrac{F(s)}{s^n}\right][s^n G(s)]$,

then $f(t) * g(t) = f^{(-n)}(t) * g^{(n)}(t)$. It follows that:

$$f(t) * g(t) = f^{(n)}(t) * g^{(-n)}(t) = f^{(-n)}(t) * g^{(n)}(t) \tag{15.4.21}$$

Although the aforementioned proof of Equation 15.4.21 is based on the one-sided Laplace transform, which applies to functions that are zero for negative time, this equation can be proved without using the Laplace transform and applies to functions that are nonzero for negative time. Equation 15.4.21 is applied to such a function in Example SE12.3.

Example 15.4.2: Application of Convolution Theorem
An unknown voltage v_{SRC} is applied at $t = 0$ to the circuit of Figure 15.4.2, with zero initial conditions. If the voltage v_1 across C_1 is known, it is required to determine i_{SRC}.

SOLUTION
From KVL around the second mesh:

$$L\frac{di_2}{dt} + \frac{1}{C_2}\int i_2 dt = v_1 \tag{15.4.22}$$

Moreover:

$$v_1 = \frac{1}{C_1}\int (i_{SRC} - i_2)dt \tag{15.4.23}$$

Taking the LTs of Equation 15.4.22 and Equation 15.4.23 and eliminating the LT of i_2:

$$I_{SRC}(s) = \left[s(C_1 + C_2)\frac{1 + s^2 L(C_1 \lozenge C_2)}{1 + s^2 LC_2} \right] V_1(s) \qquad (15.4.24)$$

where $C_1 \lozenge C_2 = \dfrac{C_1 C_2}{C_1 + C_2}$. Alternatively Equation 15.4.24 can be derived by noting that

$\dfrac{I_{SRC}(s)}{V_1(s)} = Y(s)$, where $Y(s)$ is the admittance of C_1 in parallel with the series combination of L and C_2.

If $V_1(s)$ is known analytically, then this would be a straightforward problem. $I_{SRC}(s)$ could be obtained from Equation 15.4.24 and inverted to give i_{SRC}. We are assuming, however, that $V_1(s)$ is not known analytically, and only v_1 is known experimentally. From the convolution theorem, $I_{SRC}(s)$ is the LT of the convolution of v_1 and:

$$f(t) = \mathcal{L}^{-1}\left\{ s(C_1 + C_2)\frac{1 + s^2 L(C_1 \lozenge C_2)}{1 + s^2 LC_2} \right\}. \qquad (15.4.25)$$

That is, $i_{SRC} = v_1 * f(t)$. Evaluation of $f(t)$ is left as a problem (P15.1.18).

15.5 Responses of Circuit Elements in the *s* Domain

Resistors

Taking the LT of both sides of the *v-i* relation $v = Ri$:

$$V(s) = RI(s) \qquad (15.5.1)$$

It is seen that in the *s*-domain, where the resistor voltage and current are $V(s)$ and $I(s)$, respectively, the resistor is simply represented by its resistance R.

Capacitors

Taking the LT of both sides of the *v-i* relation $v = \dfrac{1}{C}\displaystyle\int_0^t i\,dt + V_0$, where V_0 is the initial capacitor voltage (Figure 15.5.1a):

$$V(s) = \frac{1}{sC}I(s) + \frac{V_0}{s} \qquad (15.5.2)$$

Equation 15.5.2 can be represented as in Figure 15.5.1b. If $V_0 = 0$, this reduces to $1/sC$. When $V_0 \neq 0$, the *s*-domain representation includes a voltage source V_0/s whose polarity is that of a voltage drop in the direction of $I(s)$. In the sinusoidal steady state, any effects of V_0 would have died out, and $s = j\omega$, so that the *s*-domain representation of the capacitor reduces to that of the frequency domain, namely $1/j\omega C$. By analogy with $1/j\omega C$, $1/sC$ is the **s-domain impedance** of the capacitor.

Equation 15.5.2 can be rearranged as:

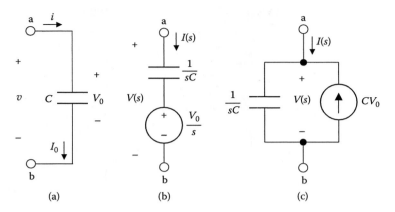

FIGURE 15.5.1
Capacitor in the s domain.

$$I(s) = sCV(s) - CV_0 \qquad (15.5.3)$$

This gives the equivalent s-domain representation of Figure 15.5.1c, which also follows from Figure 15.5.1b by source transformation. Alternatively, Equation 15.5.3 can be derived by taking the LT of the i-v relation $i = C\dfrac{dv}{dt}$.

The sources representing initial conditions are necessary because the LT of any function of time, such as v and i in Figure 15.5.1a, assumes that the function is zero for $t < 0^-$. To account for V_0 at $t = 0^-$ in the series source representation, a voltage source V_0 is inserted in series with an uncharged capacitor. As the voltage across the uncharged capacitor is zero at $t = 0^-$, the voltage that appears between terminals ab is V_0. In the s domain, V_0 is represented by V_0/s. In the parallel source representation. the voltage V_0 across the capacitor is established instantaneously by a current impulse CV_0 of the polarity indicated in Figure 15.5.1c.

Inductors

Taking the LT of both sides of the v-i relation $v = L\dfrac{di}{dt}$ for an inductor having an initial current I_0 (Figure 15.5.2a):

$$V(s) = sLI(s) - LI_0 \qquad (15.5.4)$$

Equation 15.5.4 can be represented as in Figure 15.5.2b, where sL is the s-domain impedance of the inductor. As in the case of the capacitor, an alternative representation is possible (Figure 15.5.2c) by solving for $I(s)$:

$$I(s) = \frac{1}{sL}V(s) + \frac{I_0}{s} \qquad (15.5.5)$$

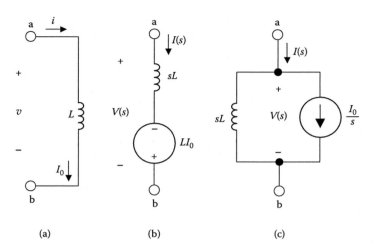

FIGURE 15.5.2
Inductor in the *s* domain.

Equation 15.5.5 can also be derived from the *i-v* relation of an inductor $i = \dfrac{1}{L}\displaystyle\int_0^t v\,dt + I_0$,
or by source transformation.

Figure 15.5.2b and Figure 15.5.2c can be interpreted in the same manner as for a capacitor. In the series source representation, a voltage impulse LI_0 of the polarity indicated in Figure 15.5.2b establishes instantaneously, through the rest of the circuit, the current I_0 in an inductor that has no initial current. In the parallel source representation, the current step I_0 accounts for the initial current through an inductor that has no initial current. Note that the circuits of Figure 15.5.1b and Figure 15.5.2c are duals, as are the circuits of Figure 15.5.1c and Figure 15.5.2b.

It should be noted that in Figure 15.5.1 and Figure 15.5.2 the source is an integral part of the circuit element in the *s* domain. Moreover, these sources can be treated just like any sources of excitation. Both of these points are illustrated by Example 15.5.1.

Example 15.5.1: Response of *RL* Circuit in the *s* and Time Domains

A voltage V_{SRC} is applied at $t = 0$ to a series *RL* circuit in which the inductor has an initial current I_0 (Figure 15.5.3a). It is desired to derive the expressions for the current in the circuit and the voltage across the inductor as functions of time.

SOLUTION

The *s*-domain representation of the circuit is shown in Figure 15.5.3b. From KVL:

$$RI(s) + sLI(s) - LI_0 = \frac{V_{SRC}}{s} \tag{15.5.6}$$

This gives:

$$I(s) = \frac{V_{SRC}}{s(R + sL)} + \frac{LI_0}{R + sL} \tag{15.5.7}$$

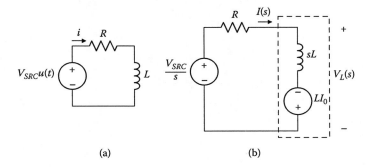

FIGURE 15.5.3
Figure for Example 15.5.1.

Equation 15.5.7 follows from superposition by applying one source at a time, with the other source set to zero. The equation can be rearranged as:

$$I(s) = \frac{1}{L} \frac{V_{SRC} + sLI_0}{s(s + 1/\tau)} \tag{15.5.8}$$

where $\tau = L/R$ is the time constant. Equation 15.5.8 can be expressed as a PFE:

$$I(s) = \frac{V_{SRC}}{sR} + \frac{I_0 - V_{SRC}/R}{s + 1/\tau} \tag{15.5.9}$$

Taking the ILT of Equation 15.5.9:

$$i = \frac{V_{SRC}}{R} + \left(I_0 - \frac{V_{SRC}}{R}\right)e^{-t/\tau} \tag{15.5.10}$$

Equation 15.5.10 is of the same form as that of Equation 13.3.2 (Chapter 13) derived earlier in terms of initial value, final value, and the time constant.

The voltage across the inductor in the s domain is:

$$V_L(s) = sLI(s) - LI_0 = \frac{V_{SRC} - RI_0}{s + 1/\tau} \tag{15.5.11}$$

Taking the ILT:

$$v_L = \left(V_{SRC} - RI_0\right)e^{-t/\tau} \tag{15.5.12}$$

Again, this result follows from Equation 13.3.2. The final value of the voltage across the inductor is zero. With an initial current I_0 in the circuit, the initial value of voltage across the inductor is $(V_{SRC} - RI_0)$.

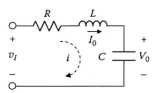

FIGURE 15.5.4
Figure for Exercise 15.5.1.

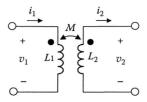

FIGURE 15.5.5
Coupled coils in the time domain.

EXERCISE 15.5.1

Given a series RLC circuit with initial current I_0 in the inductor and V_0 in the capacitor, as shown in Figure 15.5.4, represent the circuit in the s domain and derive the expression for $I(s)$.

Answer: $I(s) = \dfrac{V_I(s) + LI_0 - V_0/s}{sL + R + 1/sC}$

Magnetically Coupled Coils

The mesh equations of the basic circuit of Figure 15.5.5 are:

$$L_1 \frac{di_1}{dt} - M \frac{di_2}{dt} = v_1 \tag{15.5.13}$$

$$-M \frac{di_1}{dt} + L_2 \frac{di_2}{dt} + v_2 = 0 \tag{15.5.14}$$

Taking the LT, and including initial values I_{10} and I_{20}:

$$sL_1 I_1(s) - L_1 I_{10} - sMI_2(s) + MI_{20} = V_1(s) \tag{15.5.15}$$

$$-sMI_1(s) + MI_{10} + sL_2 I_2(s) - L_2 I_{20} + V_2(s) = 0 \tag{15.5.16}$$

Any s-domain representation of the coupled coils must satisfy Equation 15.5.15 and Equation 15.5.16. It can be readily verified that the T-circuit of Figure 15.5.6 satisfies these equations and is in accordance with the voltage source representation of the initial current through an inductor (Figure 15.5.2b). Thus, the s-domain impedances are the inductances

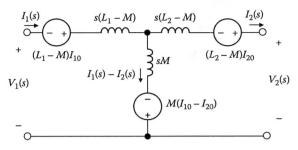

FIGURE 15.5.6
Coupled coils in the s domain.

of the T-equivalent circuit of Figure 6.2.4 (Chapter 6) multiplied by s. The initial current on the input side is I_{10} and is associated with an inductance $(L_1 - M)$, so that the source voltage accounting for the initial conditions in this inductor is a voltage rise $(L_1 - M)I_{10}$ in the direction of current; this holds true similarly on the output side. The initial current in the shunt branch is $(I_{10} - I_{20})$ and the source voltage is $M(I_{10} - I_{20})$ as indicated. Examples of the application of equivalent circuits for magnetically coupled coils with initial currents are Example SE15.3 and Example SE15.4.

EXERCISE 15.5.2
Represent the initial current in each of the three inductors in the circuit of Figure 15.5.6 by a parallel current source, as in Figure 15.5.2c. Assume that the dot markings on one coil are reversed relative to those of Figure 15.5.5.

15.6 Solution of Circuit Problems in the s Domain

It was emphasized earlier that in circuit analysis, the fundamental relations that must be satisfied are KCL, KVL, and the v-i relations of the circuit elements. When KCL is satisfied by instantaneous currents at a given node, then by taking LTs, KCL will also be satisfied in the s domain. This occurs similarly for voltage rises and drops that satisfy KVL. In the preceding section, we have seen how the v-i relations of circuit elements are represented in the s domain, including initial values. It follows that:

> **Concept:** *All circuit laws and techniques discussed previously for the frequency domain using phasor notation apply equally well to the s domain. These include: series and parallel combinations of s-domain impedances and admittances; node-voltage, mesh-current, and loop-current methods of analysis; TEC and NEC; superposition, Y-Δ transformation, etc.*

We have already applied KVL, superposition, and source transformation in the s domain in the problems of the preceding sections.

The first step in the general procedure for analyzing circuits using the LT method is to transform the circuit to the s domain, representing passive circuit elements by their s-domain impedances, initial values of voltages across capacitors and currents though inductors by appropriate sources, and independent and dependent voltage or current sources by their LTs. The circuit is then analyzed by any of the techniques described previously for phasor analysis, and the LT of the desired circuit response is derived.

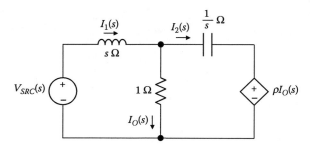

FIGURE 15.6.1
Figure for Example 15.6.1.

The essential feature of the LT method is that the circuit equations are now algebraic in s, just as they were algebraic in $j\omega$ in the case of phasor analysis. The ILT gives the response in the time domain, both transient and steady state.

Example 15.6.1: Analysis of Circuit in the s Domain

It is required to determine $I_1(s)$, $I_2(s)$, and $I_O(s)$ in Figure 15.6.1 in response to a unit voltage impulse, assuming zero initial conditions, and to interpret the behavior of the circuit.

SOLUTION
Considering $I_1(s)$ and $I_2(s)$ to be mesh currents, the mesh-current equations may be written as:

$$(s+1)I_1(s) - I_2(s) = V_{SRC}(s) \tag{15.6.1}$$

$$-I_1(s) + (1+1/s)I_2(s) = -\rho I_O(s) \tag{15.6.2}$$

where $I_O(s) = I_1(s) - I_2(s)$. Solving these equations:

$$I_1(s) = \frac{s(1-\rho)+1}{s^2(1-\rho)+s+1} V_{SRC}(s) \tag{15.6.3}$$

$$I_2(s) = \frac{s(1-\rho)}{s^2(1-\rho)+s+1} V_{SRC}(s) \tag{15.6.4}$$

$$I_O(s) = \frac{1}{s^2(1-\rho)+s+1} V_{SRC}(s) \tag{15.6.5}$$

It follows from the initial value theorem, with $V_{SRC}(s) = 1$, that:

$$i_1(0^+) = 1 \text{ A}, \quad i_2(0^+) = 1 \text{ A}, \text{ and } i_O(0^+) = 0 \tag{15.6.6}$$

To interpret these results, it must be kept in mind that KVL and KCL must be satisfied at every instant of time, including $t = 0^+$. During the application of the impulse, KVL applied to the mesh on the LHS gives:

$$\delta(t) = L\frac{di_1}{dt} + v_R$$

where v_R is the voltage across the combination of the resistor, capacitor, and dependent source. Integrating between 0^- and 0^+:

$$\int_{0^-}^{0^+} \delta(t)dt = L\int_{0^-}^{0^+} di_1 + \int_{0^-}^{0^+} v_R dt$$

The voltage across the capacitor is not forced to change, so it remains zero at $t = 0^+$; v_R will therefore be a function of i_1. Assuming that i_1 makes a finite jump between $t = 0^-$, when its value is zero, and $t = 0^+$, then:

$$\int_{0^-}^{0^+} v_R dt = 0$$

because v_R remains finite whereas the interval of integration is vanishingly small. The integration then gives $1 = i_1(0^+)$, as in Equation 15.6.6, which means that the impulse appears across the inductor, causing a step change in $i_1(t)$ from zero to:

$$\frac{1}{L}\int_{0^-}^{0^+} \delta(t)dt = 1 \text{ A}$$

The assumption of a finite jump in i_1 at $t = 0^+$ gives a consistent result and is therefore justified. As the voltage across the capacitor remains zero, this means that the voltage across the resistor is the same as that across the dependent source, that is, $i_O(0^+) \times 1 = \rho i_O(0^+)$ or:

$$i_O(0^+)\left[\rho - 1\right] = 0 \tag{15.6.7}$$

If $\rho \neq 1$, then Equation 15.6.7 can only be satisfied by having $i_O(0^+) = 0$, which makes $i_2(0^+) = i_1(0^+)$, in accordance with Equation 15.6.6. From KVL in the mesh on the LHS, with $i_O(0^+) = 0$, the voltage v_L across the inductor is zero at $t = 0^+$. To check this,

$$V_L(s) = sI_1(s) = \frac{s^2(1-\rho)+s}{s^2(1-\rho)+s+1} = 1 - \frac{1}{s^2(1-\rho)+s+1}$$

The 1 term accounts for the impulse and the term $\dfrac{1}{s^2(1-\rho)+s+1}$ is $V_L(s)$ after the impulse

is over. Applying the initial value theorem to this term gives $v_L(0^+) = 0$.

If $\rho = 1$, the voltage across the resistor is the same as that across the source, so $i_2 = 0$ for all t. Hence,

$$I_1(s) = I_0(s) = \frac{1}{s+1} \quad \text{and} \quad i_1(t) = i_0(t) = e^{-t}$$

as for R in series with L.

EXERCISE 15.6.1
Derive Equation 15.6.5 using TEC.

Switching

As mentioned in connection with the initial-value theorem (Section 15.4), the LT method is particularly useful for solving switching problems in which impulsive readjustment takes place at the time of switching, because capacitor voltages or inductor currents are forced to change at this instant. The general procedure is to represent the circuit in the s domain *after* switching occurs, that is, $t \geq 0^+$, but including the initial conditions just *before* switching takes place ($t = 0^-$). The initial-value theorem gives any desired circuit response at $t = 0^+$.

Example 15.6.2: Switched Capacitor Circuit in the s Domain

It is required to solve the problem of Example 15.4.1 entirely in the s domain.

SOLUTION
The circuit in the s domain is shown in Figure 15.6.2 for $t \geq 0^+$. Initial conditions in C_1 are represented by the current source $C_1 V_0$. The following equations may be written in terms of $V(s)$ and $I_1(s)$:

$$V(s) = \frac{R / sC_2}{R + 1 / sC_2} I_1(s) \tag{15.6.8}$$

$$I_1(s) = C_1 V_0 - sC_1 V(s) \tag{15.6.9}$$

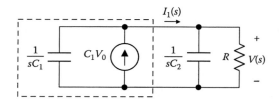

FIGURE 15.6.2
Figure for Example 15.6.2.

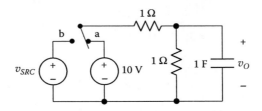

FIGURE 15.6.3
Figure for Example 15.6.3.

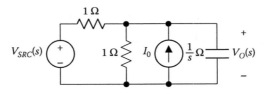

FIGURE 15.6.4
Figure for Example 15.6.3.

Solving for $I_1(s)$ gives:

$$I_1(s) = C_1 V_0 \frac{sC_2 R + 1}{s(C_1 + C_2)R + 1} \tag{15.6.10}$$

which is the same as Equation 15.4.10. Solving for $V(s)$:

$$V(s) = C_1 V_0 \frac{R}{s(C_1 + C_2)R + 1} \tag{15.6.11}$$

EXERCISE 15.6.2

Given $C_1 = C_2 = 1$ F, $R = 0.5\ \Omega$, and $V_0 = 10$ V in Example 6.3.1. Determine i_1, v, i_2, q_1, and q_2.
Answer: $i_1 = 5\delta(t) + 5e^{-t}u(t)$ A; $v = 5e^{-t}u(t)$ V; $i_2 = 5\delta(t) - 5e^{-t}u(t)$ A; $q_1 = q_2 = 5e^{-t}u(t)$ C.

Example 15.6.3: Sequential Switching of Circuit in the *s* Domain
In Figure 15.6.3, the switch is moved to position b at $t = 0$ after having been in position a for a long time. The switch is moved to position a at $t = 1$ s and back to position b at $t = 2$ s. Determine v_O during the time intervals $0 \le t \le 1$ s, 1 s $\le t \le 2$ s, and $t \ge 2$ s, given that $v_{SRC} = 15\sin tu(t)$ V.

SOLUTION
At $t = 0^-$, $v_O(0^-) = 5$ V. During the interval $0 \le t \le 1$ s, the circuit in the *s* domain is as shown in Figure 15.6.4, with $I_0 = CV_0 = 5$A and $V_{SRC}(s) = \dfrac{15}{s^2 + 1}$. From superposition:

$$V_O(s) = \frac{5}{s+2} + \frac{15}{(s+2)(s^2+1)} = \frac{5s^2+20}{(s+2)(s^2+1)} = \frac{8}{s+2} - \frac{3(s-2)}{(s^2+1)} \qquad (15.6.12)$$

Hence,

$$v_O = 8e^{-2t} - 3\cos t + 6\sin t, \ 0 \le t \le 1 \ s \qquad (15.6.13)$$

At $t = 1$ s, $v_O = 4.51$ V. During the interval $1 \ s \le t \le 2$ s, the circuit in the s domain is as shown in Figure 15.6.4, with $I_0 = 4.51\,A$ and $V_{SRC}(s) = \frac{10}{s}$. From superposition:

$$V_O(s) = \frac{4.51}{s+2} + \frac{10}{s(s+2)} = \frac{4.51s+10}{s(s+2)} = \frac{5}{s} - \frac{0.49}{s+2} \qquad (15.6.14)$$

It should be noted that when we consider the LT of the 10 V source to be $\frac{10}{s}$, it is implicitly assumed that zero time is the instant when this source is applied at $t = 1$ s. In other words, when we take the ILT of Equation 15.6.14, it is with respect to a time variable $t' = t - 1$. Thus:

$$v_O(t') = 5 - 0.49e^{-2t'}, \ 0 \le t' \le 1 \ s \qquad (15.6.15)$$

or

$$v_O(t) = 5 - 0.49e^{-2(t-1)}, \ 1 \ s \le t \le 2 \ s \qquad (15.6.16)$$

At $t = 2$ s, $v_O(t) = 4.93$ V. For $t \ge 2$ s, the circuit in the s domain is as shown in Figure 15.6.4, with $I_0 = 4.93$ and $V_{SRC}(s) = \frac{15}{s^2+1}$. From superposition:

$$V_O(s) = \frac{4.93}{s+2} + \frac{15}{(s+2)(s^2+1)} = \frac{4.93s^2+19.93}{(s+2)(s^2+1)} = \frac{7.93}{s+2} - \frac{3(s-2)}{(s^2+1)} \qquad (15.6.17)$$

As explained in connection with switching at $t = 1$ s, the ILT is with respect to $t'' = t - 2$. Hence:

$$v_O(t'') = 7.93e^{-2t''} - 3\cos t'' + 6\sin t'', \ t'' \ge 2 \ s \qquad (15.6.18)$$

or,

$$v_O(t) = 7.93e^{-2(t-2)} - 3\cos(t-2) + 6\sin t(t-2), \ t \ge 2 \ s \qquad (15.6.19)$$

Application Window 15.6.1: Fluorescent Lamp Ballast

In a fluorescent lamp, electrodes at each end of the tube are heated by an electric current so as to emit electrons. These electrons facilitate the striking of an arc between the

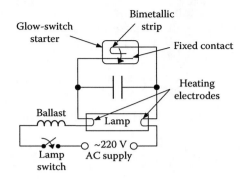

FIGURE 15.6.5
Figure for Application Window 15.6.1.

electrodes if a sufficiently high voltage is applied. Mercury in the tube is vaporized by the heat of the arc and produces ultraviolet radiation that excites the phosphor coating of the lamp, causing it to emit lower-energy radiation in the form of visible light. An inductor assembly, or **ballast**, is used for two purposes: (1) it produces, in conjunction with a capacitor, the high voltage necessary for striking the arc, and (2) once the arc is struck, it stabilizes the operating current of the fluorescent lamp.

Although fluorescent lamps can be started in several ways, a common method utilizes a glow tube filled with neon or argon gas and containing a bimetallic strip switch, as illustrated in Figure 15.6.5. The bimetallic strip is U-shaped and made up of two metals of different coefficients of thermal expansion. When heated, the different expansions of the two metals open the U, thereby closing the switch.

When the ac supply is applied to the lamp assembly, the glow switch is initially open and the lamp behaves as an open circuit. The supply voltage, applied between the bi-metallic strip and the fixed contact, is sufficient to start a discharge in the glow tube. The discharge current, limited by the ballast, heats both the electrodes of the lamp and the bimetallic strip. In a few seconds the heated bimetallic strip closes the contact in the glow switch, but this extinguishes the glow discharge, which cools the bimetallic strip and interrupts the current. The interruption of the inductive current, in conjunction with the capacitor, produces a voltage across the lamp that is high enough to strike an arc through the lamp. Once this arc is struck, the voltage across the glow switch is too low to cause a discharge, so this switch remains open. Current now passes through ballast and lamp, and is stabilized by the ballast.

The circuit in the s domain is shown in Figure 15.6.6 for $t > 0$, assuming that the glow switch opens at $t = 0$ and neglecting the initial voltage across C, which is the voltage

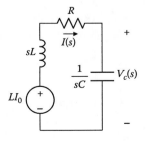

FIGURE 15.6.6
Figure for Application Window 15.6.1.

across the discharge of the glow tube. Although the ac supply is in series with the rest of the circuit, it is omitted in the figure because normally it has little effect on the arcing voltage. It follows from this figure that:

$$V_C(s) = \frac{I_0}{C} \frac{1}{s^2 + (R/L)s + 1/LC}$$

Assuming, $I_0 = 0.1$ A, $C = 0.8$ nF, $L = 0.5$ H, and $R = 50\ \Omega$, this becomes:

$$V_C(s) = \frac{1.25 \times 10^8}{s^2 + 100s + 0.25 \times 10^{10}} \cong \frac{1.25 \times 10^8}{(s + 50)^2 + 0.25 \times 10^{10}}$$

as $0.25 \times 10^{10} \gg 2500$. The ILT is: $v_C(t) = 2500e^{-50t}\sin 0.5 \times 10^5 t$. To find maximum $v_C(t)$, we set the derivative equal to zero. This gives: $t_{max} = 2 \times 10^{-5}\tan^{-1}1000 \cong 31.4\ \mu s$. As expected, t_{max} is very nearly a quarter of the period of the sine function. At t_{max}, $v_{Cmax}(t) \cong 2500$ V, which is sufficient to initiate an arc in the lamp.

15.7 Interpretations of Circuit Responses in the *s* Domain

In this section we will consider the information that can be derived from responses in the *s* domain.

Transfer Function

> **Definition:** *When a single excitation is applied to a given circuit having no initial energy storage, the transfer function H(s) is the ratio of the LT Y(s) of a designated response to the LT X(s) of the applied excitation.*

Thus:

$$H(s) = \frac{Y(s)}{X(s)} \tag{15.7.1}$$

In Example 15.6.1, for instance, $\dfrac{I_1(s)}{V_{SRC}(s)}$, $\dfrac{I_2(s)}{V_{SRC}(s)}$, and $\dfrac{I_0(s)}{V_{SRC}(s)}$ are all examples of transfer functions in the same circuit. Being the ratio of a response to an excitation, the transfer function is independent of the nature of the excitation. It depends on the circuit, on where in the circuit the excitation is applied, and on which voltage or current in the circuit is the designated response. It should also be kept in mind that:

> **Concept:** *The poles of the transfer function are the roots of the characteristic equation.*

This follows from Equation 15.3.3. With zero initial conditions, the transfer function $\dfrac{Y(s)}{F(s)}$ equals $\dfrac{1}{as^2 + bs + c}$, so that the poles of the transfer function are the roots of the characteristic equation $as^2 + bs + c = 0$.

Moreover, if the applied excitation is a unit impulse, $X(s) = 1$, $Y(s) = H(s)$, and:

$$\mathscr{L}^{-1}\left\{Y(s)\right\} = \mathscr{L}^{-1}\left\{H(s)\right\} = h(t) \tag{15.7.2}$$

It follows that:

> **Concept:** *The ILT of the transfer function is the response in the time domain to a unit impulse of excitation.*

Stability

> **Definition:** *A circuit is stable if the response to a unit impulse tends to zero as $t \to \infty$. The circuit is unstable if the response to a unit impulse increases without limit, that is, is unbounded as $t \to \infty$. The circuit is marginally stable, or metastable, if as $t \to \infty$, the response to a unit impulse does not approach zero but remains bounded.*

The response in the time domain to a unit impulse is the ILT, $h(t)$, of the transfer function $H(s)$ (Equation 15.7.2). The transfer function of lumped-parameter, LTI circuits is a proper rational function of s whose poles are the roots of the characteristic equation. The PFE of $H(s)$ therefore consists of terms of the form $\dfrac{K}{s + p_r}$ or $\dfrac{K}{\left(s + p_r\right)^m}$, where the pole at $-p_r = -\alpha_r - j\omega_r$ is, in general, complex. In the time domain, these terms have ILTs that contain terms in $e^{-\alpha_r t}$ and $t^{m-1}e^{-\alpha_r t}$, respectively. As long as $-\alpha_r < 0$, these terms vanish as $t \to \infty$, so that the impulse response tends to zero. On the other hand, if $-\alpha_r > 0$, these terms go to infinity as $t \to \infty$ and the impulse response is unbounded. It follows that:

> **Interpretation:** *If all the poles of the transfer function lie in the open left half of the s-plane, the circuit is stable. If at least one pole lies in the open right half of the s-plane, the circuit is unstable.*

The word "open" in the preceding statement excludes the imaginary axis, which is the boundary between the right and left halves of the s-plane. If the poles lie on the imaginary axis and are simple, that is, equal to $\pm j\omega$, the PFE of $H(s)$ consists of terms of the form

$$\frac{\omega}{s^2 + \omega^2} \quad \text{or} \quad \frac{s}{s^2 + \omega^2}$$

The impulse response is sinusoidal, so that the circuit is metastable. If the poles on the imaginary axis are of a higher order, the terms in the PFE of $H(s)$ consist of terms having $(s^2 + \omega^2)^n$ in the denominator. The sinusoidal functions in the impulse response will be multiplied by t^{n-1}, so that the response is unbounded and the circuit is unstable. However, if the poles on the imaginary axis are simple, that is, equal to $\pm j\omega_p$, but the excitation frequency is ω_p, the response is unbounded (Example 15.7.1).

Poles of the transfer function on the imaginary axis are those of circuits consisting of ideal, lossless inductors and capacitors. Poles in the left half of the s-plane arise because of power

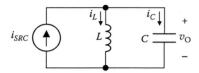

FIGURE 15.7.1
Figure for Example 15.7.1.

dissipation in the circuit, which causes the impulse response to die out as $t \to \infty$. By analogy, poles in the right half of the s-plane can be considered to be due to a negative resistance, which supplies energy that increases with time and causes an unbounded response. Poles at infinity, whether simple or multiple, do not affect the stability, as they arise from the difference in the powers of s in the numerator and denominator of the transfer function.

Example 15.7.1: Response of Lossless LC Circuit

Given a parallel LC circuit (Figure 15.7.1), with zero initial conditions, excited by a current source, it is required to determine: (a) the transfer function $\dfrac{V_O(s)}{I_{SRC}(s)}$, (b) v_O in response to a unit current impulse, and (c) v_O in response to a sinusoidal excitation of the same frequency as the poles. This example is the same as Example SE12.2 (Chapter 12) but analyzed in the s domain.

ANALYSIS

(a) It follows from the circuit in the s domain that $V_O = \dfrac{sL(1/sC)}{sL + 1/sC} I_{SRC}$, so that:

$$H(s) = \frac{V_O(s)}{I_{SRC}(s)} = \frac{1}{C} \frac{s}{s^2 + \omega_0^2} \tag{15.7.3}$$

where $\omega_0^2 = 1/LC$. $H(s)$ has a pair of simple, conjugate poles on the imaginary axis at $s = \pm j\omega_0$.

(b) If $i_{SRC} = \delta(t)$:

$$h(t) = v_O = \mathcal{L}^{-1}\{H(s)\} = \frac{1}{C} \cos \omega_0 t \tag{15.7.4}$$

This result is readily interpreted. The unit impulse instantaneously charges the capacitor to a voltage $v_O(0^+) = 1/C$ V, the energy in the capacitor being initially $\dfrac{1}{2} C v_O^2(0^+) = \dfrac{1}{2C}$. The circuit then continuously oscillates at a frequency ω_0, the amplitude of v_O being $1/C$.

$i_L = \dfrac{1}{L}\displaystyle\int_0^t v_O(t)\, dt = \omega_0 \sin \omega_0 t$. The current in the inductor has its largest magnitude when $\omega_0 t = \dfrac{n\pi}{2}$, where n is an integer. At these instants of time $v_O = 0$, so no energy is stored in the capacitor. The energy stored in the inductor is then $\frac{1}{2}LI_m^2 = \frac{1}{2}L\frac{1}{LC} = \frac{1}{2C}$, which is the same as that initially stored in the capacitor. It is seen that the energy continuously oscillates between electric energy stored in the capacitor, at $\omega_0 t = n\pi$, and magnetic energy stored in the inductor, at $\omega_0 t = \frac{n\pi}{2}$. At intermediate times, energy is stored in both the inductor and capacitor, the total energy being $1/2C$, as required by conservation of energy. As the amplitude of oscillation is bounded, the circuit is marginally stable.

(c) If $i_{SRC} = A\cos\omega t$, where $\omega \neq \omega_0$, $I_{SRC} = A\dfrac{s}{s^2 + \omega^2}$, and:

$$V_O(s) = \frac{A}{C}\frac{s}{s^2 + \omega_0^2}\frac{s}{s^2 + \omega^2} = \frac{A}{C\left(\omega_0^2 - \omega^2\right)}\left[\frac{\omega_0^2}{s^2 + \omega_0^2} - \frac{\omega^2}{s^2 + \omega^2}\right] \tag{15.7.5}$$

(Exercise 15.7.1). The ILT is:

$$v_O(t) = \frac{A}{C\left(\omega_0^2 - \omega^2\right)}\left[\omega_0\sin\omega_0 t - \omega\sin\omega t\right] \tag{15.7.6}$$

The voltage is thus a combination of the applied signal and the natural oscillation of the circuit. The response remains bounded.

If $i_{SRC} = A\cos\omega_0 t$, then:

$$V_O(s) = \frac{A}{C}\frac{s^2}{\left(s^2 + \omega_0^2\right)^2} \tag{15.7.7}$$

The PFE of $V_O(s)$ is (Exercise 15.7.1):

$$V_O(s) = \frac{A}{2C}\left[\frac{1}{s^2 + \omega_0^2} + \frac{s^2 - \omega_0^2}{\left(s^2 + \omega_0^2\right)^2}\right] \tag{15.7.8}$$

The ILT is (Table 15.3.1):

$$v_O = \frac{A}{2C}\left[\frac{1}{\omega_0}\sin\omega_0 t + t\cos\omega_0 t\right] \tag{15.7.9}$$

The response is now unbounded because the frequency of excitation is the same as that of the poles of the circuit. Note that in this case, it is not possible to solve the problem using phasor analysis because the circuit never reaches a steady state.

If two identical *LC* circuits are cascaded with isolation, so that the second circuit does not load the first, the transfer function is of the form $\dfrac{s^2}{(s^2 + \omega_0^2)^2}$. The impulse response is now unbounded because of the double pole on the imaginary axis. An impulse excites the first circuit into continuous oscillation, which provides excitation to the second circuit at its resonant frequency (Problem P15.3.33).

EXERCISE 15.7.1
Derive the PFEs of Equation 15.7.5 and Equation 15.7.8.

Sinusoidal Steady-State Response

Consider a circuit to which a sinusoidal excitation $x(t) = X_m \cos(\omega t + \theta)u(t)$ is applied. Let $y(t)$ be a designated response whose LT is $Y(s)$, and let the transfer function between excitation and response be $H(s)$, expressed as a proper rational function:

$$H(s) = \frac{Y(s)}{X(s)} = \frac{N(s)}{D(s)} \tag{15.7.10}$$

or $Y(s) = X(s) \dfrac{N(s)}{D(s)}$, where $X(s) = X_m \dfrac{s\cos\theta - w\sin\theta}{s^2 + w^2}$. Substituting for $X(s)$:

$$Y(s) = X_m \frac{s\cos\theta - \omega\sin\theta}{s^2 + \omega^2} \frac{N(s)}{D(s)} \tag{15.7.11}$$

This can be expressed as a PFE in the form:

$$Y(s) = \frac{K_1}{s - j\omega} + \frac{K_1^*}{s + j\omega} + \sum \text{terms involving roots of } D(s) \tag{15.7.12}$$

If the circuit is dissipative and we are only interested in the steady-state response as $t \to \infty$, then we can neglect the terms in the summation sign because their ILTs will have terms that include $e^{-\alpha t}$, $\alpha > 0$. Equation 15.7.12 reduces to:

$$Y_{SS}(s) = \frac{K_1}{s - j\omega} + \frac{K_1^*}{s + j\omega} \tag{15.7.13}$$

where $Y_{SS}(s)$ is the steady-state component of $Y(s)$. To determine K_1, and k_1^*, we have to refer to Equation 15.7.11 and substitute $\dfrac{N(s)}{D(s)} = H(s)$ from Equation 15.7.10. Thus:

$$Y_{SS}(s) = X_m \frac{s\cos\theta - \omega\sin\theta}{s^2 + \omega^2} H(s) = \frac{K_1}{s - j\omega} + \frac{K_1^*}{s + j\omega} \tag{15.7.14}$$

To determine K_1, we proceed in the usual manner by multiplying both sides by $(s - j\omega)$ and substituting $s = j\omega$. This gives:

$$K_1 = X_m \frac{j\omega\cos\theta - \omega\sin\theta}{j2\omega} H(j\omega) = \frac{X_m}{2} H(j\omega)\left[\cos\theta + j\sin\theta\right] = \frac{X_m}{2} H(j\omega)e^{j\theta}$$

$$= \frac{X_m}{2}\left|H(j\omega)\right|e^{j(\theta+\phi)} \tag{15.7.15}$$

where $H(j\omega) = \left|H(j\omega)\right|e^{j\phi}$. It follows from Table 15.3.1 that:

$$y_{SS}(t) = X_m\left|H(j\omega)\right|\cos(\omega t + \theta + \phi) \tag{15.7.16}$$

Interpretation: *The sinusoidal steady-state response is obtained from the transfer function by substituting $s = j\omega$, multiplying the magnitude of the excitation by $|H(j\omega)|$ and adding the phase angle of $H(j\omega)$ to that of the excitation.*

The factors of $H(j\omega)$ can be interpreted geometrically in the s plane. Consider, for example,

$$H(s) = \frac{100(s-2)}{(s+3)(s^2+2s+5)} = \frac{100(s-2)}{(s+3)(s+1+j2)(s+1-j2)}$$

The function has a zero z at $s = 2$, a pair of conjugate poles p_1 and p_2 at $s = -1 - j2$ and $s = -1 + j2$, respectively, and a pole p_3 at $s = -3$ (Figure 15.7.2). When we substitute $s = j\omega$, $H(j\omega)$ becomes:

$$H(j\omega) = \frac{100(j\omega-2)}{(j\omega+3)(j\omega+1+j2)(j\omega+1-j2)} = \frac{100(-2+j\omega)}{(3+j\omega)[1+j(\omega+2)][1+j(\omega-2)]} \tag{15.7.17}$$

$s = j\omega$ is represented by point Q at a distance $+\omega$ on the imaginary axis. Each of the factors on the RHS of Equation 15.7.17 can be represented as vectors (Figure 15.7.2). Thus,

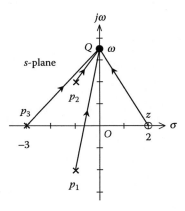

FIGURE 15.7.2
Interpretation of transfer function at real frequencies.

$(-2 + j\omega)$ is the vector zQ. It is the sum of the vector zO, which is -2, and the vector $j\omega$. Similarly, $p_3Q = p_3O + OQ = 3 + j\omega$; $p_2Q = 1 + j(\omega - 2)$; and $p_1Q = 1 + j(\omega + 2)$. These vectors can be evaluated in terms of magnitude and phase in the Argand diagram in the same way as phasors. If $\omega = 4$, for example, $zQ = 2\sqrt{5}\angle116.6°$, $p_1Q = \sqrt{37}\angle80.5°$, $p_2Q = \sqrt{5}\angle63.4°$, and $p_3Q = 5\angle53.1°$, so that:

$$H(j4) = 100\frac{2\sqrt{5}\angle116.6°}{(\sqrt{37}\angle80.5°)(\sqrt{5}\angle63.4°)(5\angle53.1°)} = 6.58\angle-80.5° \qquad (15.7.18)$$

This is the same as would be evaluated using phasors. The denominator of $H(s) = s^3 + 5s^2 + 11s + 15 \equiv (15 - 5\omega^2) + j\omega(11 - \omega^2)$, which gives for $\omega = 4$, $H(j4) = -100\dfrac{(-2 + j4)}{65 + j20}$

$= 6.58\angle-80.5°$. As ω is varied, s moves along the imaginary axis and $H(j\omega)$ assumes different values.

Interpretation of Zeros and Poles

Frequencies on the imaginary axis are those of undamped sinusoids, such as $\cos\omega t$. Terms of the form $(s^2 + \omega_z^2)$ in the numerator of a response signify zeros at $\pm j\omega_z$ on the imaginary axis and make the response zero at the frequency ω_z. Physically, the response at this frequency is generally not zero but very small due to parasitic resistances. Terms of the form $(s^2 + \omega_p^2)$ in the denominator of a response signify poles at $\pm j\omega_p$ on the imaginary axis and make the response infinite at the frequency ω_p. Physically, the response is very large at this frequency and is limited by nonlinearities and saturation in the circuit. Complex poles and zeros of the form $(-\alpha \pm j\omega)$ represent damped sinusoids of the form $e^{-\alpha t}\cos\omega t$. Formally, though not physically, they can be considered to make the response infinity or zero, respectively, at the corresponding complex frequency. They do, nevertheless, affect the response at frequencies on the imaginary axis, as illustrated in Figure 15.7.2.

EXERCISE 15.7.2
Verify that Equation 15.7.16 gives the same steady-state value of current for the series *RL*

circuit as that derived in Equation 5.2.6 (Chapter 5), using $H(s) = \dfrac{I(s)}{V_{SRC}(s)} = \dfrac{1}{sL + R}$.

Natural Response of First-Order Circuits

Figure 15.7.3 shows the *RL* circuit of Figure 12.1.1 (Chapter 12) in the *s* domain, assuming zero initial conditions. It follows from KVL that: $V_{SRC}(s) = RI(s) + sLI(s)$, which gives:

$$\frac{I(s)}{V_{SRC}(s)} = \frac{1}{R + sL} = \frac{1}{L}\frac{1}{s + 1/\tau} \qquad (15.7.19)$$

where $\tau = L/R$ is the time constant. If the applied excitation is a voltage impulse of strength *K* Vs, then $V_{SRC}(s) = K$. Substituting for $V_{SRC}(s)$ in Equation 15.7.19:

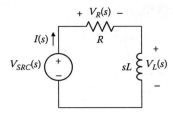

FIGURE 15.7.3
RL circuit in the s domain.

$$I(s) = \frac{K}{L}\frac{1}{s+1/\tau} \qquad (15.7.20)$$

Once $I(s)$ is known, $V_R(s)$ and $V_L(s)$ readily follow. Thus:

$$V_R(s) = RI(s) = \frac{K}{\tau}\frac{1}{s+1/\tau} \qquad (15.7.21)$$

and

$$V_L(s) = sLI(s) = K\frac{s}{s+1/\tau} \qquad (15.7.22)$$

It will be noted that all the natural responses $I(s)$, $V_R(s)$, and $V_L(s)$ have a pole at $s = -1/\tau$, which is also the root of the characteristic equation. Thus, the differential equation with zero forcing function is $L\dfrac{di}{dt} + Ri = 0$. Taking the LT, with zero initial conditions, gives: $(sL + R)I(s) = 0$. The characteristic equation is, therefore, $sL + R = 0$, whose root is $s = -1/\tau$.

> **Concept:** *All the responses of a stable first-order circuit in the s domain are characterized by a pole on the negative real axis whose magnitude is the reciprocal of the time constant.*

EXERCISE 15.7.3
If $L = 1$ H in the circuit of Figure 15.7.3, determine how the pole moves along the negative real axis as R varies between 1 Ω and 100 Ω.

Example 15.7.2: Response of First-Order Circuit in the s Domain

Consider the RC circuit shown in Figure 15.7.4. It is required to derive and interpret the transfer function $\dfrac{I(s)}{V_{SRC}(s)}$.

SOLUTION
It follows from Figure 15.7.4 that:

$$H_I(s) = \frac{I(s)}{V_{SRC}(s)} = \frac{1}{R+1/sC} = \frac{1}{R}\frac{s}{s+1/\tau} \qquad (15.7.23)$$

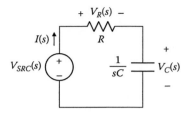

FIGURE 15.7.4
Figure for Example 15.7.2.

where $\tau = RC$. $H_I(s)$ has a zero at $s = 0$ and a pole at $s = -1/\tau$. The zero occurs at a frequency $\omega = 0$ which corresponds to dc. Under dc conditions, C behaves as an open circuit, so that the current is zero. The pole is on the negative real axis. As explained earlier, it does not occur at any physically realizable frequency. At any frequency ω, $|H_I(j\omega)| =$

$\dfrac{1}{\sqrt{R^2 + (1/\omega C)^2}}$, as obtained using phasor analysis.

If the excitation is a voltage impulse of strength K Vs, $V_{SRC}(s) = K$. Substituting for $V_{SRC}(s)$ in Equation 15.7.23:

$$I(s) = \frac{K}{R}\frac{s}{s+1/\tau} = \frac{K}{R} - \frac{K}{CR^2}\frac{1}{s+1/\tau} \tag{15.7.24}$$

From the final value theorem, $\lim_{t\to\infty} i(t) = \lim_{s\to 0} sI(s) = 0$, as to be expected because the current eventually dies out. The initial value theorem can be applied to the rational fraction part of $I(s)$ to give:

$$\lim_{t=0^+} i(t) = \lim_{s\to\infty} sI(s) = -\frac{K}{CR^2}$$

The voltage across the capacitor remains finite during the impulse, so that the voltage impulse appears across the resistor, causing a current impulse of K/R. The current impulse through the capacitor deposits a charge of K/R, which means that $v_C(0^+) = K/CR$. This causes a negative $i(t)$ to flow, whose magnitude at $t = 0^+$ is K/CR^2, in accordance with Equation 15.7.24. The ILT of Equation 15.7.24 is:

$$i(t) = \frac{K}{R}\delta(t) - \frac{K}{CR^2}e^{-t/\tau} \tag{15.7.25}$$

EXERCISE 15.7.4

Derive and interpret the transfer function $\dfrac{V_C(s)}{V_{SRC}(s)}$ in the circuit of Figure 15.7.4.

Answer: $\dfrac{V_C(s)}{V_{SRC}(s)} = \dfrac{1}{\tau}\dfrac{1}{s+1/\tau}$.

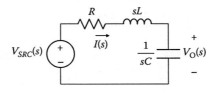

FIGURE 15.7.5
RLC circuit in the s domain.

Natural Response of Second-Order Circuits

Consider the series RLC in the s domain (Figure 15.7.5). The transfer function $\dfrac{I(s)}{V_{SRC}(s)}$ is:

$$H(s) = \frac{I(s)}{V_{SRC}(s)} = \frac{1}{R + sL + 1/sC}$$

$$= \frac{1}{L}\frac{s}{s^2 + sR/L + 1/LC} = \frac{1}{L}\frac{s}{s^2 + 2\alpha s + \omega_0^2} \tag{15.7.26}$$

where $\alpha = R/2L$ and $\omega_0 = 1/\sqrt{LC}$. The denominator of Equation 15.7.26 is the LHS of the characteristic equation of the linear differential equation derived earlier (Equation 12.2.5 Chapter 12). The poles are the roots of the characteristic equation $s_1 = -\alpha + \sqrt{\alpha^2 - \omega_0^2}$ and $s_2 = -\alpha - \sqrt{\alpha^2 - \omega_0^2}$ (Equation 12.2.8 and Equation 12.2.9), which, as explained in Section 12.2 (Chapter 12) determine the type of response. Thus, for the overdamped response, the poles s_1 and s_2 are on the negative real axis (Figure 15.7.6a). As R decreases, the poles move closer together on the real axis, until they coalesce into a double pole at $s = -\omega_0$, corresponding to critical damping $(\alpha = \omega_0)$. With further decrease in R, the poles become complex conjugates lying on the semicircle of radius ω_0. This is because the real part of the poles is

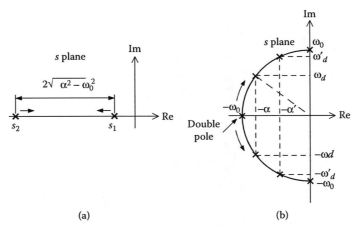

(a)   (b)

FIGURE 15.7.6
Movement of poles in the s domain.

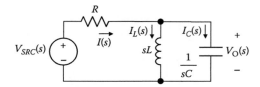

FIGURE 15.7.7
Figure for Example 15.7.3.

$-\alpha$ and their imaginary part is ω_d, where $\alpha^2 + \omega_d^2 = \omega_0^2$ (Figure 15.7.6b). The response is now underdamped. As R becomes smaller still, the poles move closer to the imaginary axis. In the limiting case of $R = 0$, $\alpha = 0$, and the poles become purely imaginary at $\pm j\omega_0$.

> **Conclusion:** *The poles of an overdamped second-order circuit lie on the negative real axis of the s-plane. A critically damped circuit has a double-pole on the real axis. The poles of an underdamped circuit occur in complex conjugate pairs.*

EXERCISE 15.7.5

Interpret the transfer function $\dfrac{I(s)}{V_{SRC}(s)}$ (Equation 15.7.26) in terms of the initial-value theorem and the final-value theorem, in response to a unit impulse, and derive the steady-state sinusoidal response.

Example 15.7.3: Response of Second-Order Circuit in the s Domain

Consider the *RLC* circuit shown in Figure 15.7.7. It is required to derive and interpret the transfer function $\dfrac{I(s)}{V_{SRC}(s)}$. This is the same bandstop circuit analyzed in Example 10.4.1 (Chapter 10).

SOLUTION

The combined impedance of L and C is $\dfrac{sL}{s^2LC+1}$. It follows that:

$$\frac{I(s)}{V_{SRC}(s)} = \frac{1}{R + sL/(s^2LC+1)}$$

which simplifies to:

$$H(s) = \frac{I(s)}{V_{SRC}(s)} = \frac{1}{R}\frac{s^2+\omega_0^2}{s^2+2\alpha_p s+\omega_0^2} \tag{15.7.27}$$

where $\alpha_p = 1/2RC$ and $\omega_0 = 1/\sqrt{LC}$. $H(s)$ has zeros at $s = \pm j\omega_0$ and has the poles of a parallel *GCL* circuit, because the circuit reduces to a parallel circuit when the source is set to zero. The zeros occur when the *LC* branch is in parallel resonance, so that its impedance is infinite, $I(s)$ is zero, and $V_O(s) = V_{SRC}(s)$.

If a sinusoidal source, $v_{SRC} = V_m \cos(\omega t + \theta)$, is applied at $t = 0$, with zero initial energy storage:

$$I(s) = \frac{V_m}{R} \frac{s^2 + \omega_0^2}{s^2 + 2\alpha_p s + \omega_0^2} \left[\frac{s\cos\theta}{s^2 + \omega^2} - \frac{\omega\sin\theta}{s^2 + \omega^2} \right] \tag{15.7.28}$$

When $\omega = \omega_0$,

$$I(s) = \frac{V_m}{R} \frac{s\cos\theta - \omega_0\sin\theta}{s^2 + 2\alpha_p s + \omega_0^2} \tag{15.7.29}$$

As the poles of the denominator have negative real parts, it follows from the final value theorem that $I_{ss}(s) = \lim_{t\to\infty} i(t) = \lim_{s\to 0} sI(s) = 0$, because of parallel resonance as argued earlier. Applying the initial-value theorem to $I(s)$ in Equation 15.7.28 or Equation 15.7.29 gives $i(0^+) = \dfrac{V_m\cos\theta}{R}$. This is to be expected since $V_{SRC}(0) = V_m\cos\theta$ and C behaves as a short circuit at $t = 0$, irrespective of whether or not ω equals ω_0.

It should be noted that whereas the steady state $I_{SS}(s) = 0$ when $\omega = \omega_0$, this is not true of $I_L(s)$ and $I_C(s)$. Thus, using Equation 15.7.29:

$$I_L(s) = \frac{1/sC}{sL + 1/sC} I(s) = \frac{V_m}{RLC} \frac{s\cos\theta - \omega_0\sin\theta}{s^2 + 2\alpha_p s + \omega_0^2} \frac{1}{s^2 + \omega_0^2} \tag{15.7.30}$$

From the initial value theorem, $\lim_{t=0^+} i_L(t) = \lim_{s\to\infty} sI_L(s) = 0$, as expected from the initial conditions. The steady-state value $I_{LSS}(s)$ cannot be obtained by applying the final-value theorem to Equation 15.7.30 because of the poles $\pm j\omega_0$ on the imaginary axis. Since $I_{SS}(s) = 0$, $I_{LSS}(s)$ cannot be obtained from the transfer function $\dfrac{I_L(s)}{I(s)}$ either. Instead, we use Equation 15.7.30, noting that the only terms of interest in the PFE of $I_L(s)$ are $\dfrac{K}{s - j\omega_0}$ and $\dfrac{K^*}{s + j\omega_0}$ because the other terms vanish as $t \to \infty$. To determine K, we multiply the RHS of Equation 15.7.30 by $(s - j\omega_0)$ and set $s = j\omega_0$. This gives $K = \dfrac{V_m}{2\omega_0 L}(\sin\theta - j\cos\theta)$, and $K^* = \dfrac{V_m}{2\omega_0 L}(\sin\theta + j\cos\theta)$. Substituting and simplifying gives:

$$I_{LSS}(s) = \frac{V_m}{\omega_0 L} \frac{\omega_0\cos\theta + s\sin\theta}{s^2 + \omega_0^2} \tag{15.7.31}$$

so that,

$$i_{LSS}(t) = \frac{V_m}{\omega_0 L} \sin\left(\omega_0 t + \theta\right) \tag{15.7.32}$$

Similarly,

$$I_C(s) = \frac{sL}{sL + 1/sC} I(s) = \frac{V_m}{R} \frac{s^3\cos\theta - s^2\omega_0\sin\theta}{s^2 + 2\alpha_p s + \omega_0^2} \frac{1}{s^2 + \omega_0^2} \tag{15.7.33}$$

From the initial-value theorem, $\lim\limits_{t=0^+} i_C(t) = \lim\limits_{s\to\infty} sI_C(s) = \dfrac{V_m \cos\theta}{R}$. To determine $i_{CSS}(t)$, we

proceed in a similar fashion considering terms $\dfrac{K'}{s - j\omega_0}$ and $\dfrac{K'^*}{s + j\omega_0}$. Multiplying the RHS

of Equation 15.7.33 by $(s - j\omega_0)$ and setting $s = j\omega_0$ gives $K' = -\dfrac{CV_m}{2}(\sin\theta - j\cos\theta)$ and

$K'^* = -\dfrac{CV_m}{2}(\sin\theta + j\cos\theta)$. Substituting and simplifying gives:

$$I_{CSS}(s) = -\omega_0 C V_m \frac{\omega_0 \cos\theta + s\sin\theta}{s^2 + \omega_0^2} \tag{15.7.34}$$

and

$$i_{CSS}(t) = -\omega_0 C V_m \sin\left(\omega_0 t + \theta\right) \tag{15.7.35}$$

The interpretation of these results is that when $v_{SRC} = V_m \cos(\omega_0 t + \theta)$ is applied at $t = 0$, with zero initial conditions, $i_L = 0$ and $v_C = 0$ at $t = 0^+$, which makes $i_C = \dfrac{V_m \cos\theta}{R}$. As

$i_{SS} = 0$, $v_O(t) = V_m \cos(\omega_0 t + \theta)$. It follows that $i_{LSS} = \dfrac{1}{L}\displaystyle\int v_O(t)dt = \dfrac{V_m}{\omega_0 L}\sin(\omega_0 t + \theta)$, ignoring

the constant of integration in the steady state, and $i_{CSS}(t) = C\dfrac{dv_O(t)}{dt} = -\omega_0 C V_m \sin(\omega_0 t + \theta)$, so that $i_{LSS}(t) + i_{CSS}(t) = 0$.

EXERCISE 15.7.6

Show that when $\omega \neq \omega_0$, the steady-state sinusoidal response $I_{SS}(s)$ in Example 15.7.3 is the same as that obtained from phasor analysis.

Answer: $H(j\omega) = \dfrac{1}{R}\dfrac{\omega_0^2 - \omega^2}{\omega_0^2 - \omega^2 + j\omega / CR}$.

EXERCISE 15.7.7

Derive $I(s)$, $I_L(s)$, and $I_C(s)$ in Example 15.7.3 when v_{SRC} is a unit impulse. Interpret the result in terms of the initial and final-value theorems.

Answer: $I(s) = \dfrac{1}{R}\left[1 - \dfrac{2\alpha_p s}{s^2 + 2\alpha_p s + \omega_0^2}\right]$; $I_L(s) = \dfrac{\omega_0^2}{R}\dfrac{1}{s^2 + 2\alpha_p s + \omega_0^2}$;

$I_C(s) = \dfrac{1}{R}\left[1 - \dfrac{2\alpha_p s + \omega_0^2}{s^2 + 2\alpha_p s + \omega_0^2}\right]$.

Summary of Main Concepts and Results

- A linear differential equation is transformed by the LT to an algebraic equation in powers of s that can be solved for the LT of the variable of the equation $Y(s)$, as in any algebraic equation. When this is done, the initial conditions $y(0^-)$ and $y^{(1)}(0^-)$ appear like applied inputs.

- The characteristic equation of a linear differential equation is a polynomial in s obtained by taking the LT of the equation with zero forcing function, zero initial conditions, and assuming a nonzero value of the variable of the equation. The LHS of the characteristic equation appears in all the responses derived from the differential equation.

- If the poles of $F(s)$ have negative real parts, except for a simple pole at the origin, if such a pole exists, then: $\lim_{t\to\infty} f(t) = \lim_{s\to 0} sF(s)$. If $F(s)$ is a proper rational function, then: $\lim_{t\to 0^+} f(t) = \lim_{s\to\infty} sF(s)$.

- According to the convolution theorem: $\mathcal{L}\{f(t)*g(t)\} = F(s)G(s)$, where $F(s) = \mathcal{L}\{f(t)\}$ and $G(s) = \mathcal{L}\{g(t)\}$.

- Circuit elements are represented in the s domain by their s-domain impedances: $R, 1/sC, sL$, and sM. Initial stored energy is represented by impulsive or step voltage or current sources of appropriate value and polarity connected in series or in parallel with the circuit element. Initial currents in a linear transformer are accounted for by an extension of the source representation for inductors using the T-equivalent circuit.

- All circuit analysis techniques that apply in the time domain or the frequency domain apply in the s domain.

- In switching problems that involve impulsive readjustments at the instant of switching, initial values in energy storage elements just before switching are used in the s-domain representation. The initial-value theorem gives circuit responses just after switching.

- When a single excitation is applied to a given circuit having no initial energy storage, the transfer function $H(s)$ is the ratio of the LT $Y(s)$ of a designated response to the LT $X(s)$ of the applied excitation.

- The poles of the transfer function are the roots of the characteristic equation and its ILT is the response in the time domain to a unit impulse of excitation.

- If all the poles of the transfer function lie in the open left half of the s-plane, the circuit is stable. If at least one pole lies in the open right half of the s-plane, the circuit is unstable.

- The sinusoidal steady-state response is obtained from the transfer function by substituting $s = j\omega$, multiplying the magnitude of the excitation by $|H(j\omega)|$, and adding the phase angle of $H(j\omega)$ to that of the excitation. As ω is varied, s moves along the imaginary axis in the s plane.

- All the responses of a stable first-order circuit in the s domain are characterized by a pole on the negative real axis whose magnitude is the reciprocal of the time constant.

- The poles of an overdamped second-order circuit lie on the negative real axis of the s-plane. A critically damped circuit has a double-pole on the real axis. The poles of an underdamped circuit occur in complex conjugate pairs.

Learning Outcomes

- Analyze circuits in the s domain and interpret the responses.

Supplementary Topics and Examples on CD

ST15.1 Derivation of the LT: Derives the LT from Fourier series and the Fourier transform.

ST15.2 Partial fraction expansion: Discusses the methods of expressing a rational fraction in terms of a partial fraction expansion.

SE15.1 Response of RC circuit to a single pulse: A pulse extending from $t = -T/2$ to $t = +T/2$ is applied to an *RC* circuit. The response is derived using: (a) the LT, and (b) convolution. It is shown in both cases how to handle a function that begins at $t < 0$.

SE15.2 Impulsive response of switched inductive circuit: Analyzes, using the LT, an inductive circuit in which inductor currents are forced to change instantaneously.

SE15.3 s-Domain equations of linear autotransformer: Derives the s-domain representation of an autotransformer having initial currents.

SE15.4 Transient in a linear transformer: Analyzes a transient in a linear transformer by three methods, one of which is based on instructive physical reasoning.

SE15.5 Response of fourth-order circuit to a single pulse: Analyzes in the s domain the response of a fourth-order *LC* circuit to a single pulse.

Problems and Exercises

P15.1   Laplace Transform

P15.1.1 Use the time-scaling property to show that $\mathcal{L}\left\{\delta(at)\right\} = \dfrac{1}{a}$. Verify by direct evaluation, with a change of the variable of integration so as to have the impulse function at the origin in the standard form.

P15.1.2 Using the division by t property and referring to the Table of Integrals in the Appendix, show that $\mathcal{L}\left\{\dfrac{\sin 5t}{t}u(t)\right\} = \dfrac{\pi}{2} - \tan^{-1}\left(\dfrac{s}{5}\right)$.

P15.1.3 Determine the LTs of the following functions: (a) $\cosh atu(t)$; (b) $\sinh atu(t)$; (c) $te^{-t}\cosh 4tu(t)$.

P15.1.4 Determine the LTs of the following functions: (a) $2e^{-2t}u(t-1)$; (b) $\cos(4t-1)u(t)$.

P15.1.5 Using the time-shift property, show that the LT of a single half sinusoid described by $f(t) = A_m \sin \omega t$, $0 \le t \le \pi/\omega$, and $f(t) = 0$ elsewhere is:

$\dfrac{\omega A_m}{s^2 + \omega^2}(1 + e^{-s\pi/\omega})$. Verify the result by direct integration using partial integration. Deduce that the LT of a full-wave rectified waveform of amplitude A_m

can be expressed as: $\dfrac{\omega A_m}{s^2 + \omega^2}\coth\left(\dfrac{\pi}{2\omega}s\right)$.

P15.1.6 From the LT of a single rectangular pulse (Equation 15.2.23), deduce that the LT of the square wave of Figure 9.2.7 (Chapter 9) can be expressed as:

$\dfrac{A_m}{s}\tanh\left(\dfrac{sT}{4}\right)$.

P15.1.7 From the LT of a single sawtooth (Exercise 15.2.4), deduce that the LT of the

sawtooth waveform of Figure 9.2.1 (Chapter 9) is: $\dfrac{A}{T}\left[\dfrac{1}{s^2} - \dfrac{Te^{-sT}}{s(1-e^{-sT})}\right]$.

P15.1.8 Show that the LT of the single reversed sawtooth pulse of amplitude A and

duration T is: $\dfrac{A}{s} - \dfrac{A}{Ts^2}(1-e^{sT})$, using the following methods: (a) direct evaluation; (b) superposition of ramp and step functions; (c) negation of the sawtooth of P15.1.7 and shifting upward by A. (d) Show that the LT of the inverted

sawtooth waveform of Figure 9.2.4(b) (Chapter 9) is: $\dfrac{A}{T}\left[-\dfrac{1}{s^2} + \dfrac{T}{s(1-e^{-sT})}\right]$.

P15.1.9 Determine the LT of the single triangular pulse of Figure P15.1.9 using the following methods: (a) expressing the function as the superposition of ramp functions; (b) integrating two rectangular pulses of appropriate amplitude, one shifted with respect to the other; (c) adding a sawtooth and a shifted inverted sawtooth; (d) direct evaluation using Equation 15.1.1.

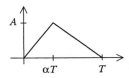

FIGURE P15.1.9

P15.1.10 Determine the LT of (a) an impulse train of strength A and period T: $A[\delta(t) + \delta(t-T) + \delta(t-2T) + \ldots]$; (b) a negative impulse train of strength A and period T, delayed by $\dfrac{T}{2}$: $-A\left[\delta(t-T/2) + \delta(t-3T/2) + \delta(t-5T/2) + \ldots\right]$. From the sum of these two LTs, deduce the LT of the square wave of Figure 9.2.7 (Chapter 9) and compare with the result of P15.1.6.

P15.1.11 The instantaneous power in a circuit is expressed in the s-domain as: $P(s) = \dfrac{s+48}{(s+1)(s+2)}$. Determine the energy delivered to the circuit between $t = 0$ and any arbitrary t.

P15.1.12 Use MATLAB's `residue` command to determine the PFE and the corresponding time function of the following polynomials: (a) $\dfrac{0.1s^4 + 10s^2 + 500}{s^4 + 20s^3 + 40s^2 + 50s + 5000}$; (b) $\dfrac{s^5 + 20s^3 + 40s + 100}{2s^5 + 8s^4 + 10s^3 + 50s^2 + 80s + 200}$.

P15.1.13 Given the differential equation: $x^{(2)}(t) + 2x^{(1)}(t) + x(t) = e^{j2t}$ with zero initial conditions at $t = 0$, determine $x(t)$ for $t \geq 0$. Note that the response to the cosine function can be obtained from that to the sine function by straight differentiation, whereas the converse is not true. Explain why. (Hint: recall that the applied signals are $(\sin 2t)u(t)$ and $(\cos 2t)u(t)$.)

P15.1.14 Given two circuits N_1 and N_2 having impulse responses $h_1(t) = 2\delta^{(1)}(t)$ and $h_2(t) = e^{-t}\sin 2t$, if the two circuits are cascaded, determine the impulse response of the combination, assuming that N_2 has a negligible loading effect on N_1, that is, the output of N_1 with N_2 connected is the same as the open-circuited output of N_1.

P15.1.15 When the excitation of a circuit having zero initial conditions is $te^{-t}u(t)$, the output is $(2 - 4t + t^2)e^{-t}u(t) - 2\cos tu(t) + 6\sin tu(t)$. Determine the transfer function and the impulse response.

P15.1.16 Prove Equation 15.2.26. Use this result to determine the LT of the following functions: (a) $(t-1)u(t-2)$; (b) $(t-2)u(t-1)$. Verify the results by expressing each function as the sum of a delayed ramp function and a delayed step function and deriving their LTs.

P15.1.17 "Helper" functions can be used to derive the LT. A simple form of a helper function is a rectangular function that is multiplied by a basic function to obtain the whole or part of the desired function. For example, the rising part of the waveform of Figure P15.1.9 can be expressed as $\dfrac{At}{\alpha T}[u(t) - u(t - \alpha T)]$, whereas the falling part can be expressed as: $\dfrac{A(T-t)}{(1-\alpha)T}[u(t-\alpha T) - u(t-T)]$. Using Equation 15.2.26, show that the sum of the LTs of these functions is the same as that derived in P15.1.9.

P15.1.18 Evaluate $f(t)$ of Equation 15.4.25, assuming $L = 1\ H$ and $C_1 = C_2 = 1\ F$.

P15.2   First-Order Circuits

P15.2.1 In Figure P15.2.1, the switch is opened at $t = 0$, after being closed for a long time. Determine v_O and the pole location of $V_O(s)$.

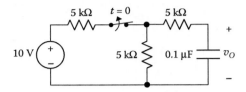

FIGURE P15.2.1

P15.2.2 In Figure P15.2.2, switch 1 has been closed for a long time, with the capacitors initially uncharged. At $t = 0$, switch 1 is opened and switch 2 is closed. Determine v_O and the pole location of $V_O(s)$.

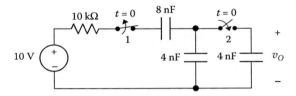

FIGURE P15.2.2

P15.2.3 In Figure P15.2.3, the switch is moved to position b at $t = 0$, after being in position a for a long time. Determine i_O and the pole location of $I_O(s)$.

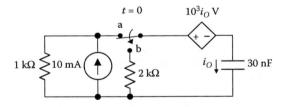

FIGURE P15.2.3

P15.2.4 In Figure P15.2.4, the switch is moved to position b at $t = 0$, after being in position a for a long time. Determine v_O and the pole location of $V_O(s)$.

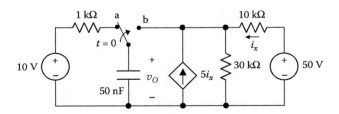

FIGURE P15.2.4

P15.2.5 In Figure P15.2.5, the switch is moved to position b at $t = 0$, after being in position a for a long time. Determine v_O and the pole location of $V_O(s)$.

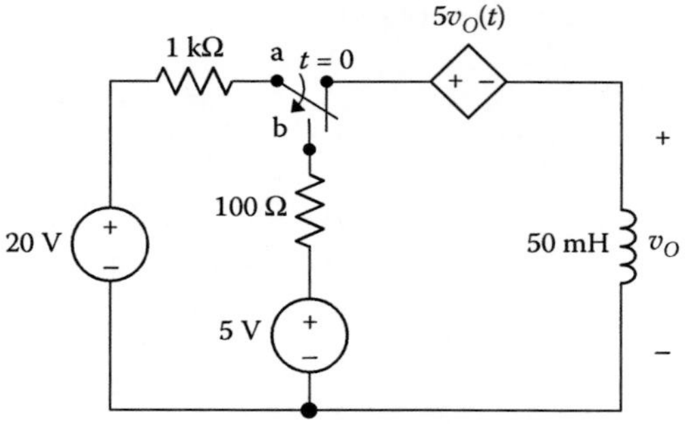

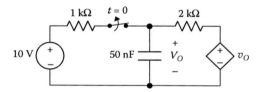

FIGURE P15.2.5

P15.2.6 In Figure P15.2.6, the switch is opened at $t = 0$, after being closed for a long time. Determine v_O and the pole location of $V_O(s)$.

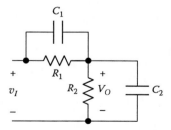

FIGURE P15.2.6

P15.2.7 Determine the current in Example 12.3.1 (Chapter 12) from the LT.

P15.3   Second-Order Circuits

P15.3.1 Figure P15.3.1 shows an attenuator commonly used in oscilloscope measurements to ensure that the input to the oscilloscope v_O is a faithful reproduction of the waveform of the applied signal v_I despite the oscilloscope input impedance represented by $C_2 R_2$. Determine the relationship between the four components that guarantees faithful reproduction of the waveform of the signal.

FIGURE P15.3.1

P15.3.2 If $C_1 = C_2 = 2\ \mu\text{F}$, $R_1 = 10\ \text{k}\Omega$, and $R_2 = 5\ \text{k}\Omega$ in Figure P15.3.1, determine the impulse response of the circuit. From the transfer function, verify the initial-value theorem, the final-value theorem, and derive the response for the sinusoidal steady state.

P15.3.3 Determine the impulse response of the circuit of Figure P15.3.3, assuming $R = 1\,\Omega$ and $C = 1\,F$. Simulate with PSpice using a narrow pulse of large amplitude. Compare with the analytical result on the same graph using Probe.

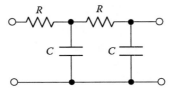

FIGURE P15.3.3

P15.3.4 Given that $R = 10\,k\Omega$ and $C = 1\,nF$ in Figure P15.3.3, from the transfer function, derive the response for the sinusoidal steady state and determine the attenuation and phase shift at $\omega = 10^5$ rad/s.

P15.3.5 At what frequency of the excitation $v_{SRC} = V_m\cos\omega t$ will the circuit of Figure P15.3.5 show unstable behavior?

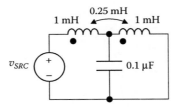

FIGURE P15.3.5

P15.3.6 Determine $V_0(s)$ and v_O in Figure P15.3.6, assuming zero initial conditions. Simulate with PSpice.

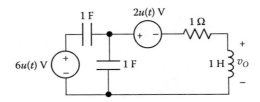

FIGURE P15.3.6

P15.3.7 The transfer function $\dfrac{V_0(s)}{V_I(s)}$ in Figure P15.3.7 has a pole at $\omega = -100 + j700$ rad/s. If $R = 500\,\Omega$, find L and C.

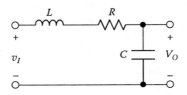

FIGURE P15.3.7

P15.3.8 Determine the transfer function of the circuit of Figure P15.3.8.

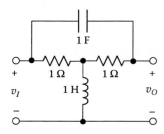

FIGURE P15.3.8

P15.3.9 Determine the transfer function $\dfrac{V_O(s)}{I_I(s)}$ in Figure P15.3.9 assuming $Z_1 = 1 + s$,

$$Z_2 = 1 + 1/s, \quad Y_1 = \frac{s}{2+s}, \quad \text{and} \quad Y_2 = 3s.$$

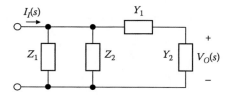

FIGURE P15.3.9

P15.3.10 Determine v_O in Figure P15.3.10 if $v_{SRC} = te^{-t}$. From the transfer function, verify the initial-value theorem, and the final-value theorem for unit impulse and step inputs, and derive the response for the sinusoidal steady state.

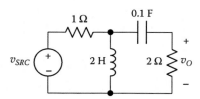

FIGURE P15.3.10

P15.3.11 Determine v_O in Figure P15.3.10 if v_{SRC} is a single rectangular pulse of 1 V amplitude and 1 s duration (Equation 15.2.23). Simulate with PSpice.

P15.3.12 Determine v_O in Figure P15.3.10 if v_{SRC} is a single half-sinusoid of 1 V amplitude and $\omega = 1$ rad/s (P15.1.5). Simulate with PSpice using two VSIN sources, one being delayed half a period with respect to the other.

P15.3.13 Determine v_O in Figure P15.3.10 if v_{SRC} is a single sinusoidal cycle, defined by: $f(t) = A_m \sin\omega t$, $0 \le t \le 2\pi/\omega$, and $f(t) = 0$, elsewhere. Assume that the amplitude of the sinusoid is 1 V and $\omega = 1$ rad/s. Simulate with PSpice using two appropriate VSIN sources.

P15.3.14 Determine v_O in Figure P15.3.10 if v_{SRC} is a single sawtooth pulse of 1 V amplitude and 1 s duration (Exercise 15.2.4). Simulate with PSpice.

P15.3.15 Determine v_O in Figure P15.3.10 if v_{SRC} is a single reversed sawtooth pulse of 1 V amplitude and 1 s duration (P15.1.8). Simulate with PSpice.

P15.3.16 Determine v_O in Figure P15.3.16 if $v_{SRC}(t) = (6 + 3t)u(t)$. Simulate with PSpice. From the transfer function, verify the initial-value theorem, the final-value theorem, and derive the response for the sinusoidal steady state.

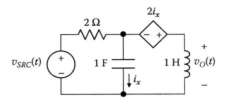

FIGURE P15.3.16

P15.3.17 Determine v_O in Figure P15.3.17 if $v_{SRC} = e^{-t}u(t)$. Simulate with PSpice using VEXP source. From the transfer function, verify the initial-value theorem, the final-value theorem, and derive the response for the sinusoidal steady state.

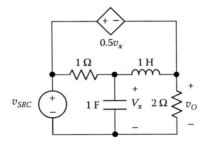

FIGURE P15.3.17

P15.3.18 Determine v_O in Figure P15.3.18 if i_{SRC} is a single rectangular pulse of 1 A amplitude and 1 s duration (Equation 15.2.23). Simulate with PSpice. From the transfer function, verify the initial-value theorem, the final-value theorem, and derive the response for the sinusoidal steady state.

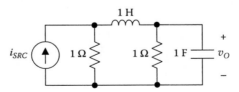

FIGURE P15.3.18

P15.3.19 Determine v_x in Figure P15.3.19 if $v_{SRC} = (t\sin t)u(t)$. From the transfer function, verify the initial-value theorem, the final-value theorem, and derive the response for the sinusoidal steady state.

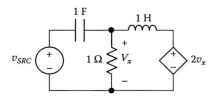

FIGURE P15.3.19

P15.3.20 Determine v_O in Figure P15.3.20 if the initial current in the inductor is 0.1 A and the initial voltage across the capacitor is zero.

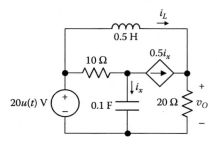

FIGURE P15.3.20

P15.3.21 Determine v_O in Figure P15.3.20 if the initial current in the inductor is zero and the initial voltage across the capacitor is 12 V, having a polarity of a voltage rise in the direction of i_x.

P15.3.22 Determine v_O in Figure P15.3.22 if the initial current in the inductor is 0.2 A and the initial voltage across the capacitor is zero.

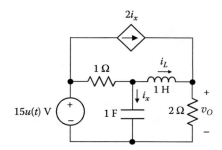

FIGURE P15.3.22

P15.3.23 Determine v_O in Figure P15.3.22 if the initial current in the inductor is zero and the initial voltage across the capacitor is 5V, the polarity being a voltage drop in the direction of i_x.

P15.3.24 Determine v_O in Figure P15.3.24 if the initial current in the inductor is 0.3 A and the initial voltage across the capacitors is zero.

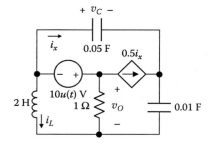

FIGURE P15.3.24

P15.3.25 Determine v_O in Figure P15.3.24 if the initial current in the inductor is zero and the initial voltage across the 0.05 F capacitor is 5 V.

P15.3.26 Determine v_O in Figure P15.3.26 if the initial current in the inductor is 0.2 A and the initial voltage across the capacitor is zero.

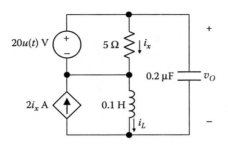

FIGURE P15.3.26

P15.3.27 Determine v_O in Figure P15.3.26 if the initial current in the inductor is zero and the initial voltage across the capacitor is 3 V and of the same polarity as v_O.

P15.3.28 Determine $v_O(t)$ in Figure P15.3.28 if the initial current in the inductor is 0.25 A in the direction of i_x and the initial voltage across the capacitor is zero.

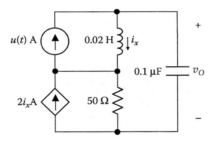

FIGURE P15.3.28

P15.3.29 Determine v_O in Figure P15.3.28 if the initial current in the inductor is zero and the initial voltage across the capacitor is 5 V and of the same polarity as v_O.

P15.3.30 In Figure P15.3.30. the switch is opened at $t = 0$ after being closed for a long time. Determine i_2 given that $i_2(0^-) = 1$ A. Verify conservation of flux linkage.

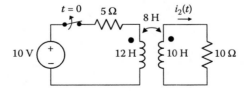

FIGURE P15.3.30

P15.3.31 In Figure P15.3.31. the switch is opened at $t = 0$ after being closed for a long time. Determine i_2. Verify conservation of flux linkage.

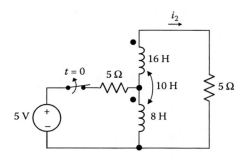

FIGURE P15.3.31

P15.3.32 Show that flux linkage is conserved at the instant of switching in the circuit of Example SE15.2 in the meshes formed by L and L_1, L and L_2, and L_2 and L_3.

P15.3.33 Consider two identical, isolated cascaded LC circuits having $L = 1$ mH and $C = 1$ μF. Simulate with PSpice using a narrow current pulse to excite the first circuit. Supply the second from a VCCS where the controlling voltage is that across the first LC circuit. Verify that the impulse response in unbounded.

P15.4   Convolution

P15.4.1 A function $f(t)$ when convolved with $(1 - e^{-2t})$ gives the function $(1 + e^{-2t} - 2e^{-t})$, $t \geq 0$. Determine $f(t)$. (Refer to Problem P12.3.1).

P15.4.2 Evaluate the functions of Problem P12.3.2 using the convolution theorem.

P15.4.3 Repeat Problem P12.3.11 in the s domain.

P15.4.4 v_{SRC} is an unknown voltage applied to the circuit of Figure P15.4.4. If $v_C = 5e^{-4t}$, determine i_{SRC}. What is $v_{SRC}(t)$?

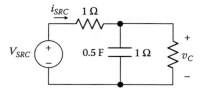

FIGURE P15.4.4

P15.4.5 v_{SRC} is an unknown voltage applied to the circuit of Figure P15.4.5. If $v_O =$
2$-3e^{-t}$, determine i_C.

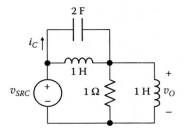

FIGURE P15.4.5

P15.4.6 Use the convolution theorem to prove that $x(t) * \delta^{(1)}(t) = \dfrac{dx(t)}{dt}$.

16

Fourier Transform

Overview

This chapter and the following two chapters are concerned with some basic signal properties and signal processing operations that are important in many applications, particularly in instrumentation, communications, and control systems. The starting point for such a discussion is the Fourier transform (FT). Although the Laplace transform has several advantages over the FT in circuit analysis, the FT is fundamental to signal analysis. Being conceptually an extension of Fourier analysis to nonperiodic signals, it utilizes the same frequency-domain representation as phasor analysis. It shares many of the operational properties of the Laplace transform but has some unique and very useful properties, which are explored in this chapter.

The FT provides a powerful tool for working in the frequency domain. This has many important applications, as explained in this chapter and the next. Among these applications is the reconstruction of medical images, as in computerized axial tomography (CAT) and magnetic resonance imaging (MRI). The usefulness of FT techniques has been greatly enhanced by digital computation, based on a rapid and efficient algorithm known as the **fast Fourier transform** (FFT) that computes the **discrete Fourier transform** (DFT). This transform is an approximation to the FT that produces a finite set of discrete-frequency spectrum values from a finite set of discrete-time values. The DFT and FFT are discussed in Section ST16.4 and Section ST16.5.

Learning Objectives

- To be familiar with:
  - FTs of functions commonly encountered in signal analysis
- To understand:
  - The basic properties of the FT, particularly duality and the convolution properties
  - The application of the FT to analyzing electric circuits
  - The interpretation and applications of Parseval's theorem
  - Some basic properties of signals, namely, causality, bandwidth, and duration-bandwidth product
  - Some basic properties of systems, namely, impulse response, causality, linearity, time invariance, memory, invertibility, and stability

16.1 Derivation of the Fourier Transform

The FT can be derived from the exponential form of the Fourier series expansion (FSE) of a periodic waveform as the period becomes infinitely large. Consider as an example the FSE of the rectangular pulse train analyzed in Example 9.2.3 (Chapter 9). As the period of the function becomes infinitely large, the waveform reduces to a single pulse of amplitude A and duration τ, extending from $t = -\tau/2$ to $t = +\tau/2$. C_n of the waveform was found to be given by:

$$C_n = A\frac{\tau}{T}\mathrm{sinc}(n\omega_0\tau/2)$$

(Equation 9.2.18, Chapter 9). As T will be made very large, and ω_0 correspondingly very small, let us consider C_nT instead of C_n. Thus:

$$C_nT = \int_{-T/2}^{T/2} f(t)e^{-jn\omega_0t}dt = A\tau\mathrm{sinc}(n\omega_0\tau/2), \quad n = 0, \pm 1, \pm 2, \ldots \qquad (16.1.1)$$

Figure 16.1.1a shows the plot of C_nT vs. $n\omega_0$ for the case of $T = 5\tau$ illustrated in Example 9.2.3 (Chapter 9). The plot is a combination of the magnitude and phase spectra of Figure 9.6.2 (Chapter 9), so that the values of C_n are positive when $\angle C_n = 0$ and are negative when $\angle C_n = 180°$. At the first zero crossing on either side of the vertical axis, $n\omega_0\tau/2 = \pm\pi$, or $n\omega_0 = \pm\dfrac{2\pi}{\tau}$, irrespective of T. When $\tau = \dfrac{T}{5}$, $\dfrac{2\pi}{\tau} = 5 \times \dfrac{2\pi}{T} = 5\omega_0$.

If τ is kept constant while T is multiplied by 4, say, to give T', $\omega_0' = \dfrac{2\pi}{T'}$ is ω_0 divided by 4. The spectral lines are now closer together (Figure 16.1.1b), and the first zero crossings occur at $\pm 20\omega_0'$ but still not at $2\pi/\tau$.

As $T \to \infty$, the function becomes aperiodic, the separation $\omega_0 = \dfrac{2\pi}{T}$ between neighboring spectral lines becomes an infinitesimal $d\omega$. The abscissa $n\omega_0$ becomes ω, so that its infinitesimal variation is $d\omega$. Equation 16.1.1 becomes:

$$C_nT = \int_{-\infty}^{\infty} f(t)e^{-j\omega t}dt = A\tau\mathrm{sinc}(\omega\tau/2) = \frac{2A}{\omega}\sin(\omega\tau/2) \qquad (16.1.2)$$

as illustrated in Figure 16.1.1c.

The integral C_nT in Equation 16.1.2 is the FT of the aperiodic time function $f(t)$ and represents a transformation from the time domain to the frequency domain. Thus:

$$\mathcal{F}\{f(t)\} = F(j\omega) = \int_{-\infty}^{\infty} f(t)e^{-j\omega t}dt \qquad (16.1.3)$$

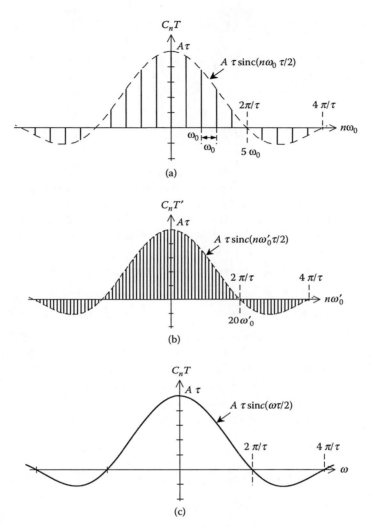

FIGURE 16.1.1
Derivation of Fourier transform.

where $F(j\omega)$ is the FT, or **frequency spectrum** of $f(t)$. The inverse transformation, from the frequency domain to the time domain, is obtained from the exponential form of the FSE:

$$f(t) = \sum_{n=-\infty}^{\infty} C_n e^{jn\omega_0 t} = \sum_{n=-\infty}^{\infty} (C_n T) e^{jn\omega_0 t} \left(\frac{1}{T}\right)$$

In the limit, as $T \to \infty$, $C_n T$ is replaced by $F(j\omega)$ and $n\omega_0$ by ω. Now,

$$\frac{1}{T} = \frac{\omega_0}{2\pi}$$

where ω_0 is the separation between adjacent spectral lines (Figure 16.1.1a). As $T \to \infty$, this separation becomes $d\omega$, as noted earlier. Hence, $\dfrac{1}{T}$ is replaced by $\dfrac{d\omega}{2\pi}$, and the summation becomes an integration, which gives:

$$\mathcal{F}^{-1}\left\{F(j\omega)\right\} = f(t) = \frac{1}{2\pi}\int_{-\infty}^{\infty} F(j\omega)e^{j\omega t}d\omega \qquad (16.1.4)$$

where $\mathcal{F}^{-1}\left\{F(j\omega)\right\}$ is the inverse Fourier transform (IFT) of $F(j\omega)$.

The following should be noted concerning the Fourier transformation:

1. The FT can be interpreted along the same line as the FSE. The frequency spectrum of a periodic function is discrete, defined at frequencies $n\omega_0$, the amplitude of the nth harmonic being $2|C_n|$. On the other hand, as $T \to \infty$, $2|C_n| = 2\dfrac{|F(j\omega)|}{T} = \dfrac{|F(j\omega)|}{\pi}d\omega$ becomes infinitesimally small, and the frequency separation ω_0 also becomes infinitesimal. Whereas a periodic function is the sum of a denumerably infinite sum of sinusoids of finite amplitude, successively separated in frequency by ω_0, the aperiodic function can be viewed as the sum of a non-denumerably infinite sum of sinusoids of infinitesimal amplitude and infinitesimal frequency separation. The amplitude of each of these components is $\dfrac{|F(j\omega)|}{\pi}d\omega$ and its phase is that of $F(j\omega)$.

2. Sufficient conditions for the existence of the FT are: (1) $\displaystyle\int_{-\infty}^{+\infty}|f(t)|dt$ is finite, that is, $|f(t)|$ is absolutely integrable, and (2) the function is piecewise continuous on every finite interval of time, as in the case of the Laplace transform, which means that the function is integrable over any finite interval. These conditions are sufficient but not necessary, so that functions for which $\displaystyle\int_{-\infty}^{+\infty}|f(t)|dt$ is infinite can have an FT. Examples are dc quantities, step functions, and sinusoidal functions. The FTs of these functions cannot be obtained directly using Equation 16.1.3 but are obtained by other means, as illustrated in the following discussion.

3. Compared to the one-sided LT of Equation 15.1.1 (Chapter 15), the FT is defined from $t = -\infty$ to $t = +\infty$ and has $s = j\omega$. The extension over all time means that functions that are not zero for $t < 0$ can have an FT, but initial conditions at $t = 0$ are not easily included. Moreover, because of the $e^{-\sigma t}$ in the LT integral, where σ is the real part of s, the LT integral converges for more functions than does the FT integral. As a consequence, functions such as $tu(t)$, for example, have an LT but not an FT.

4. The importance of the FT in signal analysis stems from the fact that it transforms a time function extending over all time to the frequency domain, thereby yielding a true representation of the frequency content of the function. Strictly speaking,

a dc or a sinusoidal signal extends from $t = -\infty$ to $t = +\infty$. To portray the frequency content of such signals through a transformation from the time domain, the transformation should include all time, as in the FT.

5. Some signal processing applications involve the FT in terms of spatial coordinates in two or three dimensions rather than time. Equation 16.1.3 applies in this case for each spatial coordinate, with the spatial coordinate replacing t.

6. The FT can be physically real. For example, if an image on a transparent film is placed at the focal point of a convex lens and illuminated by coherent light, as from a laser, the image seen at the other focal point is the two-dimensional spatial FT of the image on the transparency.

We illustrate the derivation of the FT of some basic functions, starting with a unit impulse $\delta(t)$ at the origin. It follows from Equation 16.1.3 that:

$$\mathcal{F}\{\delta(t)\} = \int_{-\infty}^{\infty} \delta(t)\, e^{-j\omega t} dt = 1 \tag{16.1.5}$$

as for the LT.

We next derive the FT of $f(t) = 1$, a constant. In this case $\int_{-\infty}^{+\infty} |f(t)| dt$ is infinite, so the FT transform cannot be obtained directly from Equation 16.1.3, but can be derived from Equation 16.1.5. According to this equation, $\mathcal{F}^{-1}\{1\} = \delta(t)$. But if we apply the inverse transform relation (Equation 16.1.4):

$$\mathcal{F}^{-1}\{1\} = \delta(t) = \frac{1}{2\pi} \int_{-\infty}^{\infty} 1 \times e^{j\omega t} d\omega \tag{16.1.6}$$

Interchanging the variables ω and t gives: $\dfrac{1}{2\pi} \int_{-\infty}^{\infty} 1 \times e^{j\omega t} dt = \delta(\omega)$. Replacing ω by $-\omega$, with $\delta(\omega) = \delta(-\omega)$:

$$\int_{-\infty}^{\infty} e^{-j\omega t} dt = 2\pi\delta(\omega) \tag{16.1.7}$$

The LHS of Equation 16.1.7 is the FT of a constant equal to 1 that extends over all time. Hence:

$$\mathcal{F}\{1\} = 2\pi\delta(\omega) \tag{16.1.8}$$

which is an impulse of strength 2π at $\omega = 0$ in the frequency domain. This to be expected because a true dc signal that extends from $t = -\infty$ to $t = +\infty$ is of zero frequency. Its representation in the frequency domain should therefore be at $\omega = 0$ only, which is an impulse.

Let us derive next the FT of $f(t) = e^{j\omega_0 t}$. From Equation 16.1.3,

$$F(j\omega) = \int_{-\infty}^{\infty} \left(e^{j\omega_0 t}\right) e^{-j\omega t} dt = \int_{-\infty}^{\infty} e^{-j(\omega - \omega_0)t} dt$$

Comparing with Equation 16.1.7, it is seen that ω in the integral on the LHS is replaced by $(\omega - \omega_0)$. Replacing ω on the RHS of Equation 16.1.7 by $(\omega - \omega_0)$ gives:

$$\mathcal{F}\left\{e^{j\omega_0 t}\right\} = 2\pi\delta(\omega - \omega_0) \qquad (16.1.9)$$

Once we have the FT of $e^{j\omega_0 t}$, the FTs of $\cos\omega_0 t$ and $\sin\omega_0 t$ readily follow. Thus:

$$\mathcal{F}\left\{\cos\omega_0 t\right\} = \mathcal{F}\left\{\frac{e^{j\omega_0 t} + e^{-j\omega_0 t}}{2}\right\} = \pi\left[\delta(\omega + \omega_0) + \delta(\omega - \omega_0)\right] \qquad (16.1.10)$$

The FT of $\cos\omega_0 t$ is, therefore, two impulse functions, one at $\omega = \omega_0$, the other at $\omega = -\omega_0$, each of strength π. As the function $\cos\omega_0 t$ extends from $t = -\infty$ to $t = +\infty$ and has a single frequency component ω_0, it is represented by impulses in the frequency domain at $\omega = \pm\omega_0$. As discussed in connection with the exponential form of Fourier series (Section 9.2, Chapter 9), the positive and negative frequencies combine to give a real function $\cos\omega_0 t$.

Similarly:

$$\mathcal{F}\left\{\sin\omega_0 t\right\} = \mathcal{F}\left\{\frac{e^{j\omega_0 t} - e^{-j\omega_0 t}}{2j}\right\} = j\pi\left[\delta(\omega + \omega_0) - \delta(\omega - \omega_0)\right] \qquad (16.1.11)$$

In this case, the impulse function at $\omega = -\omega_0$ has a phase angle of j, that is, $\pi/2$, whereas the impulse function at $\omega = \omega_0$ has a phase angle of $-\pi/2$.

Given the similarity between the LT and the FT, it may be wondered if the FT can be derived from the LT. This is, indeed, the case under certain conditions, as discussed in Section ST16.1.

As in the case of the LT, MATLAB can be used to find the FT and IFT. For example, if we enter:

```
>> syms t w
>> fourier(sin(2*t))
```

MATLAB returns

```
>> i*pi*(-Dirac(w-2)+Dirac(w+2))
```

where Dirac() denotes $\delta()$. If we then enter:

```
>> ifourier(2*sin(2*w)/w)
```

MATLAB returns

```
>> Heaviside(x+2)-Heaviside(x-2)
```

where Heaviside() denotes u() and x stands for t.

Example 16.1.1: Fourier Transform of Two Antisymmetrical Pulses

It is required to obtain the FT of the function shown in Figure 16.1.2.

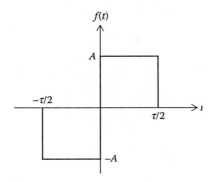

FIGURE 16.1.2
Figure for Example 16.1.1.

SOLUTION

$$\mathcal{F}\{f(t)\} = -\int_{-\tau/2}^{0} Ae^{-j\omega t}dt + \int_{0}^{\tau/2} Ae^{-j\omega t}dt = \frac{A}{j\omega}\left[e^{-j\omega t}\right]_{-\tau/2}^{0} - \frac{A}{j\omega}\left[e^{-j\omega t}\right]_{0}^{\tau/2} =$$

$$= -j\left(\frac{2A}{\omega}\right)\left(1 - \cos\frac{\omega\tau}{2}\right)$$

The FT is therefore:

$$= \left(\frac{2A}{j\omega}\right)\left(1 - \cos\frac{\omega\tau}{2}\right) \tag{16.1.12}$$

Example 16.1.2: Inverse Fourier Transform

The FT of a function is shown in Figure 16.1.3. It is required to find $f(t)$.

SOLUTION

$$\mathcal{F}^{-1}\{\mathcal{F}(j\omega)\} = \frac{1}{2\pi}\int_{-2}^{-1} 2e^{j\omega t}d\omega + \frac{1}{2\pi}\int_{-1}^{1} e^{j\omega t}d\omega \cdot \frac{1}{2\pi}\int_{1}^{2} 2e^{j\omega t}d\omega = \frac{1}{\pi jt}\left[e^{j\omega t}\right]_{-2}^{-1} + \frac{1}{2\pi jt}\left[e^{j\omega t}\right]_{-1}^{1}$$

$$+ \frac{1}{\pi jt}\left[e^{j\omega t}\right]_{1}^{2} = \frac{1}{\pi jt}\left[e^{-jt} - e^{-j2t} + e^{j2t} - e^{jt}\right] + \frac{1}{2\pi jt}\left[e^{jt} - e^{-jt}\right] = \frac{2}{\pi t}\left[\sin 2t - \sin t\right] + \frac{1}{\pi t}\left[\sin t\right]$$

$$= \frac{1}{\pi t}\left[2\sin 2t - \sin t\right].$$

EXERCISE 16.1.1

Determine FT of $e^{-a|t|}$, $a > 0$.

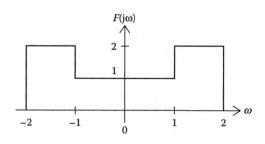

FIGURE 16.1.3
Figure for Example 16.1.2.

Answer: $\dfrac{2a}{a^2 + \omega^2}$

16.2 Some General Properties of the Fourier Transform

Real and Imaginary Parts

It should be emphasized that, according to Equation 16.1.3, $F(j\omega)$ is a complex quantity. Assuming that $f(t)$ is a real function of time:

$$F(j\omega) = \int_{-\infty}^{\infty} f(t)e^{-j\omega t}\,dt = \int_{-\infty}^{\infty} f(t)\cos\omega t\,dt - j\int_{-\infty}^{\infty} f(t)\sin\omega t\,dt$$

$$= A(\omega) + jB(\omega)$$

where $A(\omega)$ and $B(\omega)$ are the real and imaginary parts, respectively, of $F(j\omega)$, expressed as:

$$A(\omega) = \int_{-\infty}^{\infty} f(t)\cos\omega t\,dt, \text{ and } B(\omega) = -\int_{-\infty}^{\infty} f(t)\sin\omega t\,dt \qquad (16.2.1)$$

The following may be deduced from these relations:

DEDUCTIONS

1. The real part of $F(j\omega)$ is even, $A(\omega) = A(-\omega)$.
2. The imaginary part of $F(j\omega)$ is odd, $B(\omega) = -B(-\omega)$.
3. The magnitude of $F(j\omega)$, which is $\sqrt{A^2(\omega) + B^2(\omega)}$, *is even.*
4. The phase angle of $F(j\omega)$, which is $\tan^{-1}(B(\omega)/A(\omega))$, *is odd.*
5. Replacing ω *by* $-\omega$ gives the *complex conjugate* of $F(j\omega)$, that is, $F(-j\omega) = F^*(j\omega)$.

6. If $f(t)$ is even, $f(t)\cos\omega t$ is an even function of t, and $f(t)\sin\omega t$ is an odd function of t, which makes $B(\omega) = 0$. Hence, $F(j\omega)$ is real and even, with $A(\omega) =$

$$2\int_0^\infty f(t)\cos\omega t dt .$$

7. If $f(t)$ is odd, $f(t)\cos\omega t$ is an odd function of t, and $f(t)\sin\omega t$ is an even function of t, which makes $A(\omega) = 0$. Hence, $F(j\omega)$ is imaginary and odd, with

$$B(\omega) = -2\int_0^\infty f(t)\sin\omega t dt .$$

For example, $\mathcal{F}\{\cos\omega_0 t\}$ is real and even (Equation 16.1.10), whereas $\mathcal{F}\{\sin\omega_0 t\}$ is imaginary and odd (Equation 16.1.11). Similarly, the FT of the rectangular pulse of width τ, centered at the origin, is real and even (Equation 16.1.2), and the FT of two antisymmetrical pulses is imaginary and odd.

Fourier Transform at Zero Frequency

If $\omega = 0$ in Equation 16.1.3,

$$F(0) = \int_{-\infty}^\infty f(t)dt \tag{16.2.2}$$

That is, the value of FT at $\omega = 0$ is the total, net positive area subtended by $f(t)$. This area is the average, or dc component, of $f(t)$ multiplied by the interval over which $f(t) \neq 0$. Hence, if $f(t)$ has no dc component and is of finite duration, $F(0) = 0$. This can serve as a useful check on the FTs of such functions.

$F(0)$ plays an important role in the FT, as illustrated by the following examples and by the differentiation and integration properties discussed later.

Example 16.2.1: Fourier Transforms of Exponential, Signum, and Step Functions

It is required to derive the FTs of: (a) $e^{-at}u(t)$, $a > 0$; (b) a signum function, sgn(t), defined as 1 for $t > 0$ and -1 for $t < 0$; and (c) a unit step function $u(t)$.

SOLUTION

(a) From Equation 16.1.3:

$$\mathcal{F}\{e^{-at}u(t)\} = \int_0^\infty e^{-at}e^{-j\omega t}dt = \frac{1}{a + j\omega} \tag{16.2.3}$$

(b) sgn(t) = $u(t) - u(-t)$ (Figure 16.2.1). As the Fourier integral of these step functions does not converge, we consider sgn(t) as:

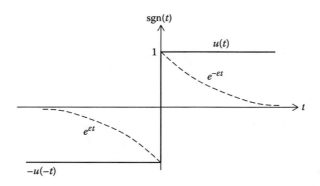

FIGURE 16.2.1
Figure for Example 16.2.1.

$$\lim_{\varepsilon \to 0}\left[e^{-\varepsilon t}u(t) - e^{\varepsilon t}u(-t)\right], \ \varepsilon > 0 \ \text{(Figure 16.2.1). Hence:}$$

$$\mathcal{F}\left\{\text{sgn}(t)\right\} = \lim_{\varepsilon \to 0}\left[\int_{0}^{\infty}e^{-\varepsilon t}e^{-j\omega t}dt - \int_{-\infty}^{0}e^{\varepsilon t}e^{-j\omega t}dt = \lim_{\varepsilon \to 0}\left[\frac{1}{\varepsilon + j\omega} - \frac{1}{\varepsilon - j\omega}\right]\right.$$

$$= \lim_{\varepsilon \to 0}\left[\frac{-2j\omega}{\varepsilon^2 + \omega^2}\right] = \frac{2}{j\omega} \tag{16.2.4}$$

(c) $u(t) = \dfrac{1}{2}\,\text{sgn}(t) + \dfrac{1}{2}$ (Figure 16.2.2). Taking the FT of both sides, and using Equation 16.1.8 and Equation 16.1.12:

$$\mathcal{F}\{u(t)\} = \pi\delta(\omega) + \frac{1}{j\omega} \tag{16.2.5}$$

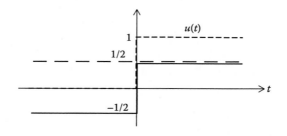

FIGURE 16.2.2
Figure for Example 16.2.1.

The following should be noted:

1. $\int_0^\infty e^{-at}dt = \dfrac{1}{a}$ and is finite for $a \neq 0$. The FT of $e^{-at}u(t)$ does not contain an impulse

   (Equation 16.2.3). If we take $\lim\limits_{a \to 0} e^{-at}u(t)$, the time function reduces to $u(t)$. Equation 16.2.3 with $a = 0$ gives an FT of $1/j\omega$, which is correct for all ω except for $\omega = 0$, where $1/j\omega$ is undefined.

2. It is generally true that functions having an infinite $\int_{-\infty}^{+\infty}|f(t)|dt$ have impulses in

   their FTs, as in the case of a constant, $u(t)$, $e^{j\omega_0 t}$, $\cos\omega_0 t$, and $\sin\omega_0 t$. The signum function is an exception because it can be considered as the sum of $u(t)$ and $-u(-t)$. In effect, the impulses of these two functions at the origin cancel out.

3. When a function has no dc component, but its $\int_{-\infty}^{+\infty}|f(t)|dt$ is infinite, it does not

   follow that $F(0) = 0$. Thus, $F(0) = 0$ in the case of $\cos\omega_0 t$ and $\sin\omega_0 t$, but $F(0) \neq 0$ in the case of $\mathrm{sgn}(t)$.

EXERCISE 16.2.1
Apply L'Hopital's rule to show that $F(0) = 0$ in Equation 16.1.12.

EXERCISE 16.2.2

The functions $e^{-at}(\cos\omega_0 t)u(t)$ and $e^{-at}\sin\omega_0 tu(t)$, $a > 0$ have a finite $\int_{-\infty}^\infty |f(t)|dt$, so their FTs do not have impulses. Show that their FTs are as given by:

$$\mathcal{F}\{e^{-at}(\cos\omega_0 t)u(t)\} = \frac{a+j\omega}{\omega_0^2 +(a+j\omega)^2}, \, a > 0 \qquad (16.2.6)$$

$$\mathcal{F}\{e^{-at}\sin\omega_0 tu(t)\} = \frac{\omega_0}{\omega_0^2 +(a+j\omega)^2}, \, a > 0 \qquad (16.2.7)$$

Use two methods: (a) from the FT of $e^{-at}e^{j\omega_0 t}u(t)$; (b) direct integration of Equation 16.1.3.

Duality

   Concept: *The symmetry between the expressions for the FT and its inverse underlies an important duality relationship:*

$$\text{If } \mathcal{F}\{f(t)\} = F(j\omega), \text{ then } \mathcal{F}\{F(jt)\} = 2\pi f(-\omega) \qquad (16.2.8)$$

PROOF

$$f(t) = \mathcal{F}^{-1}\left\{F(j\omega)\right\} = \frac{1}{2\pi}\int_{-\infty}^{\infty} F(j\omega)e^{j\omega t}d\omega$$

Multiplying both sides by 2π and interchanging the variables t and ω:

$$2\pi f(\omega) = \int_{-\infty}^{\infty} F(jt)e^{j\omega t}dt$$

Replacing ω by $-\omega$ gives Equation 16.2.8.

$F(jt)$ is the FT of $f(t)$ with ω replaced by t, and $f(-\omega)$ is $f(t)$ with t replaced by the reflected frequency $-\omega$. If $f(t)$ is an even function, then $f(-\omega) = f(\omega)$, or if $f(t)$ is an odd function, then $f(-\omega) = -f(\omega)$. Duality means that to produce a frequency spectrum $F(j\omega)$ that has the same shape as a given time function $f(t)$, the time function must have the same shape as the frequency spectrum $F(j\omega)$. We have encountered an example of duality when considering the FTs of a unit impulse and a unity constant. The FT of a unit impulse in the time domain is a unity constant in the frequency domain. To produce an impulse in the frequency domain, a constant or dc signal of $1/2\pi$ is applied in the time domain. A direct application of duality of the FT is found in MRI (Application Window 16.2.1).

Application Window 16.2.1: Slice Selection in Magnetic Resonance Imaging (MRI)

In MRI, a sequence of radio frequency (rf) magnetic pulses is applied that excites hydrogen nuclei in a longitudinal slice of the body in the presence of a strong, static, and longitudinal magnetic field. As the hydrogen nuclei revert to their quiescent state before the pulse sequence was applied the response to the rf excitation is recorded and processed, using Fourier transformations to produce an image of the slice.

In order to select a particular longitudinal slice for excitation, it is necessary to have a rectangular pulse in the frequency domain. The width of the frequency pulse determines the slice thickness, and the center frequency of the pulse determines the longitudinal distance of the slice from a reference point of the main magnet. The rectangular pulse in the frequency domain is produced by a sinc signal in the time domain, in accordance with duality of the FT. As indicated in Figure 16.1.1, the FT transform of a rectangular pulse of width τ in the time domain is a function $\text{sinc}(\omega\tau/2)$ in the frequency domain. Hence, from duality, a rectangular pulse of width τ in the frequency domain is produced by a $\text{sinc}(\omega t/2)$ signal in the time domain.

16.3 Operational Properties of the Fourier Transform

Multiplication by a Constant

If $\mathcal{F}\left\{f(t)\right\} = F(j\omega)$, then:

$$\mathcal{F}\left\{Kf(t)\right\} = KF(j\omega) \tag{16.3.1}$$

PROOF

$$\mathcal{F}\left\{Kf(t)\right\} = \int_{-\infty}^{\infty} Kf(t)e^{-j\omega t}dt = K\int_{-\infty}^{\infty} f(t)e^{-j\omega t}dt = KF(j\omega)$$

It follows that the FT of $K\delta(t)$ is K and the FT of a constant K is $2\pi K\delta(t)$.

Addition/Subtraction

If $\mathcal{F}\left\{f_1(t)\right\} = F_1(j\omega)$ and $\mathcal{F}\left\{f_2(t)\right\} = F_2(j\omega)$, then:

$$\mathcal{F}\left\{f_1(t) \pm f_2(t)\right\} = F_1(j\omega) \pm F_2(j\omega) \qquad (16.3.2)$$

PROOF

$$\mathcal{F}\left\{f_1(t) \pm f_2(t)\right\} = \int_{-\infty}^{\infty}\left[f_1(t) \pm f_2(t)\right]e^{-j\omega t}dt = \int_{-\infty}^{\infty} f_1(t)e^{-j\omega t}dt \pm \int_{-\infty}^{\infty} f_2(t)e^{-j\omega t}dt = F_1(j\omega) \pm F_2(j\omega)$$

We have already used this property and the preceding one to derive the FTs of $\cos\omega t$ and $\sin\omega t$.

Time Scaling

If $\mathcal{F}\left\{f(t)\right\} = F(j\omega)$, then:

$$\mathcal{F}\left\{f(at)\right\} = \frac{1}{|a|}F\left(\frac{j\omega}{a}\right) \qquad (16.3.3)$$

where a is a constant.

PROOF

$$\mathcal{F}\left\{f(at)\right\} = \int_{-\infty}^{\infty} f(at)e^{-j\omega t}dt$$

Let $t' = at$, where $a > 0$. Then,

$$\int_{-\infty}^{\infty} f(at)e^{-j\omega t}dt = \frac{1}{a}\int_{-\infty}^{\infty} f(t')e^{-j\omega t'/a}dt' = \frac{1}{a}F\left(\frac{j\omega}{a}\right)$$

If $a < 0$, the integration limits become interchanged, so that

$$\mathcal{F}\left\{f(at)\right\} = -\frac{1}{a}\int_{-\infty}^{\infty} f(t')e^{-j\omega t'/a}dt' = -\frac{1}{a}F\left(\frac{j\omega}{a}\right)$$

Combining these results gives Equation 16.3.3.

> **Concept:** *If a function is compressed in the time domain, it expands in the frequency domain, and conversely.*

This follows from Equation 16.3.3. For example, if the duration of the rectangular pulse of width τ, centered at the origin, is halved, its FT (Equation 16.1.2) becomes $\dfrac{A\tau}{2}\text{sinc}\left(\dfrac{\omega\tau}{4}\right)$, so that the main lobe now extends from $\omega = -\dfrac{4\pi}{\tau}$ to $+\dfrac{4\pi}{\tau}$. As a practical application, if a recording is played back at half-speed, so that its duration is doubled, its bandwidth is halved, that is, the highest frequency is now one-half of what it was at normal speed.

Time Reversal

If $\mathcal{F}\left\{f(t)\right\} = F(j\omega)$, then:

$$\mathcal{F}\left\{f(-t)\right\} = F\left(-j\omega\right) = F^{*}(j\omega) \tag{16.3.4}$$

In other words, a negation in time causes a negation in frequency. Equation 16.3.4 follows from the time-scaling property when $a = -1$. Hence:

$$\mathcal{F}\{u(-t)\} = \pi\delta(\omega) - \frac{1}{j\omega} \tag{16.3.5}$$

so that $\mathcal{F}\{\text{sgn}(t)\} = \mathcal{F}\{u(t)\} - \mathcal{F}\{u(-t)\} = \dfrac{2}{j\omega}$, as in Equation 16.2.4.

Translation in Time

If $\mathcal{F}\left\{f(t)\right\} = F(j\omega)$, then:

$$\mathcal{F}\left\{f(t-a)\right\} = e^{-j\omega a}F(j\omega) \tag{16.3.6}$$

In words, a delay of a in time reduces the phase angle of $F(j\omega)$ by ωa without changing its magnitude.

PROOF

$$\mathcal{F}\left\{f(t-a)\right\} = \int_{-\infty}^{\infty} f(t-a)e^{-j\omega t}dt$$

Substituting $t' = t - a$,

$$\int_{-\infty}^{\infty} f(t-a)e^{-j\omega t}dt = \int_{-\infty}^{\infty} f(t')e^{-j\omega(t'+a)}dt' = e^{-j\omega a}\int_{-\infty}^{\infty} f(t')e^{-j\omega t'}dt' = e^{-j\omega a}F(j\omega)$$

Example 16.3.1: Application of Translation-in-Time Property

It is required to obtain the FT of the function of Figure 16.1.2 from that of a symmetrical rectangular pulse.

SOLUTION

From Equation 16.1.2, the FT of a rectangular pulse of duration τ centered at $t = 0$ is $\dfrac{2A}{\omega}\sin(\omega\tau/2)$. When the pulse width is reduced from τ to $\dfrac{\tau}{2}$, τ is replaced by $\dfrac{\tau}{2}$, so that the FT becomes $\dfrac{2A}{\omega}\sin(\omega\tau/4)$. If the pulse is delayed by $\tau/4$, its FT becomes $\dfrac{2Ae^{-j\omega\tau/4}}{\omega}\sin(\omega\tau/4)$.

If negated and advanced by $\tau/4$, its FT becomes $-\dfrac{2Ae^{+j\omega\tau/4}}{\omega}\sin(\omega\tau/4)$. Adding the two transforms gives: $\dfrac{2A}{\omega}\sin\left(\dfrac{\omega\tau}{4}\right)(e^{-j\omega\tau/4} - e^{j\omega\tau/4}) = -j\dfrac{4A}{\omega}\sin^2\left(\dfrac{\omega\tau}{4}\right) = -j\left(\dfrac{2A}{\omega}\right)\left(1-\cos\dfrac{\omega\tau}{2}\right)$, as in Equation 16.1.12.

As an application of translation in time and duality, $\mathcal{F}\{\delta(t-a)\} = e^{-ja\omega}$ (Equation 16.3.6). It follows from Equation 16.2.8 that $\mathcal{F}\{e^{-jat}\} = 2\pi\delta(-\omega-a) = 2\pi\delta(\omega+a)$, as the impulse function is even. Setting $a = -\omega_0$ gives: $\mathcal{F}\{e^{j\omega_0 t}\} = 2\pi\delta(\omega-\omega_0)$, as in Equation 16.1.9.

An interesting and useful application of the translation-in-time property is in the derivation of the FT of a periodic signal from the exponential form of its FSE (Example 16.3.2).

Example 16.3.2: Fourier Transform of Periodic Signal

It is required to obtain the FT of a periodic signal from the exponential form of its FSE.

SOLUTION

Consider the exponential form of the FSE of a periodic signal (Equation 9.2.12, Chapter 9):

$$f(t) = \sum_{n=-\infty}^{\infty} C_n e^{jn\omega_0 t}$$

Taking the FT of both sides:

$$F(j\omega) = \sum_{n=-\infty}^{\infty} 2\pi C_n \delta(\omega-n\omega_0) \tag{16.3.7}$$

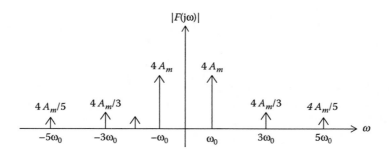

FIGURE 16.3.1
Figure for Example 16.3.2.

In words, the FT of a periodic signal is an infinite series of impulses weighted by $2\pi C_n$. As an example, consider the delayed square wave of Figure 9.2.8 (Chapter 9). From, Exercise 9.2.4: $C_n = 0$, for n even or zero, and $C_n = -\dfrac{j2A_m}{\pi n}$, for odd n. Its FT is therefore

$$F(j\omega) = \sum_{n=-\infty}^{\infty} -\frac{j4A_m}{n}\delta(\omega - n\omega_0), \quad n \text{ odd.}$$

Figure 16.3.1 shows a plot of $|F(j\omega)|$. The phase angle is $-\pi/2$ for $n > 0$, and $\pi/2$ for $n < 0$.

Note that for a cosine function, $\cos \omega t = \dfrac{1}{2}(e^{j\omega t} + e^{-j\omega t})$, which makes $C_n = 1/2$, $n = \pm 1$, and Equation 16.3.7 reduces to Equation 16.1.10. For a sine function, $\sin \omega t = \dfrac{1}{2j}(e^{j\omega t} - e^{-j\omega t})$, which makes $C_n = -j/2$ for $n = 1$ and $+j/2$ for $n = -1$. Equation 16.3.5 reduces to Equation 16.1.11. Note that $F(0) = 0$ in Figure 16.3.1, as for the sine and cosine functions.

Differentiation in Time

If $\mathcal{F}\{f(t)\} = F(j\omega)$, then:

$$\mathcal{F}\left\{\frac{df(t)}{dt}\right\} = j\omega F(j\omega) \qquad (16.3.8)$$

PROOF

$$f(t) = \mathcal{F}^{-1}\{F(j\omega)\} = \frac{1}{2\pi}\int_{-\infty}^{\infty} F(j\omega)e^{j\omega t}\,d\omega$$

Differentiating both sides with respect to t:

$$\frac{df}{dt} = \frac{j\omega}{2\pi}\int_{-\infty}^{\infty} F(j\omega)e^{j\omega t}\,d\omega$$

In other words,

$$\frac{df}{dt} = j\omega \mathcal{F}^{-1}\{F(j\omega)\}$$

Taking the FT of both sides gives Equation 16.3.8. From repeated application of Equation 16.3.8:

$$\mathcal{F}\left\{\frac{df^n(t)}{dt}\right\} = (j\omega)^n F(j\omega) \qquad (16.3.9)$$

Note that differentiation removes the dc component $\int_{-\infty}^{\infty} f(t)dt$, so that the FT derived from the differentiation property is that of $\left\{f(t) - \int_{-\infty}^{\infty} f(t)dt\right\}$, which implies that the FT derived from the differentiation property is that of $f(t)$, only when $\int_{-\infty}^{\infty} f(t)dt = 0$. Thus, in the case of $\frac{1}{2}\text{sgn}(t)$, the derivative is $\delta(t)$ and $\int_{-\infty}^{\infty} \frac{1}{2}\text{sgn}(t)dt = 0$, so that $\mathcal{F}\{\delta(t)\} = j\omega \times \frac{1}{2} \times \frac{2}{j\omega} = 1$. On the other hand, $\mathcal{F}\{\delta(t)\} \neq j\omega \mathcal{F}\{u(t)\}$, because $\int_{-\infty}^{\infty} u(t)dt \neq 0$ and is infinite.

Example 16.3.3: Application of Differentiation-in-Time Property
It is required to apply the differentiation-in-time property to $f(t)$ of Figure 16.3.2.

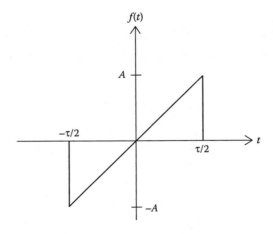

FIGURE 16.3.2
Figure for Example 16.3.3.

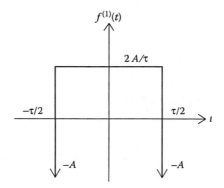

FIGURE 16.3.3
Figure for Example 16.3.3.

SOLUTION

Since $\int_{-\infty}^{\infty} f(t)dt = 0$, the differentiation-in-time property will give the FT of $f^{(1)}(t)$

unambiguously. $f(t) = \dfrac{2A}{\tau}t, -\dfrac{\tau}{2} < t < \dfrac{\tau}{2}$. Hence, $F(j\omega) = \dfrac{2A}{\tau}\int_{-\tau/2}^{\tau/2} te^{-j\omega t}dt = \dfrac{2A}{\tau}\left\{\left[\dfrac{te^{-j\omega t}}{-j\omega}\right]_{-\tau/2}^{\tau/2} + \right.$

$\left.\dfrac{1}{\omega^2}\int_{-\tau/2}^{\tau/2} e^{-j\omega t}dt\right\} = -\dfrac{2A}{j\omega}\cos(\omega\tau/2) - \dfrac{j4A}{\tau\omega^2}\sin(\omega\tau/2) = \dfrac{2A}{j\omega^2}\left[-\omega\cos(\omega\tau/2) + \dfrac{2}{\tau}\sin(\omega\tau/2)\right]$. As $f(t)$

is real and odd, $F(j\omega)$ is imaginary and odd. Moreover, substituting $\omega = 0$ gives an indeterminate value for $F(0)$. To obtain $F(0)$, we apply L'Hopital's rule and differentiate the numer-

ator and denominator with respect to ω, which gives: $\dfrac{2A}{j2\omega}\left[\dfrac{\omega\tau}{2}\sin(\omega\tau/2) - \cos(\omega\tau/2) + \right.$

$\left.\cos(\omega\tau/2)\right] = \dfrac{A\tau}{j2}\sin(\omega\tau/2)$. Substituting $\omega = 0$ gives $F(0) = 0$, as expected.

It follows that $\mathcal{F}\{f^{(1)}(t)\} = j\omega F(j\omega) = -2A\cos(\omega\tau/2) + \dfrac{4A}{\omega\tau}\sin(\omega\tau/2)$. It can be readily verified that this is the FT of $f^{(1)}(t)$ shown in Fig. 16.3.3. Using the translation-in-time property, the FT of the impulses at $t = -\tau/2$ and $t = +\tau/2$ are $-Ae^{j\omega\tau/2}$ and $-Ae^{-j\omega\tau/2}$, respectively. From

Equation 16.1.2 the FT of the pulse is $\dfrac{4A}{\omega\tau}\sin(\omega\tau/2)$. Hence, $\mathcal{F}\{f^{(1)}(t)\} = -Ae^{j\omega\tau/2} - Ae^{-j\omega\tau/2} + $

$\dfrac{4A}{\omega\tau}\sin(\omega\tau/2) - 2A\cos(\omega\tau/2) + \dfrac{4A}{\omega\tau}\sin(\omega\tau/2)$, as derived.

EXERCISE 16.3.1

Consider $e^{-at}u(t), a > 0$, whose FT is $\dfrac{1}{a + j\omega}$. As $\dfrac{d\left(e^{-at}u(t)\right)}{dt} = -ae^{-at}u(t) + \delta(t)$, take the FT of both

sides and apply the differentiation-in-time property to verify that $\mathcal{F}\left\{\dfrac{d\left(e^{-at}u(t)\right)}{dt}\right\} = \dfrac{j\omega}{a + j\omega}$.

Integration in Time

If $\mathcal{F}\{f(t)\} = F(j\omega)$, then:

$$\mathcal{F}\left\{\int_{-\infty}^{t} f(t)dt\right\} = \frac{F(j\omega)}{j\omega} + \pi F(0)\delta(\omega) \tag{16.3.10}$$

where $F(0) = \int_{-\infty}^{\infty} f(t)dt$ (Equation 16.2.2).

PROOF

$$\text{If } y(t) = \int_{-\infty}^{t} f(t)dt, \text{ then } \frac{dy}{dt} = f(t)$$

Applying the differentiation property: $j\omega\mathcal{F}\{y(t)\} = F(j\omega)$, so that:

$$\mathcal{F}\left\{\int_{-\infty}^{t} f(t)dt\right\} = \frac{F(j\omega)}{j\omega} \text{ (if } \int_{-\infty}^{\infty} f(t)dt = 0) \tag{16.3.11}$$

In view of what was said about the differentiation-in-time property, and as $f(t)$ is the derivative of $y(t)$, this implies that $\int_{-\infty}^{\infty} f(t)dt = 0$ in Equation 16.3.11, or $F(0) = 0$. When $\int_{-\infty}^{\infty} f(t)dt \neq 0$, then we should add $\pi F(0)\delta(\omega)$ as was done for $u(t)$ in Equation 16.2.5, which gives Equation 16.3.10.

We can verify Equation 16.3.10 in the case of $u(t)$, as $u(t)$ is the integral of $\delta(t)$. For the impulse function, $F(j\omega)$ and $F(0)$ are both unity. Substituting in Equation 16.3.10 gives Equation 16.2.5.

EXERCISE 16.3.2
Verify the differentiation and integration properties for $\cos\omega t$ and $\sin\omega t$. Note that multiplication or division by $j\omega$ changes an odd $F(j\omega)$ to an even $F(j\omega)$, and conversely.

Example 16.3.4: Application of Integration-In-Time Property

Given the function $g(t)$ of Figure 16.3.3, it is required to obtain the FT of its integral, the function of Figure 16.3.2, by applying the integration-in-time property.

SOLUTION
$g(t) = -A\delta(t+\tau/2) - A\delta(t-\tau/2) + 2A/\tau$, $-\tau/2 < t < \tau/2$, and $g(t) = 0$ elsewhere. Hence

$G(j\omega) = -Ae^{+j\omega\tau/2} - Ae^{-j\omega\tau/2} + \frac{4A}{\omega\tau}\sin(\omega\tau/2)$, using Equation 16.1.2 and Equation 16.3.5, or

$G(j\omega) = -2A\cos(\omega\tau/2) + \frac{4A}{\omega\tau}\sin(\omega\tau/2)$. Moreover, $\int_{-\infty}^{\infty} g(t)dt = -A-A+2A=0$. Applying

Equation 16.3.9, $\mathcal{F}\{G^{(-1)}(t)\} = \frac{G(j\omega)}{j\omega} = j\frac{2A}{\tau\omega^2}\left[\omega\tau\cos(\omega\tau/2) - 2\sin(\omega\tau/2)\right]$, as in Example 16.3.3.

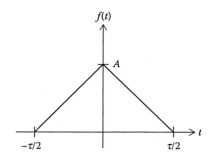

FIGURE 16.3.4
Figure for Exercise 16.3.3.

EXERCISE 16.3.3
Derive the FT of the function of Figure 16.3.4 from that of Example 16.1.1 using the integration-in-time property.

Answer: $F(j\omega) = \left(\dfrac{4A}{\omega^2\tau}\right)\left(1 - \cos\dfrac{\omega\tau}{2}\right) = \dfrac{A\tau}{2}\left(\text{sinc}\,\dfrac{\omega\tau}{4}\right)^2$.

EXERCISE 16.3.4

Consider $e^{-at}u(t)$, $a > 0$, whose FT is $\dfrac{1}{a+j\omega}$. As $\displaystyle\int_{-\infty}^{t} e^{-at}u(t) = \dfrac{1}{a}\left(1 - e^{-at}\right)u(t)$, take the FT of

both sides and apply the integration-in-time property to verify that $\mathcal{F}\left\{\displaystyle\int_{-\infty}^{0} e^{-at}u(t)\right\} =$

$\dfrac{1}{a}\left[\dfrac{1}{j\omega(a+j\omega)} + \pi\delta(\omega)\right]$.

Multiplication by *t*
If $\mathcal{F}\{f(t)\} = F(j\omega)$, then:

$$\mathcal{F}\{tf(t)\} = j\frac{d}{d\omega}F(j\omega) \tag{16.3.12}$$

Thus, multiplication by *t* corresponds to multiplying the derivative in frequency of $F(j\omega)$ by *j*.

PROOF

$$\frac{d}{d\omega}F(j\omega) = \frac{d}{d\omega}\left[\int_{-\infty}^{\infty} f(t)e^{-j\omega t}dt\right] = \int_{-\infty}^{\infty} -jtf(t)e^{-j\omega t}dt = -j\,\mathcal{F}\{tf(t)\}$$

Equation 16.3.12 then follows. Repeated application of Equation 16.3.12 gives:

$$\mathcal{F}\left\{t^n f(t)\right\} = j^n \frac{d^n}{d\omega^n} F(j\omega) \tag{16.3.13}$$

As an example of Equation 16.3.12, $\mathcal{F}\{e^{-at}u(t)\} = \dfrac{1}{a + j\omega}$ (Equation 16.2.3). Hence:

$$\mathcal{F}\left\{te^{-at}u(t)\right\} = -\frac{j \times j}{\left(a + j\omega\right)^2} = \frac{1}{(a + j\omega)^2} \tag{16.3.14}$$

$$\mathcal{F}\left\{t^2 e^{-at}u(t)\right\} = -\frac{j \times 2j}{\left(a + j\omega\right)^3} = \frac{2}{(a + j\omega)^3}$$

$$\mathcal{F}\left\{t^n e^{-at}u(t)\right\} = \frac{n!}{(a + j\omega)^{n+1}} \tag{16.3.15}$$

Translation in Frequency

If $\mathcal{F}\left\{f(t)\right\} = F(j\omega)$, then:

$$\mathcal{F}\left\{f(t)e^{j\omega_0 t}\right\} = F\left(j(\omega - \omega_0)\right) \tag{16.3.16}$$

Hence, a translation in the frequency domain is equivalent to a phase shift in the time domain.

PROOF

$$\mathcal{F}\left\{f(t)e^{j\omega_0 t}\right\} = \int_{-\infty}^{\infty} f(t)e^{j\omega_0 t}e^{-j\omega t}dt = \int_{-\infty}^{\infty} f(t)e^{-j(\omega - \omega_0)t}dt = F\left(j(\omega - \omega_0)\right)$$

Example 16.3.5: Application of Translation-in-Frequency Property

It is required to derive the FTs of: (1) $\left\{(\cos\omega_0 t)u(t)\right\}$ and (2) $\left\{(\sin\omega_0 t)u(t)\right\}$.

SOLUTION

1. We express $\cos\omega_0 t$ as $\dfrac{1}{2}(e^{j\omega_0 t} + e^{-j\omega_0 t})$. Hence, $\mathcal{F}\{(\cos\omega_0 t)u(t)\} = \mathcal{F}\left\{\dfrac{1}{2}e^{j\omega_0 t}u(t) + \right.$

$\left. \dfrac{1}{2}e^{-j\omega_0 t}u(t)\right\}$. Applying Equation 16.3.16 to Equation 16.2.5 gives:

$$\mathcal{F}\left\{(\cos\omega_0 t)u(t)\right\} = \frac{\pi}{2}\left[\delta(\omega - \omega_0) + \delta(\omega + \omega_0)\right] + \frac{j\omega}{\omega_0^2 - \omega^2} \tag{16.3.17}$$

2. We express $\sin \omega_0 t$ as $\dfrac{1}{2j}(e^{j\omega_0 t} - e^{-j\omega_0 t})$. Hence, $\mathcal{F}\{(\sin \omega_0 t)u(t)\} = \mathcal{F}\left\{\dfrac{1}{2j}e^{j\omega_0 t}u(t) - \right.$

$\left. \dfrac{1}{2j}e^{-j\omega_0 t}u(t)\right\}$. Applying Equation 16.3.16 to Equation 16.2.5 gives:

$$\mathcal{F}\{(\sin \omega_0 t)u(t)\} = \frac{\pi}{2j}\left[\delta(\omega - \omega_0) + \delta(\omega + \omega_0)\right] + \frac{\omega_0}{\omega_0^2 - \omega^2} \tag{16.3.18}$$

Convolution in Time

STATEMENT

The FT of the convolution of two time functions equals the product of their FTs:

$$\mathcal{F}\{f(t) * g(t)\} = F(j\omega)G(j\omega) \tag{16.3.19}$$

PROOF

From the discussion of Section 12.3 (Chapter 12), convolution of two functions $f(t)$ and $g(t)$ is defined as:

$$y(t) = f(t) * g(t) = \int_{-\infty}^{\infty} f(\lambda)g(t - \lambda)d\lambda \tag{16.3.20}$$

The FT of $y(t)$ is:

$$Y(j\omega) = \int_{-\infty}^{\infty}\left[\int_{-\infty}^{\infty} f(\lambda)g(t - \lambda)d\lambda\right]e^{-j\omega t}dt \tag{16.3.21}$$

Changing the order of integration and taking $f(\lambda)$ outside the integral with respect to t:

$$Y(j\omega) = \int_{-\infty}^{\infty} f(\lambda)\left[\int_{-\infty}^{\infty} g(t - \lambda)e^{-j\omega t}dt\right]d\lambda \tag{16.3.22}$$

The inner integral is the FT of the function $g(t)$ translated in time by λ. From Equation 16.3.6 this FT is $e^{-j\omega\lambda}G(j\omega)$. Taking $G(j\omega)$ outside the integral with respect to λ:

$$Y(j\omega) = G(j\omega)\int_{-\infty}^{\infty} f(\lambda)e^{-j\omega\lambda}d\lambda = G(j\omega)F(j\omega) \tag{16.3.23}$$

Equation 16.3.19 follows, bearing in mind that the ordering of the two functions on either side of the equation is immaterial because convolution, like multiplication, is commutative.

Convolution in Frequency

STATEMENT

The FT of the product of two time functions equals the convolution of their FTs divided by 2π:

$$\mathcal{F}\{f(t)g(t)\} = \frac{1}{2\pi} F(j\omega) * G(j\omega) \qquad (16.3.24)$$

PROOF

The proof exploits the symmetry between the FT and the IFT. We first express convolution of the FTs of two signals in the frequency domain by replacing time functions in Equation 16.3.21 by their corresponding FTs. Thus:

$$Y(j\omega) = F(j\omega) * G(j\omega) = \int_{-\infty}^{\infty} F(j\lambda) G(j(\omega - \lambda)) \, d\lambda \qquad (16.3.25)$$

The IFT of $Y(j\omega)$ is:

$$y(t) = \frac{1}{2\pi} \int_{-\infty}^{\infty} \left[\int_{-\infty}^{\infty} F(j\lambda) G(j(\omega - \lambda)) d\lambda \right] e^{j\omega t} d\omega \qquad (16.3.26)$$

Changing the order of integration and taking $F(j\lambda)$ outside the integral with respect to ω:

$$y(t) = \frac{1}{2\pi} \int_{-\infty}^{\infty} F(j\lambda) \left[\int_{-\infty}^{\infty} G(j(\omega - \lambda)) e^{j\omega t} d\omega \right] d\lambda \qquad (16.3.27)$$

The inner integral is the 2π times the IFT of the function $G(j\omega)$ translated in frequency by λ. From Equation 16.3.16 this integral is $2\pi e^{j\lambda t} g(t)$. Taking $g(t)$ outside the integral with respect to λ:

$$y(t) = g(t) \int_{-\infty}^{\infty} F(j\lambda) e^{jt\lambda} d\lambda = 2\pi g(t) f(t) \qquad (16.3.28)$$

Taking the FT of both sides and dividing by 2π gives Equation 16.3.24.
Table 16.3.1 summarizes the basic properties of the FT, and Table 16.3.2 lists some useful FT pairs.

16.4 Circuit Applications of the Fourier Transform

Generally speaking, the LT transform is more useful than the FT transform in circuit applications because: (1) it can easily account for initial conditions, (2) it provides a powerful tool for analyzing switched circuits, particularly those involving impulsive readjustment at the instant of switching, and (3) some functions have an LT but not an FT.

TABLE 16.3.1

Basic Properties of the Fourier Transform $F\{f(t)\} = F(\omega)$

$\mathcal{F}\{Kf(t)\}$	$KF(j\omega)$
$\mathcal{F}\{f_1(t) \pm f_2(t)\}$	$F_1(j\omega) \pm F_2(j\omega)$
$\mathcal{F}\{f(at)\}$	$\dfrac{1}{|a|}F\left(\dfrac{j\omega}{a}\right)$
$\mathcal{F}\{f(-t)\}$	$F(-j\omega) = F^*(j\omega)$
$\mathcal{F}\{f(t-a)\}$	$e^{-j\omega a}F(j\omega)$
$\mathcal{F}\left\{\dfrac{df(t)}{dt}\right\}$	$j\omega F(j\omega)$
$\mathcal{F}\left\{\dfrac{df^n(t)}{dt}\right\}$	$(j\omega)^n F(j\omega)$
$\mathcal{F}\left\{\displaystyle\int_{-\infty}^{t} f(t)dt\right\}$	$\dfrac{F(j\omega)}{j\omega} + \pi F(0)\delta(\omega)$
$\mathcal{F}\{tf(t)\}$	$j\dfrac{d}{d\omega}F(j\omega)$
$\mathcal{F}\{t^n f(t)\}$	$j^n \dfrac{d^n}{d\omega^n}F(j\omega)$
$\mathcal{F}\{f(t)e^{j\omega_0 t}\}$	$F(j(\omega-\omega_0))$
$\mathcal{F}\{F(t)\}$	$2\pi f(-\omega)$
$\mathcal{F}\{f(t) * g(t)\}$	$F(j\omega)G(j\omega)$
$\mathcal{F}\{f(t)g(t)\}$	$\dfrac{1}{2\pi}F(j\omega) * G(j\omega)$

On the other hand, the FT is useful in that it can readily handle functions defined over all time, $-\infty < t < \infty$.

The general procedure of applying the FT in circuit analysis is a generalization of the phasor method. The circuit is represented in the frequency domain, as in phasor analysis, and the usual circuit techniques applied to derive the transfer function $H(j\omega)$ as the ratio of the required response $Y(j\omega)$ to the applied excitation $X(j\omega)$, both expressed in the frequency domain (Section 10.2, Chapter 10). Thus:

$$H(j\omega) = \frac{Y(j\omega)}{X(j\omega)} \qquad (16.4.1)$$

Recall that when considering the steady-state response to a periodic input, $X(j\omega)$ was represented by the exponential form of its FSE, and $Y(j\omega)$ was derived as the response to the complex excitations represented by $+n$ and $-n$ in the FSE (Equation 9.6.4, Chapter 9). When the input is aperiodic, it is, therefore, natural to consider $X(j\omega)$ and $Y(j\omega)$ as the FTs of the excitation and response, respectively. Once $X(j\omega)$ and $H(j\omega)$ are known, $Y(j\omega)$ is

TABLE 16.3.2

Fourier Transform Pairs

$f(t)$	$F(j\omega)$	Reference Equation
$\delta(t)$	1	16.1.5
1	$2\pi\delta(\omega)$	16.1.8
$u(t)$	$\pi\delta(\omega) + \dfrac{1}{j\omega}$	16.2.5
$u(t + \tau/2) - u(t - \tau/2)$ $= \text{rect}(t/\tau)$	$\tau\,\text{sinc}(\omega\tau/2)$	16.1.2 and 16.2.8
$\dfrac{1}{2\pi}\omega_0\,\text{sinc}(\omega_0 t/2)$	$u(\omega + \omega_0/2) - u(\omega + \omega_0/2)$ $= \text{rect}(\omega/\omega_0)$	16.1.2 and 16.2.8
$\dfrac{\beta}{\pi}\,\text{sinc}(\beta t)$	$F(j\omega) = 1,\ -\beta < \omega < \beta,\ F(j\omega) = 0$ elsewhere	16.1.2 and 16.2.8
$\text{sgn}(t)$	$\dfrac{2}{j\omega}$	16.2.4
$e^{-at}u(t),\ a > 0$	$\dfrac{1}{a + j\omega}$	16.2.3
$t^n e^{-at}u(t)$	$\dfrac{n!}{(a + j\omega)^{n+1}}$	16.3.15
$e^{j\omega_0 t}$	$2\pi\delta(\omega - \omega_0)$	16.1.9
$\cos\omega_0 t$	$\pi\big[\delta(\omega + \omega_0) + \delta(\omega - \omega_0)\big]$	16.1.10
$\sin\omega_0 t$	$j\pi\big[\delta(\omega + \omega_0) - \delta(\omega - \omega_0)\big]$	16.1.11
$(\cos\omega_0 t)u(t)$	$\dfrac{\pi}{2}[\delta(\omega + \omega_0) + \delta(\omega - \omega_0)] + \dfrac{j\omega}{\omega_0^2 - \omega^2}$	16.3.17
$(\sin\omega_0 t)u(t)$	$\dfrac{\pi}{2j}[\delta(\omega + \omega_0) + \delta(\omega - \omega_0)] + \dfrac{\omega_0}{\omega_0^2 - \omega^2}$	16.3.18
$e^{-at}(\cos\omega_0 t)u(t),\ a > 0$	$\dfrac{a + j\omega}{\omega_0^2 + (a + j\omega)^2}$	16.2.6
$e^{-at}\sin\omega_0 t u(t),\ a > 0$	$\dfrac{\omega_0}{\omega_0^2 + (a + j\omega)^2}$	16.2.7

obtained as their product, and $y(t)$ as the IFT of $Y(j\omega)$. The procedure is illustrated by Example 16.4.1 and Example 16.4.2.

Example 16.4.1: Analysis of Circuit Using Fourier Transform

It is required to obtain the responses v_R and v_C in Figure 16.4.1 when v_{SRC} is: (1) $\delta(t)$, (2) $u(t)$.

SOLUTION
The transfer function is the same in both cases, namely,

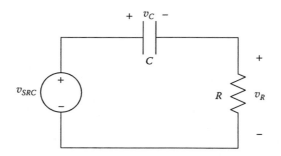

FIGURE 16.4.1
Figure for Example 16.4.1.

$$H_R(j\omega) = \frac{V_R(j\omega)}{V_{SRC}(j\omega)} = \frac{R}{R+1/j\omega C} = \frac{j\omega}{(1/\tau)+j\omega}; H_C(j\omega) = \frac{V_C(j\omega)}{V_{SRC}(j\omega)} = \frac{1/j\omega C}{R+1/j\omega C} = \frac{1}{\tau}\frac{1}{(1/\tau)+j\omega}.$$

1. $V_{SRC}(j\omega) = 1$. Hence, $V_C(j\omega) = \dfrac{1}{\tau}\dfrac{1}{(1/\tau)+j\omega}$; from Table 16.3.2, $v_C(t) = \dfrac{1}{\tau}e^{-t/\tau}u(t)$.

    $V_R(j\omega) = \dfrac{j\omega}{(1/\tau)+j\omega}$; dividing the numerator by the denominator, $V_R(j\omega) =$

    $1 - \dfrac{1/\tau}{(1/\tau)+j\omega}$, so that $v_R(t) = \delta(t) - \dfrac{1}{\tau}e^{-t/\tau}u(t)$. These are obviously the correct responses, as follows from the natural response of the circuit. Note that an initial voltage equal to the strength of the impulse divided by τ can be easily accounted for in this simple case by applying an impulse at $t = 0$ and taking the responses at $t \geq 0^+$, after the impulse is over.

2. $V_{SRC}(j\omega) = \dfrac{1}{j\omega} + \pi\delta(\omega)$. Hence, $V_R(j\omega) = \dfrac{1}{(1/\tau)+j\omega} + \dfrac{j\omega\pi\delta(\omega)}{(1/\tau)+j\omega}$. The IFT of the

    first term is $v_R(t) = e^{-t/\tau}u(t)$. The inverse of the second term is

    $\dfrac{1}{2\pi}\displaystyle\int_{-\infty}^{\infty}\dfrac{j\omega\pi\delta(\omega)}{(1/\tau)+j\omega}e^{j\omega t}d\omega = \dfrac{1}{2\pi}\displaystyle\int_{0^-}^{0^+}0\times\delta(\omega)d\omega = 0 \times 1 = 0$. As for $V_C(j\omega)$, we have:

    $V_C(j\omega) = \dfrac{1}{\tau}\dfrac{1}{(1/\tau)+j\omega}\left[\dfrac{1}{j\omega}+\pi\delta(\omega)\right] = \dfrac{1}{j\omega\tau}\dfrac{1}{(1/\tau)+j\omega} + \dfrac{\pi\delta(\omega)}{\tau}\dfrac{1}{(1/\tau)+j\omega}$. To find

    the IFT of the first term we express it in the form of partial fractions (Section 15.3, Chapter 15). Although unnecessary in this case, it is usually more convenient when finding the partial fraction expansion to substitute $s = j\omega$, determine the partial fraction expansion, and substitute back for s. The first term in terms of

    partial fractions involving s becomes: $\dfrac{1}{s\tau}\dfrac{1}{(1/\tau)+s} = \dfrac{K_1}{s} + \dfrac{K_2}{(1/\tau)+s}$, where K_1 and

    K_2 are constants to be determined. To determine K_1 we multiply both sides by s and substitute $s = 0$. This gives $K_1 = 1$. To determine K_2 we multiply both sides

    by $(1/\tau) + s$ and substitute $s = -\dfrac{1}{\tau}$, which gives $K_1 = -1$. Replacing s by $j\omega$,

$\dfrac{1}{j\omega\tau}\dfrac{1}{(1/\tau)+j\omega}=\dfrac{1}{j\omega}-\dfrac{1}{(1/\tau)+j\omega}$. As for the second term, it evaluates to zero for

all ω except $\omega=0$, for which it becomes $\pi\delta(\omega)$. Hence, $V_C(j\omega)=\dfrac{1}{j\omega}+\pi\delta(\omega)$

$-\dfrac{1}{(1/\tau)+j\omega}$. The IFT is: $v_C(t)=\left(1-e^{-t/\tau}\right)u(t)$. Both $v_R(t)$ and $v_C(t)$ are in agreement with the step response of the circuit.

There is no real advantage in obtaining the steady-state response to $\cos\omega_0 t$ using the FT because the magnitude and phase of the output with respect to the input are given by $H(j\omega)$.

EXERCISE 16.4.1

Determine v_R in Figure 16.4.1 in response to: (1) $v_{SRC}=\text{sgn}(t)$, both directly and by expressing $\text{sgn}(t)$ as $-1+2u(t)$; (2) $\cos\omega_0 t$ using the FT, and verify using phasor analysis.

Answer: (1) $2e^{-t/\tau}$; (2) $\dfrac{-\tau\omega_0}{\sqrt{1+(\tau\omega_0)^2}}\sin(\omega_0 t-\beta)$, where $\tan\beta=\tau\omega_0$.

Example 16.4.2: Response of *RL* Circuit to Exponential Input

Consider a simple series *RL* circuit (Figure 12.1.1, Chapter 12) having $R=1\ \Omega$ and $L=1$ H. It is required to determine i for all t given that $v_{SRC}=10e^{-3|t|}$ V (Figure 16.4.2).

SOLUTION

The transfer function of current in terms of applied voltage is $H(j\omega)=\dfrac{I(j\omega)}{V_{SRC}(j\omega)}=\dfrac{1}{1+j\omega}$.

From Exercise 16.1.1, $V_{SRC}(j\omega)=\dfrac{60}{9+\omega^2}$. Hence, $I(j\omega)=\dfrac{60}{(1+j\omega)(9+\omega^2)}$. To facilitate inverting $I(j\omega)$, we substitute $s=j\omega$, as in Example 16.4.1, and express the rational fraction in

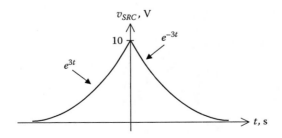

FIGURE 16.4.2
Figure for Example 16.4.2.

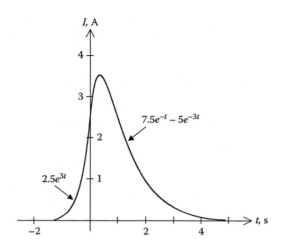

FIGURE 16.4.3
Figure for Example 16.4.2.

s in terms of partial fractions of factors of the denominator. This gives $\dfrac{60}{(1+s)(9-s^2)} =$

$\dfrac{K_1}{1+s} + \dfrac{K_2}{3+s} + \dfrac{K_3}{3-s}$. To find K_1, we multiply both sides by $(1 + s)$ and substitute $s = -1$, which gives $K_1 = 7.5$. In a similar manner, we find $K_2 = 5$ and $K_3 = 2.5$. Replacing s by $j\omega$,

it follows that: $I(j\omega) = \dfrac{7.5}{1+j\omega} - \dfrac{5}{3+j\omega} + \dfrac{2.5}{3-j\omega}$. We now have to invert each of these terms.

From Table 16.3.2, the inverse of the first two terms is $(7.5e^{-t} - 5e^{-3t})u(t)$. To invert the third term, we note that according to the time reversal property (Equation 16.3.4), $\mathcal{F}^{-1}\{F(-j\omega)\} = f(-t)$. Hence, the IFT of the last term is $2.5e^{3t}u(-t)$. It follows that: $i(t) = 2.5e^{3t}u(-t) + (7.5e^{-t} - 5e^{-3t})u(t)$, which means that $i(t) = 2.5e^{3t}$, for $t \le 0$, and $i(t) = 7.5e^{-t} - 5e^{-3t}$, for $t \ge 0$. $i(t)$ is plotted in Figure 16.4.3.

EXERCISE 16.4.2
Verify the expression for $i(t)$ in Example 16.4.2 using the convolution integral.

16.5 Parseval's Theorem

The power dissipated in a resistor R by a voltage $v(t)$ or a current $i(t)$ is, respectively, $v^2(t)/R$ and $Ri^2(t)$. When $R = 1 \ \Omega$, this power can be conveniently expressed as $f^2(t)$, where $f(t)$ could be voltage or current. The energy dissipated in the 1 Ω resistor over all time is

then $W = \displaystyle\int_{-\infty}^{\infty} f^2(t)dt$. Parseval's theorem states that:

$$\int_{-\infty}^{\infty} f^2(t)dt = \frac{1}{2\pi} \int_{-\infty}^{\infty} |F(\omega)|^2 \, d\omega \qquad (16.5.1)$$

That is, the energy dissipated in the 1 Ω resistor can be calculated either in the time domain or in the frequency domain using $|F(j\omega)|$.

PROOF

$$\int_{-\infty}^{\infty} f^2(t)dt = \int_{-\infty}^{\infty} f(t)f(t)dt = \int_{-\infty}^{\infty} f(t)\left[\frac{1}{2\pi} \int_{-\infty}^{\infty} F(j\omega)e^{j\omega t} d\omega\right] dt \qquad (16.5.2)$$

Moving $f(t)$ inside the integral with respect to ω:

$$\int_{-\infty}^{\infty} f^2(t)dt = \frac{1}{2\pi} \int_{-\infty}^{\infty} \left[\int_{-\infty}^{\infty} F(j\omega)f(t)e^{j\omega t} d\omega\right] dt \qquad (16.5.3)$$

Interchanging the order of integration and moving $F(j\omega)$ out of the integral with respect to t:

$$\int_{-\infty}^{\infty} f^2(t)dt = \frac{1}{2\pi} \int_{-\infty}^{\infty} F(j\omega)\left[\int_{-\infty}^{\infty} f(t)e^{j\omega t} dt\right] d\omega \qquad (16.5.4)$$

The inner integral is $F(-j\omega) = F^*(j\omega)$. Substituting in Equation 16.5.4 gives Equation 16.5.1.

The following should be noted concerning Parseval's theorem.

1. As $|F(j\omega)|$ is even (Section 16.2), Equation 16.5.1 could be equally expressed as:

$$\int_{-\infty}^{\infty} f^2(t)dt = \frac{1}{\pi} \int_{0}^{\infty} |F(j\omega)|^2 \, d\omega \qquad (16.5.5)$$

2. The plot of $|F(j\omega)|^2$ vs. ω is the **energy spectrum** of the signal. The energy in any frequency band from ω_1 to ω_2 is, from Equation 16.5.5:

$$W_{12} = \frac{1}{\pi} \int_{\omega_1}^{\omega_2} |F(j\omega)|^2 \, d\omega \qquad (16.5.6)$$

Note that Equation 16.5.5 was used for W_{12}, and not Equation 16.5.1, because Equation 16.5.5 accounts for positive and negative frequencies. To use Equation 16.5.1, we would have to integrate from $-\omega_2$ to $-\omega_1$ and from ω_1 to ω_2. These integrals are equal because $|F(\omega)|$ is even, which leads to Equation 16.5.6.

3. According to Equation 16.5.6, $\dfrac{|F(j\omega)|^2}{\pi}d\omega$ is the energy in an infinitesimal band of frequencies $d\omega$, so that $|F(j\omega)|^2$ is π times the energy per radian of bandwidth, or one-half the energy per hertz.

4. Some types of time functions, such as random noise, are more conveniently represented in the frequency domain than in the time domain. Parseval's theorem allows calculation of the power associated with any band of noise frequencies and hence assessment of the relative contribution of such a band to the total noise energy.

5. Because it establishes a direct relation between energy in the time domain and energy in the frequency domain, Parseval's theorem implies conservation of power and energy in the frequency domain.

6. Signals having a finite $\displaystyle\int_{-\infty}^{\infty} f^2(t)dt$ are referred to as **energy signals**. In the case of periodic signals, this integral tends to infinity. The application of Parseval's theorem to periodic signals is discussed in Section ST16.2.

Example 16.5.1: Application of Parseval's Theorem

A voltage $4e^{-2t}u(t)$ V is applied to a 20 Ω resistor. It is required to determine: (1) the total energy dissipated in both the time and frequency domains; (2) the energy associated with the frequency band $0 < f < 10$ Hz; and (3) the time interval over which an equal energy is dissipated.

SOLUTION

1. In the time domain, $W_{1\Omega} = \displaystyle\int_0^{\infty} 16e^{-4t}dt = 4$ J. The energy dissipated in the 20 Ω resistor

   by the applied voltage is $\dfrac{4}{20} = 0.2$ J. In the frequency domain, $F(j\omega) = \dfrac{4}{2+j\omega}$,

   $|F(j\omega)|^2 = \dfrac{16}{4+\omega^2}$, and $W_{1\Omega} = \dfrac{1}{\pi}\displaystyle\int_0^{\infty}\dfrac{16}{4+\omega^2}d\omega = \dfrac{16}{\pi}\left[\dfrac{1}{2}\tan^{-1}\dfrac{\omega}{2}\right]_0^{\infty} = \dfrac{16}{\pi}\times\dfrac{1}{2}\times\dfrac{\pi}{2} = 4$ J, as

   given previously.

2. The energy content of the given frequency range is $\dfrac{1}{\pi}\displaystyle\int_0^{20\pi}\dfrac{16}{4+\omega^2}d\omega = \dfrac{16}{\pi}$

   $\left[\dfrac{1}{2}\tan^{-1}\dfrac{\omega}{2}\right]_0^{20\pi} = \dfrac{8}{\pi}\tan^{-1}(10\pi) = 3.92$ J. The relative energy content over the given

   frequency range is $\dfrac{1.71}{4}\times 100 \cong 98\%$.

3. The energy dissipated in a 1 Ω resistor from 0 to t is: $\displaystyle\int_0^{t} 16e^{-4t}dt = 4(1-e^{-4t})$.

   Equating this to 3.92 J gives $t = 0.98$ s.

EXERCISE 16.5.1

Given that the current in a 5 Ω resistor has $F(\omega) = 10e^{-|\omega|}$ As, determine the total energy dissipated in the resistor (refer to the Table of Integrals in the Appendix).

Answer: $250/\pi$.

Simulation Example 16.5.2: Energies in a Lowpass Filter

The voltage $4e^{-2t}u(t)$ V of Example 16.5.1 is applied to a simple lowpass RC filter having $R = 10$ Ω and $C = 0.1$ F. It is required to determine the energy dissipated in the resistor and that supplied by the source as $t \rightarrow \infty$. Verify with PSpice.

SOLUTION

The problem can be conveniently solved in the frequency domain. $V_i(j\omega) = \dfrac{4}{2+j\omega}$, and

the transfer function of the voltage across R is: $H_R(j\omega) = \dfrac{R}{R+1/j\omega C} = \dfrac{j\omega}{1+j\omega}$.

$$V_R(j\omega) = \frac{j4\omega}{(2+j\omega)(1+j\omega)}, \text{ and } |V_R(j\omega)|^2 = \frac{16\omega^2}{(4+\omega^2)(1+\omega^2)} = \frac{16}{3}\left[-\frac{1}{1+\omega^2} + \frac{4}{4+\omega^2}\right]. \text{ It}$$

follows that $W_R = \dfrac{16}{3\pi}\displaystyle\int_0^\infty -\dfrac{1}{1+\omega^2}d\omega + \dfrac{16}{3\pi}\displaystyle\int_0^\infty \dfrac{4}{4+\omega^2}d\omega = \dfrac{16}{3\pi}\left[-\tan^{-1}\omega + 2\tan^{-1}\left(\dfrac{\omega}{2}\right)\right] =$

$\dfrac{16}{3\pi}\left[-\dfrac{\pi}{2}+\pi\right] = \dfrac{8}{3}$ J. This is the energy dissipated in a 1 Ω resistor. The energy dissipated

in the 10 Ω resistor is $8/30 = 0.267$ J.

SIMULATION

The schematic of the simulation is shown in Figure 16.5.1. A VEXP source is used, the source parameters being set as indicated for the $4e^{-2t}u(t)$ V input. The voltage

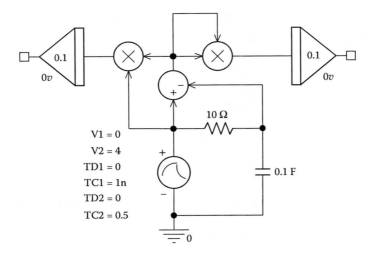

FIGURE 16.5.1
Figure for Example 16.5.2.

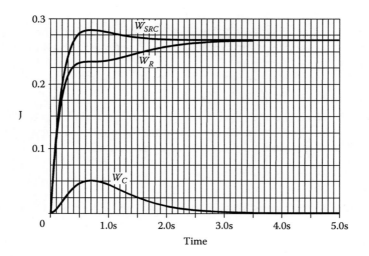

FIGURE 16.5.2
Figure for Example 16.5.2.

across R is the difference between the voltages at its two terminals, obtained using a subtracting DIFF part from the ABM library. The voltage is squared by applying it to both inputs of a multiplier MULT part from the ABM library. The multiplier output is integrated by the INTEG part from the ABM library, shown on the RHS of the figure.

Setting the gain of the integrator to 0.1, with zero initial value, gives $W_R = \dfrac{1}{R}\displaystyle\int_0^t V_R^2 dt$

which is the energy dissipated in the resistor up to time t. Another multiplier multiplies the voltage across R, which is R times the current in the circuit, by the source voltage to give the instantaneous power delivered by the source. Integrating this by the

integrator of gain 0.1 on the LHS of the figure gives $W_{SRC} = \displaystyle\int_0^t (V_{SRC})\left(\dfrac{V_R}{R}\right)dt$ which is

the power delivered by the source up to time t. Figure 16.5.2 shows the plot of the outputs of the integrators giving W_{SRC} and W_R, as well as their difference W_C, the energy in the capacitor at any time t. At large values of time, the source voltage approaches zero, the capacitor is discharged, and the energy supplied by the source is equal to that dissipated in the resistor. This energy is 0.27 J, in agreement with that calculated previously.

EXERCISE 16.5.2

The voltage $4e^{-2t}u(t)$ V is applied to the same filter of Example 16.5.2 used as a highpass filter. Using the results of Example 16.5.2, determine the energy dissipated in the resistor and that supplied by the source.

Answer: Same as in Example 16.5.2.

16.6 Basic Properties of Signals and Systems

In this section we cover several basic properties of signals and systems. Some of these properties were encountered before but were not formally or rigorously considered. It is opportune to discuss them at this point, particularly with reference to the FT.

We begin with asking what exactly constitutes a signal. Very generally, a signal is any physical entity or change that conveys information. Thus, smoke, drum beats, light flashes, a frown, or a wink are examples of signals. A narrower engineering definition of a signal is a physical, measurable quantity that conveys information, such as ultrasound waves, light traveling along an optical fiber, oven temperature, or vibrations of an engine. In electrical engineering and allied fields, a signal generally denotes something more specific:

> **Definition:** *A signal is an electrical quantity, such as voltage, charge, current, power, or energy, whose magnitude and variation with respect to time or spatial coordinates conveys information about some physical quantity.*

This assumes that non-electrical quantities have been converted to electrical quantities by appropriate transducers. Electrical signals may be *analog* or *digital*. Analog signals assume magnitudes that vary continuously over a range of values and are directly analogous to the physical quantity they represent. For example, the analog voltage signal from a microphone ideally reproduces, over a continuous range of values, the variations in air pressure that are picked up by the microphone. Digital signals, on the other hand, involve sampling and some form of digital coding of the information, as explained in Section 17.6 (Chapter 17) and are processed according to the rules of Boolean algebra.

> **Definition:** *A system is an entity that operates on one or more input signals to accomplish some function, usually by generating one or more output signals.*

The nervous system, for example, is a very complex system that receives a multitude of sensory inputs from the environment, as well as information about the internal environment of the body, and produces appropriate responses. Electric circuits are examples of electrical systems that receive one or more voltage or current inputs and produce electrical outputs that are generally useful in some way.

Impulse Response

The impulse response of a system is the output produced as a result of an applied impulse, which is approximated in practice by a large brief pulse whose duration is short compared to the smallest time constant of the system. The impulse response is characteristic of the type of system under consideration, and it completely specifies a particular system of the given type. For example, in Chapter 12 we have seen that the impulse responses of first-order systems are decaying, or saturating, exponentials. A particular impulse response would completely specify the circuit parameters of an equivalent RL or RC circuit.

The transfer function in the frequency domain, $H(j\omega)$, is also characteristic of the type of system under consideration, and it completely specifies a particular system of this type, as was discussed in Section 10.2 (Chapter 10), in connection with frequency response. It should be noted that $H(j\omega)$ is independent of any initial conditions because it is derived for a sinusoidal steady state. The same $H(j\omega)$ applies to phasor analysis, to derivation of circuit responses to periodic inputs, and to solving circuit problems using the FT.

As they both characterize a system, one in the time domain and the other in the frequency domain, the impulse response and $H(j\omega)$ are related through the FT:

Concept: *The response h(t) to a unit impulse is the IFT of H(jω).*

To prove this, Equation 16.4.1 is written as:

$$Y(j\omega) = H(j\omega)X(j\omega) \tag{16.6.1}$$

where $X(j\omega)$ is the FT of the applied input, $Y(j\omega)$ is the FT of the designated response, and $H(j\omega)$ is the corresponding transfer function. If the applied input is a unit impulse, $X(j\omega)$ = 1. Equation 16.6.1 reduces to $Y(j\omega) = H(j\omega)$. Taking the IFT of both sides gives $y(t) = h(t)$, where $y(t)$ is the response to a unit impulse and $h(t)$ is the IFT of $H(j\omega)$. The same relation applies in the case of the LT (Equation 15.7.2), as the LT and the FT of a unit impulse is unity in both cases.

Once the transfer function $H(j\omega)$ is known, the response $Y(j\omega)$ to any input $X(j\omega)$ is completely determined in the frequency domain by the product of the two FTs (Equation 16.6.1). According to the convolution-in-time property (Equation 16.3.19), the corresponding relation in the time domain is the convolution of $h(t)$ and $x(t)$:

$$y(t) = h(t) * x(t) \tag{16.6.2}$$

Equation 16.6.2 was derived in Section 12.3 (Chapter 12) by considering the response $y(t)$ in the limit as an infinite sum over a continuum of time of responses to impulses each having an infinitesimal strength that depends on $y(t)$ at a particular value of t.

Causality

Definition: *A system is causal, or nonanticipatory, if its output at the present time does not depend on future inputs.*

Thus, a system is noncausal if its output $y(t)$ at time t depends on $x(t + a)$, $a > 0$, the value of an *arbitrary* input at a later time $(t + a)$. The response of a causal system to an impulse $\delta(t)$ at $t = 0$ must be zero for $t < 0$, that is, $h(t) = 0$ for $t < 0$. The following should be noted about causality:

1. Physical systems operating in real time are causal, where real-time operation means that the system responds to inputs as they occur in time. Evidently, physical systems cannot predict future values of arbitrary inputs, and are therefore causal. On the other hand, if an input is predetermined, as in the case of a recorded input, for example, inputs for time beyond the present are available, and noncausal systems can operate on such inputs.

2. In some important applications, such as image processing, the independent variable is distance, not time, and inputs for all values of distance are available. Systems for these applications need not be causal.

3. We invoked causality in the case of the convolution integral, when we changed the upper limit of integration from ∞ to t (Equation 12.3.4, Chapter 12).

One can also speak of *causal signals*. The impulse response $h(t)$, being the response to a unit impulse $\delta(t)$ occurring at $t = 0$ is a causal signal, that is, $h(t) = 0$ for $t < 0$. By extension, any signal $f(t)$ that is zero for $t < 0$ is a causal signal. The frequency spectrum of a causal signal has the property that its real and imaginary parts are related to one another by the Hilbert transform (Section ST16.6).

Bandwidth

The bandwidth of a signal is the frequency range over which frequencies present in the signal have significant values. Although plausible, this is not a precise definition, as it does not specify a criterion for significance. In discussing frequency responses in Chapter 10, the 3-dB bandwidth was defined as the frequency range over which the signal power is not less than one-half of its maximum value. In the case of a frequency spectrum that has one main lobe and several smaller ones, as for a rectangular pulse (Figure 16.1.1c), the bandwidth may be considered to be from zero to the frequency $2\pi/\tau$ of the first zero. In this case, only the frequencies in the first lobe are considered significant, as they account for more than 90% of the energy of the signal (Example SE16.2). Nevertheless, it is generally true that:

Concept: *The faster a signal changes in time, the wider is its bandwidth.*

This may be deduced from the differentiation-in-time property of the FT by considering the FTs of a given function $f(t)$ and its derivative $f^{(1)}(t)$, which generally changes faster in time than $f(t)$. It is seen from Equation 16.3.8 that differentiating a function multiplies its $|F(\omega)|$ by ω, thereby accentuating high frequencies. Thus, the FT of a triangular pulse of duration from $t = -\tau/2$ to $t = +\tau/2$ (Figure 16.3.4) is $\dfrac{A\tau}{2}\left(\operatorname{sinc}\dfrac{\omega\tau}{4}\right)^2$ (Exercise 16.3.3), whereas the FT of a rectangular pulse of the same amplitude and duration is $A\tau\operatorname{sinc}\left(\dfrac{\omega\tau}{2}\right)$, which has a considerably higher frequency content. A function that changes particularly fast is the impulse, its amplitude going from 0 to ∞ and back to 0 over an infinitesimal interval. Its amplitude spectrum is uniform at all frequencies from zero to infinity.

Concept: *The response of physical systems is always of limited bandwidth.*

There are two main reasons for this:

1. The underlying physical processes cannot keep pace with an input that is changing at an ever-increasing rate. In a bipolar junction transistor, for example, the current gain is limited at high frequencies by the time it takes electrons to move through the base region of the transistor. In the fastest transistors, the effect may become significant at frequencies of 100 GHz or more, which is quite high but is finite nonetheless.

2. Voltage differences between terminals, and hence electric fields, are necessarily associated with input and output signals. At high enough frequencies, the displacement, or capacitive currents, of these electric fields becomes large enough to limit the response of the device. In a transformer, for example, the electric field between turns of a given winding, which is accounted for by capacitance across the winding, effectively shunts current away at high enough frequencies, thereby reducing the magnetizing current and, hence, the output of the transformer.

Duration-Bandwidth Product

Concept: *The bandwidth of a signal and its duration in time are inversely related.*

This is a kind of "uncertainty principle" that is highlighted by the FT, in that the more localized, or certain, is the duration of a signal, the wider, or more uncertain, is the bandwidth, and conversely. We have encountered several examples of this. An impulse is highly localized in time, but is of infinite bandwidth. On the other hand, a dc signal, or a sinusoidal signal, is highly localized in the frequency domain, the FTs being impulses. In the time domain, however, they extend over all time, from $t = -\infty$ to $t = +\infty$. In the case of a rectangular pulse, the narrower the pulse, that is, the smaller τ, the larger is the first zero crossing ($2\pi/\tau$), and the wider is the bandwidth, as discussed previously under time scaling.

Linearity

The general definition of linearity is in terms of superposition:

> **Definition:** *A linear system satisfies the principle of superposition, according to which if an input $x_1(t)$ produces an output $y_1(t)$, and input $x_2(t)$ produces an output $y_2(t)$, then an input $a_1 x_1(t) + a_2 x_2(t)$ produces an output $a_1 y_1(t) + a_2 y_2(t)$, where a_1 and a_2 are constants.*

The application of superposition to Ohm's law was discussed at the end of Chapter 3.

Time Invariance

A circuit was considered in Chapter 1 to be time invariant if R, L, and C, do not vary with time. The general definition of time invariance, which implies this constancy, is:

> **Definition:** *A system is time-invariant if the only effect of a shift in time of the input is to produce the same shift in time of the output.*

That is, if an input $x(t)$ produces an output $y(t)$, then shifting the input to $x(t - \tau)$ is the same as shifting the output to $y(t - \tau)$, where τ could be positive or negative. In the case of a resistor, for example, $v(t) = Ri(t)$. If the input current is delayed by τ, the output voltage becomes $Ri(t - \tau)$, which is the same as $v(t - \tau)$ if R is constant. The system is time invariant. Now consider:

$$v(t) = Ai(t) + Bt \tag{16.6.3}$$

where A and B are constants. Shifting the current by τ gives:

$$v'(t) = Ai(t - \tau) + B(t) \tag{16.6.4}$$

However, the shifted output $v(t - \tau)$ is:

$$v(t - \tau) = Ai(t - \tau) + B(t - \tau) \tag{16.6.5}$$

which is not the same as $v'(t)$. The system is not time invariant but *time varying*.

If a step voltage is applied to a thermistor (Problem P1.3.11), for example, the current increases stepwise at first, but then increases further with time afterwards, as the thermistor heats up and its resistance decreases. During the period that the resistance decreases with time, the thermistor is not time invariant. Similarly, the conductance of the membrane of the nerve cell is time varying under certain conditions, that is, it changes with time, even when the voltage across the membrane is kept constant. In an LTI electric circuit, R, L, and C, are constant with respect t, v, or i.

Memory

Definition: *A system is memoryless if its output depends only on the present input.*

It follows that a memoryless system is causal. However, a causal system has memory if its output depends, in general, on the present input as well as past inputs. A resistor is an example of a memoryless system because $v(t)$, for a given R, depends only on $i(t)$ having the same value of t. Similarly, an ideal voltage amplifier is memoryless because its output voltage $v_O(t)$ is directly proportional to the input voltage $v_I(t)$: $v_O(t) = A_v v_I(t)$, where A_v is a constant voltage gain. On the other hand, a capacitor has memory, as $v_C(t) = \dfrac{1}{C}\displaystyle\int_0^t i(\lambda)d\lambda$, assuming that the capacitor is initially uncharged and $i(\lambda)$ is applied at $\lambda = 0$. In other words, and because of the integration, $v_C(t)$ at the present value of t depends on $i(t)$ as well as on all previous values $i(\lambda)$, $0 \le \lambda \le t$. Moreover, all previous values of $i(\lambda)$ contribute to the present input with equal weight, that is, the less recent values of $i(\lambda)$ contribute to the output just as effectively as more recent values. Such a system is said to have *perfect memory*.

It is seen that the impulse response of a memoryless system, such as a resistor, is also an impulse, whereas the impulse response of a system with perfect memory, such as a capacitor, is the time integral of an impulse, that is, a step function. As discussed in Section 17.1 (Chapter 17), the extent to which the output of a circuit deviates from the input depends on system memory.

Invertibility

Definition: *A system is invertible if its input can be recovered from its output.*

Clearly, a necessary condition for invertibility is that distinct inputs should produce distinct outputs. Otherwise, if more than one input can produce the same output, the input cannot be unambiguously recovered from the output, so that the system is non-invertible. A common example is a squaring circuit described by: $y(t) = x^2(t)$. Positive and negative inputs $\pm\sqrt{x(t)}$ produce the same output, which makes the system non-invertible.

If a system is invertible, then an *inverse system* can be defined which, when cascaded with the given system, produces the original signal. This is illustrated in Figure 16.6.1. In simple cases, the inverse system is rather obvious. For example, the inverse of a voltage amplifier with a gain of 10 is a voltage attenuator, which could be a simple voltage divider,

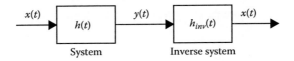

FIGURE 16.6.1
Invertibility.

having a gain of 0.1. The inverse of an integrator described by: $y(t) = A \int_{-\infty}^{t} x(\lambda)d\lambda$ is a

differentiator described by: $x(t) = \dfrac{dy(t)}{dt}$.

It is seen from Figure 16.6.1 that:

$$y(t) = x(t) * h(t) \quad \text{and} \quad [x(t) * h(t)] * h_{inv}(t) = x(t) \tag{16.6.6}$$

From the associative property of convolution (Equation 12.3.13, Chapter 12), Equation 16.6.6 can be written as:

$$x(t) * [h(t) * h_{inv}(t)] = x(t) \tag{16.6.7}$$

To satisfy Equation 16.6.7, we must have:

$$h(t) * h_{inv}(t) = \delta(t) \tag{16.6.8}$$

since convolving $x(t)$ in Equation 16.6.7 with a unit impulse gives the same function (Equation 12.4.5, Chapter 12).

The general problem of finding the inverse of a given system is not an easy one. Moreover, the inverse system may not be causal or stable (Section 17.3, Chapter 17). To illustrate stability, we can take the Laplace transform of both sides of Equation 16.6.8 to obtain:

$$H(s)H_{inv}(s) = 1 \quad \text{or} \quad H_{inv}(s) = \frac{1}{H(s)} \tag{16.6.9}$$

The poles of $H_{inv}(s)$ are the zeros of $H(s)$. For the inverse system to be stable, the zeros of the system must lie in the left half of the s plane. Inverse systems are important for deconvolution, as discussed in Section 17.3 (Chapter 17).

Stability

Stability in terms of the impulse response was discussed in Section 15.7 (Chapter 15) and related to the location of poles of the response in the s-plane. An essentially equivalent stability criterion, which, however, is not as easy to interpret in some cases, is the bounded input/bounded output (BIBO) criterion.

> **Definition:** *According to the BIBO criterion of stability, a system is stable if the magnitude of the output $y(t)$ remains bounded for any bounded magnitude of input $x(t)$.*

That is, if:

$$|x(t)| \leq M_I, \text{ then } |y(t)| \leq M_O \tag{16.6.10}$$

where M_I and M_O are some finite constants. According to this criterion, a resistor is a stable system. For if the magnitude of the current is bounded by M_I, that is, does not exceed M_I A, the magnitude of the voltage is bounded by $M_O = RM_I$ V, where both M_I and M_O are finite.

The BIBO criterion can be interpreted in terms of the integrability of the impulse response. As $y(t) = h(t)*x(t)$:

$$\left|y(t)\right| = \left|\int_{-\infty}^{\infty} h(\lambda)x(t-\lambda)d\lambda\right| \tag{16.6.11}$$

Because the magnitude of an integral is less than or equal to the integral of the magnitude of the integrand, and the magnitude of the product of two quantities equals the product of their magnitudes, Equation 16.6.11 becomes:

$$\left|y(t)\right| \leq \int_{-\infty}^{\infty} \left|h(\lambda)x(t-\lambda)d\lambda\right| = \int_{-\infty}^{\infty} \left|h(\lambda)\right|\left|x(t-\lambda)\right|d\lambda \tag{16.6.12}$$

But $\left|x(t-\lambda)\right| \leq M_I$, so that:

$$\left|y(t)\right| \leq M_I \left|\int_{-\infty}^{\infty} h(\lambda)d\lambda\right| \tag{16.6.13}$$

This means that $|y(t)|$ is bounded if the integral of $|h(t)|$ is finite:

$$\int_{-\infty}^{\infty} |h(t)|\, dt < \infty \tag{16.6.14}$$

The condition expressed by Equation 16.6.14 is that $h(t)$ be *absolutely integrable*, that is, the integral of its absolute value is finite. If the system is causal, then $h(t) = 0$ for $t < 0$, and the lower limit of integration in Equation 16.6.14 is zero.

If the integral in Equation 16.6.13 is infinite, then the output is unbounded for a bounded input and the system is unstable. A source of confusion is that in some cases, the response may be bounded for some inputs and unbounded for others. For example, if we apply a sinusoidal current to a capacitor, the capacitor voltage is sinusoidal. Both the input and output are bounded. If we apply a dc current I_{DC} to a capacitor at $t = 0$, $v_C = \dfrac{1}{C}\int_0^t I_{DC}dt = \dfrac{I_{DC}}{C}t$, which is unbounded. Such cases are said to be marginally stable, or metastable. But how do we identify such cases without having to try various inputs? As mentioned in Section 15.7 (Chapter 15), the transfer function in these cases has a simple pole on the imaginary axis in the s-plane, and the response is unbounded if the input is of the same frequency as the pole. The transfer function of a capacitor is $\dfrac{V_C(s)}{I_C(s)} = \dfrac{1}{sC}$, which has a simple pole at the origin, corresponding to zero frequency. The system is metastable, but if a dc current input is applied, which is also of zero frequency, the output voltage is unbounded. The stability criterion of Section 15.7 (Chapter 15) readily identifies the inputs that will give unbounded responses.

Summary of Main Concepts and Results

- A real function $f(t)$ has an FT $F(j\omega)$ whose magnitude spectrum is even and whose phase spectrum is odd. If $f(t)$ is even, $F(\omega)$ is real and even. If $f(t)$ is odd, $F(\omega)$ is imaginary and odd.

- A duality relation exists between the FT and the IFT: If $\mathcal{F}\{f(t)\} = F(j\omega)$, then $\mathcal{F}\{F(jt)\} = 2\pi f(-\omega)$.

- The FT of the convolution of two time functions equals the product of their FTs, and the FT of the product of two time functions equals the convolution of their FTs divided by 2π.

- The energy spectrum of a signal is a plot of $|F(j\omega)|^2$ vs. ω. According to Parseval's theorem the energy dissipated in a 1 Ω resistor over all time is $\displaystyle\int_{-\infty}^{\infty} f^2(t)dt = \frac{1}{\pi}\int_{0}^{\infty} |F(j\omega)|^2\, d\omega$.

- Parseval's theorem allows calculation of the power associated with any band of frequencies of a signal.

- The response $h(t)$ to a unit impulse is the IFT of the corresponding transfer function $H(j\omega)$.

- The faster a signal changes in time, the wider is its bandwidth.

- The bandwidth of a signal and its duration in time are inversely related.

- A system is:
  - Causal, or nonanticipatory, if its output at the present time does not depend on future inputs.
  - Linear if it satisfies the principle of superposition.
  - Time-invariant if the only effect of a shift in time of the input is to produce the same shift in time of the output.
  - Memoryless if its output depends only on the present input.
  - Invertible if its input can be recovered from its output.
  - Stable according to the BIBO criterion of stability if the magnitude of the output $y(t)$ remains bounded for any bounded magnitude of input $x(t)$.

Learning Outcomes

- Apply the Fourier transform to the analysis of signals and circuits.
- Articulate some basic properties of signals and systems.

Supplementary Topics and Examples on CD

ST16.1 Derivation of Fourier transform from Laplace transform: Shows how and under what conditions the FT can be derived from the LT.

ST16.2 Parseval's theorem for periodic inputs: Derives the average power dissipated in a 1 Ω resistor per period of a periodic voltage or current both in terms of the Fourier series expansion and the Fourier transform.

ST16.3 Parseval's theorem for the product of two functions: States and proves Parseval's theorem for the product of two functions.

ST16.4 The discrete Fourier transform: Derives the transform pair of the discrete Fourier transform.

ST16.5 The fast Fourier transform: Explains the essential features of the algorithm of the fast Fourier transform.

ST16.6 The Hilbert transform: Derives the Hilbert transform in the time and frequency domains and shows that the real and imaginary parts of the spectrum of a causal signal are related through the Hilbert transform.

SE16.1 Fourier transform of a Gaussian function: Derives the FT of a Gaussian function.

SE16.2 Energy of a rectangular pulse in time: Derives the energy in the frequency domain of a rectangular pulse in time and shows that approximately 90% of the energy is in the first lobe of the amplitude spectrum.

SE16.3 Convolution in the frequency domain: Illustrates a convolution of two functions of time that is much easier to perform in the frequency domain than in the time domain.

Problems and Exercises

P16.1   Fourier Transform and Its Properties

P16.1.1 Determine the FT of the following functions:

(a) $u(t) - u(t - 4)$.

(b) $t^2 e^{-3t} u(t)$.

(c) $|t|$. Note that $\dfrac{d\,|t|}{dt} = \mathrm{sgn}(t)$.

(d) $\dfrac{d}{dt} e^{-a|t|}$, by direct evaluation and by utilizing the result of Exercise 16.1.1.

(e) $\dfrac{2\cos(at)}{t^2 + 1}$. (Hint: use the result of Exercise 16.1.1.)

P16.1.2 Determine the IFT of the following functions:

(a) $\dfrac{5}{j\omega(j\omega + 5)}$.

(b) $\dfrac{j\omega-2}{\omega^2-5j\omega-6}$.

(c) $\dfrac{1}{\left(1+\omega^2\right)^2}$.

(d) $2u(\omega+2)-2u(\omega-2)$.

(e) $\dfrac{2\pi\delta(\omega)}{2+j\omega}+\dfrac{2}{j\omega(2+j\omega)}$.

P16.1.3 Assume that the function shown in Figure 16.1.2 is in the frequency domain. Determine $f(t)$ and verify by applying duality to Equation 16.1.12.

P16.1.4 Assume that the function shown in Figure 16.1.3 is in the time domain. Determine $F(j\omega)$ and verify by applying duality to the result of Example 16.1.2.

P16.1.5 Assume that the function shown in Figure 16.3.2 is in the frequency domain. Determine $f(t)$ and verify by applying duality to the result of Example 16.3.3.

P16.1.6 Assume that the function shown in Figure 16.3.4 is in the frequency domain. Determine $f(t)$ and verify by applying duality to the result of Exercise 16.3.3.

P16.1.7 Determine $F(j\omega)$ for: (a) $f(t)$ of Figure P16.1.7, and verify the interpretation of $F(0)$; (b) $f^{(1)}(t)$, and verify the result by applying the differentiation-in-time property to the result in (a).

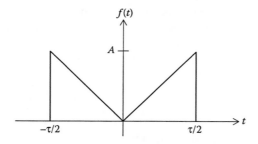

FIGURE P16.1.7

P16.1.8 Assume that the function shown in Figure P16.1.7 is in the frequency domain. Determine $f(t)$ and verify by applying duality to the result of P16.1.7.

P16.1.9 Determine $f(t)$, given $F(j\omega)$ is the single cosine pulse of Figure P16.1.9.

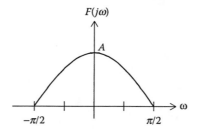

FIGURE P16.1.9

P16.1.10 Determine $F(j\omega)$ of Figure 16.3.4 as the product of the FTs of two rectangular pulses whose convolution gives $f(t)$. (Refer to Section 12.4, Chapter 12.)

P16.1.11 Determine $F(j\omega)$ of the single sinusoidal period of Figure P16.1.11.

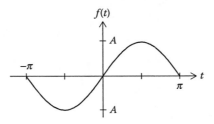

FIGURE P16.1.11

P16.1.12 Determine $F(j\omega)$ of Figure P16.1.11 by considering it as the product of a sinusoidal function and a rectangular pulse of unit amplitude extending from $-\pi$ to $+\pi$.

P16.1.13 Determine $F(j\omega)$ of the function in Figure P6.1.13 from that of its second derivative.

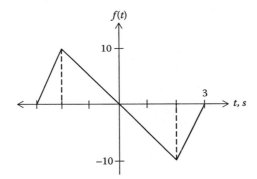

FIGURE P16.1.13

P16.1.14 Determine $F(j\omega)$ of the periodic sawtooth of Figure 9.2.1 (Chapter 9).

P16.1.15 Determine $F(j\omega)$ of the periodic sawtooth of Figure 9.2.4b (Chapter 9).

P16.1.16 Determine $F(j\omega)$ of the periodic triangular waveform of Figure 9.3.2 (Chapter 9) and verify it by applying the integration-in-time property to the square wave of Example 16.3.2.

P16.1.17 Determine $F(j\omega)$ of the half-wave rectified waveform of Figure 9.4.1a (Chapter 9), considering it as the product of a cosine function and a rectangular waveform, as in Example 9.4.1 (Chapter 9).

P16.1.18 Determine $F(j\omega)$ of the full-wave rectified waveform of Figure 9.4.1b (Chapter 9) considering it as the product of a cosine function and a square wave of zero average centered at the origin.

P16.1.19 Derive $\delta(\omega-a)*\delta(\omega-b)$ in both the time and frequency domains.

P16.1.20 Determine the FT of $f(t)=\displaystyle\sum_{n=0}^{\infty} r^{n}\delta(t-n),\ |r|<1.$

P16.1.21 Determine $F(j\omega)$ of $f(t)$ of Figure P16.1.21, where:

$$f(t) = t^2, 0 \leq t \leq 1, f(t) = (2 - t)^2, 1 \leq t \leq 2, \text{ and}$$

$$f(t) = 0 \text{ elsewhere.}$$

Verify by finding the FT of the second derivative.

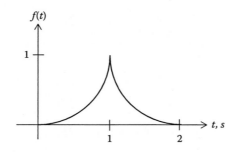

FIGURE P16.1.21

P16.1.22 Determine $F(j\omega)$ of $f(t)$ of Figure P16.1.22, where: $f(t) = t^2, -1 \leq t \leq 1, f(t) = t + 2, -2 \leq t \leq -1, f(t) = -t + 2, 1 \leq t \leq 2$, and $f(t) = 0$ elsewhere. Verify by finding the FT of the second derivative.

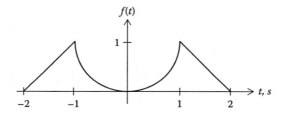

FIGURE P16.1.22

P16.1.23 Determine $F(j\omega)$ of $f(t)$ of Figure P16.1.23, where: $f(t) = 1, t \leq -1, f(t) = t^2$, $-1 \leq t \leq 0, f(t) = -t^2, 0 \leq t \leq 1, f(t) = -1, t \geq 1, f(t) = 0$ elsewhere.

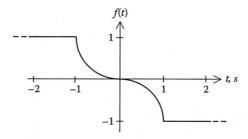

FIGURE P16.1.23

P16.1.24 Derive the IFT of $\dfrac{4}{(1+\omega^2)^2}$.

P16.1.25 The real part of $H(j\omega)$ of a causal system is $\pi\delta(\omega)$. Determine $H(j\omega)$ and $h(t)$ (refer to Section SE16.4).

P16.2 Circuit Applications

No initial energy storage is assumed in the following problems. Verify the solutions with PSpice whenever feasible.

P16.2.1 Use the FT transform to find $v_O(t)$ in Figure P16.2.1 if $v_{SRC} = \delta(t)$ Vs and interpret the result.

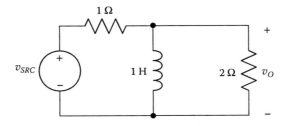

FIGURE P16.2.1

P16.2.2 Repeat P16.2.1 if $v_{SRC} = \dfrac{1}{2}$ sgn(t) V and interpret the result. How is this response related to that of P16.2.1?

P16.2.3 Repeat P16.2.1 if $v_{SRC} = u(t)$ V. How are the responses in P16.2.1 and P16.2.2 related?

P16.2.4 Repeat P16.2.1 if $v_{SRC} = e^{-t} u(t)$ V.

P16.2.5 Repeat P16.2.1 if $v_{SRC} = 2\cos 2t$.

P16.2.6 Determine $v_O(t)$ and $i_O(t)$ in Figure P16.2.6 if $v_{SRC} = 10u(t)$ V, and interpret the result.

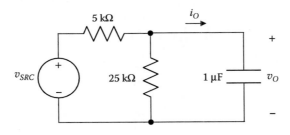

FIGURE P16.2.6

P16.2.7 Repeat P16.2.6 with $v_{SRC} = 5\text{sgn}(t)$ V, and interpret the result.

P16.2.8 Determine $v_O(t)$ and $i_O(t)$ in Figure P16.2.8 if $v_{SRC} = u(t)$, assuming: (a) $R = 2\ \Omega$; (b) $R = 2.5\ \Omega$; and (c) $R = 1.2\ \Omega$. Compare with Table 13.4.1.

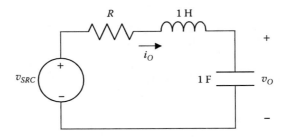

FIGURE P16.2.8

P16.2.9 Repeat P16.2.8 with $v_{SRC} = 0.5\mathrm{sgn}(t)$ V. What are the values of $v_O(t)$ and $i_O(t)$ at $t = 0^-$ and $t = 0^+$?

P16.2.10 Determine $v_O(t)$ and $i_O(t)$ in Figure P16.2.8 if $R = 2.5\ \Omega$ and $v_{SRC} = 10e^{-|t|}$ V.

P16.2.11 Repeat P16.2.10 with $R = 1.2\ \Omega$.

P16.2.12 Determine $v_O(t)$ and $i_O(t)$ in Figure P16.2.12 if v_{SRC} is the waveform of Figure 16.1.2 with $A = 10$ V and $\tau = 2$ ms. Verify that $i_O = C\dfrac{dv_O}{dt}$ in the time domain.

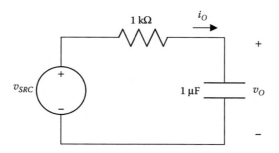

FIGURE P16.2.12

P16.2.13 Repeat P16.2.12 if v_{SRC} is the waveform of Figure 16.3.2.

P16.2.14 Determine $v_O(t)$ and $i_O(t)$ in Figure P16.2.12 if v_{SRC} is the waveform of Figure 16.3.4.

P16.2.15 Assume that $R = 1\ \Omega$ and $C = 1$ F in Figure P16.1.12. Determine $v_O(t)$ and $i_O(t)$ if v_{SRC} is the waveform of Figure P16.1.9 as a function of time, with $A = 10$ V.

P16.2.16 Repeat P16.2.15 for the waveform of Figure P16.1.11.

P16.2.17 Repeat P16.2.15 for the waveform of Figure P16.1.13.

P16.2.18 Determine $v_O(t)$ in Figure P16.2.18 if $v_{SRC} = e^{-2t}u(t)$ V.

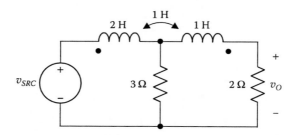

FIGURE P16.2.18

P16.3 Parseval's Theorem

P16.3.1 Use Parseval's theorem to show that $\displaystyle\int_{-\infty}^{\infty} \delta^2(t)dt$ is infinite and, therefore, undefined.

P16.3.2 If $F(j\omega) = \dfrac{1}{2 + j\omega}$, determine $W_{1\Omega}$ in both the frequency and time domains.

P16.3.3 If $f(t) = e^{-2t}(\sin 2t)u(t)$, determine $W_{1\Omega}$ in both the frequency and time domains.

P16.3.4 If $f(t) = te^{-2t}u(t)$, determine $W_{1\Omega}$ in both the frequency and time domains.

P16.3.5 If $v_{SRC} = 15e^{-|t|}$ V in Figure P16.3.5, determine the percentage of the 1 Ω energy of v_O in the frequency range $0 \le \omega \le 2$ rad/s.

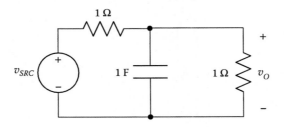

FIGURE P16.3.5

P16.3.6 If $v_{SRC} = 20e^{-2t}u(t)$ V in Figure P16.3.6, what percentage of the 1 Ω energy of v_O is in the frequency range $0 \le \omega \le 2$ rad/s if: (a) $R = 2$ Ω; (b) $R = 4$ Ω.

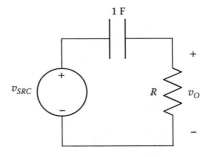

FIGURE P16.3.6

P16.3.7 Repeat P16.3.6 if $v_{SRC} = 20e^{-2|t|}$ V.

17

Basic Signal Processing Operations

Overview

The present chapter is devoted to considering some basic signal processing operations that are theoretically and practically important, with emphasis on applying what was discussed in the preceding chapter. We start with distortion, which is manifested as a difference between the input and output waveforms, and relate this to the frequency response, the impulse response, and circuit memory. In modulation, the frequency spectrum of the signal is shifted to higher frequencies. This is illustrated with a simple form of modulation — amplitude modulation (AM) — so as to highlight the important benefits of modulation.

The inverse process of convolution is deconvolution, and has the potential of undoing undesirable distortion introduced by a given system, if this distortion can be accounted for with reasonable accuracy. The difficulties of implementing deconvolution and its limitations are discussed. Another important signal processing operation is sampling of a continuous signal. A relevant question is: How often must a signal be sampled without losing any information in the signal? The answer would allow faithful reconstruction of the signal from the sampled values. It turns out that undersampling at a rate less than a critical value introduces an insidious form of distortion, known as aliasing, which cannot be removed.

Filtering and smoothing are also important signal processing operations. The object of filtering is to remove unwanted frequency components. Smoothing is a particular form of filtering, namely lowpass filtering that removes higher-frequency components in some desired manner. The chapter concludes with an explanation of the nature of digital signals and how they differ from analog signals. The advantages of digital systems, as compared to analog systems, are briefly highlighted.

Learning Objectives

- To be familiar with:
  - The benefits of modulation
  - The nature of digital signals and their advantages
- To understand:
  - How distortion of a signal by a circuit is related to phase shift and to circuit memory, as reflected in the shape of the impulse response
  - The nature, significance, and limitations of deconvolution

- Sampling of a continuous signal of time, the significance of the Nyquist frequency, and how aliasing can occur
- How functions can be smoothed by lowpass filtering or leaky integration

17.1 Distortion

A signal is distorted when its waveform is changed, which necessarily implies a change in the relative magnitudes and phases of its frequency components. Changing the size of a signal, without altering its shape, or simply delaying the signal in time, does not distort it.

Distortion is of two types: (1) **nonlinear**, or **harmonic** and (2) **linear**, or **frequency**. As its name implies, nonlinear distortion is caused by some circuit nonlinearities, which are inevitably present. The v–i characteristics of devices such as diodes and transistors, or of those having ferromagnetic elements, are inherently nonlinear. At large enough inputs, transistors saturate, which also introduces nonlinearities.

Nonlinear distortion can be illustrated by applying an input $v_I = \cos\omega_1 t + \cos\omega_2 t$ to a circuit whose output is $v_O = v_I + v_I^2$. The output can be expressed as

$$v_O = \cos\omega_1 t + \cos\omega_2 t + \left(\cos\omega_1 t + \cos\omega_2 t\right)^2$$

$$= 1 + \cos\omega_1 t + \cos\omega_2 t + \cos\left(\omega_1 + \omega_2\right)t + \cos(\omega_1 - \omega_2)t + \frac{1}{2}\cos 2\omega_1 t + \frac{1}{2}\cos 2\omega_2 t \quad (17.1.1)$$

It is seen that the nonlinearity of the square term introduces frequency components that are not present in the original signal, namely, a dc component, second harmonics, and **intermodulation products** that involve sum and difference frequencies. Although nonlinear distortion is generally objectionable, it has some useful applications, as in the mixing between a received signal and the output of a local oscillator in a superheterodyne receiver (Application Window 17.2.1).

Distortion also occurs in an LTI circuit because the presence of storage elements makes the response frequency dependent, as explained in Chapter 10. The relative magnitudes and phases of frequency components in the response differ from those in the input, which alters the shape of the waveform and distorts the signal, although no new frequency components are introduced. Consider, for example, an input $\sin\omega_0 t + 0.3\sin3\omega_0 t$ (Figure 17.1.1a). If the third harmonic is attenuated with respect to the fundamental so that output becomes $\sin\omega_0 t + 0.2\sin3\omega_0 t$ (Figure 17.1.1b), the waveform changes. This distortion, due to the variation of the relative amplitudes of the components of different frequencies, is **amplitude distortion**.

Suppose that the amplitude of the third harmonic stays the same but its phase angle is shifted by π, so that the output becomes $\sin\omega_0 t - 0.2\sin3\omega_0 t$ (Figure 17.1.1c). The waveform now looks quite different. This distortion, due to the variation of phase angles of components of different frequencies, is **phase distortion**. It is interesting to note that the human ear is largely insensitive to phase distortion.

Distortionless Delay

It is possible that phase variation with frequency does not distort the signal. Consider the case where the magnitude of the transfer function $|H(j\omega)|$ is a constant K at all frequencies, and the phase angle varies linearly with frequency as $-\omega\tau$ (Figure 17.1.2). $H(j\omega)$ is then:

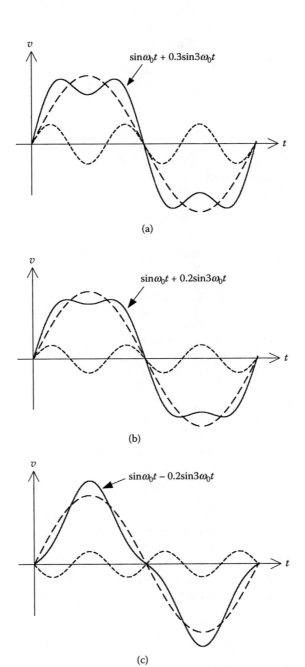

FIGURE 17.1.1
Frequency distortion.

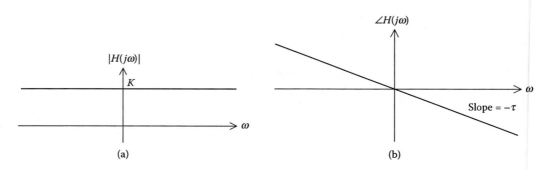

FIGURE 17.1.2
Frequency response for distortionless delay.

$$H(j\omega) = Ke^{-j\omega\tau} \tag{17.1.2}$$

If $X(j\omega)$ and $Y(j\omega)$ denote the FTs of the input and output, respectively,

$$Y(j\omega) = H(j\omega)X(j\omega) = KX(j\omega)e^{-j\omega\tau} \tag{17.1.3}$$

Applying the translation-in-time property of the FT (Equation 16.3.6, Chapter 16),

$$y(t) = Kx(t-\tau) \tag{17.1.4}$$

It is seen that the effect is simply to scale the magnitude of the function by K and to delay the function in time by τ.

This delay is also manifested in the impulse response. Taking the IFT of Equation 17.1.2:

$$h(t) = K\delta(t-\tau) \tag{17.1.5}$$

That is, the impulse response is an impulse of strength K occurring at $t = \tau$.

> **Conclusion:** *If the amplitude spectrum of the transfer function of a system is a constant and its phase spectrum is a phase lag that is directly proportional to frequency, the response is an exact replica of the input, but delayed in time.*

EXERCISE 17.1.1

A periodic input $f(t) = c_0 + \sum_{n=1}^{\infty} c_n \cos(n\omega_0 t + \theta_n)$ is applied to a system having the transfer function shown in Figure 17.1.2. Show that the function is delayed by τ .

EXERCISE 17.1.2
Show that convolving an input $x(t)$ with the impulse response of Equation 17.1.5, as discussed in Section 12.4 (Chapter 12), gives $y(t) = Kx(t-\tau)$.

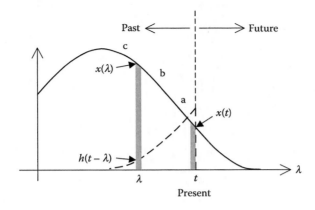

FIGURE 17.1.3
Weighting of output by impulse response.

Effect of Memory

The distortion produced by a given LTI circuit can be related to the impulse response of the circuit using the property of circuit memory (Section 16.6, Chapter 16). Figure 12.3.2b (in Chapter 12) is redrawn in Figure 17.1.3, where $\lambda = t$ denotes the present and $\lambda < t$ denotes the past. In the convolution integral, $x(t)d\lambda h(0)$ is the contribution to the response at present of the pulse $x(t)d\lambda$ occurring at the present, whereas $x(\lambda)d\lambda h(t - \lambda)$ is the contribution to the response at present of the pulse $x(\lambda)d\lambda$ that occurred at a time $(t - \lambda)$ earlier. Both these pulses are weighted by the corresponding values of the impulse response. It follows that, in determining the circuit response at present, the more recent portion of the excitation, such as ab, is weighted by the earlier part of the impulse response, whereas the less recent portion of the excitation, such as bc, is weighted by the later part of the impulse response.

As discussed in Section 16.6 (Chapter 16), the shape of the impulse response reflects circuit memory. If the circuit is memoryless, its impulse response is also an impulse. The only contribution to the response at present is the present input. The output waveform is a scaled version of the input, without distortion. If the circuit has perfect memory, the impulse response is a step function, and all past responses are equally weighted in their contribution to the present input. The deviation of the output waveform from the input waveform is maximal, and hence distortion is maximal. If the impulse response is a decaying exponential, as in Figure 17.1.3, the distortion is intermediate, the more recent inputs being more heavily weighted than less recent inputs in their contribution to the response at present.

> **Conclusions: (1)** *Circuit memory, reflected in the shape of the impulse response, affects the distortion of the input signal.* **(2)** *The impulse response of a distortionless system is an impulse of zero delay, associated with zero phase shift, or is a delayed impulse associated with a linear phase lag.*

EXERCISE 17.1.3
Consider the series RL circuit, with the applied excitation being a unit voltage step and the response being the current i. Compare how the shape of the response deviates from that of the applied excitation for various impulse responses corresponding to different values of R and L, including the limiting cases of $R = 0$ and $L = 0$.

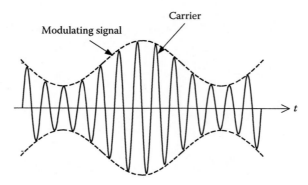

FIGURE 17.2.1
Amplitude-modulated signal.

17.2 Modulation

Definition: *Modulation is the varying of a property of a higher-frequency signal (the carrier) in accordance with the instantaneous values of a lower-frequency signal (the modulating signal).*

A simple form of modulation is amplitude modulation (AM), in which the modulating signal varies the amplitude of a sinusoidal carrier, as illustrated in Figure 17.2.1. Let the carrier, in the absence of the modulating signal $m(t)$, be $A_c \cos\omega_0 t$. When $m(t)$ is applied, the amplitude A_c is effectively modulated by $m(t)$. The resulting modulated signal is:

$$f_{AM}(t) = \left[m(t) + A_c \right] \cos\omega_c t \tag{17.2.1}$$

Substituting $\cos\omega_c t = \dfrac{1}{2}\left(e^{j\omega_c t} + e^{-j\omega_c t} \right)$,

$$f_{AM}(t) \;=\; \frac{1}{2}\left[m(t)e^{j\omega_c t} + m(t)e^{-j\omega_c t} + A_c e^{j\omega_c t} + A_c e^{-j\omega_c t} \right] \tag{17.2.2}$$

Applying the translation-in-frequency property (Equation 16.3.16, Chapter 16),

$$F_{AM}(j\omega) \;=\; \frac{1}{2}\left[M\big(j(\omega - \omega_c)\big) + M\big(j(\omega + \omega_c)\big) \right] + \pi A_c \delta(\omega - \omega_c) + \pi A_c \delta(\omega + \omega_c) \tag{17.2.3}$$

The impulse terms in Equation 17.2.3 are, of course, the FT of $A_c \cos\omega_c t$. According to Equation 17.2.3, the frequency spectrum of $m(t)$ is shifted so that it is centered about ω_c. If the magnitude spectrum of $M(j\omega)$ is as in Figure 17.2.2a, where the highest frequency is ω_{mh}, the magnitude spectrum of $F_{AM}(j\omega)$ is as in Figure 17.2.2b, where ω_c is ordinarily much larger than ω_{mh}. Whereas the bandwidth of the modulating signal is ω_{mh}, that of the modulated signal is $2\omega_{mh}$, from $(\omega_c - \omega_{mh})$ to $(\omega_c + \omega_{mh})$. The frequency range ω_c to

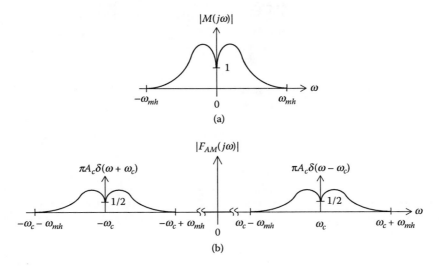

FIGURE 17.2.2
(a) Amplitude spectra of modulating signal and (b) of modulated signal.

($\omega_c + \omega_{mh}$) is the **upper sideband,** whereas the frequency range $\omega_c - \omega_{mh}$ to ω_c is the **lower sideband**.

This shifting of the spectrum of a signal to higher frequencies is characteristic of modulation and is of great practical importance because of the following main benefits:

Benefit 1: *It greatly reduces the size of the transmitting and receiving antennas.*

For efficient transmission and reception, antenna size should be at least λ /10 , where λ is the wavelength of the signal. If audio signals in the range of, say, 15 Hz to 20 kHz are to be transmitted as a wireless signal, the antenna size would have to be at least several kilometers. But if the audio signals modulate a high-frequency carrier, the antenna size is greatly reduced.

Benefit 2: *The same communication channel can be made to carry a fairly large number of modulated signals simultaneously, essentially without interference.*

The different signals modulate carriers of appropriately spaced frequencies, so that the modulated signals can be transmitted simultaneously on the same channel. Figure 17.2.3 illustrates three modulated signals carrying information in the same channel at the same time by having three carrier frequencies, so chosen that their sidebands do not overlap. This is known as **frequency-division multiplexing.** Using the same communication channel, be it a telephone channel, a mobile communication channel, or a satellite link, to

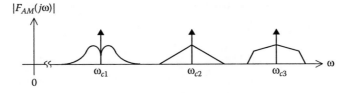

FIGURE 17.2.3
Frequency-division multiplexing.

transmit a number of signals simultaneously greatly reduces transmission cost. At the receiving end, the different modulated signals are separated by some form of bandpass filtering. Each modulated signal is then demodulated to recover the original modulating signal (Application Window 17.5.2).

> **Benefit 3:** *Depending on the modulation technique used, the modulated signal is much less vulnerable to corruption by noise.*

This depends on how noise affects the modulating signal in the modulation scheme used.

Example 17.2.1: Amplitude Modulation

Consider a simple case, where a 1 MHz carrier of amplitude A_c is modulated by a 10 kHz sinusoidal signal of amplitude kA_c, where $0 \le k \le 1$ is the **modulation factor**. It is required to determine: (1) the frequency spectrum, and (2) the 1 Ω power associated with the various frequencies.

SOLUTION

1. When $m(t) = kA_c \cos \omega_m t$, Equation 17.2.1 becomes:

$$f_{AM}(t) = A_c \left[1 + k \cos \omega_m t \right] \cos \omega_c t$$

$$= A_c \cos \omega_c t + \frac{kA_c}{2} \cos(\omega_c + \omega_m)t + \frac{kA_c}{2} \cos(\omega_c - \omega_m)t \qquad (17.2.4)$$

It follows that:

$$F_{AM}(j\omega) = \pi A_c \delta(\omega - \omega_c)t + \pi A_c \delta(\omega + \omega_c) + k \frac{\pi A_c}{2} \delta(\omega - \omega_c - \omega_m)t +$$

$$k \frac{\pi A_c}{2} \delta(\omega + \omega_c + \omega_m)t + k \frac{\pi A_c}{2} \delta(\omega - \omega_c + \omega_m)t + k \frac{\pi A_c}{2} \delta(\omega + \omega_c - \omega_m)t \qquad (17.2.5)$$

Figure 17.2.4 shows the frequency spectra of the modulating signal $M(j\omega)$, the carrier $C(j\omega)$, and the amplitude-modulated signal $F_{AM}(j\omega)$.

2. From Equation 17.2.4, the 1 Ω power associated with the carrier is $\frac{A_C^2}{2}$, and that associated with the upper or lower sidebands is $\frac{k^2 A_C^2}{8}$. Note that the power in the sidebands increases with k.

It is seen that amplitude modulation is wasteful of transmitted power. The information of interest is in the sidebands, yet the power associated with the sidebands is only a fraction of the total power transmitted. Most of the power is associated with the carrier, although the carrier does not carry any information.

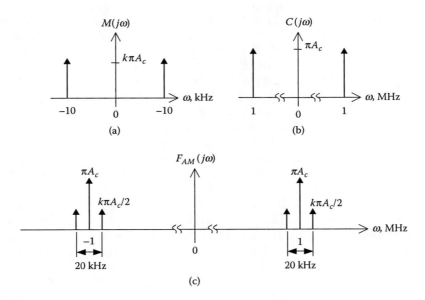

(a)

(b)

(c)

FIGURE 17.2.4
Figure for Example 17.2.1.

The power efficiency of an AM system can be improved by not transmitting the carrier because it does not carry any information, leading to a **suppressed-carrier double-sideband** AM. In **single-sideband** AM, the bandwidth is reduced by transmitting only one sideband. As the two sidebands carry the same information, no information is lost. However, in both these AM systems, more complex circuitry is required for modulating the carrier and demodulating the received signal.

Application Window 17.2.1: AM Radio Receiver

Figure 17.2.5 shows a block diagram of an AM radio receiver. The medium-wave frequency band accommodates carrier frequencies in the range 540 kHz to 1.6 MHz. The received signals may be first amplified by a radio frequency (rf) amplifier of a wide bandwidth. The signal is then mixed with a sinusoidal signal generated by a local oscillator. Tuning the receiver to a particular station of carrier frequency, say f_{c0}, sets the frequency of the oscillator to $f_{i-f} + f_{i-f}$, where f_{IF} is a fixed intermediate frequency $(i - f)$ that is usually in the range of 450 to 470 kHz. The mixing produces sum and difference frequencies, as explained in connection with Equation 17.1.1. If the desired AM signal has received frequencies in the range $(f_{c0} - f_{ch})$ to $(f_{c0} + f_{ch})$, difference frequencies in the range $(f_{i-f} - f_{ch})$ to $(f_{i-f} + f_{ch})$ are

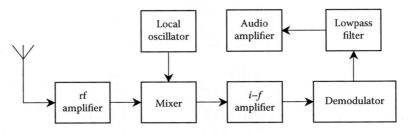

FIGURE 17.2.5
Figure for Application Window 17.2.1.

produced, in addition to other frequency components. The $i - f$ amplifier has a bandpass response, the passband being ideally from $(f_{i-f} - f_{ch})$ to $(f_{i-f} + f_{ch})$. This means that frequencies outside this range are rejected. These include other products of mixing the desired signal frequency with the local oscillator frequency, as well as the products of mixing the oscillator frequency with frequencies of undesired received signals. It is seen that the desired signal, instead of being centered around f_{c0}, where f_{c0} can be any frequency in a given frequency band, is now centered around f_{i-f}. This serves a very useful purpose, namely, allowing the main amplification and frequency selectivity of the receiver to be implemented at a fixed frequency, that is, the intermediate frequency. Otherwise, amplification and bandpass filtering will have to be performed on variable frequencies within the given frequency band, which is a much more difficult task. The shifting of the carrier frequency of a received signal to a fixed $i - f$ frequency is **superheterodyning**.

The amplified signal is then demodulated to recover the original modulating audio signal. Although a multiplier circuit can be used for this purpose, demodulation is implemented by a simpler circuit (Application Window 17.5.2). The audio signal in the output of the demodulator is easily separated from the higher intermediate frequency by simple lowpass filtering. The audio signal is then amplified and fed to the loudspeaker.

AM described earlier is a fairly simple form of modulation. Other modulation schemes are more sophisticated and offer superior performance. Examples are: (1) **frequency modulation** (FM) and **phase modulation**, whereby the modulating signal varies the frequency or phase of the carrier, respectively, rather than its amplitude, and (2) various forms of **pulse modulation**, where the carrier is a pulse train, instead of a sinusoid, and the modulating signal varies the pulse amplitude, duration, or position. The different modulation schemes are briefly discussed in Section ST17.1 and Section ST17.2.

17.3 Deconvolution

Consider a signal $x(t)$ that is affected in some undesirable way by a given system so as to produce a distorted output $y(t)$. Formally, we can describe the undesirable effect of the system by a transfer function $H(j\omega) = \dfrac{Y(j\omega)}{X(j\omega)}$, or in the time domain:

$$y(t) = x(t) * h(t) \tag{17.3.1}$$

> **Concept:** *If h(t) is known, then we can, in principle, derive an inverse system $h_{inv}(t)$ such that when y(t) is convolved with $h_{inv}(t)$ the undistorted signal x(t) is recovered:*

$$y(t) * h_{inv}(t) = x(t) \tag{17.3.2}$$

Equation 17.3.2 is a **deconvolution** operation because it basically involves the undoing of the convolution operation represented by Equation 17.3.1. As shown in Section 16.6 (Chapter 16), we should have $h(t) * h_{inv}(t) = \delta(t)$ (Equation 16.6.8, Chapter 16).

Deconvolution is important in many signal processing applications. A communication channel, such as a telephone line, can introduce frequency distortion, for example, because of its limited bandwidth. An **equalizer** circuit can be inserted in series with the telephone line to reverse the distortion in the line. If the channel introduces some time delay because of phase lag, as explained in Section 17.1, the inverse system will have to compensate this delay with an advance in time, which makes the inverse system noncausal. What is done is to make the channel and the equalizer distortionless, as discussed in Section 17.1. This gives:

$$H_{ch}(j\omega)H_{eq}(j\omega) = e^{-j\omega\tau} \tag{17.3.3}$$

where $H_{ch}(j\omega)$ and $H_{eq}(j\omega)$ are the transfer functions of the channel and equalizer, respectively, and τ is a constant delay. The required $H_{eq}(j\omega)$ is then

$$H_{eq}(j\omega) = \frac{e^{-j\omega\tau}}{H_{ch}(j\omega)} \tag{17.3.4}$$

In practice, the equalizer is designed so that its transfer function approximates that of Equation 17.3.4 as closely as possible.

A serious problem with deconvolution is the effect of noise. The basis of deconvolution is that the distorting effect of the system on the input signal is deterministic, that is, predictable, and can be used to define the transfer function of the inverse system. However, if the channel introduces random noise, the effect of this noise cannot be inverted and can have a drastic effect on the overall response of the system and its inverse. To see this, we can add noise $n(t)$ to $y(t)$ in Equation 17.3.1:

$$y(t) = x(t) * h(t) + n(t) \tag{17.3.5}$$

Taking the FT of both sides of Equation 17.3.5:

$$Y(j\omega) = X(j\omega)H(j\omega) + N(j\omega) \tag{17.3.6}$$

The FT of the inverse system is $\dfrac{1}{H(j\omega)}$ and its output $\dfrac{Y(j\omega)}{H(j\omega)} = X'(j\omega)$ is the estimated $X(j\omega)$ or its recovered FT in the presence of noise. Dividing Equation 17.3.6 by $H(j\omega)$:

$$X'(j\omega) = X(j\omega) + \frac{N(j\omega)}{H(j\omega)} \tag{17.3.7}$$

The effect of noise can be greatly accentuated at high frequencies because of the second term on the RHS of Equation 17.3.7. At high frequencies $H(j\omega)$ can be very small because of bandwidth limitations of the system, but $N(j\omega)$ can be appreciable because noise is generally broadband and has significant power over the entire frequency band of $X(j\omega)$. Dividing a significant $N(j\omega)$ by a very small $H(j\omega)$ at high frequencies can greatly amplify noise in the recovered signal. In practice, this effect is minimized by using a special filter known as a Wiener filter. Deconvolution is commonly used to obtain sharper and clearer images by compensating for aberrations introduced by optical systems of microscopes, telescopes, and cameras (Application Window 17.3.1).

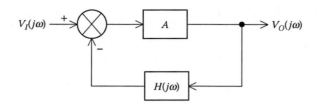

FIGURE 17.3.1
System inversion.

The inverse of a system is sometimes derived in practice by including the system in a negative feedback loop around a high-gain amplifier, as illustrated in Figure 17.3.1. The input to the amplifier is the difference between the applied input $V_I(j\omega)$ and a signal that is fed back from the output through the system to be inverted. This signal is $H(j\omega)V_O(j\omega)$, so that the input to the amplifier is $V_I(j\omega) - H(j\omega)V_O(j\omega)$. The output of the amplifier, which is $V_O(j\omega)$, is A times its input. Hence, $V_O(j\omega) = A[V_I(j\omega) - H(j\omega)V_O(j\omega)]$. Rearranging,

$$\frac{V_O(j\omega)}{V_I(j\omega)} = \frac{A}{1 + AH(j\omega)} \tag{17.3.8}$$

If A is made very large,

$$\frac{V_O(j\omega)}{V_I(j\omega)} = \frac{1}{H(j\omega)} \tag{17.3.9}$$

which is the inverse of the given system.

Another application of deconvolution is in **system identification**, which is the derivation of the transfer function of a system from its input and output. That this is essentially deconvolution is evident from Equation 17.3.1. In system identification, $h(t)$ is derived from $y(t)$ and $x(t)$, instead of the derivation of $x(t)$ from $y(t)$ and $h(t)$. Interchanging $H(j\omega)$ and $X(j\omega)$ in Equation 17.3.7:

$$H'(j\omega) = H(j\omega) + \frac{N(j\omega)}{X(j\omega)} \tag{17.3.10}$$

In this case, the effect of noise is much less serious because $x(t)$ can be chosen so that the second term in Equation 17.3.10 does not dominate at any frequency.

Example 17.3.1: Inverse of a Lowpass Response

It is required to derive the inverse of a lowpass response, $\dfrac{1}{1+j\omega}$.

SOLUTION

$H(j\omega) = \dfrac{1}{1+j\omega}$ and $H_{inv}(j\omega) = 1 + j\omega$. The impulse response corresponding to the lowpass response is, from Chapter 16, Table 16.3.2, $e^{-t}u(t)$. Using the differentiation-in-time property (Chapter 16, Table 16.3.1), the impulse response of the inverse system is $\delta(t) + \delta^{(1)}(t)$.

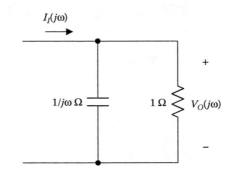

FIGURE 17.3.2
Figure for Example 17.3.1.

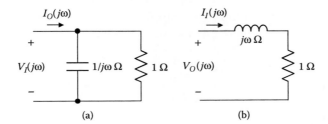

FIGURE 17.3.3
Figure for Example 17.3.1.

$H(j\omega)$ can be derived as the transfer function from a current input $I_I(j\omega)$ to the output voltage V_O $(j\omega)$ for a parallel RC circuit having $R = 1\ \Omega$ and $C = 1\ F$ (Figure 17.3.2). The inverse response can be derived from the same circuit by considering the input as a voltage $V_I(j\omega)$ and the output as the current $I_O(j\omega)$ (Figure 17.3.3a). If the inverse response is to be $\dfrac{V_O(j\omega)}{I_I(j\omega)}$, a series RL circuit can be used (Figure 17.3.3b), which is the dual of the parallel RC circuit.

EXERCISE 17.3.1

Show that if $H_{inv}(j\omega) = 1 + j\omega$ is used in the circuit of Figure 17.3.1, with very large A, the resulting transfer function is $\dfrac{V_O(j\omega)}{V_I(j\omega)} = \dfrac{1}{1 + j\omega}$.

EXERCISE 17.3.2
Show by evaluation of the convolution integral that $e^{-t}u(t) * (\delta(t) + \delta^{(1)}(t)) = \delta(t)$, in accordance with Equation 16.6.8.

Application Window 17.3.1: The Hubble Space Telescope
The Hubble Space Telescope (HST) is a famous example of the application of deconvolution. By orbiting high above Earth, the HST was intended to relay photographic images of great clarity because of elimination of blurring due to atmospheric turbulence. After it was launched in April 1990, at a total cost of almost $3 billion, it was soon discovered that the impulse response of the HST was not at all what it should be.

The impulse response of an optical system, such as a telescope or a microscope, is the **point spread function** (PSF), which is the response of the system to a point source of light. Ideally, the image of a point should be a point. In practice, the PSF is not a point but a spatial distribution of light intensity. This is because of the wave nature of light, the finite size of lenses and mirrors, and spherical aberration due to imperfections in the shapes of lenses and mirrors.

Bright, distant stars approximate a two-dimensional spatial impulse of light. The images of these stars formed by the HST were much broader and more blurred than the theoretical PSF of the HST. The problem was traced to an error made in grinding the primary mirror of the telescope to produce the required parabolic shape. An excess of about 1.3 mm of glass was removed from the face of the 1 ton, 2.4 m diameter mirror, resulting in a significant spherical aberration.

Replacing the mirror was out of the question. A team of astronauts refurbished the HST in December 1993 by replacing some cameras and computers. Because the nature of the aberration was known, deconvolution was applied to the received images so as to obtain much clearer and truly remarkable pictures of space objects.

17.4 Sampling

> **Definition:** *In sampling, a continuous function of time is expressed as a sequence of values at particular instances of time.*

Usually, sampling is done at regular intervals, as illustrated in Figure 17.4.1, where T_S is the **sampling period**, or the **sampling interval**. The function is expressed as a sequence of values ... $f(-T_S), f(0), f(T_S), f(2T_S), \ldots, f((n-1)T_S), f(nT_S), f((n+1)T_S), \ldots$

Sampling is an important operation in signal processing. It is the first step in converting a continuous-time analog signal to a digital signal (Section 17.6). Moreover, a sampled signal is the working signal in some common systems. In **pulse amplitude modulation** (PAM), for example, the carrier is a sequence of pulses of short duration, whose amplitude is varied in accordance with the instantaneous values of the modulating signal at the time of occurrence of the carrier pulses, as illustrated in Figure 17.4.2. In effect, the modulating signal is sampled at a sampling interval equal to the period of carrier pulses, as discussed in Section ST17.3.

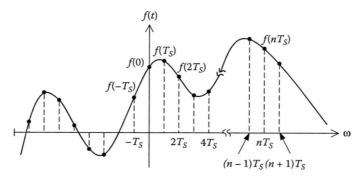

FIGURE 17.4.1
Sampling of signal.

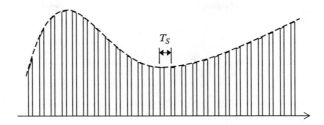

FIGURE 17.4.2
Pulse amplitude modulation.

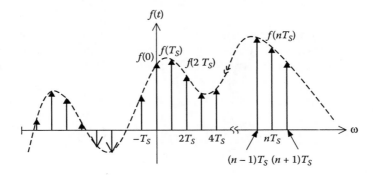

FIGURE 17.4.3
Ideal sampling.

Ideal Sampling

In ideal sampling, the signal to be sampled is multiplied by an impulse train, whose period equals the sampling interval. The continuous signal of Figure 17.4.1 is shown in Figure 17.4.3 sampled by an impulse train of period T_S. Note the similarity between Figure 17.4.1 and Figure 17.4.3. In ideal sampling, the impulse strength is modulated by the instantaneous values of the continuous signal.

The impulse train can be expressed as

$$\delta_c(t) = \sum_{n=-\infty}^{\infty} \delta(t - nT_S) \tag{17.4.1}$$

The sampled signal $f_S(t)$ is then

$$f_S(t) = f(t) \times \sum_{n=-\infty}^{\infty} \delta(t - nT_S) = \sum_{n=-\infty}^{\infty} f(nT_S)\delta(t - nT_S) \tag{17.4.2}$$

Next we will derive the FT of the sampled signal. From Equation 16.3.7 (Chapter 16), the FT of a periodic signal such as the impulse train is

$$F_c(j\omega) = \sum_{n=-\infty}^{\infty} 2\pi C_n \delta(\omega - n\omega_0)$$

where C_n for the impulse train is

$$C_n = \frac{1}{T_S} \int_{-T_S/2}^{T_S/2} \delta(t)e^{-jn\omega_0 t} dt = \frac{1}{T_S} \qquad (17.4.3)$$

assuming that an impulse occurs at the origin, as in Figure 17.4.3. It follows that

$$F_c(j\omega) = \omega_S \sum_{n=-\infty}^{\infty} \delta(\omega - n\omega_S) \qquad (17.4.4)$$

where $\omega_S = 2\pi / T_S$. In other words, the FT of an impulse train in time of period T_S is an impulse train in the frequency domain of period ω_S. Because multiplication in time is equivalent to convolution in the frequency domain (Equation 16.3.25, Chapter 16), it follows that

$$F_S(j\omega) = \frac{1}{2\pi} F(j\omega) * \left[\omega_S \sum_{n=-\infty}^{\infty} \delta(\omega - n\omega_S) \right] = \frac{1}{T_S} \sum_{n=-\infty}^{\infty} F(j\omega) * \delta(\omega - n\omega_S) \qquad (17.4.5)$$

According to Equation 12.4.4 (Chapter 12), convolution of a function with an impulse gives the same function delayed by the same delay as the impulse. Equation 17.4.5 yields

$$F_S(j\omega) = \frac{1}{T_S} \sum_{n=-\infty}^{\infty} F\big(j(\omega - n\omega_S)\big) \qquad (17.4.6)$$

The frequency spectrum of the sampled signal is illustrated in Figure 17.4.4. The frequency spectrum of the continuous-time signal $f(t)$ is assumed to be **band-limited**, that is, it does not contain frequencies above the highest frequency ω_B. $F_S(j\omega)$ is periodic, consisting of the spectrum $F(j\omega)$ reproduced at integer multiples of ω_S, that is, ... $-2\omega_S$, ω_S, 0, ω_S, $2\omega_S$, ... In effect, sampling generates harmonics of $F(j\omega)$ at integer multiples of ω_S.

Shannon's Sampling Theorem

> **Statement:** *If a band-limited signal that does not contain frequencies above f_B is sampled at a rate $f_S > 2f_B$, the signal can be uniquely recovered from its samples.*

The minimum sampling rate of $2f_B$ is the **Nyquist rate**, which means that at least two samples must be taken per period of the highest frequency component of $f(t)$. The sampling theorem can be justified with reference to Figure 17.4.4b. If $f_S > 2f_B$, then $\omega_S > 2\omega_B$, and the spectral components of $F_S(j\omega)$ shifted at integer values of ω_S do not overlap. If the sampled signal is filtered by an ideal lowpass filter of cutoff frequency $\omega_S/2$ and zero phase shift, as illustrated in Figure 17.4.5, then the function $f(t)$ can be uniquely recovered. As will be discussed in Section 17.5, such an ideal lowpass filter is physically unrealizable, but can be approximated in practice. Note that because sampling reduces the magnitude of the spectrum of the sampled signal by T_S (Equation 17.4.6), the lowpass filter must have a gain of T_S.

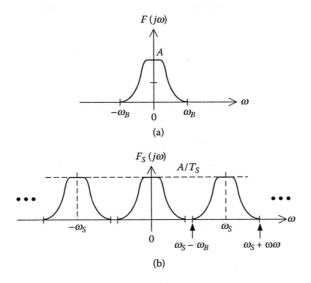

FIGURE 17.4.4
Frequency spectrum of sampled signal.

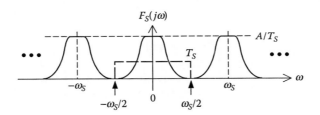

FIGURE 17.4.5
Recovery of sampled signal.

Ideal Reconstruction in the Time Domain

The ideal lowpass filtering of a sampled signal to recover the continuous signal, depicted in the frequency domain in Figure 17.4.5, can be readily illustrated in the time domain. From Chapter 16, Table 16.3.2, the FT of a rectangular pulse of unit amplitude and extending in time from $\tau/2$ to $+\tau/2$ is $\tau\text{sinc}(\omega\tau/2)$. Using the duality property of the FT, the time function whose FT is a rectangular pulse of unit amplitude and extending in frequency from $\omega_S/2$ to $+\omega_S/2$ is $\dfrac{1}{2\pi} \times \omega_S\text{sinc}(\omega_S t/2)$, which is also the impulse response of the ideal lowpass filter (Section 16.6, Chapter 16). If the lowpass filter is to have a gain of T_S, then its impulse response is

$$h_S(t) = \text{sinc}(\omega_S t/2) \tag{17.4.7}$$

$h_S(t)$ is referred to as the ideal **interpolating function**. To recover the original signal, this function must be convolved in time with the sequence of impulses representing the sampled function (Equation 17.4.2). But convolving the interpolating function with a

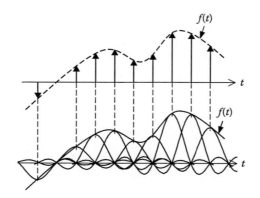

FIGURE 17.4.6
Reconstruction in time.

delayed impulse is simply equivalent to scaling the interpolation function by the strength of the impulse and delaying it by the same delay as that of the impulse. Thus,

$$f(t) = \left[\sum_{n=-\infty}^{\infty} f(nT_S)\delta(t - nT_S) \right] * \text{sinc}(\omega_S t/2) = \sum_{n=-\infty}^{\infty} f(nT_S) \, \text{sinc}\left[\omega_S(t - nT_S)/2 \right] \quad (17.4.8)$$

In words, the original function is obtained as the sum of a series of interpolating functions, each of these functions being centered at the sampling time and scaled by the sample value. This process is illustrated in Figure 17.4.6. Equation 17.4.8 can be considered a proof of the sampling theorem. An alternative proof is given in Section ST17.5.

Aliasing

What happens if $\omega_S < 2\omega_B$? As can be seen from Figure 17.4.7, neighboring, repeated spectra whose centers are separated by ω_S now overlap. This overlap of frequencies corrupts the signal so that it cannot be uniquely recovered from the sampled signal. To investigate this in more detail, consider a band-limited signal whose highest frequency is 20 kHz. Let the sampling rate be 30 kHz, so that the centers of neighboring, repeated spectra are separated by 30 kHz, with resulting overlap, as illustrated in Figure 17.4.8. Frequencies up to 15 kHz are adequately sampled with at least two samples per period whereas frequencies between 15 and 20 kHz are not. What happens is that frequencies

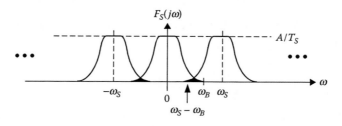

FIGURE 17.4.7
Aliasing in the frequency domain.

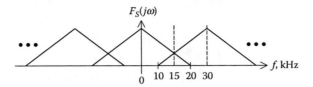

FIGURE 17.4.8
Example on aliasing.

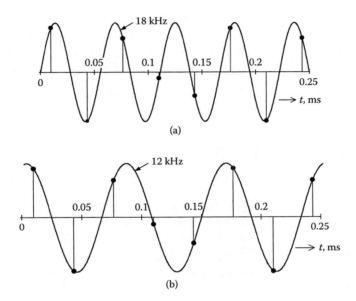

FIGURE 17.4.9
Aliasing in the time domain.

between 15 and 20 kHz are folded back at 15 KHz to lower frequencies. A frequency of $(15 + \Delta\omega)$ kHz, where $0 < \Delta\omega < 5$ kHz is inadequately sampled and is reproduced as a frequency of $(15 - \Delta\omega)$ kHz. Figure 17.4.9 illustrates this in the time domain. In Figure 17.4.9a a frequency of 18 kHz is sampled at a rate of 30 kHz, which is less than the Nyquist rate. The sampled signal is a series of values corresponding to the dots on the 18 kHz sinusoid. On average there are less than two samples per period, or one per half cycle; the positive half cycle in the middle is not sampled at all. These same sampled values in Figure 17.4.9b describe a 12 kHz sinusoid that is the 18 kHz folded back at 15 kHz. The same samples of the inadequately sampled 18 kHz signal in Figure 17.4.9a are now samples of an adequately sampled 12 kHz signal in Figure 17.4.9b. Similarly, all other frequencies between 15 and 20 kHz are folded back at 15 kHz and reproduced as lower frequencies. This phenomenon of a lower frequency taking on the identity of a higher frequency, so to speak, is **aliasing**. The aliased frequencies mix with the true frequencies to distort the signal in a way that cannot be undone.

The only way to avoid aliasing, for a given sampling rate, is to limit the bandwidth of the signal to be sampled to less than half the sampling rate. Real signals do not have a sharp frequency cutoff because higher frequencies taper off smoothly beyond a certain frequency. Antialiasing lowpass or bandpass filters are used to limit the signal bandwidth, but these filters cannot have an ideally sharp cutoff either. In practice, the sampling rate and the antialiasing filter are such that aliasing is kept to an acceptable level.

Comments on Sampling and Reconstruction

The following is a summary of the preceding discussion on sampling and reconstruction, and of related material considered in the Supplementary Topics and Examples:

1. A continuous-time, band-limited signal $f(t)$ is ideally sampled by multiplying it by a periodic impulse train of period T_S. The spectrum $F_{S\delta}(j\omega)$ of the sampled signal in the frequency domain is a periodic repetition of the spectrum $F(j\omega)$, with $F_{S\delta}(j\omega)\big|_{-\omega_B}^{\omega_B} = F(j\omega)$, where $F_{S\delta}(j\omega)\big|_{-\omega_B}^{\omega_B}$ is $F_{S\delta}(j\omega)$ for $-\omega_B < \omega < \omega_B$ and ω_B is the highest frequency in the signal. The signal $f(t)$ can be fully recovered if $\omega_S > 2\omega_B$ and the sampled signal $F_{S\delta}(j\omega)$ is applied to an ideal lowpass filter of gain T_S and cutoff frequency $\omega_S/2$. This corresponds in the time domain to convolving $f_S(t)$ with the ideal interpolating function $\text{sinc}(\omega_S t/2)$, which is the impulse response of the ideal lowpass filter.

2. If the signal $f(t)$ is sampled by multiplying it by a periodic train of pulses of amplitude A, width τ and period T_S, $F_{SP}(j\omega)$, the FT of the sampled signal, is $F_{S\delta}(j\omega)$ multiplied by $A\dfrac{\tau}{T_S}\text{sinc}(n\omega_S\tau/2)$, where $n = 0$ for $F_{S\delta}(j\omega)\big|_{-\omega_B}^{\omega_B}$. The signal $f(t)$ can be fully recovered, without distortion, if $\omega_S > 2\,\omega_B$ and the sampled signal $F_{SP}(j\omega)$ is applied to a lowpass filter of gain $T_S/A\tau$ and cutoff frequency $\omega_S/2$. Multiplying $f(t)$ by the periodic pulse train implies that the amplitude of the pulses of the sampled signal follows that of $f(t)$ over the duration of each pulse (Section ST17.3).

3. If $f(t)$ is applied to a **sample-and-hold** circuit that samples the signal at a period T_S but holds the sampled value for a duration τ, the effect is multiplying $f(t)$ by a train of periodic pulses of unit amplitude and duration τ, but with the amplitude of each sampled pulse remaining constant at the sampled value. The FT of the sampled signal is $F_{S\delta}(j\omega)$ multiplied by $\dfrac{\tau}{T_S}e^{-j\omega\tau/2}\text{sinc}(\omega\tau/2)$. The linear phase lag of $\omega\tau/2$ introduces a time delay of $\tau/2$. $F_S(j\omega)$ is somewhat distorted compared to $F_{S\delta}(j\omega)\big|_{-\omega_B}^{\omega_B}$ because of multiplication by $\text{sinc}(\omega\tau/2)$. The slightly distorted $f(t)$ can also be recovered in this case using an ideal lowpass filter of gain T_S/τ and cutoff frequency $\omega_S/2$ (Section ST17.4).

4. A sampled signal can be reconstructed by applying it to a **zero-order hold** circuit that maintains the value of a given sample until the next sample occurs. The FT of the sampled signal is $F_{S\delta}(j\omega)$ multiplied by $T_S e^{-j\omega T_S/2}\text{sinc}(\omega T_S/2)$. As in case 3, a time delay and distortion are introduced (Example SE17.1).

Application Window 17.4.1: Examples of Aliasing and Sampling

Movie cameras have a shutter speed that is usually between 25 and 30 frames/s, that is, they effectively take snapshots at this rate, which appear to the human eye to be fused and smoothly varying. Consider a spoke in the wheel of a car or a wagon that is rotating counterclockwise (CCW) and moving from right to left. If the rotation is less than 15 rev/s, corresponding to half the sampling rate of 30 frames/s, the spoke rotates less than half a turn between successive frames, and the wheel appears to rotate CCW consistent with its motion from right to left. No aliasing occurs. If the wheel's rotation is between 15 and

30 rev/s, the spoke rotates more than half a turn between successive frames, and the wheel appears to be rotating clockwise (CW) while moving from right to left. The closer the rotation is to 30 rev/s, the slower its CW rotation appears. Aliasing now occurs because the sampling rate is less than half the rotational speed. The rotational speed is effectively folded back at 15 rev/s. When the rotational speed is just less than 15 rev/s, the wheel appears to rotate CCW at this speed. When the rotational speed just exceeds 15 rev/s, the wheel appears to rotate CW at this speed. The wheel appears stationary at a rotation of 30 rev/s.

Compact disks store large amounts of data representing sound or compressed video in digital format. To store audio signals on an audio compact disk, the signals are first passed through an antialiasing bandpass filter having a bandwidth of 5 Hz to 20 kHz, required for high-quality reproduction. The filtered signal is then converted to a digital signal by an analog-to-digital converter that samples the signal at a rate of 44.1 kHz, which is more than twice the highest frequency in the filtered audio signal.

In some telephone systems, speech signals are passed through an antialiasing bandpass filter of bandwidth 200 to 3200 Hz, which is considered adequate for good speech reproduction. The filtered signal is sampled at 8 kHz, resulting in an amplitude-modulated pulse signal of this frequency (Figure 17.4.2). Because the telephone lines can support a higher frequency of transmission, the pulses from several PAM signals are interleaved in time in what is referred to as **time-division multiplexing**.

17.5 Filtering and Smoothing

> **Definition:** *A filter is a frequency-selective circuit that transmits some frequency components, essentially unaltered, while blocking other frequency components.*

Passive filters were considered in Chapter 10 and active filters will be considered in Chapter 18. We will concentrate here on the concept of an ideal filter.

> **Concept:** *Ideal filters are characterized by i) a response of unity in the passband with zero phase shift, ii) zero response in the stopband, and iii) a stepwise transition between passbands and stopbands.*

Ideal Lowpass Filter

The frequency spectrum of an ideal lowpass filter of cutoff frequency ω_{c1} is illustrated in Figure 17.5.1. $|H(j\omega)|$ is an even function when $h(t)$ is real. From Chapter 16, Table 16.3.2, the impulse response of the ideal lowpass filter is:

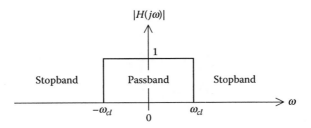

FIGURE 17.5.1
Ideal lowpass filter response.

$$h(t) = \frac{\omega_{cl}}{\pi} \, \text{sinc}(\omega_{cl}t) \qquad\qquad (17.5.1)$$

$h(t)$ is of infinite duration and is nonzero for $t < 0$, which means that it occurs before the impulse is applied at $t = 0$. This makes the ideal lowpass filter noncausal, and hence not physically realizable by a system operating in real time.

Although noncausal, an ideal lowpass filter is extremely useful as a concept for mathematical analysis and as a guide during the initial stages of a design. We used the ideal filter to recover, in the frequency domain, the original signal from a sampled signal in the absence of aliasing.

Other Ideal Filters

Filters can also be ideally highpass, bandpass, or bandstop. Figure 17.5.2a shows an ideal highpass response having an ideally sharp cutoff frequency ω_{ch}. It is seen from Figure 17.5.2b that the highpass response can be formally obtained by subtracting a lowpass response having $\omega_{cl} = \omega_{ch}$ from an ideal allpass response that has $H(j\omega) = 1$ for all ω. This was demonstrated for nonideal filters in Chapter 10.

An ideal bandpass filter has the frequency spectrum shown in Figure 17.5.3a. Formally, an ideal bandpass filter can be considered as the cascade of an ideal lowpass filter of cutoff frequency ω_{cl} and an ideal high-pass filter of cutoff frequency ω_{ch}, where $\omega_{cl} > \omega_{ch}$. Frequencies less than ω_{ch} are stopped by the highpass filter, whereas frequencies greater than ω_{cl} are stopped by the lowpass filter. Frequencies in the range ω_{ch} to ω_{cl} are passed by both filters.

An ideal bandstop filter has the frequency spectrum shown in Figure 17.5.4. Formally, an ideal bandstop response results from summing the outputs of an ideal lowpass filter and an ideal highpass filter having $\omega_{cl} < \omega_{ch}$, the inputs of the two filters being paralleled

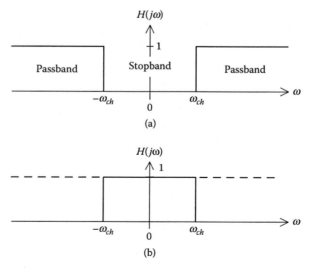

FIGURE 17.5.2
Ideal highpass filter response.

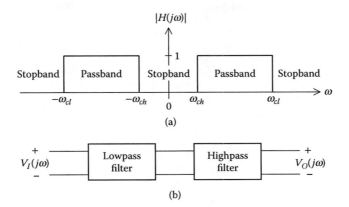

(a)

(b)

FIGURE 17.5.3
Ideal bandpass filter response.

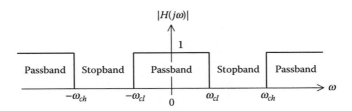

FIGURE 17.5.4
Ideal bandstop filter response.

together (see Figure 18.3.4). Alternatively, an ideal bandstop response can be formally obtained by subtracting a bandpass response from an ideal all-pass response that has $H(j\omega) = 1$ for all ω, as was demonstrated for nonideal filters in Chapter 10.

As the ideal highpass, bandpass, and bandstop filters can be derived from an ideal lowpass filter, then these filters are not physically realizable, as was argued earlier for a lowpass filter.

Smoothing

Before discussing smoothing, we should have some criterion for judging how "smooth" is a given time-varying function. A commonly used criterion is to look for sudden jumps, or discontinuities, in the value of a given function or its higher derivatives. The higher the order of the derivative that first shows a discontinuity, the smoother is the function. This is illustrated in Figure 17.5.5. The square wave is itself discontinuous (Figure 17.5.5a). The integral of a square wave of zero average is the triangular wave of Figure 17.5.5b. This function is continuous, but its first derivative, the square wave, is discontinuous. Hence, the triangular wave is smoother than the square wave. Similarly, the integral of the triangular wave of Figure 17.5.5b is the parabolic wave of Figure 17.5.5c. This function and its first derivative are continuous, but its second derivative, the square wave, is discontinuous. The parabolic wave is therefore smoother than the triangular wave, and is smoother than the square wave.

We encountered this criterion in Section 9.5 (Chapter 9), when discussing the rate of attenuation of harmonics of a periodic function. It was argued that the higher the order of the derivative that is discontinuous, the smaller is the harmonic content. This implies

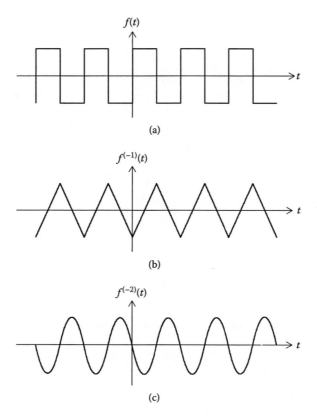

FIGURE 17.5.5
Smoothness of functions.

that the smoother the function, the less rapidly it changes in time and the narrower is its bandwidth (Section 16.6, Chapter 16).

The preceding discussion suggests the following:

> **Concept:** *Integration makes a function smoother, whereas differentiation makes it less smooth.*

This follows from the integration-in-time (Equation 16.3.11, Chapter 16) and the differentiation-in-time (Equation 16.3.8, Chapter 16) properties. When a function is integrated, its magnitude spectrum is divided by ω, which means that the high-frequency components are attenuated with respect to the low-frequency components. The bandwidth is reduced, the function changes less rapidly with time, and becomes smoother. The converse is true of differentiation. In fact, differentiation is avoided as much as possible in practice because the accentuation of high frequencies introduces noise.

Leaky Integration

Signals are commonly smoothed using a lowpass filter in what is essentially a form of imperfect, or leaky, integration. An ideal capacitor is a perfect integrator because $v_C = \dfrac{1}{C} \int i_C dt$. A resistor R connected in parallel with C provides a leakage path for the

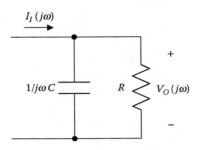

FIGURE 17.5.6
Leaky integration.

charge on the capacitor (Figure 17.5.6). The transfer function from input current to output voltage is:

$$H(j\omega) = \frac{V_O(j\omega)}{I_I(j\omega)} = \frac{R}{1 + j\omega CR} \qquad (17.5.2)$$

which is a lowpass characteristic (Equation 10.1.1, Chapter 10). If $R \to \infty$, $H(j\omega) = \frac{1}{j\omega C}$, which corresponds to perfect integration. With finite R, Equation 17.5.2 represents imperfect, or leaky, integration. In the time domain, if an impulse is applied to a perfect integrator, the impulse response is: $h_{int}(t) = \int_{-\infty}^{\infty} \delta(t)dt = u(t)$. For a leaky integrator, the IFT of $H(j\omega)$ in Equation 17.5.2 is, from Table 16.3.2, Chapter 16, $h_{lk\,int} = e^{-t/t}u(t)$, where $\tau = RC$, assuming the multiplying factor to be unity.

The smoothing effect of the RC circuit of Figure 17.5.6 is simulated in Example 17.5.1 and illustrated in Figure 17.5.7. Rectangular current pulses of 100 mA amplitude, 0.1 ms duration, and 1 ms period are applied, with $R = 1\ k\Omega$ and $C = 3\ \mu F$, which makes $\tau = 3$ ms.

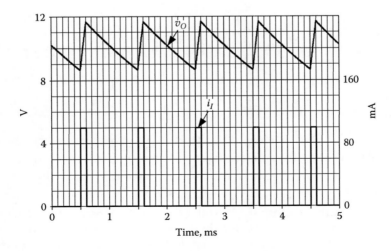

FIGURE 17.5.7
Smoothing operation.

Although the current input i_I is pulsating, the output voltage v_O is much smoother, varying between 8.65 and 11.66 V about an average value of 10.09 V.

If R is omitted, v_O increases progressively as each current pulse during the on period pumps additional charge into the capacitor, in accordance with the integrating property of C. The effect of R, which normally represents a load, is to leak some charge, that is, discharge C during the off period, between successive pulses, so that a steady state is reached. In this steady state, the average voltage across C is such that the charge gained by C during the on period is equal to the charge lost during the off period. An equivalent interpretation of the smoothing effect is in terms of the lowpass response. The reactance of the RC combination decreases with frequency. With a current input, the high-frequency components of the output voltage are smaller at high frequencies than at low frequencies. The reduced high frequency in the output voltage means a smoother waveform.

A direct application of this smoothing operation is the capacitor-input filter, which is commonly used in conjunction with a rectifier circuit to supply a dc load of relatively low power from an ac mains supply (Application Window 17.5.1). In this case τ is made as large as practicable. Another application is the demodulation of an AM wave, in which case τ is chosen so that the modulating signal is adequately recovered (Application Window 17.5.2). If the pulses in Figure 17.5.7 occur at a varying rate, the RC circuit can be used to provide a signal that is indicative of the pulse rate (Application Window 17.5.3).

Simulation Example 17.5.1: Smoothing Effect of RC Circuit

The schematic is shown in Figure 17.5.8, which also indicates the parameters of the source IPULSE. The simulation is run for 25 ms, with data collection starting at 20 ms, so as to show a steady state.

The maximum and minimum voltages V_{Omax} and V_{Omin} can be readily calculated and compared with simulated values. During the on period, C charges from V_{Omin} toward a maximum of $(0.1A) \times (1\ k\Omega) = 100$ V. At $t = 0.1$ ms, v_O reaches V_{Omax}. From Equation 13.3.2 (Chapter 13),

$$V_{Omax} = 100 + (V_{Omin} - 100)e^{-0.1/3} \tag{17.5.3}$$

where $RC = 3$ ms. During the off period, Equation 13.3.2 (Chapter 13) gives

$$V_{Omin} = V_{Omax}e^{-0.9/3} \tag{17.5.4}$$

Solving for V_{Omax} and V_{Omin} gives $V_{Omax} = 100\dfrac{1-e^{-0.1/3}}{1-e^{-1/3}} = 11.56$ V and $V_{Omin} = 8.57$ V. The

corresponding values from the simulation are: $V_{Omax} = 11.68$ V and $V_{Omin} = 8.66$ V. Assum-

ing a linear variation of v_O between V_{Omax} and V_{Omin}, $V_{Oavg} = \dfrac{V_{Omax} + V_{Omin}}{2} = 11.07$ V. Using

the command $AVGX(V_O, 1m)$ assuming VO is labeled VO, gives an average value from the simulation of 10.09 V. The discrepancy is due to V_O being exponential rather than linear.

EXERCISE 17.5.1

Assume that in simulation Example 17.5.1, the capacitor charges and discharges linearly. Express the average charge gained during the on period and the average charge lost during the off period in terms of V_O, the average output voltage, taking into account the trapezoidal shape of the input current pulse. By equating these two quantities, show that $V_O = 10.1$ V. Note that during charging, part of the input current flows through R.

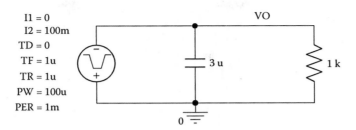

FIGURE 17.5.8
Figure for Example 17.5.1.

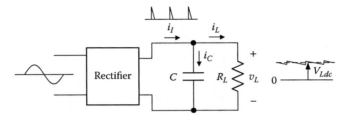

FIGURE 17.5.9
Figure for Application Window 17.5.1.

Application Window 17.5.1: Capacitor Input Filter

A common requirement of electronic equipment that is supplied from the ac mains is to convert the ac supply to a dc output. The first step is to use a rectifier circuit that converts the ac supply to a unidirectional one. The next step is to use some kind of filter that smoothes the unidirectional output of the rectifier circuit. In the case of relatively low-power loads, smoothing is achieved by a capacitor connected across the input of an IC regulator that provides a regulated and almost true dc supply that does not change with variations of the load current or the supply voltage, within specified limits. The capacitor in parallel with the load represented by R_L constitutes a capacitor input filter (Figure 17.5.9). This type of filter is popular because of the availability of low-cost electrolytic capacitors having a large capacitance in a relatively small volume.

For large enough C, the rectifier applies current pulses to the parallel combination of C and R_L. The shape of the current pulses is almost a right triangle, and their frequency is twice that of the AC supply, assuming a full-wave rectifier. The output voltage is very similar to that of Figure 17.5.7. C is typically such that $\omega C R_L$ is at least 50, where ω is the radian frequency of the supply. The resulting peak-to-peak ripple, $V_{Omax} - V_{Omin}$, as a percentage of the average (or dc) value normally does not exceed 5%.

EXERCISE 17.5.2

Consider a pulsating inflow of instantaneous value F_I m^3/s to a water tank of area A m^2. The instantaneous outflow from the tank is F_O m^3/s, where F_O can be assumed to be directly proportional to the instantaneous head of water H in the tank, that is, $F_O = KH$,

where K is a constant. Show that H satisfies the differential equation: $A\dfrac{dH}{dt} + KH = F_I$,

which is exactly analogous to the differential equation $C\dfrac{dv_O}{dt} + \dfrac{v_O}{R} = i_I$ satisfied by the RC leaky integrator of Figure 17.5.6.

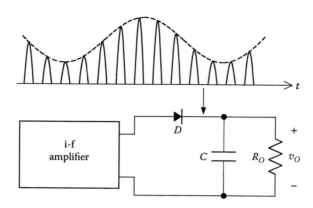

FIGURE 17.5.10
Figure for Application Window 17.5.2.

Application Window 17.5.2: AM Demodulator

The RC leaky integrator is used for demodulating the amplitude-modulated, intermediate frequency of a superheterodyne receiver (Application Window 17.2.1). The output of the i-f amplifier is applied to the RC circuit through a diode D, which allows conduction in one direction only, thereby eliminating the negative part of the waveform (Figure 17.5.10). To recover the modulating signal, v_O should follow the envelope of the waveform, which places some constraints on the time constant RC. In the absence of a modulating signal, v_O should remain nearly constant at the peak value of the waveform. This is similar to the requirement of the capacitor-input filter and implies that $\omega_{i\text{-}f}RC \gg 2\pi$, or $RC \gg 1/f_{i\text{-}f}$, where $f_{i\text{-}f}$ is the intermediate frequency. On the other hand, to follow the envelope faithfully, the phase lag due to the RC circuit should be small. From Equation 17.5.2, the phase lag at the highest frequency ω_{mh} in the modulating signal is $\tan^{-1}\omega_{mh}RC$. For a small angle, $\tan^{-1}\omega_{mh}CR \cong \omega_{mh}CR$, and this should be small compared to 2π. In other words, $RC \ll 1/f_{mh}$. Combining this with the previous inequality,

$$1/f_{mh} \gg RC \gg 1/f_{i\text{-}f} \qquad (17.5.5)$$

In an AM radio, $f_{i\text{-}f}$ is 450 to 470 kHz and f_{mh} does not exceed 5 kHz. A reasonable choice for RC that equally satisfies both sides of this inequality is the geometric mean

$$RC = \frac{1}{\sqrt{f_{i\text{-}f}f_{mh}}} \cong 2\times10^{-5}\ \text{s}.$$

Before v_O is applied to the audio amplifier, the dc component is removed by a sufficiently large series capacitor that does not significantly affect the low-frequency components of the recovered signal. This dc component is not present in the original sound-modulating signal, but is introduced by using the diode to eliminate the negative part of the AM input. v_O also contains a small i-f component because $\omega_{IF}CR$ is finite. If objectionable, this component is practically eliminated by a small shunt capacitor connected across the input of the audio amplifier.

Application Window 17.5.3: Estimating Pulse Rate

The RC leaky integrator can also be used for determining the pulse rate of a train of identical pulses of varying rate. The principle is conveniently illustrated by approximating the input current pulses by impulses of appropriate strength (Figure 17.5.11). The voltage

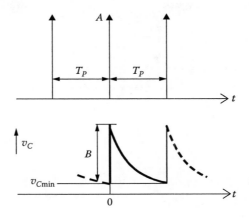

FIGURE 17.5.11
Figure for Application Window 17.5.3.

across the *RC* circuit is the sum of the impulse responses to each pulse, as shown in the lower part of Figure 17.5.11.

Let the train of impulses be applied for a sufficiently long time so that a steady state is reached, and assume that an impulse occurs at $t = 0$. If the effect of previous impulses at $t = 0^-$ is $v_{C\,min}$, then $v_C(0^+) = v_{C\,min} + B$, where $B = \dfrac{A}{C}$ and A is the impulse strength. Hence,

$$v_C(t) = (v_{C\,min} + B)e^{-t/\tau}, \quad t \geq 0^+ \tag{17.5.6}$$

At $t = T_P$, $v_C(T_P) = v_{C\,min}$ because of the assumption of a steady state. Substituting in Equation 17.5.6 and solving for $v_{C\,min}$ gives: $v_{Cmin} = \dfrac{Be^{-T_P/\varepsilon}}{1-e^{-T_P/\tau}}$. Substituting for $v_{C\,min}$ in Equation 17.5.6:

$$v_C(t) = \frac{Be^{-t/\varepsilon}}{1-e^{-T_P/\tau}}, \quad t \geq 0^+ \tag{17.5.7}$$

The average of $v_C(t)$ is:

$$v_{Cavg} = \frac{B}{(1-e^{T_P/\tau})} \int_0^{T_P} e^{-t/\tau}dt = \frac{B\tau}{T_P} = B\tau f_P \tag{17.5.8}$$

where $f_P = \dfrac{1}{T_P}$ is the pulse rate. It is seen that v_{Cavg} is directly proportional to f_P. If the pulse rate varies over a time interval that is comparable or larger than T_P, v_{Cavg} will vary accordingly.

17.6 Analog-to-Digital Conversion

The conversion of a continuous analog signal to a digital signal involves the following steps:

1. *Sampling:* The analog signal is sampled at a regular rate as described in Section 17.4.
2. *Quantization:* The range of values of the output is divided into a number of discrete levels, and the value of each sample is expressed as a number equal to that of the nearest, or lower, discrete level.

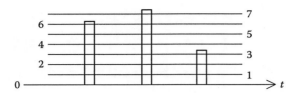

FIGURE 17.6.1
Quantization.

3. *Coding:* The number corresponding to the value of each sample is coded in the form of some digital code, normally a binary code in which a number is expressed as a string of 1s and 0s.

To illustrate these steps, suppose that the range of values of a signal is divided into 8 levels, including the 0 level, as illustrated in Figure 17.6.1. The magnitude of the first sample is between levels 6 and 7, so it is assigned the number 6. The magnitude of the second sample slightly exceeds level 7, so it is assigned the maximum level of 7. The magnitude of the third sample is between levels 3 and 4, and is assigned the value of 3. In binary code, these numbers are expressed in the binary number system in which only the binary digits, or **bits**, 0 and 1 are used. As we have 8 levels 0 to 7 and $2^3 = 8$, we need 3 bits to represent a number in the range 0 to 7. The number is expressed as the sum of 3 terms, where each term is a bit multiplied by 2 raised to a power equal to the position of the bit in the number, starting from a power of zero for the least significant position. The number 4, for example, is represented as 100 because $1 \times 2^2 + 0 \times 2^1 + 0 \times 2^0 = 4$. The three values 6, 7, and 3 are represented by the sequence of 3 binary numbers 110, 111, and 011.

In practice, many more bits are used, typically 32 or 64 bits. These bits are generally represented in digital circuits by voltage levels. A binary digit of 0 is represented by a low voltage level that is close to zero, whereas a binary digit of 1, or logic 1, is represented by a higher voltage level that may be close to, say, +5 V. The 1 and 0 bits are identified with True and False statements of Boolean logic, so that digital signals are processed according to the rules of Boolean algebra.

Digital-to-analog conversion can be readily performed with operational amplifiers and resistive networks, as described in Example SE18.9 (Chapter 18). An analog-to-digital converter (ADC) can be constructed using a digital-to-analog converter (DAC) connected in a feedback loop, so as to perform the inverse operation, in a manner analogous to the scheme of Figure 17.3.1. The basic arrangement is illustrated in Figure 17.6.2. A clock generator is applied to an up/down counter, whose count increments or decrements by one with each clock pulse, depending on whether the voltage of the control line is high or low. The output of the counter is fed to a DAC, whose output v_O is connected to one input of the comparator (Section 18.5, Chapter 18), the other input being connected to the analog input v_I. The output of the comparator controls the direction of counting of the up/down counter.

If $v_O < v_I$, the comparator output is high and causes the counter to count clock pulses upwards; that is, as an increasing number. The input to the DAC increases and v_O increases until it just exceeds v_I. The output of the comparator then goes low, causing the counter to count downwards; that is, in a decreasing number of counts. v_I is generally time varying, so the feedback to the comparator keeps v_O close to v_I, and v_O is said to *track* v_I. Because

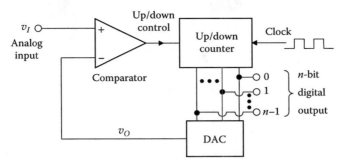

FIGURE 17.6.2
Analog-to-digital converter.

v_O is the output of a DAC, the input to the DAC (the counter output) must be the digital equivalent of v_I.

Before ending this section, it is appropriate to highlight the advantages of digital systems, which can be summarized as follows:

1. *Accuracy:* The accuracy of a digital system is determined by the number of bits used to represent the various quantities. In a 16-bit system, for example, there are 2^{16}, or 65,536, levels. The corresponding accuracy is $\dfrac{100}{2^{16}} = 0.0015\%$. Analog systems, on the other hand, seldom achieve an accuracy better than 0.1%. It should be noted, however, that in digital computations errors may accumulate, so that the overall accuracy may not be as high as suggested by the number of bits used.

2. *Reliability:* Because only two voltage levels are involved, corresponding to 0 and 1, electronic devices such as transistors need only switch between two states — fully on or fully off. Noise immunity is high because misreading a 0 level as a 1 level, or conversely, can occur with only relatively large noise inputs. In analog circuits, on the other hand, transistors operate on a continuum of levels over a given range, so that errors and corruption by noise are more likely to occur.

3. *Versatility:* Digital systems are much more versatile than analog systems in several respects:

   a. Processing in analog systems is performed by special-purpose units, such as adders, integrators, amplifiers, etc. In combinational digital circuits, whose outputs depend only on present inputs, all Boolean logic can theoretically be performed by only one type of logic gate (NAND or NOR). In sequential digital circuits, whose outputs depend on present as well as past inputs and which therefore incorporate some memory elements, only a single type of memory element (a bistable) need, in theory, be used.

   b. The processing in digital systems can be controlled by software. In other words, an operation, or the parameters of a given operation, can be altered, not by changing or adjusting hardware, as in analog systems, but by programming, as in the case of digital computers, digital filters, or computer-controlled communications and control systems.

    c. Some operations, such as data storage, data compression, and error detection and correction are implemented much more conveniently and effectively in digital systems than in analog systems.

Summary of Main Concepts and Results

- Distortion of a signal could be due to nonlinearities or due to the frequency response of an LTI circuit.

- The impulse response of a distortionless LTI circuit is an impulse of zero delay associated with zero phase shift, or is a delayed impulse associated with a linear phase lag.

- The extent of the distortion produced by an LTI circuit is related to the shape of the impulse response. The briefer the impulse response, the smaller the distortion.

- Modulation shifts the frequency spectrum of the modulating signal to higher frequencies. The main benefits are: (1) a great reduction in the size of transmitting and receiving antennas, (2) transmission of more than one modulated signal on the same communication channel, and (3) reduced corruption of the modulating signal by noise.

- In principle, deconvolution can be used to undo the distortion of a signal by a given system. Noise can have a serious effect on the recovery of a distorted signal by deconvolution.

- If a band-limited signal that does not contain frequencies above f_B is sampled at a rate $f_S > 2f_B$, the signal can be uniquely recovered from its samples.

- When $f_S < 2f_B$, frequencies in the range between $f_S/2$ and f_B are folded back at the frequency $f_S/2$ and appear as frequencies less than $f_S/2$, thereby resulting in aliasing. The aliased frequencies mix with the true frequencies to distort the signal in a way that cannot be undone.

- Ideal filters are characterized by: i) a response of unity in the passband and zero phase shift, ii) zero response in the stopband, and iii) a stepwise transition between passbands and stopbands.

- Ideal filters are not realizable by physical systems operating in real time.

- Integration makes a function smoother, whereas differentiation makes it less smooth.

- Compared to analog systems, digital systems are more accurate, reliable, and versatile.

Learning Outcomes

- Articulate the basic features of distortion, modulation, deconvolution, sampling, filtering and smoothing, and digitizing

Supplementary Topics and Examples on CD

ST17.1 Angle modulation: Explains phase modulation (PM) and frequency modulation (FM), and derives the expression for a frequency-modulated signal.

ST17.2 Pulse modulation: Briefly explains the various types of pulse modulation, namely, pulse amplitude, pulse width, pulse duration, pulse position, pulse code, and delta modulation.

ST17.3 Nonideal sampling: Considers the case when the sampling signal is a pulse of finite width rather than an ideal impulse function. The analysis also applies to PAM.

ST17.4 Flat-topped amplitude modulation: Analyzes the case of applying the modulating signal to a sample-and-hold circuit that produces flat-topped sample pulses.

ST17.5 Sampling theorem: Gives an alternative proof of the sampling theorem.

ST17.6 Sampling theorem in the frequency domain: Presents and proves the sampling theorem in the frequency domain.

SE17.1 Practical reconstruction with zero-order hold: Discusses the reconstruction of a sampled signal when the last sample value is held constant until the next sample occurs.

SE17.2 Nyquist rate and reconstruction of bandpass signals: Derives the Nyquist rate for a bandpass rather than a baseband signal and determines the transfer function and the impulse response of the ideal interpolation filter.

SE17.3 Lowpass filtering of sampled signal: Shows that if a sampled signal is applied to a lowpass filter of cutoff frequency equal to one half of the sampling frequency, the filter output equals the original continuous signal at integer multiples of the sampling period, irrespective of the bandwidth of the original signal.

Problems and Exercises

P17.1   Distortion

P17.1.1 A voltage $v = \cos 4,000t + 2\sin 8,000t$ V is applied across a 1 Ω resistor whose v–i relation is: $i = v + 0.1v^2$. What are the frequency components present in the current?

P17.1.2 A voltage $v = \cos\omega_0 t + 2\sin 2\omega_0 t$ is applied to a circuit that gives an output $v_O = v + v^2$. Determine the FSE of the output. What is the amplitude and phase distortion of the fundamental and the second harmonic?

P17.1.3 The transfer function of an amplifier is expressed as $\dfrac{500}{(\omega + j10^4)(\omega + j5 \times 10^4)}$.

Determine the amplitude and phase distortion, with respect to very low frequencies, at: (a) $\omega = 2 \times 10^4$ rad/s; (b) $\omega = 10^5$ rad/s.

P17.1.4 Is it possible to have a system whose phase spectrum is $+\omega\tau$ instead of $-\omega\tau$?

P17.1.5 If a constant phase angle θ is added to the phase spectrum of a distortionless delay system, does the system remain distortionless for all θ?

P17.1.6 In passing through a system, the phase angle of a signal is reduced at a rate of 10^{-2} degrees/Hz. By how much is the waveform delayed in time?

P17.2 Modulation

P17.2.1 A 5 KHz sinusoidal signal amplitude modulates a 1.5 MHz carrier of 10 V amplitude, the modulation factor being 50%. Express the modulated signal in both the frequency and time domains and determine the 1 Ω power associated with the carrier and each sideband.

P17.2.2 A music signal has frequency components from 10 Hz to 30 kHz. If this signal amplitude modulates a 1 MHz carrier, what is the range of frequencies for the upper and lower sidebands?

P17.2.3 Assuming double-sideband transmission, and a maximum bandwidth of 5 kHz for the modulating signal, how many stations can the AM medium-wave band (530 kHz to 1.6 MHz) accommodate in a given locality?

P17.2.4 A channel of 120 KHz bandwidth is to be used to transmit a number of different audio signals of 4 kHz bandwidth using AM and frequency-division multiplexing. How many signals can be transmitted using double-sideband modulation and single-sideband modulation?

P17.2.5 Suppose that in an AM receiver the modulated signal is multiplied by the output of an oscillator that has the same frequency as the carrier. What is the resulting spectrum? How can the modulating signal be recovered? What is the main practical difficulty with such a system? This type of demodulation is **coherent demodulation**. It provides an alternative to a superheterodyne receiver known as a **synchrodyne** receiver.

P17.2.6 A radar signal consists of pulses of sinusoids of frequency 1 GHz, the pulse duration being 1 μs and the pulse period 1 ms. The signal may be considered as a carrier that is modulated by a rectangular wave whose amplitude varies between 1 and 0. Determine the magnitude spectrum of the modulated signal.

P17.2.7 Figure P17.2.7 shows the block diagram of a balanced modulator. The same carrier is modulated by two identical AM modulators, by $m(t)$ in one modulator and by $-m(t)$ in the other. The outputs of the two modulators are then subtracted to produce the output modulated signal $f_{AM}(t)$. Show that $f_{AM}(t)$ is a double-sideband AM signal with the carrier suppressed.

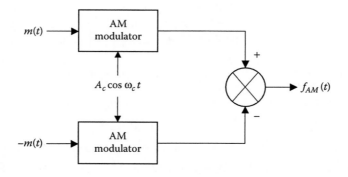

FIGURE P17.2.7

P17.3 Deconvolution

P17.3.1 If $H(j\omega) = 2 + j\left(\dfrac{\omega}{2} - \dfrac{2}{\omega}\right)$, determine $H_{inv}(j\omega)$ and $h_{inv}(t)$.

P17.3.2 If $H(j\omega) = e^{1+j\omega}(1 + j\omega)$, determine $H_{inv}(j\omega)$ and $h_{inv}(t)$.

P17.3.3 The output of a system is $2e^{-2t}u(t)$ when the input is $e^{-t}u(t)$. Determine $H(j\omega)$ and $h(t)$.

P17.3.4 Consider that $H(j\omega) = \dfrac{1}{j\omega}$, corresponding to perfect integration in the scheme of Figure 17.3.1. Show that if A is very large, $\dfrac{V_O(j\omega)}{V_I(j\omega)} = j\omega$, corresponding to perfect differentiation.

P17.4 Sampling

P17.4.1 Given $x(t) = \text{sinc}(400\pi t)$. (a) What is the Nyquist rate? (b) If the signal is sampled at half the Nyquist rate, what is the frequency around which frequencies are folded back?

P17.4.2 Given $x(t) = \dfrac{\omega_B}{2\pi}\,\text{sinc}^2(\omega_B t/2)$, sketch $X(j\omega)$ and $X_S(j\omega)$ for sampling intervals $\pi/2\omega_B$, $2\pi/3\omega_B$, π/ω_B, and $2\pi/\omega_B$. Which sampling interval corresponds to the Nyquist rate? For which sampling interval does aliasing occur?

P17.4.3 Suppose that the sampling interval in P17.4.2 is increased to $4\pi/\omega_B$. What frequencies are still adequately sampled? What is the frequency around which frequencies are folded back?

P17.4.4 Consider the bandpass signal of Figure P17.4.4. (a) What is the Nyquist rate that would result in a spectrum of the sampled signal without gaps and overlap? Determine the sampling rate and sketch the resulting spectrum.

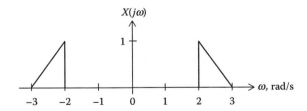

FIGURE P17.4.4

P17.4.5 Determine the impulse response of the ideal interpolation filter that can be used to reconstruct the signal in P17.4.4.

P17.4.6 Sketch the frequency spectrum of the signal $\cos 4\pi t$ when sampled at sampling intervals of, $\dfrac{1}{8s}$, $\dfrac{1}{2s}$, and $\dfrac{3}{4s}$. For which intervals does aliasing occur?

P17.4.7 A signal $x(t) = \cos(600\pi t)$ is sampled at a rate $\omega_S = 400\pi$. The sampled signal is then filtered with an ideal, lowpass filter having a gain of $1/200$ and a cutoff frequency of 400π. Determine the resulting time signal and sketch its frequency spectrum.

P17.5 Filtering and Smoothing

P17.5.1 A signal $x(t) = \dfrac{\sin\tau t}{\pi t}$ is applied to an ideal lowpass filter of cutoff frequency ω_{cl}. What should be the relation between τ and ω_{cl} if the signal is to be recovered without distortion?

P17.5.2 A signal $x(t) = e^{-4t}u(t)$ is applied to an ideal lowpass filter of cutoff frequency ω_{cl}. Determine ω_{cl} such that $W_{1\Omega}$ of the output signal is 90% that of the input signal.

P17.5.3 Two signals, $x_1(t) = \cos(400\pi t)$ and $x_2(t) = \cos(800\pi t)$, are multiplied together. If the resulting signal is passed through an ideal high-pass filter of cutoff frequency $1000\pi t$, what is the frequency of the output of the filter?

P17.5.4 A periodic rectangular train (Figure 9.2.5, Chapter 9) having $\tau = 1$ ms and $T = 4$ ms is passed through the following types of ideal filters: (a) a lowpass filter of 1,200 Hz cutoff frequency, (b) a highpass filter of 4,300 Hz cutoff frequency; (c) a bandpass filter of passband from 2,400 to 3,600 Hz, and (d) a bandstop filter having cutoff frequencies as in case c. Determine the frequencies in the output in each case.

P17.5.5 Determine the leaky integrator circuit that is the dual of the parallel RC circuit of Figure 17.5.6.

P17.5.6 A practical application of the dual circuit of P17.5.5 is the inductor-input filter. The output of a full-wave rectifier is, in this case, a *voltage* waveform as illustrated in Figure 9.4.1b (Chapter 9), provided the inductance is large enough to ensure that current flows continuously through the inductor. Explain the filtering effect in terms of the smoothing effect of integration.

18

Signal Processing Using Operational Amplifiers

Overview

An operational amplifier, or op amp for short, is a high-gain voltage amplifier, designed to amplify signals in the frequency range from dc to a specified upper frequency. Operational amplifiers are so called because they were initially introduced to perform mathematical operations — such as addition, subtraction, differentiation, and integration — in analog computers that were commonly used to solve differential equations, before digital computers prevailed. Nowadays, op amps of high performance and low cost are widely available in integrated-circuit (IC) form, which makes them an important building block in a variety of signal-processing applications. A number of these applications are considered in this chapter.

The most important feature of linear op amp circuits is the use of negative feedback to trade-off the high gain of the op amp for some desirable feature, thereby allowing such operations as precision amplification and integration. Op amps are at the heart of active filters of various types, whose performance is superior in many respects to that of passive filters, particularly in their allowing high Q values to be obtained without the use of inductors. An important switching application of op amps is in comparators which, as their name implies, are used to compare two voltage levels.

Learning Objectives

- To be familiar with:
  - The broad range of applications of op amps in signal processing
- To understand:
  - The characteristics of an ideal operational amplifier and the nature of imperfections in practical amplifiers
  - The essential features of the noninverting and inverting operational amplifier configurations and how virtual ground in an inverting configuration can be utilized in a number of important practical applications
  - Practical considerations in the design of op amp circuits and the usefulness of some circuits such as the unity-gain amplifier and the instrumentation amplifier

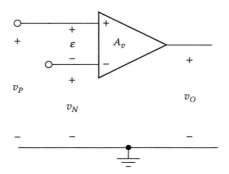

FIGURE 18.1.1
Operational amplifier.

- Configurations and features of some basic types of active filters, including low-pass, highpass, bandpass, bandstop, allpass, and universal filters, of both the single op amp type and biquads
- The use of an op amp as a comparator and the advantages of introducing positive feedback

18.1 Ideal Operational Amplifier and Basic Configurations

Definition: *An operational amplifier, or op amp, is a device whose output voltage is directly proportional to the difference between the voltages applied to its two input terminals.*

The op amp has an output terminal and two input terminals: a noninverting terminal denoted by the (+) sign, and an inverting terminal denoted by the (−) sign, as illustrated on the symbol for an op amp in Figure 18.1.1. According to the preceding definition,

$$v_o = A_v(v_p - v_N) = A_v\varepsilon \tag{18.1.1}$$

where A_v is the voltage gain of the amplifier. The following should be noted concerning this definition:

1. Because the output is determined by the difference between the two inputs, the op amp is an example of a **differential amplifier.**
2. The two terminals are designated as noninverting and inverting because v_O is of the same polarity as the input v_p applied to the noninverting terminal, and is of opposite polarity to the input v_N applied to the inverting terminal.

Ideal Operational Amplifier

This is a very useful idealization that can serve as an initial design step and which is commonly invoked to illustrate some important circuit concepts. Many practical op amps approach the ideal in several respects, so the concept is not as far fetched as it may seem.

Concept: *An ideal op amp has the following properties:*

1. The output is an ideal voltage source of voltage $v_O = A_v(v_P - v_N)$, with $A_v \to \infty$.
2. Like an ideal voltage source, the amplifier has zero output resistance and can deliver any output voltage or current, at any frequency.
3. Both inputs behave as open circuits.
4. The amplifier is free from other imperfections that are mentioned below, such as common-mode response, slew rate, noise, and drift.

There are two basic op amp configurations — noninverting and inverting — that are discussed next.

Noninverting Configuration

Consider an ideal op amp connected as shown in Figure 18.1.2. It is desired to derive the relation between the output voltage v_O and the input voltage v_I. Recall that $v_O = A_v(v_P - v_N)$, with $A_v \to \infty$. Although the op amp is ideal, it is useful to consider A_v as finite, then let $A_v \to \infty$. From Figure 18.1.2, $v_I = v_P$ and

$$v_N = \frac{R_r}{R_r + R_f} v_O = \beta v_O$$

where

$$\beta = \frac{R_r}{R_r + R_f}$$

Substituting for v_P and v_N in Equation 18.1.1 and rearranging:

$$\frac{v_O}{v_I} = \frac{A_v}{1 + \beta A_v} \tag{18.1.2}$$

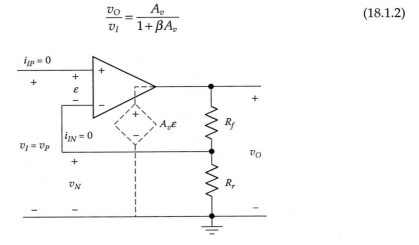

FIGURE 18.1.2
Noninverting configuration.

Because v_O is in phase with v_I, the configuration is noninverting. If $A_v \to \infty$,

$$\frac{v_O}{v_I} = \frac{1}{\beta} = 1 + \frac{R_f}{R_r} \tag{18.1.3}$$

The voltage gain $\dfrac{v_O}{v_I}$ cannot be less than unity. If $R_r = 1\ k\Omega$ and $R_f = 4\ k\Omega$, for example, $\beta = \dfrac{1}{5}$

and $\dfrac{v_O}{v_I} = 5$. A_v for a general purpose, low-frequency op amp is typically about 10^5. Sub-

stituting for β and A_v in Equation 18.1.2 gives $\dfrac{v_O}{v_I} = 4.99975$.

A basic concept that governs the behavior of op amp circuits is illustrated by the foregoing analysis and may be stated as:

> **Concept:** *If there is a circuit connection between the output terminal and the inverting terminal, and the gain of the op amp is very large, the differential input is vanishingly small for any finite output, so that the two input terminals are virtually at the same voltage.*

The justification follows from Equation 18.1.1 because $\varepsilon = \dfrac{v_O}{A_v}$. With v_O finite, then as

$A_v \to \infty$, $\varepsilon = (v_P - v_N) \to 0$, and $v_P = v_N$. If $|v_O| = 15$ V and $A_v = 10^5$, then $|\varepsilon| = 0.15$ mV. Hence, as v_I varies in Figure 18.1.2, v_O varies so as to maintain $v_N = v_P$. That is, the voltage of the inverting terminal follows that of the noninverting terminal. Equation 18.1.3 readily

follows from this condition. Thus, $v_N = \dfrac{R_r}{R_r + R_f} v_O = v_P = v_I$, or $\dfrac{v_O}{v_I} = \dfrac{R_r + R_f}{R_r}$.

Having $v_P = v_N$ is tantamount to having a *virtual short-circuit* between the two input terminals, without being actually connected together. This greatly simplifies the analysis of op amp circuits.

EXERCISE 18.1.1

Determine V in Figure 18.1.3 so that no current flows in the 2 kΩ resistor.
Answer: 10 V.

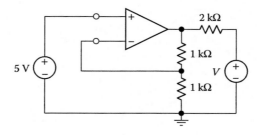

FIGURE 18.1.3
Figure for Exercise 18.1.1.

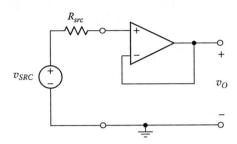

FIGURE 18.1.4
Unity-gain amplifier.

Unity-Gain Amplifier

An important special case of a noninverting configuration is the **unity-gain amplifier**, also known as a **voltage follower**, in which $R_f = 0$, so that the output is directly connected to the inverting input (Figure 18.1.4). R_r across the output of the op amp becomes redundant and is removed. With $\beta = 1$, the overall gain is very nearly unity when A_v is large (Equation 18.1.3).

The unity-gain amplifier is extremely useful for isolating a source from a load. Isolation is desirable in many cases for the following reasons:

1. Changes in the load do not affect the source current and hence the source voltage.

2. When source and load resistances are appreciably mismatched, the signal at the load can be severely attenuated, and the power delivered to the load is limited.

   In Figure 18.1.5a, for example, the voltage v_O across the load is only $\dfrac{0.1}{1.2} \times 6 = 0.5$ V compared to an open-circuit source voltage of 6 V, and the power delivered to the load is 2.5 mW. With a unity-gain amplifier connected between the source and load (Figure 18.1.5b), the source current is ideally zero, independently of load variations. The voltage across the load is now 6 V because the amplifier appears as a voltage source, and the power delivered to the load is 360 mW, assuming that the op amp is capable of delivering the required power.

3. If there is a mismatch between the frequency dependence of the source and load impedances, then the voltage across the load varies with frequency. Connecting a unity gain amplifier having a high input impedance and a low output impedance overcomes this problem and makes the voltage across the load virtually independent of frequency over a specified frequency range.

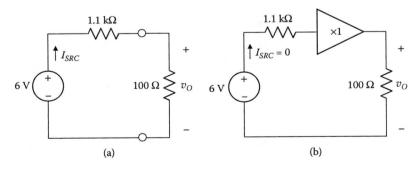

FIGURE 18.1.5
Load isolation.

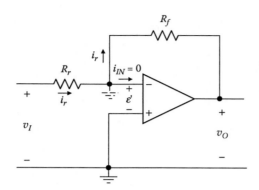

FIGURE 18.1.6
Inverting configuration.

Inverting Configuration

The basic inverting configuration is shown in Figure 18.1.6. The noninverting input is connected to the common reference, which is usually grounded. To analyze the circuit, KVL at the input gives:

$$v_I = R_r i_r + \varepsilon' \tag{18.1.4}$$

Note that i_r passes through R_f because no current flows into the input terminal of an ideal op amp. From KVL at the output,

$$\varepsilon' = R_f i_r + v_O \tag{18.1.5}$$

For the op amp,

$$v_O = -A_v \varepsilon' \tag{18.1.6}$$

where $\varepsilon' = v_N - v_P = -\varepsilon$. Substituting for ε' from Equation 18.1.5 in Equation 18.1.4 and Equation 18.1.6, eliminating i_r between the resulting equations, and rearranging gives:

$$\frac{v_O}{v_I} = -\frac{R_f}{R_r}\frac{\beta A_v}{1+\beta A_v} \tag{18.1.7}$$

where $\beta = \dfrac{R_r}{R_r + R_f}$, as for the noninverting configuration. If $A_v \to \infty$,

$$\frac{v_O}{v_I} = -\frac{R_f}{R_r} \tag{18.1.8}$$

Because v_O is in antiphase with v_I, the configuration is inverting. Unlike the noninverting configuration, the magnitude of the gain could be less than unity.

In accordance with the previous discussion, a circuit connection exists between the output and the inverting terminal through R_f. If $A_v \to \infty$, then $v_P = v_N$. But because $v_P = 0$

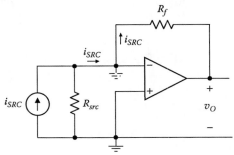

FIGURE 18.1.7
Current-source-to-voltage-source converter.

in this case, $v_N = 0$. The inverting input is a **virtual ground**: it is at ground potential without being actually connected to ground. This means that the current $i_r = \dfrac{v_I - 0}{R_r}$ is determined solely by the input circuit. Because $i_{IN} = 0$, i_r is forced to flow through R_f. Hence, $v_O = 0 - R_f i_r$, and Equation 18.1.8 follows.

The fact that the inverting input of the op amp in the inverting configuration is a virtual ground is highly important and leads to a host of useful applications, the underlying concept being the following.

> **Concept:** *When the inverting input is a virtual ground, the input current is determined solely by the input circuit. The input current is forced to flow through the circuit element connected between the inverting input and the output, irrespective of the value or nature of this element.*

In other words, the circuit element connected between the inverting terminal and output sees a current source whose value is the input current. Next we will consider a number of applications based on the virtual ground feature.

Current-Source-to-Voltage-Source Converter

The nonideal current source in Figure 18.1.7 is ideally terminated with a short circuit. No current flows through R_{src} because the voltage across it is zero. i_{SRC} flows through R_f, resulting in an ideal voltage source output of $v_O = -R_f i_{SRC}$.

EXERCISE 18.1.2

Determine i_O in Figure 18.1.8.

Answer: 2.5 mA.

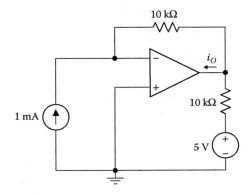

FIGURE 18.1.8
Figure for Exercise 18.1.2

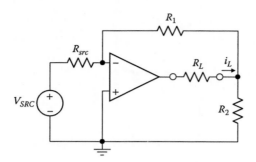

FIGURE 18.1.9
Figure for Exercise 18.1.3.

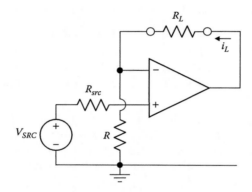

FIGURE 18.1.10
Figure for Exercise 18.1.4.

EXERCISE 18.1.3

Show that in Figure 18.1.9, $i_L = -\dfrac{v_{SRC}}{R_{src}}\left(1 + \dfrac{R_1}{R_2}\right)$. Note that i_L is, in this case, larger than the current drawn from the source.

EXERCISE 18.1.4

Show that in Figure 18.1.10, $i_L = \dfrac{v_{SRC}}{R}$. Note that the voltage source is ideally terminated with an open circuit, and the current in R_L is independent of R_L, as if R_L is connected to an ideal current source.

Perfect Integrator

If R_f is replaced by a capacitor (Figure 18.1.11), i_r still equals $\dfrac{v_I}{R_r}$ and

$$v_O = -\frac{1}{C_f}\int_0^t \frac{v_I}{R_r}\,dt + v_O(0) \qquad (18.1.9)$$

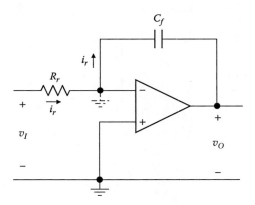

FIGURE 18.1.11
Perfect integrator.

If $v_O(0) = 0$, v_O is $\dfrac{-1}{R_r C_f}$ times the time integral of v_I. Ideally, the circuit acts as a perfect integrator.

Perfect Differentiator

If R_r is replaced by a capacitor (Figure 18.1.12), $i_r = C_r \dfrac{dv_I}{dt}$ and

$$v_O = -R_f C_r \frac{dv_I}{dt} \qquad (18.1.10)$$

Ideally, the circuit acts as a perfect differentiator. The same result can be achieved by replacing R_f in Figure 18.1.6 with an ideal inductor. However, capacitors are preferred to inductors because they can be closer to the ideal than inductors, and are less bulky and expensive.

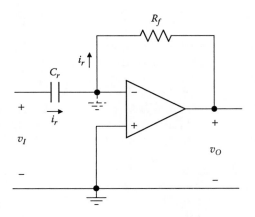

FIGURE 18.1.12
Perfect differentiator.

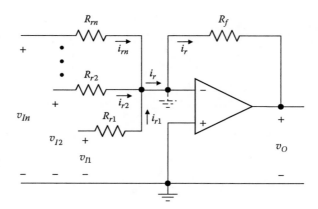

FIGURE 18.1.13
Adder.

Adder

The virtual ground can act as a summing point, leading to an adding circuit (Figure 18.1.13). The current $i_r = i_{r1} + i_{r2} + \ldots + i_{r\,n}$. It follows that:

$$v_O = -R_f \left(\frac{v_{I1}}{R_{r1}} + \frac{v_{I2}}{R_{r2}} + \ldots + \frac{v_{In}}{R_{rn}} \right) \qquad (18.1.11)$$

In words, the output is the weighted sum of the inputs, with sign inversion. If all the input resistances are equal to R_r,

$$v_O = -\frac{R_f}{R_r} \left(v_{I1} + v_{I2} + \ldots + v_{In} \right) \qquad (18.1.12)$$

An important feature of this adding circuit is that the virtual ground prevents interaction between the input circuits, thereby effectively isolating them from one another. A change in v_{I1} or R_{r1}, for example, affects only i_{r1} and has no effect on the other input circuits (Section ST18.2).

EXERCISE 18.1.5
Show that the circuit of Figure 18.1.14 is a noninverting integrator in which C_f is multiplied by the gain of the inverting amplifier. Note that the noninverting input of the integrator is now a virtual ground.

Before ending this section, we wish to answer an important and relevant question that is suggested by Equation 18.1.3 and Equation 18.1.8, namely, what is the wisdom of starting with an op amp of high voltage gain, such as 10^5, and ending with an amplifier circuit having a much smaller voltage gain, say 5 or 100? The answer is that the gain A_v of the op amp is not precisely defined and is subject to variations due to the following: (1) changes in internal components because of environmental factors or "aging," (2) changes in supply voltages, or (3) replacing the op amp by another op amp of the same type because such op amps are never identical in every respect but differ due to manufacturing tolerances. When the op amp is connected as in Figure 18.1.2 or Figure 18.1.6, with A_v very large, the gain is no longer determined by A_v but by resistance ratios, which can be

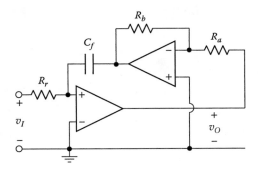

FIGURE 18.1.14
Figure for Exercise 18.1.5.

accurate to better than 0.01% and highly stable. The result is precision amplification of a specified value. This underscores a concept of great practical importance.

> **Concept:** *Op amp circuits can be designed to trade off the high gain of the op amp for some desirable characteristics of the circuit.*

High gain in IC op amps is quite inexpensive and can be advantageously traded off, not only for precision and stability of output, as already illustrated, but for other desirable characteristics such as reduced distortion, wider bandwidth, increased input impedance, or reduced output resistance. The general circuit configuration that achieves these benefits is **negative feedback**. As its name implies, negative feedback results when a signal is fed back from the output to the input in such a manner as to oppose the change in input. In both the noninverting and inverting configurations, this is accomplished by the connection from the output to the inverting input. Section ST18.1 discusses the two op amp configurations in terms of negative feedback and explains its benefits.

18.2 Op Amp Imperfections and Their Effects

Finite Gain

In the case of amplifiers, the effect of finite gain is generally not very significant, as long as A_v is large, as illustrated by the reduction in gain cited earlier, from 5.0 to 4.99975. However, near-perfect integration down to very low frequencies requires a very high value of gain (Section ST18.3). This can be readily inferred from the transfer function of a perfect integrator, whose magnitude is $1/\omega$. As $\omega \to 0$, this magnitude tends to infinity.

Amplifier Saturation

According to item 3 of the concept of an ideal op amp (Section 18.1), the input power to the amplifier is zero. Yet according to item 2, the output power delivered by the op amp is finite. This may look like a violation of conservation of energy. In fact, at least one dc supply is required for proper operation of the op amp, which leads to conservation of energy when the dc supplies are included. In practice, two dc supplies, $+V_{CC}$ and $-V_{CC}$,

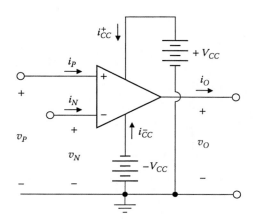

FIGURE 18.2.1
Bias supplies and currents of an op amp.

are normally used, as illustrated in Figure 18.2.1. The supplies usually have the same magnitude of voltage, in the range of a few volts to 20 V. The dc supplies of opposite polarity allow the output voltage to swing positively or negatively with respect to the common, zero reference. Moreover, the amplifier as a whole obeys KCL, so that:

$$i_P + i_N + i_{CC}^- + i_{CC}^+ - i_O = 0 \qquad (18.2.1)$$

A practical implication of the need for dc supplies is that the output voltage cannot become more positive than $+V_{CC}$, nor more negative than $-V_{CC}$. As the magnitude of the differential input ε increases, the output *saturates*, that is, stops increasing, or increases at a very slow rate, as illustrated by the transfer characteristic in Figure 18.2.2. In the positive or negative saturation regions, the incremental or small-signal, voltage gain of the amplifier, which is equal to the slope of the transfer characteristic, becomes very small. Hence, the normal operating range as an amplifier is in the nonsaturation region of the transfer characteristic. This region is not quite linear but can be approximated by a linear range ab of slope A_v. As discussed in Section ST18.1, distortion is reduced by a factor of $1 + \beta A_v$ by the negative feedback. This is appreciable when A_v is large, which is the case as long as the op amp is not in saturation. Note that the linear range of differential inputs to a typical op amp is usually

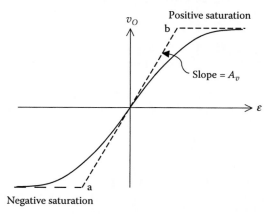

FIGURE 18.2.2
Transfer characteristics of op amp.

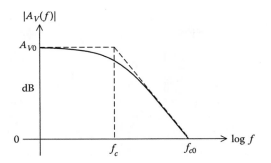

FIGURE 18.2.3
Frequency response of op amp.

quite small. In the case considered earlier, if saturation occurs at ±15 V and $A_v = 10^5$, the differential input range of the linear part of the input–output characteristic is only ±0.15 mV.

Frequency Response

Practical op amps do not amplify equally all frequencies of the input signal, which means that the gain A_v is frequency dependent, as illustrated in Figure 18.2.3. It may be assumed that

$$A_v(f) = \frac{A_{v0}}{1 + jf/f_c} \qquad (18.2.2)$$

where f_c is the 3-dB cutoff frequency. At high-enough frequencies, A_v is relatively small, which limits the maximum frequency at which the amplifier or integrator still has acceptable performance. The effective bandwidth of the op amp circuit is increased by the negative feedback by a factor of $1 + \beta A_{v0}$, the same factor by which the gain is reduced (Section ST18.1), so that the gain bandwidth product remains the same.

Output Resistance

The output of an op amp further deviates from that of an ideal source by having a finite output resistance. Figure 18.2.4 shows an op amp with a dependent voltage source $A_v \varepsilon$ at

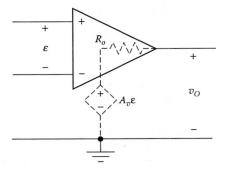

FIGURE 18.2.4
Output resistance of op amp.

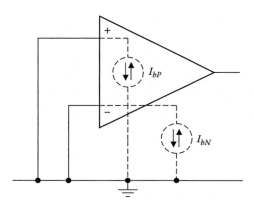

FIGURE 18.2.5
Input bias currents of op amp.

the output in series with an output resistance R_o, which is usually in the range of a few hundred ohms. The output resistance is defined *dynamically* under small signal conditions, that is, as the ratio of the change in the output voltage to a small change in the output current, with the source voltage remaining constant. Manufacturer data sheets usually specify a short-circuit current for maximum input. This depends on the large-signal effective output resistance at maximum input and not on the small-signal output resistance. Again, negative feedback reduces R_o by a factor of $1 + \beta A_{v0}$ in both the inverting and noninverting configurations (Section ST18.1).

Input Impedances and Bias Currents

In general, a small current flows through each op amp input, even when these inputs are shorted to the common reference (Figure 18.2.5). These currents are the bias currents of the input transistor stages and are represented by two current sources I_{bP} and I_{bN} of the same polarity. Depending on the type of input stage, these currents may actually flow in or out of the input terminals, their magnitude ranging between a few nanoamperes and less than a picoampere. Because the two input currents are not exactly equal, the **input bias current** I_{ib} is the average of I_{bP} and I_{bN}, $I_{ib} = \frac{1}{2}(I_{bP} + I_{bN})$, whereas the **input offset current** I_{io} is the magnitude of their difference, $I_{io} = |I_{bP} - I_{bN}|$. In amplifier circuits, it is not the absolute values of these currents that are important, but their drift with temperature or with time. The steady component of bias currants can, at least in principle, be compensated, but the random component cannot, which sets a limit to the stability of amplifier output (Section ST18.2). Moreover, input bias current has a drastic effect on integrators (Section ST18.3).

In addition to the input bias currents, there is an input current due to the finite impedances at the amplifier inputs. Three impedances Z_1, Z_2, and Z_3 are identified in Figure 18.2.6, connected as shown. They give rise to three impedances: Z_{id}, the **differential input impedance** that effectively appears between the two inputs, in addition to Z_{PG} and Z_{NG}, the effective impedance between the noninverting and inverting inputs and the common reference, respectively. As in the case of output resistance, these impedances are defined dynamically under small-signal conditions, that is, as the ratio of a small change in the voltage across the impedance to the resulting change in the current through the impedance. The impedances are largely resistive in nature, with a small, parallel, capacitive component. Z_{id} is generally much smaller than Z_{PG} or Z_{NG} and may vary from a few tens of kilo-ohms to thousands of megohms, depending on the type of op amp. If the two input stages

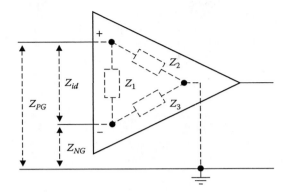

FIGURE 18.2.6
Input impedances of op amp.

of the op amp are well matched, Z_{PG} and Z_{NG} are almost equal. Their parallel combination, with the op amp terminals shorted together, is the **common-mode impedance**. The effect of negative feedback is to increase Z_{id} in the noninverting configuration by the factor $1 + \beta A_{v0}$ (Section ST18.1). In the inverting configuration, the input impedance is essentially that connected between the input terminal and virtual ground.

In an ideal op amp, both inputs behave as open circuits, so that all input impedances are assumed infinite, and bias currents are assumed zero.

Input Offset Voltage

When the two inputs of an op amp are shorted together and connected to the common reference, with the two V_{CC} supplies having equal magnitude, the output of the amplifier is not exactly zero, as it should be, because of imperfect matching of circuit components associated with the two inputs. This effect could be simulated by connecting a voltage source V_{io} in series with either input of an ideal op amp (Figure 18.2.7), the polarity of V_{io} depending on the op amp. V_{io} is the **input offset voltage** and is generally in the range of several millivolts to less than a millivolt, depending on the input stage. Because V_{io} may be sufficient to saturate the output, practical op amps usually have external terminals for zeroing the output.

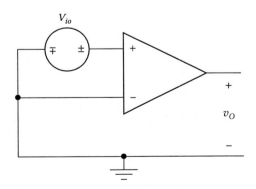

FIGURE 18.2.7
Input offset voltage.

As in the case of input bias current, it is not the absolute value of input offset voltage that is important in many applications, but its drift with temperature or with time (Section ST18.2). Moreover, input offset voltage has a drastic effect on integrators (Section ST18.3).

EXERCISE 18.2.1

A nonideal op amp has input bias currents of 1 and 1.2 nA, an input offset voltage of 1 mV, a voltage gain of 10^5, and voltage supplies of +15 V and –15 V. Determine: (a) the input bias current; (b) the input offset current; (c) if the two inputs of the amplifier are shorted together, without compensating for the input offset voltage, will the amplifier saturate?

Answer: (a) 1.1 nA; (b) 0.2 nA; (c) yes, because $10^3 \times 10^5 = 100 > 15$.

EXERCISE 18.2.2

Show that the unity-gain frequency f_{c0} in Figure 18.2.3 is given by: $f_{c0} = f_c\sqrt{A_{v0}^2 - 1} \cong A_{v0}f_c$. If an op amp has $A_{v0} = 10^5$ and $f_{c0} = 1$ MHz, determine f_c.

Answer: 10 Hz.

Common-Mode Rejection

If the two inputs of an op amp are shorted together, and a voltage v_{cm} is applied between the shorted inputs and the common reference, then ideally $v_O = 0$ because the differential input is zero. However, in a practical amplifier, there will be a finite output voltage, or **common mode response**.

Consider an op amp that is ideal except for finite gain, with inputs applied as shown in Figure 18.2.8. It is seen that

$$v_P = v_{cm} + \frac{v_d}{2} \quad \text{and} \quad v_N = v_{cm} - \frac{v_d}{2} \tag{18.2.3}$$

The differential input is $v_d = v_P - v_N$, whereas the common-mode input is $v_{cm} = \frac{1}{2}(v_P + v_N)$, the average of v_P and v_N.

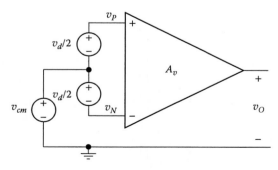

FIGURE 18.2.8
Differential and common-mode inputs.

An ideal op amp has equal voltage gains from each input to the output. In a practical op amp, imperfections in the differential amplifier stages result in a common-mode response, which can be accounted for by assuming different voltage gains between the two inputs and the output. Let the voltage gains from the noninverting and inverting inputs of the amplifier be denoted as A_{vP} and A_{vN}, respectively. The output voltage may be expressed as: $v_O = A_{vP}v_P - A_{vN}v_N$. Substituting for v_P and v_N from Equation 18.2.3 gives:

$$v_O = (A_{vP} - A_{vN})v_{cm} + \frac{A_{vP} + A_{vN}}{2}v_d = v_{Ocm} + v_{Od} \qquad (18.2.4)$$

where $v_{Ocm} = (A_{vP} - A_{vN})v_{cm}$, and $v_{Od} = \dfrac{A_{vP} + A_{vN}}{2}v_d$ are the components of output due to the common-mode and differential inputs, respectively. If $A_{vP} = A_{vN} = A_v$, then $v_{Ocm} = 0$, and $v_{Od} = A_v v_d = A_v(v_P - v_N)$, as has been assumed until now.

The common-mode gain is the ratio of the common mode output to the common-mode input. That is, $A_{vcm} = \dfrac{v_{Ocm}}{v_{cm}} = A_{vP} - A_{vN}$. The **common-mode rejection ratio** (CMRR) is the magnitude of the ratio of the differential gain A_v to the common-mode gain:

$$\text{CMRR} = \left| \frac{A_v}{A_{vcm}} \right| \qquad (18.2.5)$$

The **common-mode rejection** (CMR) is the CMRR expressed in dB; CMR = $20\log_{10}$CMRR, or the CMRR itself is sometimes expressed in dB. In general, the CMRR varies with frequency and with the value of the common-mode input.

Let us define a differential signal v_{dcm} that, if amplified by the differential gain A_v, produces the same output voltage as the common-mode output: $A_v v_{dcm} = A_{vcm}v_{cm}$. It follows that $\left| \dfrac{A_v}{A_{vcm}} \right| = \left| \dfrac{v_{cm}}{v_{dcm}} \right| = \text{CMRR}$. Thus, if the CMRR is 10^5, a common-mode signal of 2 V produces the same output voltage as a differential signal of $\dfrac{2\text{ V}}{10^5} = 20\ \mu\text{V}$.

The definition of v_{dcm} allows a convenient representation of the common-mode signal as a voltage source $\dfrac{v_{cm}}{\text{CMRR}}$ in series with one of the inputs, just like the input offset voltage, which also depends on dissymmetry between the circuitry associated with the two inputs. The polarity of this source is also indefinite, depending on the particular op amp.

The common-mode input may be appreciable in the noninverting configuration but is negligible in the inverting configuration, in which the voltages of the inputs are essentially zero with respect to the common reference. The effect of the CMRR on the output of a unity-gain amplifier is examined in Example 18.2.1.

Example 18.2.1: Unity-Gain Amplifier Having Finite Gain and CMRR

Determine the output of a unity-gain amplifier assuming the op amp is ideal except for finite A_v and CMRR.

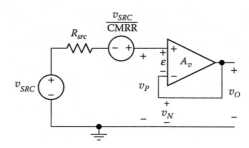

FIGURE 18.2.9
Figure for Example 18.2.1.

SOLUTION

The amplifier is shown in Figure 18.2.9 with a voltage source $\dfrac{v_{SRC}}{\text{CMRR}}$ in series with the noninverting input. The common-mode input v_{cm} is considered equal to the input signal, because $v_P \cong v_N \cong v_{SRC}$. It follows that $v_P = v_{SRC} \pm \dfrac{v_{SRC}}{\text{CMRR}}$, $v_N = v_O$, $\varepsilon = v_P - v_N$, and $v_O = A_v \varepsilon$. Eliminating ε, v_P, and v_N, gives:

$$\frac{v_O}{v_{SRC}} = \frac{1 \pm 1 / \text{CMRR}}{1 + 1 / A_v}$$

If, for example, $A_v = 20{,}000$ and $\text{CMRR} = 10^5$, $\dfrac{v_O}{v_{SRC}} = 0.99996$ or 0.99994, depending on whether the sign in the numerator is plus or minus, respectively.

A common-mode *response* v_{Ocm} is not only because of A_{vp} not being equal to A_{vN} but could arise from a common-mode input that causes a *differential* input due to impedance unbalance at the inputs of the op amp. This is illustrated in Figure 18.2.10, where the input impedances are shown as Z_1, Z_2, and Z_3, as in Figure 18.2.6. Also shown are source impedances Z_{src1} and Z_{src2}, which could be due to the differential voltage sources in Figure 18.2.8, which are set to zero in the figure. Clearly, if $Z_{src1} = Z_{src2}$ and $Z_2 = Z_3$, it follows from symmetry that $v_P = v_N$, so that the differential input due to v_{cm} is zero. More generally, $v_P = v_N$ if a bridge balance condition is satisfied, that is, $\dfrac{Z_{src1}}{Z_{src2}} = \dfrac{Z_2}{Z_3}$. Otherwise, a differential

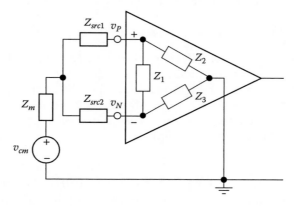

FIGURE 18.2.10
Effect of impedance mismatch.

input arises because of v_{cm}. Example SE18.3 illustrates a practical example. In general, Equation 18.2.4 may be expressed as:

$$v_O = v_{Ocm} + v_{Od} = A_{vcm}v_{cm} + A_v v_d \qquad (18.2.6)$$

where A_{vcm} and A_v are related by the CMRR (Equation 18.2.5).

EXERCISE 18.2.3

Determine the common-mode response, assuming that in Figure 18.2.10 $Z_m = 10$ kΩ, $Z_{src1} = Z_{src2} = 10$ kΩ, $Z_1 = 1$ MΩ, $Z_2 = 10$ MΩ, $Z_3 = 10.01$ MΩ, $V_{cm} = 1$ V, and $A_v = 10^5$.
Answer: 97.6 mV.

Slew Rate

The unity-gain amplifier may be used to measure the **slew rate** (SR) of an op amp. The amplifier in Figure 18.1.4 is connected to a specified load, and a square wave of a relatively large, specified amplitude is applied to the input. Ideally, the amplifier output should

follow the input. In fact, the output is trapezoidal in shape, rising and falling at rates $\dfrac{v_O}{t_1}$

and $\dfrac{v_O}{t_2}$, as illustrated in Figure 18.2.11a. The smaller of these two rates is the SR of the

amplifier and is usually expressed in volts per microsecond.
    The SR is a large-signal frequency response limitation that is distinct from the small-signal frequency response limitation. If the frequency variation of A_v is that of a single time constant (Equation 18.2.2), the output response to a square-wave input will have exponential rise and fall, without delays. The SR is mainly due to the fact that a rapid

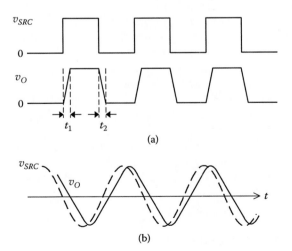

(a)

(b)

FIGURE 18.2.11
Slew rate of op amp.

change in output requires an equally rapid charging or discharging of capacitors associated with the amplifier. These could be inherent in the transistor amplifier stages or could have been added to shape the frequency response of the amplifier in some desired manner.

Rapid charging and discharging of capacitors require large currents because $\dfrac{dv_C}{dt} = \dfrac{i}{C}$. But in a practical amplifier, i is limited, which in turn limits the rate of change of capacitor voltages, and hence the rate of change of output voltage.

A limited SR produces nonlinear distortion. If a small-amplitude sinusoidal input of high enough frequency is applied to the amplifier, the effect of a limited, small-signal frequency response is to reduce the output sinusoid in amplitude and delay it in phase, but the output waveform is still sinusoidal. However, if a large-amplitude sinusoidal input of high-enough frequency is applied, the SR limitation of the amplifier prevents the output from following the fast rise and fall of the sinusoid, resulting in a more triangular-shaped output waveform (Figure 18.2.11b).

When an op amp is used in an amplifier circuit, then depending on the type of op amp, the SR of the output is generally better with the inverting than the noninverting configuration.

Noise and Drift

Noise refers to spurious, unwanted signals that are random in nature and therefore uncorrelated with the input signal. Drift is often regarded as low-frequency noise, and "random" implies a zero average over a sufficiently long period that depends on the frequency components of the noise that are being averaged.

Basically, noise arises from imperfections in components, including transistors, as well as from the discrete and statistical nature of current at the atomic level. In op amps, noise is accounted for by adding a noise voltage source in series with V_{io} and noise current sources in parallel with I_{bP} and I_{bN}.

Drift is manifested in op amps as slow changes in input offset voltage and input bias currents. It could be due to internally generated changes in the amplifier or to external factors, such as changes in temperature, humidity, dc supply voltages, or the parameters of external circuit components. Although unpredictable, drift due to these causes is usually in a given sense and may not average to zero. Manufacturer data sheets quote a figure for the temperature variation of input offset voltage, input bias currents, and input offset current. It is important to note that, whereas steady values of input offset voltage and input bias currents may be compensated by means of appropriate external circuitry, it is not possible to compensate the drift components because of their unpredictability. Hence, these components are usually more significant than the absolute values and set a limit to the output stability of the amplifier (Section ST18.2).

Spurious signals may also arise mainly because of inductive or capacitive pickup from installation wiring, or from nearby equipment, due to coupling through magnetic or electric fields (Example SE18.4). Table 18.2.1 compares the noninverting and inverting configurations.

Example 18.2.2: Difference Amplifier

It is required to determine the output of the difference amplifier shown in Figure 18.2.12.

TABLE 18.2.1

Comparison of the Noninverting and Inverting Configurations

	Noninverting	Inverting
Phase inversion	No	Yes
Magnitude of gain	≥1	Any
Feedback factor, $\beta_f = \dfrac{R_r}{R_r + R_f}$	≤1	< 1
Virtual ground	No	Yes
Input offset voltage and noise voltage relative to input signal	×1	$\times \dfrac{R_r + R_f}{R_f}$
Input impedance	High	Relatively low
Common mode response	Present	Negligible
Slew rate limitation	Present	Generally less pronounced

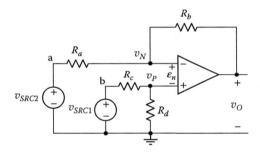

FIGURE 18.2.12
Figure for Example 18.2.2.

SOLUTION

The input v_P at the noninverting input is simply the output of a voltage divider:

$v_P = \dfrac{R_d}{R_c + R_d} v_{SRC1}$. Applying KCL at the inverting input: $\dfrac{v_{SRC2} - v_N}{R_a} = \dfrac{v_N - v_O}{R_b}$. In an ideal

op amp, $v_P = v_N$. Substituting for v_N and eliminating v_P gives:

$$v_O = -\frac{R_b}{R_a} v_{SRC2} + \frac{1 + R_b / R_a}{1 + R_c / R_d} v_{SRC1} \qquad (18.2.7)$$

If $\dfrac{R_b}{R_a} = \dfrac{R_d}{R_c} = k$, Equation 18.2.7 reduces to:

$$v_O = k(v_{SRC1} - v_{SRC2}) \qquad (18.2.8)$$

The circuit can therefore be used for directly subtracting one signal from another. Close matching of resistance ratios can be achieved when the resistances are included with the op amp on the same IC chip. Subtraction can also be performed indirectly by inverting one signal and adding it to the other signal.

It is instructive to derive Equation 18.2.7 from a modification of the relation for an inverting amplifier that has $v_{SRC1} = 0$ in Figure 18.2.12. The output of the amplifier under these conditions is simply $v_O' = -\dfrac{R_b}{R_a} v_{SRC2}$. With v_{SRC1} applied, $v_P = \dfrac{R_d}{R_c + R_d} v_{SRC1}$. Because $v_N = v_P$ in the ideal case, v_N must change from zero, under virtual ground conditions, to the new v_P. Considering R_a and R_b as a voltage divider with v_{SRC2} applied to one end and v_O at the other,

$$v_N = \frac{R_b}{R_a + R_b} v_{SRC2} + \frac{R_a}{R_a + R_b} v_O \qquad (18.2.9)$$

In order for v_N to change by $\dfrac{R_d}{R_c + R_d} v_{SRC1}$, v_O must change by an amount v_O'' such that $\dfrac{R_a}{R_a + R_b} v_O'' = \dfrac{R_d}{R_c + R_d} v_{SRC1}$. Hence, $v_O'' = \dfrac{R_d}{R_c + R_d} \dfrac{R_a + R_b}{R_a} v_{SRC1}$. Adding v_O' and v_O'' gives Equation 18.2.7.

EXERCISE 18.2.4

Using the same reasoning as in the preceding paragraph, show that if an input v_{SRC3} is added in series with R_d in Figure 18.2.12, the output is given by:

$$v_O = -\frac{R_b}{R_a} v_{SRC2} + \frac{R_d}{R_c + R_d} \frac{R_a + R_b}{R_a} v_{SRC1} + \frac{R_c}{R_c + R_d} \frac{R_a + R_b}{R_a} v_{SRC3}$$

Instrumentation Amplifier

Some of the concepts discussed earlier are illustrated by the instrumentation amplifier shown in Figure 18.2.13. It consists of two stages, the first stage having two op amps in a noninverting configuration, and connected so as to provide a differential input, with their feedback resistors essentially combined so as to provide a differential output. The second stage is a difference amplifier. Because the input source is applied to the non-inverting input, the input impedance is high. Moreover, as is shown later, the overall gain is determined by the value of a single resistance R_1. To minimize the common-mode response, the resistors having the same subscript number should be carefully matched. Because of these features, the instrumentation amplifier, or IA as it is often referred to, is available in IC form.

If ideal op amps are assumed, the differential gain of the first stage can be very simply derived by noting that, because of the virtual short at the inputs of an ideal op amp, $v_{P1} = v_{N1} = v_{SRC1}$ and $v_{P2} = v_{N2} = v_{SRC2}$. This means that the voltage $v_{N1} - v_{N2}$ across R_1 equals the differential input voltage $v_{SRC1} - v_{SRC2}$, so that the current in R_1 is $\dfrac{v_{SRC1} - v_{SRC2}}{R_1}$. This same current flows in the two resistances R_2. Hence, $v_{O1} - v_{O2} = \dfrac{v_{SRC1} - v_{SRC2}}{R_1}(R_1 + 2R_2)$. The differential gain of the first stage is therefore,

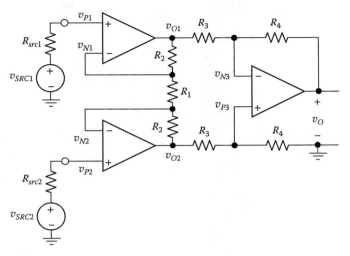

FIGURE 18.2.13
Instrumentation amplifier.

$$\frac{v_{O1} - v_{O2}}{v_{SRC1} - v_{SRC2}} = 1 + 2\frac{R_2}{R_1} \tag{18.2.10}$$

In order to obtain the overall gain, this differential gain has to be multiplied by that of the difference amplifier. It follows from Equation 18.2.7 that $\dfrac{v_O}{v_{O2} - v_{O1}} = \dfrac{R_4}{R_3}$. Multiplying this by Equation 18.2.10 gives:

$$\frac{v_O}{v_{SRC2} - v_{SRC1}} = \left(1 + 2\frac{R_2}{R_1}\right)\frac{R_4}{R_3} \tag{18.2.11}$$

It is seen that the gain can be varied by varying a single resistance R_1, which does not have to be matched to any other resistance.

It is clear from the symmetry in Figure 18.2.13 that the effects of input offset voltages in the two input amplifiers cancel out at the differential output, as long as the two halves are identical. Similarly, the common-mode response at this output cancels out, which makes the common-mode response at the output of the IA dependent only on the common-mode response of the second stage. Also, the effects of the input bias currents of the inverting inputs of the two input amplifiers cancel out at the differential output if the source resistances R_{src1} and R_{src2} are equal. It is seen that, as long as the two halves of the IA are properly matched, the IA is a high-performance amplifier.

Application Window 18.2.1: Bridge Amplifier

An instrumentation amplifier may be used to amplify the output of a transducer bridge. Figure 18.2.14 shows a two-element strain-gauge transducer arranged so that when the resistance of one element increases by αR_o as a result of strain in a given direction, the resistance of the other element decreases by the same amount, where R_o is the nominal resistance of the elements in the rest position. The upper two resistors are dummy strain

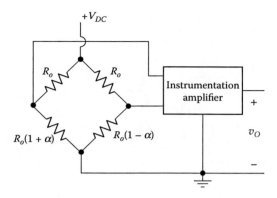

FIGURE 18.2.14
Figure for Application Window 18.2.1.

gauges that are identical to the active strain gauges but are not subjected to any strain. Such a bridge configuration has two advantages: (1) as the resistances of the lower, active strain-gage elements change in opposite directions, the bridge output for a given strain is essentially doubled, and (2) because the four elements have the same resistance at rest and are made of the same material, they are equally affected by changes in ambient temperature, so that the bridge output is independent of these changes.

18.3 First-Order Active Filters

Op amps may be used with RC circuits to construct active filters. RC rather than RL circuits are used for this purpose because capacitors are less bulky and expensive than inductors and are closer to an ideal circuit element. Op amps in active filters provide gain, isolation, load drive, and a higher Q for improved performance. Active filters are extensively used for frequencies up to a few megahertz, limited by the frequency response of op amps. For higher frequencies, passive filters are generally preferred. The simplest forms of active filters are the first-order filters discussed in this section.

Lowpass Filter

The transfer function of the ideal integrator of Figure 18.1.11 is $\dfrac{1}{sC_f R_r}$. The gain $\dfrac{1}{\omega C_f R_r}$ is high at low frequencies because of the large reactance of the capacitor. To have a lowpass characteristic, the response must become flat at low frequencies. This can be achieved by connecting a resistance R_f in parallel with C_f (Figure 18.3.1), so that at low frequencies, as the reactance of the capacitance becomes very large, the gain of the amplifier becomes essentially independent of frequency. The transfer function is:

$$H_l(s) = -\frac{R_f \parallel (1/sC_f)}{R_r} = -\frac{R_f}{R_r}\frac{\omega_{cL}}{s + \omega_{cL}} \qquad (18.3.1)$$

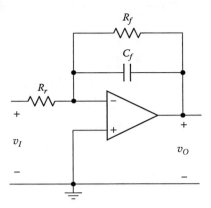

FIGURE 18.3.1
First-order, lowpass active filter.

where $\omega_d = \dfrac{1}{C_f R_f}$. The transfer function of Equation 18.3.1 is that of a first-order lowpass

response having a cutoff frequency ω_d and a gain of $-\dfrac{R_f}{R_r}$ in the passband.

EXERCISE 18.3.1

A filter of the type of Figure 18.3.1 is required to have a passband gain of −10 and a
3-dB cutoff frequency of 1 krad/s, using a 0.1 µF capacitor. Determine R_r and R_f.
Answer: $R_r = 1$ kΩ, $R_f = 10$ kΩ.

Highpass Filter

A highpass filter may be obtained by replacing the resistors in the lowpass filter by
capacitors and the capacitor by a resistor, as may be readily verified from Equation 18.3.1.
An alternative to using two capacitors is to start with an ideal differentiator (Figure
18.1.12), whose transfer function is of the form $sC_r R_f$. The gain $\omega C_r R_f$ is high at high
frequencies because of the small reactance of the capacitor. To have a highpass characteristic, the response must become flat at high frequencies. This can be achieved by connecting
a resistance R_r in series with C_r (Figure 18.3.2), so that at high frequencies, as the reactance
of the capacitance becomes very small, the gain of the amplifier becomes essentially
independent of frequency. The transfer function is:

$$H_h(s) = -\frac{R_f}{R_r + 1/sC_r} = -\frac{R_f}{R_r}\frac{s}{s + \omega_{ch}} \qquad (18.3.2)$$

where $\omega_{ch} = \dfrac{1}{C_r R_r}$. The transfer function of Equation 18.3.2 is that of a first-order highpass

response having a cutoff frequency ω_{ch} and a gain of $-\dfrac{R_f}{R_r}$ in the passband.

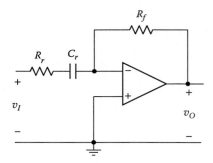

FIGURE 18.3.2
First-order, highpass active filter.

As explained in Section 17.5 (Chapter 17), a first-order, lowpass filter can be combined with a first-order, highpass filter to obtain a bandpass or a bandstop response. When the upper 3-dB cutoff frequency of this response is much larger than the lower 3-dB cutoff frequency, the filter is **broadband**, as illustrated by Example 18.3.1 and Example 18.3.2.

EXERCISE 18.3.2

A filter of the type of Figure 18.3.2 is required to have a passband gain of −10, a 3-dB cutoff frequency of 10 krad/s, and an input impedance of magnitude 10 kΩ at 20 krad/s. Determine C_r, R_r and R_f.

Answer: $C_r = 5\sqrt{5}$ nF, $R_r = 20/\sqrt{5}$ kΩ, $R_f = 200/\sqrt{5}$ Ω.

Design Example 18.3.1: Broadband Bandpass Filter

It is required to design a broadband bandpass filter in the audio frequency range, 20 Hz to 15 kHz, having a passband gain of 5, using 50 nF capacitors and op amps in the inverting configuration.

SOLUTION

The overall transfer function is the product of the two transfer functions given by Equation 18.3.1 and Equation 18.3.2:

$$H_{bp}(s) = \left[-\frac{R_{fl}}{R_{rl}} \frac{\omega_{c2}}{s + \omega_{c2}} \right] \left[-\frac{R_{fh}}{R_{rh}} \frac{s}{s + \omega_{c1}} \right]$$

$$= K \frac{\omega_{c2}s}{s^2 + (\omega_{c1} + \omega_{c2})s + \omega_0^2} \qquad (18.3.3)$$

where $K = \dfrac{R_{fl}R_{fh}}{R_{rl}R_{rh}}$ is the product of the bandpass gains of the two filter, $\omega_{c1} = \dfrac{1}{C_{rh}R_{rh}}$ is

the cutoff frequency of the highpass filter, $\omega_{c2} = \dfrac{1}{C_{rl}R_{rl}}$ is the cutoff frequency of the

lowpass filter, and $\omega_0 = \sqrt{\omega_{c1}\omega_{c2}}$ is the center frequency of the passband. For a bandpass response, $\omega_{c2} > \omega_{c1}$.

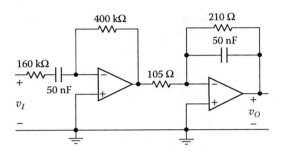

FIGURE 18.3.3
Figure for Example 18.3.1.

It follows from Equation 18.3.3 that:

$$\left|H_{bp}(j\omega)\right| = K\frac{\omega_{c2}\omega}{\sqrt{\left(\omega_0^2 - \omega^2\right) + \omega^2(\omega_{c1} + \omega_{c2})^2}}$$

The maximum value of this magnitude occurs when $\omega = \omega_0$ and is equal to $K\dfrac{\omega_{c2}}{\omega_{c1} + \omega_{c2}}$.
For a broadband filter, $\omega_{c2} \gg \omega_{c1}$, which makes ω_{c1} and ω_{c2} independent of one another, so that each filter can be designed separately.

For the highpass filter, $R_{rh} = \dfrac{1}{\omega_{c1}C_{rh}} = \dfrac{1}{2\pi \times 20(50 \text{ nF})} \cong 160 \text{ k}\Omega$. For the lowpass filter,

$R_{fl} = \dfrac{1}{\omega_{c2}C_{fl}} = \dfrac{1}{2\pi \times 15{,}000(50 \text{ nF})} \cong 210 \ \Omega$. $K = 5 = \dfrac{R_{fl}R_{fh}}{R_{rl}R_{rh}}$. We may choose $R_{fh} = 400 \text{ k}\Omega$,

so $R = 105 \ \Omega$. The filter circuit is shown in Figure 18.3.3. The highpass filter precedes the lowpass filter so as to have a high input impedance.

Design Example 18.3.2: Broadband Bandstop Filter
It is required to design a broadband bandstop filter in the frequency range 125 rad/s to 100 krad/s, having a passband gain of 10, using op amps in the inverting configuration. The filter is to have an input impedance of at least 25 kΩ.

SOLUTION
To implement a bandstop filter, the input is applied to the paralleled inputs of a low-pass filter and a highpass filter, and the outputs of these filters are summed in an adder

to obtain the bandstop response (Figure 18.3.4). If $\omega_d = \dfrac{1}{C_{rl}R_{rl}}$ is the cutoff frequency of

the lowpass filter, and $\omega_{ch} = \dfrac{1}{R_{rh}C_{rh}}$ is the cutoff frequency of the highpass filter, then for

a bandstop response, $\omega_{cl} < \omega_{ch}$. For simplicity, it is assumed that the resistors in each filter are equal so that each filter will have unity gain in its passband. The transfer function is:

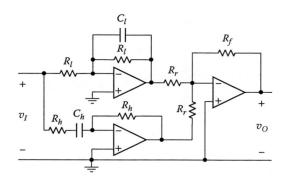

FIGURE 18.3.4
Figure for Example 18.3.2.

$$H_{bs}(s) = -\frac{R_f}{R_r}\left[-\frac{\omega_{cl}}{s+\omega_{cl}} - \frac{s}{s+\omega_{ch}}\right]$$

$$= \frac{R_f}{R_r}\frac{s^2 + 2\omega_{cl}s + \omega_{cl}\omega_{ch}}{s^2 + (\omega_{cl} + \omega_{ch})s + \omega_{cl}\omega_{ch}} \tag{18.3.4}$$

Again, it is assumed that $\omega_{ch} \gg \omega_{cl}$ so that the two cutoff frequencies are independent of one another, and the two filters can be designed separately. The minimum input impedance is $R_l \parallel R_h$ at high frequencies. We may choose $R_l = R_h = 50$ kΩ.

For the highpass filter, $C_h = \dfrac{1}{10^5 \times 50 \times 10^3} \equiv 0.2$ nF. For the lowpass filter, $C_l = \dfrac{1}{125 \times 50 \times 10^3} = 0.16$ μF.

EXERCISE 18.3.3

Show that the minimum gain in the stopband is, from Equation 18.3.4, $\dfrac{R_f}{R_r}\dfrac{2}{1+\omega_{ch}/\omega_{cl}}$ and occurs at $\omega_0 = \sqrt{\omega_{cl}\omega_{ch}}$.

EXERCISE 18.3.4

Use MATLAB or PSpice to obtain the Bode magnitude plots in Design Example 18.3.1 and Design Example 18.3.2 and verify that they correspond to the desired responses. Determine the minimum gain in the stopband.
Answer: −32.1 dB.

First-order active filters may be cascaded just like passive filters (Section 10.6, Chapter 10), with the op amps providing isolation between cascaded stages. The same disadvantage still applies, namely, the reduction of the 3-dB bandwidth as the number of cascaded stages is increased.

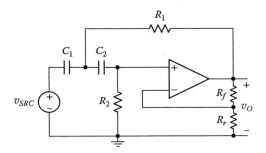

FIGURE 18.4.1
Highpass noninverting filter.

18.4 Second-Order Active Filters

Using op amps in second-order filters allows increasing the Q to any desired value. Butterworth filters of any order can therefore be implemented using first-order and second-order active filters. Two general classes of second-order active filters are discussed in this section: filters based on a single op amp and those using more than one op amp.

Single Op Amp Filters

Single op amp filters may be noninverting or inverting. An instructive derivation of these types of filters is given in Section ST18.4 and Section ST18.5, respectively, based on feedback relations and circuit manipulation techniques for obtaining a common terminal between input and output.

A highpass noninverting filter is illustrated in Figure 18.4.1. From Section ST18.4, the transfer function is:

$$\frac{V_o}{V_{src}} = A \frac{s^2}{s^2 + s\left[\frac{1}{R_2}\left(\frac{1}{C_1} + \frac{1}{C_2}\right) + \frac{1}{C_1 R_1}(1-A)\right] + \omega_c^2} \tag{18.4.1}$$

where $A = \left(1 + \frac{R_f}{R_r}\right)$, $\omega_c = 1/\sqrt{C_1 C_2 R_1 R_2}$, $Q = \frac{\sqrt{C_1 C_2 R_1 R_2}}{(C_1 + C_2)R_1 + C_2 R_2(1-A)}$, and $s = j\omega$ at real

frequencies. As $s \to \infty$, the capacitors behave as short circuits, and the circuit reduces to a noninverting amplifier of gain A. As $s \to 0$, the capacitors behave as open circuits. The amplifier is isolated from the input source, with its input connected to ground, and the output is zero.

If $A = 1$, and the filter is normalized by having $C_1 = C_2 = 1$ F and $\omega_c = 1$ rad/s, Equation 18.4.1 becomes:

$$\frac{V_o}{V_{src}} = \frac{s^2}{s^2 + \frac{2}{R_2}s + 1}, \text{ with } R_1 R_2 = 1 \tag{18.4.2}$$

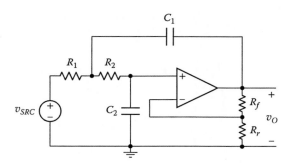

FIGURE 18.4.2
Lowpass noninverting filter.

Design Example 18.4.1: Second-Order, Highpass Butterworth Filter

It is required to design a unity-gain, second-order, highpass Butterworth filter having a 3-dB cutoff frequency of 500 Hz, using 0.5 µF capacitors.

SOLUTION

We will illustrate two design procedures; one is based on normalized responses and scaling, whereas the other is not. Starting with the latter, it follows from the discussion in Chapter 10 that to obtain a Butterworth response of the second order requires $\xi = 1/\sqrt{2}$ or $Q = 1/\sqrt{2}$. From Equation 18.4.1, with $A = 1$, $\omega_c = 1/\sqrt{C_1 C_2 R_1 R_2}$, and $Q =$

$\dfrac{\sqrt{C_1 C_2 R_1 R_2}}{(C_1 + C_2) R_1} = \dfrac{1}{\omega_c (C_1 + C_2) R_1}$. Substituting for Q, ω_c, C_1 and C_2, gives $R_1 = \dfrac{1000\sqrt{2}}{\pi} =$

$450.16\,\Omega$. From the expression for ω_c, $R_2 = \dfrac{1}{\omega_c^2 C_1 C_2 R_1} = \dfrac{2000\sqrt{2}}{\pi} = 900.32\,\Omega$.

The second procedure, which is the standard design procedure, is to refer to the normalized second-order Butterworth polynomial. From Table 10.6.1 (Chapter 10), this polynomial is: $B_2(s) = s^2 + \sqrt{2}s + 1$. Comparing with the denominator of $H(s)$ in Equation 18.4.2 gives $R_2 = \dfrac{2}{\sqrt{2}}\,\Omega$, so that $R_1 = R_2 = \dfrac{1}{\sqrt{2}}\,\Omega$. To move the cutoff frequency from 1 rad/s to $2\pi \times 500$ Hz requires a frequency scale factor $k_f = 1000\pi$. To use capacitors of 0.5 µF requires a magnitude scale factor $k_m = \dfrac{1}{1000\pi \times 0.5 \times 10^{-6}} = \dfrac{2000}{\pi}$ (Equation 10.5.14, Chapter 10). Resistances are multiplied by k_m (Equation 10.5.12, Chapter 10), so that $R_1 = \dfrac{1000\sqrt{2}}{\pi} = 450.16\ \Omega$ and $R_2 = \dfrac{2000\sqrt{2}}{\pi} = 900.32\ \Omega$.

A lowpass filter can be derived from the highpass filter by replacing every resistor of the filter circuit, that is, excluding R_r and R_f, by a capacitor, and conversely. This is equivalent to replacing every R_k in the transfer function by $1/sC_k$, and every C_m by $1/sR_m$ (Figure 18.4.2). From Equation 18.4.1, the transfer function is:

$$\frac{V_o}{V_{src}} = \frac{A}{C_1 C_2 R_1 R_2} \frac{1}{s^2 + s\left[\frac{1}{C_1}\left(\frac{1}{R_1} + \frac{1}{R_2}\right) + \frac{1}{C_2 R_2}(1-A)\right] + \omega_c^2} \tag{18.4.3}$$

where $\omega_c = 1/\sqrt{C_1 C_2 R_1 R_2}$ and $Q = \dfrac{\sqrt{C_1 C_2 R_1 R_2}}{C_2(R_1 + R_2) + C_1 R_1(1-A)}$. As $s \to 0$, the capacitors behave as open circuits and the circuit reduces to a noninverting configuration of gain A. As $s \to \infty$, the capacitors behave as short circuits. C_2 effectively grounds the input of the op amp, so the output is zero.

If $A = 1$, and the filter is normalized by having $R_1 = R_2 = 1\ \Omega$, and $\omega_c = 1$ rad/s. Equation 18.4.3 becomes:

$$\frac{V_o}{V_{src}} = \frac{1}{s^2 + \dfrac{2}{C_1}s + 1}, \text{ with } C_1 C_2 = 1 \tag{18.4.4}$$

Design Example 18.4.2: Third-Order, Lowpass Butterworth Filter

It is required to design a third-order, lowpass Butterworth filter having a gain of 3 and a 3-dB cutoff frequency of 2000 rad/s, using 1 kΩ resistors in the filter sections.

SOLUTION

Referring to Table 10.6.1 (Chapter 10), the normalized third-order Butterworth polynomial is $(s^2 + s + 1)(s + 1)$. The filter is implemented as a cascade of second-order and first-order filters. Comparing $s^2 + s + 1$ with the denominator of $H(s)$ in Equation 18.4.2 gives $C_1 = 2$ F, so that $C_2 = 0.5$ F. To move the cutoff frequency to 2000 rad/s requires $k_f = 2000$. To use 1 kΩ resistors requires $k_m = 1000$. It follows that $C_1 = \dfrac{2\ \text{F}}{1000 \times 2000} = 1\ \mu\text{F}$

and $C_2 = \dfrac{C_1}{4} = 0.25\ \mu\text{F}$ (Equation 10.5.14, Chapter 10).

The first-order filter has the same k_f and k_m. Hence, $C = \dfrac{1\ \text{F}}{1000 \times 2000} = 0.5\ \mu\text{F}$. A nominating lowpass filter may be implemented using a simple RC lowpass filter and a noninverting amplifier having a gain of 3. The filter circuit is shown in Figure 18.4.3.

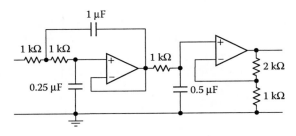

FIGURE 18.4.3
Figure for Example 18.4.2.

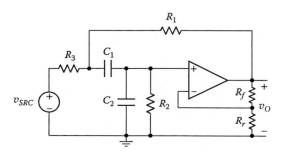

FIGURE 18.4.4
Bandpass noninverting filter.

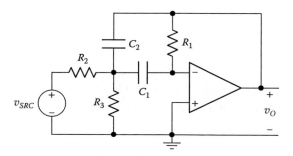

FIGURE 18.4.5
Bandpass inverting filter.

A bandpass filter is shown in Figure 18.4.4. From Table 10.5.1 (Chapter 10), the coefficient of s in the denominator of the transfer function of a band pass filter is ω_0/Q, and the normalized gain is unity when the numerator is also ω_0/Q. The transfer function of the circuit of Figure 18.4.4, derived in Section ST18.4, is:

$$\frac{V_o}{V_{src}} = \frac{A}{C_1 R_3} \frac{s}{s^2 + s\left[\frac{1}{(R_1 \parallel R_3)}\left(\frac{1}{C_1} + \frac{1}{C_2}\right) + \frac{1}{C_2}\left(\frac{1}{R_2} - \frac{A}{R_1}\right)\right] + \omega_0^2} \tag{18.4.5}$$

where the center frequency is $\omega_0 = 1/\sqrt{C_1 C_2 (R_1 \parallel R_3) R_2}$, the gain at this frequency is

$\dfrac{QA}{\omega_0 C_1 R_3}$, and $Q = \dfrac{\sqrt{C_1 C_2 (R_1 \parallel R_3) R_2}}{R_2(C_1 + C_2) + C_1(R_1 \parallel R_3)[1 - AR_2/R_1]}$. That the circuit behaves as a

bandpass can be readily checked. As $s \to 0$, C_1 isolates the input of the op amp from the source, and the output is zero. As $s \to \infty$, C_2 grounds the input of the op amp, and the output is again zero.

Another class of single op amp filters is the inverting type. A bandpass filter of this type is illustrated in Figure 18.4.5. From Section ST 18.5, the transfer function is:

$$\frac{V_o}{V_{src}} = -\frac{1}{R_2 C_2} \frac{s}{s^2 + \frac{1}{R_1}\left(\frac{1}{C_1} + \frac{1}{C_2}\right)s + \omega_0^2} \tag{18.4.6}$$

where $\omega_0 = 1/\sqrt{C_1 C_2 R_1 (R_2 \parallel R_3)}$, $\dfrac{\omega_0}{Q} = \dfrac{1}{R_1}\left(\dfrac{1}{C_1}+\dfrac{1}{C_2}\right)$ = BW is the 3-dB bandwidth. The

gain at the center frequency, obtained by setting $s = j\omega_0$, is $K = -\dfrac{R_1 C_1}{R_2(C_1 + C_2)}$.

If the filter is normalized by having $C_1 = C_2 = 1$ F and $\omega_o = 1$ rad/s, Equation 18.4.6 becomes:

$$\frac{V_o}{V_{src}} = -\frac{1}{R_2}\frac{s}{s^2 + \dfrac{2}{R_1}s + 1} \;, \text{ with } R_1(R_2 \parallel R_3) = 1 \qquad (18.4.7)$$

where $\dfrac{1}{Q} = \text{BW} = \dfrac{2}{R_1}$ and $K = -\dfrac{Q}{R_2}$. It follows that $R_1 = 2Q$, $R_2 = Q/K$, and $R_3 = \dfrac{Q}{2Q^2 + K}$.

Once the center frequency ω_0, the passband gain K, and BW or Q are specified, R_1, R_2, and R_3 can be determined, as illustrated by the following example.

Design and Simulation Example 18.4.3: Second-Order Bandpass Filter

It is required to design a bandpass filter having a center frequency of the passband of 20 krad/s, $Q = 10$, and a passband gain of -4, using 0.05 μF capacitors.

SOLUTION
Using the relations derived in connection with Equation 18.4.7, $R_1 = 2Q = 20\ \Omega$, $R_2 =$

$\dfrac{10}{4} = 2.5\ \Omega$, and $R_3 = \dfrac{10}{2\times(10)^2 - 4} = \dfrac{10}{196}\ \Omega$. The frequency scale factor k_f is 20,000. To have

$C = 0.05$ μF requires a magnitude scale factor $k_m = \dfrac{10^6}{20{,}000\times 0.05} = 1000$. Hence, $R_1 = (20\ \Omega)\times$

$1000 = 20$ kΩ, $R_2 = (2.5\ \Omega)\times 1000 = 2.5$ kΩ, and $R_3 = \dfrac{10}{196}\times 1000 = 51\ \Omega$. The bandwidth

is $\dfrac{\omega_o}{Q} = \dfrac{2}{R_1}k_f = 2{,}000$ rad/s.

SIMULATION
An ideal op amp is used that is available in PSpice as part OPAMP in the ANALOG library. It has only three terminals and default power supplies of ±15 V, which can be changed in the Property Editor spreadsheet of the OPAMP. The circuit of Figure 18.4.5 is entered in the schematic with a VAC of 1V applied. When the simulation is run as AC Sweep/Noise, the magnitude plot of Figure 18.4.6 is obtained. Using the Probe cursor, it can be readily verified that the magnitude of the output, the center frequency, the Q, and the bandwidth are in accordance with the figures already mentioned.

If in Figure 18.4.5, R_2 is replaced by a capacitor C_3, the resulting filter becomes an inverting highpass filter having a transfer function (Problem P18.3.5):

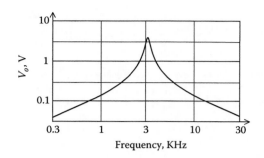

FIGURE 18.4.6
Figure for Example 18.4.3.

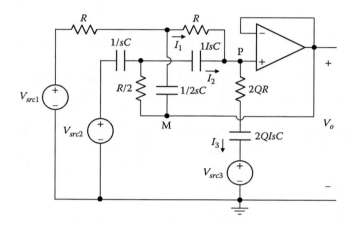

FIGURE 18.4.7
Universal filter.

TABLE 18.4.1

Input Combinations and Responses of
Universal Filter

V_{src1}	V_{src2}	V_{src3}	Response
V_{src}	0	0	Lowpass
0	V_{src}	0	Highpass
0	0	V_{src}	Bandpass
V_{src}	V_{src}	0	Bandstop
V_{src}	V_{src}	V_{src}	Allpass

$$\frac{V_o}{V_{src}} = -\frac{C_3}{C_2} \frac{s^2}{s^2 + \frac{1}{R_1}\left(\frac{1}{C_1} + \frac{1}{C_2} + \frac{C_3}{C_1 C_2}\right)s + \omega_c^2} \tag{18.4.8}$$

where $\omega_c = 1/\sqrt{C_1 C_2 R_1 R_3}$. Replacing capacitors by resistors, and conversely, in the highpass filter results in an inverting lowpass filter (Problem P18.3.5).

As a last example of a single op amp filter, a universal filter (Figure 18.4.7) is considered that is capable of providing all five of the basic filter responses (Table 18.4.1). The filter has three input sources, the output being given as:

$$V_o = \frac{s^2 V_{src2} - (\omega_0/Q)s V_{src3} + \omega_0^2 V_{src1}}{s^2 + (\omega_0/Q)s + \omega_0^2} \tag{18.4.9}$$

where $\omega_0 = 1/RC$ and Q has a chosen value. By suitable combinations of inputs, any of the five filter responses could be obtained.

Equation 18.4.9 may be derived in a creative and instructive way, assuming an ideal op amp. Node M is connected to an ideal voltage source V_o representing the amplifier output (Figure 18.4.8). Node P is also at a voltage V_o with respect to the common reference because it is the input of a unity-gain amplifier. According to the substitution theorem (Section 4.2, Chapter 4), the branch between node P and the common reference may be replaced by an ideal voltage source V_o, so that nodes P and M may be connected together to the same source V_o. The currents I_1 and I_2 become the output short-circuit currents of the two T-sections, having inputs $(V_{src1} - V_o)$ and $(V_{src2} - V_o)$, respectively. The current I_1 is given by:

$$\frac{I_1}{V_{src1} - V_o} = \frac{1}{R + R \parallel (1/sC)} \frac{1/2sC}{R + 1/2sC} = \frac{1}{2R} \frac{\omega_0}{s + \omega_0} \tag{18.4.10}$$

where $\omega_0 = 1/RC$. The current I_2 may be determined in the same way, or by replacing every R in Equation 18.4.10 by $1/sC$ and every C by $1/sR$, to obtain:

$$\frac{I_2}{V_{src2} - V_o} = \frac{C}{2} \frac{s^2}{s + \omega_0} \tag{18.4.11}$$

From Figure 18.4.7, I_3 is given by: $I_3 = \dfrac{V_o - V_{src3}}{2QR + 2Q/sC}$ or:

$$\frac{I_3}{V_{src3} - V_o} = \frac{1}{2QR} \frac{s}{s + \omega_o} \tag{18.4.12}$$

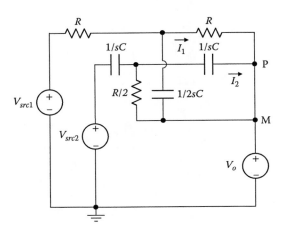

FIGURE 18.4.8
Transformed universal filter.

Substituting for the currents from Equation 18.4.10 to Equation 18.4.12 in the equation $I_3 = I_1 + I_2$, and collecting terms gives Equation 18.4.9.

Two-Integrator Loop Biquad Filters

Consider a highpass transfer function of the form:

$$\frac{V_{oh}}{V_{src}} = K \frac{s^2}{s^2 + (\omega_0/Q)s + \omega_0^2} \tag{18.4.13}$$

Equation 18.4.13 can be rearranged as:

$$V_{oh} = KV_{src} + \frac{1}{Q}\left(-\frac{\omega_0}{s}\right)V_{oh} - \left(-\frac{\omega_0}{s}\right)^2 V_{oh} \tag{18.4.14}$$

Equation 18.4.14 may be represented by the block diagram of Figure 18.4.9. The second and third terms on the RHS of Equation 18.4.14 represent successive integrations of V_{oh}, with sign inversion. These are then added to KV_{src} to give V_{oh}. The output of the first integrator, being V_{oh} multiplied by $1/s$, is a bandpass response, whereas the output of the second integrator, being V_{oh} multiplied by $1/s^2$, is a lowpass response. Hence, all three types of responses are available simultaneously. The implementation of Figure 18.4.9 is shown in Figure 18.4.10, where a difference amplifier is used to sum the three terms on

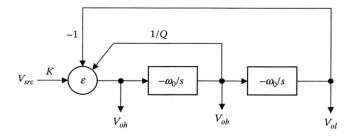

FIGURE 18.4.9
Block diagram of a biquad filter.

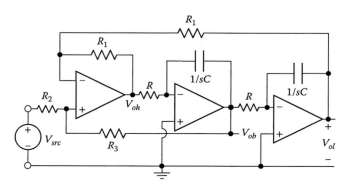

FIGURE 18.4.10
KHN biquad.

the RHS of Equation 18.4.14. The circuit is known as the **Kerwin-Huelsman-Newcomb** (KHN) biquad. Using the result of Exercise 18.2.4, with $v_O \equiv V_{oh}$, $v_{SRC1} \equiv V_{src}$, $v_{SRC2} \equiv V_{ol}$, $v_{SRC3} \equiv V_{ob}$, $R_a = R_b = R_1$, $R_c = R_2$, and $R_d = R_3$, we obtain:

$$V_{oh} = \frac{2R_3}{R_2 + R_3} V_{src} + \frac{2R_2}{R_2 + R_3} V_{ob} - V_{ol} \tag{18.4.15}$$

Comparing the RHS of Equation 18.4.14 and Equation 18.4.15, term by term, it is seen

that $\dfrac{1}{Q} = \dfrac{2R_2}{R_2 + R_3}$ and $K = \dfrac{2R_3}{R_2 + R_3}$. It follows that:

$$\frac{R_3}{R_2} = 2Q - 1 \quad \text{and} \quad K = \frac{1}{Q}\frac{R_3}{R_2} = 2 - \frac{1}{Q} \tag{18.4.16}$$

From Equation 18.4.13, the passband gain of the highpass and lowpass filters is K. For the bandpass filter, the coefficient of s in the numerator is $\omega_0 K$. The maximum gain is

therefore $-\omega_0 K \dfrac{Q}{\omega_0}$, which gives KQ or $1 - 2Q$.

EXERCISE 18.4.1

Design a KHN bandpass filter having a center frequency of the passband of 10 krad/s and $Q = 20$, using 0.01 μF capacitors, with $R_1 = R_2 = 10$ kΩ, and determine the maximum gain.

Answer: $R = 10$ kΩ; $R_3 = 390$ kΩ; 39.

EXERCISE 18.4.2

Show that if the V_{oh}, V_{ob}, and V_{ol} outputs of a KHN filter are added in an inverting adder having input resistances R_H, R_B, and R_L, respectively, the feedback resistance being R_f, a universal response is obtained as follows:

$$\frac{V_o}{V_{src}} = -K\frac{(R_f/R_H)s^2 - (R_f/R_B)\omega_0 s + (R_f/R_L)\omega_0^2}{s^2 + (\omega_0/Q)s + \omega_0^2}.$$

Where it is desirable to have a virtual ground in all the amplifiers, the second inverting integrator may be replaced with a noninverting integrator (Figure 18.1.14) and the summation implemented at the virtual-ground input of the first integrator. The resulting circuit, shown in Figure 18.4.11, is known as the **Åkerberg-Mossberg biquad**. It is seen that R_3 in Figure 18.4.10 is now in parallel with C of the first integrator. With the resistor values

shown, it follows that $-V_{ob} = \left(\dfrac{V_{src}}{R_2} + \dfrac{V_{ol}}{R}\right)\dfrac{R_3}{1 + sCR_3}$ and $V_{ol} = \dfrac{V_{ob}}{sCR}$. These relations give

$$\frac{V_{ol}}{V_{src}} = -\frac{1}{C^2 RR_2}\frac{1}{s^2 + \dfrac{s}{CR_3} + \dfrac{1}{C^2 R^2}} \tag{18.4.17}$$

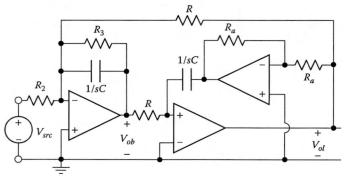

FIGURE 18.4.11
Åkerberg-Mossberg biquad.

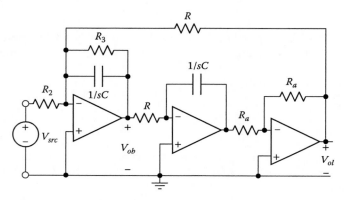

FIGURE 18.4.12
Tow-Thomas biquad.

and
$$\frac{V_{ob}}{V_{src}} = -\frac{1}{CR_2}\frac{s}{s^2 + \dfrac{s}{CR_3} + \dfrac{1}{C^2R^2}}$$
(18.4.18)

Because $\dfrac{\omega_0}{Q} = \dfrac{1}{CR_3}$, where $\omega_0 = 1/RC$, it follows that $R_3 = QR$. The passband gain of the lowpass filter, as $s \to 0$ is $-K = -R/R_2$. The maximum gain of the bandpass filter is $-\dfrac{1}{CR_2} \times \dfrac{Q}{\omega_0} = -\dfrac{1}{CR_2} \times CR_3 = -\dfrac{R_3}{R_2}$. As for the KHN filter, $KQ = R_3/R_2$. A highpass response is no longer available.

An alternative that leads to all amplifiers being in the inverting configuration is to invert the output of the second integrator in Figure 18.4.11 by means of a unity gain inverter, leading to the **Tow-Thomas biquad** filter shown in Figure 18.4.12. It is clear that the same relations for the Åkerberg-Mossberg filter apply in this case.

Sensitivity

An important practical consideration in selecting a suitable filter for a particular application is the sensitivity of a filter characteristic, such as Q or ω_0, to variations in circuit parameters, such as values of resistances, capacitances, and op amp gain. The sensitivity of a quantity y to variations in another quantity x is generally defined as:

$$S_x^y = \frac{\Delta y/y}{\Delta x/x}, \text{ which in the limit is: } S_x^y = \frac{\partial y}{\partial x}\frac{x}{y} = \frac{\partial(\log y)}{\partial(\log x)} \qquad (18.4.19)$$

Basically, S_x^y is the fractional change in y per fractional change in x. For example, in RC filters, ω_0 is given by an expression of the general form $\dfrac{1}{\sqrt{C_a C_b R_a R_b}}$. The sensitivity of ω_0 to any of these parameters, say C_a, is derived from Equation 18.4.19. Thus, $\dfrac{\partial \omega_0}{\partial C_a} = -\dfrac{\omega_0}{2C_a}$, which gives $S_{C_a}^{\omega_0} = -\dfrac{1}{2}$. This means that a 1% increase in C_a deceases ω_0 by 0.5%. Sensitivity analysis can become quite involved algebraically and is most conveniently performed using computer programs.

It can be shown that the variation of S_A^Q is of the order of $\dfrac{Q^2}{A}$ for single op amp active filters and $\dfrac{Q}{A}$ for biquad filters, where A is the gain of the op amp. Hence, biquad filters are generally preferred for high-Q applications (Problem P18.3.19 and Problem P18.3.20).

EXERCISE 18.4.3
The definition of sensitivity may also be applied to a transfer function to obtain a **function sensitivity**. If $H(s) = \dfrac{N(s)}{D(s)}$, show that: $S_x^{H(s)} = x \left[\dfrac{1}{N(s)}\dfrac{\partial N(s)}{\partial x} - \dfrac{1}{D(s)}\dfrac{\partial D(s)}{\partial x} \right]$, where x is some element in the circuit implementation of $H(s)$.

Analog filters are being increasingly implemented as switched-capacitor IC circuits, the object being to replace resistors by capacitors and transistor switches. The advantages are improved accuracy in implementing the frequency response, coupled with good linearity and dynamic range. The improved accuracy of frequency response results from the filter parameters being determined by capacitance ratios. These can be set in integrated circuits to a precision as high as 0.1%, compared to a variation as large as 20% for RC time constants. Switched-capacitor filters are explained in Section ST18.6.

18.5 Comparators

A **comparator circuit** compares two voltages and gives an output that depends on the sign of the difference between the compared voltages. Evidently, an op amp can be used for this purpose. In Figure 18.5.1, v_I is applied to the noninverting input of an op amp, with the inverting input connected to a reference voltage V_{ref} derived from a potential divider across the $+V_{CC}$ supply. The op amp is driven to positive saturation if $v_I > V_{ref}$, and to negative saturation if $v_I < V_{ref}$. It changes state when v_I passes V_{ref} in either direction.

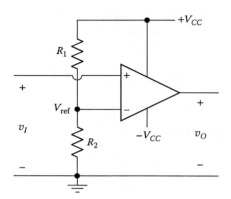

FIGURE 18.5.1
Op amp comparator.

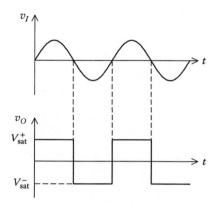

FIGURE 18.5.2
Zero-crossing detection.

The input level at which the output changes is the **threshold**, and is nominally V_{ref} in Figure 18.5.1.

Comparators may be used as **zero-crossing detectors**. If the inverting input is connected to the common reference, so $V_{ref} = 0$, the op amp output changes between the two saturation limits every time v_I passes through zero (Figure 18.5.2).

Although simple, the circuit of Figure 18.5.1 suffers from some limitations. As it crosses the threshold, the op amp passes through the linear region, where the gain may be very high. An op amp that is not properly frequency compensated will almost certainly break into oscillations when operating in the high-gain region between the saturation limits. If the transition between the saturation limits is fast enough, the oscillations are hardly noticeable. However, if v_I changes relatively slowly, the oscillation in the output may be objectionable. Another limitation is encountered if v_I is corrupted by noise, in which case several spurious crossings of the threshold may occur. These limitations are overcome in the regenerative comparator.

Schmitt Trigger

In the regenerative comparator, or **Schmitt trigger**, positive feedback is applied between the output of the op amp and its noninverting input (Figure 18.5.3a). If v_I is zero, it follows from superposition that:

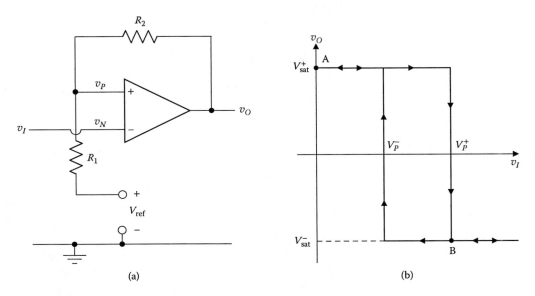

FIGURE 18.5.3
Schmitt trigger circuit (a) and characteristic (b).

$$v_P = \frac{R_1}{R_1 + R_2} v_O + \frac{R_2}{R_1 + R_2} V_{ref} \tag{18.5.1}$$

Assume, for the sake of argument, that V_{ref} is positive and that R_1 and R_2 are such that the RHS of Equation 18.5.1 is positive even when v_O is at its negative saturation limit. Then, if $v_N = v_I = 0$ and $v_P > 0$, as assumed in Equation 18.5.1. The op amp is in positive saturation because $v_P > v_N$. The operating point on a v_O vs. v_I plot will be A in Figure 18.5.3b. From Equation 18.5.1, with $v_O = V_{sat}^+$, the voltage V_P^+ of the noninverting input is:

$$V_P^+ = \frac{R_1}{R_1 + R_2} V_{sat}^+ + \frac{R_2}{R_1 + R_2} V_{ref} \tag{18.5.2}$$

To switch the op amp to negative saturation, v_I should increase in the positive direction so that it exceeds the upper threshold V_P^+ (Equation 18.5.2). The operating point now moves to B in Figure 18.5.3b. When v_I is more positive than V_P^+, v_O remains at V_{sat}^-; v_P assumes a new value V_P^- given by:

$$V_P^- = \frac{R_1}{R_1 + R_2} V_{sat}^- + \frac{R_2}{R_1 + R_2} V_{ref} \tag{18.5.3}$$

To switch the op amp back to positive saturation, v_I should decrease to less than the lower threshold V_P^-, where $V_P^- < V_P^+$.

The inequality of thresholds, depending on whether the output is at V_{sat}^+ or V_{sat}^-, is described as **hysteresis**. From Figure 18.5.3b, the width of the hysteresis loop is

$$V_P^+ - V_P^- = \frac{R_1}{R_1 + R_2} (V_{sat}^+ - V_{sat}^-) \tag{18.5.4}$$

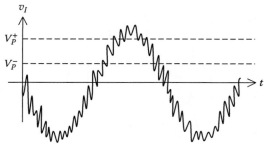

FIGURE 18.5.4
Noise in the presence of hysteresis.

The hysteresis loop is centered at $\frac{1}{2}(V_P^+ + V_P^-) = \frac{R_2}{R_1 + R_2} V_{\text{ref}}$.

Hysteresis can be useful. In the case of the noisy signal of Figure 18.5.4, as the input increases past V_P^+, the output switches to V_{sat}^- and stays at this level despite fluctuations of the input around V_P^+. The output will not switch back to V_{sat}^+ until the input drops below V_P^-. The noise will not cause multiple transitions in the vicinity of the threshold, although it may affect the instant of switching. This is acceptable if one is interested in the number of crossings of the threshold in a given interval, and not in the exact time of crossings.

Once a threshold is crossed and the output begins to change, the change in output augments the differential input of the op amp, through positive feedback, which speeds up the change in output. As a result, the transition is faster and becomes independent of the rate of change of the external input v_I.

Design Example 18.5.1: Schmitt Trigger

It is required to design a Schmitt trigger that operates from ±5 V supplies and has a hysteresis loop 1 V wide, centered at 2.5 V.

SOLUTION

$V_P^+ - V_P^- = 1$ V and $\frac{1}{2}(V_P^+ + V_P^-) = 2.5$ V. This gives: $V_P^+ = 3$ V and $V_P^- = 2$ V. From Equation

18.5.4, $1 = \frac{R_1}{R_1 + R_2} \times 10$, so $R_2 = 9R_1$. It follows that $\frac{R_2}{R_1 + R_2} = 1 - \frac{R_1}{R_1 + R_2} = 0.9$. Substituting

in Equation 18.5.2, $V_{\text{ref}} = 2.78$ V. If R_1 is arbitrarily chosen to be 1 kΩ, then $R_2 = 9$ kΩ.

EXERCISE 18.5.1

Repeat Example 18.5.1, assuming that the hysteresis loop is to be centered at the origin. Trace the hysteresis loop as v_I is varied.

Answer: $R_1 = 1$ kΩ, $R_2 = 9$ kΩ, $V_{\text{ref}} = 0$.

Summary of Main Concepts and Results

- An ideal op amp is a differential-input, single-ended output, voltage amplifier that is characterized by infinite differential gain, an ideal voltage source output, open-circuit inputs, and freedom from imperfections.

- The two basic op amp configurations are noninverting and inverting. The most salient feature of the noninverting configuration is high input impedance, whereas the most salient feature of the inverting configuration is a virtual ground. This allows important practical applications such as integrators, differentiators, and adders.
- The unity-gain amplifier, or voltage follower, is useful for isolation purposes and for increased power drive to loads when the source impedance is relatively high.
- The common-mode response of an op amp arises from: (1) unequal gains from each of the two inputs to the output, and (2) imbalance between source impedances and the impedances between each op amp input and the common reference.
- The slew rate of an op amp is a large-signal frequency limitation on how fast the output signal can change. As a result, large, fast signals are distorted.
- Op amps in active filters provide gain, isolation, load drive, and high Q.
- Second-order lowpass, highpass, bandpass, and bandstop active filters having any desired Q may be constructed using operational amplifiers in the noninverting or inverting configuration. Single op amp filters are possible as well as filters based on a three-integrator loop.
- A comparator circuit compares two voltages and gives an output that depends on the sign of the difference between the compared voltages.
- In a Schmitt trigger, positive feedback is added to a comparator so as to increase the speed of response and introduce hysteresis.

Learning Outcomes

- Analyze circuits involving op amps
- Design some basic types of op amp circuits, including amplifiers, integrators, active filters, and comparators

Supplementary Topics and Examples on CD

ST18.1 Negative feedback: Explains negative feedback and its advantages, and discusses the noninverting and inverting op amp configurations in terms of negative feedback and signal-flow diagrams.

ST18.2 Effects of input bias currents and offset voltage: Analyzes the effects of these imperfections and demonstrates how drift in these quantities sets a limit to performance. Inverting and noninverting adders are compared, taking into account input offset voltage, input bias currents, and finite A_v.

ST18.3 Integrators: Investigates the effects of input offset voltage, input bias currents, and finite A_v on the performance of an integrator and deduces some design guidelines and constraints. The performance in the frequency domain is related to that in the time domain.

ST18.4 Derivation of noninverting single op amp filters: Derives the transfer functions of these types of filters.

ST18.5 Derivation of inverting single op amp filters: Derives the transfer functions of these types of filters.

ST18.6 Switched-capacitor filters: Explains the principle of operation of switched-capacitor filters.

SE18.1 Simulation of large feedback resistor: A T-circuit is used to simulate a large R_f in an inverting configuration. The performance of the circuit is compared with that using a single resistor for R_f when finite A_v, input bias current, and input offset voltage are taken into consideration.

SE18.2 Input impedance and output resistance of noninverting configuration: The input impedance and output resistance of the noninverting configuration are determined taking into consideration the finite resistances of the feedback voltage divider.

SE18.3 ECG recording: Gives an example of ECG recording that emphasizes the common-mode response of the amplifier due to pick-up from the installation wiring.

SE18.4 The μA741 in PSpice: Obtains the frequency response of the μA741, its bias currents at 25 and 100°C, the input offset voltage, A_{v0}, and the CMRR.

SE18.5 Conversion of inverting op amp circuit to a circuit with unity-gain amplifier: Illustrates how an inverting configuration having a floating source can be converted to a circuit having a unity-gain amplifier and a grounded source.

SE18.6 Relocation of source in an op amp circuit: Derives the modified transfer function when a source is relocated inside a three-terminal circuit connected to an amplifier.

SE18.7 Notch filter: Derives a second-order, high-Q, op amp notch filter from a twin-T passive notch filter.

SE18.8 Digital-to-analog converter: Discusses DAC conversion using an op amp.

Problems and Exercises

P18.1   Ideal Op Amp Circuits

In the following problems, op amps are assumed ideal except as may be indicated.

P18.1.1 Consider the circuit of Figure 18.1.10 with an ammeter that can indicate a current of up to 100 μA connected in series with R_L. It is desired that the 100 μA current be read as 10 V applied to the noninverting terminal. Determine suitable values of R and R_L.

P18.1.2 In Figure P18.1.2 a T-circuit is used at the input of the integrator. Determine the integration time constant.

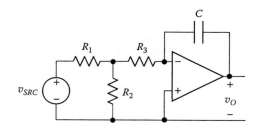

FIGURE P18.1.2

P18.1.3 The circuit of Figure P18.1.3 uses matched resistor pairs $R_1, R_2, ..., R_n$ to provide the average of the applied inputs. Show that: $v_O = -(a_1v_1 + a_2v_2 + ... + a_nv_n)$,

where $a_k = \dfrac{G_k}{G_1 + G_2 + ... + G_n}$, $G_k = 1/R_k$, and $a_1 + a_2 + ... + a_n = 1$.

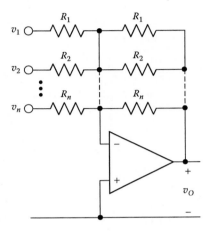

FIGURE P18.1.3

P18.1.4 Show that in Figure P18.1.4, $C_{in} = \dfrac{C}{\alpha}$. What is the purpose of the first amplifier?

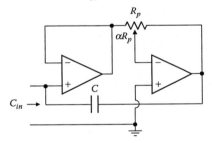

FIGURE P18.1.4

P18.1.5 Consider the difference amplifier circuit of Figure 18.2.12 with $R_a = R_c = R$ and $R_b = R_d = kR$. Determine the input resistance seen by the following sources: (a) v_{SR1}; (b) v_{SRC2}; (c) a source connected between nodes a and b; and (d) a source connected between nodes a and b joined together and common reference.

P18.1.6 Determine i_O in Figure P18.1.6. Simulate with PSpice.

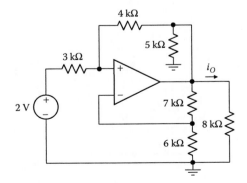

FIGURE P18.1.6

P18.1.7 Modify the circuit of the inverting adder so as to obtain a summing integrator and differentiator, and verify the output relations.

P18.1.8 Determine R_{in} in Figure P18.1.8.

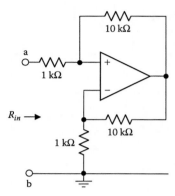

FIGURE P18.1.8

P18.1.9 Figure P18.1.9 illustrates a circuit that may be used with a strain gauge bridge. Show that $v_O = \dfrac{R}{R_o}\dfrac{\alpha}{(1+\alpha)(1+R_o/R)+1}V_{DC}$. Note that if $\dfrac{R_o}{R}\ll 1$ and $\alpha\ll 1$, then $v_O = \dfrac{\alpha R}{R_o}V_{DC}$.

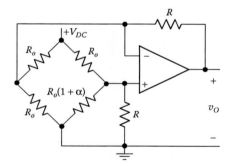

FIGURE P18.1.9

P18.1.10 Show that i_L in Figure P18.1.10 is given by: $i_L = (v_2 - v_1)/R_2$. Note that the load sees a current source whose value is determined by the differential input $(v_2 - v_1)$.

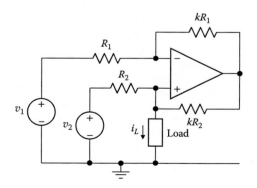

FIGURE P18.1.10

P18.1.11 Consider the circuit of Figure P18.1.11. Using Equation 18.2.3 and Equation 18.2.6,

show that: $\text{CMR} = 20 \log_{10} \dfrac{1 + \dfrac{1}{2}\left(\dfrac{R_a}{R_b} + \dfrac{R_c}{R_d}\right)}{\left(\dfrac{R_a}{R_b} - \dfrac{R_c}{R_d}\right)}$. If $\dfrac{R_a}{R_b} = \dfrac{R_c}{R_d} = k$, the CMR → ∞. Show

that if k is large and the resistors have a small tolerance δ, the worst-case CMR is

approximately $20 \log_{10}\left(\dfrac{1}{4\delta}\right)$. Calculate the worst-case CMR when the resistors

have a ±2% tolerance. Simulate with PSpice.

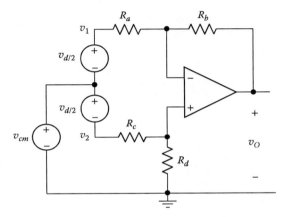

FIGURE P18.1.11

P18.1.12 Show that in Figure P18.1.12, $v_{O1} = -\dfrac{R_2}{R_1} v_2$ and $v_O = R_5\left(\dfrac{R_2}{R_1 R_3} v_2 - \dfrac{v_1}{R_4}\right)$. Note

that if $\dfrac{R_2}{R_1} = \dfrac{R_3}{R_4}$, $v_O = \dfrac{R_5}{R_3}(v_2 - v_1)$. The circuit acts as a difference amplifier but

can tolerate a relatively large common mode input. If $R_1 = 10R_2$, and the
saturation limits are ±12 V, what is the largest value of v_2 that makes v_{O1} reach
the saturation limit?

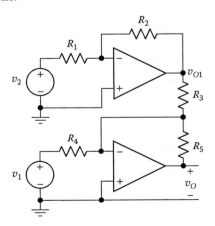

FIGURE P18.1.12

P18.1.13 In Figure P18.1.13, R may be used to reduce the current I_{SRC} drawn from the source. (a) Determine R for zero I_{SRC}. (b) Determine the maximum V_{SRC} if the saturation limits of the second op amp are ± 12 V.

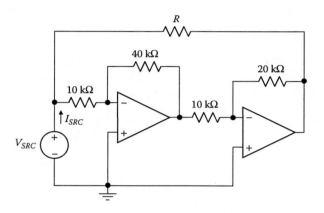

FIGURE P18.1.13

P18.1.14 Determine v_O/v_{SRC} in Figure P18.1.14. Simulate with PSpice, using appropriate resistor values.

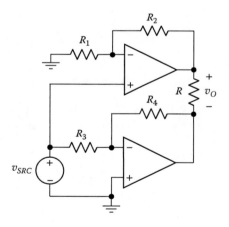

FIGURE P18.1.14

P18.1.15 Determine v_O in Figure P18.1.15. Simulate with PSpice, using appropriate resistor values.

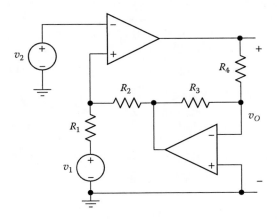

FIGURE P18.1.15

P18.1.16 Determine v_O in Figure P18.1.16. Simulate with PSpice.

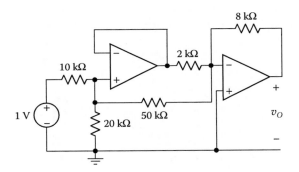

FIGURE P18.1.16

P18.1.17 In Figure P18.1.17, the switch is closed at $t = 0$ with no initial energy storage in the capacitor and inductor. Determine v_O for $t \geq 0$. Simulate with PSpice.

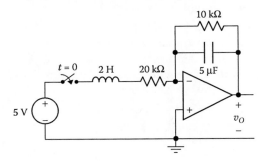

FIGURE P18.1.17

P18.1.18 An integrator has an input resistor of 10 kΩ and an integrating capacitor of 0.1 µF. A 1 MΩ resistor is connected in parallel with this capacitor to prevent saturation due to input offset voltage and bias current. (a) If a sinusoidal input is applied, determine the frequency at which the amplitude of the output deviates from the ideal by 1%. What is the phase deviation? (b) If a ramp voltage is applied, determine the time at which the output deviates from the ideal by 1%. Simulate with PSpice.

P18.1.19 Show that if a resistor R_f is placed in series with the integrating capacitor in

the simple integrator circuit, $v_O = -\dfrac{R_f}{R_r}v_I - \dfrac{1}{R_rC}\int v_I dt$. If the input is a square

pulse train of a 4 ms period and amplitude that varies between 0 and 1 V,

sketch the output, assuming $\dfrac{R_f}{R_r} = 2$ and an integration time constant of 10 ms.

Simulate with PSpice and interpret the result.

P18.1.20 Show that V_o in Figure P18.1.20 is given by: $V_o = -\dfrac{1}{sC_fR_r}\dfrac{sC_fR_f}{1+sC_fR_f}$

$\cdot\dfrac{sC_rR_r}{1+sC_rR_r}V_{src}$. Note that if $R_fC_f = R_rC_r = \tau$, then $V_o = -\dfrac{1}{sC_fR_r}\left(\dfrac{s\tau}{1+s\tau}\right)^2 V_{src}$.

Deduce that the circuit behaves as a band pass filter for frequencies in the neighborhood of $\omega = 1/\tau$, as a differentiator at low frequencies, and as an integrator at high frequencies. Simulate with PSpice. If the input offset voltage

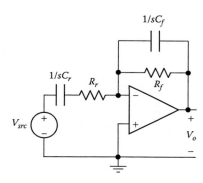

FIGURE P18.1.20

is 2 mV, the input bias current at the inverting input is 10 nA and $R_f = 1$ MΩ, what is the worst-case bias in the output?

P18.1.21 The circuit of Figure P18.1.21 is referred to as a "charge amplifier" that may be used with capacitive transducers, where v_{src}, C_{src}, and R_{src} represent the transducer voltage, capacitance, and resistance, respectively. R provides a dc path for the input bias current and should be large, compared to the reactance of C at the lowest operating frequency.

A change in either v_{SRC} or C_{src} changes the charge on C_{src}. This charge is transmitted to C, thereby causing a change in the output voltage. Ignoring R,

show that: $\Delta v_O = -\dfrac{\Delta C_{src}}{C}v_{SRC} - \dfrac{C_{src}}{C}\Delta v_{SRC}$. Note that this relation is not affected

by a shunt capacitance at the virtual ground, which means that a long shielded cable may be used between the transducer and amplifier.

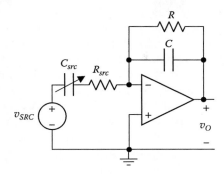

FIGURE P18.1.21

P18.1.22 The circuit of Figure P18.1.22 provides a double integration, using a single op amp. Show that: $\dfrac{V_o}{V_{src}} = -\dfrac{1}{(s\tau)^2}$, where $\tau = RC$.

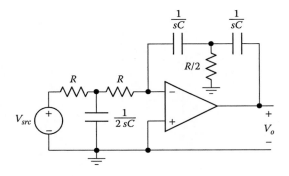

FIGURE P18.1.22

P18.1.23 The circuit of Figure P18.1.23 provides an adjustable response that could be a lag or a lead. Show that: $\dfrac{V_o}{V_{src}} = -\dfrac{R_{p2}}{R_{p1}} \dfrac{1+s\tau_1}{1+s\tau_{p1}} \dfrac{1+s\tau_{p2}}{1+s\tau_2}$, where $\tau_1 = R_1 C_1$, $\tau_2 = R_2 C_2$, $\tau_{p1} = C_1[R_1 + R_{p1}\alpha_1(1-\alpha_1)]$, and $\tau_{p2} = C_2[R_2 + R_{p2}\alpha_2(1-\alpha_2)]$.

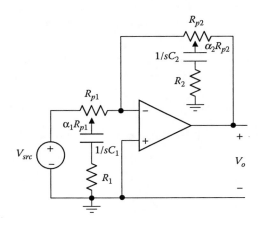

FIGURE P18.1.23

P18.1.24 This problem and the next deal with op amp AC amplifiers that use a single dc voltage supply $+V_{CC}$. The noninverting input is dc biased by means of resistors to $+V_{CC}/2$. The feedback to the inverting input ensures that in the absence of an applied signal, the inverting input and the output are also at a dc voltage of $+V_{CC}/2$. dc blocking capacitors of negligible reactance at the lowest operating frequency are connected in series with the input and output for dc isolation of the amplifier. A dc path must always be provided between the inputs of the op amp and common reference for the input bias currents.

Figure P18.1.24 illustrates an inverting ac amplifier circuit that is supplied from a single dc bias supply. Neglecting the reactances of C_r and C_o, show that the voltage gain is $-R_f/R_r$, and the inverting input is a virtual ground as far as the ac signal is concerned. If $R_L = 2$ kΩ, $R_r = 20$ kΩ, and the lowest operating frequency is 1 kHz, determine appropriate values for C_r and C_o such that the reactance at this frequency is one tenth of the associated resistance.

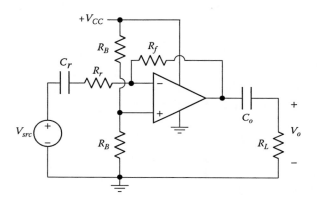

FIGURE P18.1.24

P18.1.25 Figure P18.1.25 illustrates a noninverting ac amplifier circuit that is supplied from a single dc bias supply. The bias to the noninverting input is applied through R_A so as to increase the input resistance. In order that the bias to the inverting input be provided through feedback from the output, it is necessary to include C_r and C_A to isolate this input from all dc voltages other than the op amp output.

Assuming that R_A and R_B are much larger than R_r, and that the reactances of all capacitors are negligible compared with the associated resistances, show that at the signal frequency $V_o = \dfrac{1}{\beta}\dfrac{1}{1+1/\beta A_v}V_{src}$, where $\beta = \dfrac{R_r}{R_r+R_f}$. Show that the input resistance seen by the source is $R_A(1 + \beta A_v)$.

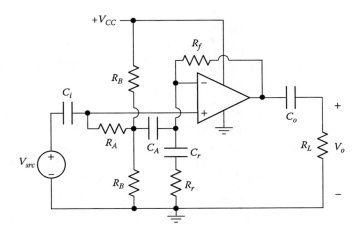

FIGURE P18.1.25

P18.2 Nonideal Op Amp Circuits

P18.2.1 Consider the inverting configuration using an op amp that is ideal except for finite A_v and a finite differential input resistance R_{id}. Show that $\dfrac{v_O}{v_{SRC}} =$

$-\dfrac{R_f}{R_r}\dfrac{1}{1+1/\beta'A_v}$, where $\dfrac{1}{\beta'}=1+\dfrac{R_f}{R_r\parallel R_{id}}$.

P18.2.2 Consider the inverting configuration connected to a load R_L and using an op amp that is ideal except for finite A_v and a finite output resistance R_o. Show

that: $\dfrac{v_O}{v_{SRC}}=-\dfrac{R_f}{R_r}\dfrac{1}{1+1/\beta A_v'}$, where $A_v'=-\dfrac{v_O}{\varepsilon_n}$ is the modified ratio of the out-

put voltage to the differential input and is given by: $A_v' = \dfrac{R_f \parallel R_L}{R_o+R_f \parallel R_L}A_v -$

$\cdot\dfrac{R_o \parallel R_L}{R_f+R_o \parallel R_L}\cong\dfrac{R_L}{R_o+R_L}A_v$. (Note: The solution to this problem illustrates an instructive application of the substitution theorem.)

P18.2.3 An op amp having $A_v(f)=\dfrac{10^4}{1+jf/10}$ and $R_o=200\ \Omega$ is used in an inverting configuration at 10 kHz, with $R_r = R_f = 10\ \mathrm{k\Omega}$ and $R_L = 1\ \mathrm{k\Omega}$. Using the results of P18.2.2, determine v_O/v_{SRC}. Verify with PSpice. Simulate the op amp using a VCVS with added resistor and capacitor.

P18.2.4 Given a unity-gain amplifier connected to a source of resistance $R_{src} = 10$ kΩ and a load resistance of $R_L = 1$ kΩ, the op amp has $A_v = 10^5$, $R_{id} = 200$ kΩ, and $R_o = 200$ Ω. Show that the voltage gain is $A =$

$$\frac{R_L(R_o + A_v R_{id})}{R_L(R_o + A_v R_{id}) + (R_L + R_o)(R_{id} + R_{src})}$$. Assuming the numerical values given,

determine $S_{A_v}^A$.

P18.2.5 When the inputs to a differential amplifier are $(\sin t + \cos t)$ mV and $(\sin t - \cos t)$ mV, the output is $(20{,}000\cos t - \sin t)$ mV. Determine the CMRR.

P18.2.6 An op amp having $A_v = 10^5$ and CMRR = 1000 is used in a noninverting configuration with $R_r = R_f = 10$ kΩ and $V_{SRC} = 1$ V. Determine the percentage error due to finite A_v and that due to the common-mode response.

P18.2.7 An op amp has a differential gain of 10^5 and a CMR of 80 dB. If $v_P = 10$ μV, $v_N = -10$ μV, and $v_{cm} = 0.1$ V, determine A_{cm} and the output voltage. Simulate with PSpice.

P18.2.8 The SR of an amplifier is 2 V/μs. What is the maximum frequency for an undistorted sinusoidal output voltage of (a) 1 V peak, and (b) 5 V peak.

P18.2.9 An op amp having an SR of 10 V/μs is to be used in a noninverting configuration of voltage gain 2 and a pulse train input of 0 to 1V pulses. Assuming equal SR limitation on the leading and trailing edges of a pulse, what is the minimum pulse width for which a fully amplified pulse output is obtained?

P18.2.10 This problem illustrates noise calculations. Consider an op amp in the inverting configuration having $R_r = 1$ kΩ and $R_f = 10$ kΩ, so that the input resistance is R_r and the gain is 10. The total $(\text{rms})^2$ noise voltage per unit bandwidth is given by

$$v_n^2 = 4kTR_r + v_{ni}^2 + (R_r i_{ni})^2$$

where k is Boltzmann's constant and T is the absolute temperature. The first term on the RHS is due to the input resistance R_r and the second and third terms are due, respectively, to the noise voltage and current of the op amp. Assuming $4kT = 0.3 \times 10^{-20}$ VA/Hz, $v_{ni} = 20$ nV/$\sqrt{\text{Hz}}$, $i_{ni} = 10$ pA/$\sqrt{\text{Hz}}$, and that the bandwidth of the amplifier is 1 kHz, determine: (a) the input noise

power given by: $P_n = \dfrac{v_n^2}{R_r} \times (\text{bandwidth})$; (b) the rms noise voltage at the input,

$\sqrt{P_n R_r}$; and (c) the rms noise voltage at the output.

P18.2.11 Consider the circuit of Figure P18.2.11 with the effects of input offset voltage and bias currents included. Determine v_O when: (a) S_1 and S_2 are closed; (b) S_1 is open and S_2 is closed; (c) S_1 is closed and S_2 is open; and (d) S_1 and S_2 are open.

P18.2.12 The circuit of Figure P18.2.12 may be used to compensate for the effects of input offset voltage and bias currents in the inverting configuration. The two resistors of resistance R_a are used in order to have a good sensitivity of zero adjustment of output voltage by means of the potentiometer. Normally, R_3 is much larger,

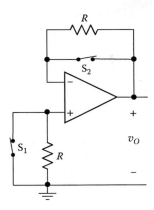

FIGURE P18.2.11

say 20 times, than R_2 and is also large compared to $(R_a/2 + R_p/4)$. Under these conditions, show that suitable guidelines for choosing the resistances in the compensating circuit are: (1) $R_1 + R_2 = R_r \parallel R_f$ and (2) $\dfrac{|V_p|}{R_3} R_2$ is slightly larger than $|V_{io}| + (R_r \parallel R_f)I_{ib}$, where $|V_p| \cong \dfrac{|V_{CC}| R_a}{R_a + R_p/2}$ is the largest magnitude of the voltage at the slider of the potentiometer with respect to ground. Assuming $R_r = 10$ kΩ, $R_f = 100$ kΩ, the maximum input offset voltage to be 5 mV, the maximum input bias current to be 1 μA, and the voltage supplies to be ±10 V, determine suitable values for the compensating circuit.

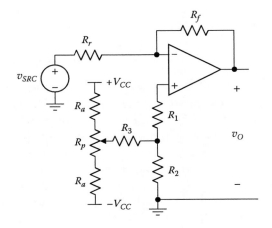

FIGURE P18.2.12

P18.2.13 Consider an inverting amplifier having $R_r = 0$ and $A_v(f)$ given by Equation 18.2.2. Show that for $f \ll f_{c0}$ and A_{v0} is large, Z_{in} at the inverting input approximately equals $\dfrac{R_f}{A_{v0}} + j\dfrac{f}{f_{c0}}R_f$.

P18.2.14 Show that if in Figure P18.1.10, $k = 1$, v_2 and R_2 are removed and the load is a resistor R_3, Z_{in} between the inverting input and the common reference is approximately $\dfrac{R_1}{A_{v0}} - \dfrac{R_1 R_3}{R_2 + R_3} + jf\dfrac{R_1}{f_{c0}}$. The circuit may be used to simulate an inductor of very high Q because the sum of the two real terms can be made arbitrarily small.

P18.3 Active Filters

The solutions of P18.3.1 to P18.3.3 are considerably facilitated by applying the substitution theorem to the output of the op amp. Verify all filter designs by simulation with PSpice.

P18.3.1 Verify Equation 18.4.1 by evaluating V_o/V_{src} in Figure 18.4.1.

P18.3.2 Verify Equation 18.4.3 by evaluating V_o/V_{src} in Figure 18.4.2.

P18.3.3 Verify Equation 18.4.5 by evaluating V_o/V_{src} in Figure 18.4.4.

P18.3.4 Verify Equation 18.4.6 by evaluating V_o/V_{src} in Figure 18.4.5.

P18.3.5 Derive the inverting lowpass filter as mentioned in connection with Equation 18.4.8 and verify its transfer function.

P18.3.6 The circuit of Problem P18.1.20 is used as a bandpass filter with $R_r = R_f = 10\ \text{k}\Omega$ and $C_r = C_f = 0.01\ \mu\text{F}$. Determine the center frequency, the Q, the gain, and the input impedance at the center frequency. Simulate with PSpice.

P18.3.7 Design a wideband bandpass filter, using first-order filters. The filter is to have a gain of 20 dB in the passband and 3-dB cutoff frequencies in the audio range at 100 Hz and 10 kHz. 0.01 μF is to be used and the input impedance of the filter is to be very high.

P18.3.8 Show that the slope at the 3-dB frequency is: (a) $-10n$ dB/decade for an nth-order lowpass Butterworth filter; (b) $-\dfrac{20n(2^{1/n} - 1)}{2^{1/n}}$ dB/decade for n cascaded, identical, first-order, lowpass filters. Compare the slopes for $n = 1, 2, 3, 10,$ and 50.

P18.3.9 Design a notch filter based on Figure 18.4.7 to reject the power frequency of 50 Hz with a Q of 50. Use $C = 0.1\ \mu\text{F}$. Simulate with PSpice.

P18.3.10 Design a broadband bandpass Butterworth filter having a lower 3-dB cutoff frequency of 10 kHz and an upper 3-dB cutoff frequency of 80 kHz. The passband gain should be 20 dB and both lowpass and highpass filters are to be of the third-order, the second-order circuits using single op amps in the noninverting configuration. Use 0.1 μF capacitors in the highpass filter and 10 kΩ resistors in the lowpass filter. Determine the gain at 4 kHz and at 200 kHz. Verify your design with PSpice.

P18.3.11 Repeat DP18.3.10 using second-order Butterworth filters, instead of third-order, based on single op amps in the inverting configuration. The filter is to have a midband gain of unity.

P18.3.12 A bandpass filter is required having a center frequency of 100 kHz and a Q of 20. Design a suitable filter using a single-op amp noninverting filter, as in Figure 18.4.4, having 1 kΩ resistors and capacitors of equal value C. Determine C, A, and the maximum gain.

P18.3.13 Implement the filter of P18.3.12 using a single-op amp inverting filter, as in Figure 18.4.5, having 1 nF capacitors and $R_2 = R_3 = R$. Determine R, R_1, and the maximum gain.

P18.3.14 Implement the filter of Problem P18.3.12 using a KHN biquad having 1 nF capacitors and $R_1 = R_3 = 10$ kΩ. Determine R, R_2, and the maximum gain.

P18.3.15 Implement the filter of P18.3.12 using a Tow-Thomas biquad having 1 nF capacitors and $R_2 = 10$ kΩ. Determine R, R_3, and the maximum gain.

P18.3.16 Show that if the Tow-Thomas biquad filter is modified as shown in Figure P18.3.16, a universal filter results, having the transfer function:

$$\frac{V_o}{V_{src}} = -\frac{\left(\dfrac{C_1}{C}\right)s^2 + \dfrac{1}{C}\left(\dfrac{1}{R_1} - \dfrac{R_a}{RR_3}\right)s + \dfrac{1}{C^2RR_2}}{s^2 + \dfrac{1}{QCR}s + \dfrac{1}{C^2R^2}}$$

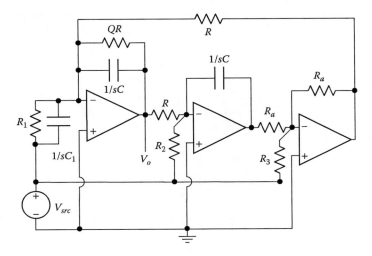

FIGURE P18.3.16

P18.3.17 Implement the transfer function $\dfrac{V_o}{V_i} = \dfrac{10^4 + 10s}{s^2 + 5s + 10^4}$ using a universal filter of the KHN type (Exercise 18.4.2), with 0.1 μF capacitors and 10 kΩ resistors whenever possible.

P18.3.18 Implement the transfer function of P18.3.17 using the universal filter of Figure P18.3.16, with 0.1 μF capacitors and 10 kΩ resistors whenever possible.

P18.3.19 For the single-op amp low-pass noninverting filter (Equation 18.4.3), determine the sensitivities of Q to the passive components and to amplifier gain. Note that the choice of components affects the sensitivities. In particular, show that if $C_1 = C_2$ and $A = 2$, $S_A^Q = 2Q^2$.

P18.3.20 For the KHN biquad, determine the sensitivities of: (a) Q to the passive components and (b) gain to Q.

P18.3.21 Verify the following properties of the sensitivity function: (a) $S_x^{ky} = S_{kx}^y = S_x^y$, where

k is a constant; (b) $S_{1/x}^y = S_x^{1/y} = -S_x^y$; (c) $S_x^{y_1 y_2 y_3} = S_x^{y_1} + S_x^{y_2} + S_x^{y_3}$; (d) $S_x^{y^n} = nS_x^y$, where

n is a constant; (e) $S_x^{y_1/y_2} = S_x^{y_1} - S_x^{y_2}$; (f) $S_x^{y_1+y_2} = \dfrac{y_1 S_x^{y_1} + y_2 S_x^{y_2}}{y_1 + y_2}$; (g) if $y = f_1(u)$ and

$u = f_2(x)$, $S_x^y = S_u^y S_x^u$; and (h) if $y = |y| \angle\theta$, then $S_x^y = S_x^{|y|} + j\theta S_x^\theta$.

P18.4 Miscellaneous

P18.4.1 Design a circuit that solves two simultaneous equations using two difference amplifiers. Consider the equations: $a_1 x + a_2 y = A$ and $b_1 x + b_2 y = B$. Express

these as: $x = \dfrac{1}{a_1}A - \dfrac{a_2}{a_1}y$ and $y = \dfrac{1}{b_1}B - \dfrac{b_2}{b_1}x$. Represent each of these equations

by a difference amplifier in which the resistances are adjusted to correspond to the expressions involving the coefficients. Because the output of either difference amplifier represents the value of a variable, it can be used to provide one of the inputs to the other amplifier, as shown in Figure P18.4.1. For given A, B, and resistance values, the outputs represent x and y.

Choose appropriate values of resistances for solving the equations $2x + 5y = 20$ and $5x - 2y = 21$. Verify your circuit by simulating with PSpice. Note that, in some cases, it may be necessary to modify the supply voltages of the op amp to avoid amplifier saturation.

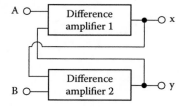

FIGURE P18.4.1

P18.4.2 Design a circuit that solves a linear, second-order differential equation, based on op amp integrators. Consider the equation:

$$\frac{d^2 x}{dt^2} + a\frac{dx}{dt} + bx = y$$

with initial conditions $x(0)$ and $x^{(1)}(0)$. Assume that $x(t)$ is available as the

output of an integrator 2. The input to this integrator is then $\dfrac{dx}{dt}$. If the given

equation is rearranged as $\dfrac{d^2 x}{dt^2} = -a\dfrac{dx}{dt} - bx + y$ and integrated, it gives:

$$\frac{dx}{dt} = \int_0^t \left(-a\frac{dx}{dt} - bx + y\right)dt + x^{(1)}(0)$$

This equation may be implemented as in Figure P18.4.2a. The input to the inverting integrator 1 should be the negation of the quantity in brackets. The integration time constant RC is assumed to be unity, and $x^{(1)}(0)$ is introduced across C as a voltage source in series with a switch that is opened at $t = 0$. Note that the polarity of the source gives the correct initial value of $x^{(1)}(0)$. Once we have $\frac{dx}{dt}$, we can integrate it to obtain:

$$x = \int_0^t \frac{dx}{dt}dt + v(0)$$

(Figure P18.4.2b). The output of the inverting integrator should be inverted to obtain x.

Design a circuit for solving the differential equation:

$$\frac{d^2v}{dt^2} + 4\frac{dv}{dt} + 2v = 10\cos 2t$$

for $t \geq 0$, with $v(0) = 1$ V and $v^{(1)}(0) = 2$ V/s.

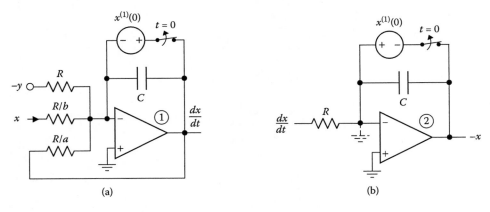

(a)   (b)

FIGURE P18.4.2

19

Electric Circuit Analogs of Nonelectrical Systems

Overview

Electric circuit analogs provide a useful description of nonelectrical systems, in that they yield additional insight into the behavior of these systems, and allow application of the powerful analytical tools and computer aids that are available for the analysis and design of electrical systems. Sophisticated representations, such as linear graphs and bond graphs, have been developed for the unified modeling of engineering systems, both electrical and nonelectrical. This chapter is not concerned with such methods but deals instead with the basic derivation and illustration of electric circuit analogs of mechanical, fluid, and thermal systems as well as the ionic system of the cell membrane.

The basis for all analogies is a common mathematical description between electrical and nonelectrical systems. Once a given nonelectrical system is described in mathematical terms, it becomes relatively straightforward to derive the electrical analog of the given system. Based on these analogies, certain general rules can be inferred that allow the derivation of the equivalent electric circuit without having to formulate the mathematical relations. This approach is applied first to mechanical systems to establish the electrical analogs of basic mechanical quantities, namely, force and displacement. These analogies can then be naturally extended to other quantities that derive from these such as velocity, mass, stiffness, etc. The electrical analog of fluid systems is essentially an adaptation of that of mechanical systems because of the natural correspondence between the variables in the two systems.

The analogy between thermal and electrical systems is well established and is commonly invoked in thermal analysis. After all, there is some correspondence between thermal and electrical properties, as good conductors of heat are generally good conductors of electricity, and conversely. In ionic systems, the flow of ions under the influence of an electric field constitutes a current, so the system is basically electrical in nature. However, ions also flow under the influence of a concentration gradient, which has to be accounted for in the equivalent electric circuit. It may be noted that current carriers in semiconductors, that is, electrons and holes, also flow under the influence of both electric fields and concentration gradients. A close analogy therefore exists between ionic and semiconductor systems.

Learning Objectives

- To be familiar with:

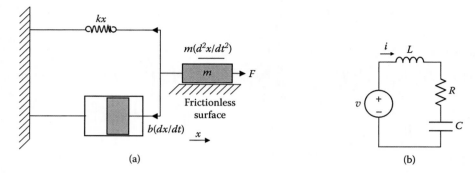

FIGURE 19.1.1
Single-mass system.

- • The general concept of representing nonelectrical systems by means of equiv-
    alent electric circuits
- • To understand:
  - • The basis for deriving the electric circuit that is equivalent to a given mechan-
      ical, fluid, thermal, or ionic system
  - • How circuit concepts and techniques of analysis can be applied to investigating
      the behavior of nonelectrical systems

19.1 Mechanical Systems

Single Mass

Consider the system shown in Figure 19.1.1a consisting of a mass m that is free to move
on a frictionless surface under the influence of a force F. This force is opposed by the
inertia of m, by a force kx due to a linear spring, and by a viscous damping force $b\dfrac{dx}{dt}$
due to a linear dashpot, where x is the displacement from a specified reference, or rest
point, k is the stiffness constant of the spring, and b is the damping constant. Neglecting
the mass of the spring, dashpot, and connections, the equation of motion of m may be
written as:

$$m\frac{d^2x}{dt^2}+b\frac{dx}{dt}+kx=F \tag{19.1.1}$$

To derive the equivalent electric circuit, consider a series RLC circuit excited by a voltage
source v (Figure 19.1.1b). From KVL:

$$L\frac{di}{dt}+Ri+\frac{1}{C}\int idt=v \tag{19.1.2}$$

TABLE 19.1.1

Analogous Mechanical and Electrical Quantities

Mechanical Quantity	Electrical Quantity
Force (N)	Voltage or emf (V)
Displacement (m)	Charge (C)
Velocity (m/s)	Current (A)
Mass or inertia (kg)	Inductance (H)
Damping constant (N/m/s)	Resistance (Ω)
Elastance (m/N) (1/stiffness)	Capacitance (F)
Momentum (kgm/s) (mass ×velocity)	Flux linkage (Vs) (inductance ×current)
Kinetic energy (J)	Magnetic energy stored in inductor (J)
$\left(\dfrac{1}{2}mv^2\right)$	$\left(\dfrac{1}{2}Li^2\right)$
Potential energy stored in spring (J)	Electric energy stored in capacitor (J)
$\left(\dfrac{1}{2}kx^2\right)$	$\left(\dfrac{1}{2}\dfrac{q^2}{C}\right)$
Force aiding motion	Voltage rise
Force opposing motion	Voltage drop
In-line, or parallel, forces convergent at a point	Voltages in series
Balance of forces at a point	KVL
Relative displacement	KCL

If we express the LHS of Equation 19.1.2 in terms of the charge $q = \int i\,dt$ rather than i, the equation becomes:

$$L\frac{d^2q}{dt^2} + R\frac{dq}{dt} + \frac{q}{C} = v \tag{19.1.3}$$

which is identical in form to Equation 19.1.1, leading to the analogy between mechanical and electrical quantities summarized in Table 19.1.1. Note that consistent SI units are used in this table for each set of electrical and mechanical quantities. Although translational mechanical quantities are shown in the table, the analogy clearly applies to the corresponding rotational mechanical quantities.

Intuitively, electromotive force (emf) or voltage, which accelerates current carriers in a conductor, is analogous to mechanical force, which accelerates a mass. Current is thus analogous to velocity, and charge to displacement. The remaining analogies readily follow from the analogy between these basic quantities.

Based on the analogies of Table 19.1.1, the equivalent electric circuit may be written by inspection. Basically, all the forces in Figure 19.1.1a are considered to act at the center of mass of m. They are, therefore, convergent at this point and sum to zero algebraically. In the equivalent electric circuit, they are represented as voltages around a loop. Force F, causing motion, is considered a voltage rise. The remaining force opposing motion are considered voltage drops.

It may be noted that an alternative analogy between electrical and mechanical systems is possible in which force is analogous to current and velocity to voltage. The electric circuit derived in this manner is the dual of that based on the analogy of Table 19.1.1.

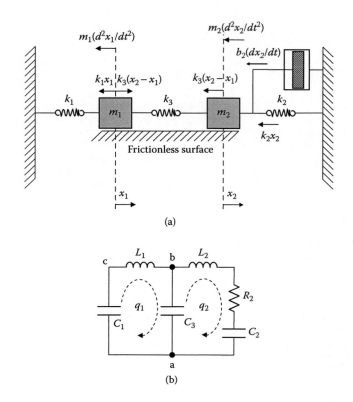

FIGURE 19.1.2
Coupled masses.

EXERCISE 19.1.1

What is the mechanical analog of an *LC* circuit that is set into oscillation because of an initial charge on the capacitor?

Answer: A mass coupled to a spring, as in Figure 19.1.1a without the dashpot, with the mass given an initial displacement.

Coupled Masses

A more complex system is illustrated in Figure 19.1.2a. As there are two displacement variables at the two ends of spring 3, the equivalent circuit has two meshes coupled by a capacitor C_3 (Figure 19.1.2b). The length of spring 3 is $x_2 - x_1$, so that the force developed by the spring is $k_3(x_2 - x_1)$. This force tends to move m_1 in the positive *x*-direction and m_2 in the negative *x*-direction.

To write the equations of motion, consider both masses moving in the positive *x*-direction, with $x_2 > x_1$. There will be a force $k_3(x_2 - x_1)$ moving m_1 in the positive *x*-direction, opposed by k_1x_1, due to spring 1, and $m_1 \dfrac{d^2x_1}{dt^2}$ due to the inertia of m_1. As in Equation 19.1.1, the equation of motion of m_1 is:

$$m_1 \frac{d^2 x_1}{dt^2} + k_1 x_1 = k_3(x_2 - x_1) \tag{19.1.4}$$

Considering m_2, all the forces shown in Figure 19.1.2a oppose the motion. Hence,

$$m_2 \frac{d^2 x_2}{dt^2} + b_2 \frac{dx_2}{dt} + k_2 x_2 + k_3(x_2 - x_1) = 0 \tag{19.1.5}$$

In terms of q_1 and q_2, the mesh equations in the circuit of Figure 19.1.2b are:

$$L_1 \frac{d^2 q_1}{dt^2} + C_1 q_1 - C_3(q_2 - q_1) = 0 \tag{19.1.6}$$

$$L_2 \frac{d^2 q_2}{dt^2} + R_2 \frac{dq_2}{dt} + C_2 q_2 + C_3(q_2 - q_1) = 0 \tag{19.1.7}$$

Equation 19.1.6 and Equation 19.1.7 are in agreement with Equation 19.1.4 and Equation 19.1.5, respectively.

Suppose that for $t < 0$, m_1 is displaced a distance X_0 to the right, while m_2 is held at its rest position. Let both masses be released at $t = 0$. When the masses are released, x_2 increases $\left(\frac{dx_2}{dt} > 0 \right)$ because of compression in spring 3. Simultaneously, x_1 decreases $\left(\frac{dx_1}{dt} < 0 \right)$ because of compression in spring 3 and tension in spring 1. The corresponding initial conditions in the equivalent electric circuit are: zero initial energy storage in C_2, L_1, and L_2, an initial charge Q_0 on C_3 (node b positive with respect to node a), and $-Q_0$ on C_1 (node c negative with respect to node a). This can be justified by noting that when spring 1 is in tension, x_1 is positive, and q_1 in Figure 19.1.2b is positive, which makes node c negative with respect to node a. Similarly, when spring 3 is in compression, x_2 and hence q_2, tend to increase, so that node b is positive with respect to node a.

Vibrations

Steady vibrations are a good example of periodic phenomena. Figure 19.1.1 may be considered as a basic vibration isolation system, where m represents a vibrating mass, and the spring and dashpot represent stiffness and damping of the support, respectively. The vibrating mass could be a reciprocating internal combustion engine, or an electric motor that is not perfectly balanced. The vibration effectively produces a force F that, under steady-state conditions, is periodic. To reduce the effect of the vibration on the surroundings, the force F_T transmitted to the surroundings through the spring and dashpot should be small. The problem in the steady state may be analyzed by considering F to be sinusoidal, that is, $F = F_m \cos \omega t$, and using phasor notation.

In the equivalent electric circuit, the force F_T corresponds to the sum of the voltages across R and C. In terms of phasors, the transmittivity ratio (TR) is given by:

$$TR = \frac{\mathbf{F_T}}{\mathbf{F}} = \frac{R + 1/j\omega C}{\omega L + R + 1/j\omega C} = \frac{1 + j\omega CR}{1 - \omega^2 LC + j\omega CR} \tag{19.1.8}$$

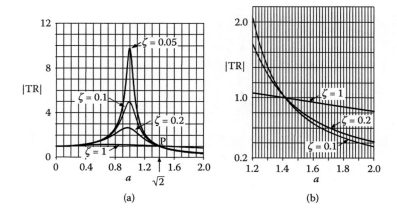

FIGURE 19.1.3
Variation of transmittivity with frequency.

It follows that:

$$|TR| = \sqrt{\frac{1+(\omega CR)^2}{(1-\omega^2 LC)^2 + (\omega CR)^2}} \tag{19.1.9}$$

Let $a = \dfrac{\omega}{\omega_n}$, where $\omega_n = \dfrac{1}{\sqrt{LC}} = \sqrt{\dfrac{k}{m}}$ is the resonant frequency of the system, and

$\zeta = \dfrac{\omega_n CR}{2} = \dfrac{\omega_n b}{2k}$ is the damping parameter defined by Equation 12.2.7 (Chapter 12).
Equation 19.1.9 may be expressed as

$$|TR| = \sqrt{\frac{1+(2\zeta a)^2}{(1-a^2)^2 + (2\zeta a)^2}} \tag{19.1.10}$$

$|TR|$ is plotted in Figure 19.1.3 as a function of a for various ζ. It is seen that $|TR| = 1$ for $a = \sqrt{2}$, irrespective of ζ. In order to reduce the vibrations transmitted to the surroundings, $|TR|$ should be small, which means that a should be as large as practicable. a is of course smallest for the frequency of the fundamental of the periodic function F. In the case of vibration associated with rotary or reciprocating motion, the frequency of the fundamental corresponds to the speed of rotation or the periodicity of the reciprocating motion, respectively. For a given frequency of fundamental and a given m, k should be chosen to make $a \geq 3$. It will be noted that for $a > \sqrt{2}$, $|TR|$ is smaller, the smaller is the damping. For example, if $a = 4$,

$$|TR| = \sqrt{\frac{1+64\zeta^2}{225+64\zeta^2}} \tag{19.1.11}$$

TABLE 19.2.1

Analogous Fluid and Electrical Quantities

Fluid Quantity	Electrical Quantity
Pressure difference (N/m², or pascal)	Voltage or emf (V)
Volume (m³)	Charge (C)
Flow rate (m³/s)	Current (A)
Inertance (Ns²/m⁵)	Inductance (H)
Fluid resistance (Ns/m⁵)	Resistance (Ω)
Compliance or fluid capacitance (m⁵/N)	Capacitance (F)
Conservation of flow (incompressible liquid)	KCL

$|TR| = \dfrac{1}{a^2-1} = 0.067$ if $\xi=0$, $|TR| = 0.072$ if $\xi=0.05$, and $|TR| = 0.085$ if $\xi=0.1$.

In practice, some damping is needed because during start-up, as opposed to steady-state conditions, the frequency is low, and the system generally passes through resonance ($a = 1$).

Damping reduces $|TR|$ at resonance. Thus, although $|TR|$ is reduced by nearly 21% if ξ is reduced from 0.1 to 0.015, with $a = 4$, the resonance peak is multiplied by about 6.5.

The preceding conclusions readily follow from consideration of the equivalent electric circuit. For a given L, the larger C, the smaller is ω_n, and the smaller is the voltage drop across C and R as a fraction of the applied voltage. This fraction is not sensitive to the value of R, as long as $\omega CR \ll 1$.

EXERCISE 19.1.2

What is the electrical equivalent of an ideal reduction gear of ratio n, that is, one that reduces a rotational speed N at its input to N/n at its output?

Answer: An ideal step-up transformer of turns ratio n, as it reduces a current i at the input to i/n at the output.

19.2 Fluid Systems

The analogy between fluid and electrical systems (Table 19.2.1) is basically an adaptation of that between mechanical and electrical systems (Table 19.1.1). Pressure difference is analogous to voltage, volume to charge, and flow rate to current. An incremental compliance, or fluid capacitance, that is analogous to electric capacitance is generally defined as:

$$C_F = \frac{d\Lambda}{d(\Delta p)} \tag{19.2.1}$$

where Λ is the volume and Δp is the pressure difference. Analogous to $v = L\dfrac{di}{dt}$, we can write:

$$\Delta p = L_F \frac{d\Phi}{dt} \tag{19.2.2}$$

where Φ is the flow rate. L_F in Equation 19.2.2 defines, in general, an incremental **inertance**, analogous to inductance. It can be readily shown that in the case of a fluid-filled tube of uniform cross-section, the inertance of the fluid is given by:

$$L_F = \frac{\rho l}{A} \tag{19.2.3}$$

where ρ is the density of the fluid, l is the length of the tube, and A is its cross-sectional area (Exercise 19.2.2).

Power dissipation is associated with fluid flow in conduits or openings because of internal friction between elements of the fluid moving at different velocities. Analogous to electric resistance, an incremental fluid resistance is generally defined as:

$$R_F = \frac{d(\Delta p)}{d\Phi} \tag{19.2.4}$$

For laminar flow in a long, circular pipe of length l and diameter d, the fluid resistance is, from Poiseuille's law:

$$R = \frac{128 \eta l}{\pi d^4} \tag{19.2.5}$$

where η is the fluid viscosity. C, L, and R are constant only if the quantities defining them are linearly related. They can also be considered constant for small variations in these quantities.

EXERCISE 19.2.1

Given a tank of cross-sectional area A that is filled with a liquid of density ρ to a height h, from the relation between the pressure at the bottom of the tank and the volume of liquid in the tank, determine the fluid capacitance of the system.

Answer: $\dfrac{A}{\rho g}$ (g is acceleration due to gravity. Note that the fluid capacitance of the system is not equal to the volume of liquid; see Problem P19.2.4.)

EXERCISE 19.2.2

Derive Equation 19.2.3 by considering the equation of motion ($F = ma$) of a column of liquid of length l, area A, and density ρ subjected to a pressure difference Δp.

Liquid-Filled Tube

As an example of a fluid system, consider a liquid-filled flexible tube that is open at one end and connected at the other end to a pressure sensor that consists of a cylindrical

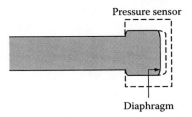

FIGURE 19.2.1
Liquid-filled catheter.

chamber whose closed end is a flexible diaphragm (Figure 19.2.1). When periodic pressure pulsations are applied to the open end, the pulsations are transmitted through the liquid column in the tube to the pressure sensor. The pressure pulsations applied to the pressure sensor are affected by the dynamics of the tube and sensor chamber. It is desired to represent these dynamics by means of an equivalent electric circuit so as to determine how the pressure at the sensor is related to that applied to the open end.

Consider first the chamber of the pressure sensor. The applied pressure causes the liquid in the chamber to move and deflect the sensor's diaphragm. The movement is opposed by three forces due to: (1) the inertia the liquid in the chamber, (2) frictional forces, and (3) the compliance of the diaphragm. It follows from the discussion of the single-mass system of Figure 19.1.1 that the equivalent electric circuit of the chamber consists of L_{FS}, R_{FS}, and C_{DS} (Figure 19.2.2a), where L_{FS} and R_{FS} are given by Equation 19.2.3 and Equation

19.2.5, respectively, and $C_{DS} = \dfrac{d\Lambda_{\text{chamber}}}{dp_{\text{chamber}}}$.

In a similar manner, a length Δx of the tube may be represented by an inductance L_{FX} in series with a resistance R_{FX} and a shunt capacitance C_{CX}, connected as in Figure 19.2.2b, where C_{CX} represents the compliance of a length Δx of the tube wall. Note that L_{FX} and R_{FX} are in series because the pressure differences due to inertia and friction of the fluid in the tube add together along the length of the tube; C_{CX} is in shunt because the pressure difference acting on the tube wall is between the inside of the tube and the outside. The outside pressure is represented in the equivalent electric circuit by the voltage of the common reference, which may be assumed to be zero. Strictly speaking, the three elements L_{FX}, R_{FX}, and C_{CX} are distributed along the length of the tube, so that the equivalent electric circuit is that of a transmission line. However, if the tube is relatively short, the compliance of its wall may be neglected compared to that of the diaphragm, in which case the tube by be represented by two lumped elements L_{FC} and R_{FC} in series. Here, L_{FC} is the inertance of the fluid in the tube, and is given by Equation 19.2.3, whereas R_{FC} is the resistance of the fluid in the tube, and is given by Equation 19.2.5. Note that $L_{FC} = L_{Fx} L / \Delta x$ and $R_{FC} = R_{Fx} L / \Delta x$. Moreover, L_{FS}

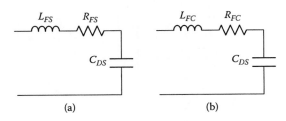

(a)   (b)

FIGURE 19.2.2
Equivalent circuits of sensor chamber (a) and catheter (b).

and R_{FS} of the chamber, which appear in series with L_{FC} and R_{FC}, may be neglected in comparison, leading to the equivalent electric circuit of Figure 19.2.2b.

As the pressure pulsations are assumed periodic, we may consider the pressure response to be a sinusoidally varying input and apply phasor analysis. Noting that the pressure response corresponds to the ratio of the voltage across the capacitor to the applied voltage:

$$\frac{\mathbf{P_S}}{\mathbf{P}} = \frac{1/j\omega C_{DS}}{\omega L_{FC} + R_{FC} + 1/j\omega C_{DS}} = \frac{1}{1 - \omega^2 L_{FC} C_{DS} + j\omega C_{DS} R_{FC}} \tag{19.2.6}$$

It follows that

$$\left|\frac{\mathbf{P_S}}{\mathbf{P}}\right| = \frac{1}{\sqrt{(1 - \omega^2 L_{FC} C_{DS})^2 + (\omega C_{DS} R_{FC})^2}} \tag{19.2.7}$$

Let $a = \dfrac{\omega}{\omega_n}$, where $\omega_n = \dfrac{1}{\sqrt{L_{FC} C_{DS}}} = \dfrac{d}{2}\sqrt{\dfrac{\pi}{\rho \, l C_{DS}}}$ is the resonant frequency of the system

and $\zeta = \dfrac{\omega_n C_{DS} R_{FC}}{2} = \dfrac{32\eta}{d^3}\sqrt{\dfrac{l C_{DS}}{\pi\rho}}$ is the damping ratio. Then Equation 19.2.7 may be

expressed as

$$\left|\frac{\mathbf{P_S}}{\mathbf{P}}\right| = \frac{1}{\sqrt{(1 - a^2)^2 + (2\zeta a)^2}} \tag{19.2.8}$$

Equation 19.2.8 is plotted in Figure 19.2.3 for various values of ζ. Resonance occurs at $a = 1$, the resonance peak increasing with decreasing ζ. If the pressure at the sensor is to be a faithful replica of the applied pressure, the frequency of the highest-order, significant harmonic should be well below the resonant frequency. This is in contrast to the vibration support considered in Section 19.1.1, where the operating frequency should be well above resonance.

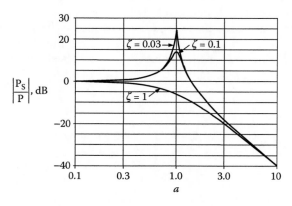

FIGURE 19.2.3
Frequency response of catheter and chamber.

What is the effect of an air bubble in the sensor chamber? Such a bubble has an appreciable compliance because air is compressible. The compliance of the air bubble adds to C_{DS}, thereby reducing ω_n and increasing ζ (Exercise 19.2.3) Unless the applied frequency is very much smaller than the resonant frequency, the air bubble degrades performance.

EXERCISE 19.2.3

Given a 1 m long tube of 1 mm internal diameter filled with saline $(\rho = 1 \text{ kg/l})$, determine f_n and ζ, assuming the compliance of the diaphragm to be $2 \times 10^{-15} \text{ m}^5/\text{N}$ and the viscosity of water to be 0.0007 Pas at 37°C. Calculate the new values of f_n and ζ when the sensor chamber contains an air bubble of volume $5 \times 10^{-3} \text{ cm}^3$, assuming that $\Delta\Lambda/\Delta p$ for air is $1 \text{ cm}^3/1/ 100$ Pa.

Answer: 99.7 Hz, 0.018; 19.6 Hz, 0.091.

19.3 Thermal Systems

Consider a body of mass m that is initially at thermal equilibrium with its surroundings. At $t = 0$, heat is transferred to the body at a constant rate I_{HI} W. As the body stores heat, its temperature rises, and heat is dissipated to the surroundings. It is desired to determine the temperature rise ΔT of the body for $t \geq 0$.

While the temperature of the body is rising,

$$I_{HI} = I_{HD} + I_{HS} \tag{19.3.1}$$

where I_{HD} is the rate at which heat is dissipated to the surroundings, and I_{HS} is the rate at which heat is stored in the body. Assuming that I_{HD} is directly proportional to ΔT:

$$I_{HD} = \frac{\Delta T}{\Theta} \tag{19.3.2}$$

where $1/\Theta$ is the constant of proportionality.

The quantity of heat H_S that must be stored in the body to raise its temperature ΔT degrees is

$$H_S = c \times m \times \Delta T \tag{19.3.3}$$

where c is the specific heat of the body, defined as the heat that must be transferred to a unit mass of a given substance to raise its temperature by 1K.

If heat is stored in the body at a rate $\dfrac{dH_S}{dt}$, the temperature of the body rises at a rate $\dfrac{d(\Delta T)}{dt}$, where, by differentiating both sides of Equation 19.3.3:

$$I_{HS} = \frac{dH_S}{dt} = cm\frac{d(\Delta T)}{dt} \qquad (19.3.4)$$

Substituting from Equation 19.3.2 and Equation 19.3.4 in Equation 19.3.1,

$$cm\frac{d(\Delta T)}{dt} + \frac{\Delta T}{\Theta} = I_{HI} \qquad (19.3.5)$$

Equation 19.3.5 is exactly analogous to

$$C\frac{dv}{dt} + \frac{v}{R} = i_I \qquad (19.3.6)$$

which is the equation for the voltage v across a parallel RC circuit excited by a current source i_I. The thermal equivalent circuit is shown in Figure 19.3.1, where $R_T = \Theta$ and the thermal capacitance C_T is

$$C_T = cm \qquad (19.3.7)$$

The specific heat of most materials is nearly constant at or above room temperature.

Comparison between Equation 19.3.5 and Equation 19.3.6 leads to the analogy between thermal and electrical systems summarized in Table 19.3.1.

Heat flows down a temperature gradient, just as current flows down an electric potential gradient. Temperature difference is thus analogous to voltage, heat flow rate to current, and the quantity of heat to charge. Table 19.3.1 does not include a thermal quantity that corresponds to inductance. This is because no significant physical phenomenon that corresponds to thermal inductance has been identified in simple thermal systems of interest in engineering.

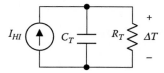

FIGURE 19.3.1
Thermal equivalent circuit.

TABLE 19.3.1

Analogous Thermal and Electrical Quantities

Thermal Quantity	Electrical Quantity
Temperature difference (K)	Voltage or emf (V)
Quantity of heat (J)	Charge (C)
Heat flow rate (W)	Current (A)
Thermal resistance (K/W)	Resistance (Ω)
Thermal capacitance (J/K)	Capacitance (F)

Thermal resistance Θ is defined as

$$\Theta = \frac{d(\Delta T)}{dI_H} \tag{19.3.8}$$

If the temperature difference ΔT is directly proportional to I_H, as assumed earlier, Θ is constant. In actual fact, Θ is not, in general, constant. Heat may be transferred by conduction, convection, and radiation. Conduction is the dominant mechanism of heat transfer in solids and fluids at moderate temperatures. The heat flow rate by conduction due to a temperature difference ΔT is:

$$I_{Hcond} = k_{cond} \Delta T \tag{19.3.9}$$

where $k_{cond} = 1/\Theta_{cond}$ is the thermal conductance due to conduction. For a homogeneous, uniform body of length l and area A:

$$k_{cond} = \frac{\sigma_T A}{l} \tag{19.3.10}$$

where σ_T is the thermal conductivity. Equation 19.3.10 is analogous to Equation 2.4.14 (Chapter 2) in the electrical case. For most substances, there is a high correlation between thermal and electrical conductivities. Thermal conductivities of gases generally increase with temperature, whereas thermal conductivities of solids and liquids may increase, decrease, or stay substantially constant with temperature.

In moving fluids, heat may be transferred by convection, through bulk movement of parts of the fluid. In a heat sink, for example, air adjacent to the heat sink is heated, becomes less dense, and rises, carrying heat with it. It is replaced by colder air that is heated in turn. That is why air should be free to move upward in the vicinity of a heat sink. Heat transfer by convection may be increased by forcing air, or a cooling fluid such as water or oil, past a heated surface.

Heat flow rate by convection may be expressed as

$$I_{Hconv} = k_{conv} A \Delta T \tag{19.3.11}$$

where A is the contact area between a heated surface and the moving fluid, and k_{conv} is the convection heat transfer coefficient (W/m^{2}K). The corresponding thermal resistance is

$$\Theta_{conv} = \frac{1}{k_{conv} A} \tag{19.3.12}$$

k_{conv} depends strongly on the cooling fluid, its velocity, and flow geometry,

Heat may be also transferred from one body to another by radiation, in accordance with the Stefan-Boltzmann law,

$$I_{Hrad} = k_{rad} \left(T_1^4 - T_2^4 \right) \tag{19.3.13}$$

where T_1 is the temperature of the hotter body, T_2 is the temperature of the colder body, and k_{rad} is the radiation heat transfer constant (W/K^4). k_{rad} depends on the geometry and properties of the radiating and absorbing surfaces. For a given material and geometry, k_{rad} is maximized by painting the radiating surface black. According to Equation 19.3.13, heat transfer by radiation is highly nonlinear and cannot be properly defined in terms of a constant thermal resistance. Where heat is transferred by more than one mechanism, the corresponding heat flow rates add together.

As a simple application, consider a 10 Ω resistor that is connected to a 10 V dc source. The power dissipated in the resistor as heat is 10 W. The steady-state temperature of the resistor is such that this rate of heat generation equals the rate of heat dissipation to the surroundings through conduction, convection, and radiation; that is, $I_{HI} = I_{HD}$ in Equation 19.3.1. The larger k_{cond}, k_{conv}, and k_{rad}, the smaller is the temperature rise for a given I_{HD}. The thermal capacitance of the resistor plays no role in determining the steady-state temperature rise because in the steady state,

$$\frac{dT}{dt} = 0,$$

so $I_{HS} = 0$ (Equation 19.3.4). In the circuit of Figure 19.3.1, I_{HI} is a dc source of 10 W, whereas I_{HD} and I_{HS} are the currents in R_T and C_T, respectively.

Consider next the opposite extreme case, where the voltage is a surge of very short duration, say 10 µs, so that no significant amount of heat is transferred to the surroundings during this time. The temperature rise of the resistor under these conditions is entirely determined by its thermal capacitance. If the amplitude of the voltage rise is V, the thermal capacitance of the resistor is C_T, and the maximum allowable temperature rise is 100°C, then the quantity of heat generated by the voltage pulse is

$$\frac{V^2}{10}10^{-5} = 10^{-6}V^2 \text{ J}$$

where V is in volts. From Equation 19.3.3 and Equation 19.3.7, the temperature rise is related to the quantity of heat stored by the relation:

$$\Delta T = \frac{10^{-6}V^2(\text{J})}{C_T \text{ (J/K)}}$$

If $C_T = 0.16$ J/°C and $\Delta T = 100$°C, then $V = 4$ kV. In the circuit of Figure 19.3.1, I_{HI} is a heat impulse that flows through C_T only.

Heat Sink

Heat sinks are commonly used with semiconductor devices in order to limit the maximum steady-state temperature of a device that dissipates substantial power. For Si devices, this maximum temperature is in the range of 150 to 200°C, beyond which the semiconductor device is irreversibly damaged.

A typical thermal equivalent circuit of a power bipolar junction transistor is shown in Figure 19.3.2. The current source P_D represents the power dissipated in the transistor. Four nodes are identified as the junction (J), the case (C), the heat sink (S), and the ambient

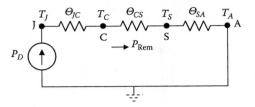

FIGURE 19.3.2
Thermal resistances associated with a heat sink.

environment (A). A thermal resistance Θ connects adjacent nodes. Under steady-state conditions, heat is generated at a rate of P_D and removed at an equal rate of P_{Rem}. It follows that

$$P_D = P_{Rem} = \frac{T_J - T_A}{\Theta_{JA}} = \frac{T_J - T_A}{\Theta_{JC} + \Theta_{CS} + \Theta_{SA}} \qquad (19.3.14)$$

For a given $T_J - T_A$, the lower the total thermal resistance, the higher is the allowed P_D, and conversely. If the transistor is mounted in free air, Θ_{JC} is unchanged because it depends on the transistor, but $\Theta_{CA} = \Theta_{CS} + \Theta_{SA}$, without a heat sink, is generally much larger than $\Theta_{CS} + \Theta_{SA}$ with an efficient heat sink. The allowed P_D is therefore much lower.

EXERCISE 19.3.1

A transistor has $\Theta_{JC} = 2.5°C/W$, $\Theta_{CS} = 0.4°C/W$, and $\Theta_{SA} = 3.6°C/W$. (1) If the transistor dissipates 25 W at $T_A = 25°C$, what is T_J? (2) When mounted without a heat sink, $\Theta_{CA} = 27.5°C/W$. What is P_D at $T_A = 25°C$, if $T_J = 150°C$?

Answer: (1) 187.5°C; (2) 4.17 W.

19.4 Cell Membrane

Structure

The basic structure of the membrane that encloses living cells is essentially that of a bimolecular layer of phospholipid molecules that is about 50 to 80 Å thick (Figure 19.4.1). Protein molecules are found on either side of the membrane or extending across it. These proteins serve many different functions, one of which is to form ionic channels, or pores, through which ions can pass.

Ionic Distribution and Flow

Although many ionic species are present in the intracellular and extracellular media, the ionic species present in the largest concentrations are Na^+, K^+, and Cl^-. In almost all cases, these ions are distributed as illustrated in Figure 19.4.2. Na^+ and Cl^- are more concentrated in the extracellular medium, whereas K^+ is more concentrated in the intracellular medium. This medium also contains indiffusible anions $[A^-]$ in the form of proteins, amino acids, and other charged molecules. Many types of uncharged molecules are also present.

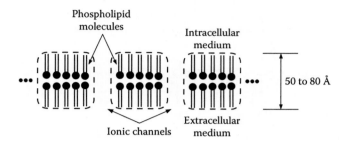

FIGURE 19.4.1
Basic structure of a cell membrane.

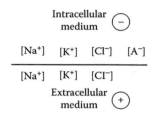

FIGURE 19.4.2
Ionic distribution across the cell membrane.

In almost all cases, the inside of the cell is at a voltage of −10 to −100 mV with respect to the outside. This resting membrane potential, as it is called, depends on the type of cell.

> **Concept:** *Freely moving ions in solution flow from regions of high potential energy to regions of low potential energy.*

Freely moving ions in solutions are subject to two types of driving force:

1. Voltage gradients, which drive positively charged ions (cations) from positively charged regions to negatively charged regions, and which drive negatively charged ions (anions) in the opposite direction. In both cases *ions are driven down an electric potential energy gradient.*

2. Concentration gradients, which drive ions from regions of higher concentration to regions of lower concentration. In accordance with the preceding concept, a type of potential energy, known as **chemical potential energy**, is associated with concentration.

If we consider these gradients for Na⁺, K⁺, and Cl, we find that in the case of K⁺ and Cl, the two types of gradient are in opposition, though this does not mean they are equal, whereas in the case of Na⁺ they both drive these ions inward (Figure 19.4.3). Evidently, under steady-state conditions, ionic concentrations do not vary with time. If Na⁺ flows inward under the influence of both the voltage and concentration gradients, then there must be some mechanism that drives them outward at the same rate; otherwise, the intracellular concentration of Na⁺ will change with time. This mechanism is Na⁺ ionic "pump" that actively pumps Na⁺ outward. The flow of Na⁺ due to this pumping action is *active*, in that energy must be expended in overcoming the two potential energy gradients. In contrast, the flow under the influence of these gradients, depicted in Figure 19.4.3, is *passive*. It turns out that this same Na⁺ pump also pumps K⁺ in the opposite direction

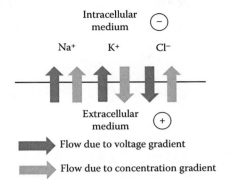

FIGURE 19.4.3
Ionic flow across the cell membrane.

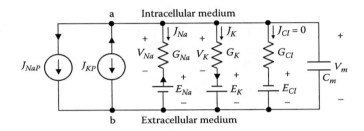

FIGURE 19.4.4
Equivalent electric circuit.

but at roughly 2/3 the rate for Na⁺. It is usually assumed that Cl⁻ is not actively pumped, which means that its rate of flow under the influence of the voltage gradient is equal and opposite to that under the influence of the concentration gradient.

Equivalent Circuit

The electrical state of the cell membrane can be represented as in Figure 19.4.4. Each of the three ionic species is represented by a branch consisting of a conductance and a battery. The assigned positive direction of voltage is as shown, so that V_m, the voltage across the membrane, has a negative numerical value. Because of the assigned positive direction of V_m, the assigned positive direction of current is outward, as indicated for the labeled currents in the figure. The arrow in a given branch indicates the actual direction of current in that branch. The current sources J_{NaP} and J_{KP} represent, respectively, the current per unit area due to the ionic pump that pumps Na⁺ outward and K⁺ inward. The conductances per unit area are due to the ionic channels through which ions pass and represent current density of the given ion per unit voltage across the membrane. These conductances are generally nonlinear. The membrane has a capacitance C_m per unit area because it is essentially an insulating structure between two aqueous electrolytic solutions. It can, therefore, separate electric charges and hence store electric energy. Because of the thinness of the membrane, C_m has a relatively large value of about 1 μF/cm².

To appreciate the significance of the batteries in each ionic branch, let us consider the Na⁺ branch first. KVL for this branch is: $V_{Na} + E_{Na} - V_m = 0$. This equation may be written as:

$$E_{Na} - V_m = -V_{Na} = -\frac{J_{Na}}{G_{Na}} \qquad (19.4.1)$$

where each term in Equation 19.4.1, including the sign, has a positive numerical value.

As J_{Na} has a negative numerical value, $-\dfrac{J_{Na}}{G_{Na}}$ represents a positive voltage *drop* across G_{Na} due to $-J_{Na}$. It is equal to the sum of two positive voltage *rises* $-V_m$ and E_{Na}, where $-V_m$ has a positive numerical value and represents the voltage gradient driving Na⁺ inward. E_{Na} is also a positive quantity that represents the electrical equivalent of the concentration gradient. Both of these gradients drive Na⁺ inward (Figure 19.4.3). E_{Na} is given by the Nernst equation for Na⁺ (Section ST19.1) as follows:

$$E_{Na} = \frac{kT}{|q|} \ln \frac{[Na^+]_o}{[Na^+]_i} \qquad (19.4.2)$$

where k is Boltzman's constant in J/K, T is the absolute temperature, $|q|$ is the magnitude of the electronic charge, and $[Na^+]_o$ and $[Na^+]_i$ are the concentrations of Na⁺ in the extracellular and intracellular media, respectively.

As for the K⁺ branch, KVL for this branch is: $V_K + E_K - V_m = 0$. In terms of quantities that have positive numerical values, this equation becomes

$$-E_K - (-V_m) = V_K = \frac{J_K}{G_K} \qquad (19.4.3)$$

Here, J_K has a positive numerical value, so that $\dfrac{J_K}{G_K}$ represents a positive voltage *drop* across G_K due to J_K. It is now equal to the difference of two positive voltage *rises* $-E_K$ and $-V_m$. As in the case of Na⁺, $-V_m$ represents the voltage gradient driving K⁺ inward. $-E_K$ is a positive quantity that represents the electrical equivalent of the concentration gradient, which drives K⁺ outward. $-E_K > -V_m$ because in the steady state, the net passive outflow of K⁺ is equal and opposite to the net inflow due to the ionic pump. E_K is given by the Nernst equation for K⁺ (Section ST19.1) as follows:

$$E_K = \frac{kT}{|q|} \ln \frac{[K^+]_o}{[K^+]_i} \qquad (19.4.4)$$

where $[K^+]_o$ and $[K^+]_i$ are the concentrations of K⁺ in the extracellular and intracellular media, respectively.

As Cl⁻ is not pumped, $J_{Cl} = 0$. E_{Cl} is given by the Nernst equation for Cl (Section ST19.1):

$$E_{Cl} = V_m = -\frac{kT}{|q|} \ln \frac{[Cl^-]_o}{[Cl^-]_i} \qquad (19.4.5)$$

where $[Cl^-]_o$ and $[Cl^-]_i$ are the concentrations of Cl⁻ in the extracellular and intracellular media, respectively. The negative sign in Equation 19.4.5 is due to the negative charge of Cl⁻.

KCL is, of course, satisfied by the circuit of Figure 19.4.4 but is further constrained by the need to conserve the current due to each ionic species. With $J_{Cl} = 0$ and no current though C_m under steady-state conditions,

$$J_{NaP} + J_{Na} = 0 \quad \text{and} \quad J_{KP} + J_K = 0 \tag{19.4.5}$$

For small changes v_m around the resting membrane potential V_m,

$$C_m \frac{dv_m}{dt} + G_m v_m = 0 \tag{19.4.6}$$

where $G_m = G_{Na} + G_K + G_{Cl}$.

It may be noted that in actual fact E_{Na}, E_K, and E_{Cl} are not independent of J_{NaP} and J_{KP}. If these current sources become zero, the concentrations of Na^+, K^+, and Cl will change, which changes E_{Na}, E_K, and E_{Cl}. However, both in the steady state and under transient conditions of short duration, these battery voltages can be considered constant.

Summary of Main Concepts and Results

- In mechanical systems, force is analogous to voltage and velocity to current, which leads to analogies between various derived quantities. The balance of forces at a point is analogous to KVL.
- In fluid systems, pressure difference or head is analogous to voltage and flow rate is analogous to current
- In thermal systems, temperature difference is analogous to voltage and heat flow rate is analogous to current.
- In ionic systems, ionic currents may be due to both voltage gradients and concentration gradients. Both of these driving forces have to be accounted for, in general, in equivalent electric circuits.

Learning Outcomes

- Be able to derive the equivalent electric circuit of representative mechanical, fluid, thermal, and ionic systems

Supplementary Topics and Examples on CD

ST19.1.1 Equilibrium voltage: Derives the expressions for chemical potential, electrochemical potential, and equilibrium voltage.

SE19.1 *Heat sink selection:* Shows how a heat sink for a power transistor can be selected from specified thermal parameters of the transistor and heat sink.

Problems and Exercises

P19.1   Mechanical Systems

P19.1.1 A mass of moment of inertia J is suspended at the end of a shaft of torsional stiffness k (Figure P19.1.1). Derive the equivalent electric circuit and determine the frequency of oscillation, neglecting friction.

FIGURE P19.1.1

P19.1.2 A railway car of mass m moving at a velocity u_0 along a level track strikes a spring-loaded buffer of stiffness k and velocity damping constant b. Derive the equivalent electric circuit. If the damping is critical, determine: (1) the variation of u with t, and (2) the maximum compression of the buffer.

P19.1.3 A mass m is suspended at the end of two springs in series (Figure P19.1.3). If F denotes the force acting on the mass, derive the equivalent electric circuit and determine the frequency of oscillation, neglecting friction.

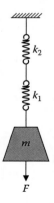

FIGURE P19.1.3

P19.1.4 Repeat P19.1.3 assuming the springs are in parallel.

P19.1.5 A baseball weighing 50 g is thrown horizontally by a pitcher at 30 m/s and is hit by a batter, causing it to travel at 40 m/s at an angle of 45° with respect to the horizontal. Assuming that the bat exerts an impulsive force on the ball, derive the equivalent electric circuit and determine the strength of the impulse.

P19.1.6 A mass of 2 kg moving horizontally at a velocity of 5 m/s impacts a stationary mass of 10 kg resting on a frictionless horizontal surface, and sticks to it, so that the two masses move together thereafter. Derive the equivalent electric circuit and determine: (1) the final velocity of the two masses, (2) the strength of the impulse exerted by the stationary mass on the moving mass, and (3) the energy lost.

P19.2 Fluid Systems

P19.2.1 Consider a number of pipes radiating from a common junction. If all the pipes carry an incompressible liquid, what relation between the rates of flow in the pipes corresponds to KCL?

P19.2.2 Consider the hydraulic jack shown in Figure P19.2.2, where a force F_1 is applied to a piston of area A_1 to lift a weight by means of a force F_2 applied to a piston of area A_2. Assume that the forces F_1 and F_2 move through distances x_1 and x_2, respectively. Neglecting losses and assuming the liquid is incompressible, it follows from conservation of energy that $F_1 x_1 = F_2 x_2$. (1) If both sides of this relation are differentiated with respect to time, what is the electrical analog of the new relation? (2) Deduce from the constancy of volume of liquid that

$$\frac{F_1}{A_1} = \frac{F_2}{A_2},$$ which means that the pressure in the liquid is the same everywhere.

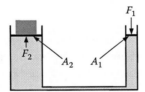

FIGURE P19.2.2

P19.2.3 Consider a tank of area A which is being filled by a liquid flow F_I and emptied by a flow F_O, where $F_I > F_O$, as illustrated in Figure P19.2.3. Assuming that F_O is directly proportional to the head H, derive the electrical analog of the hydraulic system, assuming the motion of liquid in the tank is small enough for inertance to be negligible.

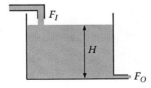

FIGURE P19.2.3

P19.2.4 The tank on the left in Figure P19.2.4 supplies the tank on the right at a constant head H_I. The valve is opened at $t = 0$, with the tank on the right initially empty. Assuming that the flow between the tanks is small, so that the effect of inertance can be neglected, derive the equivalent electric circuit and determine H as a function of time.

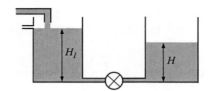

FIGURE P19.2.4

P19.2.5 A u-tube of diameter a is filled with a liquid of density ρ, the total length of the liquid in the tube being l. If the liquid is displaced as illustrated in Figure P19.2.5, derive the equivalent electric circuit and determine the frequency of oscillation. Neglect all frictional losses.

FIGURE P19.2.5

P19.3 Thermal Systems

P19.3.1 A power transistor in a TO-3 case has $\Theta_{JC} = 1.8°C/W$ and $\Theta_{CA} = 25°C/W$ when mounted in free air. Determine the case and junction temperatures when the transistor dissipates 5 W at an ambient temperature of 30°C, with the transistor: (1) mounted in free air; (2) used with a heat sink having $\Theta_{CS} = 0.5°C/W$ and $\Theta_{SA} = 2.5°C/W$.

P19.3.2 Transistor manufacturers' data sheets usually give a power derating curve, illustrated in Figure P19.3.2. The curve indicates a maximum power dissipation $P_{Dmax}(T_{Cref})$ at a reference case temperature T_{Cref} and a linear derating curve up to a maximum junction temperature T_{Jmax}. The slope of this line is $-1/\Theta_{JC}$.

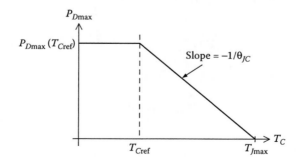

FIGURE P19.3.2

A transistor has P_{Dmax} = 160 W at $T_{C(Ref)}$ = 40°C has T_{Jmax} = 120°C. Determine the maximum allowed power dissipation at an ambient temperature of 80°C if θ_{CS} = 0.3°C/W and θ_{SA} = 0.2°C/W.

P19.3.3 A transistor has T_{Jmax} = 120°C and P_{Dmax} = 5 W at $T_{C(Ref)}$ = 25°C in free air. Determine θ_{JC} and the maximum allowed power dissipation when the case temperature is 80°C.

P19.3.4 For the transistor of P19.3.3, if θ_{CS} = 0.5°C/W and θ_{SA} = 5°C/W, what is the maximum allowed power dissipation at an ambient temperature of 40°C?

P19.3.5 A concentric cylinder of inner diameter a and outer diameter b conveys steam at a temperature of 125°C, the outside temperature being 25°C. Show that the thermal resistance per unit length is given by $\dfrac{\ln(b/a)}{2\pi\sigma_T}$, where σ_T is the thermal conductivity of the cylinder wall. If σ_T = 4 W/°Cm, b = 4 cm and a = 2 cm, determine the heat flow rate per meter.

FIGURE P19.3.5

P19.3.6 A small sphere of 0.5 mm diameter, representing a thermocouple junction, has a density of 8 g/cm^3 and thermal capacitance per unit mass of 500 J/kgK. The sphere, initially at a temperature of 25°C, is inserted in a gas stream at 200°C. If the thermal resistance between the gas stream and the sphere is $1/(hA_s)$, where A_s is the area of the sphere and h is 400 W/Km^2, derive the equivalent electric circuit and determine the thermal time constant and the time it takes the sphere to reach a temperature of 195°C.

P19.4 Cell Membrane

P19.4.1 Two NaCl solutions of different concentrations are separated by a membrane that is permeable to Na$^+$ but not to Cl$^-$. (1) What is the equivalent circuit? (2) How would the equivalent circuit change if the membrane were permeable to Cl$^-$ but not to Na$^+$?

P19.4.2 Consider a membrane having a capacitance of 1 μF/cm^2. (1) What is the charge per unit area that results in a voltage of 100 mV across the membrane? (2) If the magnitude of the electronic charge is 1.6×10^{-19} C, how many monovalent ions per unit area would produce this charge? (3) If ionic concentrations on either side of the membrane are approximately 100 mM, what is the concentration of monovalent ions per cubic centimeter^3? (Note that the number of ions per liter of a 1-M solution equals Avogadro's number, i.e., 6.02×10^{23} ions per liter.) (4) What thickness would be represented by the ions required in (2)? It is to be concluded that the charge that produces the voltage across the cell membrane is confined to the immediate vicinity of the membrane on either side. Beyond this immediate vicinity, any small volume of solution is electro-neutral.

P19.4.3 Consider the ionic system of Figure P19.4.3 having a membrane that is perme-able to Na^+ and Cl^-, but not to A^-. No ionic pump is present. Deduce from a qualitative argument that under equilibrium conditions, side 1 will be at a negative voltage with respect to side 2. This system is said to be in **Donan equilibrium**.

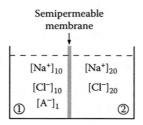

Semipermeable
membrane

$[Na^+]_{10}$ $[Na^+]_{20}$

$[Cl^-]_{10}$ $[Cl^-]_{20}$

① $[A^-]_1$ ②

FIGURE P19.4.3

P19.4.4 To analyze the system of Figure P19.4.3 at equilibrium, write the Nernst equa-tions for Na^+ and Cl, and the electroneutrality condition for both sides. Let

$$k = \frac{[Na^+]_{2o}}{[Na^+]_{1o}} = \frac{[Cl^-]_{1o}}{[Cl^-]_{2o}}.$$ Show that $k = \sqrt{1 - \frac{[A^-]_1}{[Na^+]_{1o}}} < 1.$ Deduce that: (1) side 1 is

at a negative voltage with respect to side 2, and (2) the total ionic concentration on side 1 is greater than on side 2. Note that because of the unequal total ionic concentrations on both sides, the system is not in osmotic equilibrium. To apply the Nernst equations for Na^+ and Cl^-, the system should, strictly speaking, be in osmotic equilibrium. This can be achieved by adding an appropriate con-centration of uncharged molecules to side 2.

P19.4.5 Repeat P19.4.3 and P19.4.4 assuming the indiffusible ions are positively charged.

P19.4.6 Consider the system of Figure 19.4.3. Let $J_{NaP} = 300$ nA/cm², $E_{Na} = 50_m V$, $G_{Na} = 2.5$ µS/cm², $E_K = -90$ mV, and $G_K = 60$ µS/cm². Determine V_m and E_K.

P19.4.7 Consider an ionic system of n ionic species, each being pumped, so that the jth ion is represented by an outward current source J_{jp} and a passive branch consisting of a conductance G_j in series with a battery E_j whose assigned positive direction is the same as V_m in Figure 19.4.3. Show that:

$$V_m = \frac{-\sum_{j=1}^{n} J_{jP} + \sum_{j=1}^{n} E_j G_j}{\sum_{j=1}^{n} G_j}.$$

Appendix: Reference Material

Trigonometric Relations

$$\sin(\alpha \pm \beta) = \sin\alpha\cos\beta \pm \cos\alpha\sin\beta$$

$$\cos(\alpha \pm \beta) = \cos\alpha\cos\beta \mp \sin\alpha\sin\beta$$

$$\tan(\alpha \pm \beta) = \frac{\tan\alpha \pm \tan\beta}{1 \mp \tan\alpha\tan\beta}$$

$$\sin 2\alpha = 2\sin\alpha\cos\alpha$$

$$\cos 2\alpha = 2\cos^2\alpha - 1 = 1 - 2\sin^2\alpha$$

$$\tan 2\alpha = \frac{2\tan\alpha}{1 - \tan^2\alpha}$$

$$\sin\alpha + \sin\beta = 2\sin\frac{\alpha+\beta}{2}\cos\frac{\alpha-\beta}{2}$$

$$\sin\alpha - \sin\beta = 2\cos\left(\frac{\alpha+\beta}{2}\right)\sin\left(\frac{\alpha-\beta}{2}\right)$$

$$\cos\alpha + \cos\beta = 2\cos\left(\frac{\alpha+\beta}{2}\right)\cos\left(\frac{\alpha-\beta}{2}\right)$$

$$\cos\alpha - \cos\beta = -2\sin\left(\frac{\alpha+\beta}{2}\right)\sin\left(\frac{\alpha-\beta}{2}\right)$$

$$2\sin\alpha\sin\beta = \cos(\alpha-\beta) - \cos(\alpha+\beta)$$

$$2\cos\alpha\cos\beta = \cos(\alpha-\beta) + \cos(\alpha+\beta)$$

$$2\sin\alpha\cos\beta = \sin(\alpha+\beta) + \sin(\alpha-\beta)$$

$$2\cos\alpha\sin\beta = \sin(\alpha+\beta) - \sin(\alpha-\beta)$$

$$\cos^2\alpha = \frac{1}{2} + \frac{1}{2}\cos 2\alpha$$

$$\sin^2\alpha = \frac{1}{2} - \frac{1}{2}\cos 2\alpha$$

$$\cos^3\alpha = \frac{1}{4}\left(3\cos\alpha + \cos 3\alpha\right)$$

$$\sin^3\alpha = \frac{1}{4}\left(3\sin\alpha - \sin 3\alpha\right)$$

$$\cos^4\alpha = \frac{1}{8}\left(3 + 4\cos 2\alpha + \cos 4\alpha\right)$$

$$\sin^4\alpha = \frac{1}{8}\left(3 - 4\cos 2\alpha + \cos 4\alpha\right)$$

TRIANGLE RULES

Let a, b, and c be the lengths of the sides of a triangle, and α, β, and γ be, respectively, the angles opposite these sides. Then,

Sine rule: $$\frac{a}{\sin\alpha} = \frac{b}{\sin\beta} = \frac{c}{\sin\gamma}$$

Cosine rule: $a^2 = b^2 + c^2 - 2bc\cos\alpha$

$$b^2 = a^2 + c^2 - 2ac\cos\beta$$

$$c^2 = a^2 + b^2 - 2ab\cos\lambda$$

SMALL ANGLES

If α is a small angle expressed in radians, then $\sin\alpha \cong \alpha$, $\tan\alpha \cong \alpha$, and $\cos\alpha \cong 1 - \alpha^2/2$.

Useful Mathematical Relations

INTEGRATION BY PARTS

$$\int u\left(\frac{dv}{dx}\right)dx = uv - \int v\left(\frac{du}{dx}\right)dx$$

L'HOSPITAL'S RULE

A function $\dfrac{f(x)}{x}$ that is indeterminate (that is $\dfrac{0}{0}$), when $x = 0$, can be evaluated by differentiating the numerator and denominator with respect to x any number of times until a finite answer is obtained.

Table of Integrals

$$\int xe^{ax}dx = \frac{e^{ax}}{a^2}(ax - 1)$$

$$\int x^2 e^{ax}dx = \frac{e^{ax}}{a^3}(a^2x^2 - 2ax + 2)$$

$$\int_0^\infty x^{1/2}e^{-x}dx = \frac{\pi^{1/2}}{2}$$

$$\int x\sin axdx = \frac{1}{a^2}\sin ax - \frac{x}{a}\cos ax$$

$$\int x\cos axdx = \frac{1}{a^2}\cos ax + \frac{x}{a}\sin ax$$

$$\int e^{ax}\sin bxdx = \frac{e^{ax}}{a^2 + b^2}(a\sin bx - b\cos bx)$$

$$\int e^{ax}\cos bxdx = \frac{e^{ax}}{a^2 + b^2}(a\cos bx + b\sin bx)$$

$$\int \frac{1}{x^2 + a^2}dx = \frac{1}{a}\tan^{-1}\frac{x}{a}$$

$$\int \frac{1}{ax^2 - x}dx = \ln\left(1 - \frac{1}{ax}\right)$$

$$\int \frac{1}{(x^2 + a^2)^2}dx = \frac{1}{2a^2}\left(\frac{x}{x^2 + a^2} + \frac{1}{a}\tan^{-1}\frac{x}{a}\right)$$

$$\int \sin ax \sin bxdx = \frac{\sin(a-b)x}{2(a-b)} - \frac{\sin(a+b)x}{2(a+b)}, \quad a^2 \neq b^2$$

$$\int \cos ax \cos bxdx = \frac{\sin(a-b)x}{2(a-b)} + \frac{\sin(a+b)x}{2(a+b)}, \quad a^2 \neq b^2$$

$$\int \sin ax \cos bxdx = -\frac{\cos(a-b)x}{2(a-b)} - \frac{\cos(a+b)x}{2(a+b)}, \quad a^2 \neq b^2$$

$$\int \sin^2 axdx = \frac{x}{2} - \frac{\sin 2ax}{4a}$$

$$\int \cos^2 axdx = \frac{x}{2} + \frac{\sin 2ax}{4a}$$

$$\int_0^\infty \frac{a}{a^2 + x^2}dx = \begin{cases} \frac{\pi}{2}, & a > 0 \\ 0, & a = 0 \\ -\frac{\pi}{2}, & a < 0 \end{cases}$$

$$\int_0^\infty \frac{\sin ax}{x}dx = \begin{cases} \frac{\pi}{2}, & a > 0 \\ -\frac{\pi}{2}, & a < 0 \end{cases}$$

$$\int x^2 \sin axdx = \frac{2x}{a^2}\sin ax - \frac{a^2x^2 - 2}{a^3}\cos ax$$

$$\int x^2 \cos axdx = \frac{2x}{a^2}\cos ax + \frac{a^2x^2 - 2}{a^3}\sin ax$$

$$\int e^{ax} \sin^2 bx \, dx = \frac{e^{ax}}{a^2 + 4b^2} \left[(a \sin bx - 2b \cos bx) \sin bx + \frac{2b^2}{a} \right]$$

$$\int e^{ax} \cos^2 bx \, dx = \frac{e^{ax}}{a^2 + 4b^2} \left[(a \cos bx + 2b \sin bx) \cos bx + \frac{2b^2}{a} \right]$$

$$\int_{-\infty}^{\infty} e^{-x^2/2\sigma^2} \, dx = \sigma \sqrt{2\pi} \, , \, \sigma > 0$$

$$\int_{-\infty}^{\infty} x^2 e^{-x^2/2\sigma^2} \, dx = \sigma^3 \sqrt{2\pi} \, , \, \sigma > 0$$

Bibliography

Alexander, C.K. and Sadiku, M.N.O., *Fundamentals of Electric Circuits*, 3rd ed., McGraw-Hill, Boston, MA, 2006.

Bellman, R., Kalaba, R.E., and Lockett, J.A., *Numerical Inversion of the Laplace Transform*, Elsevier, New York, 1966.

Biran, A. and Breiner, M., *Matlab® for Engineers*, 2nd ed., Prentice Hall, Harlow, U.K., 1999.

Brown, F.T., *Engineering System Dynamics: A Unified Graph-Centered Approach*, Marcel Dekker, New York, 2001.

Cavallo, A., Setola, R., and Vasca, F., *Using Matlab®*, Prentice Hall, London, 1996.

Conway, J.B., *Functions of One Complex Variable*, Springer-Verlag, New York, 1973.

Davis, A.M., *Linear Circuit Analysis*, PWS, Boston, MA, 1998.

DeCarlo, R.A. and Lin, P.M., *Linear Circuit Analysis*, 2nd ed., Oxford University Press, New York, 2001.

Eide, A.R., Jenison, R.D., Mashaw, L.H., and Northup, L.L., *Engineering Fundamentals and Problem Solving*, 4th ed., McGraw-Hill, Boston, MA, 2002.

Feinberg, B.N., *Applied Clinical Engineering*, Prentice Hall, Englewood Cliffs, NJ, 1986.

Fogiel, M., *The Electric Circuits Problem Solver*, Research and Education Association, Piscataway, NJ, 1998.

Fogler, H.S. and LeBlanc, S., *Strategies for Creative Problem Solving*, Prentice Hall PTR, Upper Saddle River, NJ, 1995.

Girod, B., Rabenstein, R., and Stenger, A., *Signals and Systems*, John Wiley & Sons, Chichester, U.K., 2001.

Gray, B.F., *Elementary Engineering Systems*, Longman Group Limited, London, 1974.

Haykin, S. and Van Veen, B., *Signals and Systems*, 2nd ed., John Wiley & Sons, New York, 2003.

Hayt, W.H., Durbin, S.M., and Kemmerly, J.E., *Engineering Circuit Analysis*, 7th ed., McGraw-Hill, Boston, MA, 2006.

Herniter, M.E., *Schematic Capture with Cadence PSpice*, 2nd ed., Prentice Hall, Upper Saddle River, NJ, 2003.

Hsu, H.P., *Schaum's Outline of Theory and Problems of Signals and Systems*, McGraw-Hill, New York, 1995.

Huelsman, L.P., *Active and Passive Analog Filter Design*, McGraw-Hill, New York, 1993.

Incropera, F.P. and Dewitt, D.D., *Fundamentals of Heat and Mass Transfer*, 5th ed., John Wiley & Sons, New York, 2002.

Inman, D.J., *Engineering Vibration*, 2nd ed., Prentice Hall, Upper Saddle River, NJ, 2001.

Irwin, J.D. and Kerns, D.V., *Introduction to Electrical Engineering*, Prentice Hall, Englewood Cliffs, NJ, 1995.

Jackson, L.B., *Signals, Systems, and Transforms*, Addison-Wesley, Reading, MA, 1991.

Kaplan, W., *Advanced Calculus*, 5th ed., Addison-Wesley, Redwood City, CA, 2003.

Keown, J., *OrCAD PSpice and Circuit Analysis*, 4th ed., Prentice Hall, Upper Saddle River, NJ, 2003.

Kreyszig, E., *Advanced Engineering Mathematics*, 9th ed., John Wiley & Sons, New York, 2005.

Macfadyen, K.A., *Small Transformers and Inductors*, Chapman and Hall, London, 1953.

Marshall, S.V. and Skitek, G.G., *Electromagnetic Concepts and Applications*, 3rd ed., Prentice Hall, Englewood Cliffs, NJ, 1990.

Nilsson, J.W. and Riedel, S.A., *Electric Circuits*, 7th ed., revised printing, Prentice Hall, Upper Saddle River, NJ, 2005.

Palm, W.J., III, *Matlab® for Engineering Applications*, WCB McGraw-Hill, Boston, MA, 1999.

Philips, C.L., Parr, J.M., and Riskin, E.A., *Signals, Systems, and Transforms*, 3rd ed., Prentice Hall, Upper Saddle River, NJ, 2003.

Poularakis, A.D. and Seely, S., *Signals and Systems*, 2nd ed., Krieger, Malabar, FL, 1994.

Research and Education Association, *The Electric Circuits Problem Solver*, Research and Education Association, Piscataway, NJ, 1998.

Rowell, D. and Wormley, D.N., *System Dynamics: An Introduction*, Prentice Hall, Upper Saddle River, NJ, 1997.

Van Valkenburg, M.E., *Network Analysis*, 3rd ed., Prentice Hall, Englewood Cliffs, NJ, 1974.

Webster, J.G., Ed., *Medical Instrumentation: Application and Design*, 3rd ed., John Wiley & Sons, New York, 1997.

Index